MARIA GEORGIADOU

Constantin Carathéodory

*Mathematics and Politics
in Turbulent Times*

Springer

A hardcover edition of this book is available with the same title
with the ISBN 3-540-44258-8

Cataloging-in-Publication Data applied for

Bibliographic information published by Die Deutsche Bibliothek
Die Deutsche Bibliothek lists this publication in the Deutsche Nationalbibliografie;
detailed bibliographic data is available in the Internet at http://dnb.ddb.de

Mathematics Subject Classification (2000):
01 A 70, 01 A 60, 01 A 73, 01 A 74

ISBN 3-540-20352-4 Springer-Verlag Berlin Heidelberg New York

With 87 illustrations

Springer-Verlag is a part of Springer Science+Business Media
springeronline.com

© Springer-Verlag Berlin Heidelberg 2004
Printed in Germany

Typeset in Word by the author and edited by PublicationService Gisela Koch, Wiesenbach
using a modified Springer LATEX macro-package.
Cover design: design & production GmbH, Heidelberg
Printed on acid-free paper   41/3142Ko - 5 4 3 2 1 0

Foreword

The life and times of Constantin Carathéodory (1873–1950), a brilliant German mathematician of Greek descent, should surely interest mathematicians, German and Greek academics, historians of science as well as many others. The author of this documentary biography has researched with great care a large part of Carathéodory's correspondence, papers and other relevant documents and archives, and has constructed a narrative that is driven by Carathéodory's own words and thoughts. She has provided excellent personal, historical and mathematical background information, as well as lean and precise comments and interpretation. The result is stunningly effective. Carathéodory and his time spring to life: his mathematics and science, his love of history, his politics, his unshakeable belief in the power of the German intellectual tradition, his emotional attachment to Greece, his setbacks, his lost causes, and his disappointments. We see the world changing around him, sometimes too fast. But we also see his own efforts to make changes, for he was enormously influential both in the academic world and more broadly in science and education politics. The dramatic events of the first half of the twentieth century changed the world and seeing them in this gripping narrative through the actions and thoughts of a leading intellectual, mathematician and academic provides a unique and fascinating perspective.

Carathéodory did not decide to go into mathematics until he was twenty-seven, after he had studied engineering at the Military School of Belgium and worked for a few years as an engineer. He was born in 1873 in Berlin, where his family was residing while his father was the diplomatic representative of the Ottoman Empire in Germany. He studied mathematics at first in Berlin, attracted to the lectures of H. A. Schwarz on function theory. He then went to Göttingen, receiving a doctoral degree in 1904 with a dissertation entitled "On the discontinuous solutions in the calculus of variations". He was close to Klein and Hilbert in Göttingen but his dissertation was signed by Minkowski. His first teaching position was in Göttingen, where he stayed until 1908. He subsequently moved to Bonn, Hanover and Breslau before returning to Göttingen in 1913 as full professor. In 1918 he went to Berlin and stayed till the end of 1919. His meteoric rise in the German academic world was, of course, a direct recognition of his many seminal contributions in function theory, the theory of measure, first-order partial differential equations, the calculus of variations and the foundations of thermodynamics.

His mathematical interests were deeply rooted in geometry and mechanics, but Carathéodory was above all a powerful analyst who understood and appreciated

the importance of simplicity, naturalness and elegance in mathematics. He actively
sought these qualities in his work. He was also very interested in the origin and
historical evolution of important mathematical ideas and methods, which are often
lost in more efficient and abstract formulations. His work in function theory is closely
tied to the geometry of conformal mapping, elegant and unexpected uses of Schwarz's
lemma, the various variational forms of the Riemann mapping theorem and many
results that are part of harmonic analysis today. He published his beautiful lecture
notes on conformal mapping in 1932 and, very late in his life, he wrote an elegant
treatise on complex function theory that appeared in 1950.

Function theory must have been his first and enduring love. It suited his mathematical style very well and he must have liked lecturing on it. The calculus of variations
must not have been very far behind, however, as his favorite area in mathematics.
He had a very specific point of view here: that the elaboration of the equivalence
between the Huygens principle and Fermat's principle in Optics is in fact central to
the mathematical foundations of the calculus of variations. In Carathéodory's work
this idea leads to a far-reaching generalization and extension of the Hamilton–Jacobi
theory, in which the level surfaces of the action, or travel time, that are solutions
of the Hamilton–Jacobi equation, are also generated by families of extremals of
the associated Lagrange functional. His work is presented in his famous 1935 book
Variationsrechnung und partielle Differentialgleichungen erster Ordnung (Calculus
of Variations and Partial Differential Equations of the First Order), translated into
English in 1965. This is still one of the best books in the area even though the modern
theory of control and dynamical systems is often oriented toward the inclusion of
stochastic effects, which entered much more recently.

The way that the Hamilton–Jacobi theory emerges in high-frequency asymptotics
in wave propagation was surely known to Carathéodory but it is not discussed in his
book. It played a very important role in the development of quantum mechanics,
which Carathéodory followed closely, and before that in optics. The mathematical
foundations of high-frequency asymptotics that connect Hamilton–Jacobi theory to
wave propagation did not emerge until many years after Carathéodory's work, in the
sixties and seventies, so he could not deal with it mathematically even if he understood
it physically. It is interesting, however, that his intensely geometrical point of view
is becoming increasingly important today, in connection with level-set methods for
the efficient numerical computation of travel times for image reconstruction.

Carathéodory began his work on measure theory around 1913 and in the following five years, until the publication of his book *Vorlesungen über reelle Funktionen*
(Lectures on Real Functions) in 1918, made several fundamental contributions that
solidified his standing as a leading mathematician of his generation in Germany
and around the world. His theory of outer measure, and the Carathéodory extension
theorem, are now part of the standard material taught in every graduate-level real
variables course. For forty years, until the sixties, his real variables book was one
of the best sources for learning measure and integration theory. What makes it look
a bit strange today is the absence of basic Hilbert space theory and functional analysis, which did not emerge until the thirties. Another field that did not exist before
the thirties is the measure-theoretic foundation of probability theory, established by

Kolmogorov in 1933, which provided a major thrust for the further development of general measure theory. For a mathematician with Carathéodory's command of classical analysis (function theory, calculus of variations) it is somewhat surprising that at age forty his research interests turned into a more abstract direction, measure and integration theory.

His elegant mathematical formulation of thermodynamics attracted the attention of several physicists but remained somehow outside the mainstream. The physicist Max Born, a lifelong friend of Carathéodory, appreciated and promoted it but mathematical foundations of established physical fields seem to penetrate slowly and with difficulty into the world of physics. It is interesting to see how his involvement with applications of mathematics, thermodynamics in particular, as well as his background in engineering influenced his appointments and promotions. It appears that they did play a role but not a big one. It is also interesting to see how the German mathematics community viewed applied mathematics just before the First World War. It is only a partial view, of course, but it seems to be mainly as a way to make the teaching of mathematics more effective. Carathéodory was less interested in applications than in the way mathematics interacts intimately with physics at a basic level, as in the general theory of relativity and in quantum mechanics. An exception is his very detailed study of aberrations in optical instruments in the late thirties.

Carathéodory's life changed dramatically in the fall of 1919 when he accepted a mandate from E. Venizelos, the prime minister of Greece, to organize a new university in Smyrna. He left Berlin at the end of that year and spent the next four years in Athens and Smyrna before returning to Germany as professor of mathematics in Munich. He remained there for the rest of his life, resuming his brilliant career almost as if nothing had happened. But a great deal had happened, and it had touched Carathéodory in a very personal way, even if it did not diminish his self-confidence and his creative drive. Accepting Venizelos' calling had clearly been an emotional decision, not a rational one, for Carathéodory belonged to an extended family of distinguished diplomats and knew at first hand the complex and unstable political scene in Greece at the time of the Versailles Peace Conference. By the fall of 1920, when Venizelos was voted out of office, Carathéodory was unable to recruit even close former colleagues to the new university and was at odds with the royalist government in Athens as a Venizelist (a liberal). He clung to the noble cause of building a university, a civilizing mission if ever there was one. In the words of Arnold Toynbee, a war correspondent for the *Manchester Guardian* in Asia Minor during this period, Carathéodory the scientist-humanist looked like fish out of water. The background and events of this period are presented with such clarity and precision in this biography that even readers familiar with the complicated and little-known historical details will find the narrative fascinating. Focusing on Carathéodory's doomed efforts to build a university in Smyrna is special, personal and even tragic. But it is also a rare glimpse at the end of an era in Europe.

Remarkably, Carathéodory was able to continue his research in mathematics through the turbulent period in Smyrna and Athens, and no doubt keep his mental balance by doing so. After returning to Germany as professor in Munich he received many honors and much international recognition. In 1928 he became the first Visiting

Lecturer of the American Mathematical Society and visited and lectured at several American universities, including Harvard, the University of Texas at Austin, Stanford and Berkeley. He was considered for a permanent appointment at Harvard but he did not receive an offer. He did, however, receive an offer of a professorship at Stanford, in early 1929, which he did not accept. The American West was beautiful and attractive but at the other end of the world for a cosmopolitan European like Carathéodory. His academic life in Munich was fulfilling and productive and there was, of course, no hint that a storm was about to grip Germany and Europe.

Carathéodory's relations with Greece took another interesting turn when Venizelos returned to power and appointed him, in early 1930, adviser to the Government for the reorganization of Greek universities. He took a leave of absence from Munich for about two years to conduct an in-depth study of universities in Greece. His prestige was enormous, but even an apolitical academic of his stature, who surely acted in the best interests of Greek universities, could not overcome the political divisions that plagued the Greek university system. He eventually terminated his formal relations with Greece in 1932, and from then on his academic position remained fixed in Munich until his retirement in 1938 at the age of sixty-five.

The arrival of national socialism in March 1933 took many academics in Germany by surprise, especially the speed with which racial and political discrimination set in, as well as the loss of personal freedom and the obligation to demonstrate compatibility with and adherence to the new regime. Carathéodory had been born in Germany, studied mathematics in Germany, became a distinguished professor in several German universities, but he was not a German. He was an ethnic Greek, a Christian, and was able to function in Nazi Germany as long as he did not openly oppose the regime even if he did so privately. His behavior and actions during the Nazi period, meticulously researched and accurately reported in this biography, were exactly like those of many German academics of his generation. He actively and energetically helped his Jewish friends and colleagues who became persecuted undesirables in their own country almost overnight, and sometimes, in later years, intervened on their behalf through more powerful colleagues who were closer to the regime. Persecution was not unknown to Carathéodory who came from a family of ethnic Greeks that had risen to prominent positions in the Ottoman Empire. At any moment, events out of their control could, and did many times, turn the state against them with unpredictable intensity. He no doubt understood very well that his academic status and prestige could hardly protect him should his loyalty to the state be questioned, especially since he was not German. Where he crossed the line, however – and the author is very careful here – is in allowing himself to be used by the Nazis in their international propaganda. When he, as a distinguished academic, represented German mathematics at international meetings that the state allowed him to attend, he was being used. During the war he continued to help colleagues outside Germany who now were not only losing their positions and their property but also were increasingly in danger of their lives. However, neither he nor other engaged academics could do much, and there is no indication at what time Carathéodory became aware of the Holocaust, or what his reaction was to the invasion and occupation of Greece by the Germans in April 1941.

Carathéodory tried to re-establish his contacts outside Germany after the war but he quickly found that the world had changed completely. His colleagues in America helped him generously but not enthusiastically. When he tried to differentiate himself from other Germans as an ethnic Greek living in Germany he discovered that the American occupying forces did not understand the difference. When he tried to move to Greece with his family he was rebuffed by the Greeks. The only escape he had was in mathematics, and that is when he wrote his book on function theory. He was diagnosed with an enlarged prostate in the spring of 1947 and died in February 1950.

Carathéodory is a towering figure in the world of mathematics. His work on first order partial differential equations and the calculus of variations is still cited and studied today, as are his contributions in real and complex analysis. He lived in complicated and turbulent times, and had roots in many cultures and traditions, most of which are fading memories today. He was a cultivated and engaged intellectual, and this biography is a fitting tribute to his life and work.

George Papanicolaou

Preface

Όπου και να ταξιδέψω, η Ελλάδα με πληγώνει.
Wherever I travel, Greece wounds me.

G. Seferis, 1936

This book treats elements of the life and work of Constantin Carathéodory, a world-famous mathematician of the first half of the 20th century, who emerged from the Göttingen school of mathematics and whose main contribution to his science is the modern-classical treatment of the calculus of variations. Though a cosmopolitan, Carathéodory lived within two cultural environments, the Greek and the German, and my version of his story is based for the most part on his activity within these two cultures.

The presentation has a documentary character. Biographical details of Carathéodory's life are interwoven with a presentation of his scientific work and its reception.

The first chapter, on Carathéodory's origin and formative years, covers the period 1873–1908. It traces the family's roots back to the Greek mercantile bourgeoisie and the Greek urban elite of Constantinople, after the Greek War of Independence and the formation of the Greek national state. It then covers his formative years in secondary and higher military education in Brussels, his views about the Graeco-Turkish War of 1897, his first engineering work for the British colonial service in Egypt and his decision to pursue a mathematical career. The chapter concludes with a survey of his mathematical studies in Berlin and Göttingen.

The second chapter tells of Carathéodory's academic career in Bonn, Hannover, Breslau, Göttingen and Berlin from 1908 up to the Versailles Peace Treaty. The way that appointments to academic posts were influenced and controlled by prominent figures emerges from a description of Carathéodory's appointments to the respective universities in the German Reich.

The third chapter tells of Carathéodory's services to the Greek national goal of a Great Greece by means of organising a new university in Smyrna, on territory that had been ceded to Greece by the victors of World War I. The goal of a Great Greece had been pursued and realised by the Greek Prime Minister and leader of the Liberals, Eleutherios Venizelos, in the years 1919–1922. Carathéodory's views on the refugee settlement in Greece after the Asia Minor disaster in September 1922 are contained in his report to Henry Morgenthau, Sr., which is presented in this chapter.

The fourth chapter gives an account of Carathéodory's scientific work and his educational activities in Greece, Germany and the United States in the years 1924–1933, starting with his appointment to the chair of mathematics at the University of Munich in 1924. It includes his contact with the Nobel Prize winner Robert Andrew Millikan in connection with the proposed foundation of a scientific institute associated with the University of Athens and with American aid to post-war Europe.

Carathéodory's scientific contacts with philosophers of science, mathematicians and physicists in Germany, Greece and the United States are described, as well as his academic tour in the States and his visits to Harvard, Stanford and Berkeley in 1928. It further presents the conflict concerning the foundation of mathematics that ended in the so-called *Annalenstreit* (*Annalen* quarrel) and the dissolution of that journal's editorial board according to his own proposal of a way out of the crisis. Finally, it shows, how Carathéodory's concept for the reorganisation of the Greek universities in the years 1930–1932 led to his dismissal from the post of Commissioner of the Greek Government.

The fifth chapter considers the troubled period of the growth of national socialism and the Second World War. It exposes the conditions at the universities of Nazi Germany and, particularly, the persecutions under the racial laws of the years 1933, 1935 and 1937, which deprived Carathéodory of the majority of his friends and colleagues. It further deals with the developments in Greece during these years marked by Metaxas's dictatorship, which lasted from August 1936 until some days before the German invasion of Greece in the spring of 1941. It describes Carathéodory's activities during the Nazi era, his membership in German delegations sent to international mathematical events, his appointment to the Carl Schurz Professorship at the University of Wisconsin and his designation to Papal Academician. This chapter also demonstrates Carathéodory's efforts to save human lives and cultural goods. In addition, it focuses on his retirement in 1938 and the six-year dispute over his successor. Carathéodory's denouncement by the lecturers of the University of Munich, the report of the Rosenberg authority on his activities and his assessment by the Nazis are included as documents. The chapter concludes with the decision of the Bavarian Academy to publish Carathéodory's collected works towards the end of the war, his increasing isolation from his international friends, his contacts with persons belonging to the Nazi party and, above all, with the president of the German Union of Mathematicians, Wilhelm Süss.

The sixth chapter, on his final years, describes Carathéodory's mediating role between Germany and the USA for the recovery of mathematics. It reveals his unrealised intention to return to a damaged Greece, proceeding towards civil war. Finally, it presents his post-war academic contacts and the dissolution of his library after his death.

Some manuscripts and prints of Carathéodory's mathematical papers are kept in the Bavarian Academy of Sciences in Munich, but his personal archive is lost and his correspondence, consisting for the most part of letters to friends and colleagues, is scattered all over the world. Part of this correspondence can be traced in libraries and archives, mainly in Germany and the USA and but also in Israel, Denmark and Hungary. The extensive search for sources of information was extremely difficult: much material has been destroyed by war, or by its owners when Germany occupied their countries, or it was per force left behind in Germany by those who emigrated during the Nazi period and it is now lost; some archives or collections have not yet been opened up, or are only partly accessible, or are kept by relatives who have not yet decided how to handle the material in their possession; other archives are not ordered, or, even worse, have disappeared altogether; a few persons possessing sources of information are not willing to respond to enquiries; the archives of the Belgian Military School were lost during both world wars; thus, only Carathéodory's autobiographical entries in various university questionnaires and his *Autobiographical Notes* up to 1908, which he had prepared for the Austrian Academy of Sciences, give information about his military training; original Greek documents concerning Carathéodory's engineering activities and his major Greek educational projects, except for the minutes of the General Assembly of Samos of the year 1897 and the minutes of sessions of the School of Natural and Mathematical Sciences of the University of Athens, are non-existent; of relevance for the assessment of Carathéodory's engagement in Greek educational policies are documents already presented in two publications of 1956 and 1962; the archive of the Ionian University kept at the University of Athens at least up until 1971 and including Carathéodory's letters (without registration number) to the High Commissioner of Ionia and to persons living in Athens in the years 1919–1920 has vanished. Much of the material used in this study has been collected over several years and has not been previously published. Oral history has been included only where the stories seemed plausible and showed consistency with the documents.

Thus far, biographical data on Carathéodory could only be found in his incomplete *Autobiographical Notes*, in obituaries or sparse entries in dictionaries. This is his first comprehensive biography. There has been no serious research concerning Carathéodory's personality. Authors of recent publications and speakers at hastily organised events do not treat him properly: his Greek work is presented out of its historical context, while his activities in Germany and the USA are variously ignored, and an attempt is made to cultivate a myth around his name. This book is an attempt at a fair biography beyond myth and monumentalisation.

The mathematical parts of the book are mainly based on Carathéodory's own work as this appears in his books, his collected papers and his correspondence. Apart from his major contributions, lesser ones are also included to show the broad scope of his mathematical agenda. The extent of Carathéodory's influence on later mathematicians, and the question of whether he could be said to have established a particular school of mathematics, is left open. In the absence of any extensive studies of the succeeding mathematical generation, it may yet be too early for such an assessment. In this respect, the most that can be done at the moment is to provide a list of Carathéodory's students with their biographical data.

Maria Georgiadou

Acknowledgements

My acknowledgements and thanks are due to the following specialists who, out of pure interest in the subject, have contributed to the completion of this project:

Dr. Caesar Alexopoulos, emeritus professor of physics at Athens University, for his communications; Pavlos Xynidakis, theologian, for books and archival information from Crete; Vasiliki Konti, mathematician, for research in Athenian educational institutions and the Academy of Athens; Dr. George Papanicolaou, mathematics professor at Stanford University, and Dr. Peter Duren, mathematics professor at the University of Michigan, for their comments, remarks, suggestions and aid in mathematical terminology; Dr. Demetrios Christodoulou, mathematics professor at Princeton University, for his enlightening explanations of the theory of general relativity; Dr. Wendell Fleming, professor at the Division of Applied Mathematics, Brown University, for his report on my manuscript; Dr. Constantine Dafermos, professor at the Division of Applied Mathematics, Brown University, for his comments and general support of this project; above all, Dr. Freddy Litten, researcher at the Institut für Geschichte der Naturwissenschaften, LMU München for his always prompt response to my queries as well as for material, control of information and suggested bibliography;

to the following academics who allowed me the opportunity to present my views on Carathéodory and open them up to discussion:

Professor Dr. Klaus Kreiser, Chair of Turkish Language, History and Culture at Bamberg University; Priv. Doz. Dr. Reinhard Heydenreuter, Archive Director of the Bavarian Academy of Sciences; Dr. Konrad Clewing, editor of the *Südost-Forschungen*; Méropi Anastassiadou-Dumont, PhD, chargée de recherche au CNRS, Institut Français d'Etudes Anatoliennes; May Davie, PhD, editor of *Chronos*;

to the following scholars who kindly responded to my enquiries:

Ernest D. Courant for insight into his father's correspondence with Carathéodory; Dr. Antal Varga, Bolyai Institute, József Attila University, Szeged for information, letters, suggested bibliography and a photograph; Dr. Kurt Ramskov, professor at the Department of Mathematics of the University of Copenhagen, for discussions and letters by B. Jessen; Dr. Dirk van Dalen, curator of the Brouwer archive and professor of philosophy at the University of Utrecht, for Brouwer's and other letters; Molly Greene, assistant professor in History and the Program in Hellenic Studies at Princeton University, for suggested bibliography and useful discussions; Mike Mahoney, emeritus professor of the history of science at Princeton University, for atmospheric information on post-war Munich, where he attended Perron's lectures; Dr. h.c. mult. John Argyris, emeritus professor of the University of Stuttgart, for his impressions of Germany during the Nazi era; Dr. Klaus von Dohnanyi and Dr. Winfried Meyer, biographer of Hans von Dohnanyi, for checking a reported story which had allegedly happened in the year 1942; Dr. Johann Strauss, Département d'études turques, Université Marc-Bloch, Strasbourg for information about Carathéodory's library; Dr. Yeşim Işil Ülman, Medical History and Ethics, Faculty of Medicine, Istanbul University, for information about the medical activity of Carathéodory's grandfather in Constantinople; Dr. Nuran Yıldırım, Turcologist and medical historian, Professor at the Faculty of Medicine, Istanbul University, for information about Carathéodory's doctoral student Nazim Terzioğlu; Dr. Ali Onur, architectural historian, for photographs of the Ionian-University building in Izmir; Dr. Necmi Ülker, professor at Ege Üniversitesi, Izmir, for information about the unpublished Turkish PhD thesis of Engin Berber; Dr. Rolf-D. Philips for a photograph of Carathéodory's house in Bonn; Mrs. Lieve Lettany, Projectcoördinator, "Musea in Mechelen", and Karl-Friedrich Koch, production editor of this book, for their research in the history of a Rubens triptychon in a church at Mechelen; Professor Leonidas Kamarinopoulos, president of the Greek Atomic Energy Committee for Behnke's account of Carathéodory's life and work; Vicki Lynn Hill, PhD, mathematics instructor at the American University, Washington DC, for photocopies of John Horváth's letter to her and of documents from the Central Archive of the Academy of Sciences of the former German Democratic Republic; Rena Fatsea, PhD, Assistant Professor, School of Architecture, University of Thessaly, for her recommendation of Sturdza's geanological dictionary and information about Greek dynasties and Dr. Nikolaos Chryssidis, scholar of Russian history at Southern Conecticut State University, for helpful discussions about the Phanariots and the Oecumanical Patriarchate;

to my friends Angela Kakavoutis, Deutsche Schule Thessaloniki, and Fotini Pelteki for double-checking information from libraries in Thessaloniki.

Carathéodory's daughter, Mrs. Despina Rodopoulos-Carathéodory, deserves a special mention, for two interviews and for supplying me with published material concerning her father, a poem of 1918 from a female student of Carathéodory in Göttingen, an unpublished letter of 1922 from the General Secretary of the YMCA to Carathéodory, a photocopy of Carathéodory's guest-book in Munich, and also for the photographs she set at my disposal for publication.

My sincere appreciation goes to the following scholars, who knew many persons mentioned in this book, either personally or through their own research, and have shown willingness to help me by suggestions, information and material:

Christos Landros, director of the Archives of the Nomarchy of Samos, General State Archives; Emmanuel Trajas, Κοβεντάρειος Δημοτική Βιβλιοθήκη Κοζάνης; Professor Adrianos Melissinos, student of Hondros, Department of Physics and Astronomy, University of Rochester, New York; Professor Dr. Kristie Macrakis, College of Arts and Letters, Department of History, Michigan State University; Professor Peter Havas, College of Arts and Sciences, Temple University; the late Professor Dr. Achilleas Papapetrou, Directeur de Recherche, CNRS, Paris and head of the Laboratoire de physique théorique at the Henri Poincaré Institute; Professor M. Roilos, Department of Physics, University of Patras; Professor Leonidas Resvanis, Director of the Physics Laboratory, University of Athens; Professor Josef Ventura, Department of Physics, University of Crete; Professor Eutychios Bitsakis, Physics Department, Nuclear and Particle Physics Section, University of Athens; Dr. Leonhard Weigand, Carathéodory's last doctoral student; Professor Werner Romberg, Carathéodory's and Sommerfeld's student from 1929 to 1932; Dr. Michael Eckert, Institut für Geschichte der Naturwissenschaften, LMU München; Dr. Wilhelm Füßl, Leiter der Archive, Deutsches Museum; Professor Dr. Hans Josef Pesch, Universität Bayreuth; Dr. Klaus-Dieter Reinsch, Zentrum Mathematik der Technischen Universität München; Professor Dr. Dr. h.c. Walter Benz, Hansische Universität Hamburg; Dr. Gerhard Betsch, Mathematisches Institut, Universität Tübingen; Professor Dr. Günther Frei, Laval University, Canada; Professor John Stachel, Boston University.

For their special interest in the cultural aspects of this project I would like to thank:

the architect Ralph Schweitzer, Bonn; Mark Graf Fugger von Babenhausen, director of the museum in Babenhausen; Sir Michael Llewellyn Smith, former ambassador of Great Britain in Athens and author of the book *Ionian Vision*; Pater Stephan Kessler, SJ, Institut für Biblische und Historische Theologie, Universität Freiburg; Father Olivier Raquez, Rector of the Pontificio Collegio Pio Romeno; Dr. Albert Rauch, Ostkirchliches Institut Regensburg; the Reverend Fathers Paul Buhagiar, SJ and Professor Michalis Roussos, SJ.

I am also indebted to the following, whose commitment to their work in archives, libraries and authorities has been a great encouragement for the completion of this book:

Dr. Norbert Becker, Universitätsarchiv Stuttgart; Dr. Brigitte Uhlemann, Philosophisches Archiv der Universität Konstanz; Frau Zeipel, Stadtarchiv und Stadthistorische Bibliothek Bonn; Dr. W. Schultze, Archive, Humboldt University of Berlin; Rafael Weiser, Department of Manuscripts and Archives, the Jewish National and University Library; Birgit Schaper and Christine Weidlich, Universitäts- und Landesbibliothek Bonn, Handschriften- und Rara-Abteilung; Dr. Marion Kazemi, Archiv zur Geschichte der Max-Planck-Gesellschaft; A. Sawada, Akademie der Wissenschaften zu Göttingen; Peter Bardehle, Niedersächsisches Hauptstaatsarchiv Hannover; Dr. K. Volkert, Pädagogische Hochschule Heidelberg; Dr. Keßler, Universitätsarchiv, Ruprecht-Karls-Universität Heidelberg; Stefanie Golach, Senckenbergische Bibliothek der Johann Wolfgang Goethe-Universität; Dr. Rita Seidel, Universitätsbibliothek Hannover und technische Informationsbibliothek; Dr. Heidelies Wittig-Sorg und Frau Wunderlich, Staatsarchiv Hamburg; Priv. Doz. Dr. Bernd Dörflinger, Kant-Studien-Redaktion, Johannes Gutenberg-Universität Mainz; Erika Rasthofcr, Institut für Zeitgeschichte, München; Priv. Doz. Dr. Reinhard Heydenreuter, Archiv der Bayerischen Akademie der Wissenschaften; Dr. Andrea Schwarz, Bayerisches Hauptstaatsarchiv; Bärbel Mund, Abteilung für Handschriften und seltene Drucke, Niedersächsische Staats- und Universitätsbibliothek; Sabine Happ, Archiv, Rheinische Friedrich-Wilhelms-Universität Bonn; Dagmar Kicherer and Dr. Dieter Speck, Universitätsarchiv Freiburg; Gabriele Stefanski, Dieter Lange, Roland Klein and Frank Dannenberg, Staatsbibliothek zu Berlin, Preußischer Kulturbesitz; Dr. Petra Blödorn-Meyer, Staats- und Universitätsbibliothek Hamburg Carl von Ossietzky; Jochen Stollberg, Archivzentrum Universität Frankfurt am Main; Dr. Ulrich Hunger, Universitätsarchiv Georg-August-Universität Göttingen; Gisela Sprenger, Universitätsbibliothek Würzburg; Katrin Schuch, Bibliothek des Mathematischen Seminars, Universität Frankfurt am Main; Petra Schwarz, Bibliothek des Mathematischen Instituts, Universität Erlangen; Stefanie Golath, Senckenbergische Bibliothek der Johann Wolfgang Goethe-Universität; Dr. John Moore, Universitätsarchiv Bamberg; Dr. Eva-Marie Felschow, Universitätsbibliothek, Universität Giessen; Prof. Dr. Laetitia Böhm, and Dr. Wolfgang Smolka, Universitätsarchiv der LMU München; H. Scheiner, Bibliothek, Mathematisches Institut der LMU München; Frau Deppe, Sekretariat, Akademie der Wissenschaften zu Göttingen, Frau Bilewski, Deutsche Dienststelle für die Benachrichtigung der nächsten Angehörigen von Gefallenen der ehemaligen deutschen Wehrmacht; Friedrich, Staatsarchiv Nürnberg; Jürgen Matthes, Bibliothek des Mathematischen Instituts der Universität Göttingen; Dr. Winfried Hagenmaier, Universitätsbibliothek Freiburg; Frau Reich, Institut für Geschichte der Naturwissenschaften, Mathematik und Technik, Universität Hamburg; Dr. Martina Haggenmüller, Staatsarchiv Augsburg; Frau Wolff, Herr Meentz, Frau Blumberg, Deutsches Bundesarchiv; Staatsanwalt Wacker, Zentrale Stelle der Landesjustizverwaltungen; Frau Kalies, Der Bundesbeauftragte für die Unterlagen des Staatssicherheitsdienstes der ehemaligen Deutschen Demokratischen Republik; Ze'ev Rosenkranz, Curator of The Albert Einstein Archives, Judith Levy and Dina Carter, assistants to the Bern Dibner Curator, the Jewish National and University Library; Dr. Ada Baccari and Margaret Evangelisti, Accademia Nazionale dei Lincei; Adrian Allan, University of Liverpool Archives; Felicity Pors, Niels Bohr Archive; Seira Airas, Technical University of Helsinki; Dr. Leonard Smołka, Archivum, Uniwersytet Wrocławski; Milena Hermanova, Central Library of the Faculty of Mathematics and Physics, Charles University, Prague; Pentti Kauranen, Finnish Academy of Science and Letters; Christiane Demeulenaere-Douyère, Service des archives, Institut de France, Académie des sciences, Paris; Dr. Richard Boijen, Documentation Center, Koninklijk Museum van het leger en de Krijgsgeschiedenis, Brussels; Dr. Beat Glaus, Wissenschaftshistorische Sammlungen der ETH-Bibliothek Zürich; Simon Wiesenthal, Dokumentationszentrum des Bundes jüdischer Verfolgter des Naziregimes, Wien; DDr. Fischer, Bundesministerium für Inneres, Republik Österreich; HR Dr. Manfred Fink, Österreichisches Staatsarchiv; Dr. Rosemary Moravec and Christa Bader, Österreichische Nationalbibliothek; Erich Jiresch, Universitätsarchiv, Technische Universität Wien; Dr. Kurt Mühlberger, Archiv der Universität Wien; Dr. Gugler, Universitätsbibliothek Wien; Erich Reiter, Bibliothek, Österreichische Akademie der Wissenschaften; Dr. Gerhart Marckhgott, Oberösterreichisches Landesarchiv; Dr. Elisabeth Klamper, Stiftung Dokumentationsarchiv des Österreichischen Widerstandes; Lois Beattie, Institute Archives, MIT; Shelley Erwin, California Institute of Technology Archives; Dr. Raimund Goerler, Bertha Ihnat and Tamar Galed, University Archives, The Ohio State University; Lisa Richmond, Library, St. John's College, Annapolis; Brian Sullivan, Harvard University Archives; Teresa Mora, New York University Archives; Ralph Elder, The Center for American History, The University of Texas at Austin; Carol Hutchins, Library, Courant Institute, New York University; William Roberts, University Archive, University of California, Berkeley; Martin Hacket, University Archives, University of Pennsylvania; David Farrel, The Bancroft Library, Berkeley; Heather Briston, Tom Hyry, Karen Jania, and Leigh Jasmer, Bentley Historical Library, University of Michigan; Sheila Cummins, Harlan Hatcher Graduate Library, University of Michigan; Art Carpenter, Loyola University Archive, New Orleans; Steven Tomlinson, Department of Special Collections and Western Manuscripts, and Colin Harris, Bodleian Library, Oxford; Barbara Luszczynska, Mathematics Library, Washington University; Barbara Cain, North Carolina State Archives; Margaret Kimball, Archives, Stanford University and Ryan Max Steinberg and Polly Armstrong, Department of Special Collections, Stanford University Libraries; Carole Prietto, Archives, Washington University; Tom Owen, University Archives and Records Center, University of Louisville; Terry Abraham, Special Collections and Archives, University of Idaho; Lisa Coats and James Fein, Archives, Institute for Advanced Study, Princeton University; Christine Turner, Archives, Mudd Library, Princeton University; Donald Skemer, Rare Books and Manuscript Collection, Firestone Library, Princeton University; Winifred Okamitsu, Archive, the Historical Society of Princeton; Leverett Smith, Jr., Department of English, NC Wesleyan College; Mary Hickey, Academic Library, St. Mary's University; Robin Witmore, Lick Observatory; Bob Donahue, Mt. Wilson Observatory; Judy Larson and Maria Dora Guerra, Our Lady of the Lake University Library; J. Frank Cook and Steve Masar, University of Wisconsin-Madison Archives; Barbie McConnell, Mathematics Library, University of Wisconsin-Madison; Jane Turner, Archives, University of Victoria Libraries; Susan Storch, Archives, University of Oregon Library; Professor James Rovnyak, Mathematics Department, University of Virginia; Jill Fatzer, Library, University of New Orleans; Jill Jackson, Special Collections and Archives, University of Texas at San Antonio; Talar Kizirian, Harvard University Archives; Patrick Quinn, Archives, Northwestern University; Katherine Markee, Special Collections Library, Purdue University; Diane Kaplan, Manuscripts and Archives, Yale University Library; Walter Gerald Heverly, Archives of Scientific Philosophy, University of Pittsburgh Libraries; Kathleen Whalen, Bryn Mawr College Library; Gugliotta, Archives, The University of New Mexico General Library; Elaine Engst, Rare and Manuscript Collections, Kroch Library, Cornell University; R. Todd Crumley, Duke University Archives; Dr. Elisabeth Leedham-Green, Cambridge University Archives; Janice Sabec, Trinity University Archives; Karen Gillum, Olin Science Library, Colby College; John Weeren, Seeley G. Mudd Manuscript Library, Princeton University; Christopher Rooney and Martin Hackett, University of Pennsylvania Archives; Meg Spencer, Cornell Library of Science and Engineering, Swarthmore College; Mary Sampson, Archive, The Royal Society London; E. Poulle, Académie internationale d'histoire des sciences, Paris; Donald L. Singer, Modern Military Records, National Archives, Maryland.

My sincere thanks go to the following authorities, archives, libraries, academies and museums, who granted me insight into their holdings and/or permission for publication:

Harvard University Archives; California Institute of Technology Archives; Archives of the University of California at Berkeley; The Center for American History of the University of Texas at Austin; Department of Special Collections of Stanford University Libraries; New York University Archives; The Ohio State University Archives; University of Wisconsin-Madison Archives; Special Collections and Archives of the University of Texas at San Antonio; Archives of the Institute for Advanced Study at Princeton; Archiv der Rheinischen Friedrich-Wilhelms-Universität Bonn; Archiv der Georg-August-Universität Göttingen; Universitätsarchiv Freiburg; Archiv der Ruprecht-Karls-Universität Heidelberg; Philosophisches Archiv der Universität Konstanz; Archiv der Humboldt-Universität zu Berlin; Archiv der Ludwig-Maximilians-Universität München; Archiv der Technischen Universität Wien; University of Athens Archives; University of Wrocław Archives; University of Liverpool Archives; Brouwer Archive; Archives of Scientific Philosophy at the University of Pittsburgh Libraries; Manuscript Division of The Library of Congress; The Historical Society of Princeton Archive; Abteilung Handschriften, Alte Drucke und Rara der Universitätsbibliothek Freiburg im Breisgau; Handschriften- und Rara-Abteilung der Universitäts- und Landesbibliothek Bonn; Abteilung für Handschriften und seltene Drucke der Niedersächsischen Staats- und Universitätsbibliothek Göttingen; Handschriften- und Inkunabelabteilung der Bayerischen Staatsbibliothek München; Handschriftenabteilung der Staatsbibliothek zu Berlin – Preußischer Kulturbesitz; Bibliothek des Mathematischen Instituts der Universität Göttingen; Archive des Deutschen Museums; Historical Archive of Samos; US National Archives and Records Administration; Staatsarchiv Hamburg; Niedersächsisches Hauptstaatsarchiv Hannover; Staatsarchiv Augsburg; Bayerisches Hauptstaatsarchiv; Archiv des Instituts für Zeitgeschichte in München; Amtsgericht München; Bundesarchiv in Berlin; Zentrale Stelle der Landesjustizverwaltungen in Ludwigsburg; Deutsche Dienststelle für die Benachrichtigung der nächsten Angehörigen von Gefallenen der ehemaligen deutschen Wehrmacht; Der Bundesbeauftragte für die Unterlagen des Staatssicherheitsdienstes der ehemaligen Deutschen Demokratischen Republik; Bundesministerium für Inneres der Republik Österreich; Dokumentationszentrum des Bundes jüdischer Verfolgter des Naziregimes in Wien; Wissenschaftshistorische Sammlungen der ETH-Bibliothek Zürich; Department of Manuscripts and Archives of the Jewish National and University Library; Archiv der Bayerischen Akademie der Wissenschaften; Archiv der Akademie der Wissenschaften zu Göttingen; Archiv der Österreichischen Akademie der Wissenschaften; Benakis Museum; Deutsches Museum; Architectural Museum of Wrocław; Museum of Mechelen; Academy of Athens.

I am grateful to my husband Sokratis Georgiadis for moral and financial support and for his photographs; to my son Spyridon for practical help with classification of archival material and linguistic corrections; to the Program in Hellenic Studies of Princeton University for a Stanley Seeger Visiting Fellowship in support of this research; to the copy editors Dr. Victoria Wicks and especially Chris Weeks for their conscientious, detailed and critical copy editing; to the production editor Karl-Friedrich Koch for the setting and layout of this book; to the Springer publishers and the mathematics editorial in Heidelberg, especially to Dr. Catriona Byrne, senior editor in mathematics, and Susanne Denskus, editorial assistant, for their persistence in the completion of this project and their reliable and substantial support all the way long; finally, to the many anonymous reviewers for their critical view of my work.

Maria Georgiadou

Contents

CHAPTER 3

The Asia-Minor Project

CHAPTER 4

A Scholar of World Reputation

CHAPTER 5

National Socialism and War

CHAPTER 6

The Final Years

CHAPTER I

Origin and Formative Years

1.1

From Chios to Livorno and Marseille

Constantin Carathéodory was a cosmopolitan of Greek origin, whose intellectual profile was moulded at the cross roads of European and Oriental culture. On his mother's side he stemmed from the Greek bourgeois culture of the Diaspora, which originated in the modern Greek settlements[1] (παροικίες) and then spread, in the 18th and early 19th centuries to important trade centres of the Ottoman Empire around the Mediterranean and also to European capitals, especially Vienna and Paris.

Based on trade and international commercial shipping, this culture was characterised by its openness and cosmopolitanism. Its moving force, the Greek bourgeoisie, was familiar with Western tradition and manners and promoted Western education thus aiding the transfer of Western culture and way of life into the Orient. The so-called Modern Greek Enlightenment (Νεοελληνικός Διαφωτισμός)[2] of the years 1750–1821 sprang out of the Greek Diaspora culture. The Greek intellectuals of the Enlightenment promoted social and national liberation in the period leading to the Greek Revolution through the use of scientific reasoning and also by adopting the characteristic style, way of expression and ideas of the European Enlightenment. These intellectuals had studied at the universities of Europe and had a broad educational and cultural background.

Despite its European origin, the Modern Greek Enlightenment was strikingly different in being rooted in the idea of a transfer of knowledge rather than a self-generated development of science. Constrained by, or maybe also because of, respect for old traditions, especially those of Aristotle and, conversely, those of the Eastern church, the Greek philosophers of the Enlightenment tried to find a compromise between a scientifically founded knowledge of nature and a metaphysically founded or specific Christian cosmology. This ambivalence hindered the essential adoption of enlightened ideas; the transformation to the Greek context led instead to their attenuation. Despite all attempts to develop a Greek alternative discourse – extensive philosophical and scientific texts by Greek intellectuals of the Enlightenment bear witness to this – the daring intellectual balancing act did not succeed in the end. The Diaspora produced an additional insurmountable difficulty, namely the absence of nationally organised and institutionalised knowledge.

The Petrocochinos, Constantin's family on his mother's side, were without exception businessmen and traders belonging to the ethnically mixed Chiot upper class, who used their relative political freedom and the autonomous status of the island of Chios under Ottoman rule in order to promote their financial interests. Many travellers have recounted their impressions of the famous production of mastic and the abundance of fruit, silk and linen textiles in this "paradise of the Orient" and admired the wealth and luxury, as well as the high educational background, of its people.[3] The prosperity of the Petrocochino family was so great that they even possessed their own church, the church of St. Basil.[4]

By 1822, when Chios was invaded and laid waste by the Turks, the island comprised over sixty-six villages and local markets, over three hundred monasteries and six hundred churches. The Turkish repression was brutal: six male members of the Petrocochino family were held hostage in the fort and hanged by the Turks. Another one died of hardship in the fort before he could be executed, while another was released for a ransom to the Americans and was able to reach Boston in 1824. It is not known how other members of the Petrocochino family escaped the massacre. They may have been set free in exchange for a ransom and fled the island, like other Greek families, on Greek ships from Psara. Somewhat later, we find in Rio de Janeiro a flourishing Greek settlement, whose leading members included the names of the Chiot refugees, Petrocochino and Rodocanachis, to whom the Carathéodorys were related by marriage.[5]

After their flight, Carathéodory's great-grandparents, then newly married, settled first in Livorno, and some years later in Marseille. The small Greek settlement of Livorno consisted of wealthy Greeks, who provided grants and scholarships for their young countrymen, graduates of the schools of Athens, Ioannina and Chios, to study in Italy, especially in Pisa. The community also gave material support to the Greek War of Independence. By 1816 Chiots such as Michael Petrocochino and Michael Rodocanachis were trustees of the Oriental Church in Livorno. Their families had gained social prestige through their ownership of real estate and their luxurious life-style, by becoming part of a powerful social network and, significantly, by expending time and money on philanthropic and cultural activities. The Rodocanachis's businesses, combining international trade with banking and insurance activities, dominated Livorno's financial and commercial life throughout the 19th century. Carathéodory mentions Livorno in a letter to Professor Kalitsounakis[6] as a place regularly visited by Greek traders before 1821. He wrote that Livorno had a greater significance than Marseille for Greek merchants during the period of the Continental System.[7] The merchants were better able to trade there with grain from Russia, since they could exploit fluctuations in the French customs tariffs on account of Livorno's proximity to France. Greek merchants who had settled in French Mediterranean harbours, of which Marseille was the most famous, were active in the foreign trade of the Ottoman Empire. France played a crucial role in this trade in the 18th century but there came a point when the French began to fear the loss of their share in the Levant trade to the Greek merchants of Marseille and to the Greek trading fleet.[8]

Later, when the French Revolution and the Napoleonic Wars brought the trading activities of the French to a standstill, the Greek sea traders filled the gap. This activity led to a Greek "stranglehold"[9] on a large part of the Ottoman sea trade with

Europe and to a decisive presence of the Greek fleet in the Mediterranean, which was later to the benefit of the Greek War of Independence that broke out in 1821. The Greek trading fleet was to a great extent superior to the Ottoman fleet, being equipped with cannons and enjoying the exclusive privilege of being allowed to transport ammunition.

Many Greek refugees of the Turkish persecution, who had settled down in Marseille, returned to Greece as soon as she gained her independence through the London Treaty of 3 February 1830. Others, among them the Petrocochino family, remained in Marseille and strengthened the Greek settlement there. This settlement, and the one in Alexandria, had been founded in the same urban Mediterranean centres of the Greek colonies of Antiquity. Their spectacular financial growth in the 1840s was due to the transport of grain from southern Europe and the eastern Mediterranean to western Europe. This was the time of the creation of the great cargo, ship-insurance, and ship-owner companies.

Carathéodory remembers that the house belonging to the Petrocochino family in Marseille "was in summer the meeting point for numerous descendants scattered all over the world so that over many years [he] had the opportunity of playing with small male or female cousins who felt at home in Alexandria or Liverpool, in Odessa or Trieste, in Smyrna or London, in Braila or Paris, in Vienna or Constantinople and who were dragged along on the summer journey by their parents."[10]

1.2
The Carathéodorys in the Ottoman Empire

On his father's side Carathéodory belonged to a family rooted in the Phanariot tradition. The Phanariot society grew out of the mixing of the westernised Greek trade-bourgeoisie with the remaining core of the former Byzantine nobility that had occurred around the end of the 16th century. The name Phanariots (Φαναριώτες) comes from the northern section of Constantinople on the west coast of the Golden Horn where they lived and where the Phanar (φανός), or lighthouse, illuminated the imperial landing stage.

The Phanariots were scholars, diplomats and administrators in the Sultan's service, who had studied in the West and were the leading bearers of education among the Sultan's subjects in the Ottoman Empire. They held three vital systems of power in their hands: the Patriarchate; the posts of the Grand Dragoman of the Sublime Porte and the Grand Dragoman of the Fleet (the dragomans were, loosely, interpreters and foreign affairs advisors and the Sublime Porte is the name given to the government of the Ottoman Empire); and the thrones of the Danube Principalities of Moldavia and Wallachia. Their power reached its peak in the 18th century, the so-called Phanariot Epoch.[11] Essentially, they enabled the transfer of the European Enlightenment into the Balkans until at least 1790, but after the French Revolution they either mistrusted, or even opposed, the new ideas. Under Ottoman rule, the Phanariots developed into the most powerful intellectual presence of the Greeks and ensured the survival of Greek culture and the particular position of the Greeks between Orient and Occident. But at the same time, they grew into the notorious caste of technicians of power integrated in the Ottoman system of rule. Their diplomatic skills, their flexibility and their adaptability produced a peculiar character blend of *râya* (non-Muslim subjects of the Sultan), despot and gentleman. Their mobility was guaranteed by an impressive multilinguality; their political experience was accompanied by a broad education; their wealth, influence and authority were secured by risky intrigues.

Despite all these common characteristics, the Phanariots did not form a compact aristocracy, did not possess a collective consciousness nor a coherent ideology. They moved with the times, became increasingly numerous and shared in the melting pot of all the political and ideological movements of their age. Balkan nationalism, especially the Greek Revolution, on the one hand and political interests of the Great Powers on the other, hindered the realisation of their plans for a shared rule with the Ottomans in the Balkans and the Near East. By participating in the Greek War of Independence, they lost their numerous privileges, their titles, the great Dragomanies and the thrones of the Danube Hegemonies. They soon became embroiled in conflicts with the local administration and the leaders of the rebels in insurgent Greece, but managed in the end to take on a significant political role. After the foundation of the Greek state and at the moment when the Great Idea (*Μεγάλη Ιδέα*)[12] of the Greeks emerged as the ruling national ideology, the reputation of the Phanariots was historically restored and they were even celebrated as preachers of this Great Idea. It was, of course, those Phanariots who had settled in Athens and took on leading positions in the political, social and intellectual life of the young state, who contributed to this development.

Although much is known about the social and political role of the Phanariots and the Greek ruling class in the period 1711–1821, the Greeks who remained subjects of the Sultan after the Greek War of Independence have been little researched. It is remarkable that until the Balkan Wars (1912–1913) the majority of Greek people continued to live in territory ruled by the Ottomans. In addition, despite Ottoman reprisals against the Greeks after the Greek insurrection on the Peloponnese, the Ottoman Greeks, especially in the period 1860–1913, achieved a significant role in the financial, social and political affairs of the Empire. Some of these Ottoman Greeks occupied offices of influence within the Ottoman government in matters of foreign policy. Some diplomats and higher civil servants of Greek descent were even able to impose their will on vital affairs of the Ottoman Empire.[13] The Carathéodorys were a dynasty of Christian Ottoman Greeks who belonged to the upper stratum of the powerful urban elite of Constantinople. As Neo-Phanariots or Second Phanar they revived the Phanariot tradition by taking up state positions in the Sultan's service as doctors, administrators, politicians and diplomats when the city came under the financial and cultural influence of the West and Sultan Mahmud II was putting forward his West-oriented reform programme in the Empire. As members of this elite they worked for Ottoman purposes and supported the establishment.

Counter to the separatist activities of the Balkan peoples, including the Greeks, the Neo-Phanariots opposed the dismembering of the Ottoman Empire almost up until the Balkan Wars. They believed that a social, economic and political flourishing of the Greeks would only be possible if they remained connected with a modern and cosmopolitan Ottoman Empire friendly disposed towards the West. Within their national-religious community, the so-called *Rum millet*, or "Greek" *millet*, the Carathéodorys actively participated in the Mixed Council of the Oecumenical Patriarchate and the National Assemblies, an administrative body with jurisdiction over schools, hospitals and philanthropic establishments of the Greeks, and the Greek Philological Society of Constantinople (*Ελληνικός Φιλολογικός Σύλλογος Κωνσταντινουπόλεως*), a literary society propagating secular education among the Greeks of the Empire and gradually encouraging the idea of a Greek national ideology.[14]

1.3
Stephanos Carathéodory, the Father

Unlike engineering studies, which were promoted by the Ottomans in the first half of the 19th century, the study of European philosophy, law and political sciences remained blocked within the Ottoman Empire. These disciplines were considered a potential threat to the existing structures of power. A change occurred for the first time after the Paris peace settlement (1856) at the end of the Crimean War. The Western powers had supported the Ottomans in order to stop the Russian advance into the Dardanelles, but demanded in return a modernisation of the Empire. The wave of reform policies following the peace settlement gave rise to an apparatus in the Ottoman administration that was given the task of organising a state on modern foundations so as to enable it to consolidate its extended relations to Europe. The Ottoman delegations in Europe gained the status of embassies in 1834. Between 1840 and 1912 at least fifteen Greek diplomats were authorised by the Ottomans to represent the Sublime Porte in the West. Ten of Carathéodory's relatives were among them and headed legations in Athens, Vienna, London, Turin, Rome, Berlin, The Hague, Washington and Brussels. In his *Autobiographical Notes* Carathéodory remarked: "I have spent a little longer than might appear necessary on the Greeks who were active in the Turkish diplomatic service. But the whole atmosphere in which I have grown up would not be comprehensible if one did not know of the existence of these men, whose exceptional qualities were developed through the most peculiar interplay of unique historical events."[15]

Stephanos Carathéodory, the father of our subject, belonged to those Ottoman Greeks who served the Sublime Porte as diplomats. He had studied law in Berlin under the supervision of his uncle Ioannis Aristarchis, the Ottoman ambassador in Berlin. In 1860, he obtained his doctorate with a dissertation on international river law[16] and later became the secretary at the Ottoman embassies in Berlin, Stockholm and Vienna. In the period 1866–1871 he replaced the ambassador of the Ottoman Empire in St. Petersburg and directly afterwards was appointed to the post of the first secretary of the Ottoman Legation in Berlin where his son Constantin was born on 13 September 1873. Constantin was christened in the Russian Orthodox church in Berlin[17] with Ioannis Aristarchis as his godfather. Stephanos led a very intense social

The infant Constantin.
Courtesy of Mrs. Rodopoulos-Carathéodory.

life that grew as the years went by and well went beyond the narrow limits required of a diplomatic engagement. As his son writes in his *Autobiographical Notes*

in all the most significant cities of Europe from Portugal to Russia and from Sweden to Sicily he had relations and friends with whom he corresponded extensively. Among them were diplomats such as Earl Brandenburg, Earl Alvensleben, Prince Ouroussow, Lord Vivian, or the French envoy Bonrée, musicians such as Massenet or Paderewski, music historians such as Gevaert or Bourgault-Ducoudray, artists such as Paul Meyerheim, Emile Wanters or Constantin Meunier, international law experts such as the Swiss Alphonse Rivier, the Dutch Asser, or the Russian F. Martens. In addition, there were also Greeks, Albanians, Turks, Armenians, Egyptians, in a word, the whole contact with the Orient.[18]

Many of his contacts were honorary members of the Greek Philological Society of Constantinople. Constantin would get to know all the politicians and artists who came to their house. His cultural education was due to his father, who took him on travels everywhere in Europe, excepting only Spain and Russia, mostly to visit churches and museums.

Stephanos spent the year 1874–1875 in Constantinople on leave, granted for a personal service to the Sultan. Then, for the last quarter of the 19th century he was the Ottoman Ambassador in Brussels. He was dismissed from his post there in 1900 because of his failure to prevent the publication of a libel against the Sultan and he remained in Brussels until the end of his life.[19] Carathéodory's *Autobiographical Notes* contain no mention of his father's dismissal from the Sultan's service.

Stephanos Carathéodory around 1900.
Courtesy of Mrs. Rodopoulos-Carathéodory.

1.4

Early Years in Belgium

Constantin's formative years, then, were in Brussels. At the age of six, with his sister just four years old, he was bereaved of his mother Despina, daughter of Antonios Petrocochino. Despina died at the age of 28 of pneumonia in Cannes. Both children were then brought up by their grandmother Petrocochino, who, as Constantin mentions, kept house for their father "with the meticulous care typical of every Chiot housewife" right up to her death.[20] A German maid was also employed to look after the children and to teach them German. In Brussels, Constantin's school education and military training were conducted in French, which he spoke with the same fluency as he spoke Greek.

In 1881, Carathéodory was sent to the private school of Vanderstock, where he remained for only two years. He then accompanied his father to Berlin, where they enjoyed the hospitality of Gustav Richter[21] in the house at 6 Bellevue Street, which the painter had inherited from his father-in-law, the composer Meyerbeer.[22] The outstanding Berlin artist, whose favourite occupation was painting the portraits of the high society of his time, had also painted the portrait of Constantin's mother, Despina.

Despina Petrocochino-Carathéodory.
Courtesy of Mrs. Rodopoulos-Carathéodory.

Constantin at the age of 15.
Courtesy of Mrs. Rodopoulos-Carathéodory.

Constantin with his father on graduation day.
Courtesy of Mrs. Rodopoulos-Carathéodory.

The Carathéodory family spent the winter months of the years 1883–1884 and 1884–1885 on the Italian riviera in Bordighera and San Remo. After returning to Brussels, Constantin was sent for a year to a secondary school where no Latin was taught. This lack of Latin was to become significant to him and he later made up for it by starting to learn Latin at the age of 40. His love for mathematics began to develop in the geometry lessons. He subsequently attended the *Athénée Royal d'Ixelles* grammar school in the years 1886–1891. In the general competition for all Belgian grammar schools, he was twice awarded the first prize in mathematics, consisting of Jules Verne's books that still exist in the remains of his library. After his final examinations he entered the *École Militaire de Belgique* as *élève étranger* in the *57ᵉ promotion d'artillerie et du génie* (foreign student of the 57th year of entry to the school of artillery and engineering) which he attended in the years 1891–1895 and he also attended the *École d'Application* from 1893 to 1896.

*Constantin with relatives in Constantinople in 1896. From left to right: Stephanos,
Euphrosyne's brother; Pavlos, Euphrosyne's brother; Alexander, Euphrosyne's cousin;
Constantin. Courtesy of Mrs. Rodopoulos-Carathéodory.*

The Belgian Military School was modelled on the *École Polytechnique* in Paris.
Its first commander, Chapelié, had stressed the scientific aspect of training, appointing
the best scholars he could find, such as Dandelin, the Belgian statistician Adolphe
Quetelet and the chemist Stas as lecturers, and outstanding results were achieved
by the students. Carathéodory attached great significance to the foundation of the
École Polytechnique during the French Revolution, particularly because Monge had
lectured there and breathed new life into the study of geometry.[23] One of the graduates
of the *École Polytechnique* was Carathéodory's great uncle Konstantin Carathéodory,
an engineer who specialised in bridge construction. Konstantin, as head of Public
Works in the Ottoman Empire, represented it in the European Commission of Danube
(CED = *Commission Européenne du Danube*),[24] which at the end of the Crimean
War (1853–1856) was given the task of administering the navigation of the lower
Danube and its three branches flowing into the Black Sea.

Carathéodory received a thorough technical and mathematical training at the Belgian Military School under the direction of General Leman, who later commanded the defence Liège at the beginning of the First World War. Quartered in barracks, and following a strict programme of drilling and riding, Carathéodory attended lectures and took examinations. He considered the infinitesimal calculus course, based on the work of Charles-François Sturm, to be antiquated. By contrast, he enjoyed the teaching of Chomé in descriptive geometry, since Chomé presented the subject in the spirit of Monge, namely integrating it with the theory of surfaces. Carathéodory learned to use geometry as a kind of game, with the help of which he could approach different problems. In the lectures on mechanics he learned how to tackle the most difficult problems skilfully. Charles Lagrange, one of the best teachers in his eyes, held lectures in probability calculus, astronomy and geodesy, while his younger brother Edouard lectured on thermodynamics, a subject that immediately caught the attention of Carathéodory and never left him. He became close friends with the Lagrange brothers and their family. The purely technical instruction given by highly experienced officers of the engineer corps was also of high quality. Carathéodory's rich experience at the school had a considerable influence on his scientific career.

By specialising as an engineer, Carathéodory followed a family tradition that went back to the first half of the 19th century, when the rulers of Muslim countries actively engaged in promoting the transfer of knowledge, especially in engineering, to the Ottoman Empire. It was for this reason that members of the Carathéodory family, such as cousin James Aristarchis, great uncle Konstantin Carathéodory and uncle Telemachos Carathéodory, had studied engineering in various European capitals.

1.5
The Graeco-Turkish War of 1897

Shortly after Carathéodory finished his studies at the military school, the Graeco-Turkish War of 1897 broke out. Although a young military man, Carathéodory did not participate in this war. However, he did not hesitate to state his views on the events. It was the first time that he expressed himself in matters of Greek policies.

In the last four decades of the 19th century, and especially in the final decade, Greece had experienced a strengthening of the parliamentary system, financial growth, rationalisation of the administration, extension of the infrastructure and considerable cultural achievements. The driving force was the Greek bourgeoisie, who had given up their exclusive orientation towards international trade in order to invest at home in industry, mining and finances. Politically, reform was pushed through by the liberal prime minister Charilaos Trikoupis, but its financing resulted in heavy debts for the state and finally in the state bankruptcy of 1893.

The rise of the bourgeoisie in the final decades of the century was followed by a strengthening of national consciousness, avidly embraced by all sections of society, from the King and the prosperous classes to the general populace, and generated a mood of nationalism that aspired to Greek irredentism either in Crete or Epirus, in Macedonia or Thrace. But any attempt to realise these ambitions was bound to fail, being based on a completely wrong estimate of the balance of power in the region combined with an underestimate of the power of the Ottoman army.

The outbreak in 1869 of rebellion on Crete, partly fomented by the secret Greek nationalist society *Ethniki Etairia* (*Εθνική Εταιρεία*), appeared to present Greece with an opportunity to annex the island. By the beginning of 1897, large consignments of arms were sent to Crete from Greece. On 21 January the Greek fleet was mobilised and in early February Greek troops landed on the island and its union with Greece was proclaimed. The following month, however, the European powers, fearing that disturbances would spread to the Balkans, imposed a blockade on Greece to prevent assistance being sent from the mainland to the island.

On 3 [15] April 1897, the Ottoman Empire declared war on Greece because of frontier violations, and two days later the Greek Ambassador in Constantinople, Nikolaos Mavrocordato, left Turkey. After a four-week expedition in Thessaly, commanded by Crown Prince Konstantin, the Graeco-Turkish War ended with the Greek army suffering a devastating defeat. Inadequately prepared for war, the Greeks had been overwhelmed by a Turkish army which had recently been reorganised under German supervision. Greece had no option but put her fate in the hands of the Great Powers who mediated an armistice agreement. Greece had to yield some territory in Thessaly to the Ottoman Empire, withdraw her troops from Crete, pay the Ottomans an indemnity and accept an international financial commission that would control Greek finances.

At the beginning of the war, Carathéodory wrote a letter from Athens to his cousin Stephanos, son of Alexander Carathéodory Pasha. Albeit sceptical, he tried to appear confident about future victory for the Greeks.

Dearest Stephanos,

the news is not so good today. A great battle was to take place yesterday, seemingly decisive for the fate of Thessaly. Our own people fight like lions, but what for? However, I hope that the final result will turn out in our favour. I believe I told you in Br[ussels] that the Turks were going to gain some victories in the beginning. I had not believed that they were going to endanger us greatly. But from a political point of view, I see things in a different way today. Criminal people spread rumours about the royal family in order to achieve their extremely egoistic goals. They say that G[eorge, second oldest son of King George I] is a coward, that he had disobeyed the commands of Sachtouris [commander of the fleet]. They say that N[ikolaos, son of the King] is frightened, they tell … stories about everyone! If the Turks cross the borders to Thessaly, I am sure that the nation will be ready to make sacrifices to chase them away and that it will achieve it. On the other hand, I am not sure about the fate of the local government. The willingness to make sacrifices is comforting. In residences there is a lack of servants. I know of people who were freed from military service and could not enlist but, because the government needed them, they volunteered under a false name and moved up to the borders. I know of families who have five children at the borders, etc. The same 'clique' who slander the r[oyal] f[amily], do the same as regards the fleet and claim that it is good for nothing, that three torpedo boats are already useless, since they have been damaged on their way to Crete. They say that the ships had no precise destination. Finally, they spread the news that the Turks prepare to land in Piraeus! […]

I kiss you, etc. K. [C. Carathéodory].

I ask you to show this letter at home. You might remember, that I have shown you a bridge on the map to be blown up; they have tried to blow it up but I do not know whether they were successful.[25]

Carathéodory was asking his cousin to show the letter to his father, Stephanos, the then Ottoman ambassador in Brussels! Two days after Carathéodory's letter, Turkish troops entered Larisa, which had been abandoned by the Greeks.

Later, on 28 April [10 May], when the Greek retreat was a fact and defeat more than obvious, Carathéodory revised his views of the royal family in a second letter to his cousin.[26] He considered the war the greatest disaster for Greece and the entire Hellenism, maybe the "coup de grâce". He wrote that political corruption in Greece had drawn the royal family into its current, something he could not have known, since he was living outside the country. Carathéodory felt compassion only for Sophia, third daughter of Kaiser Friedrich III and wife of Crown Prince Konstantin, and he realised that she was enjoying the people's sympathy. Carathéodory considered everyone to be responsible for the war and to have failed when the country was in crisis. In his opinion, the country lacked political personalities with the necessary will, intelligence and patriotism. Greece had been driven to war completely unprepared and the inexplicable apathy of the armed forces, especially that of the fleet, evoked the suspicion that the war had been undertaken for purposes of foreign policy.

It is true that Russia had used that war to expand its influence in Greece. It is further noteworthy that the Turcophile Kaiser Wilhelm II, despite the fact that he had been brother-in-law to Crown Prince Konstantin since 1889, adopted the sharpest tone against Greece, while England and Italy expressed their sympathy for the Greeks. Carathéodory believed that only the presence of the Turks in Athens or, even better, the installation of a scaffold in the centre of the Greek capital would force politicians to reflect on the situation. He favoured the resignation of the King's sons Nikolaos and George from the officers' corps and the dismissal of the crown prince and army commander-in-chief Konstantin from the general staff. With these views he expressed the mood of the Greeks at that time. Even the royalist Ion Dragoumis stated that the country had become victim of a foreign pseudo-policy under the leadership of a King who himself was not a Greek. Also the royalist prime minister, Dimitrios Rallis, whom Carathéodory reproached for having great ambitions and small political weight, spoke of an attack against the nation and the armed forces and accused King George and his sons of treachery. The national consensus, finding the cause of the Greek defeat in dishonesty and treason, was reflected in the pages of the satirical journal *Romios* (Ρωμηός = Modern Greek), issues of which accumulated in Carathéodory's personal library.

While Carathéodory was showing concern for the fate of Greece, other members of his family were still serving the Sublime Porte. When, in 1898, the candidacy for the post of the future Governor General of Crete, then under military control of the Great Powers, was being discussed, Prince George, second son of the King of Greece, was proposed as commissar by Russia, Great Britain, France and Italy. The Porte on the other hand, favoured the candidacy of Alexander Carathéodory Pasha, Carathéodory's future father-in-law, who had already failed once in this office. In the end, the Turkish troops left the island, Crete was made an international protectorate and an autonomous government under Prince George was established. Finally, the island was ceded to Greece under the Treaty of London (1913), thus concluding the first Balkan War.

I.6
With the British Colonial Service in Egypt

In 1897, Carathéodory spent his time in Athens and other European cities such as Trieste, Semmering, Vienna, Dresden, Brussels, London and Paris. His name is mentioned in the minutes of the General Assembly of Samos[27] with respect to the debate of December 1897 on the road construction of the island. Carathéodory himself reports that he was helping his cousin Ioannis Aristarchis, alias James Bey, chief engineer of the archipelago in the service of the Sublime Porte, to design the road network. Aristarchis had studied at Collège Sainte-Barbe in Paris, at Oxford University and at Zurich Polytechnic. While in the service of the *Divan* (the Ottoman State Council) he had constructed roads in the islands of Lesbos, Chios, Ikaria and Rhodes. The minutes of the General Assembly of Samos contain the additional information that Carathéodory and the local authorities were bound by a provisional contract but the commission was finally given to one of his competitors. Carathéodory himself mentions in his *Autobiographical Notes* that this project was not realised because of the Graeco-Turkish War of 1897.

But Carathéodory experienced more luck in London that year, when he was offered a job as engineer in the British colonial service. The following year he travelled to Egypt to work on the building of the Assiut dam, an 833 metre long and 12.5 metre high construction. He remained there until April 1900, suffering an unfavourable climate and poor hygienic conditions. At the site of the first Aswan (Assuan) dam, which was being erected at the same time, 200 000 people were diseased with Rift-valley fever, malaria and bilharzia. Reservoirs used as parts of the irrigation system were a breeding ground for viruses and parasites. From September to January flooding brought almost every building activity to a standstill. Carathéodory used these interruptions to his engineering work to devote himself to something that was more and more alluring to him, the study of mathematics. In later years, he repeatedly explained that he owed the greatest part of his mathematical knowledge to Camille Jordan's *Cours d'Analyse*.[28] The mathematician Heinrich Behnke reports[29] that he himself experienced how, in 1918, Carathéodory, then professor in Göttingen, recommended the *Cours d'Analyse* to demanding young students of mathematics. Carathéodory's own study in Egypt was complemented by the *Analytische Geometrie der Kegelschnitte* (Analytical Geometry of Conic Sections)[30] by George Salmon and Wilhelm Fiedler, which was considered to be the best book on analytic geometry in Germany, and this too he recommended to students attending his course in projective geometry in Göttingen in the summer semester of 1917.

In Egypt, Carathéodory had the opportunity of meeting renowned Englishmen. In Assiut, a city with no European society, his social circle consisted of the project leader, Stephens, and another four or five colleagues. The most talented among them was a descendant of Admiral Lord Samuel Hood, commander of the British fleet at the beginning of the French Revolution and one of the first patrons of Nelson. Carathéodory's friend Hood had pursued a career as an engineer. The particular geography of upper Egypt, a narrow strip 1000 km long and not much more than 12 km wide, meant that there was little space for habitation, so this small company of engineers often received house guests, who stayed with them for a few days. In this way, Carathéodory made the acquaintance of Sir Benjamin Baker, one of the most famous engineers of his time, architect of the Aswan barrage completed in 1902 and designer, with J. Fowler, of the magnificent steel railway bridge across the Firth of Forth in Scotland. Carathéodory also met the archaeologist and historian Archibald Henry Sayce, one of the founders of Assyriology; they met again at Queen's College in Oxford many years later. Carathéodory learned from him about the German archaeologist Heinrich Schliemann, a friend of Sayce.

In February 1900, Carathéodory visited the building site of the Aswan dam, and the Nile island Philae with its sanctuary consisting of four temples – still above the watermark – at which pilgrims gathered in ancient times to worship the mysterious goddess Isis. For a few days he also stayed in Luxor, the site with the great Amon temple on the east bank of Nile, 725 km south of Cairo. He spent a night in the house of the archaeologist Howard Carter, still then completely unknown, who was head of the local police near Madinat-Habu (temple of the dead), the "most beautiful Egyptian temple" in Carathéodory's opinion, built during the reign of Ramses III in an already sacred area. Later, Carter carried out excavations in the necropolis of Thebes, where he discovered among others the royal tombs of Amenhotep I, Hatshepsut and Thutmose IV and, in 1922, the tomb of Tutankhamen, which was the richest and most complete burial site in Egypt. In a footnote to his *Autobiographical Notes*, Carathéodory described the accidental discovery of the first royal burial chamber carved in the rock:

After a downpour in this country, in which it rains briefly only once every two or three years, the horse of Howard Carter slipped and Carter realised that the earth did not consist of rocks at this place of some square meters. He asked his superior authority to allow him to proceed to excavations, was given three English pounds (around 63 Reich marks) towards it and in this way he reached the grave of Mentuhotep II, which, according to Baedeker's Egypt, 8th edit., 1929, p. 323, is still called by the Arabs Bab-el-Hassan, i.e. the Porte of the Horse. It is in the forecourt of the temple of Kings Mentuhotep II and III, which was excavated only later (1905–07).[31]

While in Egypt, Carathéodory carried out measurements at the main entrance of the Cheops pyramid. His results are in the article *Nouvelles mesures du mur sud de la grande galerie de la grande pyramide de Cheops* (New Measurements of the South Wall of the Great Gallery of the Great Pyramid of Cheops),[32] written in Berlin on 6 December 1900 and published in Brussels in 1901. Comparing his own measurements with those of his contemporaries, the British pyramidologist C. Piazzi Smith, who saw in the great pyramid divinely inspired prophecies and codes confirming Biblical prophecy, and Sir William Matthew Flinders Petrie, the British archaeologist and Egyptologist and founder of the scientific excavation methods in Egypt, he found only slight differences. In his paper he says that he had carried out the measurements, at the request of Charles Lagrange, and with the aid of Etienne Zigada, on 2 April 1900. Carathéodory's work confirmed the results of Smith and Petrie who, as he remarked, had contributed to making the grand pyramid known in a definite and exact way by utilising the modern means of science and technology.

NOUVELLES MESURES DU MUR SUD

DE LA

GRANDE GALERIE DE LA GRANDE PYRAMIDE DE CHEOPS

PAR

C. CARATHEODORY

ANCIEN ÉLÈVE DE L'ÉCOLE MILITAIRE DE BELGIQUE

BRUXELLES

HAYEZ, IMPRIMEUR DE L'ACADÉMIE ROYALE DE BELGIQUE

rue de Louvain, 112

1901

Carathéodory's paper on the measurements of the great pyramid.
Courtesy of the Bavarian Academy of Sciences, Archive.

During his two-year stay in Egypt, Carathéodory had the opportunity of meeting relatives on his mother's side living in Alexandria and to become fairly well acquainted with the local Greek society. The Greek presence in Alexandria goes back to the 16th century. In the 19th century, Greeks in Alexandria financially supported the Greek War of Independence and some of them were members of the Friendly Society (*Φιλική Εταιρεία*).[33] The Greek settlement of Alexandria experienced a significant increase in numbers when, towards the middle of 19th century, numerous Greeks settled in the city. They were part of an influx of foreigners, among them Jews, Armenians and Central Europeans of various nationalities, who reached Alexandria after Mehmed Ali Pasha decided to extend the Egyptian cotton trade. Cairo, Port Said and other centres of cotton cultivation and packing also attracted Greeks as workers, tradesmen and businessmen. Some of the most well-known Greek families established their enterprises in Egypt at that time and became distinguished as traders of oil, tobacco, grain and cotton. The demand for Egyptian cotton increased significantly in European markets after the decline in availability of American cotton as a consequence of the American civil war. Towards the end of the 19th century, as much as half the cotton crop and the cleansing, packing and exporting of cotton was in Greek hands and the same held for many other branches of Egyptian domestic trade.

An additional significant factor for the flourishing of Greek settlements lay in the beginning of work on the Suez Canal. Egyptian, Syrian and European workers and engineers were engaged in the construction of the canal, two-thirds of the latter being Greeks. Not only did the Greeks comprise the most numerous community in Egyptian urban centres, but they were also the group holding a major part of the trade and stock market activities in Egypt up to the beginning of World War I.

During the 19th century, the Greek "national benefactors" (*εθνικοί ευεργέτες*) began to play an increasingly important role. These were members of the Greek settlements who financed the foundation of schools and the building of stadiums, museums, churches and orphanages in the Greek state, primarily in their places of birth but also in their places of residence. However, the only recipients of their charity were Greeks or those foreigners who intended to study the Greek language. The national benefactor Emmanuel Benakis[34] held a prominent position among the wealthy Greeks of Alexandria. He was a cosmopolitan millionaire who had made his fortune in the Egyptian cotton trade, mostly with England, running the largest enterprise of Greek cotton brokers in Egypt. He was the president of the Greek community of Alexandria, the most wealthy, exclusive and elitist foreign community of a city with both the most emphatic Greek character in the whole Middle East and the most intense cosmopolitan features. Benakis's trade activities determined his ideology and his political proximity to the future premier of Greece, Eleutherios Venizelos,[35] who, in 1910, put Benakis in charge of the newly founded Ministry of National Economy. Benakis's daughter, the famous writer of children stories and publicist Penelope S. Delta, was not only a close friend of Carathéodory but also distantly related: her husband's mother was Carathéodory's father-in-law's sister.

One of the most significant Greek poets of modern times, the best known and most translated, was Konstantinos Kavafis who was born in Alexandria in 1863, the same year in which the company Choremis, Benakis & Co. was founded in order to exploit the boom in Egyptian cotton. Kavafis was in fact distantly related to Carathéodory's father and was living in Alexandria when Carathéodory visited the city. There is much in common between the two men. Intellectually, Kavafis shared Carathéodory's preference for history based on extensive study of sources and very wide reading. The poet used history as his poetic raw material, while the mathematician used history as pre-existing knowledge. Both men were precise, laconic and sober in their expression; both used irony; both pursued a kind of linguistic eclecticism, the poet based on demotic Greek (δημοτική), the mathematician, at least in his scientific writings, on purified language (καθαρεύουσα). Both also worked for the British, Kavafis in the Egyptian Irrigation Office, Carathéodory on the construction of the Assiut dam, and both men had lived for a while in Constantinople. While Carathéodory was in Athens in the spring of 1932 at the invitation of Venizelos, Kavafis paid his final visit to the Greek capital, but it seems that they did not meet, and a year later Kavafis died in Alexandria. Despite their similarities, and the fact that they were distantly related, it seems that the paths of these two men never crossed, neither in Alexandria nor in Athens. Carathéodory might well have known of Kavafis, of course, but at that time the reception of Kavafis's poetry was extremely controversial and Carathéodory seems to have been rather conservative in his preferences in the field of art.

Although in the early 19th century the Greeks were under Ottoman rule and, after the War of Independence, the Greek Kingdom was subjected to political control by the Great Powers, the Greeks of the settlements considered themselves a constituent part of the colonial rule in the Orient and were advocates of the Great Idea. Egypt served them as a starting point for their expansion into the Near East, since it harboured the Greek trade fleet. Carathéodory appealed to "the wealthy members of the middle class of Greek Egypt," as he called them, to take the first place in the emerging industry, which would necessarily fall into European hands, since, as he asserted, the Egyptians themselves knew nothing but the cultivation and irrigation of the ground. Carathéodory expressed this exhortation at the end of his book *Egypt* (*Η Αίγυπτος*),[36] which he wrote in Berlin. The work contained information on the country's geography, including a description of the Nile and its properties, on the climate, watering system and agriculture, on the political, military and cultural history of Egypt, on the archaeology, architecture, landscape, population and conditions of life, religion and every-day culture, flora and fauna, diseases, social divisions, colonialism, technical achievements, such as the Suez Canal or the Assiut dam, administration, European foreigners and, especially, the Greek settlements in the country. Carathéodory's library contained the first edition of the seminal *Description de l'Egypte*, published on Napoleon's instructions in Paris in the years 1809–1822. Carathéodory was probably inspired by this family heirloom, since, in his own book, he dealt with similar subjects to those of the earlier publication.

Egypt was published in 1901 as the 14th volume of a series edited by the Society for the Promotion of Useful Books (Σύλλογος προς διάδοσιν ωφελίμων βιβλίων). Leading members of the society included Stephanos Deltas, husband of Penelope Delta, Markos Dragoumis, politician and Carathéodory's friend, Georg von Streit, Carathéodory's brother-in-law, and also included many poets, such as Kostis Palamas,

Aristomenis Provelengios and Georgios Drosinis. The society was founded in Athens by Dimitrios Vikelas in May 1899 following the example of French, English and Swiss societies. It set the goal of using education to help the Greek people regain their national pride after their defeat in the Graeco-Turkish War of 1897. Markos Dragoumis, who had studied in Paris, had been a member of the Greek diplomatic corps for thirty years and was remarkably active in support of the Greek national movement of Macedonia at the beginning of the 20th century. He published the *Bulletin d'Orient* in French to inform Europeans about the struggle for Macedonia. Dimitrios Vikelas, who had studied in London, maintained close contacts with the members of the Greek community there; it was in London that he began his literary activity. In 1872 Vikelas moved to Paris and four years later settled in Athens. He was instrumental in securing Athens as the home for the first modern Olympic Games in 1896.

1.7
Studies in Berlin

Markos Dragoumis and Dimitrios Vikelas, Carathéodory's older friends, criticised his decision in 1900 to give up his engineering profession in order to study mathematics. They believed he was turning his back on a promising and secure career in order to pursue a dream. Carathéodory's family were equally outraged, since they thought that his studies in mathematics would, at best, lead merely to a teaching career in a secondary school and nothing more. Carathéodory himself was not convinced that his plan would come to a happy conclusion, but he could not resist the persistent idea that only his unbridled preoccupation with mathematics would give meaning to his life. He was supported in this decision only by his great uncle, Alexander Carathéodory Pasha, an Ottoman Greek who had served the Sultan in the post of the foreign minister. John Horváth recollects, from what he has learned from his teacher Fejér, that Carathéodory went on vacation to Europe and visited his great uncle in Berlin during the academic year 1899–1900. Since Carathéodory "was always interested in mathematics and Hermann Amandus Schwarz (the leading mathematician in Berlin of the period) conducted a seminar, he profited from his stay to attend the lecture given there. It was Fejér who was giving a lecture: he presented a beautiful, simple proof of a result of Schwarz, according to which the triangle with the smallest perimeter inscribed in a given triangle is the one which joins the points in which the heights meet the opposite sides. [...] Now the legend has it that Carathéodory there and then decided to give up engineering (he could afford it) and became a professional mathematician."[37] Carathéodory himself says that, for his studies, he chose Berlin, where he could go underground and live for himself. He wanted to avoid Paris, where he had many relatives and friends. Berlin had the advantage that it was not completely unknown to him since, as we have seen, his father Stephanos had made good friends there, not only with members of the diplomatic community, but also with painters, musicians and scholars.

In his study, Carathéodory kept a large engraving of a signed portrait of Alexander von Humboldt as a precious memento of that time. This had been personally presented to Carathéodory's father by the famous scholar. The idea of establishing an important mathematics research centre at the University of Berlin in the late 1820's had been actively pursued by Alexander von Humboldt.[38]

At the turn of the century, the Friedrich-Wilhelm University in Berlin was one of the most respected centres of learning. Friedrich Althoff, department head at the Prussian Ministry of Education and Cultural Affairs from 1882 and a controversial personality because of his challenge to the autonomy of the universities, had found significant scientists for this university with the help of personal contacts and the participation of private investors in scientific projects. At that time, scientists in Germany held themselves in high esteem and members of the academic community conceived their task as requiring dedication to the fatherland and to science. Germany had attained a leading position in astronomy through work at the Königsberg, Gotha and Potsdam observatories. Natural sciences were advancing with the inventions of glass bulbs, batteries, microscopes and telescopes. Helmholtz discovered the law of conservation of power, Röntgen discovered X-rays, acoustic behaviour was explained, the nature of heat as motion was analysed, and the basic laws of electricity were established. In mathematics, names such as Gauss, Weber, Kronecker and Weierstrass took the place of Leibniz and Lagrange.[39] The University of Berlin itself employed renowned scientists, who, alongside their duties as university professors, pursued the establishment of public scientific institutions. Among these eminent scientists we find the chemist Emil Fischer, the physician Robert Koch, the historian Theodor Mommsen and the physicist Max Planck – all Nobel Prize winners.

Carathéodory began his studies in the summer semester of 1900. He was classified as an advanced student and he participated in the colloquium of Hermann Amandus Schwarz, which was held twice a month. He maintained a close friendship with Schwarz and shared his love of geometry. Following one of Schwarz's lectures on the foundations of projective geometry, Carathéodory read the book by Reye and much of the work of Steiner and von Staudt,[40] which he later recommended to his students in Göttingen in the summer semester of 1917. After the sessions, the participants of the colloquium gathered together in a public house on Friedrich-strasse under the arch of the urban railway, where they would read from the collected treatises of Schwarz. There, Carathéodory gained basic knowledge in the theory of functions, with the help of the fundamental theorem of conformal mapping and the triangle functions of Schwarz, and he realised that the best understanding of general theories presupposes the best knowledge of special examples. In the *Deutsches Biographisches Jahrbuch* (German Biographical Yearbook) of 1927,[41] Carathéodory remarked that the works of Schwarz were imbued with a peculiar power and naturalness, which only a completely independent and original thinker could have. The strength of Schwarz's talent was to be revealed most of all in geometry. His merit, according to Carathéodory, was in having shown that many of the main (unproved) results of Riemann could be established with the help of methods based on principles set up by Weierstrass. In this way, Schwarz had managed to bridge the gap between those mathematicians who criticised Riemann sharply and those who admired Riemann's results and their coherence and could recognise nothing but pettiness in the critique. Carathéodory found the theory that Schwarz created for establishing Riemann's conclusions in a rigorous manner to be particularly elegant, interesting and instructive, even if somewhat intricate.[42]

The Schwarz colloquiums offered Carathéodory the opportunity of getting acquainted with promising young mathematicians who were studying at various German universities or trying to publish their first independent works. This opportunity was of even greater value to his scientific career than the colloquiums themselves. Within the first weeks of his arrival in Berlin he had already become close friends with Erhard Schmidt and Leopold Fejér, and in Schwarz's colloquium he got to know Friedrich Hartogs, Paul Koebe, Emil Hilb, the Swiss G. Dumas and the Americans Oliver Kellog and Max Mason. In 1902, Kellog was to become the first of Hilbert's students, with a doctoral dissertation on integral equations, and later a professor of mathematics at Harvard, where he met Carathéodory again in 1928. A life-time friendship developed between Carathéodory and Erhard Schmidt, who came from the Baltic. Behnke remarks that Schmidt had a comparable social background to Carathéodory's and that he was always open to discussions about mathematical problems.[43] In Born's words, he was an aristocrat by looks and by conviction with great charm, sparkling humour, a clear mind, lively intellectual curiosity and the attitude of an artist towards mathematics.[44] Both Schmidt and Carathéodory adored their professor. According to Schmidt, Schwarz united the three necessary sources of the human spirit for mathematical work, these being geometric experience, the art of logic and the ability of critical abstraction.[45]

Carathéodory compiled a programme of studies for himself. Parallel to the pure mathematics lectures, he became an assiduous visitor to the Max Planck lectures, especially on mechanics and Maxwell's theory, and to Julius Bauschinger's lectures on celestial mechanics. The latter was one of the last representatives of classical astronomy, whose research dealt with the determination of the orbits of planets and comets. Carathéodory also attended lectures on logic held by the philosopher Carl Fr. Stumpf, as well as seminars or lectures held by Ferdinand Georg Frobenius, Immanuel Lazarus Fuchs, Kurt Hensel and Georg Hettner. Frobenius, professor at the University of Berlin from 1892, had developed, together with Issai Schur, the theory of group characters and the representation theory of abstract groups, which were used later in the mathematical formulation of quantum theory. He is described by Biermann as an extremely pugnacious and aggressive person inclined to attacks and, in his aversion against persons and situations, as a gross man but as an excellent researcher.[46] Carathéodory had a very limited personal relationship with Frobenius, although he attended his lectures which he described as so perfect and straightforward that they did not allow the assumption of the existence of unresolved problems. Fuchs, who succeeded Kummer in Berlin in 1884, is considered the founder of the modern theory of differential equations and is known for his work on the theory of functions. According to Gray,[47] his work can be seen as an attempt to impose upon the underdeveloped world of differential equations the conceptual order of the emerging theory of complex functions. He is credited as being the first to apply the theory of functions to linear differential equations of any order, with rational coefficients. Biermann quotes L. Königsberger and L. Heffter in describing Fuchs as "indecisive and anxious, procrastinating and easily swayed, yet humorous and unselfishly kind."[48] Carathéodory did not approach Fuchs, who had become rector and cared little for his rather elementary lectures. Also other students characterise Fuchs as a non-brilliant lecturer, rarely well prepared, but always certain of the essential points of his argument. Fuchs, Schwarz and Frobenius were full professors. They were assisted by the associate professors K. Hensel, J. Knoblauch and G. Hettner and the honorary professor R. Lehmann-Filhés. Hensel, at the university since 1892, had worked in number theory. He made his mark with his *Neue Begründung der Theorie der algebraischen Zahlen* (New Foundation of the Theory of Algebraic Numbers)[49] in 1899. In 1901, he took over as the editor of Crelle's *Journal für die reine und angewandte Mathematik* (Journal for Pure and Applied Mathematics), replacing Fuchs, who was succeeded by Schottky at the University of Berlin, while he himself replaced Schottky at the University of Marburg. Hettner, a professor at the Technical University of Charlottenburg since 1894, lectured at Berlin University on probability calculus. Knoblauch lectured on differential equations and the application of elliptic functions. The astronomer Lehmann-Filhés held mathematical lectures for the newcomers.

Carathéodory attended lectures in what one of Fuchs's students[50] described as a small and crowded room, poorly ventilated, stuffy and hot in the summer days but full of meaning and inspiration for the earnest and thoughtful student. Through the Greek algebraist and geometer, Cyparissos Stephanos, professor at the University of Athens since 1884, Carathéodory got to know St. Jolles, professor of geometry at the Technical University of Charlottenburg. In his house, he also met Adolph Kneser for the first time, one of the founders of the Berlin Mathematical Society in autumn 1901. Kneser had come to Berlin from Dorpat (now Tartu) in Estonia, where Erhard Schmidt had been among his students. From 1901 to 1905 Kneser taught at the *Bergakademie* in Berlin. The Berlin Mathematical Society, which gave form to the loosely organised mathematical life of Berlin, appealed more to young students than to the older Berlin professors. In Jolles's house, Carathéodory also met the famous surface theoretician, Julius Weingarten, professor at the Technical University of Charlottenburg and founding member and first chairman of the Berlin Mathematical Society, who could recall a lot of things from Berlin in the 1850s as well as the beginnings of Weierstrass' mathematics. Weingarten was in contact with August Leopold Crelle, who had founded his journal in 1826.

Carathéodory also had friendly contacts and scientific co-operation with four other Greeks, all younger than him, who were living in Berlin at that time. Georgios Ioakimoglou, a future lecturer of pharmacology at Berlin University, Periklis Vizoukidis, a future jurist and professor at the Law Faculty of the University of Thessaloniki, the physician Aristotelis Siniossoglou and, finally, Ioannis Kalitsounakis, a member of the teaching staff at the Seminar for Oriental Languages[51] of Berlin University and a future professor of economy at the University of Athens. Decades later, on 17 June 1965, at a ceremony in honour of Ioakimoglou, Kalitsounakis spoke of the Berlin years before World War I. Germany, he said, was experiencing a period of social and scientific blossoming at that time. Teaching staff and students were enjoying the diversity of possibilities that the big city offered them. Unfortunately, the war put an end to this with the destruction of life as well as the academic environment.[52]

1.8
The German University

What Kalitsounakis described were in fact the features of the German university in the Kaiserreich up to World War I. Science was autonomous, the significance of scientific research was increasing and full professors enjoyed high social reputation and undisputed rank. They were the heads of their disciplines and self-regulating within a science that meant anything else but "application". The attitude towards those high schools that taught applications was condescending to the point that to be engaged in such work represented a "fall from grace". The university did not require a prescribed course of study for the students but provided an opportunity to either study a discipline intensively or to follow a general education. The former could lead to brilliant academic success while the latter was often a preparation for a future career in the civil service. While the education provided by the university could not be described as humanistic, it was nevertheless broader than a narrow specialist training. The qualifications acquired by many of the students secured their entrance to a social class respected by the rest of the society and acknowledged as the upper class. The universities enjoyed a harmonious relationship with the state and the wider society. Socially, they were part of traditionally educated middle-class intellectuals, the so- called *Bildungsbürgertum*. Politically, universities were formally neutral (socialists were tolerated at the university), but the claim that university professors were apolitical was of course a lie. In general, they supported the Kaiser, and some became members of parliament or political publishers. As a group they held on devotedly to a primitive nationalism.[53]

1.9
Friends in Göttingen

At the beginning of the century, the reputation as the leading German centre of mathematics passed from Berlin to Göttingen. The Georg-August University in Göttingen already had an impressive tradition in mathematics, primarily as the result of the contributions of Carl Friedrich Gauss, Peter Gustav Lejeune Dirichlet and Bernhard Riemann in the 19th century, but at the beginning of the 20th century it acquired modern intellectual life and attracted an international public.[54] Carathéodory's friend Erhard Schmidt travelled to Göttingen as early as 1901. When he returned to Berlin to celebrate Christmas, he described his experiences in Göttingen so enthusiastically that Carathéodory decided to follow him there. Carathéodory left for Göttingen in the summer semester of 1902 having written *La Géométrie synthétique*,[55] his paper on synthetic geometry, before he left Berlin.

At the beginning of his time in Göttingen, Carathéodory associated almost exclusively with Erhard Schmidt and his Baltic friends, particularly with the up-and-coming Sanskrit and Buddha researcher Alexander Wilhelm Baron Stael von Holstein, consultant to the Chinese government in educational matters in 1921. Carathéodory lived in a boarding house at Nicolausbergerweg 49, traditionally inhabited by young scholars and university teachers. He enjoyed the company of the Hungarian

With Fejér in Göttingen.
Courtesy of Mrs. Rodopoulos-Carathéodory.

Leopold Fejér, who, after gaining his doctorate in Budapest, spent the winter of 1902–1903 at the same address. In the summer of 1903, Carathéodory began to establish closer contact with other mathematicians and physicists. These included the brilliant and eccentric Ernst Zermelo,[56] a nervous, solitary man[57] who was Planck's assistant for theoretical physics in the years 1894–1899 and also lecturer in mathematics at Göttingen. There was also Max Abraham, a theoretical physicist who had obtained his habilitation in Göttingen with research in mathematical physics in 1900 and was lecturing on the electromagnetic theory of light. He was the author of a textbook on electromagnetism, one of the first systematic presentations of Maxwell's theory in German, and he would later contribute to the development of wireless telegraphy. The future Nobel Prize winner Max Born, who admired Carathéodory and considered him his mathematical friend was also there, as were Johann Oswald Müller, with former studies in Leipzig, Zurich and Paris, Conrad Müller, alias Thorschreiber, who possessed extensive mathematical knowledge and a sound mathematical judgement but showed no productivity[58] and Ludwig Otto Blumenthal, Hilbert's first doctoral student, whose doctorate was awarded in 1898, "a gentle, fun-loving, sociable young man who spoke and read a number of languages and was interested in literature, his-

The house at Nicolausbergerweg (today Nikolausberger Weg) 49.
Photograph: S. Georgiadis.

tory and theology as well as mathematics and physics".[59] Blumenthal had studied under Edouard Borel and Camille Jordan in Paris and qualified as a university lecturer in Göttingen in 1901. Two further friends were Grace Chisholm-Young, the first woman mathematician in Germany to be awarded a doctoral degree, whose son would later express the wish to write his doctoral thesis under Carathéodory's supervision in the summer semester of 1928,[60] and Felix Bernstein, the future holder of the Chair of Statistics that was to be created by Klein. All of them belonged to the Göttingen Mathematical Society, to which William Henry Young, who had lived in Göttingen in the period 1899–1908, referred in a report on university mathematics in England and Germany.

The German professors have instituted at Göttingen and elsewhere a Mathematical Society of their own, meeting one evening in the week, to which the professor, Privatdozents and a few advanced students have access. The current mathematical literature is, as far as possible, divided up among the members for perusal, and subsequently to report to the Society as to the contents. Free criticism and suggestion is allowed, and in particular any references to other writers, ancient or modern, in which the subjects treated of in the society occur, are welcomed.[61]

This circle of similar-aged students and lecturers was a stage for free and intense scientific exchange and was for Carathéodory of greater value than the occasional opportunity he had to come into contact with major celebrities. Students flocked to Göttingen from Germany, from the rest of Europe and also from afar afield as India, America and Japan. Carathéodory himself described the town as the "seat of an international congress of mathematicians permanently in session".[62] According to Behnke,[63] Carathéodory's exchange with the young mathematicians took place in an atmosphere corresponding to his intellectual expectations. He liked to discuss problems informally within small groups, to find an approach and, if possible, immediately to work out a solution while sitting at the marble table of a café. He was very happy to be able to lecture on his findings as soon as he was able to. It is evident, Behnke remarks, that such a man fitted splendidly into the Göttingen atmosphere. His conspicuous aristocratic nature, his Hellenic aura that aroused admiration and his convincing manner contributed to his popularity. He bore himself with great confidence and had a command of many languages. Reid describes Carathéodory as tall (in fact, just 1.70 m = 5 ft 7 in), impressive, and aristocratic-looking, although unfortunately cross-eyed.[64] He belonged to the most important Göttingen group of young learned men from all faculties, which included Ernst Zermelo, Johann and Conrad Müller, Max Abraham and Erhard Schmidt. These scholars followed unwritten rules in their social life, although they were not a club or a society. They met at the *Schwarzer Bär* (Black Bear) in the old town of Göttingen for Bavarian beer and conversation and had the habit of "treating the unwelcome intruder in such a polite but icy way that he would never come again." One had to be witty or original to be accepted in that group. Students could only attend as occasional guests of one of the members, which was the case with Born, who had been introduced to the company by Carathéodory himself.[65] Born had met Carathéodory for the first time during a visit to Minkowski.

There was only one other guest who later was to be my great friend and of considerable influence on my scientific development. He had a very un-German face, fine and noble features, a long nose and high forehead, and a faint foreign accent. He was introduced as Dr Constantin Carathéodory; I had already heard of this brilliant young Greek who was about to be admitted as lecturer in mathematics. He amused us by telling stories about his work as railway engineer in Egypt.[66]

I.10
Connections with Klein and Hilbert

It seems that Carathéodory had contact not only with Minkowski but also with the other Göttingen celebrities, Felix Klein and David Hilbert, who held chairs in the mathematical faculty of the Georg-August University. Hilbert had been offered and had accepted a full professorship on Klein's initiative in 1895. In the summer of 1902, he rejected an offer to succeed Fuchs, who had died in April, but used this opportunity to improve the chances for the deployment of mathematics in Göttingen, where, in harmonious co-operation with Klein, he was having spectacular success. In addition, he was preparing the appointment of Minkowski, who somewhat earlier had given a lecture in Göttingen entitled *Über die Körper konstanter Breite* (On Bodies of Constant Width).[67] Minkowski had succeeded Hilbert at the University of Königsberg in 1895 (and had been lecturing at the University of Zurich since 1896).

On 24 June 1902, Klein and Hilbert jointly petitioned Althoff for the establishment of a new chair in pure mathematics at Georg-August University; the request was successful and Hermann Minkowski was appointed to the post in the autumn of 1903. In addition, Friedrich Schilling's associate professorship for graphic exercises and mathematical instruments was turned into a full professorship for applied mathematics in 1904. It was the first of its kind to be established at a German university. The chair was occupied by the 48-year-old Carl Runge, who was a professor at the Technical University of Hannover at that time. Runge is known for the numerical procedure for solving initial-value problems in the case of ordinary differential equations, which he developed with Kutta. In the same year, the 29-year-old Ludwig Prandtl, one of the founders of modern hydrodynamics and aerodynamics, obtained an associate professorship for technical physics in Göttingen.

Carathéodory was bound to Klein by a special relationship. Behnke asserts that, due to his sophisticated manners and his great geometrical background, Carathéodory was getting on very well with Klein and had in fact become Klein's intimate friend. He often had the opportunity of discussing with Klein the Monge's school and the development of geometry in France, since Klein was eager to learn about Carathéodory's experiences at the Belgian Military School. Carathéodory felt close to Klein from the very moment of his arrival in Göttingen and denied the commonly shared view that Klein was unapproachable. Their friendship became closer over the years and lasted right up to Klein's death. Carathéodory speaks of him with admiration, albeit not uncritically. In his youth, he writes, Klein was able to guess the solutions to the most difficult mathematical questions and he was exceptionally productive. He was the main editor of the *Mathematische Annalen*, the style of which he determined for decades, and he became full professor at the young age of 22. At the time Carathéodory got to know him, however, Klein was suffering from the consequences of a disease of the nerves that had stricken him at the age of 33 and put an abrupt end to his academic career. Klein was at that time occupied with three projects: the Encyclopaedia of Mathematical Sciences and their Applications (*Encyklopädie der mathematischen Wissenschaften mit Einschluß ihrer Anwendungen*), an enterprise seeking to present all important developments in mathematics and its disciplines of immediate application; the application of advanced pure mathematics to technology, an effort that met opposition from engineers and their lack of understanding but was at least able to produce the world-famous Prandtl Institute for air-stream research; finally, the reform of mathematical instruction at secondary schools.

Carathéodory characterised Klein's viewpoint regarding this reform as romantic and doctrinaire, because Klein had not taken into account the impossibility of training the great number of highly qualified teachers needed to prevent teaching from becoming trivial. But apart from this, in Carathéodory's opinion, Klein's efforts were emerging from the most broad-minded synthesis in the conception of mathematics. This conception found expression in Klein's last lectures on the history of mathematics, which he gave for several semesters from autumn 1914, initially at the university and subsequently in his own house, and which Carathéodory attended despite being himself a professor at that time. He considered Klein to be one of the most brilliant and conscientious teachers and as a person able to recognise the common ground of the remotest problems and to make this awareness accessible and credible to his audience. In the closing remarks Carathéodory made later in the *Encyklopädie*,[68] he identified Klein's ability and the tenacity with which he looked into the smallest detail as evidence of Klein's strength and greatness, qualities which had enabled him to influence the development of the *Encyklopädie*.

In addition to Carathéodory's remarks, Klein's activity included co-operation with industry, the chairmanship of ICMI, the International Commission for Mathematical Instruction (IMUK = *Internationale Mathematische Unterrichtskommission*), plans for the integration of universities and technical universities as well as the presidency of the German Union of Mathematicians (DMV = *Deutsche Mathematiker-Vereinigung*) in 1897, 1903 and 1908. His popular attempt to formalise the applications of mathematical thought as an independent discipline of teaching and research appeared to lack depth and was not of interest to pure mathematicians[69] such as Carathéodory. This lack of depth was implied by Carathéodory, who reproached Klein for impatience in unquestionably asserting predicates, of whose correctness he was convinced, instead of establishing them by a logical, indisputable proof. According to Carathéodory, the essence of mathematics was to be found in the correct implementation of the art of putting forward such proofs, something that Klein would not accept. In this refusal, Carathéodory recognised the reason why Klein could not sense the hidden beauties of Euclid's *Elements* and why he viewed Euclid, as opposed to Archimedes, as a boring schoolmaster.[70]

Indeed, contrary to Carathéodory and certainly also to Hilbert, Klein – expressing the classical view – considered Euclid's axioms to be "evident truths" or "not arbitrary, but sensible predicates".[71] He objected to the theory of axiomatics and opposed Hilbert's deductive (formal) system: "The researcher himself works in mathematics as in every other science by no means in this strict deductive way, but he uses his fantasy considerably and based on heuristic means he proceeds inductively."[72] A formalist interpretation of mathematics had been presented through Hilbert's *Grundlagen der Geometrie* (Foundations of Geometry)[73] in 1899. Through his axiomatics, Hilbert renounced an ontological foundation of mathematics. Mathematical expressions were freed from semantics and presented no other property than the syntactic. The existence of mathematical objects was guaranteed only by the consistency of the theories in which they were described. Carathéodory's critique of Klein unfolded against that background. Carathéodory was indeed using and propagating Hilbert's axiomatics: he introduced, for instance, his Munich lecture on analytical mechanics on 2 November 1925 as follows:

Our experiences from the outside world are summarised in a small number of predicates, the axioms, that fit the image which we have of the outside world anyway. These axioms are not, of course, allowed to be self-contradictory and, additionally, they must be made in such a way as to enable us (without new dependence on experience) through only purely mathematical conclusions to design an image out of them that would do justice to the facts that we aim to master. This method possesses the advantage that, if new phenomena, not belonging to the image formed in this way, should be taken into account, only the axioms have to be subjected to close and critical examination until we manage to resolve all contradictions.[74]

In his address on the occasion of the 50th anniversary of Hilbert's doctorate,[75] Carathéodory argued that the way in which Hilbert improved the axiomatic method and defeated the respectable Euclid with Euclid's own arguments had made the greatest impression on Hilbert's fellow mathematicians. By this statement, he was expressing a view described by Max Steck, later a candidate for Carathéodory's chair, as typically "non-German". Steck's book about *Das Hauptproblem der Mathematik*

(The Main Problem of Mathematics)[76] was strongly polemic against the view that Euclid's definitions are only superfluous accessories. Steck attributed this view to modern formalism, which under Hilbert's leadership had given up the "German line" in mathematics, whereas, as regards content, Greeks and the "German line" had focused on the way of thinking of which Euclid's definitions were substantial parts.[77]

It seems likely, however, that Carathéodory was able to penetrate Klein's reserve, at least on particular points and confessed that on one occasion he had been able to influence the old Privy Councillor. Because of a misunderstanding, the relationship between the two men had momentarily deteriorated. In December 1906, Carathéodory was staying in the Grand Hotel in Paris, where he had attended the funeral of a family member. On Christmas eve, he informed Klein about an article by Jules Huret in the culture section of the Parisian *Le Figaro* of 11 December which referred personally to Klein. "These very unfavourable and totally wrong claims of this gentleman" he wrote "have touched me all the more in a painfully embarrassing manner because much information contained in this article comes from a discussion that I had had with him during a party at Brendel's."[78] The journalist had written that the professors of Göttingen University were paid partly by the state according to their reputation and partly by students attending their lectures. Their income fluctuated between 5 and 40 marks per semester. Professors whose lectures were in great demand could earn a lot of money in this way. Huret gave the example of Klein as the most well known mathematician in Germany, who was criticised for choosing to lecture to the numerous first-year students in order to make more profit. Of course, it was to the advantage of the beginners, he wrote, to be taught by a man in control of his subject and lecturing with admirable ease, such as Klein. On the other hand, the older students were missing out on Klein's lectures to their disadvantage. Huret continued his article by praising the conditions for students and staff at Göttingen University.[79] Klein answered Carathéodory two days after receiving his message.[80] He demanded that Carathéodory tell him which information had been at Huret's disposal. Klein accused Huret of a total lack of responsibility; as a motive for Huret's "insults" against him he believed he recognised the "lure of a piquant reading" for the French readership. He announced to Carathéodory that he had already written to Painlevé, one of his French academician friends, and was hoping that the latter would be able to "find the means and the way to correct the matter". There was a certain tension in French-German relations, not only because of the traditional rivalry between these two nations, but also because a strong pole of mathematics was formed in France around Poincaré, who was in competition with Klein.

Carathéodory, who had obviously provoked this incident with his stories to Huret, and shared some responsibility for the release of the article, was apparently trying to deny his involvement in the affair. Strangely, Klein's over-sensitive reaction to the accusations did not lead to a break-up with Carathéodory. Despite the fact that Klein was feared because of his dictatorial behaviour and had the name of being able to destroy the careers even of talented scientists if they got in his way, his relation with Carathéodory, apart from this incident, remained in the long term free from problems or worries.

Carathéodory developed another kind of relationship with Hilbert, not only because of Hilbert's openness and naturalness in associating even with younger scientists, but especially because there was a kind of deep intellectual affinity between them. Behnke remarks[81] that Hilbert and Klein complemented each other and Zassenhaus[82] asserts that Hilbert, unlike Klein, tried to purify the applications of mathematics but his approach lacked applicability and found no interest with physicists in the pre-computer era. Hilbert was never well prepared for his lectures. Carathéodory found them not so polished as those of Klein, and hardly to be compared with the self-contained lectures of Frobenius but, due to the abundance of Hilbert's unique ideas, the most original and beautiful he had ever attended. In his opinion, new scientists became bound to Hilbert and were attracted to his way of thinking because of this fact. These observations are confirmed by Weyl,[83] who attributed Hilbert's enormous influence and peculiar charm to his optimism, his spiritual passion, his unshakeable faith in the supreme value of sciences, his firm confidence in the power of reason to find simple and clear answers to simple and clear questions, his directness and his industriousness.

A total of sixty-nine students authored valuable dissertations under Hilbert's supervision. Among them there were many Anglo-Saxons who subsequently played a considerable role in the development of American mathematics. Hilbert's students also included women – studies for them had been introduced on Althoff's instructions – despite all opposition from the cool, very conservative and hierarchically structured Göttingen society. This society, in which "carefully observed distinctions of rank cut the professors off from the docents and advanced students"[84] was later described by Einstein as a "small circle of self-important and usually petty and narrow-minded scholars".[85] These distinctions do not seem to have reduced the students' respect for Hilbert. It would be no problem to find eighty to ninety or even more of Hilbert's students and friends, Carathéodory wrote to Weyl on 17 April 1928,[86] who would be willing to commission a bust of Hilbert from a Hamburg sculptor for the Göttingen Institute.

In the speech that Carathéodory later prepared for Hilbert's funeral, he stressed that Hilbert believed persistently in the mission of Göttingen University and pursued science policies with this awareness.[87] It is not without significance that Carathéodory attributed to Hilbert the belief in a mission. Carathéodory himself repeatedly used the notion of mission, or vocation, with regard to his own efforts aiming at the organisation of the Greek universities in the 1920s and 1930s. Also, Hilbert's method, "to frontally attack all difficulties he met", to which Carathéodory explicitly referred, corresponded absolutely to his own way of action.

On 18 August 1900, Hilbert, as chairman of the DMV, gave a speech at the second International Congress of Mathematicians (ICM) in Paris on Mathematical Problems, which according to Carathéodory, marked the turning point in his scientific career. He formulated twenty-three problems belonging to all areas of mathematics, explained why they deserved interest, and estimated the degree of their difficulty. One of them was the further development of the calculus of variations, to which Carathéodory would dedicate a great part of his life. Hilbert had started lecturing on the calculus of variations for the first time in his career during the winter semester of 1899–1900 and he made significant contributions. According to Weyl,[88] the influence of Hilbert's ideas "upon the whole trend of the modern development of the calculus of variations" was "of a more indirect character [than that in the theory of conformal mapping and of minimal surfaces], but of considerable vigor".

Weyl went on to refer to the names of Carathéodory, Lebesgue and Tonelli in Europe and of O. Bolza and M. Morse in the USA as the main contributors to this development. In the English and American specialist literature, Carathéodory is mentioned together with Hilbert, Adolf Kneser, Hans Hahn, William Osgood and Gilbert Bliss as one of the most significant researchers in the modern-classical formulation of the calculus of variations in the early 20th century, which had begun with the textbooks of Kneser (1900) and Bolza (1904).[89]

I.II

Doctorate: Discontinuous Solutions in the Calculus of Variations

When Carathéodory started to become occupied with the calculus of variations, the subject was already in the air. Zermelo's doctoral thesis, for example, carried the title *Untersuchungen zur Variationsrechnung* (Studies in the Calculus of Variations).[90] Through A. Kneser's *Lehrbuch der Variationsrechnung* (Textbook of the Calculus of Variations),[91] the Weierstrass theory on the calculus of variations became accessible to a broader public and was completed in various fundamental points. Hilbert had introduced "direct methods", whose characteristic feature was a peculiarly direct attack on problems, unfettered by algorithms; he always went back to the questions in their original simplicity.[92]

However, Carathéodory was not simply following a "school". He was convinced of the "peculiar appeal" of the calculus of variations, which

has to do on one hand with the fact that it originates in problems which are among the oldest and the most beautiful that mathematics has ever faced, in problems, the significance of which every layman can grasp, but furthermore also and above all, with the fact that since Lagrange it stands at the centre point of mechanics and that an always repeated experience has demonstrated that in the final analysis the mathematical core of almost all theories of physics could be traced back to the form of variational problems.[93]

In 1903 Carathéodory joined the DMV. He spent the Easter of that year in Isthmia, near Corinth, where he conceived the idea for a paper that would become his "first independent mathematical work". He stayed in the house of his uncle Telemachos Carathéodory, an engineer who three years earlier had become director of the company responsible for maintenance works on the Corinth Canal.[94] Carathéodory concluded his paper in Göttingen with the title *Zur geometrischen Deutung der Charakteristiken einer partiellen Differentialgleichung erster Ordnung mit zwei Veränderlichen* (On the Geometric Interpretation of the Characteristics of a Partial First-Order Differential Equation with Two Variables)[95] in February 1904.

In December, the 24-year-old Hans Hahn,[96] who had just obtained his doctoral degree with a dissertation submitted to G. von Escherich in Vienna, gave a speech on Escherich's theory of the second variation in the case of a Lagrange problem.[97] Hahn

Ueber die

diskontinuirlichen Lösungen

in der Variationsrechnung.

Inaugural-Dissertation,

zur

Erlangung der Doktorwürde

der

hohen philosophischen Fakultät der Georg-Augusts-Universität

zu Göttingen

vorgelegt von

Constantin Et. Carathéodory
aus Konstantinopel.

Göttingen 1904.
Druck der Dieterich'schen Univ.-Buchdruckerei.
(W. Fr. Kaestner,)

Carathéodory's doctoral thesis On the Discontinuous Solutions in the Calculus
of Variations. *Courtesy of the Bavarian Academy of Sciences, Archive.*

referred to an exception in this kind of problem which von Escherich was not able to deal with. This fact came as a surprise to Carathéodory, who took it as an opportunity to construct a simple, geometrically clear example which exhibited the anomaly. The problem he faced read: "The points of a semi-spherical globe that surrounds a lamp are projected through the lamp centrally onto the floor. A curve of a given length has to be drawn between two points on the globe, such that the shadow of the curve on the floor would be as long or as short as possible."[98] Carathéodory could not get this problem out of his mind anymore. In Berlin, where he went to celebrate the 200th Colloquium of H. A. Schwarz on 22 January 1904, he calculated the Weierstrass E function[99] of his problem within a few hours in a cafe on Potsdamer Platz. In the following weeks he outlined the elements of a work that was to become his doctoral thesis *Über die diskontinuierlichen Lösungen in der Variationsrechnung* (On the Discontinuous Solutions in the Calculus of Variations).[100]

Carathéodory spent the Easter vacation in 1904 together with his father in Brussels. As he had always done, he used it also this time to accelerate his work and was soon in the position to submit his dissertation. He chose to submit it to Hermann Minkowski rather than Klein or Hilbert, of whom he was in awe. In the subsidiary subjects of applied mathematics and astronomy he was examined by Klein and also by the "genius" astronomer Karl Schwarzschild,[101] professor and director of the Göttingen observatory, and of the same age as Carathéodory.

In his memorial speech for Schwarzschild in 1916, Einstein would find the effortless mastering of mathematical research methods and the ease with which Schwarzschild discovered the substantial aspects of an astronomic or physical question to be especially striking in the astronomer's theoretical work. Einstein asserted that such significant mathematical knowledge combined with so much sense of reality and such adaptable thought seldom existed and he attributed Schwarzschild's impulsiveness for valuable theoretical work in various areas where mathematical difficulties scared others off more to the artistic pleasure of inventing a mathematical idea than to the longing for the cognition of hidden connections in nature.[102] This artistic pleasure was also what Carathéodory enjoyed in mathematics and it is worth mentioning that in 1949 he recommended that Sommerfeld reply to Einstein, who was complaining that the work had appeared to him hard at times, with the verse:

> *Die Lust des Schaffens in heiteren Stunden*
> *Haben die Dilettanten erfunden.*
> *Die Qual des Schaffens in nie sich genügen,*
> *Das ist das wahre Künstlervergnügen*

> (The lust of creation in cheerful moments
> Has been invented by the amateur
> The pain of creation never accomplished
> This is the artist's real bonheur)[103]

Among Carathéodory's papers in the Bavarian Academy of Sciences there is a chronological list of his mathematical works in his own handwriting. On the history of his dissertation he remarked: "Berlin, Café Josty 22.1.04, Göttingen, Brussels. Day of oral examination 13.07.04, printed in September". He submitted his work under

Q·F·F·F·Q·S

AUSPICIIS · ET · AUCTORITATE

AUGUSTISSIMI · POTENTISSIMI · PRINCIPIS · AC · DOMINI

WILHELMI II

IMPERATORIS · GERMANORUM · BORUSSIAE · REGIS

DOMINI · NOSTRI · LONGE · CLEMENTISSIMI

RECTORE · ACADEMIAE · GEORGIAE · AUGUSTAE · MAGNIFICENTISSIMO

ALBERTO

REGIO · BORUSSIAE · PRINCIPE · CELSISSIMO · DUCATUS · BRUNSVICENSIS · SUMMO · MODERATORE

PRORECTORE · MAGNIFICO

VICTORE · EHRENBERG

IURIS · UTRIUSQUE · DOCTORE · ET · PROFESSORE · PUBLICO · ORDINARIO
ORDINIS · REGII · AQUILAE · RUBRAE · QUARTAE · CLASSI · ADSCRIPTO

EGO

GUILELMUS · FLEISCHMANN

PHILOSOPHIAE · DOCTOR · ARTIUM · LIBERALIUM · MAGISTER · PROFESSOR · PUBLICUS · ORDINARIUS

ORDINIS · PHILOSOPHORUM · H · T · DECANUS · ET · PROMOTOR · LEGITIME · CONSTITUTUS

VIRUM · NOBILISSIMUM · DOCTISSIMUM

CONSTANTINUM · CARATHÉODORY

BEROLINENSEM

QUI · DISSERTATIONE · EDITA · REPERIENDI · SOLLERTIA · CONSPICUA

„ÜBER DIE DISKONTINUIRLICHEN LÖSUNGEN IN DER VARIATIONSRECHNUNG"

ET · EXAMINE · SUPERATO · SCIENTIAM · MATHESEOS · ASTRONOMIAE · ET · MATHESEOS · APPLICATAE

MAGNA · CUM · LAUDE · COMPROBAVIT

DIE · I · MENSIS · OCTOBRIS · A · MCMIV

PHILOSOPHIAE · DOCTOREM · ET · ARTIUM · LIBERALIUM · MAGISTRUM

CREAVI

EIUSQUE · REI · HAS · LITTERAS · TESTES

SIGILLO · ORDINIS · PHILOSOPHORUM

MUNIRI · IUSSI

Carathéodory's doctoral title. Courtesy of the University Archive Bonn.
Photograph: S. Georgiadis.

the name "Constantin Et. Carathéodory from Constantinople" and was awarded the doctoral degree on 1 October 1904.

Carathéodory dedicated his dissertation to his father and, in the acknowledgements, he mentioned those persons whose lectures and seminars he had attended at the universities of Berlin and Göttingen: Bauschinger, Frobenius, Fuchs, Hensel, Hettner, Planck, Schwarz, Stumpf, as well as Abraham, Blumenthal, Hilbert, the phenomenologist Edmund Husserl, associate professor of philosophy in Göttingen since 1901, Klein, Minkowski, Zermelo and Schwarzschild. Carathéodory especially acknowledged the role of H. A. Schwarz for the scientific foundations gained under his instruction and he also thanked Minkowski and Zermelo for the multiple stimulus that he had experienced in their company. At that time Minkowski was showing interest in classical treatises on special relativity and electromagnetic theory,[104] whereas Zermelo and H. Hahn were writing the article *Weiterentwicklung der Variationsrechnung in den letzten Jahren* (New Developments in the Calculus of Variations in Recent Years) for Klein's *Encyklopädie*. This article mentioned Carathéodory's dissertation for the first time, in fact before it had been released, though indirectly, "on the basis of oral communications".[105]

In his doctoral thesis, Carathéodory went beyond the classical framework of the calculus of variations as defined by Weierstrass and Hilbert, both in his formulation of the problem and in his method. The awareness that in the calculus of variations the requirements for differentiability set on the integrand of a problem do not guarantee that the dimension number of the continuum covered by the allowed extremals cannot change disjointedly, led Carathéodory to introduce the notion of a class of an arc of extremals.[106] He worked out a comprehensive theory of discontinuous solutions, proved the existence of a field of broken extremals (the solutions) in the vicinity of a corner point, worked out the theory of conjugate points, and established the necessary and sufficient conditions for the existence of a discontinuous solution with several break points. Earlier, only the so-called Weierstrass–Erdmann corner conditions were known, that is the necessary conditions for an extremum, additional to the Euler equation, specified at points at which the extremal has a corner. Carathéodory showed that the theory which holds for continuously bent curves could also be applied to broken curves.

Until then, nobody had really dared to engage in the so-called discontinuous solutions, in the case of which the usual presuppositions are no longer satisfied throughout.[107]

Familiar with the history of his subject, Carathéodory drew on the ideas of Huygens und Johann I. Bernoulli. He introduced the concept of the indicatrix of his variational problem, whose thorough definition and properties he presented only later in his habilitation. He also examined the discontinuous solutions in the case of isoperimetric problems, the simplest example thereof being the circle's property of having the greatest area of all convex curves with the same circumference. The term "isoperimetric problem" has been extended to mean any problem in the calculus of variations in which a function is to be made a maximum or a minimum, subject to an auxiliary condition, called the isoperimetric condition. Carathéodory dealt especially with the above mentioned clear geometrical example of the hemisphere. He formulated it as follows: "A hemisphere of radius one is projected from the centre onto a tangential plane which is parallel to the equator. It is required to connect two points of the plane through a curve of a given length in such a way that the prototype of this curve on the sphere obtains a maximum or a minimum length."[108] In this way a broken line resulted as the solution.

Perron characterises Carathéodory as "a master at getting to the heart of the matter and finding valuable approaches for an exact treatment of the raised problem. In special recognition of this fact, Alfred Pringsheim, when he was still handing out souvenirs to his loyal followers before leaving the Third Reich, presented to our Carathéodory a precious piece of jewellery from his library, today an extremely rare print from the year 1700, containing a letter in Latin from Jakob Bernoulli to his brother Johann with an appendix on the solution of the isoperimetric problem. The recipient was highly delighted with this gift and his joy was even intensified when he read the dedication, in which Pringsheim imaginatively described him as Isopéri-maître incomparable [incomparable master of isoperimetry]."[109] The print, dating in fact from 1701, was apparently Jakob Bernoulli's *Analysis magni problematis isoperimetrici* mentioned by Carathéodory in his paper *Basel und der Beginn der Variationsrechnung* (Basel and the Beginning of the Calculus of Variations)[110] in connection with the history of Johann Bernoulli's treatises on isoperimetric problems. The paper was published in the Festschrift on the occasion of the 60th birthday of the Swiss professor Dr. Andreas Speiser.

Carathéodory's dissertation was a significant contribution to the calculus of variations. Throughout his life he concerned himself successfully with variational problems in the most general case of m-dimensional surfaces in n-dimensional space by introducing "geodesic fields", which ought to be solutions of the Euler–Lagrange equation. Until then, results had been achieved only for the case $m = n - 1$.

As was to be proved later, this inaugural work showed points of contact with Jacques Hadamard's considerations presented at the IV ICM in Rome in 1908. In addition, Carathéodory found Hadamard's viewpoint, that the calculus of variations was only "a chapter of functional calculus still in its beginnings", to be modern and this modern way of looking at things to have been emphasised for some years by Hilbert, and particularly by Minkowski, in their Göttingen lectures.[111]

I.12

The Third International Congress
of Mathematicians

From 8 to 13 August 1904 Carathéodory attended an international congress of mathematicians for the first time. It was the third ICM held in the university town of Heidelberg. An admission ticket of 20 marks entitled the bearer to participate in sessions, festivities and the banquet in the city ballroom, to visit the exhibition of mathematical models, apparatus and publications of the preceding decade in the hall of the university museum, and to obtain the Festschrift and the congress files. The exhibition was opened with three general lectures including one on Leibniz's original calculating machine displayed by Carl Runge. The programme also included a trip on the river Neckar sponsored by the town of Heidelberg, an evening of entertainment organised by the DMV and excursions in the surroundings of Heidelberg. A special committee organised the reception and entertainment of the ladies who accompanied the participants and another committee endeavoured to find lodgings for the congress members according to their wishes, either in an inn or in a private house. Not a single participant was female. Jacobi's centennial birthday was celebrated by a special ceremony, since it coincided with the year of the congress. Following the opening address by Heinrich Weber, Leo Königsberger, a professor of mathematics at Heidelberg and a friend of Fuchs, spoke about Jacobi. A transcript of his speech was presented to all participants as a gift at the end of the congress.[112] The participants could purchase a reduced price biography of Jacobi published by B. G. Teubner, the famous publisher and bookseller in Leipzig, who was a member of the organising committee. This committee had decided on four plenary addresses in English, French, German and Italian, and on the publication of the lectures in the language in which they were given.[113]

In Heidelberg, Carathéodory had the opportunity of becoming acquainted with more than a hundred mathematicians. In his *Autobiographical Notes*, he mentions Paul Painlevé, a professor in Paris and, since 1900, member of the Académie Française, Lorenz Leonard Lindelöf and his son Ernst, professor in Helsingfors from 1903, Adolf Mayer, member of the editorial board of the *Mathematische Annalen* from 1873 and honorary professor at the University of Leipzig from 1881, and Sir Alfred George Greenhill, Klein's future deputy in the principal committee of the ICMI. Carathéodory does not, however, mention Paul Tannery, who held the French post of *Directeur des Manufactures de l'Etat*, a mathematician who had applied himself to historical and philosophical mathematical problems of Ancient Greece and who spoke at the congress on the history of mathematics. However, he does refer to the connection between the Frenchman and his own future father-in-law, Alexander Carathéodory Pasha: "Alexander Carathéodory was an orientalist: he had prepared the text and the French translation of an unknown Arabic mathematical work of the 12th century. In this respect, he had a lively correspondence with Paul Tannery but he could not decide whether to publish the historical introduction to the book, because, as a Christian and at the same time being a high dignitary of the Sultan, he should exercise some restraint on the judgement of a Muslim scholar, though not justified from the purely scientific point of view."[114]

After leaving Heidelberg, Carathéodory travelled together with H. Hahn and W. Wirtinger, professor in Vienna, through Munich to Achensee. There, Wirtinger introduced him to Otto Stolz, a famous professor in Innsbruck who had produced significant work on analysis. Stolz died a year later and Hahn deputised for him in the academic year 1905–1906. Achensee is mentioned by Carathéodory, in a later biographical note he wrote on Wirtinger,[115] as the place where Wirtinger improved the theory of capillary waves during his summer vacation.

I.13

A Visit to Edinburgh

Accompanying his father Stephanos to a conference of the Institute of International Law (*Institut de Droit International*), Carathéodory visited Edinburgh in the autumn of 1904. On their trip, father and son were accompanied by Constantin's sister, Ioulia, together with her husband Georg von Streit, a reputable jurist, member of the Permanent Arbitration Tribunal in The Hague in 1900, expert in international law and professor of jurisprudence at the University of Athens. Like his father-in-law, he also participated in the conference of the Institute of International Law. Carathéodory had the privilege "of being invited to all festivities, to visit the castle of Lord Rosebery and to enjoy the beauties of the incomparable city in the marvellous autumn weather."[116] Archibald Philip Primrose, fifth Earl of Rosebery, had played a significant role in diplomatic activities concerned with southern Europe. The highly educated politician and author, while Foreign Minister in 1886, agreed with the representatives of the Great Powers that there was no legal reason for a war led by

Ioulia Streit. Courtesy of Mrs. Rodopoulos-Carathéodory.

Greece against Turkey. After Gladstone's resignation in 1894, Lord Rosebery was invited to form a government but was obliged to resign the following year, in 1895. A year later he also gave up the leadership of the Liberal Party. Diametrically opposed to his adversary Gladstone, Rosebery believed that a condemnation of the Ottoman Empire for the massacre of Armenians could only draw Great Britain into war, which he himself wanted to avoid at all costs.

I.14
Habilitation in Göttingen

Under the German system, a candidate has to write a post-doctorate thesis ('habilitation') in order to be appointed to a university lectureship. Carathéodory ascribed an almost destiny-like significance that was to determine the course of his life when, at Klein's suggestion in 1904, he determined to stay on in Göttingen for his habilitation. Up to that point he had not thought for a moment of remaining in Germany. What finally persuaded him to stay was the open, internationally oriented scientific work and the expectation of a promising academic career. After being awarded his doctoral degree he had decided to work in Greece at a military or a naval school; he did not aspire to a professorship at the University of Athens. But even this rather modest goal could not be fulfilled. He was offered only a job as a teacher of the Greek language at a school in the provinces.[117] Consequently, he returned to Germany full of disappointment.

During the Christmas holidays of 1904, Carathéodory used his stay in Brussels to work on his habilitation. He was able to complete it in January 1905. On 5 March 1905, he submitted it to Göttingen University with the title *Über die starken Maxima und Minima bei einfachen Integralen* (On the Strong Maxima and Minima in the Case of Simple Integrals),[118] thus gaining the authorisation to lecture during the 10th semester of his studies. The formal arrangement was due to Hilbert's proposal to the mathematical faculty to revoke the stipulated period between doctorate and habilitation by an official *ad hoc* decision. The dean of the Philosophical Faculty, W. Fleischmann, gave the following reasons for accepting Hilbert's proposal:

On the grounds of the expert reports and also through the performance of Herr Carathéodory in the colloquium and in the demonstration lecture, the faculty is convinced that both the scientific and the personal abilities of the named correspond throughout to the demands that fall on an academic teacher. His scientific works reveal him as an especially successful researcher in the area of the calculus of variations and at the same time as a skilful geometer. His personality is that of a mature and well-modelled mathematician, who has a good command of modern function theory and its diverse branches and applications.[119]

Carathéodory's exceptional performance confirmed the brilliant educational policy followed by Hilbert and Klein. His demonstration lecture marking the start of Carathéodory's academic career was entitled "Über Länge und Oberfläche" (On Length and Area).[120] Forty-five years later, on 16 December 1949, shortly before the end of his teaching career, Carathéodory gave his final lecture with the same title at the Munich Mathematical Colloquium.[121] Carathéodory revised his habilitation in spring 1905 for publication in the *Mathematische Annalen*.[122]

1.15
Lecturer in Göttingen

When he returned to Göttingen after the Easter vacations of 1905, which he had spent with his father and sister in Isthmia enjoying the hospitality of his dear uncle Telemachos, Carathéodory deputised for the very ill professor Brendel at Klein's request. Brendel was teaching actuarial mathematics, which had been introduced in the winter semester of 1895–1896. Carathéodory's first lecture was on kinetics and he remained as lecturer in Göttingen until Easter 1908.[123]

In the summer of 1905, Caratheodory fell when his horse stumbled and fractured his right clavicle. He then trained himself to write all certificates of attendance for his students with his left hand until he had recovered from the after-effects of the accident.[124] This effort of self-discipline and will to be on top of his form is evident from another incident related by one of his students in Munich in the years 1930–1932: Werner Romberg[125] tells us that when teaching Carathéodory would wipe his nose with a handkerchief held in his left hand so as not to interrupt his writing on the board with his right hand. And his daughter tells that when her father was ill, he would still come to breakfast and give the impression of a healthy and happy person so as not to cause distress or embarrassment to his wife or children.

Carathéodory reveals himself as a man who put an effort into hiding his emotions and one who attempted to exercise his abilities to the maximum. The second characteristic was a tradition in his father's family. There is evidence, however, that emotional outbursts by his relatives were not rare.

At the time Carathéodory became lecturer, mathematics was blossoming in Germany, through academic institutions and teaching, through scientific and educational societies and through literary enterprises. Mathematical seminars and institutes, their reading rooms and specialised libraries, large public libraries, collections of models, the use of projectors and cinematographs in lectures, further training and vacation courses for teachers, study plans and "tips" for students, doctoral degrees, prize questions, and so on were provided for teaching and research activity at schools and universities. There were also numerous student and learned societies, such as mathematical student associations, the Naturalists' Assembly (*Naturforscherversammlung*) and its mathematical section the DMV, various regional and local mathematical societies, "circles", "meetings", the Association for the Promotion of Instruction in Mathematics and Natural Sciences (*Verein zur Förderung des mathematischen und naturwissenschaftlichen Unterrichts*), the Göttingen Union for the Promotion of Applied Physics and Mathematics (*Göttinger Vereinigung zur Förderung der Angewandten Physik und Mathematik*), academies of sciences and the Cartel of German Academies (*Kartell der deutschen Akademien*), the Association of Scientific Academies of all Civilised Nations (*Assoziation der wissenschaftlichen Akademien aller Kulturländer*), the ICM and the ICMI, all of which served to promote scientific research. There were also various mathematical journals, such as the Yearbook on the Progress of Mathematics (*Jahrbuch über die Fortschritte der Mathematik*), the *Encyklopädie*, as well as publications of mathematical collected works and other published material helping to disseminate mathematical knowledge among the specialised public and serving to create links between various mathematical areas.[126] As Hilbert remarked:

For what a delight it is to be a mathematician today, when mathematics is sprouting up everywhere and the sprouts blossoming; when it is being brought to bear on applications to the natural sciences as well as the philosophical direction; and it is on the point of re-conquering its former central position.[127]

Göttingen mathematicians were inclined to neighbouring disciplines, where they perceived opportunities for applications of their mathematical theories. That is why the foundation of the two new institutes in physics in December 1905, the one in applied electricity, the other in applied mathematics and mechanics, could be directly ascribed to Klein's influence. Of Carathéodory's friends, those with a keen interest in physics were Zermelo, who declared his doubts about the consequences of statistical mechanics, Max Abraham who worked out a seriously discussed theory of the electron, Gustav Herglotz and Max Born, who investigated how the concept of the rigid body could be replaced in that theory, and Minkowski, who realised that the competing electrodynamical theories proposed by both Einstein and Lorentz must lead to a four-dimensional, non-Euclidean, space–time continuum. In his *Autobiographical Notes*, Carathéodory mentions further examples of the crossing of physics and mathematics in Göttingen:

Not only did applied mathematics gain a bright representative with Runge, and technical mechanics alike with L. Prandtl, but also G. Herglotz, P. Koebe and O. Toeplitz habilitated there. Runge, who was born in Havana and had an English mother, possessed many qualities that came from that origin. The way in which he also handled mechanics was astonishing. When, for example, the Wright brothers made their first flight experiment, he was able to quite precisely estimate the power of the engine, information about which was kept secret, with the help of models made from scraps of paper which he loaded with a pin and left fall in glide. This ability of his impressed me the most then. Besides, he was not only a brilliant experimental physicist, whose measurements of spectral lines were epoch-making, but also a first-rate pure mathematician, whose name would be perpetually linked to the approximation of analytic functions by polynomials.[128]

In 1908, under the direction of Prandtl, the Society for the Study of Motor-Powered Flight (*Motorluftschiff-Studiengesellschaft*) built the world's first sinuous-circuit, return-flow wind tunnel, with a wind velocity of $10 \, \mathrm{m \, s^{-1}}$ and a power of 34 hp. The high efficiency of its design, the incorporation of vanes at the corners to alter the flow, and the use of strategically positioned screens and honeycombs to smooth the airflow, made the Göttingen tunnel a standard to copy. In his tunnel, Prandtl tested a variety of aerofoils, streamlined bodies, and aircraft components. He also measured pressure distributions over rotating propeller blades for the first time.[129] Just before the end of World War II, the President of the DMV would tell him that research in aeronautics would have been unthinkable without his fundamental works on the theory of wings and boundary layers.[130]

Toeplitz went to Göttingen in 1906 and was a member of the narrow circle of Hilbert's students. He worked on integral equations and equations with infinitely many variables. His subjects overlapped with Carathéodory's during the pre-war years, but Carathéodory seems not to have had close contact with him, since the time they were both at Göttingen had been short[131] and when Carathéodory returned to Göttingen as professor in 1913, Toeplitz left to take up an appointment in Kiel. Later, in 1928 Toeplitz succeeded Carathéodory's colleague Eduard Study in Bonn.

Ernst Zermelo, who was a lecturer in Göttingen at that time, became Carathéodory's life-long friend. Zermelo's paper of 1904, that proves that every set can become well-ordered (*Beweis, daß jede Menge wohlgeordnet werden kann*),[132] had immediately provoked a lively and controversial discussion about the procedure used for the proof, which, according to Carathéodory, could not be surpassed with respect to simplicity, shortness and classical elegance.[133] This procedure was based on the axiom of choice[134] enabling a specific kind of mathematical construction. Zermelo lay the foundations for a polemical confrontation with his critics in his new paper of 1908, *Neuer Beweis für die Möglichkeit einer Wohlordnung* (A New Proof for the Possibility of Well-Ordering).[135] His second paper of 1908, *Untersuchungen über die Mengenlehre I* (Studies of the Foundations of Set Theory I),[136] is the first axiomatisation of set theory. There, he proposed seven axioms, namely the axiom of extensionality, the axiom of elementary sets, the axiom of separation, the power set axiom, the axiom of union, the axiom of choice, and the axiom of infinity.

In a letter to Zermelo on 30 August 1907,[137] Carathéodory refers to these two papers of 1908, which he had failed to check through. Apologising to Zermelo for his omission he said that he had been dedicating much time to his father. Carathéodory congratulated Zermelo on his "stabilisation": "still, it is even much better than the B story, which one could not know how it would develop, and you are relieved of a few pecuniary worries. It will also get much better with time, that is, within the foreseeable future, as I have definitely heard." Carathéodory was apparently informed of the fact that the lecturer's scholarship which Zermelo had been granted for the previous five years had not been further extended. A ministerial decree signed by Althoff on 20 August 1907 authorised Zermelo to lecture on "mathematical logic and related subjects" with effect from the winter semester of 1907. The teaching commission by which mathematical logic was included for the first time in the programme of a German university would be remunerated with 1800 marks per year. That sum was raised to 3000 marks in April 1909 and indeed relieved Zermelo from financial worries.[138]

Carathéodory might have used the expression "B story" either to describe the Klein–Born conflict that forced Born to leave Göttingen at the end of the summer semester of 1906, or Blumenthal's position in Göttingen, which was comparable to that of Zermelo until the former took up an appointment in 1905 as professor at the Technical University of Aachen. Born's "insult" against Klein had been his refusal to compete for the academic prize on The Stability of the Elastic Line as Klein had suggested to him. Just like his friends from Breslau and also from the *Schwarzer Bär*, he also felt a strong dislike for Klein's predilection for applied mathematics but he knew at the same time that having Klein as an enemy would mean the end of his mathematical career. Toeplitz, Hellinger, Carathéodory and Schmidt had advised him to give way.[139]

As Carathéodory told Königsberger,[140] he was occupied with algebraic problems which were more remote to him than the mechanical ones; he had collected material for a book on the calculus of variations; he had previously "worked out the invariants for any variational problem, which in the theory of curved surfaces pass over to angles, total curvature and surface." He was not sure whether he ought to publish it or not because it seemed to him "that it was not much good to anyone." On 22 July 1907, Carathéodory declined an invitation to visit Königsberger in Heidelberg, because he had decided to spend most of the vacation with his father in Brussels. He had not seen Stephanos, who was not "entirely well", for almost a year.[141] In a later message, he informed Königsberger[142] of the "great misfortune" of losing his father.

Carathéodory spent the first three weeks of November 1907 in a Göttingen clinic because of an infectious fever. Now he had to cope with double the amount of work, not only to regain lost time, but also because he had added extra hours to the second part of his series of lectures on differential and integral calculus.[143]

CHAPTER 2

Academic Career in Germany

2.1

Habilitation (again) in Bonn

In his letter of 30 August 1907 to Zermelo, Carathéodory spoke about the "uncertain" prospect of going to Bonn. He did not know whether he would like to leave Göttingen: "Now that you and Weyl remain, I would almost prefer G[öttingen] again". Hermann Weyl, one of Hilbert's students, was working at that time on the theory of integral equations and the eigenvalue problems for differential equations.

Earlier that same year, Eduard Study had asked the Philosophical Faculty of the Rhine Friedrich-Wilhelm University, Bonn, for Carathéodory to be appointed there. The case presented for Carathéodory's appointment as Philipp Furtwängler's successor was that Carathéodory had mastered both pure mathematics as a lecturer in Göttingen and applied mathematics as a former engineer in Egypt and therefore united both qualities in one person.[1] Furtwängler and Study were probably known to Carathéodory from the 1904 International Congress of Mathematicians in Heidelberg. Furtwängler had made contributions to geodesy and the theory of algebraic numbers; Study's works on projective and differential geometry were held to be authoritative.

On 29 January 1908, Carathéodory asked the Dean of the Philosophical Faculty in Bonn, Brinkmann, for his permission to become a lecturer there.[2] Two years earlier, Erhard Schmidt had submitted his own habilitation on the theory of linear and non-linear integral equations to the same faculty. On the basis of confidential information from Elster, a senior civil servant in Berlin, Carathéodory considered that he was guaranteed a teaching commission in applied mathematics. Consequently, he asked the faculty to include him on the teaching staff. On 31 January, Study submitted Carathéodory's request for habilitation, together with the applicant's doctoral degree and twelve treatises, to Dean Brinkmann and asked him to allow Carathéodory to lecture on any subject of mathematics, as he used to do in Göttingen, and to fix a date in the current semester for the inaugural lecture. Study proposed to the Dean that the choice of the subject and the timing should be left to Carathéodory and believed that the faculty would be valuably enriched by Carathéodory's appointment.[3] Carathéodory's request for authorisation to teach was passed on by Dean Brinkmann on 13 February 1908 to be decided by all members of the faculty. Brinkmann himself was in favour of a positive outcome and he referred to a custom followed in similar

The house at Venusbergweg 32 (today 43) in Bonn.
Photograph: Dr. Rolf-D. Philips.

previous cases, in which all habilitation achievements, such as the habilitation itself, the demonstration lecture and the inaugural lecture, had been waived. Thirty faculty members demonstrated agreement with their signatures.[4] A week later, the dean announced to the rector of the University that the Philosophical Faculty had accepted Carathéodory without demanding a new habilitation.[5]

The reason this "brilliant Greek"[6] had chosen Bonn was that he saw it as a stepping stone to his academic career. Since the end of the 19th century, as we have remarked earlier, academics in Germany were regarded as belonging to the so-called *Bildungsbürgertum*, well-paid and respected members of an established order, possessing a significant social standing and with a civil-service mentality. Carathéodory, keen to become a member of such a social class, was appointed to the Rhine Friedrich-Wilhelm University on 1 April 1908. Although now officially appointed in Bonn, Carathéodory attended the International Congress of Mathematicians at the Palazzo Corsini in Rome as "Dr. phil., Brussels", his teaching activity in Bonn starting properly in the summer semester of 1908.

In June 1908, Carathéodory was granted leave of absence for the summer semester, since he had to stay in Constantinople "for several months" because of urgent family business.[7] In Constantinople he enjoyed the company of Born, who recalled:

In spring 1908 I undertook a journey to Greece [...]. The first plan for this journey had sprung from an accidental meeting with Carathéodory in Berlin who told me that he intended to visit his relatives in Athens and Constantinople and asked me to join him. I took a small Austrian steamer from Trieste. After a pleasant journey (I was almost the only passenger on the boat) along the Dalmatian coast and a short stop at Corfu we steamed through the gulf of Corinth and entered the canal, which is a narrow and deep cut through the rock. And when we reached the other end which opens to the blue Aegean Sea there standing at the quayside was my friend Carathéodory in his usual black attire, a cigar in his mouth. But he told me to my great disappointment that he could not come with me (I think his uncle, who was director of the canal, was ill) but would join me later in Athens. So I had to spend most of the time without my friend on whom I had counted as an expert guide. But I hardly needed a guide; thanks to my classical education the whole journey was almost a home- coming; all was well-known, names and views familiar from pictures and descriptions, and everything beautiful beyond description, for Greek air and light can only be experienced. [...] [In Constantinople] from Carathéodory I had introductions to some of his relatives and was invited to parties. In this way I saw something of the life of these Greco-Turkish families. They were rather grand people with lovely dark features, in particular the women, and they received me with great kindness.[8]

2.2

Axiomatic Foundation of Thermodynamics

In Xirokrini (Kuru Çesme in Turkish) on the Bosphorus, Carathéodory worked from July to September 1908 on his *Untersuchungen über die Grundlagen der Thermo-dynamik* (Studies in the Foundations of Thermodynamics),[9] which he completed in Bonn in November that year. In the meantime, "in acknowledgement of his scientific performance" he was awarded "the rating 'professor'" by ministerial decree on 3 October 1908[10] after lecturing in Bonn for almost two months in total.

Thermodynamics enjoys a special place in Carathéodory's work. This area of theoretical physics had attracted him since his time at the Belgian Military School. His *Studies in the Foundations of Thermodynamics*, he once said, was the only work that he had thought through from the beginning to the end before writing it down.[11] With this work he endeavoured to lay down the axiomatic foundations of thermodynamics. "One can derive the whole theory", he wrote, "without assuming the existence of heat, i.e. of a physical quantity that differs from the usual mechanical quantities." He reformulated the first and second theorems of thermodynamics as the first and second axioms. He then introduced two assumptions that differed from the traditional approach. First, heat was reduced to mechanical work and, second, the system was treated mainly as adiabatic instead of as a cyclic process. The first axiom, $\varepsilon' - \varepsilon + A = 0$, where $\varepsilon' - \varepsilon$ denotes the difference between the final and initial values of the energy and A the external work done by the system, was formulated as follows:

In every adiabatic change of state the change in energy increased by the external work A equals zero,

and the second axiom read:

In any neighbourhood of any given initial state there are states which cannot be approximated arbitrarily through adiabatic changes of state.

Carathéodory described the quasi-static adiabatic changes of the state of the system by the curves of a Pfaff equation with *n* deformation co-ordinates. A. Landé wrote in his *Axiomatische Begründung der Thermodynamik durch Carathéodory* (Axiomatic Foundation of Thermodynamics by Carathéodory),[12] that "the work needed here to eliminate all unimportant elements and to give thermodynamics a more geometric foundation equal to that of mechanics has been achieved by Carathéodory and indeed in a very extensive and abstract way". He also noted that Carathéodory's achievement lay in the fact

that he had reduced the existence of an integral denominator for the expression $dQ = dU - dA$ and thus the existence of the entropy function to an axiom simpler than Thomson's principle and Clausius' principle. Using the theory of Pfaff equations which he even completed with a fundamental theorem, he presented the proof of the existence of the integral denominator in a particularly simple form and he further showed the mathematical reasons why the integral denominator should have a factor able to split off, which is only a function of empirical temperature, independent of particular body properties and which represents the universal 'absolute temperature'.

In other words, by stating the second law in terms of the inaccessibility of certain states by adiabatic paths and by using a mathematical theorem,[13] Carathéodory was able to infer the existence of an entropy function and an integrating factor connected with Kelvin temperature.

Ten years later, Carathéodory referred to thermodynamics in his inaugural speech at the Prussian Academy of Sciences:

One should not forget that in problems involving an interaction of pure mathematics and natural sciences, the mathematician, who feels satisfied as soon as the questions posed by his mind are answered, is more frequently the one who takes than the one who gives. To take one of a number of examples: one can ask oneself the question, how should a theory of phenomenological thermodynamics be constructed, so that the calculations only use directly measurable quantities such as volume, pressure and the chemical composition of bodies. The resulting theory is incontestable from a logical point of view and satisfies the mathematician completely, since, starting only from the observed facts, it is established from a small number of assumptions. And yet these are exactly the merits that, from the naturalist's general view, make it of little use, not only because temperature appears as a derived quantity, but, most of all, because, through the smooth walls of the structure that so elaborately and artistically fit together, one can establish no passage between the world of visible and perceptible matter and the world of atoms.[14]

The Philosophical Faculty of Berlin University drew attention to Carathéodory's approach to thermodynamics in their proposal to the Ministry of Education regarding Frobenius' successor in 1917: "Finally, let us stress another contribution to thermodynamics, highly appreciated by authoritative physicists, in which a new axiomatic derivation of this theory is developed that makes no use of the assumption of the existence of heat as a physical quantity, differing from the usual mechanical quantities."[15]

Max Planck, Max Born and Arnold Sommerfeld indeed showed a great interest in this work and made it known to their specialist circles, thereby helping Carathéodory become a respected figure among German physicists. Born, for example, wrote:

He [Carathéodory] and I discussed among other things the peculiar fact that the quite abstract science of thermodynamics was built on technical concepts, namely on "heat machines". Could that not be avoided? Some years later, Carathéodory found a new, absolutely strict, direct approach [to the theory]. He published it in a quite general and abstract way in the *Mathematische Annalen*; this work was hardly noticed. Fifteen years later, I attempted to make this theory more popular through a simpler account in the *Physikalische Zeitschrift*, but without success. Only now after fifty years, are textbooks using this simple and clear approach being published.[16]

In his introduction to the *Kritische Betrachtungen zur traditionellen Darstellung der Thermodynamik* (Critical Considerations of the Traditional Representation of Thermodynamics) Born wrote:

In order to facilitate the study of Carathéodory's treatise, I have made an attempt here to present its basic ideas in an entirely simple way and explain them to specialists. I know that thermodynamics exercises a strong intellectual attraction in the form in which it has been created by the great masters and that it is firmly embodied in the consciousness of physicists. Nevertheless, the new reasoning may gain friends; and although it lacks those wonderful ideas that in an almost marvellous way lead from the simple facts of experience to the basic principles, it is more transparent and employs just "normal" mathematics, which everyone has learned.[17]

At the end of his paper Born continued:

The cardinal point of Carathéodory's theory is the understanding that a more general formulation of the experience principle about the impossibility of certain processes is sufficient, without new physical ideas and in the simplest way, to arrive at the second main theorem with the help of the previously proved theorem on Pfaff's differential expressions.[18]

From *Café Métropole* in Brussels, Carathéodory wrote to Born on 9 October 1907[19] that he was able to define the concepts "amount of heat" and "reversibility" more precisely than usual and that he could prove that the equation $d\varepsilon = t\,dy - p\,dV$ could be derived from the energy principle and the Carnot principle. He asked Born to let him know whether he was satisfied with that thermodynamic version or whether he still had "other objections". Such a question led Born to the conviction that he himself had taken part in the development of Carathéodory's thermodynamics.[20]

Referring to the summer of 1907, which he spent in Cambridge, Born wrote in his memoirs:

Apart from my experimental course with Searle, I had nothing to do, and I spent most of my time lying in a punt or canoe on the Cam and reading Gibbs, whom I now began to understand. From this sprang an essential piece of progress in thermodynamics – not by myself, but by my friend Carathéodory. I tried hard to understand the classical foundations of the two theorems, as given by Clausius and Kelvin; they seemed to me wonderful, like a miracle produced by a magician's wand, but I could not find the logical and mathematical root of these marvellous results. A month later I visited Carathéodory in Brussels where he was staying with his father, the Turkish ambassador, and told him about my worries. I expressed the conviction that a theorem expressible in mathematical terms, namely the existence of a function of state like entropy, with definite properties, must have a proof using mathematical arguments which for their part are based on physical assumptions or experiences but clearly distinguished from these. Carathéodory saw my point at once and began to study the question. The result was his brilliant paper, published in *Mathematische Annalen*, which I consider the best and clearest presentation of thermodynamics. I tried to popularize it in a series of articles which appeared in *Physikalische Zeitschrift*. But only a few of my colleagues accepted this method, amongst them R. H. Fowler, one of the foremost experts in this field. Fowler and I intended, a few years ago, to write a little book on this subject in order to make it better known in the English-speaking world, when he suddenly died. That will, I suppose, be the end of it, until somebody re-discovers and improves the method. [21]

On 12 February 1921, some time before publishing his attempt to popularise Cara-théodory's theory, Born wrote to Einstein: "Firstly, I have written an account of Carathéodory's thermodynamics that will appear some time in the *Phys. Z.* I am really curious to hear, what you will say about it. Carathéodory himself, to whom I have sent the printed sheet in Smyrna, found the account to his liking." [22] On 24 March 1949, Carathéodory responded to Born again to thank him for a book he had received. He did not mention the title but it was apparently Born's *Natural Philosophy of Cause and Chance*. "Your presentation of thermodynamics is superb. That you have not forgotten me through all these long years and you have cited me in so friendly a way is wonderful." [23] Carathéodory was mentioned in the book as Born's mathematical friend who had shown that the laws of thermodynamics were connected with the properties of the Pfaff equations and who had established the so-called Cara-théodory principle, namely that adiabatically inaccessible states exist in any vicinity of a given state. [24] Despite all efforts, Born's attempts to popularise Carathéodory's foundation of thermodynamics did not have the expected effect of displacing the classical, and in his opinion ponderous and mathematically inscrutable, method.

From the mathematical point of view, the procedure of inferring the existence of an entropy function on the basis of the Clausius principle is indeed unsatisfactory and that is the reason for Carathéodory's axiomatic treatment. However, in his *Rational Thermodynamics*, [25] C. Truesdell harshly criticised Carathéodory's treatment of thermodynamics, which he himself described finally as "an axiomatization for axiomatization's sake, not attempting to solve any physical problem beyond those already solved informally". He went on to say that "Carathéodory's greatest legacy of disaster is his attempt to define heat in terms of work and energy. Any such attempt must fail, since it can neither come to grips with the idea of heat supply as something corresponding in thermodynamics to force in ordinary mechanics nor distinguish between supplies of heat at different temperatures."

Planck's attitude towards Carathéodory's treatment is also interesting. On the one hand, he did not accept its claim to be a fundamental theory because of the statistical character of the second theorem but, on the other hand, he was impressed by Carathéodory's mathematical deduction and, like Carathéodory himself, he refrained from introducing cyclic processes and ideal gases. [26]

Carathéodory's paper *Über die Bestimmung der Energie und der absoluten Temperatur mit Hilfe von reversiblen Prozessen* (On the Definition of Energy and Absolute Temperature with the Aid of Reversible Processes) [27] was preceded by a conversation with Max Planck in Munich in December 1924. Unlike Planck who, according to his own statement, found it increasingly difficult to quickly leave a matter in which he was engaged and to quickly seize it again in a more favourable moment, [28] Cara-théodory was able to return to his previous efforts to establish thermodynamics. With the aid of the most elementary means, he tried to prove that – provided that reversible thermodynamic processes could be realised on their own in nature – the scale of absolute temperature could be defined, although the location of the absolute zero on this scale would remain undefined. In other words, the zero point of the absolute temperature scale could be fixed only with the aid of an irreversible process. His method consisted of the construction of thought experiments on which the axioms were applied. The axioms themselves had to satisfy "the observed things" and logical conclusions to be deduced from the given premises. Sixteen years earlier, he had pursued this direction in his *Studies in the Foundations of Thermodynamics* and had concluded that "if the value of entropy has not remained constant in some change of state, then no adiabatic change of state can be found that could transfer the observed system from its final to its initial state. Every change of state in which the entropy value varies is 'irreversible' ".[29]

In the prologue to the 8th edition of his *Vorlesungen über Thermodynamik* (Lectures on Thermodynamics),[30] Planck mentioned the doubts he had expressed at a session of the Physics-Mathematics Class of the Prussian Academy of Sciences[31] about the value of the principle used by Carathéodory as a starting point for his reasoning: "In any proximity of any given initial state there are states which cannot be approximated by an adiabatic change of state."[32] Planck argued that the statement containing Carathéodory's principle was not valid for all systems for which the quasi-ergodic hypothesis applies and that, although Carathéodory's principle dealt with the non-feasibility of certain neighbouring states, it did not present any feature through which the attainable neighbouring states could be distinguished from the unattainable ones. These considerations had incited Planck to develop a foundation of the second heat theorem on the basis of the Thomson principle of the impossibility of a *perpetuum mobile* of the second kind, which, to his mind, possessed the advantages of Carathéodory's train of thought without being subject to any doubts.

2.3
Marriage, a Family Affair

On 3 February 1909, Carathéodory asked for permission to conclude his lectures on technical mechanics before the end of the term, namely on 12 February, in order to undertake a journey.[33] On 18 February, in the municipality of Xirokrini, Constantinople, he married Euphrosyne Carathéodory, his aunt and eleven years his junior. The matrimonial ceremony was performed by the parish priest, Ioannis, according to the rites of the Eastern Orthodox Church. Carathéodory's sister, Ioulia Streit, and Euphrosyne's brother, Pavlos Carathéodory, were the witnesses.[34]

By taking a relative as his wife, Constantin was following a tradition of intermarriage within the neo-Phanariot circles, a strategy adopted by the Carathéodorys to further the growth of their political influence and financial power, and which helped to strengthen the bonds between the members of a privileged social group; the birth of many children ensured the numerical expansion of the Constantinople Greek elite. Euphrosyne was one of seven children, two sons and five daughters, of Alexander Carathéodory Pasha and Cassandra Mousouros.

Euphrosyne as a bride.
Courtesy of Mrs. Rodopoulos-Carathéodory.

Alexander was a renowned polymath who was said to speak sixteen European, African and Asiatic languages.[35] As deputy foreign minister of the Ottoman Empire, he had participated in the Constantinople Conference,[36] in which the Great Powers attempted to mediate a reform programme for the Christian insurgent regions of the Balkans (Bosnia, Herzegovina and Bulgaria). Alexander's international fame, however, is due to his representing the Empire as the first Ottoman plenipotentiary at the Berlin Congress convened in 1878 on the insistence of Great Britain and Austria-Hungary by the Austrian foreign minister, Count Gyula Andrássy,[37] to revise the preliminary Peace Treaty of San Stefano (Yesilköy in Turkish). The Treaty of San Stefano had concluded the Russo-Turkish war of 1877–1878 by creating a Greater Bulgaria extending from the Aegean to the Black Sea, and Russia was supposed to have set up the principality within two years.

The outcome of the 1878 congress was that Russia was denied the means to extend its naval power, and its international position was weakened, while the Ottoman Empire retained its status as a European power and thus the interests of Great Britain were satisfied; further, it allowed Austria-Hungary to occupy Bosnia and Herzegovina, thereby increasing its influence in the Balkans. However, by failing adequately to consider the aspirations of the Balkan peoples, the congress sowed the seeds for future crises in the Balkans. The British diplomat Edwin Pears writes that Alexander, "a Greek of quite exceptional ability, [...] struggled to preserve the interests of his master as stoutly as any man could do. He was constantly and rudely snubbed by Bismarck, who told him in so many words that he was there to accept what the Powers dictated." Pears mentions further that, with one exception, Alexander was "the only Christian ever allowed to be Minister of Foreign Affairs, the tradition being that a Ghiaour should occupy that office whenever territory had to be ceded, the evident intention being that any odium connected with such surrender should not fall upon a Believer."[38] The Ottoman Empire departed from the congress having lost two-fifths of its territory and one-fifth of its population.[39]

During his office as *Vali* (Governor General) of Crete in that same year (1878), Alexander was appointed Foreign Minister by the Sublime Porte. From 1885 to 1894 he ruled over Christians and Muslims as the Prince of Samos. Despite his endeavours to modernise the island through measures to improve health, education, the penal system and administration, he used the force of the Ottoman army to suppress riots that broke out in resistance to his attempts to combat phylloxera, which was destroying the famous vineyards of Samos. Bringing in the Ottoman army was a violation of the autonomy of the island and Alexander was relieved of his post in 1894. The Porte sent him back to Crete as Governor General in May 1895, in an attempt at a reconciliation with the Cretan Greeks.

It was in August that same year that Alexander Pasha received a visit from his nephew in Chania. There, Constantin Carathéodory met the future Prime Minister of Greece, Venizelos, for the first time.[40] Venizelos had entered the political scene six years earlier, when he called for a boycott of the elections and the General Assembly on the island, protesting against the suspension of representative institutions by Sultan Abdülhamid II. Despite the initial enthusiasm that the Pasha's arrival evoked among the Christian population, Alexander failed in the end because the administrative apparatus resisted his reforms which aimed at restoring the profitability of public finances and also because he himself proved incapable of dealing with the demands of the Christian population. Venizelos had refused to participate in the Ottoman administration of Crete despite offers made to him by the Pasha. Moreover, he even opposed the autonomy of the island under the guarantee of the Great Powers and supported instead Crete's incorporation into Greece. Constantin Carathéodory was obviously very sympathetic towards Venizelos's political convictions whereas Alexander Pasha, saying that he had daughters to marry off, showed little interest in Greek nationalism.[41] Carathéodory's meeting with Venizelos in Chania in the late summer of 1895 had marked the start of a new friendship based on mutual respect. Years later, Carathéodory, recounting their acquaintance, told Penelope Delta[42] of his admiration for Venizelos's intelligence and humanism. Through his subsequent connection with Venizelos, Carathéodory placed himself in the service of Greek nationalism and liberalism long before the Great Idea was adopted by Ottoman Greeks like his great uncle, who at that time still envisaged an oriental Graeco-Ottoman confederation.

Euphrosyne. Photograph dated May 1906 with dedication to Constantin.
Courtesy of Mrs. Rodopoulos-Carathéodory.

A year after his appointment as Governor General of Crete, the Pasha was recalled to Constantinople. Until his death in 1906 he carried out the tasks of First Translator of His Imperial Majesty the Sultan, Abdülhamid II, and supervised the work of the Ottoman Foreign Ministry. Abdülhamid, who trusted Alexander and appreciated him as not only the most brilliant diplomat in the service of the Ottoman Empire but also one of the most intelligent in Europe, gave him an apartment in Yildiz.[43]

The Ottoman poet E. Fazy describes Carathéodory Pasha as

by far the first man of the Orient. Why did fate paralyse this brain in Istanbul, in the sterile atmosphere of Muslim offices, this brain which, otherwise, in an active environment, would have been able to change the character of things, restore his race and save Hellenism?

Carathéodory belongs to a great Phanariot family. In him lies the substance of several scholars. He unites his knowledge of about ten languages, which he speaks and writes fluently, with that of mathematical sciences. The competent amateurs are researching into his French version of Nasreddin Eltoussi's *Treatise on Trigonometry*. He is a doctor of law from the University of Paris. Being of a superior, modest and firm character, he possesses, they say, beneath an air of timidity, full of moral courage, the Christian faith of a Pascal. In the Orient, politics and its necessities corrupt nearly everything. But one can have no doubts about the sincerity of a Carathéodory Pasha.

After having been irreparably wrong to propose to the Sultan the institution of a responsible and ... autonomous ministry, Carathéodory had to go through diverse tasks, which he did not deserve: he was even sent to govern Crete!

Alexander Carathéodory Pasha. Sketch from T. Tsonidis's book
To Γένος Καραθεοδωρή (The Carathéodory Family) (Orestias 1989).
Courtesy of Mrs. Rodopoulos-Carathéodory.

Now, he has his apartments in the Palace and bears the title of Translator to His Majesty. (Nevertheless, he insists on preferring to stay at the Sublime Porte ...) There is no awkward document or annoying complaint that is not submitted to him. And, in situations of danger, this ghiaour is asked for advice. But because it is God's will, he is wasting away and, until his death, he will be wasting away in the common misery, in which all the Nazarene servants of the Islam are declining.[44]

Euphrosyne's mother, Cassandra Mousouros, was descended from an old-established and powerful family who had been able, soon after the Phanariots' disaster of 1821, to regain access to both the Danube Principalities and to Constantinople.

Euphrosyne's grandmother, Loukia Mavrocordato-Carathéodory, was a direct relation of the Great Dragoman Mavrocordato *ex aporriton* (εξ απορρήτων),[45] who became a role model for all Phanariots by his invention of the doctrine of co-operation between subjects and rulers within the Ottoman Empire.[46] Loukia was also a cousin of Alexander Mavrocordato, one of the most significant politicians of the new Greek state. He had a clear political leaning towards Great Britain and, as President of the first Greek National Assembly, he had proclaimed the manifesto of Greek independence in 1822. Through her marriage to Stephanos Carathéodory, Loukia Mavrocordato, a highly educated, socially and intellectually active figure, became the link between the Phanar and the Second Phanar traditions.

Euphrosyne's grandfather and Loukia's husband, who was said to be able to read the Bible in sixteen languages, had studied medicine in Pisa, just like his nephew Constantin, Constantin Carathéodory's grandfather. Constantin had acquired the additional qualification of eye-specialist in London. Both Stephanos and Constantin were physicians to Sultan Mahmud II and his son and successor to the throne, Abdülmecid.[47] Both of them lectured in Turkish and French for many years at the Imperial School of Medicine, a higher secular school, which Sultan Mahmud II had founded in Constantinople in 1827 to train doctors for the new army. In his *Autobiographical Notes*, Constantin Carathéodory emphasises the definite European character of Mahmud's reforms and the modernisation measures accompanying them that were taken against the initial reaction of the *Ulema* (Islamic scholars of law). Among the measures, he mentions an order that was to become the basis of the modern Turkish medical service and the isolation wards which his great uncle and his grandfather set up in their successful attempt to combat the plague that was depleting the population of Constantinople in 1836.

Constantin Carathéodory's marriage to his aunt was dictated by his desire for social advancement through his connection to Phanar's nobility. Through his father he had secured connections in the West; through his father-in-law he acquired connections in the East. Of course he could have resisted the temptation of extending his political power and he could have just concentrated on building up his obviously very promising mathematical career in Germany. But he decided otherwise, thus meeting the expectations of his family and also making up for their disappointment in his choice of profession. From now on his life would follow twin tracks, dedicated to mathematics and politics, to the Orient and to the West. Constantin took his wife to Germany and they started a family there, but he only completely turned his back on the Orient after the Greek national disaster of 1922.

The timing of Constantin's marriage had nothing to do with political developments in the Orient, such as the Young Turks' revolution, which had broken out some months before. After all, members of the Carathéodory family continued to live on in Constantinople. It was only connected with his intention to prove to his relatives that, having secured the professorship, a brilliant career was awaiting him and he would be in a position to care for a family of his own.

In his obituary of Carathéodory, Heinrich Tietze underlines the trust that existed between Constantin and Euphrosyne right from the start of their marriage:

Anyone who knew his wife, with her admirable spirited optimism, may well feel that [even after the catastrophe of Smyrna] she did not despair. She was a remote relative of her husband, whom he married before taking up his professorship in Hannover. [...] And he travelled with [his wife] who certainly spoke French but not a word of German and who attended an opera for the first time during a short stopover in Vienna on their way from her country to Hannover. But she had followed him with confidence into the foreign conditions.[48]

Perron describes Euphrosyne as a woman with a lively sunny temperament, who soon felt at home in Germany and who always shared joy and suffering with her husband.[49] She spoke French, Greek and Turkish at the time she got married, but she soon learned German and later English during the couple's visit to the United States in 1928. Members of the Greek community in Munich remember her as a very discrete, hospitable, warm and kind person, absolutely dedicated to her family.

2.4
First Professorship in Hannover

Carathéodory remained in Bonn for only a very short period. While he was there, he associated with Erhard Schmidt who, however, attained a full professorship in Zurich in 1908, and with Eduard Study, with whom he worked on the isoperimetric problem in early 1909. The paper *Zwei Beweise des Satzes, dass der Kreis unter allen Figuren gleichen Umfanges den größten Inhalt hat* (Two Proofs of the Theorem that among all Figures with the Same Perimeter the Circle has the Most Content)[50] sprang out of this co-operation. Both proofs were purely geometric but differed from each other. Carathéodory used approximations through curves, Study through polygons. Each one signed his own work separately.

On 20 April 1909, the board of trustees of the University of Bonn informed the philosophical faculty that Carathéodory had been appointed "budgeted" *(etatsmäßiger)* professor to the Royal Technical University of Hannover. He was to occupy the Chair of Higher Mathematics as Paul Stäckel's successor.[51] That same day, Carathéodory also informed the faculty accordingly,[52] which in turn proposed Gerhard Hessenberg, professor at the Academy of Agriculture in Bonn from 1907, as his successor.[53] Hessenberg was appointed only a year later to the Royal Technical University of Breslau, where he would later co-operate with Carathéodory.

On 9 March 1908, more than a year before Carathéodory's appointment in Hannover, the Minister of Education had asked the Department for General Sciences of the Royal Technical University of Hannover to submit proposals for a successor to Stäckel, who had accepted an appointment to Karlsruhe.[54] As a result, on 26 March 1908, the head of department presented a list with the names of Friedrich Karl Wieghardt, already professor at the Technical University of Hannover, Georg Hamel, professor at the Technical University of Brünn (today Brno), and Heinrich Timerding, professor at the University of Pressburg (today Bratislava), taking into account "the condition that the professor to be appointed, besides having fully proved teaching skills and a comprehensive knowledge of mathematics, should also have a clear understanding of its application to engineering." The arguments for the names on the list were that Wieghardt already had lecturing experience, possessed a lively and fresh mind and a gift for teaching, and his works, directly gathered for the most part from practical experience and requiring great mathematical skill, dealt mainly with mechanics. Hamel had a similar history; his mathematical works also mainly concerned mechanics and he had been particularly recommended by some of the professors in Karlsruhe. Timerding was competent in analysis and had occupied himself in detail with the practical problem of a ship's motion. Much was expected of him.[55]

In a letter to the rector on 22 April 1908,[56] the minister supported the continuation of Wieghardt's teaching commission and even an increase in Wieghardt's remuneration in return for his good performance, but he instructed the Department for General Sciences to submit additional proposals for Stäckel's replacement and to express their opinion on Fricke (Braunschweig), Carathéodory (Bonn) and Maurer (Tübingen) without getting in touch with them. The head of department replied on 4 May 1908[57] that Fricke would have been on the original list had they not had doubts about his work, which was not so closely related to engineering as that of Wieghardt. If Wieghardt were to have been elected to the chair, then Carathéodory's name would probably have been raised as a possible successor to him. Carathéodory was felt to be a kind of traitor to the engineering profession: "This gentleman was earlier a civil engineer but then he devoted himself to pure mathematics and he finds satisfaction in the occupation with problems of an abstract nature. He is warmly recommended by his former colleagues from Göttingen; but it should at least be questionable whether he would again return to engineering subjects with particular affection; in any case, he would firstly have to produce the proof of his qualifications through a fairly long teaching activity." Maurer was recommended neither for the one nor for the other position. Though found to be competent and well-versed in analysis, he was not considered, on the basis of his knowledge, as being capable of contributing to teaching at a technical university.

The department declared that they were not in a position to submit other proposals and the matter remained in the air until the minister demanded a statement from the department and the submission of new proposals.[58] On 27 February, the department informed the rector and the senate that, by a majority of four to one, they insisted on the previous list and on the candidates' names in the same order.[59] Only Dr. Kiepert, a close friend of Klein, voted against this list. In support of their vote, the department added the argument that Wieghardt owed his popularity among the students to particularly skilful teaching and stimulating lectures. Members of staff had attached the greatest importance to this, since they considered that, apart from rare exceptions, students at a technical university lacked initial interest in mathematics and this interest had to be instilled. According to their arguments, Wieghardt stimulated the students to be independent and he put all his energy into the mathematical treatment of difficult problems in engineering mechanics. Finally, the majority of staff stressed that despite Felix Klein's expressed view that Wieghardt lacked the universality of mathematical thought, they continued to support his candidacy, since Wieghardt's special field was of equal interest to both mathematicians and engineers. Also Wieghardt's appointment would in no way do harm to the guiding principles of mathematical instruction at technical universities as they had been established by Klein at the *Lehrerfeier* (Teachers' Day) at the *Deutsches Museum* on 1 October 1908. Moreover, in their opinion, not only the comments of the mathematicians but also those of the engineers had to be taken into account when considering the appointment to mathematical chairs at technical universities. Before writing the first report on 26 March 1908, the head of department had consulted engineering colleagues and received favourable information about Wieghardt's teaching activities. They thought that the gentlemen named for the second and third places fulfilled the same prerequisites as Wieghard, albeit to a lesser degree. The assessment in favour of Wieghardt was based on the experience of the university's own staff and not on the information given by other specialists and therefore was regarded as considerably more reliable.

On 3 March 1909, Kiepert submitted a special recommendation to the minister.[60] Although he acknowledged Wieghardt's good qualities as expressed in the report from the department, he said that he would propose Carl Friedrich August

Gutzmer, professor of applied mathematics at the University of Halle, and Carathéo-dory, lecturer in applied mathematics at the University of Bonn, for the first place on the list. The three names proposed by the majority of the teaching staff would follow in the list after these two. Kiepert supported his recommendation with the argument that professorships for higher mathematics ought to be occupied by mathematicians, who admittedly should have a lively interest in the engineering applications of math-ematics, but who would basically pursue mathematics for mathematics' own sake. He added that Klein's reservations regarding Wieghardt could not simply be disre-garded, since Wieghardt had been Klein's assistant and therefore Klein knew his work intimately. Kiepert argued that he had asked for the opinion of many colleagues with regard to the occupation of the chair. Klein had written to him that although Gutzmer was scientifically inferior to Carathéodory, he was an excellent teacher and an organ-iser of the first rank. Since it was not at all sure whether Gutzmer would accept the professorship, Kiepert expressed the view that Carathéodory's appointment would also greatly profit the university. He then gave an account of various assessments by Carathéodory's academic teachers: he was "an excellent, gifted mathematician" and his most significant works were "met with lively approval in the circles of col-leagues." Kiepert added the opinion of Schwarz and Friedrich Albert Wangerin, a full professor at the University of Halle since 1905, as well as that of Klein, Runge, Blumenthal and Stäckel. A telegram from Schwarz praised Carathéodory for his "excellent scientific activity", his "background in engineering" and his "excellent characteristics." Wangerin's letter presented Carathéodory as qualified for a profes-sorship because of a number of competent published works. Klein added that he would even prefer Carathéodory to Hamel on the basis of Carathéodory's general interests and his superiority in engineering and he referred to Carathéodory's civil engineering activity at the Assiut dam. Runge underlined Carathéodory's suitability for the professorship, since he was "a fine head, independent and original as a math-ematician, with a lively interest in applied mathematics, although his works lie in the areas of pure mathematics and mathematical physics. As a man he is very broadly educated, speaks English and French just as well as German, has seen much of the world and has moved in many circles."[61]

In his review to the minister, Kiepert also referred to Stäckel, the former holder of the chair, who in May 1908 had named Carathéodory as the desired candidate for his professorship. He continued by saying that Blumenthal had reported to him from Aachen that Carathéodory was preparing two publications testifying to his interest in applied mathematics, namely "a strict formulation of the second main theorem of thermodynamics, which pleased Schwarzschild very much, and [...] a strict derivation of the surface equation from the general elastic equations." Kiepert attached to his review Carathéodory's curriculum vitae and a list of the latter's scientific works and lectures. The publications mentioned by Blumenthal were apparently Carathéodory's *Studies in the Foundations of Thermodynamics* and a paper *Über den Variabilitäts-bereich der Fourierschen Konstanten von positiven harmonischen Funktionen* (On the Variability Area of Fourier Constants of Positive Harmonic Functions).[62]

The minister informed the University on 6 April 1909 that the decision had been made in favour of Carathéodory, who would begin lecturing in Hannover in the coming term.[63] After short negotiations, Carathéodory was appointed on 16 April with effect from 1 April 1909. A salary providing for a pension and including the statutory housing benefit of 660 marks was fixed at 4500 marks *per annum*. In addition, Carathéodory would be paid an amount for lecturing and taking examinations that was determined by law.[64] On 7 June 1909, in the Rector's office and in front of the Rector of the Technical University, Dr. Ost, and the transcript writer, he took the following oath: "I swear to God, the Omnipotent and the Omniscient, that I will be subservient, loyal and obedient to his Majesty Wilhelm, King of Prussia, my most powerful master, and I will faithfully fulfil the duties of my office to the best of my knowledge and consciousness. I will also conscientiously uphold the constitution, so help me God."[65]

Thus, Carathéodory owed his first professorship basically to Klein, who established direct contact with the ministry and insisted on his favourite candidate, disregarding the will of the appointment committee of the Hannover Technical University. Within the university, Kiepert supported Klein's wish and thus contributed to the success of the operation. This *modus operandi*, which did nothing for the autonomy of the university, was not unusual. Carathéodory, as may be expected, was not against it and, indeed, only a while later adopted the same method himself to favour the appointment of his successor.

Carathéodory did not in fact stay in Hannover for long. After a period of only three terms, he left the Technical University to accept an appointment in Breslau (today Wrocław). In the meantime, he had become a father for the first time; his son Stephanos was born on 7 November 1909. Carathéodory must have already known of his appointment in Breslau in the spring of 1910, when he contacted Felix Klein to discuss his own successor in Hannover. On 14 March 1910,[66] he wrote to Klein that Conrad Müller, former librarian of the mathematical reading room in Göttingen and at that time a librarian in Hannover, was his favourite for the job. He held a high opinion of Müller's mathematical performance (he had contributed to vol. IV of Klein's *Encyklopädie*), his knowledge in applied mechanics, in which he was just as good as Wieghardt although on a completely different level, and his positive personality. "The more I think about this idea, which initially seemed to me a bit paradoxical, the more I find that it has a lot to offer and it must be considered seriously." Carathéodory reckoned with the opposition of his colleagues to his proposal in favour of Müller, since "also personal motives play a powerful part." As alternatives, he named Wilhelm Ahrens, Paul Koebe, Georg Hamel and Hans Hahn and informed Klein that Kiepert was in the picture. He asked Klein for his advice, which he promised to follow. To Klein, the idea of Müller's appointment seemed realisable,[67] although – as Carathéodory wrote – it would be difficult for him to part with Müller and the latter's appointment would be a major loss for the university library.

In the end Conrad Müller was given the professorship at the Technical University. In the autumn of 1910, Müller was assisted in running the classes on mathematical exercises by his friend Georg Prange, who travelled in from a village on the Deister hills (to the southwest of Hannover) once a week.[68] And in 1941, when Prange died, Conrad and Martha Müller visited Carathéodory in Munich, probably to seek his advice regarding Prange's successor, who should also be "a qualified mathematician, having connections not only with the real applications of mathematics but especially with problems of the theory of electricity and at least be close to theoretical physics through the area of his work."[69]

Carathéodory's involvement in matters concerning educational policies had started as soon as he acquired his professorship in Bonn.[70] From then on, he started building a career for himself, moving from one university to another to improve his income and his status and began to acquire managerial skills.

2.5
Professor at the Royal Technical University of Breslau

In March 1910, Carathéodory heard in Berlin that Steinitz seemed to have the greatest chance of obtaining the professorship for mathematics in Breslau. He informed Klein[71] accordingly and remarked that "Dehn was much less abstract and therefore much more qualified". Carathéodory added that during the previous years Dehn had had a great deal of influence over the students in Münster. He could, however, very well understand the reasons that led the ministry to support Steinitz.

Ernst Steinitz, a Jew from Upper Silesia who for many years had to "struggle along on very little to eat",[72] had obtained his habilitation in 1897 and had been an associate professor at the Technical University of Berlin-Charlottenburg since 1903. His pioneer work of 1910, *Algebraische Theorie der Körper* (Algebraic Theory of Bodies), in which he gave the first abstract definition of a field, established him as the founder of abstract algebra. As it turned out, he was indeed appointed to the Technical University of Breslau that year but again only as an associate professor. Being a pure mathematician, Steinitz was not considered particularly suitable for a technical university. His appointment later to a full professorship at Kiel University in 1920 must have been a great relief to him.

Max Dehn had begun his career as a student of Hilbert's producing work on the foundations of geometry[73] and obtained his habilitation in Münster in 1901. In 1910, he produced a paper on the topology of three-dimensional space.[74] With the help of group-theoretical methods, he proved that the trefoil knot was not isotopic with a circle in that space. In 1911, Dehn became an associate professor in Kiel. Carathéodory expressed his joy to Hilbert[75] about Dehn's appointment to Kiel. His favourite candidate was "a passionate humanist, incessantly endeavouring to exactly and open-mindedly understand human beings and reality. He was convinced that the occupation with spiritual things made people happy and liberated them from arrogance and bias, fear, hatred and greed."[76] Dehn's suitability for a technical university is sustained by his remark to Fraenkel that "young mathematicians ought not to concern themselves with such abstract subjects but treat more concrete problems instead."[77] In 1913, when Carathéodory left Breslau three years after his appointment there, Dehn succeeded him. He was, however, also considered for a professorship in Zurich. On 27 June 1913, Frobenius replied to a confidential request from the President of the Zurich Polytechnic regarding the occupation of a chair in Zurich:

"I would also put *Weyl* in first place. His recently published book about Riemann surfaces is being praised a lot. By the way, it is said that he has the prospect of being appointed to Breslau in place of Carathéodory. [...] To summarise my opinion once more: if you appoint Weyl, you will make an excellent and indisputable choice. But if he is not available any more, then take Dehn or Hellinger."[78]

After negotiations with the ministry, Carathéodory was transferred to the Royal Technical University of Breslau as a budgeted professor for higher mathematics on 26 March 1910. He had to start his work there on 1 October.[79] In Breslau, Carathéodory endeavoured to organise the mathematical instruction in accordance with Klein's guidelines; to this end he called on the help of the Ministry of Spiritual, Medical and Instruction Affairs, the Rector Rudolf Schenk and the Department of General Sciences of the Technical University. The ministry inquired on 27 December 1910 about the necessity of permanent assistant positions for the two chairs for higher mathematics, which, wherever possible, were expected to devote an equal number of hours to the supervision of exercises as to their lectures. Carathéodory and Steinitz explained the reasons for the employment of assistants in their reply of 12 January 1911: an assistant was indispensable in helping the students to directly apply what they had learned from the lectures and in modelling instruction to the individual needs of the students. Freed from this obligation, the professor would be able to cope with extensive independent work and keep it up to date. Carathéodory and Steinitz advocated the employment of a permanent assistant who would cost the university less in the end than the minimum twelve-hour weekly commission. They were against the employment of an honorary assistant, who would not be allowed to work beyond the obligatory hours. A letter dated 30 November 1912 proves that the ministry had accepted Carathéodory and Steinitz's proposal, which was also supported by the Department of General Sciences. A permanent assistant was engaged, but because of the limited budget, his work came to an end in 1914.[80]

Carathéodory was optimistic about the prospects of his activity in Breslau. "Contact with the other colleagues", he wrote to Hilbert on 2 February 1911,[81] "promises to be very stimulating. We even revived a technical-mathematical colloquium that started off very well. In addition, the physics colloquium at the university is also very interesting for the mathematician and I learn many things there." In another place he wrote:

Minkowski's collected works arrived yesterday and I would like to immediately thank you most warmly for this friendly gesture. I was very happy to leaf through them and to find all kinds of things which reminded me of my Göttingen time. But I did not expect that through a mere compilation of Minkowski's works his significance would increase to such a great extent, even for me, who knew him well and appreciated him highly. On every page one opens, there are fine, beautiful, deep thoughts. There is, for example, the work about convex polyhedrons and the other about the motion of a rigid body in a fluid,[82] both of which I did not know and whose elegance have made a great impression on me. I mostly regret that the first volume means little to me. Maybe, I will have the opportunity to become familiar with its subjects later.

The first volume contained a series of papers on the theory of quadratic forms and another one on the geometry of numbers. As Hilbert wrote in his memorial speech for his friend,[83] Minkowski's pure mathematical investigations of quadratic forms were generalised and culminated in the *Geometrie der Zahlen* (Geometry of Numbers). This geometry had led Minkowski to work on convex bodies and on ways in which figures of a given shape can be placed within a given figure. Minkowski's collected works were published by Hilbert with the help of Andreas Speiser and Hermann Weyl in two volumes in Leipzig in 1911. Hilbert had set himself this task directly after Minkowski's death,[84] and, for the preparation of the edition he had read more than ninety of Minkowski's letters to him since their common time together as students.[85]

Carathéodory announced to Hilbert later[86] that he much valued the fact that Erhard Schmidt would live in the same city as him again. Schmidt had acquired a full professorship in Breslau in 1911. Probably in Breslau Carathéodory also got to know Emil Toeplitz, Toeplitz's father.[87] Carathéodory was in continuous communication with the physicists and he had promised them that he would present a paper on the common work of Sommerfeld and Fräulein Runge in one of the winter sessions. He found the scientific circle in Breslau "very pleasant" and described his own activity at the Technical University with the same words, this time in a letter to Hilbert on 5 May 1912,[88] "particularly because we can allow the students to make considerable progress."

Carathéodory had to adjust his programme to meet the needs of the engineers at the Technical University of Breslau. At the Department of Mechanical and Electrical Engineering, he belonged to a number of committees including, with Steinitz, the committee for the pre-diploma examination in Higher Mathematics and, with Mann, the committee for the main diploma examination in the Mathematical Treatment of Technical Problems. Hessenberg and Mann taught mechanics, descriptive geometry and graphic statics.[89] Documents from 1912 show that Carathéodory also held the office of Vice-Rector at the Technical University.[90]

As regards his mathematical work during his years in Breslau, which we shall look at in the next section, Carathéodory mainly concentrated on classical function theory, i.e. the Picard theorem, its extension by Landau, the study of functions that map uniformly bounded domains onto the unit disc, the possible simplest proof of the Riemann mapping theorem and the theory of prime ends.

2.6
Theory of Functions

2.6.1
The Picard Theorem

Soon after his habilitation, Carathéodory devoted himself to the theory of functions. Henri Poincaré's nephew, Pierre Léon Boutroux, who was in Göttingen for a few days shortly after the summer semester in 1905, talked to Carathéodory about his efforts to simplify Borel's proof of the Picard theorem.[91] Carathéodory wrote:

Boutroux had noticed that this proof was successful only because in the case of conformal mappings there was a remarkable rigidity, which, by the way, he was not able to put into formulae. Boutroux's discovery did not let me rest and six weeks later I was able to prove Landau's sharpening of the Picard theorem in a few lines by using the theorem which is today called the lemma of Schwarz. I produced this theorem with the help of Poisson's integral; only through Erhard Schmidt, whom I had informed of my findings, did I learn not only that the theorem already exists in the work of Schwarz, but also that it can be gained by absolutely elementary means. Indeed, the proof, which Schmidt informed me about, cannot be improved. Thus, I gained a further field of activity apart from the calculus of variations.[92]

Perron remarks that Carathéodory's "début" in the theory of analytic functions "was a note in volume 141 of the Paris *Comptes Rendus*,[93] small in size, but completely great in content." Perron continued

A year earlier, Landau had been able to extend the Picard theorem through the exciting discovery that the convergence radius of a power series

$$a_0 + a_1 z + \ldots (a_0 \neq 0, \ a_0 \neq 1; \ a_1 \neq 0),$$

which omits the values 0 and 1 within its circle of convergence, can at most equal a limit depending only on a_0 and a_1. However, the limit found by Landau was not the best possible. Carathéodory found then what is today called the Landau radius;[94] this is equal to

$$\frac{|\tau(a_0) - \overline{\tau(a_0)}|}{|a_1 \tau'(a_0)|},$$

where $\tau(z)$ denotes the inverse of the elliptic modular function, which maps the upper z-half plane onto the circular triangle of the τ-plane with the vertices $0, 1, \infty$, whose three angles equal 0. This was a great discovery, which revealed at the same time that the so-called elementary methods for the proof of the Picard theorem are not entirely appropriate and that Picard's original proof with the aid of the modular function, which at first appeared far-fetched, was in fact natural. In the same short note, however, Carathéodory had developed a further brilliant idea. The above mentioned triangle with angles 0, in union with its mirror images, covers the whole upper τ-half plane schlicht, and the performance of the function $\tau(z)$ in the Picard–Landau context of ideas is to a great part based exactly on this.

Now Carathéodory asks[95] why shouldn't something analogous be established by other Schwarz triangle functions, where the angles are not 0, but $\frac{\pi}{m}, \frac{\pi}{n}, \frac{\pi}{p}$ with integers m, n, p, for which $\frac{1}{m} + \frac{1}{n} + \frac{1}{p} < 1$? In this case, instead of the logarithmic points we arrive at m-, n-, p-sheeted branching points. And, indeed, he arrived at the following surprising theorem: if the order of every 0-place is divisible by m, the order of every 1-place by n and the order of every pole by p in the case of a regular function $f(z)$, except for the pole, in the circle $|z| < R$ with a Taylor series expansion

$$f(z) = a_0 + a_1 z + \ldots (a_0 \neq 0, a_0 \neq 1; a_1 \neq 0),$$

then

$$R \leq \frac{1}{|a_1 \psi'(a_0)|}$$

where ψ denotes an appropriately normalised triangle function with the angles $\frac{\pi}{m}, \frac{\pi}{n}, \frac{\pi}{p}$. This limit is again the best possible.[96]

On 15 January 1908, Carathéodory thanked Friedrich Schottky[97] for showing interest in his "C. R. note" and for mentioning him in such a way that he had by no means expected. He expressed self-criticism regarding an error at the end of his "C. R. note", the existence of which Schottky had indicated: "As you will have certainly seen from the text of my work", Carathéodory wrote, "I have rather occupied myself with the general theory of functions, but never with automorphic functions in detail. Of course,

I certainly knew that the task of mapping originating from *three* circles should be solved and that it leads to the mapping of two circular hexagons onto each other, but I had the wrong idea that this could be achieved by the quotient of *ordinary* hypergeometric functions. This is all the more inexcusable, since a little reflection would have put it right."

By "C. R. note" Carathéodory meant his 1905 note in the *Comptes Rendus*,[98] which he had concluded with the remark: "Thus, the case in which for $|x| = 1$ the function $y = f(x)$ is subject to taking only the values in the plane of y at the exterior of three circles, all of them being exterior to each other, will lead us to consider the quotient of two hypergeometric functions anew."[99]

In the *Beiträge zur Konvergenz von Funktionenfolgen* (Contributions to the Convergence of Power Sequences),[100] which emerged from the correspondence between Carathéodory and Edmund Landau in the winter of 1910–1911, a particular theorem was formulated and proved for the first time. It contained as special cases all theorems of a chain, which successively, from always fewer assumptions, provided the proof that a certain expression represents an analytic function. This theorem reads:

Let the analytic functions $f_1(x), f_2(x), \ldots, f_n(x), \ldots$ be regular for $|x| \leq 1$. Let a and b be two different complex constants such that each of the functions $f_n(x)$ omits both values a and b. Let $\lim_{n=\infty} f_n(x)$ exist for infinitely many points that have at least one accumulation point in the interior of the unit circle. Then, for all x of the domain $|x| < 1$ there exists $\lim_{n=\infty} f_n(x) = f(x)$. Further, $f(x)$ is regular for $|x| < 1$. Finally, if $0 < \theta < 1$ and θ is constant, then $\lim_{n=\infty} f_n(x) = f(x)$ uniformly for $x \leq \theta$.[101]

In this case it was implicitly proved that the regular functions with the above properties form a normal (according to Montel's definition) family.[102]

As Carathéodory and Landau showed, the above theorem can replace the assumption of local boundedness in Vitali's theorem. They mentioned the contribution from Paul Isaak Bernays, then a student in Göttingen, to the direct transition from the so-called theorem IV to theorem V as a new proof of Landau's auxiliary theorem. Theorem IV reads:

Let the function $F(x) = a_0(x) + a_1 x + \ldots + a_m x^m + \ldots$ be regular, $\neq 0$ and $\neq 1$ for $|x| < 1$. Let $0 < \Theta < 1$. In the case of a constant but arbitrarily large ω and a constant but arbitrarily small positive ε, let the number a_0 satisfy the conditions $|a_0| \geq \varepsilon$, $|a_0 - 1| \geq \varepsilon$, $|a_0| \leq \omega$. Then, there exists a number $\Psi = \Psi(\Theta, \omega, \varepsilon)$ dependent only on $\Theta, \omega, \varepsilon$ (not on $a_0, a_1, a_2, \ldots$) such that $|F(x)| \leq \Psi$ for $|x| \leq \Theta$.

Landau had shown that the ε boundedness can be omitted, i.e. he showed theorem V:

Let the function $F(x) = a_0(x) + a_1 x + \ldots + a_m x^m + \ldots$ be regular, $\neq 0$ and $\neq 1$ for $|x| < 1$. Let $0 < \Theta < 1$ and $|a_0| \leq \omega$. Then, there exists a number $\Phi = \Phi(\Theta, \omega)$ dependent only on Θ and ω (not on $a_0, a_1, a_2, \ldots$) such that $|F(x)| \leq \Phi$ for $|x| \leq \Theta$.

Carathéodory, who had proved Bernays' theorem with the help of both his results from the *Untersuchungen über die konformen Abbildungen von festen und veränderlichen Gebieten* (Studies in Conformal Mappings of Constant and Variable Domains)[103] and the function Ω appearing in his common paper with Landau, contacted

Bernays on 23 September 1912.[104] He asked for Bernays' proof, which greatly in-
terested him, and presented his own: "Let $f(x)^{\frac{1}{k}}, (1 - f(x))^{\frac{1}{l}}, f(x)^{-\frac{1}{m}}$ be regular
everywhere, thus also finite in the circle $|x| < 1$. Secondly let $|f(0)| < M$ and $\theta < 1$.
Then, there are two numbers $A(M, \theta)$ and $B(M, \theta)$ such that for $m > A$, $|f(x)| < B$
is always true if $|x| < \theta$." Shortly thereafter, in a second undated letter to Bernays,[105]
Carathéodory referred to his "first work (1905) on the subject", in order to prove
the theorem which Bernays had mentioned in his letter: "however, the method is the
same, instead of a point β, it suffices to take a circle C of radius ε around a_0." In his
letter, Carathéodory presented "a much more general and nicer theorem":

Let $f(x)$ for $|x| < 1$ be meromorphic and for $0 < |x| < 1$ let $f(x)^{\frac{1}{m}}, (f(x) - 1)^{\frac{1}{n}}, f(x)^{-\frac{1}{p}}$
be regular everywhere and also finite $\left(\frac{1}{m} + \frac{1}{n} + \frac{1}{p} < 1\right)$. If δ is any positive quantity, one can
find a quantity ε dependent only on δ, m, n, p, such that for $|x| < \varepsilon$ either $|f(x) - f(0)| < \delta$
or $\left|\frac{1}{f(x)} - \frac{1}{f(0)}\right| < \delta$ is true. [...] One can surely get rid also of m, n, p, in the same way as
a_0.

The Picard theorem occupied Carathéodory until the 1920s. In a paper *Über eine Verallgemeinerung der Picardschen Sätze* (On a Generalisation of Picard Theorems),[106]
he proved purely geometrically and "therefore very transparently" that both Picard
theorems could be considered the natural corollary of a more general and much
more complicated theorem. "The theorem of Picard, that an analytic function, in
every neighbourhood of one of its essential singular points, can leave out at most
two values, is derived in every modern textbook of function theory. Lesser known
is the other result by the same author, according to which two single-valued functions $f(z)$ and $g(z)$, between which there is an algebraic relation of genus $p > 1$,
cannot possess any isolated essential singular points."[107] Carathéodory's reasoning
was quite similar to his proof in the paper *Sur le théorème général de M. Picard*
(On the General Picard Theorem)[108] published in 1912. He had then reformulated
the two Picard theorems: "if R denotes the Riemann surface of the elliptic modular
function or any closed algebraic Riemann surface of genus $p > 1$, the boundary of
the domain G cannot contain any isolated points in which $f(z)$ possesses an essential
singularity."[109] In his 1920 paper, $f(z)$ denoted a single-valued analytic function in
a domain G of the z plane and R the Riemann surface of the multivalued analytic
function $\varphi(u)$, or part of it. Carathéodory gave a condition for R, which satisfied the
generalisation and from which the same property of $f(z)$ could be derived.

2.6.2

Coefficient Problems

In his 1916 paper *Über ein Problem des Herrn Carathéodory* (On a Problem of Herr
Carathéodory),[110] Friedrich Riesz proved that the necessary and sufficient condition
for the series

$$1 + (a_1 + ib_1)z + \ldots + (a_n + ib_n)z_n + \ldots$$

to be continued in such a way as to converge inside the unit disc, and for the real part
of the function that is represented by it to be positive everywhere in the unit disc,
was equivalent to the one presented by Carathéodory in his 1907 paper *Über den*

Variabilitätsbereich der Koeffizienten von Potenzreihen, die gegebene Werte nicht annehmen (On the Variability Area of Coefficients of Power Series that do not Assume Given Values),[111] as well as that of Toeplitz in his 1911 paper *Über die Fouriersche Entwickelung positiver Funktionen* (On the Fourier Development of Positive Functions).[112]

Direct proofs of the agreement between Carathéodory and Toeplitz's condition were given by Carathéodory in his 1911 paper *Über den Variabilitätsbereich der Fourierschen Konstanten von positiven harmonischen Funktionen*,[113] by E. Fischer,[114] by I. Schur[115] and by Frobenius.[116] Their condition has a semi-algebraic character, containing an infinite number of algebraic conditions. The class C of functions

$$f(z) = 1 + \sum_{n=1}^{\infty} c_n z^n$$

that are regular in the disc $|z| < 1$ and have a positive real part there is named after Carathéodory. Carathéodory also determined the precise set of values of the system of coefficients $\{c_1, \ldots c_n\}$, $n \geq 1$ on the class C (Carathéodory's lemma).[117]

In the introduction to his 1911 paper, Carathéodory referred to Toeplitz's 1911 contribution and 1910 paper *Zur Theorie der quadratischen Formen von unendlich vielen Veränderlichen* (On the Theory of Quadratic Forms of Infinitely Many Variables)[118] as follows: "I had worked out a parametric representation for the boundary of the [convex] body K_{2n}, which completely masters this boundary through trigonometric polynomials. Now, Herr Toeplitz managed to discover a very simple representation of this boundary in the form of determinants, by which all the theorems found can be brought to an algebraic form."

The Carathéodory–Toeplitz theorem states that the set of values of the system $\{c_1, \ldots c_n\}$, $n \geq 1$, on C is the closed convex bounded set K_n of points of the n-dimensional complex Euclidean space at which the determinants

$$\Delta k = \begin{vmatrix} 2 & c_1 & \ldots & c_k \\ \overline{c}_1 & 2 & \ldots & c_{k-1} \\ \cdot\cdot\cdot\cdot\cdot\cdot\cdot\cdot\cdot\cdot\cdot\cdot\cdot\cdot\cdot\cdot\cdot\cdot \\ \cdot\cdot\cdot\cdot\cdot\cdot\cdot\cdot\cdot\cdot\cdot\cdot\cdot\cdot\cdot\cdot\cdot\cdot \\ \overline{c}_k & \overline{c}_{k-1} & \ldots & 2 \end{vmatrix}, \ 1 \leq k \leq n$$

are either all positive or positive up to some number, beyond which they are all zero.

Carathéodory's questioning and "sharp analytic solutions" for the relations between the coefficients of an infinite Fourier series and the values of the represented function as well as Toeplitz's "elegant algebraic solutions" also occupied Fejér in Budapest in his 1916 paper *Über trigonometrische Polynome* (On Trigonometric Polynomials).[119]

In the early 1920s, Szegö studied the distribution of eigenvalues of finite Toeplitz forms belonging to a function of class $L(-\pi, \pi)$ and introduced orthogonal polynomials on the unit circle with respect to a weight function.[120] Since then, the subject has been further studied, especially by American, Russian and Scandinavian mathematicians.[121]

A problem analogous with the Landau theorem was treated and solved by Carathéodory in a common work with Fejér *Über den Zusammenhang der Extremen von harmonischen Funktionen mit ihren Koeffizienten und über den Picard–Landauschen Satz* (On the Connection of Extreme [Values] of Harmonic Functions with their Coefficients and on the Picard–Landau Theorem)[122] in 1911. Carathéodory and Fejér showed that another class of power series bounded in the unit disc, $|f(z)| \le M$, can be reduced to the class mentioned and formulated in such a way as to produce what is now called the Carathéodory–Fejér theorem:

Let $P(z) = c_0 + c_1 z + \ldots + c_{n-1} z^{n-1}$ be a given polynomial $\neq 0$. There exists a unique rational function $R(z) = R(z, c_0, \ldots, c_{n-1})$ of the form

$$R(z) = \lambda \left(\bar{a}_{n-1} + \bar{a}_{n-2} + \ldots + \bar{a}_0 z^{n-1} \right) \left(a_0 + a_1 z + \ldots + a_{n-1} z^{n-1} \right)^{-1} \quad \lambda > 0$$

regular in the unit disc and having $c_0, \ldots, c_{n-1}$ as the first n coefficients of its MacLaurin expansion. This function, and only this, realises the minimum value of $M_f = \sup_{|z|<1} |f(z)|$ in the class of all regular functions $f(z)$ in the unit disc of the form $f(z) = P(z) + a_n z^n + \ldots$ and this minimum value is $\lambda = \lambda(c_0, \ldots, c_{n-1})$.

The number $\lambda(c_0, \ldots, c_{n-1})$ is equal to the largest positive root of the following equation of degree $2n$:

$$\begin{vmatrix} -\lambda & 0 & \ldots & 0 & c_0 & c_1 & \ldots & c_{n-1} \\ 0 & -\lambda & \ldots & 0 & 0 & c_0 & \ldots & c_{n-1} \\ \cdot & \cdot & \ldots & \cdot & \cdot & \cdot & \ldots & \cdot \\ 0 & 0 & \ldots & -\lambda & 0 & 0 & \ldots & c_0 \\ \bar{c}_0 & 0 & \ldots & 0 & -\lambda & 0 & \ldots & 0 \\ \bar{c}_1 & \bar{c}_0 & \ldots & 0 & 0 & -\lambda & \ldots & 0 \\ \cdot & \cdot & \ldots & \cdot & \cdot & \cdot & \ldots & \cdot \\ \bar{c}_{n-1} & \bar{c}_{n-2} & \ldots & \bar{c}_0 & 0 & 0 & \ldots & -\lambda \end{vmatrix} = 0$$

If $c_0, \ldots, c_{n-1}$ are real, then $\lambda(c_0, \ldots, c_{n-1})$ is the largest of the absolute values of the roots of the following equation of degree n:[123]

$$\begin{vmatrix} -\lambda & 0 & \ldots & 0 & c_0 \\ 0 & -\lambda & \ldots & c_0 & c_1 \\ \cdot & \cdot & \ldots & \cdot & \cdot \\ \cdot & \cdot & \ldots & \cdot & \cdot \\ c_0 & c_1 & \ldots & c_{n-1} & -\lambda \end{vmatrix} = 0$$

2.6.3
The Schwarz Lemma

Carathéodory engaged with the Schwarz lemma for the first time when he was a lecturer in Göttingen. There, he and Fejér published their first joint article[124] in which they introduced what was to become known as the "Blaschke product."[125]

In his 1928 work *Über die Winkelderivierten von beschränkten analytischen Funktionen* (On the Angle Derivatives of Bounded Analytic Functions),[126] Carathéodory generalised G. Julia's theorem:[127]

$$|1 - f(z)|^2 (1 - |f(z)|^2)^{-1} \le a |1 - z|^2 (1 - |z|^2)^{-1}$$

by also allowing the real positive number a to be infinite. He proved that under the only condition that the analytic function $\omega = f(z)$ is regular and of modulus less than unity in the disc $|z| < 1$, the $\lim_{z=1}(1 - f(z))(1 - z)^{-1}$ converges uniformly in every triangle A, which lies in the unit disc and possesses an angle at the point $z = 1$. Then, for a sequence of points $z_1, z_2, \ldots$, converging to the point $z = 1$ and lying inside a triangle A, there exists the $\lim_{n=\infty}(1 - f(z_n))(1 - z_n)^{-1}$, which is either infinite or a real positive number a_0. In the second case, it is additionally true that $\lim_{n=\infty} f'(z_n) = a_0$ and the function $f(z)$ possesses an angle derivative at the point $z = 1$, which equals a_0. Finally, relation

$$0 < |1 - f(z)|^2 (1 - |f(z)|^2)^{-1} \leq a|1 - z|^2 (1 - |z|^2)^{-1}$$

is likewise satisfied for $a = a_0$ in the entire unit disc. Carathéodory used the generalised theorem to study the behaviour of conformal representation in certain boundary points of the considered domain.

Carathéodory engaged again with generalisations in 1936: at a session of the Athens Academy on 14 May he spoke on "A Completion of the Schwarz Lemma" (*Μία συμπλήρωσις του λήμματος του Schwarz*);[128] on 27 November he delivered an address on "Bounded Analytic Functions" to the American Mathematical Society at Lawrence, Kansas which appeared in the Bulletin of the AMS as *A Generalization of Schwarz's Lemma*[129] in April the following year. He then started a correspondence with Szegö, who was teaching at Washington University in St. Louis,[130] on this matter:

Let the analytic function $f(z)$ be regular, for $|z| < 1$ unimodular bounded, and let $f(0) = 0$ and $f(z)$ non-linear. Then, for all functions of this family is uniformly

(1)
$$\left| \frac{f(z_2) - f(z_1)}{z_2 - z_1} \right| < 1$$

a) when $z_1 = 0$ (the usual Schwarz lemma)
b) when *both* z_1 and z_2 lie within the closed circle $|z| \leq \sqrt{2} - 1$.

In a certain sense, this theorem cannot be improved: in every domain G within the unit circle, containing points for which $|z| > \sqrt{2} - 1$, there are pairs of points z_1, z_2 for which the maximum of the quotient of differences (1) is > 1.

The following theorem is also true: one can assign a closed neighbourhood $G(z_1)$ of the centre $z = 0$ to each point z_1 inside the unit circle, so that (1) will be correct for all functions of the given family with $z_2 < G(z_0)$.

But while the theorem given in the beginning is not much more difficult to prove than the Schwarz lemma itself, I can not survey the form of the domains $G(z_0)$. I suppose that I have calculated very awkwardly and that you could do it better! Would you look at this thing a little bit?

Of course, I got a bit annoyed because of your last remark in your paper of the *Zentralblatt*. After all, it is natural that I can do such things elegantly after having done nothing else in my life from 1905 to 1914 than to combine the lemma of Schwarz with conformal mappings. These things were then so new that Koebe, for example, needed my help in order to prove his distortion theorem. It's true that my works of that time fell so much into oblivion that either Lindelöf is cited, who formulated the same thing as a general principle several years later than me (1908), or even Rogosinski, who has proved many of my results anew. Therefore, I have intentionally chosen the formulation in the beginning of my work in the *Monatshefte*, which does not completely agree with the rendition given in your review.[131]

ΑΚΑΔΗΜΙΑ ΑΘΗΝΩΝ

ΣΥΝΕΔΡΙΑ ΤΗΣ 14 ΜΑΪΟΥ 1936

C. CARATHÉODORY

EINE VERSCHÄRFUNG DES SCHWARZSCHEN LEMMA'S

Κ. ΚΑΡΑΘΕΟΔΩΡΗ. — ΜΙΑ ΣΥΜΠΛΗΡΩΣΙΣ ΤΟΥ ΛΗΜΜΑΤΟΣ ΤΟΥ SCHWARZ

Ἀνάτυπον ἐκ τῶν Πρακτικῶν τῆς Ἀκαδημίας Ἀθηνῶν, 11, 1936, σ. 276

Extrait des Praktika de l'Académie d'Athènes, 11, 1936, p. 276

(Séance du 14 mai 1936)

Carathéodory's speech on A Completion of the Schwarz Lemma.
Reprint from the Minutes of the Academy of Athens.
Courtesy of the Bavarian Academy of Sciences, Archive.

In a new letter to Szegö, Carathéodory presented the geometric meaning of the theorem stated in his paper *A Generalization of Schwarz's Lemma*, whereby he mentioned the bicircular curves of 4th order, which he had examined in his paper *Über beschränkte Funktionen, die in einem Paar von vorgeschriebenen Punkten gleiche Werte annehmen* (On Bounded Functions which Assume Equal Values in a Pair of Given Points). [132]

The theorem which I have recently mentioned is not at all as difficult to prove as I had believed. One should only pull the right corner of the blanket which wraps it up. Strangely enough, there are again bicircular curves of 4th order that play the main role. The result reads:

For h real and $0 < h < 1$ the unit circle will be split up through the curve

$$4h^2(x^2 + y^2) = \left[(1 - h^2)(1 - x^2 - y^2) - 2hx\right]^2$$

into two simply connected domains A and B. For all unimodular bounded functions $f(z)$, which are regular in the unit circle and vanish for $z = 0$, the maximum of

$$\left| \frac{f(z) - f(h)}{z - h} \right|$$

is equal to one, as long as z lies in $A + C$ [C is the boundary of A] and is accessed only for the functions $e^{i\theta}z$. This maximum is always > 1 if z lies somewhere in B. For $h \to 0$ the theorem goes over to the usual lemma of Schwarz continuously. [133]

In a later letter, Carathéodory explained his method to Szegö and characterised it as "more elementary, if not shorter", than Szegö's. He finished his letter with a promise: "I will look at your note in the Bulletin; I have not thought of your or Fejér's investigations, which, unfortunately, I do not know, but rather of Rogosinski and of certain Frenchmen. [134] I just think that, in the case of your calculations, it may be better to just use the Stieltjes integral instead of trigonometric polynomials." [135]

On 10 December 1936, Carathéodory contacted Szegö anew to send him

a letter of Rogosinski which, by chance, has arrived today and contains some of his new results. You will see that these are of more formal nature and that I was not all so wrong. You would have observed yourself that, in order to determine the maximum of the absolute value in the formula

$$\frac{f(z_2) - f(z_1)}{z_2 - z_1} = \frac{a(1 - \bar{z}_1 z_2) - (z_2 - a\bar{a}z_1)\varphi(z_2)}{1 - \bar{z}_1 z_2 - \bar{a}(z_2 - z_1)\varphi(z_2)},$$

one is allowed to put $\varphi(z_2) = 1$ and, in addition, one can write

$$\frac{a(1 - \bar{z}_1 z_2) - (z_2 - a\bar{a}z_1)}{1 - \bar{z}_1 z_2 - \bar{a}(z_2 - z_1)} = \frac{a - u}{1 - \bar{a}u}, \quad u = \frac{z_2}{1 - \bar{z}_1 z_2 + \bar{a}z_1}$$

that is, one can end up in exactly the same way as you do. However, one finds the less elegant condition $|z_1| + |z_2| \le |1 - \bar{z}_1 z_2|$ which, subsequently, has to be reduced to your own. But the proof can hardly be made more beautiful and elegant. [136]

In *A Generalization of Schwarz's Lemma*, Carathéodory stressed that whenever the relation $|z_1 + z_2| + |z_1 z_2| \le 1$ is satisfied, "we need not state that the points z_1 and z_2 do not lie outside the unit circle. This most elegant form of the inequality $[|z_1| + |z_2| \le |1 - \bar{z}_1 z_2|]$ was pointed out to me by *Szegö*; it shows at first sight the symmetry in z_1 and z_2 of the condition obtained." [137]

Carathéodory remarked to Szegö on 16 December 1936:

The method of your earlier letter is *principally* highly interesting. Only that you took a sledge hammer to crack a nut. However, I very much like the new problems you mention and I am sure that you are going to develop a beautiful theory out of these things. In order to conclude from $|z_1| + |z_2| \leq |1 - \bar{z}_1 z_2|$ that $|z_1 + z_2| + z_1 z_2 \leq 1$ and the reverse are true, it is probably easiest and quickest not to use the identity you mention but to *derive* it.[138]

In his 1941 paper *Über das Maximum des absoluten Betrages des Differenzenquotienten für unimodular beschränkte Funktionen* (On the Maximum of the Absolute Value of the Quotient of Differences for Unimodular Bounded Functions),[139] Carathéodory calculated the maximum $M(z_1, z_2)$ of $|(f(z_2) - f(z_1))(z_2 - z_1)^{-1}|$, where $f(z)$ is any regular and unimodular bounded analytic function in the unit disc $|z| < 1$, which vanishes at the origin $z = 0$, and z_1, z_2 are two different from each other points in the unit disc, and proved that

$$M(z_1, z_2) = \cosh(\tau - \sigma) \text{ if } \tau > \sigma \text{ or } M(z_1, z_2) = 1 \text{ if } \tau \leq \sigma,$$

where $\cosh \sigma = |(1 - \bar{z}_1 z_2)(z_2 - z_1)^{-1}|$ and $\cosh \tau = (|z_1| + |z_2|)|z_2 - z_1|^{-1}$.

2.6.4
Conformal Mapping

2.6.4.1
Existence Theorems

In a letter dated 22 July 1907 to Königsberger, Carathéodory wrote that Koebe had been able "to prove in a strict way the great uniformisation theorem of automorphic functions."[140] In an account of Koebe's works,[141] Bieberbach mentions that with the help of a *Verzerrungssatz* (distortion theorem)[142] appearing in Koebe's work of 1907,[143] the solution of the Riemann–Schottky mapping problem and the Schwarz–Poincaré–Klein uniformisation question[144] can be obtained.

In his 1936 course in the theory of functions at the University of Wisconsin, Carathéodory referred to Koebe's distortion theorem as follows:

As early as 1883 it was known to Poincaré and Klein that the curves given by the power series $w = a_0 + a_1 z + a_2 z^2 + \ldots$ and its continuations can be represented by a single parameter, i.e. by

$$z = \varphi(t), \quad w = \psi(t)$$

and these functions are single power series that are either:

1) regular inside $|z| < 1$,
2) entire functions, or
3) meromorphic functions on the whole sphere.

Several proofs have been given. We shall examine one given by Koebe at the suggestion of C. Carathéodory.

Theorem: *If the family $\{f(z)\}$ is meromorphic for $|z| < 1$ and $\neq p$ and q, and if $f(z') \neq f(z'')$ for all $f(z)$ when $z' \neq z''$, then $\{f(z)\}$ is a normal function, i.e. $|f(z)| < R$.*

Choose p, q as a, ∞; also let $f(0) = 0$. Then $w = f(z)$ covers a simply connected domain D not containing a or ∞ but containing 0. Construct a circle about 0 containing a on its interior. Then there is at least one point P of this circle not in D since if this were not the case, the boundary D would be divided and hence D would not be simply connected. Thus $f(z) \neq a, P,$ and ∞. Applying the lemma of Schwarz, we find $|f(z)| < z$ whence the function is bounded uniformly and the theorem is proved.

The Koebe Constant

Let $f(z)$ be regular in $|z| < 1$ and let $f(z)$ be different from a and ∞ in that circle. Putting

$$\varphi(z) = \log[f(z) - a]$$

it can be shown that $|\varphi'(0)| \leq 4$. This is known as the Koebe constant. For the best complete discussion see Conformal Representation, C. Carathéodory, page 48 for a proof by Erhard Schmidt. A second proof will be found in R. Nevanlinna, *Eindeutige analytische Funktionen*, Berlin, 1936.[145]

In 1912 the *Göttinger Nachrichten* published Koebe's paper *Über eine neue Methode der konformen Abbildung und Uniformisierung* (On a New Method of Conformal Mapping and Uniformisation)[146] in which Koebe sketched a method of conformal mapping of a simply connected schlicht domain onto the unit disc.

Koebe presented his *Schmiegungsverfahren* (osculation method), an iteration process, later in 1914 in a work[147] dedicated to H. A. Schwarz on his 50th doctoral anniversary. This work outlined the square-root algorithm in detail and somewhat systematised and simplified Carathéodory's method based on linear and square-root mappings. Carathéodory's contribution to Schwarz's Festschrift[148] contained the presentation of his own method, the explicit formulation of the convergence theorem and its direct proof.[149] Carathéodory and Koebe share the reputation for proving the Riemann mapping theorem through repeated square-root mappings.[150]

A general convergence theorem presented by Carathéodory in a paper *On Dirichlet's Problem*,[151] had been announced to Szegö:

This winter I was more interested in potential theory and I found a general convergence theorem, of which I informed F. Riesz to learn whether he knew it. He wrote to me yesterday that he liked the theorem very much. In its simplest form it reads: let G be a domain in n-dimensional space for which the Dirichlet's problem is soluble. Let $L_j f (j = 1, 2, \ldots)$ be infinitely many linear operators with the following properties:

1) If $f(P)$ is continuous in the closed domain $\overline{G}$, the same is true for $L_j f$
2) If $f \geq 0$ in G, the same is true for $L_j f$
3) If U is harmonic in G and continuous in $\overline{G}$, then $L_j U \equiv U$
4) Every operator $L_j f$ appears in the series $L_1 f, L_2 f \ldots$ infinitely often.
5) a sphere K_P with centre P which lies with its boundary in G is assigned to every interior point P of G, and likewise an index j_P, so that if $f(Q)$ is on the boundary of $K_P \leq f(P)$ and is not $\equiv f(P)$, the value of $L_{j_P} f$ in the point P is really smaller than $f(P)$.

If now $f_1(P)$ is continuous in $\overline{G}$ and one puts $f_{K+1} = L_K f_K$, then f_K converges towards a harmonic function with the same boundary values, as f_1 does.

Condition 5) is necessary, since otherwise all $L_j f$ could equal the identity. Condition 4) could be avoided but then one has to demand other limiting properties for the $L_j f$. The above theorem contains the proof of convergence for almost all classical procedures (Schwarz, Neumann, Poincaré, Lebesgue, Kellogg).[152]

Nearly all the solutions to Dirichlet's problem, which had been proposed since the time of Schwarz and Poincaré and which were based on a convergent sequence of continuous operators, could be treated as special cases of that theorem. However, to include the solution proposed by Kellogg, condition 5 of Carathéodory's theorem had to be slightly revised. In his paper, Carathéodory endeavoured "to give a very elementary treatment of the principal properties of harmonic functions culminating in the existence proof of Dirichlet's problem devised by O. Perron and very much simplified by T. Radó and F. Riesz."[153]

Conformal mapping can be employed for solving the Dirichlet problem for domains in R^2, since in R^2 the theory of harmonic functions is closely related to the theory of analytic functions of a complex variable.[154] On the other hand, Dirichlet's principle together with methods developed in potential theory for problems concerning values on the boundary and methods taken exclusively from the theory of functions are mentioned by Carathéodory[155] as three distinct methods of attack used to prove the fundamental theorem of conformal mapping.

In his *Bemerkungen zu den Existenztheoremen der konformen Abbildung* (Remarks on the Existence Theorem of Conformal Mapping),[156] Carathéodory stated that although many proofs, shorter and better than the one given by Schwarz, had been proposed in rapid succession for the main theorem of conformal mapping, only Fejér and Riesz had returned to Riemann's basic idea and combined the solution of the problem of conformal mapping with the solution of a variational problem. Thereby, they chose a problem which certainly had a solution and not a problem difficult to handle, as was the Dirichlet principle. Fejér and Riesz's proof of how to obtain the Riemann mapping function was presented by Radó in just one page.[157] Carathéodory showed that through a minor modification in the choice of the variational problem the Fejér–Riesz proof could be considerably simplified.

2.6.4.2

Variable Domains

On 2 February 1911, Carathéodory informed Hilbert[158] that he had thought about a theorem on conformal mappings which, if it were really original, would be appropriate for the *Mathematische Annalen*. He had proved the theorem and believed that it would have many applications. He asked Hilbert to inform him through Alfréd Haar, a Hilbert student known for his work on functional analysis, if it would be worth documenting both the theorem and the proof. "It is about the following", he wrote:

Let $G_1, G_2, \ldots$ be infinitely many domains in the u plane, all of which contain the point $u = 0$ in their interior and all of which lie inside a fixed circle (the last [we assume] only for the sake of simplicity; one can set up much more general conditions instead of these). For $u = 0$, I call the domain K a *kernel* of a sequence of domains if it has the following properties: (a) it contains the point $u = 0$ in its *interior*; (b) it is the largest domain for which every domain K' that lies with its boundary in the interior of K also lies, from a certain n onwards, in the interior of every G_j. By the way, in the limiting case, K can consist only of the point $u = 0$. I say now: "the sequence $G_1, G_2, \ldots$ converges to K if K is also the kernel of every arbitrary subsequence $G_{n_1}, G_{n_2} \ldots$" Now I map the domains G_j onto the unit circle, so that $u = 0$ corresponds to the centre of the circle and the derivative $f_j'(0)$ of the function $f_j(z)$ that induces the mapping is real.

The theorem which I can prove is the following: *The necessary and sufficient condition that the series $f_j(z)$ converges to an analytic function is that the sequence of domains G_j converges to its kernel. The limit function $f(z)$ produces the mapping of the kernel.*
A first application of this theorem is, for instance, the proof of continuity of the conformal mapping as a function of its boundary, even if the boundary is a non-analytic curve and the Cauchy theorem cannot be applied. [...] [In this case], I can also prove the uniformity of convergence under very general assumptions.

What Carathéodory stressed in his letter to Hilbert is now known as Carathéodory's theorem on conformal mappings with variable boundaries. According to Duren, the convergence theorem is "*the* theorem that connects geometric behaviour with analytic behaviour of a sequence of mapping functions", and Carathéodory "certainly was one of the true pioneers of geometric function theory."[159]

Carathéodory sent his manuscript to Hilbert together with a letter, probably written in spring 1911.[160] He was not entirely satisfied with the presentation of his work: "Certainly, the leading thought is so elementary everywhere that, it seems to me, it cannot be simplified. The difficulty is purely editorial; namely, to set up a definite proof of the most general Riemann surfaces imaginable that are simply connected, without using many words." As he remarked, he had systematically used Hilbert's "method of choice" in his work.

In both his letters, Carathéodory apparently referred to his *Untersuchungen über die konformen Abbildungen von festen und veränderlichen Gebieten*, which he dated Breslau, 17 July 11, and was published in the *Mathematische Annalen* of 1912. He wrote there: "On the contrary, we would proceed using the theory of functions and, with relatively simple means, we will prove not solely the possibility of a conformal mapping of a domain but we will gain the mapping function itself through a recurrent procedure which in every step demands only the solution of first- or second-degree equations." He used the Schwarz lemma to prove the uniqueness of conformal mapping of a simply connected domain onto the unit disc when the mapping function has a non-vanishing derivative at $z = 0$. He further showed the existence and uniqueness of the kernel K of a sequence of domains and gave its definition. A kernel should "consist of only the point $u = 0$ if there exists no circle which has $u = 0$ as its centre and which lies in the interior of all G_n from a certain n onwards. In any other case, K should be the largest domain with the property that every closed domain $\overline{H}$, which contains the point $u = 0$ as internal point and, together with its boundary, lies in the interior of K, also lies in the interior of all G_n from a certain n onwards."[161] The sequence $G_1, G_2, \ldots$ is said to converge to its kernel K if K is also the kernel of each infinite subsequence of $G_1, G_2, \ldots$. Carathéodory proved that the sequence $f_1, f_2, \ldots$ converges to the limit function $f(z)$, which produces the mapping of K onto $|z| < 1$ if and only if the sequence $G_1, G_2, \ldots$ converges to its kernel K.

In the third chapter of his paper, Carathéodory introduced what Shields defines[162] as bounded, simply connected domains G such that every neighbourhood of every boundary point contains points that can be bound to ∞ by a path that avoids the closure of G. These domains, now known as Carathéodory domains, have proved to be useful in polynomial approximation. Every Carathéodory domain can be represented as the kernel of a decreasing convergent sequence of simply connected domains $\{G_n\}$: $\overline{G} \subset G_{n+1} \subset \overline{G}_{n+1} \subset G_n, n = 1, 2, \ldots$, and every domain G for which there exists such a sequence is a Carathéodory domain.

A year after Carathéodory's work appeared in the *Mathematische Annalen*, Bieberbach, using a definition of the kernel of a sequence of domains that was equivalent to Carathéodory's definition, generalised Carathéodory's theorem and proved that it held not only for mappings to a disc but also to any other either simply or multiply connected domain. He reproached Carathéodory for adhering to the Schwarz lemma, which had hindered him from presenting a generalised formulation.[163] In 1924, Joseph Leonard Walsh proved that a function f that is continuous on a closed domain bounded by an analytic Jordan curve and is holomorphic in the interior can be uniformly approximated by polynomials on the closed domain.[164] Two years later he proved this for arbitrary Jordan domains. In the introduction to his paper *Über die Entwicklung einer analytischen Funktion nach Polynomen* (On the Expansion of an Analytic Function in Series of Polynomials),[165] he stated that Carathéodory had suggested that Courant's theorem should be applied to this problem. Allen Shields presents the proof as follows: First, he considers the special case in which the kernel G of the sequence $\{G_n\}$ equals D (the open unit disc). He then replaces f by f_r, where $f_r = f(rz)$ and $0 < r < 1$. Then, f_r is holomorphic in the neighbourhood of a closed unit disc and, consequently, it may be approximated uniformly on the closed disc by polynomials. Finally, f_r is uniformly close to f on the closed disc, and so f can be approximated uniformly by polynomials.[166] Courant's theorem[167] provides for a similar line of argument if larger discs $\{|z| < 1/r\}$ are replaced by larger Jordan domains shrinking down to G. Although Carathéodory had apparently realised the use of Courant's theorem in this case, he obviously suggested the subject to Walsh for publication. Walsh's signature is in Carathéodory's guest-book next to the date 13 March 1926, so perhaps they had discussed this point during that visit.

On 30 March 1913, Carathéodory showed Felix Hausdorff[168] how his own definition of kernel could harmonise with Hausdorff's definition. He proposed the following formulation: "The kernel is the union of $\{0 + \underline{L}^0\}$, where $\underline{L}^0$ denotes the connected domain of L that contains 0 in its interior, if it exists."[169] In Carathéodory's words, Hausdorff's notions $L, M, \overline{L}, \overline{M}, \underline{L}, \underline{M}$, would "certainly provide for the derivation a long series of nice things".

Weyl, who was engaged with similar studies in Göttingen, stresses Hausdorff's significance in topology: "Hilbert defines a two-dimensional manifold by means of neighborhoods, and requires that a class of 'admissible' one-to-one mappings of a neighborhood upon Jordan domains in an x, y-plane be designated, any two of which are connected by continuous transformations. When I gave a course on Riemann surfaces at Göttingen in 1912, I consulted Hilbert's paper and noticed that the neighborhoods themselves could be used to characterize this class. The ensuing definition was given its final touch by F. Hausdorff; the Hausdorff axioms have become a byword in topology."[170] In his book *Grundzüge der Mengenlehre* (Basic Features of Set Theory),[171] Hausdorff had developed topological spaces from a set of axioms.

2.6.4.3
Mapping of the Boundary

In the letter of spring 1911 to Hilbert, mentioned above, Carathéodory wrote that he still had to reformulate his presentation throughout a second work, but it would be better to do that only after the submitted paper was published. The second work implicitly contained a method by which all results of Schwarz relating to boundaries could be deduced from the last chapter of his first work. "Then, both works together form a purely function-theoretical foundation of the theorems regarding conformal mapping, at least for the simplest case of simply connected schlicht domains." Carathéodory reported to Hilbert, apparently in the spring of 1912,[172] that the edited manuscript of his first work concerning the *Untersuchungen über die konformen Abbildungen von festen und veränderlichen Gebieten* had become much longer than initially planned:

> Not only have I worked out the conditions for the continuity of the mapping function in the case of a change in the boundary of the domains, but I have also done the same for arbitrary simply connected Riemann surfaces. This has allowed me to provide a purely function theoretical-method to prove the *existence* of the conformal mapping of a domain (without the aid of the Green function and of the boundary-value problem of $\Delta u = 0$). In this method, the mapping function of the given domain is approximated by rational functions whose coefficients are determined by the solution of linear and quadratic equations in a recurrent way. [...]
>
> Besides this first work, which exclusively concerns the mapping of the *interior* of a domain onto the *interior* of a circle, and consequently does not take the boundaries into account, I have written a second work, which is about the mapping of boundaries onto each other, in which I have proved that in the case of conformal mappings of the interior of the domain, which is bounded by an arbitrary Jordan curve, onto the interior of a circle, the Jordan curve and the circumference correspond to one another uniquely and continuously. *Inter alia* the boundary-value problem of $\Delta u = 0$ is thereby solved for arbitrary Jordan curves, whereas, as it seems to me, the conditions of Poincaré's sweeping out process[173] along the boundary have been indispensable for this problem (and also for the conformal mapping) up to now.

Carathéodory announced to Hilbert his intention of having his second work, "methodically completely different from the first one", published in the *Acta Mathematica*. Fejér, who visited Breslau for a few days, was of the opinion that it would be more advantageous if both works were published in the same journal. Carathéodory wanted to hear Hilbert's opinion on that matter before sending him his first manuscript so that he could change his introduction if need be. The second work was his paper *Über die gegenseitige Beziehung der Ränder bei der konformen Abbildung des Inneren einer Jordanschen Kurve auf einen Kreis* (On the Reciprocal Relation of the Boundaries in the Case of Conformal Mapping of the Interior of a Jordan Curve onto a Circle),[174] which he completed over Easter 1912. With this, he had attempted "to show in the simplest way that the boundaries of both constructs turn into each other uniquely and continuously" or, in other words, if G is a Jordan domain, then the Riemann mapping function from D (the open unit disc) to G extends to be a homeomorphism between the closed domains. Carathéodory enclosed this work together with a second manuscript in a letter to Hilbert on 5 May 1912. If Hilbert wished to accept these for publication, Carathéodory would send the drawings of the figures directly to Blumenthal.[175]

Riemann had shown in his dissertation called *Grundlagen für eine allgemeine Theorie der Functionen einer veränderlichen complexen Größe* (Foundations for a General Theory of Functions of a Complex Variable),[176] that an analytic function $\omega = f(z)$, for every z for which the derivative does not vanish, produces a conformal mapping. According to the Riemann mapping theorem, every simply connected domain G with at least two boundary points, i.e. which does not comprise the whole plane, can be mapped by an analytic function conformally onto the unit disc. In modern proofs one uses sequences of holomorphic functions which converge uniformly in the interior but can diverge on the boundaries. On the other hand, Schwarz had already shown with his reflection theorem that if the boundary of a domain consists of finitely many real analytic curves which intersect at angles different from zero, then the domain can be mapped onto the unit disc biholomorphically. Paul Painlevé and Paraf separately generalised the conditions, but only Carathéodory worked out the final theorem and proved it. In the second manuscript (a third work) he sent to Hilbert, he answered the question of what happens if the boundary of the domain to be mapped cannot be conceived as a Jordan curve by developing "a complete theory of the boundary of the most general simply connected domains". His intention was "to be absolutely strict." "You know", he wrote to Hilbert, "how often, if one leaves *Brouwer* aside, one sins in this part of mathematics" and he was of the opinion "that the most significant *simplest* and at the same time most necessary basic notions have to be created first." "You will see", he continued, "what it is about if you care to check through the introduction, pages 21–28 of the manuscript and the wording of theorems I–XIII. In addition, I would be extremely obliged to you if you would tell me whether you are pleased with my terminology, which, by the way, I have discussed thoroughly with Schmidt and Hessenberg. Can one use especially the words "Kette von Ausschnitten" [chain of cuts] and "Primende" [prime end]? But a chain has already been introduced in the abstract set theory in another sense (Dedekind),[177] and no notion of a product is assigned to the notion of divisibility, which I explain on p. 24.!"[178]

The second manuscript, for which Carathéodory gave the date 1 May 1912, was published as well, under the title *Über die Begrenzung einfach zusammenhängender Gebiete* (On the Boundary of Simply Connected Domains).[179] In the printed paper he used the expression *Kette von Querschnitten* (chain of cross-sections) instead of *Kette von Ausschnitten* (chain of cuts). Riemann had used the term cross-section in his dissertation to define the simple connectivity: *"Eine zusammenhängende Fläche heißt, wenn sie durch jeden Querschnitt in Stücke zerfällt, eine einfach zusammenhängende"* (a connected surface is called simply connected if it is divided into pieces by every cross-section).[180] In his letter to Hilbert, Carathéodory reported further about "the most peculiar result of the whole work [...], namely that exactly *four* types of prime ends exist." He had often corresponded with Study, who was occupied with similar problems, but Carathéodory found Study's considerations to be very complicated since, as he said, Study did not have the notion of prime ends. Carathéodory wrote in his paper that Study had drawn his attention to the division of the points of a prime end into main points and secondary points. Carathéodory had learned of Study's parallel research in September 1911, when he presented his solution for the conformal mapping of the interior of a Jordan curve onto a circle and for the boundary of simply connected domains[181] to the *Naturforscherversammlung* in Karlsruhe. He was allowed to read Study's manuscript concerning *Konforme Abbildung einfach zusammenhängender Bereiche* (Conformal Mappings of Simply Connected Domains),[182] which prompted him to write the third chapter of his work, in which he geometrically examined the possible form of the prime ends. In the introduction to his work he gives the definition of prime ends as constructs which (in analogy with the prime numbers) possess no proper divisors, which in the simplest case consist of one point and which can be defined purely set-theoretically. "Not only do they allow us to master completely the problem of conformal mapping in the neighbourhood of the boundaries of domains that are to be mapped onto each other, but we are also hardly able to do without them primarily for the description of this boundary." Carathéodory explained that the prime ends, for which Erhard Schmidt had proposed a definition, build "in a way a substitute for the points of the boundary". Carathéodory's main result reads:

By the conformal representation of the interior of any simply connected simple domain onto the interior of a circle, every set of points converging to a prime end of the domain passes over to a set of points which converge to a point on the circumference, and vice versa. The prime ends of the domain and the points of the circle correspond to each other one-to-one.[183]

Thus, Carathéodory extended the conformal mapping of a domain onto a disc continuously to the boundary. Carathéodory's extension theorem is stated as follows. The map $f : E \to G$ can be extended to a topological map from $\overline{E}$ onto $\overline{G}$ if and only if the boundary of G is a closed Jordan curve. With regard to the mapping of a simply connected domain onto the unit disc, Lindelöf showed in a treatise of 1915 that the mapping remains bijective and continuous on the boundary when this consists of a simple curve. He also treated the more subtle questions related to multiply accessible and to inaccessible boundary points, questions that had been investigated by Carathéodory.[184]

On 30 March 1913 Carathéodory sent his paper *Über die Begrenzung einfach zusammenhängender Gebiete* to Felix Hausdorff[185] with the remark: "Recently, I was able to prove that a domain consisting only of prime ends of the second kind [...] cannot exist. But in these questions that is almost all one can do and the silly thing is that there is no general method to proceed here systematically."

Decades later, Lars V. Ahlfors evaluated Carathéodory's relevant performance: "His understanding of Riemann's mapping theorem was far ahead of his contemporaries', and his feat of promoting an obscure remark of H. A. Schwarz to become the famous Schwarz's lemma was as important as it was generous. In the same area, I can personally never forget how impressed I was, and still am, by his invention of prime ends."[186] L. Young writes about Carathéodory's "famous theory of prime ends" as follows.

Imagine a rectangle, with the lower left-hand corner at the origin O of the x, y plane, and let the sides parallel to the y axis (the vertical sides) be of unit length. Denote by D the domain obtained by removing from the interior of this rectangle a sequence of vertical segments of heights 1/2, whose lower extremities are points of the x axis, tending to O from the right. We refer to these vertical segments as hurdles, and we term pockets the parts of D between successive hurdles, and below the line of height 1/2. How in this domain D is it possible to approach the point O? – Clearly, we must go deeper and deeper into an infinite succession of the pockets: the point O can then be a limit point of our path. However, the whole segment OY of the figure will then also consist of such limit points, since the path cannot proceed from one pocket to the next without exceeding somewhere in between the height 1/2. Thus from the point of view of approach from the interior of D the whole segment OY counts as the terminal extremity: its points form a single unit, it constitutes what Carathéodory calls a prime end. Of this new entity Carathéodory was able to prove that, when a domain is conformally mapped onto the unit disc, the map extends to the boundary continuously, in such manner that each prime end becomes a point of the circumference. This beautiful extension of the Riemann mapping theorem applies to extremely general simply-connected domains. The conformal mapping of domains of the type of D above turns out to be also a useful tool for certain counter-examples. If we alter D to make it triangular – with the right-hand vertical side squashed into a single point, and if we replace the hurdles by interlocking triangular 'teeth', it becomes the domain in a little gem of a note by Littlewood 'On the conformal representation of the mouth of a crocodile'.[187]

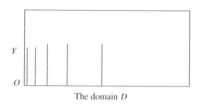

The domain D

Perron wrote in 1950 that "this theory has still not gone down in textbooks. In the main, one contents oneself with boundaries, which are Jordan curves or at least consist of accessible points alone, and refers to the original work in Math. Annal. 73 for the general case. By the way, also Carathéodory himself does without the general case in his 1932 released Cambridge Tract *Conformal Representation*, as in the footnote on page 47 of the mentioned work, but at the same time without mentioning his own work, as well."[188]

2.6.5

Normal Families

In his *Remark on a Theorem of Osgood concerning convergent Series of analytic Functions*,[189] Carathéodory extended the theorem of Osgood[190] to hold for meromorphic functions, provided that the limiting function $f(z)$, to which a sequence of meromorphic functions $f_n(z)$ converges in a region R, is finite at every point of this region, and proposed a proof for an approach by rational functions to meromorphic functions.

Let $\varphi(z) = cz + c_0 + c_1 z^{-1} + c_2 z^{-2} + \ldots, c > 0$ be an analytic function, regular and univalent for $|z| > 1$, which maps $|z| > 1$ conformally onto a simply connected domain in the complex x plane (containing $x = \infty$ as an interior point), preserving the point at infinity and the direction therein. The boundary of the domain is assumed to be a continuum consisting of a finite number of rectifiable Jordan arcs. Then, according to a theorem by Osgood and Carathéodory, the function $\varphi(z)$ is continuous in $|z| \geq 1$ and furnishes a one-to-one and continuous correspondence between the unit circle $|z| = 1$ and the boundary of the domain.[191] Carathéodory himself formulated this theorem as follows: "If one Jordan domain is transformed conformally into another, then the transformation is one-to-one and continuous in the closed domain, and the two frontiers are described in the same sense by a moving point on one and the corresponding point on the other."[192]

Radó would write with admiration about Carathéodory's treatment of the transformation of one Jordan domain into another: "Let R be a bounded simply connected Jordan region in the $\omega = u + iv$ plane and let R_* be a region of the same type in the $z = x + iy$ plane. Then, there exists a function $\omega = f(z), z \in R_*$, with the following properties. (i) $f(z)$ is continuous in R_* and analytic in R_*^0. (ii) The formula $\omega = f(z), z \in R_*$, defines a homeomorphism from R_* onto R. A beautiful presentation of this subject may be found in Carathéodory."[193]

Carathéodory admits that a communication to him from Radó revealed that the proved generalisation of Osgood's theorem was much more trivial than he had initially assumed: namely, under the assumption that the meromorphic functions $f_n(z)$ converge in a domain G, it is solely required to prove the existence of a subdomain H of G, in which either the sequence of $f_n(z)$ or the sequence of $1/f_n(z)$ would be uniformly bounded.[194]

In his paper *Stetige Konvergenz und normale Familien von Funktionen* (Continuous Convergence and Normal Families of Functions),[195] Carathéodory showed that the convergence of a sequence of analytic functions can be continuous at only those points at which the fluctuations of the various functions of the sequence satisfy a certain condition, namely the vanishing of the limiting oscillation at a convergence point of the sequence, worked out convergence properties of meromorphic functions and gave the definition of continuous convergence for meromorphic functions.[196] In his paper, Carathéodory mentions that Ostrowski had characterised normal families by a theorem in which he used a notion similar in content with the limiting oscillation and extremely suitable for the systematic representation of the theory of these families. In Carathéodory's opinion, this led to a handier definition of normal families than that by Montel.

Perron remarks that "Carathéodory proposed a definition, completely different in wording and somewhat more general in nature, to which he kept in his later works and is also to be found in the two-volumed work *Funktionentheorie*, which was released shortly after his death, and by which readers familiar only with the easily remembered Montel definition, could easily become somewhat confused."[197]

In the notes to his 1932 book *Conformal Representation*, Carathéodory mentions Paul Montel's *Leçons sur les familles normales de fonctions analytiques et leurs applications* (Lessons on Normal Families of Analytic Functions and their Applications),[198] in which Montel gave his theorem on conditions for normality of a family of holomorphic functions. Montel introduced the notion of a "normal family" in function theory in 1912.[199] By "normal family" he means a set M of holomorphic functions in a domain D of the complex plane, if, out of every sequence S of M, one can select a subsequence which converges uniformly in every compact part of *D*. This notion can be generalised to apply to meromorphic functions. Montel had shown that a uniformly limited set of holomorphic functions always builds a normal family. Carathéodory, Bieberbach, Julia and other researchers have used this construction to either simplify procedures of proof or achieve new results in the theory of conformal representation.

In a letter to Zermelo on 14 June 1935 Carathéodory wrote about the history of the notion of "normal family":

The word and the notion "normal family" comes from Montel, who had shaped it around 1904. This notion has emerged from a further development of the Weierstrass double-series theorem stemming from Stieltjes (around 1895).[200] If one notes that for all *analytic* functions $f(z)$, which are regular for $|z| < 1$ and satisfy the condition $|f(z)| < 1$ there, all coefficients of the power series $a_0 + a_1 z + \ldots = f(z)$ are uniformly limited, it follows that from every set $\{f(z)\}$ of such functions one can choose a uniformly convergent sequence on every circle $|z| \leq r < 1$. This led Montel to give the name "normal families" to all sets of functions which possess an analogous property. So, one was able to show that all functions which are regular in a domain *G* and are $\neq 0, 1$ constitute a normal family; the Picard theorem follows on from here easily. The notion of the limiting oscillation which allows us to speak of families that are normal in a *point* comes from me.[201]

2.6.6
Functions of Several Variables

Perron remarks that "the special difficulties, with which the study of analytic functions of several variables is faced, also appealed to Carathéodory".[202] The title of Carathéodory's first work in this little-researched area at that time, which Behnke described as a work pointing the way ahead,[203] reads *Über das Schwarzsche Lemma bei analytischen Funktionen von zwei komplexen Veränderlichen* (On the Schwarz Lemma in the Case of Analytic Functions of two Complex Variables)[204] and his main result, "a peculiar analogue to the Schwarz lemma" (Perron), is as follows: "Through the bounded analytic functions of two complex variables, which are regular in a given four-dimensional domain, a metric is imposed on this domain with the help of a distance function."[205] Based on this metric, he showed that it is impossible to map analytically the interior of a bicylinder onto the interior of the hypersphere.

In his paper *Über die Geometrie der analytischen Abbildungen, die durch analytische Funktionen von zwei Veränderlichen vermittelt werden* (On the Geometry of Analytic Mappings which are Induced by Analytic Functions of Two Variables),[206] he showed "that the wish to transfer the lemma of Schwarz also to analytic functions of several variables leads one to assign to every $2n$-dimensional domain a metric characteristic of it"[207] and studied this metric exclusively for domains of four or more dimensions, the *Kreiskörper* (circular bodies). Blaschke's work *Zur Geometrie der Funktionen zweier komplexer Veränderlicher I: Die Gruppen der Kreiskörper* (On the Geometry of Functions of Two Complex Variables I: The Group of Circular Bodies)[208] simplifies and completes Carathéodory's studies into these domains through geometric considerations. Blaschke mentioned therein that with a series of lectures held in Hamburg in July 1926 as well as with two of his works, Carathéodory had achieved significant progress in the general theory of functions of two complex variables and together with Behnke and Wirtinger he had brought new life to a subject that had been forgotten for almost twenty years. Kritikos writes that this branch remained largely unexplored ('lived as a vegetable') despite a most significant work by Poincaré in 1907 and two nice investigations by Reinhardt in 1921.[209] In his note *Über eine spezielle Metrik, die in der Theorie der analytischen Funktionen auftritt* (On a Special Metric which Appears in the Theory of Analytic Functions),[210] Carathéodory considered a metric relevant also for other point sets. In later works, he attempted to find analogies between the function theory of several variables and classical function theory.

Carathéodory's contribution to the Schottky issue of Crelle's journal, dedicated to Schottky on his 80th birthday, was *Ein dem Vitalischen analoger Satz für analytische Funktionen von mehreren Veränderlichen* (A Theorem Analogous to that of Vitali for Analytic Functions of Several Variables).[211] Carathéodory presents his work as follows. Vitali's theorem says that a sequence of functions $f_n(z)$, which belong to a normal family in a domain G, converges in this domain continuously towards an analytic function $f(z)$ if the convergence points of $f_n(z)$ possess an accumulation point in the interior of G. This theorem is based on the fact that the zeros of a non-identically vanishing analytic function lie isolated. But this is not true for analytic functions of several variables, so that the attempt to transfer Vitali's theorem to these functions appears to be impossible. Carathéodory was able to show that it is possible to construct countable sequences of points $P_1, P_2, \ldots$ in the $2n$-dimensional space of complex numbers $(z_1, \ldots, z_n)$, which converge towards a point O and are so irregularly distributed that every function $f(z_1, \ldots, z_n)$ analytic in O and vanishing in every point P_ν, must identically equal zero. Then, the analogue to the Vitali theorem would also be true for such sequences of points.

Kritikos maintains that the reverberations caused by Carathéodory's investigations into the mappings of $2n$-dimensional space by analytic functions of $n \geq 2$ complex variables spread so quickly that within six years (1926–1932) the relevant bibliography was enriched with more studies than those published in the previous three decades. And in 1932, at the International Congress of Mathematicians held in Zurich from 4 to 12 September, in one of the so-called general speeches, Carathéodory, with his erudition, sharp scientific conception and widely acknowledged well elaborated presentation, spoke about the main progress and the still unsolved problems of this branch of the theory.[212] His contribution was included in the congress proceedings under the heading *Über die analytischen Abbildungen von mehrdimensionalen Räumen* (On the Analytic Mappings of Multi-Dimensional Spaces).[213] There, he highlighted the first prospects for the application of the concept of the Riemann surface to spaces of n complex variables. According to Carathéodory, Weyl had produced a comprehensive theory in his work on *Die Idee der Riemannschen Fläche* (The Idea of the Riemann Surface). He defined a Riemann surface R as any topological two-dimensional set of points that possesses the following three properties: (1) it can be triangulated; (2) every point of R possesses surroundings that can be mapped conformally onto the interior of a circle; (3) the conformal mappings of two overlapping surroundings are consistent with each other. Radó had shown that the consequence of properties (2) and (3) is the possibility of triangulation; but the proof he gave makes the possibility of transferring this theorem to multi-dimensional constructs appear very doubtful.

In his work *Über die Abbildungen, die durch Systeme von analytischen Funktionen von mehreren Veränderlichen erzeugt werden* (About Mappings Produced by Systems of Analytic Functions of Several Variables)[214] Carathéodory systematically transferred the theorems of conformal representations on the plane to analytic representations of $2n$-dimensional domains on one another by systems of n analytic functions of n variables. The main theorem of his theory reads as follows. "Let there be given a sequence of topological transformations S_k, through which the domains G_k are converted into certain domains G'_k. Let the sequence of domains G_k converge to a non-empty kernel G with the basic point O. Additionally, at the point O, both the transformation functions and their first partial derivatives should converge. If all these conditions are satisfied, then the sequence of representations S_k converges continuously in G towards a non-degenerate transformation S if and only if the sequence of domains G'_k converges towards a non-empty kernel G' with the basic point $O' = SO$. In this case, the limiting transformation S converts the kernel G into G'." In a final section about the hypersphere and the polycylinder he proved that these two domains cannot be mapped onto each other.

2.7
Elementary Radiation Theory

In Breslau, Carathéodory became a father for the second time. His daughter Despina was born on 13 October 1912. Some weeks later, in November 1912, Carathéodory gave a lecture on Hilbert's "wonderful work" on the elementary radiation theory at the Breslau physics colloquium, which he informed Hilbert about on 12 December 1912.[215] In Hilbert's own words, the elementary radiation theory was "that phenomenological part of radiation theory which is based directly on the concepts of emission and absorption and culminates in Kirchhoff's theorems on the relation between emission and absorption."[216] Hilbert had directed his attention away from the kinetic gas theory to the elementary radiation theory, in which the concepts also led directly to integral equations.[217] Reacting to Carathéodory's lecture, the physicist Ernst Pringsheim, described by Born as "a quiet thinker, elegant in manners and attire, cautious and reserved in his statements, modest and unobtrusive",[218] raised objections to Hilbert's work.

O. Lummer and E. Pringsheim, who together obtained the spectral distribution (relative amount of energy associated with each wavelength interval of the emitted radiation) of blackbody (perfect absorber and emitter of radiation) radiation for several absolute temperatures in 1900,[219] were the main representatives of physics in Breslau. The two friends were then endeavouring to test Planck's law experimentally and their measurements contributed to its final acknowledgement. In Carathéodory's opinion, Pringsheim's objections to Hilbert's work were at first so generally formulated that he was not at all able "to understand what it was all about; I was always told 'that this is not what we understand under temperature radiation' but I was not given an exact definition of temperature radiation", he wrote to Hilbert. However, Carathéodory finally realised how grave the objections were and informed Hilbert about the "whole matter". He gave the definitions of temperature radiation and luminescence as well as the conditions of the problem as follows.

I. The radiation of matter in an adiabatically isolated container could be called temperature radiation if the three positive functions of the frequency of oscillation $n, q(n), \eta(n), \alpha(n)$, assigned to every material volume element were unambiguously determined through the thermodynamic co-ordinates of the volume element. Carathéodory referred here to his own work about the foundations of thermodynamics; if, in addition, these three functions depended on the state of radiation, i.e. the intensity distribution of the various colours, one would speak of luminescence. Temperature radiation was assumed to be available and the state of matter, but not necessarily the state of radiation, was assumed to be stationary.

II. Energy is composed of the adherent energy (that is the thermodynamic energy of matter itself), which spreads through heat conductivity and whose density H is a function of only the thermodynamic co-ordinates and is independent of the frequency of oscillation n, and of the radiating energy, which spreads with the velocity of light and depends on n.

III. Calculation of the density of radiation energy according to Hilbert's formula shows that this quantity does not depend on time, so that the state of radiation is stationary as well, and this holds for every colour.

IV. If O is any closed surface considered during the time interval Δt and if, further, $[\Delta i]_{n_1}^{n_2}$ denotes the radiation energy Σ, which lies in the part $n_1 < n < n_2$ of the spectrum and in Δt passes through the surface O from the outside to the inside, $[\Delta k]_{n_1}^{n_2}$ denotes the relative expression for the energy flowing from the inside to the outside and $[\Delta e]_{n_1}^{n_2}$ and $[\Delta a]_{n_1}^{n_2}$ the emitted and absorbed energy during Δt inside O respectively, then equation $0 = [\Delta i - \Delta k + \Delta e - \Delta a]_{n_1}^{n_2}$ ought to be satisfied, since the state of radiation is stationary. Planck attempted to show that $\Delta i - \Delta k$ equals zero by starting from a homogeneous medium for which such an equation surely existed. If it is possible to prove it "directly and generally", one would obtain $[\Delta e - \Delta a]_{n_1}^{n_2} = 0$, which would lead to Hilbert's "integral equation 8 p. 7 of the Gött. Ber."[220]

V. If the problem was put as Carathéodory had put it, then loss and gain of heat were to be understood for the adherent energy of matter. In this case, one ought to equate the whole emitted energy (integrated over n) with the whole absorbed energy. Then, instead of Hilbert's integral equation, one would obtain the integral equation

$$\int_0^\infty dn \left(\eta - \frac{\alpha}{4\pi q^2} \int K \eta_1 dv_1 \right) = 0.$$

Carathéodory closed his letter with the remark that the leading idea of Pringsheim's proof could be useful and that notice should be taken of the physicists' point of view: "Maybe one could achieve the aim by pursuing it in an analogous way to that of E. Pringsheim[221] [...] who varies the material composition of the medium and keeps the temperature constant. There is much to say about Pringsheim's derivation; I have drawn his attention to it. But maybe the leading idea of his proof is useful after all. However, the greatest difficulty surely lies in the fact that the variations of the three functions q, η, α are not independent of each other and that in any case your equation

(12) $\left[1 - \frac{1}{4\pi} \int \frac{a_1}{q_1^2} K \, dv_1 = 0 \right]$ must always be satisfied. It seems to me that, in any

case, the point of view of the physicists here deserves consideration and I hope that I have now made myself clear."

In the winter of 1912, Hilbert gave a lecture on the *Begründung der elementaren Strahlungstheorie* (Foundation of the Elementary Radiation Theory) in Munich and he published a paper with the same title in 1913. He added to it "an axiomatic treatment of the elementary theory of radiation" after "his attention was drawn to the interesting and substantial treatise by E. Pringsheim".[222]

Hilbert sent his corrections, together with the printed text of his axiomatic treatment of the elementary theory of radiation, to Carathéodory who replied on 4 April 1913:[223] "The matter seems to me to be completely resolved now and should satisfy every physicist." However, for Carathéodory "a basic question" remained unanswered. It concerned Hilbert's third axiom, according to which there existed materials with an absorption coefficient a and refraction capacity such that the quotient α/q^2 became equal to an arbitrarily given function of the wavelength λ. How could that be brought into agreement with the fact, Carathéodory asked, "that, in nature, a *finite* (though very great) number of parameters p exists"? To resolve the question, Carathéodory suggested: "one should show that Kirchhoff's law is approximated with an increasing number of parameters. In this case, I think of a theorem that would be somewhat analogous to the following: If $f(x)$ is continuous in the interval $\overline{0 \, 2\pi}$ and $|[f(x+h) - f(x)]/h| < 1$, then it follows from the vanishing of the first n Fourier coefficients of $f(x)$ that in the whole interval $|f(x)| < 1$, where $\lim_{n=\infty} M_n = 0$. Thus, if one is contented with restricted precision, then the vanishing of the function follows from the vanishing of the first n Fourier coefficients." Carathéodory proved the above theorem in a few lines of his letter. He closed it with the remark: "If one represents especially a continuous function by a trigonometric polynomial with a given number of terms (as is, for instance, the case when using the Michelson apparatus), then the exactness of the representation depends only on the maximum of $|f'(x)|$. It is peculiar that this remark, which justifies even more the use of harmonic analysers, is not more widespread." Carathéodory wrote that he had also communicated his remark to Ernst Hellinger, who was in Breslau at that time and who told him that he had found $M_n < 1/n^{2/3}$. Hellinger, jointly with Toeplitz, had proved a significant theorem on symmetric transformations in Hilbert spaces.[224] Toeplitz made connections between Carathéodory's work on coefficient problems[225] and Hilbert's theory of quadratic forms in his 1910 paper.[226]

It is striking that in his communications on the elementary theory of radiation, Hilbert referred to Planck and Pringsheim but he did not mention Carathéodory, who had drawn his attention to the work of both. Also in his *Bemerkungen zur Begründung der elementaren Strahlungstheorie* (Remarks on the Foundation of the Elementary Theory of Radiation), Hilbert withdrew his objections to Planck's proof of Kirchhoff's laws, which he had expressed in the addendum to his *Begründung der elementaren Strahlungstheorie*: "In his textbook on the theory of heat radiation, to which I owe the stimulus for this entire study, M. Planck attempts to prove the Kirchhoff theorems without axiom 1 or without further axioms equivalent to it; according to my above observations this proof must contain a loophole; I see the same one in the explanations in § 26 on p. 27; the attempted spectral splitting of the equilibrium condition appears to me not sufficiently motivated and, in any case, not permitted in the generality necessary later for the proof of Kirchhoff's theorems."[227] But now Hilbert admitted: "The procedure by M. Planck is in no way – as I believed earlier – contra to my observations, since assumptions equivalent to my axiom 2 are to be seen in the physical reflections of M. Planck that refer to the second heat theorem". In this case, Hilbert referred to Planck's textbook on the *Theorie der Wärmestrahlung* (Theory of Heat Radiation).[228] In contrast to the reconciliation with Planck, the Hilbert–Pringsheim controversy continued.

2.8
Venizelos Calls Carathéodory to Greece

In Greece the Liberal Party (*Κόμμα των Φιλελευθέρων*) won the elections of December 1910. The new prime minister, Eleutherios Venizelos, was now empowered to implement his programme of domestic reform and economic and political modernisation combined with the aggressive pursuit of the union of "unredeemed" Greeks into a single state.

Venizelos's first government (1910–1915) placed special emphasis on educational reform. Primary education became compulsory and the responsibility for its costs was transferred from the communities to the state. Two hundred primary schools were founded and primary-school teachers were sent to Western Europe for further studies. In 1910, the Educational Group (*Εκπαιδευτικός Όμιλος*) was created to implement the reform, especially with respect to the question of language. Over time it demonstrated its ability to present the demotic Greek language as both the symbol of a subversive discourse and the realisation of a political will. The educational theorists and scholars of language and literature, Alexandros Delmouzos, Dimitris Glinos and Manolis Triantafyllidis, became its mainstays. Each of them headed the Educational Group for a fairly long period, resulting in rather different emphases in its policies. The educational reform, as conceived by all three of them, bore the ideals of middle-class liberalism. Under their leadership, the Educational Group caused teachers to become more aware of professional, social, political and educational matters. They raised awareness of the Greek language among teachers and significantly contributed to the formulation of syntax and grammar rules for the spoken language. These latter were recorded later by Manolis Triantafyllidis in his *Modern Greek Grammar* (*Νεοελληνική Γραμματική*), in 1941. But at the beginning of the 20th century, the question of language, a delicate question burdened with previous philological-political struggles, became the focus of conflict between reform supporters and reform opponents. The reformers were defeated and the "purified" language was thus established as the official language in the constitution of 1911.

Penelope Delta, Frankfurt 1912.
Courtesy of the Historical Archives, Benakis Museum.

Triantafyllidis was the one of the three Educational Group personalities who was a friend of Carathéodory. Penelope Delta also belonged to the Educational Group from 1910, when she started writing short stories which were published in its journal.

In the years 1910–1911, despite sporadic endeavours to obtain some sort of autonomy, the university sector in Greece depended absolutely on the Ministry of Education and Religious Affairs. Also at that time, purges of teaching staff were carried out on the government's initiative; a common desire of both the ministry and the university was to limit the number of secondary-school graduates moving up to the university. This was attempted by the introduction of a law to impose tuition fees and entrance examinations. However, the law "Concerning the Completion and Organisation of the University" of 1911, albeit part of the reform, did not in fact result in any change in the orientation and aims of the university. The new organisation was so incomplete, and the state interventions so frequent, that a full and detailed statement of the many modifications was demanded.

Georgios Papandreou, the future Minister of Education of the Liberal Government, was studying law, social and political sciences in Berlin at that time. His testimony gives an idea of the educational conditions in Germany as compared with those in Greece:

The more I saw that in the absolutist Germany the freedom of thought and speech was absolute; that university professors were teaching philosophy and natural sciences with no religious reservation; that some professors of political economy were even declaring aloud from the Chair of the University of Berlin, the capital of Prussia, that they were genuine students of Marx; and that, on the contrary, in Greece with the democratic traditions and the democratic regime there was such a lack of freedom, such a spiritual retrogression, such a demagogic terrorism, [. . .] the more my heart was filled with sorrow and indignation.

Papandreou surely included Carathéodory among the representatives of the new world, who would do their best for the rebirth of Greece because they had "clear ideas, deep convictions, courage to express their opinion, belief and boldness."[229]

Besides the foundation of a School of Fine Arts, the integration of the Capodistrian University of Athens into the National University, and the establishment of new chairs and laboratories, the law of 1911 regulated the occupation of vacant chairs. Up to then, the relevant procedures followed the principle of self-service and were opportunities for political favours. For the first time, decisions on new appointments were transferred to a committee. Venizelos proposed Carathéodory as a member of the committee responsible for the appointments to chairs of the Faculty of Natural Sciences and Mathematics. One of the committee's tasks was to decide on problems that had been caused by a senseless purge at the university, as Carathéodory would later remark, in 1935, in a memorandum to the Greek Prime Minister, Panayis Tsaldaris.[230] Other members of the committee were Theodoros Skoufos, professor of geology and palaeontology of Athens University since 1906, and Leonidas Arapidis, graduate of the Zurich Polytechnic and scientific consultant to the Greek industry.

The appointments to the Faculty of Natural Sciences and Mathematics in 1911 were carried out in an exemplary manner. (This was not always the case: breaches of the rules led to the intervention of the courts and parliament in corresponding procedures at the Faculty of Medicine.) The 29-year-old Dimitrios Hondros was offered a full professorship for physics. Hondros had held a scholarship awarded to capable and cultured Greeks of Macedonia by the Bavarian foundation of the Viennese-Greek Baron Konstantinos Velios. The condition of the grant was that the recipients should study in Munich and then return to offer their services to Greece. Accompanied by the politician Dragoumis, Hondros began his study in Germany in 1905. He was Sommerfeld's student in Munich and then Carathéodory's student in Göttingen in 1907–1909. Two years before his appointment to Athens University, and having just been awarded his doctoral title for his work *Über elektromagnetische Drahtwellen* (On Electromagnetic Wire Waves), Hondros had expressed doubts about the possibility of an appointment in Greece. In a letter to Sommerfeld dated 24 December 1909 he described the situation in Athens as follows: "I am now hoping to become an assistant at the Chemical Laboratory. Since there is no free position at the Physics Institute, I have to be content with it, whether I want to or not. The salary would be about 100–100 Fr. [*sic*]. But all at once the political situation has become gloomy and everything here is so closely related to politics that my appointment becomes quite questionable."[231] At the request of a Greek professor at Athens University, who considered Hondros quite incapable of applying for a professorship, Sommerfeld wrote to the Greek Prime Minister Venizelos and recommended Hondros. He concluded his letter with the wish "to see our common science represented at that place, which once was the mother of all sciences."[232]

As soon as he was appointed, Hondros started his lectures with an account of Einstein's theory of relativity, thus meeting a general demand from Greek scientists for the introduction of modern physics to the university. Relativity was of course a favourite subject of discussion. Even Venizelos showed an interest in it and it is said that, among friends, he had expressed his regret for having missed the opportunity of obtaining the relevant information from Carathéodory, who had otherwise made statements on everything imaginable.

On Carathéodory's insistence, Georgios Remoundos was also appointed as a professor for mathematical analysis to the Faculty of Natural Sciences and Mathematics. With a scholarship granted to him by the Greek government for the years 1901 to 1906, Remoundos had studied at the *École Normale Supérieure* in Paris under Borel and Picard. His first work, produced during his studies, treated the Picard theorem and its generalisations. His contributions are mentioned by P. Montel and O. Blumenthal.

2.9
Carathéodory Succeeds Klein in Göttingen

On 6 February 1913, the Minister of Spiritual and Instruction Affairs announced to the rector and the senate of the Royal Technical University of Breslau that Carathéodory would probably accept an appointment to the University of Göttingen on 1 April 1913.[233] Six days later, the Department of General Sciences submitted a list of four candidates to succeed Carathéodory. Zermelo, full professor at the University of Zurich since 1910, was first on the list, Issai Schur, who had made excellent contributions to analysis, was in second place along with Max Dehn and Hermann Weyl, whose studies on the asymptotic distribution of eigenvalues of integral equations gave him the reputation of being one of the best mathematicians of his time, was third.

Carathéodory's appointment to full professor at the Philosophical Faculty of the University of Göttingen by "His Imperial Majesty, the King" was decided on 24 February and announced to him on 3 March. Thus, he was informed that he had obtained Klein's chair "with the obligation to teach the mathematical disciplines in co-operation with the other faculty professors through lectures and exercises in full completeness." At the same time, the minister appointed him co-director of the Physical-Mathematical Seminar.[234] His annual salary, providing for a pension, would be 7200 marks, plus a housing benefit of 720 marks and an additional amount for lecturing and for other activities to do with his university position.[235]

In a letter dated 4 April 1913,[236] Carathéodory informed Hilbert that he would visit him in Göttingen around 20 April; his family would follow the week before Whitsun. Before he left Breslau, Rector Schenk organised a farewell party for Carathéodory, to which he invited all professors, lecturers, assistants, junior university teachers in charge of practical or supplementary classes, the legal adviser and the librarian of the Technical University.[237]

In Göttingen, Carathéodory joined colleagues Hilbert and Landau. Reid's account of Klein's succession mentions Carathéodory's adaptability to the Göttingen atmosphere: He "returned to relieve Klein after the older man had suffered another breakdown in 1911. The easy, cultured Greek, whose family motto was "No Effort Too Much", had become a mathematician in the traditional Göttingen style."[238] But there were surely additional reasons for that arrangement. Carathéodory himself mentioned in his *Autobiographical Notes* that his friendly relations with Klein lasted until the latter's death. Moreover, Klein was convinced of Carathéodory's

"Exercise book of the schoolboy Costias Carathéodory":
Carathéodory's self-ironical entry on the cover of his Latin vocabulary book;
probably Göttingen 1913.
Courtesy of the Bavarian Academy of Sciences, Archive.

scientific performance before he asked him to habilitate in Göttingen. Georg Faber supports the view that almost no other mathematician in Germany with such an extensive overview of the entire area of pure and applied mathematics could have been appointed to succeed Klein.[239] Carathéodory shared with Klein his interest in educational innovation. Finally, it should be noted that Klein was a passionate lover of Greek antiquity and Carathéodory offered him the opportunity to discuss subjects of ancient Greek mathematics and culture. Carathéodory himself, now already past 40 years of age, started to learn both Ancient Greek and Latin with enthusiasm.

That Carathéodory was not going to stay permanently at Göttingen University is evident from an application he submitted from Göttingen to the Breslau building authorities on 26 November 1913. He intended to install an elevator in his Breslau house at Scharnhorststr. 30,[240] probably to help his son, who was ill with an infantile paralysis. And this is an indication that he intended to return to Breslau.

When Carathéodory arrived in Göttingen, a seven-day Wolfskehl congress on the kinetic theory of matter and electricity, organised by Hilbert, was in progress. A prize had been established at Paul Wolfskehl's bequest after his death in 1906 for the mathematician who could supply a proof of Fermat's last theorem or contradict it through a counter example. The interest from this prize money was used to bring famous scientists to Göttingen for a series of lectures. On this occasion, the theoretical physicist Hendrik Antoon Lorentz, who had retired from the University of Leiden the year before, described the development of the kinetic theory of gases. Hilbert had worked out a new mathematical method for the solution of problems in that theory. Sommerfeld reported on Hilbert's work and Peter Debye, professor of theoretical physics at the University of Utrecht and former assistant to Sommerfeld in Munich, presented a theory of heat conduction for the first time. Further speakers were Planck, Nernst, Smoluchowski, Kammerlingh Onnes and Keesom. The Wolfskehl congress and the approval for the foundation of the *Modellversuchsanstalt der Göttinger Universität* (Institute of Test Modelling of Göttingen University) under Prandtl, which followed in the same year, marked a renewal of interest in physics in Göttingen, which would reach its peak after World War I, through the work of Max Born, James Franck and Robert Wichard Pohl.

Physical chemistry at Göttingen was co-ordinated by Gustav Tammann and technical physics by Carl Runge, Theodor von Kármán and others. In 1903, Tammann was appointed to the recently founded chair of inorganic chemistry where he remained until his retirement in 1930. He became known as the founder of metallurgy, contributing fundamental research in metallography, melting and crystallisation processes. One of his assistants, P. Kyropoulos, was later appointed to the University of Smyrna by Carathéodory. Runge had drawn attention to the need for reform in the mathematical instruction of physics students in a report about *The Mathematical Training of the Physicist in the University* which he presented at the International Congress of Mathematicians in Cambridge (22–28 August 1912). He thought the main danger was in the widening gap between physicists and pure mathematicians, the latter suffering from over-specialisation and cutting themselves off from natural philosophy and experimental science.[241] Kármán, like Prandtl, was one of the founders of modern air-stream physics. The Swiss physicist Paul Scherrer also became a student at Göttingen in 1913.[242]

This was a time of great activity and an opportunity for Carathéodory to associate with major mathematicians. Klein was endeavouring to promote the internationalisation of mathematical instruction. Weyl, who had earlier attended Carathéodory's lectures in Göttingen, published his monograph *Die Idee der Riemannschen Fläche* with the intention "to develop the basic ideas of Riemann theory on the algebraic functions and their integrals in such a way that concepts and theorems of analysis situs [topology] involved in the theory will also be treated sufficiently according to modern demands for strictness", which he dedicated to Klein "in gratitude and admiration".[243] The number of Hilbert's students was continually increasing and Hilbert himself was attempting an axiomatisation of physics.[244] Prange worked on his dissertation on *Die Hamilton–Jacobische Theorie für Doppelintegrale* (The Hamilton–Jacobi Theory for Double Integrals) with which he gained his doctoral degree under Hilbert's supervision in 1915. His attempt to extend the Hamilton–Jacobi theory for the case of double integrals was mentioned by Carathéodory in the paper *Über die geometrische Behandlung der Extrema von Doppelintegralen* (On the Geometric Treatment of the Extremes of Double Integrals).[245] In that work Carathéodory obtained the Hamilton–Jacobi results without the prior fixing of boundary values for the sought solution and he also defined the concept of a cross-section. Together with Courant, Carathéodory held a seminar on the calculus of variations in the winter semester of 1913.[246] Meanwhile, Landau, with the advantage of inherited fortunes from his parents and his father-in-law Paul Ehrlich, the physician, chemotherapist and Nobel Prize winner of 1908, had been able to build his palace-like residence in Göttingen and house his huge scientific library there. Except for his tireless mathematical research and conscientious lecturing, Landau cultivated two hobbies, namely stamp collecting and passionate reading of detective stories, and showed contempt for everything even remotely practical. His primary research focused on analytic number theory and especially on the distribution of prime numbers.

2.10
On the Editorial Board of the Mathematische Annalen

Some time after his arrival in Göttingen, Carathéodory was proposed for membership of the editorial board of the *Mathematische Annalen*. On 22 December 1913, he wrote to Klein[247] that he would be pleased to accept Klein's "honourable proposal", but he suggested a delay in announcing the acceptance. On the same day, he wrote to Hilbert, who was pressing for Carathéodory to become an *Annalen* board member quickly:[248] "After having repeatedly spoken with Landau and also with Klein about your recent proposal and having thought everything over, I principally agree to accept this honourable proposal of yours. The only point which has still to be considered is whether this honour has not arrived too early after my appointment in Göttingen and whether it would be appropriate to wait before I join the editorial board of the *Annalen*. Basically, this point is not very important but I ask you to consider it before you make any decision."[249] Carathéodory remarks that the foundation of the *Mathematische Annalen* reflected a split among German mathematicians that had been caused by a petty criticism against methods deployed by Riemann in his proofs. In the light of criticism from the Weierstrass school, those methods had been recognised as not entirely capable of providing a proof.[250] The *Mathematische Annalen* had been founded by Alfred Clebsch and Carl Neumann in 1868 and aimed at being "open to all original works of a purely scientific content, which would promote the discipline of mathematics itself or its scientific applications in any possible way".[251] Klein, who, together with Paul Gordan, Adolf Mayer and Karl von der Mühll, joined the editorial board in 1873, the year of Carathéodory's birth, took on the publication in 1876 and had a considerable impact on the journal's orientation. In 1897, Hilbert was accepted as a board member and became one of the editors from volume 55 (1902) onwards. Carathéodory and the Dutch mathematician Luitzen Egbertus Jan Brouwer appear as co-editors together with O. Hölder, C. Neumann and Max Noether in volume 76 (1915). Klein remained one of the main editors alongside Hilbert, Einstein and Blumenthal up to volume 92 (1924). From then on, Carathéodory took up his position and Klein was mentioned as co-editor up to volume 94 (1925).

On 22 June 1925 Felix Klein died. On 1 July, Carathéodory wrote to Hilbert that he found Blumenthal's obituary of Klein for the *Mathematischen Annalen* quite impossible as regards both content and style. Carathéodory felt that Klein's mathematical performance had to be described in a longer article, by Weyl. He thought there could be a "short obituary with a maximum of twenty lines, but in which every word should have an effect", centred around Klein's connection with the *Annalen*. Carathéodory asked Hilbert himself to write such an obituary, for Blumenthal would otherwise feel insulted. Finally, Carathéodory protested about not having been informed of Klein's death earlier and therefore not having been able to attend the funeral.[252] Carathéodory himself published an obituary in the *München-Augsburger Abendzeitung* of 3 July 1925; he also wrote the death notice in the name of the editorial board of the *Annalen* which appeared in volume 95 (1925):

Felix Klein has parted from us. In him we lose a man whose name has stood at the head of the editorial board of this journal for almost fifty years, but whose spirit will live on in the *Annalen* as long as this journal exists.

After Clebsch's death and while Neumann was the editor-in-chief, it was mainly Klein who kept our journal alive. A few years later, he joined the main editorial board, in which he worked with all his power. He withdrew hardly ten months ago, when this power started to wane, because he did not want to be, merely on account of his name, in charge of a post for which he could not fully hold responsibility anymore. In its present form, the *Mathematische Annalen* is substantially Klein's creation. The guidelines which the founders of this journal had in mind became a firm tradition through him. If the *Mathematische Annalen* has included equally, as far as was possible, all mathematical disciplines that are currently active, then this is due to Klein. He made sure that the various mathematical trends were represented in the editorial board and that the members of the board worked alongside him with equal rights. As always in his life, he never paid any attention to himself, but he constantly kept an eye simply on the aim to be achieved.[253]

This last sentence almost coincided with Hilbert's praise of Klein on the latter's 60th birthday on 25 April 1909: "You were never interested in personal advantage, nor were your actions directed towards the interests of another party, but rather continuously in pursuit of the cause." [254]

Carathéodory himself remained co-editor of the journal until the so-called *Annalenstreit* (*Annalen* quarrel) of 1928, namely the controversy regarding the foundations of mathematics, which ended in the dissolution of the old *Annalen* editorial board and the emergence of a new one under Hilbert's sole command. Carathéodory worked closely with Klein and even advised him on tactics. He asked him, for instance, from La Haye, Holland, on 1 September 1919[255] "not to accept any work by [von] Schouten for the *Annalen* without previously asking Brouwer. The work that Schouten wants to have published now has a long and very complicated background. Brouwer is part of it and since he sits on our editorial board, I think it is necessary to listen to his opinion as well. It would be best if we could keep our fingers *entirely* out of this affair between two Dutchmen." There is a draft of a letter from Klein to Schouten (13 March 1920), whose paper had received a negative evaluation from Brouwer, mentioning that on joining the editorial board of the *Annalen*, Brouwer had reserved, especially, the right to decide about Dutch papers.

Carathéodory was also in permanent contact with Hilbert as regards the *Annalen*. They discussed not only their mathematical concerns but also matters of everyday life. Particular emphasis was put on social relations among the colleagues who formed a small exclusive society. Carathéodory wrote to Hilbert on 14 April 1926[256] that Blumenthal would be 50 years old on 20 June. He and Sommerfeld had thought of a birthday present for Blumenthal in the name of the editorial board "for the great pains he had taken during his twenty years on the board and for the qualities he brought to the journal". Sommerfeld had thought of a bust of a mathematician and Carathéodory suggested that every member contribute 20 marks. [257] Meanwhile, Carathéodory had been thinking about a different possible birthday gift and on 17 April 1926 he wrote to Hilbert[258] that he and Sommerfeld were expecting to hear from Hopf whether Blumenthal possessed Jacobi's texts. He knew an excellent bookbinder and he himself offered to have a good cover made and a dedication printed at the beginning of the book. He announced that he would forward the proposal to Bieberbach, Bohr, Brouwer, Dyck, Einstein and Kármán. He asked Hilbert to inform Hölder, whom Carathéodory hardly knew. The Jacobi idea, however, was dropped. Carathéodory's letter to Einstein on 4 October 1926[259] reveals that Blumenthal received "Gauss's works in a very nice cover" as a birthday present. This time, 40 marks were to be paid by each contributor. Carathéodory had enclosed a payment slip and Einstein transferred the money to him.[260] From Kobe, Japan, Kármán sent Carathéodory 21 yen and remarked: "I am already looking forward to imagining you going from one bank to the other in Munich in order to cash [21 yen]. We will thus see, whether Munich is really a cosmopolitan city."[261]

During his time on the editorial board, Carathéodory was sometimes faced with extremely delicate problems. This was especially the case in the war years, when the production of the *Annalen* became considerably difficult. Carathéodory's correspondence with Georg Pólya, on two works sent to the *Annalen* for publication by Robert Jentzsch and Pólya, gives an impression of the complicated situation. Georg Pólya, who had gained his doctorate at the University of Budapest, spent 1913 in Göttingen and directly afterwards, through the mediation of Hurwitz, went to Zurich, where he became an assistant professor. There, he wrote numerous works in various areas of mathematics. In 1918, he published papers on series, number theory, combinatorial analysis and electoral systems. Carathéodory wrote to him on 7 February 1917:[262] "I have just received the manuscript of an interesting work by Jentzsch, in which he announces a work on the same subject by you. But the present difficulties in printing the *Math. Ann.* are so serious that I can hardly advise you and Jentzsch to have these works printed there. A whole series of manuscripts, which are not yet even set, lies there since June of this year, and Teubner can continue setting again only in May. I have also informed Jentzsch; maybe, you could get in touch with him and find a way out. Please, write to tell me what you have decided as soon as possible. Of course, we would gladly welcome the work, if it is not too long and if you are willing to accept the unavoidable delay." Exactly two months later, Carathéodory informed Pólya[263] that he had received his work and would let it be printed immediately after Jentzsch's work. The *Annalen* were not printed in 1917. But volume 78 of 1918 does indeed contain two papers with the same title by Jentzsch and Pólya.[264]

When, in 1944, Carathéodory forwarded Popoff's paper *Sur une propriété des extrémales et le théorème de Jacobi* (On a Property of the Extremals and the Theorem of Jacobi) to Behnke for publication in the *Annalen*, without knowing that van der Waerden had rejected it,[265] he admitted that he had treated the work with more laxity than in other cases, since he could not apply to Popoff, a Bulgarian, the strict standards he would apply to a Greek.[266]

Much later, Carathéodory's obituary signed by the editorial board of the *Mathematische Annalen* in 1950, emphasised his contribution during the war years: "He was active for more than four decades as a colleague of the Mathematische Annalen. During the difficult period of World War I, he took care of the edition. His initiative and his confident advice were invaluable to the board. He helped us keep the gates open to the world in critical times. He was still contributing to the editing of the most recent issues."[267]

2.11

War

In the period between 1900 and 1910 the bourgeois world to which both Born and Carathéodory belonged was convinced that no European war would ever happen again. Military service was taken for granted as an unavoidable meaningless exercise with all the discomfort that might accompany it but with the expectation that it would later be reduced or abolished. In 1907 Carathéodory wished Born would soon free himself from the military, underlining the word soon. "You are obviously riding *too* well",[268] he wrote, probably referring to an incident during a riding lesson which had outraged Born's lieutenant: Born had been the only one of the soldiers not thrown off his horse when jumping a hurdle.[269]

But war broke out in the summer of 1914 and in October that year, many scholars and intellectuals signed the Appeal to the Cultural World (*Aufruf an die Kulturwelt*), which was published by the Reich government. That call was intended to refute

The house at Friedländerweg 31. Photograph: S. Georgiadis.

what the Germans called propaganda by their enemies. Instead, it damaged the international reputation of German science. Among those who signed the appeal were famous personalities from Carathéodory's intellectual environment, such as Felix Klein, Emil Fischer (professor of chemistry in Berlin and tutor of Ioakimoglou), Adolf von Harnack (General Director of the Royal Library in Berlin), Walter Nernst, Max Planck and Richard Willstätter (all of whom were professors of physics in Berlin), Wilhelm Wien (professor of physics in Würzburg) and Paul Ehrlich. Carathéodory himself did not sign it and it was maybe his non-German citizenship that protected him from being accused of treason. Hilbert, however, was not excused for not signing, since he was not only a German, but, even more, a Prussian. Klein's name was crossed off the list of members of the French Academy and he was himself aware of the consequences of the predictable German isolation but, despite that, he insisted on the development and improvement of mathematical instruction in military education.[270] Carathéodory also did not sign the later Declaration of Intellectuals of July 1915 that sought to confer a mantle of moral justification for Germany's expansionist policies.

Born remembers that

in 1914 there was a patriotic outburst of foolish enthusiasm in all countries. We had it in Göttingen in full measure: flags and marching and singing. Troops marched through streets lined by people throwing flowers. Flags everywhere – in the streets and on the trains carrying soldiers to the front. The barracks behind our house was the centre of excitement, for the regiment had to march in a few day's time. Many of our young friends had to join up, among

Constantin and Euphrosyne in front of their house in Göttingen in 1915.
Courtesy of Mrs. Rodopoulos-Carathéodory.

them my colleague Richard Courant. He was Sergeant-Major in the reserve army and got a commission on the outbreak of war, like many other Jews who would never have become officers in peace-time. [...]

The patriotic frenzy was coupled with wild rumours and a spy hunt: wells were said to be poisoned, the horses of the regiment paralysed, bridges blown up. All foreigners were rounded up and put into custody, [...]

The newspapers were full of pep articles.[271]

A victim of this hysteria was Carathéodory's wife Euphrosyne, who was accused of having poisoned her husband's colleagues and their wives with cakes at a reception in their house in Göttingen at Friedländerweg 31. The outcome of this case is unknown because the files no longer exist.[272]

In the entire German Reich 13.2 million men were recruited for the war. The political leadership of the Reich was degraded, whereas the *3. Oberste Heeresleitung* (Third Superior Army Command Staff) under Paul von Hindenburg and Erich Ludendorff was upgraded in August 1916 to a leading authority with the traits of a military dictatorship. The self-awareness of a German *Sonderweg* (special way) fused with the social Darwinist idea of the survival of the fittest. Inhibitions towards violence gradually fell. The economy was set up according to the dictation of military demands. With devastating consequences for their health, women and adolescents filled the gap left by the men in the war economy. Hundreds of thousands of deported foreign workers were forced into slave-labour. The war expenditure increased to the enormous amount of 162 billion marks, those engaged in the war economy increased their profit by several hundred percent. Public corporations contributed to the war through taxes of merely 21 billion marks. During the war, the national product decreased to 60% of that in 1913. The standard of living deteriorated rapidly. In 1917, only one third of the necessary minimum number of calories was available per person and the black market flourished. The upper financial bourgeoisie was blossoming, whereas the classically educated middle classes were financially weakened. The real income of the higher civil servants decreased by 60% on average. The income of employees was reduced by a quarter and not unusually to a half. Workers experienced an impoverishment comparable to that of the early industrialisation. In addition deficiency diseases spread.[273]

In the years 1915–1917, the army and navy command staff supported war projects run by the University of Göttingen, such as the construction of a new wind tunnel, considerably bigger than the one built in 1908. At the beginning of the war, Max Abraham was forced to leave his professorship of rational mechanics at the University of Milan, Italy, and returned to Göttingen. He worked in Göttingen on the theory of radio transmission for the war authorities. Runge and Courant, together with Peter Debye and his assistant, Paul Scherrer, worked in the late summer of 1915 on the development of earth telegraphy, a project conceived for military application. The Debye–Scherrer method was discovered in 1917 for the establishment of the order of atoms within a crystal through measurement of the bending of X rays on crystal powder. The brilliant scientific work of the Göttingen mathematicians was the incentive behind Rockefeller's later investment of hundreds of thousands of marks for the building of university institutes, despite their involvement in war research. It was Courant who mediated the American connection.

In Ebertal-Göttingen, a large prisoner of war camp was set up and supervised by the university and the university library opened an issue desk at the camp. Lecturers taken as prisoners of war established a camp university providing lectures and seminars. Since the prisoners were allowed to receive their scientific literature by mail, they were able to set up a camp library which, by the end of war, remarkably contained over ten thousand volumes. Carl Runge had responsibility for imprisoned foreigners and after the war he successfully contributed to restoring contact with English natural scientists.[274]

Carathéodory spent the war years in Göttingen and tried to survive as best he could with his family. The majority of students and lecturers served in the military. In a postcard that Carathéodory sent to Zermelo on 31 March 1915[275] to invite him to Göttingen he excused himself for not having written earlier, "because, up to now, we have had a large billeting of soldiers and we did not know for how long they would stay". Feeling totally isolated, Carathéodory wrote to Fejér on 2 July 1916:

I think of you very often, dear Fejér, and I am pleased that you still intend to visit us. It is actually terrible that the end of the war cannot be predicted; now I am entirely cut off from my sister and other friends and relatives in Athens. Fortunately, I have so much to do that this time passes by still tolerably for me.[276]

But he was able to spend September 1917 with his sister in St. Moritz, where she was staying.[277] She had left Greece with her husband, Georg von Streit, who had accompanied the King of Greece, Konstantin, in exile. Konstantin had departed from Greece on the Allies' demand, without formally abdicating.

2.12

Famine

The so-called turnip winter of 1917 was a metonymy for the worst famine of the war. The food shortage is mentioned in a letter from Carathéodory to Hilbert:[278] "Three days ago, I was in the small hotel of the Gauss tower for a night (the next day, we went to Münden). We were the sole guests and the calmness up there was very agreeable. Yesterday, Heis was here and advised us to go to Warnemünde; it is said that good food in sufficient amounts could be obtained there and W. has the advantage that my wife could also relax a bit". Despina recounts that her father kept two goats in their house yard to supply the children with fresh milk. Purchases on the black market and storage in the house had become a sort of mania and a kind of sport employing tricks, if we are to believe Klaus Mann. "Frau Carathéodory understands how to panic-buy". A female student explained that if she found no eggs, she would buy coffee cans. For Carathéodory himself she made clear that he was dreaming of chops and was longing for luxury. Frau Hilbert was trying to "supply her David with calories", while Frau Runge endeavoured to plant vegetables in her garden.[279]

2.13

Insipid Mathematics

Carathéodory calculated for Hilbert the motion of a fluid by the aid of Gauss's theorem. He sent the calculation on a postcard to Hilbert, who was staying in Flims near Chur, Switzerland, on 12 August 1916.[280] On 15 September 1916, Carathéodory recommended Poincaré's lectures on celestial mechanics to Klein:

It seems that your question could be solved most elegantly if you introduced canonical variables. Let $x_1, x_2, \ldots, x_{3n}$ be the co-ordinates of the n bodies, $m_1, \ldots, m_{3n}$ their masses $(m_1 = m_2 = m_3, m_4 = m_5 = m_6, \ldots)$, $y_k \equiv m_k \dot{x}_k$ the impulses. Then, if one puts

$$H = \sum \frac{y_k^2}{2m_k} + \sum \left(\frac{-m_1 m_4}{z_{14}} + \frac{-m_1 m_8}{z_{18}} + \ldots \right),$$

one obtains

(1)
$$\dot{x}_k = \frac{\partial H}{\partial y_k} \qquad \dot{y}_k = -\frac{\partial H}{\partial x_k}.$$

The centres of gravity and the surface theorems are consequently to be derived in exactly the way you have used with the other notation*. *But now the energy theorem is indeed a consequence of*

$$\frac{\partial H}{\partial t} = 0,$$

since from (1) follows only

$$\dot{H} = \frac{\partial H}{\partial t}.$$

Compare with Hamel's old seminar lecture!

With best regards
Yours obediently
C. Carathéodory

*See for example Poincaré
Leçons de Mécanique Céleste I p. 25 [281]

Student notes reveal that in the summer semesters of 1915–1918 Carathéodory lectured successively on first-order partial differential equations, conformal representation, projective geometry and analytical geometry. His students had the impression that, just like Hilbert, he lectured in a confusing way and was a casual lecturer who, moreover, took no account of his audience, even if that included Leon Lichtenstein, as they said. At that time, Lichtenstein, a Polish-Jewish mathematician, was lecturer at the Technical University of Berlin-Charlottenburg and was a consultant to the Julius Springer publishing company in Berlin. At the end of 1917, Springer announced the establishment of the *Mathematische Zeitschrift*, whose first issue was published in January 1918. The editor was Lichtenstein. [282]

2.14
"German Science and its Importance"

By the end of the war, the Göttingen Mathematical Society was dispersed. The reading room had great gaps in its collection and the construction of the mathematical institutes had to be postponed. Almost all German editors of scientific journals and books were pulling out. [283] Carathéodory, however, assessed the losses of German science during the war as not particularly great, rather the opposite was the case. In an article published on 12 April 1929 in the *Deutsche Allgemeine Zeitung* under the title *Deutsches Wissen und seine Geltung* (German Science and its Importance), [284] he asserted:

A defeat of the exact natural sciences and mathematics which would have been caused by the war and the post-war period is out of the question. Moreover, a kindly Providence has allowed a new recognition of German science in the whole world, especially during the past decade, through sensational *discoveries*. On this occasion, *Einstein's relativity theory* must be mentioned first, the significance of which has rather to be sought in its basic thoughts, which have determined an entirely new attitude towards nature, than in the complicated and still not fully complete apparatus of forms, [...]

However, I see a more profound reason why scientific research in Germany has been touched so little by the rigours of these times, especially in the *constitution of our universities*, which connects life and scientific research most fortunately. [...]

The *younger* scientific *generation* has not suffered that much either, as was initially feared: the first generations of students who returned from the battlefield were more hard-working, eager to learn, more mature than in normal times, yet they felt the urge to get back to real life as soon as possible and only a few individuals among them were destined to deploy their abilities in pure science. But immediately in the following years, a whole series of brilliant talents revealed themselves, so that one could be assured of the fact that scientific work in the disciplines with which I sympathise will still continue to develop favourably in Germany.

The *respect for German science* plays a much greater role as regards the relationships with foreign countries than is known to the broad public.

2.15
Einstein Contacts Carathéodory

Albert Einstein, then professor at Zurich Polytechnic, returned to Germany in March 1914. As a full member of the Physics-Mathematics Class of the Prussian Academy of Sciences in Berlin, to which he had been appointed in 1913, he had the right to give lectures at the University of Berlin. He gave two at the Berlin Academy itself called *Über die allgemeine Relativitätstheorie* (On the General Theory of Relativity) on 11 and 25 November 1915. After years of effort he was able to formulate the general theory of relativity, i.e. a field theory of gravitation, according to which gravitational waves and consequently also the corresponding elementary particles, the "gravitons", should exist. His performance was rewarded by his election to Chairman of the German Physical Society (*Deutsche Physikalische Gesellschaft*) on 5 May 1916.

Einstein included Carathéodory in his efforts to generalise the theory of gravitation. He had met the Göttingen mathematicians in the summer of 1915 when he held six two-hour lectures on the general relativity theory there. Hilbert had presented *Die Grundlagen der Physik* (The Foundations of Physics) claiming that he would deduce a new system of equations simultaneously containing the solution to the problems of Einstein and Mie.[285] Einstein got in touch with Carathéodory to ask him for the derivation of the Hamilton–Jacobi relation in a letter on 6 September 1916.[286] Therein, he himself presented his "simple considerations" of how the Hamilton–Jacobi equation could be derived. Einstein ended his letter with the words: "Naturally, the Jacobi transformation is still in no way proved by this. But for this, the formal, less transparent proof as given by Appell suffices for me". In the same letter, Einstein asked Carathéodory to reflect on "the problem of the closed time-lines a bit more", since therein lay "the nucleus of the still unresolved part of the space–time problem". He further added: "Of course, I do not imagine that these trivialities are in any way original or new. They are just the things that give me the feeling of being familiar with the object". The Hamilton–Jacobi theorem was employed by Einstein in his paper *Zum Quantensatz von Sommerfeld und Epstein* (On the Quantum Theorem of Sommerfeld and Epstein), which was presented to the German Physical Society on 11 May 1917.

Einstein went on to request from Carathéodory an exposition of the canonical transformations and particularly of the solution of the problem of closed time-lines:[287] "I find your derivation wonderful", he wrote to Carathéodory, probably referring to the derivation of the Hamilton–Jacobi relation, and he recommended its publication in the *Annalen der Physik*,

since physicists usually know nothing of this subject, as was the case with me as well. With my letter, I must have appeared to you like a Berliner who had just discovered the Grunewald [a pine forest in Berlin] and asked whether people have ever been to it yet. If you would like to make the effort to explain to me the canonical transformations, you will find a grateful and conscientious listener. But, if you solve the problem of the closed time-lines, I shall place myself before you with hands folded [in reverence]. ... Behind this is something worthy of the sweat of the best of us.

Carathéodory replied on 16 December 1916.[288] The theory of canonical substitutions, he wrote, was a consequence of the transformation which had led to the following equation:

$$\sum y_k \, dx_k = d\Omega + \sum A_j \, da_j + H dt,$$

where $d\Omega$ is a total differential and the A_j are independent of t. Every time when such an equation is satisfied, the functions

$$x_k = \overline{x}_k(a_1 \ldots a_{2n}, t), \qquad y_k = \overline{y}_k(a_1 \ldots a_{2n}, t)$$

would be solutions of the canonical differential equations

$$\frac{\partial}{\partial t} \sum_k \overline{y}_k \frac{\partial \overline{x}_k}{\partial a_j} = \frac{\partial^2 \overline{\Omega}}{\partial a_j \partial t},$$

$$\frac{\partial}{\partial a_j} \sum_k \overline{y}_k \frac{\partial \overline{x}_k}{\partial t} = \frac{\partial^2 \overline{\Omega}}{\partial a_j \partial t} + \frac{\partial \overline{H}}{\partial a_j}.$$

Carathéodory suggested to Einstein the relevant representation in Whittaker's *Analytical Dynamics*.[289]

The problem of canonical transformations reappeared in 1922, when Carathéodory wrote the fifth chapter, on the calculus of variations, of the book *Die Differential-und Integralgleichungen der Mechanik und Physik* (The Differential and Integral Equations of Mechanics and Physics).[290] The term "closed time-lines" refers to the geodesic lines corresponding to light rays or paths of free particles (not influenced by other masses) that are closed in the model of a static universe, which Einstein introduced in 1917.[291] The term was hinted at by Einstein himself in the formal foundation of general relativity that he submitted to the Prussian Academy of Sciences at the end of 1914: "Therefore, one can think *a priori* of a motion of a point by which the four-dimensional trajectory curve of the point would be an almost closed one."[292] According to D. Christodoulou,[293] the term "closed time-lines" can be interpreted as follows:

"First, the date of the letter places it one year after Einstein's formulation of the general theory of relativity. It clearly refers to what we nowadays call "closed time-like curves". The notion of space–time introduced by general relativity, namely that of a manifold endowed with a structure analogous to that of a Riemannian manifold, possessing curvature, had opened up possibilities which did not exist for the linear structure of the Minkowski space–time of special relativity. Similarly, a Riemannian manifold had possibilities which were unthinkable in the Euclidean framework of a linear structure for space. Examples related to the issue in question are the Riemannian notion of a compact space without boundary and the existence of closed geodesics. In the case of space–time geometry, it must have been quickly realised that an example of a space–time containing closed time-like curves can be constructed out of Minkowski space–time, by simply considering the domain in Minkowski space–time bounded by two parallel space-like hyperplanes and imposing an identification of the boundary hyperplanes. This produces a manifold which has the topology of the product of a circle with 3-dimensional Euclidean space. Being everywhere flat, it is also a solution of Einstein's equations of general relativity in the absence of matter. However, it is clearly an artificial example and its physical consistency is by no means guaranteed. By this we mean that it is not a priori clear whether the equations embodying the basic physical laws, notably Maxwell's equations of electromagnetic theory, possess general solutions in the given space–time framework. In the case of the artificial example just mentioned, this question is equivalent to the question of the existence of general time-periodic solutions of the Maxwell equations in the original Minkowskian framework and the answer is, as we now know, in the negative. Perhaps Einstein, by saying "If you solve the problem of the closed time-like lines I shall place myself before you with hands folded in reverence" in his letter to Caratheodory, is actually proposing the investigation of this problem: namely the existence or non-existence of space–times which possess closed time-like curves and in which Maxwell's equations are well posed."

2.16
The Theory of Relativity in its Historical Context

The significance of relativity theory for Carathéodory emerges directly from his first lecture in the winter semester 1926–1927 at the University of Munich when he put it into its historical context in an extended review of the history of mathematics and physics:[294]

Introduction. If someone asked me what the most distinctive feature of our culture is, in contrast to previous great cultures such as the Egyptian, the Assyrian or the Aegean, or in contrast to those cultures that live on independently of us today, as in India or China, then I would not set this difference so much on external features that immediately hit every traveller in the eye, but rather say: What characterises our culture mostly is the *belief that we can calculate in advance external events with the aid of mathematics.* The fact that such great human masses were able to live without this belief shows that this is not that natural. But for us, who – as a consequence of this attitude – were able to register the greatest successes, this belief forms the roots of our thinking.

Also, the man in the street – not only the educated mathematician – knows, or imagines, that the course of heavenly bodies can be calculated in advance, as well as the most daring bridges, the biggest dynamo machines and the most complicated chemical processes. The expert shares this point of view by and large, but he knows better where the limits of human potentialities are to be drawn and where the powers of mathematics cease.

Now, if we reveal the reasons for this way of thinking (or if we only want to justify it to ourselves), we have to postulate from the beginning that a mathematical description of inanimate nature, whose results agree with the material of observation, is possible. The solution to this problem is the main aim of that science which is today described as *mathematical physics* and which the English more succinctly call *natural philosophy.* Because, if one thinks over the matter more exactly and frees oneself from the usual classification of sciences, *geometry* must also be accepted among those objects which belong to a mathematical description of the objective world.

From this point of view, it is not surprising that people who, in *ancient times*, set up the principle that there are natural laws and, to be precise, *mathematical* natural laws, have engaged themselves particularly with *geometry* and brought this science to a perfection that can only be exceeded with difficulty today.

The second chapter of mathematical physics is *mechanics*, whose beginnings have likewise to be sought in antiquity. One, of course, has to be well educated in order to acknowledge the role that *Aristotle* has played in the development of mechanics. But, one only needs to open *Archimedes'* works to see that he has treated problems of mechanics with methods which have not yet lost their validity. Anyhow, almost 1900 years had to pass since Archimedes until mechanics acquired a new face through the great men of the 17th century, *Galileo, Kepler, Huygens* and *Newton*.

Since *Newton* and *Leibniz*, infinitesimal calculus has been employed in the service of the natural sciences, and mechanics – especially celestial mechanics – made gigantic progress during the 18th century.

In the 19th century, mathematical physics was rapidly enriched through new chapters: *optics, heat conductivity, potential theory, theory of electricity* and finally *thermodynamics* (1850). A proud picture of the universe emerged, which seemed to be final. A rather good overview of science at that time is given by the famous book by *G. Kirchhoff* in four volumes. Then, the general opinion prevailed that one could complete the painting through new brush-strokes here and there, but the frame in which the whole had been built seemed to be unshakeably stable.

I am talking roughly of the year 1880. Shortly before, *Maxwell's theory* of electricity was born, which seemed to be the icing on the cake, yet it contained the first seeds of decay.

In its original form, Maxwell's theory postulated the existence of a totally rigid and motionless body in space, called the ether, and experiments were devised in order to measure the motion of our earth in relation to the ether. Since the motion of the earth constantly changes, according to the usual notions, one expected to find a relative velocity at least during part of the year. But, the experiment that we will discuss later proceeded as if the earth was constantly stationary in the sea of the ether.

The first mathematical interpretation, with the help of a modified Maxwell theory, was given by the Dutch physicist *H. A. Lorentz*, but some years later, *Einstein* came and gave a much simpler solution to the problem, which, however, turned our traditional and biased opinions about space and time upside down. The theory of relativity, the purpose of this lecture, was born.

By elaborating on his ideas, *Einstein* found that they were extremely suitable for tackling a problem which had resisted all previous attempts at a solution and for solving it: namely the *gravitational problem*. If one thinks over the matter, one has to admit that the idea of *action at a distance*, as postulated by Newton's law, is much more difficult to bear for the human mind than the most daring combination of thoughts by Einstein. After all, the horror of action at a distance was the reason why *Faraday* and *Maxwell* liberated the theory of electricity from this concept.

Einstein did not content himself with building up his theory mathematically, but he was able to draw [from it] some consequences that could be confirmed experimentally.

From the point of view of a mathematician, the construction of Einstein's ideas is one of the most beautiful ever. Even if it did not have any physical background it would be worth going through.

Still, there are renowned physicists and some philosophers at present – they are of course in the minority – who reject Einstein's ideas (misunderstandings). On the other hand, there are even more people who, the more they think they have the right to extol Einstein's ideas, the less they know about them. There are entire libraries with popular Einstein literature with the most bizarre titles (e.g. The Theory of Relativity for the Tired Businessman); one has to warn against these books (the more popular, the less understandable).

When Einstein arrived in New York, there was such a crowd standing shoulder to shoulder that he needed hours to disembark. Later, he once said: 'They wanted to see a man whose thoughts no-one can understand.' But, we shall try to really understand the theory of relativity.

In 1934, Carathéodory himself attempted to popularise the theory of special relativity competently. For the *Great Greek Encyclopaedia* he wrote an extensive article entitled *Space–Time* (Χωρόχρονος).[295] He concluded it with the remark that

C. Carathéodory

Relativitätstheorie

1. Vorl. 2. 11. 26.

Einleitung. Wenn mich jemand fragen würde, welches das markanteste Kennzeichen unserer Kultur ist, gegenüber den vorangegangenen grossen Kulturen, der Ägyptischen, Assyrischen oder Egäischen, oder gegenüber den Kulturen, (gegenwärtig) die heute noch unabhängig von uns fortleben, wie in Indien oder in China, so würde ich diesen Unterschied nicht so sehr auf äussere Merkmale verlegen, die jedem Reisenden sofort in die Augen fallen sondern sagen: was unsere Kultur am meisten charakterisiert ist der Glaube, dass wir mit Hülfe der Mathematik das äussere Geschehen im Voraus berechnen können. Die Tatsache, dass so grosse Menschenmassen ohne diesen Glauben leben können, zeigt dass er nicht ganz selbstver-

Carathéodory's manuscript for his lecture on the "Theory of Relativity".
Courtesy of the Bavarian Academy of Sciences, Archive.

a more exact study of various, rather simple problems has shown the existence of difficulties which threaten to topple the whole construction. In the meantime, *Einstein* had continued his research on the mutual attraction of bodies and had just found out that the theory which we developed can only be considered as the first approximation to a much more general theoretical construction called the general theory of relativity. In this theory, the space–time transformations which are used are not strictly linear any more; but the paths of light are not linear either, because they are somewhat bent through the attraction of the bodies. [...] It may suffice to mention that, in many cases, the approximation given to us by the theory of special relativity, whose basic characteristics we have described, is so exact that it offers us a satisfactory picture of reality and the difficulties mentioned do not exist anymore in the general theory of relativity.

2.17
Functions of Real Variables

In Göttingen, Carathéodory changed the main direction of his research. He began to study real functions and, with Lebesgue, was responsible for laying down the first and fundamental steps in the modern development of the theory of measure. This extends the quantitative concepts of Euclidean geometry to general subsets of q-dimensional Euclidean space. In order to keep the discussion of the mathematical development of Carathéodory's work together, we shall deal with it here and return to the Carathéodory's professional work in Göttingen again in Sect. 2.18.

2.17.1
Theory of Measure

Carathéodory presented his formal theory of measure in the 1914 paper *Über das lineare Maß von Punktmengen – eine Verallgemeinerung des Längebegriffs* (On the Linear Measure of Point Sets – A Generalisation of the Notion of Length),[296] in which he defined k-dimensional measure for subsets of q-dimensional Euclidean space through the following five basic properties. For $k = 1$ his method gives linear measure and for $k = q$ Lebesgue measure.

I A number $\mu * A$, which is either zero, or positive, or infinite ($0 \leq \mu * A \leq \infty$) and which is called the outer measure of A, is uniquely assigned to every point set A of q-dimensional Euclidean space.

II $\mu * B \leq \mu * A$ for every subset B of A.

III $\mu * A \leq \mu * A_1 + \mu * A_2 + \ldots$ if A is the union of a sequence of finite or countably infinite many sets $A_1, A_2, \ldots$.
This inequality makes sense only if the right-hand sum converges.

IV $\mu * (A_1 + A_2) = \mu * A_1 + \mu * A_2$ if A_1 and A_2 are two sets whose distance is positive.

V The outer measure $\mu * A$ of any point set is the lower limit of the measure μB of all measurable sets B_n for which $A < B_n$ and $\mu B_n \leq \mu * A + (1/n)$ (i.e. which contain A as a subset).
A set A is called measurable (with regard to $\mu*$) if the equation $\mu * (W) = \mu * (AW) + \mu * (W - AW)$ is true.[297]

Subsequently, with the help of certain sets which satisfied a further condition, Carathéodory defined for every set an inner measure as the upper limit of the outer measures of all measurable sets contained in *A*. Perron presents Carathéodory's method as follows:

If *A* is a point set in R_q, let us cover it with at most countably infinite many convex sets U_1, U_2, whose diameters d_1, d_2 are smaller than a number ρ. The lower limit $\sum d_\nu$ for all these coverings is a function of ρ, which increases with decreasing ρ, consequently striving for a limiting value (eventually ∞) for $\rho \to 0$. It will be called the *linear* outer measure of *A* and it will be shown that this measure satisfies the axioms for measure functions. Especially, if *A* is a curve, it is further shown that it is measurable and its measure equals the usual arc length (∞, if it is not rectifiable). A simpler notion, more general than that formerly used, of surface area for surfaces in space is also completely analogously defined. And the notion of *k*-dimensional measure for any set point in R_q ($k < q$) turned out generally. Thus, the *k*-dimensional diameter simply appears in the place of the above used diameter *d* of a convex point set *U* and can be understood as follows: let one project *U* perpendicularly on a *k*-dimensional hyperplane; the projection is a convex set of a certain *k*-dimensional content, which, in addition, depends on the position of the hyperplane in R_q. Then, the *k*-dimensional diameter of *U* will be defined as the upper limit of these contents for all possible positions of the hyperplane."[298]

In his *Beiträge zu Caratheodorys Meßbarkeitstheorie* (Contributions to Carathéodory's Theory of Measurability),[299] Artur Rosenthal answered the still open question of an axiomatic definition of the inner measure, independent of the outer measure, and defined the outer measure through a system of four axioms, whose independence he proved. His reduced system of axioms for the definition of the outer measure corresponded to that of Carathéodory. He also showed that the asymmetry between the theories of inner and outer measures cannot be avoided.[300]

In 1918, Hausdorff gave a similar definition to that of Carathéodory, which is somewhat more convenient and which has become the "standard" definition of *k*-dimensional measure.[301] The idea of *k*-dimensional measure is an essential part of what is now called "geometric measure theory."[302]

Carathéodory's linear measure has been accepted in mathematical literature as an adequate generalisation of the notion of length. Only in 1945 did A. P. Morse and J. F. Randolph[303] publish a paper on the *Gillespie measure*,[304] which, however, was based on Carathéodory's purely formal theory of measurability.

2.17.2
One-to-One Mapping

In his Berlin paper *Über den Wiederkehrsatz von Poincaré* (On Poincaré's Recurrence Theorem)[305] Carathéodory gave a surprisingly short measure-theoretical proof of one of the basic theorems in the general theory of dynamical systems with an invariant measure. The so-called re-ergodic hypothesis was first established by Poincaré. Carathéodory chose this subject only because, as he wrote, "through a small modification of Poincaré's train of thought one could simplify his proof to a considerable degree and could argue in a few lines."

Some years later he wrote in his paper *Some Applications of the Lebesgue Integral in Geometry*[306] about Poincaré's theorem "that there is the probability *one* that in a steady motion of an incompressible liquid the path of a given molecule does not return to any neighbourhood of the place where the molecule was located at the time $t = 0$. At the time when Poincaré introduced this theorem (1890), it was impossible to understand the meaning of it, and Poincaré's proof was of course inaccurate. But twelve years later, Borel and Lebesgue invented the new theory of measure by which the very proof of Poincaré was saved. It is now possible to give a much shorter proof of Poincaré's theorem of return than the original one."

To prove this theorem, Carathéodory considered a dynamical system $f(p, t)$ ($p = P$ for $t = 0$) in a metric space R; he then called the measure invariant with respect to the system $f(p, t)$ if for any μ-measurable set $A \subset R$, $\mu f(A, t) = \mu A$, $-\infty < t < +\infty$. Assuming the measure of the whole space R to be finite, Carathéodory proved that

(1) if $\mu A = m > 0$, then values t can be found, $|t| \geq 1$, such that $\mu[A \cdot f(A, t)] > 0$, where $A \cdot f(A, t)$ is the set of points belonging simultaneously to the sets A and $f(A, t)$;

(2) if in a space R with a countable base, $\mu R = 1$ for the invariant measure μ, then almost-all points $p \in R$ return an infinite number of times to a neighbourhood of the initial state.[307]

2.17.3
Carathéodory's Books on Real Functions

In Göttingen, Carathéodory started to prepare his book *Vorlesungen über reelle Funktionen* (Lectures on Real Functions),[308] which he dedicated to his friends E. Schmidt and E. Zermelo. He had consulted the *Cours d'Analyse* by Camille Jordan, the *Cours d'Analyse* by Ch.-J. de la Vallée Poussin and works by Baire, Hausdorff, Lindelöf and Young, as well as Lebesgue's treatise *Sur l'intégration des fonctions discontinues* (On the Integration of Discontinuous Functions).[309] Carathéodory's book, representing both a completion of the development begun around 1900 by Borel and Lebesgue and the beginning of the modern axiomatisation of this field,[310] was based on the lectures he had given during the summer semester of 1914, but it was only published four years later. It is very detailed and can be read easily, since it starts with the basic properties of real numbers and proceeds to the theories of point sets and limits, which have their beginning in Cantor's theory of sets. Subsequently, Carathéodory methodically introduces the reader to the basic properties of the function of n variables, to the theory of measurability and measure of point sets in n-dimensional space and to one of its applications, namely the definite and indefinite integral and the derivative of the latter (all these according to Lebesgue)[311] and, finally, to the specialisation and application of these theories to known problems of mathematical analysis for functions of one or more variables. He himself described his book as follows:

I have used the past year to write a book (unfortunately too thick) about real functions which is now being printed. It is no great deed but I believe that it will be useful, since Lebesgue's theory of integration will be presented in an easily understandable manner for the first time. Also, through systematic occupation with the subject, some minor things have emerged such as, for instance, the reversal of Scheeffer's theorem. L. Scheeffer showed that if, for example, the top-right derivative of a continuous function $f(x)$ is finite everywhere except at most in countably infinite many points, there is no second continuous function $f_1(x)$ (except for $f_1(x)$ + const.) which possesses the same top-right derivative. If now $D^+f(x)$ equals $\pm\infty$ in a *non*-countable p.s. [point set], there will be functions $f_1(x)$ such that $D^+f_1 \equiv D^+f$ and $f_1 - f$ is not constant. De la Vallée Poussin and others have generalised Scheeffer's theorem and not noticed that these generalisations are *illusory*.[312]

This book of "fundamental significance" was one of the reasons why Hilbert, Klein, Landau and Runge proposed Carathéodory as a corresponding member of the Mathematical-Physical Class of the Göttingen Academy of Sciences (*Gesellschaft der Wissenschaften zu Göttingen*) on 22 February 1919.[313] Its importance can scarcely be overemphasised, notes Fleming. "Until the 1950s (and perhaps even after) this was the best source from which young mathematicians could learn real analysis."[314] This view is also affirmed by Schmidt's dedication to Carathéodory on his 70th birthday:

At the turn of the century Lebesgue had reformed the integral calculus for the contents of a point set with unexpected success on the basis of the new definition of a concept stemming from Borel and described as measure. Carathéodory's book about real functions with a far-reaching effect followed this development. Here, for the interpretation of the concept of measure and measurability, the old procedure of interval covering was overtaken by the definition of these concepts through some simple requirements of axiomatic character. From this point of view, simple and beautiful ways leading far into new territory, were derived from the earlier, often difficult, development. In recent years, Carathéodory, progressing in this direction, has detached the entire formation of concepts based on the integral calculus, such as union of sets, average, measure, measurability and, finally, also the function defined on a set of points, from their association with regions of space or sets of points, and has raised them in their sublimated form into the ether of axiomatics. The freedom, purity and clarity ruling there secure the same rebirth for the integral calculus, which geometry has experienced through Hilbert and which also arithmetic and algebra have experienced in the last decades.[315]

Carathéodory's 1918 book *Vorlesungen über Reelle Funktionen* was published without major alterations for the second time in 1927 and the publisher Teubner called for a third edition. But as Carathéodory started to prepare this third edition, he noticed that he had to revise his text. Kritikos remarks that

in the meantime, however, the theory of functions had advanced greatly. A very lively new branch, general analysis, had sprung from an initial idea of the Frenchman Fréchet,[316] in which the examined spaces [...] are completely abstract sets of objects of our thought, for which only a few of the familiar basic properties of number spaces are each time assumed, and the relevant conclusions are subsequently drawn. The notion of the integral had also been extended to such spaces. Thus, Lebesgue's theory, which had constituted the central subject of Carathéodory's textbook, could be incorporated into more general theories. Therefore, Carathéodory decided to take account of these developments for the third edition of his book. But while working on this, he discovered that one could proceed even further in the way of these generalisations and abstractions, since what mattered most were not the properties of the elements over which the integration would extend, but the properties one would demand from the act of integration itself. In this way, a new series of Carathéodory's original papers emerged, of which I will only mention the first: Draft for an Algebraic Theory of the Notion of the Integral (1938). (*Entwurf für eine Algebraisierung des Integralbegriffs*)[317] To accommodate these in his book without damaging its pedagogic value, Carathéodory thought of splitting it into three volumes.[318]

The first volume was released under the title *Reelle Funktionen, Bd. I, Zahlen, Punkt-mengen Funktionen* (Real Functions, Vol. 1, Numbers, Point Sets, Functions)[319] in 1939 and contained the material which had been treated in the first 230 pages of his 1918 book. It was condensed so as to include recent results and the order and methods were presented so as to allow later generalisations if required. Kritikos, Boerner and R. Steuerwald helped Carathéodory to deal with the corrections of his new manuscript.

Referring to the "nice and interesting separata" he had received from Carathéodory early in 1938, Rosenthal announced to Carathéodory: "I will immediately set about reading them more precisely. I have very little to report about me. You may have already heard from Hartogs or Liebmann that I have undertaken the arrangement and publication of the 2nd volume of Hahn's real functions (on the basis of the unpublished manuscript); this really takes a lot of trouble and I am almost exclusively busy doing it; I have already written more than 100 pages; but the arrangement will take a long time." [320] Apparently Carathéodory's "interesting mathematical announcements" concerned his *Entwurf für eine Algebraisierung des Integralbegriffs*. Rosenthal found Carathéodory's general axiomatic definition of sets "very amusing and, of course, surely new" and Carathéodory's system of axioms "natural and comprehensible".

Blaschke established that, since Riemann, a far-reaching clarification of the notion of the integral had been achieved for sufficiently simple areas of integration. "But it is different for surface integrals and similar integrals over relatively general point sets. Many attempts have been made [...] in this direction, for example by Carathéodory. But for Carathéodory's integrals, [...] neither Gauss's integral theorem nor that of Stokes could be proved until now." [321]

The *Entwurf* was followed by the *Bemerkungen zur Axiomatik der Somentheorie* (Remarks on the Axiomatics of the Theory of Somas).[322] On 26 July 1938, Carathéodory sent this paper to Børge Jessen, a representative of the Danish school of analysis, with the remark that Bochner had drawn his attention to three Danish works by Jessen on the same subject. However, it was then too late for Carathéodory to cite Jessen, and he apologised: "It doesn't matter, since I have many new results which I will write down soon and I will use this opportunity to study and cite your works. You would do me a great favour, if you could send me these Danish works, which, if I am not mistaken, appeared in *Matem. Tidsskrift*." [323] Carathéodory presented his "new results" in a paper on *Die Homomorphien von Somen und die Multiplikation von Inhaltsfunktionen* (Homomorphisms of Somas and Multiplication of Content Functions).[324] There he mentions Jessen as follows: "In a complicated case Herr B. Jessen has treated an absolutely similar mapping as a transfer principle and was able to make many applications of this principle. These results should not surprise us, since it will turn out that every content function, whose measure set possesses a countable base, can be mapped homomorphically and measure-preserving onto a set of somas, which is also homomorphic to and preserves the measure of the usual Lebesgue content on the straight line." Carathéodory continued his letter to Jessen using the terminology of the *Entwurf* in order to give Jessen an idea of the main theorem[325] appearing in *Die Homomorphien*: "perhaps, it may interest you that the measure-preserving mapping of the torus Q_w[326] onto a distance presents only a special case of an entirely general theorem. With the notation of my last work, it is proved like this: if $\mu * A$ is a content function, which is declared to be any complete set of somas, then one can map this set homomorphically and measure-preserving onto a set of somas, in which the zero somas of the first set are being mapped onto the empty set of the second. For the Lebesgue linear measure $m * a$ this means, that all point sets, which differ only by one zero quantity can be considered a single soma. Then, one can map the thus reduced measure functions $\mu * A$ and $m * a$ isomorphically and measure-preserving onto each other, as long as the measure set of $\mu * A$ can be gained from a countable base. It is very remarkable that the absolutely abstract measure functions and the Stieltjes integrals are not considerably more general than the Lebesgue linear content. Not everything is as simple as I write it to you but only roughly so." According to Carathéodory, the Jessen torus is a non-trivial example that extends beyond the theory of Lebesgue measure.

Jessen outlined a reply to Carathéodory. His works in *Matematisk Tidsskrift*, he wrote, did not reveal any connection with Carathéodory's measure theory. "The three already released works of the series treat the measure and integral in abstract sets and contain, apart from the cited works in your first footnote, objectively nothing or only little new. In the further works of the series, I intend, among others, to treat Wiener's measure theory in product spaces and in differential space; it will here concern things which, as far as I know, have not been treated systematically. [...] Your theory and especially the information about the isomorphism of measure functions were of great interest to me. In another context, I have felt the usual conception of set functions to be too narrow."[327]

Carathéodory continued his engagement with measure theory with contributions dedicated to Perron,[328] Herglotz[329] and Knopp[330] on their 60th birthdays and ended this series of papers with his *Bemerkungen zum Ergodensatz von G. Birkhoff* (Remarks on Birkhoff's Ergodic Theorem).[331] He noted that, in earlier proofs of the Principal Ergodic Theorem, it was necessary to assume that the homomorphism $A' = \tau A$ is an isomorphism and the inverse transformation $\tau^{-1} A$ exists; however, a trick due to H. R. Pitt,[332] which could be further simplified, enabled a proof of the theorem under more general conditions.

2.17.4
The Book on Algebraic Theory of Measure and Integration

Of the two further volumes on the theory of real functions, which Carathéodory had heralded in the preface to his *Reelle Funktionen, I*, the Real Functions, vol. II was to be released in 1943. But the publishing house of Teubner was destroyed during the bombardments of Leipzig and Carathéodory took the decision to revise the material prepared for these volumes. He managed to secure Rosenthal and Steuerwald's co-operation for the corrections of his manuscript and announced its publication later to Born, on 24 March 1949: "I also have another book on measure and integral in Boolean spaces. It had already been printed but was destroyed by fire in Leipzig in 1943. I have rewritten it. It will also be published by Birkhäuser."[333]

Carathéodory died a year later and Rosenthal, Steuerwald and Finsler edited "the final [work] from the pen of this great scholar". It was published, under the title *Mass und Integral und ihre Algebraisierung* (Measure and Integral and their Algebraisation),[334] by Birkhäuser in 1956.

The work treated a general theory of abstract entities, which Carathéodory had named somas from the Greek word σῶμα meaning body. Carathéodory explained that his only excuse for introducing the word "soma" was that the elements of a Boolean algebra which he treated in his book had to be given a special name. Since he was obliged to consider certain sets of such elements as themselves elements of a set, this special name had to serve the same purpose as the word "point" in the theory of abstract spaces.[335] The somas were defined through a few simple demands, namely to have the properties enjoyed by both subsets of a fixed, arbitrary set of elements and by the figures of elementary geometry. The notion of the Lebesgue, Stieltjes, etc. integrals emerged as an application of this theory.

2.17.5
Correspondence with Radó on Area Theory

On 10 December 1948, Carathéodory asked Radó for some reprints of his work on the area of surfaces. Some months before, he had had a conversation about that topic with Besicovitch, some of whose papers he had also studied.[336] Radó replied to Carathéodory six days later:

Dear Professor Caratheodory:

[...]

We have now with us, at the Ohio State University, L. Cesari as a visiting professor. He made some very brilliant contributions to the theory of the Lebesgue area. He showed me one of your letters in which you make some comments on the isoperimetric problem for surfaces. I have published some time ago a paper on this subject (The isoperimetric inequality and the Lebesgue definition of surface area, Transactions of the American Mathematical Society, vol. 61, 1947, pp. 530–555). I am sending you a reprint of this paper under separate cover. However, I should like to make a few simple remarks at this time. Let U be the unit sphere $x^2 + y^2 + z^2 = 1$, and let a surface S be defined by equations

$$S: x = x(p), \quad y = y(p), \quad z = z(p),$$

where p varies on U, and the equations just written define a topological transformation. Let D be the set the points (x, y, z) interior to S, and let $|S|, |D|$ denote the three-dimensional Lebesgue measure of S, D, respectively. If $|S| > 0$, then let us say that S is a "heavy" surface. For a heavy surface, we may consider two concepts of "enclosed volume", namely

$$V_i = |D| \quad \text{(interior volume)},$$
$$V_e = |D| + |S| \quad \text{(exterior volume)}.$$

The Besicovitch example shows that V_e may be very large and at the same time $A_L(S)$, the Lebesgue area of S, may be very small.[337] On the other hand, as it follows from my paper quoted above, we have always

$$V_i(S)^2 \leq A_L(S)^3/36\pi$$

In other words, the isoperimetric inequality holds if we use the "interior volume". As a matter of fact, in my paper I prove the isoperimetric inequality for the most general closed continuous surface, using the Lebesgue area and an appropriate concept of "enclosed volume".

Perhaps you know that I have published a book (Length and Area, American Mathematical Society, vol. 30). If you feel that mail from America reaches you safely, I shall be very happy to send you a copy of that book with my compliments. In that book, quite a few important and difficult problems were left unsolved. Practically all those problems have been solved since by brilliant younger mathematicians, including Cesari. This means that there is now available a complete and very beautiful theory for applications and generalizations.[338]

In his paper *The Isoperimetric Inequality and the Lebesgue definition of surface area*,[339] Radó studied the Lebesgue area $A_L(S)$ from the point of view of fitness relative to the spatial isoperimetric inequality

$$V(S)^2 \leq A(S)^3/36\pi$$

$A(S)$ denoting the area of a closed surface S in Euclidean three-dimensional space and $V(S)$ the volume enclosed by S. Radó confirmed that the Lebesgue area is expected to yield a sharper isoperimetric inequality. He obtained results leading to simplifications and improvements concerning the most familiar plane isoperimetric inequality

$$a(C) \leq l(S)^2/4\pi$$

$a(S)$ denoting the area enclosed by a closed curve C in an Euclidean plane and $l(C)$ the length of C.

Carathéodory replied on 2 January 1949:

My dear Radó,

Have many thanks for your kind letter. I thank you also for your sending me your paper on the isoperimetric inequality. You should look also into the very beautiful contribution on that subject, which Erhard Schmidt has published a few months ago (Mathem. Nachrichten, Berlin, Vol. I, pp. 81–157).[340] Schmidt has succeeded in proving the inequality for quite general closed point sets in the spherical, Euclidean and Lobatsewkian space of n-dimensions with one and the same method. He uses though for the content of the surface the definition of Minkowski.

Two days ago I was visited by L. C. Young and we have spoken at length of you and your work. I am very much interested in all these problems and should greatly appreciate it if you would send me your book about length and area. Mail from America comes quite safely into Germany; so I have received some months ago the very interesting book of A. Rosenthal. [...]

Forward please my greetings to L. Cesari whose very interesting letter has been brought to me just at this moment.[341]

On 10 January Radó sent his book to Carathéodory together with the following letter:

Dear Professor Caratheodory:

Thanks for your letter of January 2, 1949. I am sending you today a copy of my book on Length and Area.[342] The first Chapter of the book is a general introduction, and each one of the main parts end[s] with a special chapter which gives a summary as well as historical references. The manuscript of the book was closed [finished] in December 1945. Since that time, all the major problems left open in the book have been solved, and thus the theory is now essentially complete. The most important fact, in my opinion is the development of two-dimensional concepts of bounded variation and absolute continuity (called eBV and eAC in the book),[343] which have now been thoroughly tested in many important and difficult situations.

Since I last saw you (at Harvard, I believe), I studied a great deal of topology and real variables, in connection with my work in surface area theory and Calculus of Variations. [...]

Sincerely yours,
T. Rado
92 Walhalla Road,
Columbus 2, Ohio, USA[344]

In the meantime, Carathéodory had received Radó's paper on the isoperimetric inequality and wanted to express his agreement with Besicovitch's assertion:

Munich February 1st 1949
Josephinum: Schönfeldstr. 16

Dear Radó,

Have many thanks for your kind letter, as well as for your most important paper which I received only a couple of days ago. It is exceedingly well written and I have read it from cover to cover with the greatest interest. Still I cannot understand why you give such preeminence to the isoperimetric inequality. The problem is to check if the assertion of Besicovitch that the Lebesgue area is inadequate for certain surfaces is correct. Now, it is possible to simplify his example in such a way as to show that he is quite right.

The choice of the parameters is most important; even if you map bicontinuously a Jordan curve of finite length onto the unit circle it may happen that some zero sets on the circle are images of point sets on the Jordan curve whose linear measure is positive. It is then impossible to calculate the length of the curve by means of a Lebesgue integral involving the parameter u. This is also the case if you choose the parameter properly.

Now the Besicovitch example consists of two parts

1. an open polyhedron A and
2. a perfect totally discontinuous point set B.

Besicovitch chooses B with a non vanishing volume but this latter supposition is quite irrelevant. Take B in the plane $z = 0$ so that its area $\mu B = 1$. Build up the polyhedron A in the half space $z > 0$ so that its area $\mu A < 1/2$. It is evident that for any proper definition of the area of the closed surface $A + B$ you must have $\mu(A + B) \geq 1$ and it is very easy to fulfil this requirement. For it is possible to map $A + B$ bicontinuously onto the surface of the closed sphere and choose the mapping as well as the parameters u, v in such a way that the area of any measurable point set S lying on $A + B$ be calculated as an ordinary Lebesgue $\int\int$ which satisfies the following property: Call A_1, B_1, S_1 the point sets of the sphere corresponding to A, B and S. Then $\int_{A_1} \int_{S_1}$ is equal to the ordinary area of the point set taken on the polyhedron and $\int_{B_1} \int_{S_1}$ the area of the BS, which is lying on $z = 0$ and whose area is also measured in the ordinary way.

Yet by the definition of the Lebesgue area you have $L(A + B) = \mu A < 1/2$ and the property of the lower semicontinuity does *not* hold. I cannot see any argument that would lead to the elimination of the ominous points as you have tried to do in treating the isoperimetric inequality.

Yours very sincerely
C. Carathéodory[345]

Radó's book reached Carathéodory at the beginning of February and filled him with interest.

Munich February 5th 1949
Josephinum: Schönfeldstr. 16

Dear Radó,

The day before yesterday I received your book. I could not imagine that it was such a bulky affair and would not have asked for it if I knew. But I shall pursue it with the greatest interest and I am much indebted to you for this splendid gift; have many, many thanks.

To the letter which I sent you some days ago I should like to add some obvious remarks.

1. You will have noticed that the postulatum I made use from is much weaker than the "projection postulate".

2. Take your example (I.1.11 of your book, case 1d). Call B the Osgood curve itself: then you have $R = R^0 + B$. The Lebesgue areas of R and R^0 exist and you have $LR = LR^0$. Now, if I am not mistaken, you have by V.2.69 $LB = 0$. In this case there is no difficulty: you have $LR = LR^0 + LB$.

3. A quite similar and perhaps more instructive example is the following. Take $Q: 0 \le x \le 1$, $0 \le y \le 1$ i.e. the unit square and call B a totally disconnected pointset whose superficial *measure* $mB = 1/2$ and which is lying in Q. Put $Q = B + C$. The Lebesgue areas for Q and C are immediately to be calculated and you have $L(B + C) = 1, LC = 1/2$. Now either the Lebesgue area LB has no meaning at all or according to Besicovitch you must put $LB = 0$. In either case the Lebesgue area is not a "measure function".

4. The Plateau Problem: Take in the plane $z = 0$ a finite number of circles $C_1, \dots, C_m$ lying outside one another. Call $|C_k|$ the area of the circle C_k. For any surface having the border $C_1 + \dots + C_m$ the area is not smaller than $|C_1| + \dots + |C_m|$. This is no more the case if you identify the border of the surface with a pointset B like those we have considered under 3. The Plateau Problem is meaningless for such a border.[346]

2.18
Doctoral Students in Göttingen

To return to Göttingen in the difficult times of the First World War, among Carathéodory's students are two doctoral students worthy of mention who made major contributions to mathematics.

First, Rademacher, who had been enlisted into the army in 1914, finished his dissertation on *Eindeutige Abbildungen und Meßbarkeit* (Unique Mappings and Measurability)[347] under Carathéodory's supervision in 1916. That work, treating some delicate problems of the differentiation and integration of real-valued functions of real variables was contemporary with similar considerations by A. Denjoy in France and G. C.-Young in England. Rademacher also published a paper *Über die Eindeutigkeit im Kleinen und im Grossen stetiger Abbildungen von Gebieten* (On the Local and Global Uniqueness of Continuous Mappings of Domains)[348] together with Carathéodory.

We should also mention Paul Finsler, who gained his doctorate with his famous thesis *Über Kurven und Flächen in allgemeinen Räumen* (On Curves and Surfaces in General Spaces) in 1918. Finsler generalised Riemann's classical differential geometry for n-dimensional spaces. These generalised spaces, in which Minkowski's geometry holds locally, were named after him in 1934, when Élie Cartan wrote his book *Les espaces de Finsler* (Finsler's Spaces).[349]

2.19
Succeeded by Erich Hecke in Göttingen

Having survived the war years in Göttingen, it was time for Carathéodory to leave for Berlin, where he carried with him the ideas and outlook which had been so inspired by Klein. L. C. Young writes:

In World War I and in the difficult years that followed, Göttingen declined and languished, rendered all the more vulnerable by its universality. Even so, it was characteristic of this universality, that the successor of Felix Klein in 1913 was Constantin Carathéodory. [...] Carathéodory stayed a relatively short time, leaving for Berlin in 1918 [...], but he carried the spirit of Göttingen, the keen forward-looking interests of Minkowski together with the insight and traditions of Gauss, of Riemann, of Felix Klein. In the matter of traditions, he went even further back and became thoroughly familiar with the works of Euler, parts of which he edited, and with those of the Bernoulli's, Huygens, Leibniz, not forgetting the mathematicians of Ancient Greece, all of whom he could quote in connection with some extremely modern and up-to-date idea. By Hilbert he was perhaps less influenced, but that was because he liked to have his own ways of doing things, usually with some 'Kunstgriff' of his own devising.[350]

In Göttingen Carathéodory was replaced by Erich Hecke. Hecke had gained his doctoral degree under Hilbert with a work on *Höhere Modulfunktionen und ihre Anwendungen auf die Zahlentheorie* (Higher Modular Functions and their Applications to the Theory of Numbers) and obtained his habilitation in Göttingen in 1912, when he was also Hilbert's mathematical assistant. Hecke became an associate professor in Basel in 1915 and later a full professor there. In 1918 he was appointed to Göttingen to replace Carathéodory, but remained there for only a year, leaving in 1919 when he was appointed to Hamburg. He was also a member of the editorial board of the *Mathematische Annalen*. Hecke's research concerned multiply periodical functions and hyperelliptic constructs, Dedekind zeta functions of the algebraic number sets and their generalisations and, finally, special automorphic functions, especially elliptic modular functions.

2.20
Professor in Berlin

Carathéodory accepted the appointment in Berlin on 1 April 1918. He had been invited to the University of Berlin by Schmidt, who himself had succeeded Hermann Amandus Schwarz as a full professor there on 1 October 1917. "With the obligation to cover the whole area of mathematics in lectures and exercises", Carathéodory was offered the chair of Georg Frobenius, who had died in August 1917. However, at his own request, Carathéodory did not take up his appointment as full professor for mathematics until 1 October 1918, and he also became joint head of the mathematical seminar.[351]

Carathéodory's appointment to the Frobenius chair was preceded by complicated negotiations. Arguing that Frobenius himself had referred to a mathematician called Issai Schur as his successor, Schwarz and Schottky wanted to put Schur first on the list. In addition, Schottky credited Schur with astuteness, exceptional knowledge of mathematical research in all areas including geometry and, not least, with an absolutely noble character.[352] Schur had gone to Berlin to study mathematics at about the age of twenty. Frobenius soon took notice of the unusually talented student and made sure that much of Schur's work on group theory was published in the *Session Reports of the Prussian Academy of Sciences*. Both Frobenius and his student analysed the properties of finite groups[353] and were among the well-known representatives of the Berlin algebraic tradition of the beginnings of the century. Despite this fact, even many years after his habilitation in 1903, Schur held the extremely underprivileged and unpaid position of an assistant professor at Berlin University. Fraenkel remarked[354] that an appointment for Schur was a compelling necessity, since Schur was a Russian Jew and, as such, not naturalised according to the Prussian administration's practice. Before the "closing of the gates", Fraenkel continued, "things changed through Schur's appointment as an associate professor in Bonn (1913), which had been initiated by E. Study; in this way, he became a civil servant and automatically a Prussian citizen. By the way, he was called back to Berlin in 1916."

The Philosophical Faculty, to which all mathematical chairs belonged, had already decided on 11 October 1917 to put Schur and Carathéodory in joint first place.[355] Planck opposed any change to this decision with the argument that the faculty ought to insist on the objective requirements and leave personal considerations aside. The decision was indeed obeyed, despite the fact that in a session of the appointment committee five days later, Schottky declared that he considered Schur more significant than Carathéodory; in his opinion, Schur was the stricter, Carathéodory the more elegant.[356] Schmidt, on the other hand, stated that if the admittedly very difficult comparison of the candidates' significance could not be avoided, he would not consider Carathéodory less significant than Schur.[357] Finally, Schwarz and Schottky agreed with him.

There were also different opinions and fierce controversy regarding the third candidate on the list. Schmidt's favourites were H.Weyl (Zurich), Erich Hecke (Basel), G. Herglotz (Leipzig) and A. Kneser (Breslau).[358] At first, the committee placed Weyl and Hecke on the list, *ex aequo*. Schottky had doubts about Hecke. Planck preferred Kneser instead of Hecke, particularly since, in his opinion, Kneser had exercised "a stimulating and beneficial influence" on physics in Breslau. Finally, Weyl was placed second as the only candidate, despite the fact that Schmidt would have liked to have put Herglotz next to him.

The text containing the decision of the appointment committee was written by Schmidt on 22 October 1917 and addressed to the State Minister of Spiritual, Medical and Instruction Affairs, Friedrich Schmidt-Ott. Schmidt praised Carathéodory:

> Only a few purely geometric works such as the isoperimetric property of the circle and the geometric interpretation of the characteristics of first-order partial differential equations derive from Carathéodory's pen, and yet all his analytic works are penetrated by the spirit of geometry. He touches every problem geometrically and uses his extraordinary spatial fantasy as a powerful aid. He has extensive and deep geometric knowledge which reaches as far as applied mathematics. After all, he himself claimed once that descriptive geometry is that mathematical discipline of which he has the best command. Therefore, Carathéodory is beyond comparison the most suitable personality to fill the currently perceptible gap at our university in the teaching of geometry.
>
> As a teacher, he is especially qualified for, and reliable in, the beautiful task of the teacher, the stimulus and instruction of beginners to independent scientific work, by virtue of his versatility and adaptability and his wealth of ideas.[359]

The text ends with the remark that the performance of the candidates in first place on the list was not comparable because of their difference in the directions they followed. Both of them were the most suitable personalities to fill in two completely

DIE KURVEN MIT BESCHRÄNKTEN BIEGUNGEN

VON

C. CARATHÉODORY

SONDERAUSGABE AUS DEN SITZUNGSBERICHTEN
DER PREUSSISCHEN AKADEMIE DER WISSENSCHAFTEN
PHYS.-MATH. KLASSE. 1933. III

BERLIN 1933
VERLAG DER AKADEMIE DER WISSENSCHAFTEN
IN KOMMISSION BEI WALTER DE GRUYTER U. CO.

Carathéodory's paper on Curves with Bounded Curvatures.
Special edition from the Session Reports of the Prussian Academy of Sciences.
Courtesy of the Bavarian Academy of Sciences, Archive.

different and considerable gaps in research and teaching at the university. Therefore, the faculty would request the appointment of both, if conditions allowed it. The request was signed by Schmidt, Rubens, Schwarz, Planck, Wehnelt, Schottky, Cohn and Nernst.

2.21

Geometry

Then let us consider this, I said, as one point settled. In the second place let us examine whether the science bordering on arithmetic concerns us.
What is this? Do you mean geometry? He said.
Exactly, I replied.
So far as it bears to military matters, he said, it obviously concerns us.
[...]
But for these purposes, I observed, a trifling knowledge of geometry and calculations would suffice; what we have to consider is whether a more thorough and advanced study of the subject tends to facilitate contemplation of the Idea of the Good. [...] Well, even those who are only slightly conversant with geometry will not dispute us in saying that this science holds a position the very opposite from that implied in the language of those who practice it.
How so? he asked.
They speak, I gather, in an exceedingly ridiculous and poverty- stricken way. For they fashion all their arguments as though they were engaged in business and had some practical end in view, speaking of squaring and producing and adding and so on, whereas in reality, I fancy, the study is pursued wholly for the sake of knowledge.

Thus Plato in *Republic* vii, 526c–527b.[360] Schmidt's text of 22 October 1917 also focuses on gemetry and its applications and reflects the strong preoccupation with the subject among professors in Berlin. Even during the procedure that had led to Schmidt's appointment in 1917 and after Hilbert rejected a position in Berlin, Carl Heinrich Becker,[361] *Vortragender Rat* (Speaker Councillor) at the Prussian Ministry of Education, had strong doubts as to whether he ought to name the proposed mathematicians Erhard Schmidt and Issai Schur to the minister. He wrote to Klein: "I have the impression that professorships for mathematics in Berlin expose a certain one-sidedness and I ask myself if I should rather propose to the minister that he make sure that at least one of the professorships for mathematics here should be oriented somewhat more towards the geometric or, anyway, towards the applied side."[362]

Carathéodory himself dealt with the risks of specialisation and the extraordinary significance of geometry as the connecting medium of all mathematical areas in the paper *Die Bedeutung des Erlanger Programms* (The Significance of the Erlangen Programme),[363] which he wrote in Berlin a short time after his appointment there: "For the first time, a tendency that later became decisive in all of Klein's works came up in the Erlangen Programme and revealed the connection between remote areas, thus creating new fruitful possibilities for research. Through it, Klein contributed more than anyone else in mathematics to overcoming the risks of science fragmentation caused by a much too great specialisation."

The significance of groups for geometry was established in the Erlangen Programme in the form of Klein's synthesis of geometry. Klein conceived geometry as the study of the properties of a space that are invariant under a given group of

ΑΚΑΔΗΜΙΑ ΑΘΗΝΩΝ

ΣΥΝΕΔΡΙΑ ΤΗΣ 6 ΝΟΕΜΒΡΙΟΥ 1930

C. CARATHÉODORY

ÜBER FLÄCHEN
DEREN KRÜMMUNG ALLGEMEIN BESCHRÄNKT IST

Κ. ΚΑΡΑΘΕΟΔΩΡΗ. — ΠΕΡΙ ΕΠΙΦΑΝΕΙΩΝ ΜΕ ΠΕΠΕΡΑΣΜΕΝΗΝ ΓΕΝΙΚΗΝ ΚΑΜΠΥΛΟΤΗΤΑ

'Απόσπασμα ἐκ τῶν Πρακτικῶν τῆς 'Ακαδημίας 'Αθηνῶν, **5,** 1930, σ. 345

Extrait des Praktika de l'Académïe d'Athènes, 5, 1930, p. 345

(Séance du 6 novembre 1930)

Carathéodory's paper On Surfaces whose Curvature is Generally Bounded.
Reprint from the Minutes of the Academy of Athens.
Courtesy of the Bavarian Academy of Sciences, Archive.

transformations. The Programme was published when he took up his appointment in Erlangen in 1872 with an inaugural lecture in which Klein set forth the views contained in the Programme. Carathéodory explains that a group of space transformations can be considered as a group of transformations of algebraic figures that are characterised by finitely many 'co-ordinates' and can therefore be conceived as elements of space. This group of transformations generates,

according to *Klein's* principle, a certain geometry of these figures. Now it can happen that several geometries constructed in this way possess the same group and therefore agree with one another. As a result [...], a uniform foundation was created for the establishment of new transference principles. Out of this idea, which also proved later to be fruitful in many works, *Klein* made a series of significant applications. [...] Later, *Klein* repeatedly stressed that the ideas of the Erlangen Programme can be considered as the highest classification principle for mechanics as well. At first he showed how one can treat the mechanics of the rigid body from this point of view. But then, the theory of relativity and Einstein's new theory of gravitation gave him a new opportunity to examine the fundamental role played right here by the group, just in the sense of his Erlangen Programme. [...] One sees how the original scope of *Klein's* ideas widened through the addition of problems which did not even exist when the ideas were emerging and for which science was not even mature; and this is exactly a touchstone for the far-reaching consequences of the progress achieved through the Erlangen Programme.[364]

Carathéodory later engaged himself in geometrical studies during his time in Munich. Hilbert, who had earlier fallen ill with pernicious anaemia, was now recovering and encouraged by this fact, Carathéodory promised him in April 1926 a corrected version of the paper *Über Flächen mit lauter geschlossenen geodätischen Linien und konjugierten Gegenpunkten* (On Surfaces with Nothing but Closed Geodesic Lines and Conjugate Opposite Points):[365] "I will soon send you a small note", he announced to Hilbert,[366] "which contains only a small remark. Maybe, you will take a look at it, since it deals with an old problem of yours (*Surfaces with Nothing but Closed Geodesic Lines*) and to be precise, in that form which you had suggested back then to Paul Funk as a dissertation topic. Namely, it has been established that the question of Liouville surfaces[367] leads to an Abel integral equation and that the line element can be completely given by elementary integrations. The matter is so trivial that I was not prepared to believe that is was new. But it really seems to be the case." In his paper, Carathéodory remarked: "My study touches upon a piece of work by *P. Stäckel* that *P. Funk* has continued. *Stäckel* has, namely, pointed to the possibility that all geodesic lines on a Liouville surface can be closed; Funk[368] had given a proof for the existence of such surfaces. We will now show directly that the line element of Liouville surfaces, which possess not only the properties claimed by Stäckel and Funk, but also the additional property that to every point of the surfaces an opposite point [...] can be ascribed, can be given in closed form."[369] Carathéodory proceeded to show that, of all considered Liouville surfaces with opposite points, the sphere is the only one which is closed. This work was destined for the mathematical seminar of the newly founded (1919) University of Hamburg which Carathéodory visited often as an honorary guest during the first years of his stay in Munich.[370] On Blaschke's initiative, renowned researchers were invited to the Hamburg Mathematical Seminar to bring the mathematicians there into contact with recent developments in mathematics in addition to differential geometry, algebra and number theory, which were particular topics cultivated in Hamburg. In the 1920s, the invited lecturers included D. Hilbert, H. Hjelmsler, H. Tietze, W. Wirtinger, H. Bohr and Carathéodory.[371]

In his *Einfache Bemerkungen über Nabelpunktskurven* (Simple Remarks about Curves of Umbilic Points),[372] which emerged about ten years later, Carathéodory examined the main properties of surfaces with continuous lines of umbilic points, rediscovering "with very clear means" the results of Monge, who had studied these surfaces in his famous book *Application de l'analyse à la Géométrie* (Application of Analysis to Geometry).[373] Carathéodory assumed the existence of at least two umbilic points on an analytic ovaloid in E_3. Hans Hamburger solved this problem in a very complicated way,[374] whereas Gerrit Bol found a simpler solution.[375] Blaschke[376] reported on it as a response to Bol's[377] communication.

Blaschke frequently mentions Carathéodory with respect to problems and theorems in differential geometry concerning the extremals of closed spatial curves and the theory of surfaces in the large.[378]

2.22
Supervision of Students

The Schwarz, Frobenius and Schottky era in Berlin ended when Carathéodory was appointed there, and a short transitory period (1918–1919), marked by Carathéodory and Schmidt, began for mathematics in Berlin. Although both friends had studied in Berlin, they were strongly influenced by the intellectual atmosphere of Göttingen. Carathéodory began his teaching in Berlin with a lecture on the calculus of variations. In the following semester he lectured on projective geometry and mechanics, but did not have the chance to lecture on advanced mechanics.

Together with Schmidt, Carathéodory supervised or reviewed the following works completed at the university: Erich Bessel-Hagen's dissertation *Über eine Art singulärer Punkte der einfachen Variationsprobleme in der Ebene* (On a Category of Singular Points of Simple Problems of the Calculus of Variations in the Plane),[379] Hans Rademacher's habilitation *Über partielle und totale Differenzierbarkeit von Funktionen mehrerer Variabeln und über die Transformation der Doppelintegrale* (On Partial and Total Differentiability of Functions of Several Variables and about the Transformation of Double Integrals),[380] and Hans Hamburger's habilitation *Über eine Verallgemeinerung des Stieltjesschen Momentenproblems* (On a Generalisation of the Stieltjes Moment Problem).[381] In Carathéodory's opinion, Bessel-Hagen's work represented the first great progress in the theory of discontinuous solutions of the calculus of variations after Carathéodory's own works of 1904–1905. Rademacher became an assistant professor and remained in Berlin until 1922, when he was called to the recently created University of Hamburg as an associate professor; in April 1925 he went on to Breslau as a full professor. In 1922, Hamburger got Schur's associate professorship, and he accepted an appointment to Cologne two years later. Both Carathéodory and Schmidt supported the approval of Robert Remak's application for habilitation, which was, however, rejected by Schwarz, Schottky, Planck and Nernst among others. Remak, a specialist in the geometry of numbers and known for his works on mathematical economy,[382] had gained his doctorate under Frobenius in 1911 and had the reputation of being a critical and obstinate person.

2.23
Applied Mathematics as a Consequence of War

After the war Carathéodory participated in activities aimed at boosting the subject of applied mathematics. On 10 June 1918, following Schmidt's initiative, the Philosophical Faculty asked the Minister of Education for the creation of a chair for applied mathematics which would satisfy "the unexpected need for practically and theoretically trained mathematicians, which has appeared as a consequence of the war".[383] The person to be employed for whom only a full professorship would be appropriate, ought to guarantee the creation of a new centre at the University which, apart from instruction in descriptive geometry, would most of all favour lecturing and research in the methods of numerical, graphic and instrumental calculation and in other disciplines necessary for practical life.

The ministry rejected that request on 7 January 1919. The relevant committee, consisting of Schmidt, Carathéodory, and Planck among others, repeated their demand six months later, on 5 July 1919, and this time they also submitted a list of candidates for the post with the names of Carl Runge (Göttingen), Gerhard Hessenberg (Breslau) and Richard Edler von Mises (Dresden), in that order. The first two candidates had emerged from the Berlin school, whereas von Mises, who later belonged to the nucleus of the Vienna circle of logical positivism in the 1920s, had followed a different career. At the outbreak of the war, von Mises had volunteered for the Imperial and Royal Squadrons of the Austro-Hungarian army, but was in action for only a short time. He was mainly given organisational tasks, held lectures about flying for officers and directed the construction and testing of a major 600 hp military aeroplane. After the war, he obtained a teaching commission for applied mathematics and, especially, for aeronautics at the University of Hamburg. When his name was added to the committee's list, he had already accepted an offer to be the chair for the theory of strength of materials, hydrodynamics and aerodynamics at the Technical University of Dresden.

At the end of July 1919, the Philosophical Faculty was asked to take steps to fill the vacant post of associate professorship for pure mathematics following Theodor Knopp's departure for Königsberg (now Kaliningrad). The appointing committee used the opportunity to place the discussion on applied mathematics on the agenda once more. Its members stressed that an appointment that covered both pure and applied mathematics was needed and they asked the ministry to offer the associate professorship which, however, would acquire the rank of a full professorship *ad personam*, to one of the candidates named in their request of 5 July 1919. Since it did not seem likely that Runge, who at that time was already 63 years old, would accept the offer, the faculty proposed Hessenberg, followed by von Mises. They did not name a third candidate, since they did not find anyone "suitable". The ministry agreed on this solution but Hessenberg preferred to accept an appointment in Tübingen. Thus,

Richard von Mises took on this post on 1 April 1920, assuming that it would be changed into a scheduled full professorship as soon as possible.

In the same year, an institute for applied mathematics was founded, with Richard von Mises as its director, and twenty-two students participated in its first year of seminars. The tasks of the institute were the instruction of students in all areas of applied mathematics and their training in numerical, graphical and instrumental processes in higher mathematics, the promotion and carrying out of scientific research work and, occasionally, the receipt of commissions for technical-industrial projects. Von Mises founded the *Zeitschrift für angewandte Mathematik und Mechanik* (Journal for Applied Mathematics and Mechanics) in 1921, for which he served as editor for the following twelve years.

2.24
Collapse of Former Politics

Carathéodory's Berlin environment was anything but happy. During the war, the number of students at the university dropped drastically. Directly afterwards, the daily life of the city was troubled with price inflation, social want, revolution and putsch. In a letter Carathéodory sent to Hilbert from Halle on 10 October 1918,[384] he complained about the difficulty in finding an appropriate apartment in Berlin. He settled down in the modest guest house Ludwig at Markgrafenstr. 33, which at least had the benefit of being favourably situated. Carathéodory reported to Hilbert what he knew about the anticipated end of the war and asked him to be discrete: "Finally, after all the war will be over in the next weeks. According to concurrent news that I have heard from different sides, it seems that all the conditions will be accepted. The situation in the West pushes for that. It is even said, for example, that demands will be made for a neutral power (Holland?) to administer the occupied areas up until the peace agreement in order to create a very extended neutral zone. The question is, how will the people bear the collapse of the former politics. Please, do not tell anyone that I have written you this. From my wife I hear that in Göttingen people still live in a totally different world."

Four years after the *Aufruf an die Kulturwelt* (see 2.11), on 4 October 1918, the new Reich leadership requested of the American President Wilson a cease fire and the start of peace negotiations based on his fourteen points. To accept it, Wilson demanded, among other things, the evacuation of all areas occupied by the Central Powers and an end to the submarine war, which was indeed ordered on 20 October. The October Reforms of 1918 introduced by the new Reich Chancellor, Prince Max of Baden, though representing a first step towards a parliamentary system of government, could do nothing to halt the "collapse of former politics."

From the summer of 1918, with millions of dead or seriously disabled in the war, together with reports of the refusal of soldiers at the front to obey orders, the decay of war morale and total exhaustion,[385] the attitude of the German people began to change. It was recognised that the Western front was in danger of collapsing and the populace lost its blind trust in German military superiority. But the general feeling of weariness with the war was not apparent among the nations's elite. In the final year of the war the numbers of university scholars in the right-wing Fatherland Party (*Vaterlandspartei*), for example, by far exceeded the few who desperately favoured peace through negotiations. Wilhelm II and the military command were not at all prepared to submit to the new Reich government supported by parliament. The end came quite suddenly. Within a few days, the revolutionary movement spread across Germany. After mutiny in the deep-sea fleet and the extension of the unrest to Berlin and to the larger cities of Bavaria, the Kaiser abdicated and fled the country to the Netherlands, early in the morning of 9 November 1918. In the afternoon of the same day, the German Republic was proclaimed, and two days later the ceasefire was signed in the forest of Compiègne.

Shortly afterwards, Karl Liebknecht, leader of the Spartacus Union, proclaimed the Free Socialist Republic. The Social Democrats accepted the "mandate of the revolution" for governmental authority and became the leading political force in Germany. They, however, co-operated with the old, largely discredited, military establishment and the imperial bureaucracy in an attempt to ward off attacks from the radical revolutionaries and to cope with the consequences of losing the war, with the aim of "saving the German people from civil war and famine".

On 17 November 1918, Carathéodory wrote to Klein[386] about the November revolution that overthrew the monarchy. His subject was the revolution days in Berlin and his primary concern was the fate of the Royal Library. He had experienced the events "most closely; from such a small distance the whole movement made a rather harmless impression. However, a few volumes of the Schlömilch Journal[387] were shot in the seminar library (12 bullets penetrated the room in total) and the Royal Library is affected much worse. No books were damaged there, but, of course, as I hear, doors, furniture and a big chandelier, which Friedrich der Große [Frederick the Great] had donated, were. However, the whole fight seems to have been a tilt at windmills, since no one saw who it was that had shot from the buildings. I am not fully unconcerned about the coming developments."

Carathéodory exhibited the typical "apolitical" behaviour of German university professors: they were dismayed but not prepared to view themselves critically. Indeed, the "coming developments" confirmed Carathéodory's worst fears. An unstable situation emerged at the end of 1918, mostly out of fundamental differences concerning military policy within the government and the bloody suppression of the Spartacus uprising in Berlin by the "free corps", which culminated in the murder of Rosa Luxemburg and Karl Liebknecht in January 1919. A National Assembly was elected on 19 January, with women entitled to vote and stand for election for the first time. In the constituent assembly, obliged to convene in Weimar because of the unstable situation in Berlin, the Social-Democratic Party of Germany (SPD = *Sozialdemokratische Partei Deutschlands*) formed the so-called Weimar Coalition with the democratic middle-class parties, including the Centre (*Zentrum*), and the German Democratic Party (DDP = *Deutsche Demokratische Partei*), which commanded a three-quarters majority. The constitution of what was to be called the Weimar Republic was strongly influenced by the liberal and democratic tradition of 1848, which did not correspond to the existing social reality. Many democrats fully expected that "Prussian presumption, junkerdom, predominance of the aristocracy, civil servant cliques and the military [would be] over once and for all".[388] However, the new Republic was constantly under attack by monarchists who wanted to restore the Reich, by communists who envisioned a soviet republic, and by National Socialists who demanded the rearming of Germany. The generality of professors of German universities, who were undergoing a crisis of self-confidence in the face of the economic and cultural changes of the era, had nothing but contempt for the Weimar Republic that turned into hatred when inflation slashed their salaries and reduced government appropriations for higher education. With social and economic decline, they became a disturbed and dissatisfied class. They attacked the Republic from their university chairs and contributed thus to the cultivation of an antidemocratic attitude among their students.

Carathéodory was living alone at that time in Berlin. His wife and the two children had remained behind in Göttingen. However, in October 1918 the family travelled to the "extremely ugly city of Halle",[389] which had been recommended to them for the medical treatment of Carathéodory's son Stephanos, who had been suffering from poliomyelitis since he was five. But in Halle, the boy fell ill with pneumonia and his father visited him there twice.

2.25
Member of the Prussian Academy of Sciences

Carathéodory was proposed for membership of the Prussian Academy of Sciences on 7 November 1918. His impressive list of sponsors were Albert Einstein, Adolf Engler, Theodor Liebisch, Gustav Müller, Johannes Orth, Albrecht Penck, Max Planck, Heinrich Rubens, Erhard Schmidt, Friedrich Schottky, Hermann Amandus Schwarz and Hermann Struve. There was a discussion as to whether the approval of the new government was required to confirm their proposal, but there was no common agreement on the matter. Carathéodory was appointed a full member on 12 December that year.[390] On 3 July 1919, together with Schmidt, he was introduced to the Academy during a public session. In his inaugural speech, Carathéodory attempted to present his feelings for Germany through a cultural recourse to his own family history.

My origin, my early education and also my first training indicate different countries and cultural circles, and therefore I would like to tell you first why I do not entirely feel like a stranger in this country. Even purely externally, Berlin is the place of my birth, but even more valuable for me is the fact that from my very youth I have gained impressions that did not make it difficult to find a homeland here in this country for the last two decades. In my parents' house, even if it was far away, German history and literature and, even more, German science and art did not remain foreign to me, since through many personal relations the web that was once formed was further spun again and again.

[...] A picture of *Alexander v. Humboldt* dedicated by his own hand more than sixty years ago, which I still retain with pride in my office, was kept in our house. And through him, a tradition remained alive which almost unconsciously led me [...] to the place where that old prince in European intellectual life had summed up his life's work.[391]

In his response to Carathéodory's and Schmidt's inaugural speeches, Max Planck, Secretary of the Physics-Mathematics Class, compared Schmidt and Carathéodory, in

SITZUNGSBERICHTE

1919.
XXXIII.

DER PREUSSISCHEN

AKADEMIE DER WISSENSCHAFTEN.

Öffentliche Sitzung zur Feier des LEIBNIZischen Jahrestages vom
3. Juli.

Antrittsreden

der HH. SCHMIDT und CARATHÉODORY

und Erwiderung

des Hrn. PLANCK,

Sekretars der phys.-math. Klasse.

Carathéodory's and Schmidt's inaugural speeches at the Prussian Academy of Sciences and Planck's response. Courtesy of the Bavarian Academy of Sciences, Archive.

order to show that they complemented each other "in a splendid way". Carathéodory's "treasure trove of ideas enriched by much activity abroad and in different directions" was the counterpart of Schmidt's "specific German education." [392]

2.26

Supporting Brouwer's Candidacy

Carathéodory tried to arrange his own successor in Berlin. He and Schmidt supported the candidacy of Luitzen Egbertus Jan Brouwer, the Dutch logician who since 1912 had been the professor for set theory, function theory and axiomatics at the University of Amsterdam. In the years 1911–1914, Brouwer had mastered certain very difficult topological problems connected with the concept of dimension. In 1912 he developed the so-called fixed-point theorem or Brouwer's theorem, stating that in any continuous transformation of all points in a circle or on a sphere, at least one point must remain unchanged. But already before that, and again in the years 1918–1930, he rejected everything in "classical" mathematics and "classical" (Aristotelian) logic that was based on a formalist existence and not on construction. Brouwer had distanced himself from Hilbert's formalism already with his doctoral dissertation in 1907. The two ways of thought differ in one considerable consequence. Hilbert's approach justified the so-called existence proofs by which the existence of a certain number, or a mathematical truth, follows from the fact that the assumption of the opposite leads to a contradiction (proof by *reductio ad absurdum*). But Brouwer demanded that one ought to accept the existence of a mathematical construct only if a procedure was given by which this could really be constructed. In this way, he founded the philosophy of mathematical Intuitionism, according to which mathematics is a construction of the human consciousness, the source and seat of all knowledge.[393] The intuitionist-formalist debate centred on the requirement for a reliable foundation of mathematics, which Brouwer saw in a constructive definition of a set, thereby rejecting the axiom of comprehension as a foundation of set theory and the logical principle of the excluded middle as part of a mathematical proof.[394] The School of Intuitionism starts from the problem of avoiding antinomies in set theory and attempts to solve it by activating Kronecker and Poincaré's mathematical ideas, as well as Kant's philosophical theory. Intuitionism dominated the discourse on the foundations of mathematics throughout the 1920s and exercised influence on the later philosophy of language. In 1917, at the beginning of the formalism-intuitionism discourse, the *Mathematische Annalen* opened its doors again for work on the foundation of mathematics. As Peckhaus writes,[395] Hilbert had in mind an interdisciplinary centre of basic research in Göttingen that would decisively promote the dialogue between philosophy and mathematics. It was apparently as a result of this, that Hilbert offered Brouwer an appointment to Göttingen in 1918 which the latter however rejected. It was now Carathéodory and Schmidt's turn with their offer for Berlin. On 19 December 1919, the committee in charge of Carathéodory's appointment adopted their proposal and argued in favour of Brouwer, underlining the originality of Brouwer's methods and his very extensive and profound knowledge not only in pure but also in applied mathematics.

Some months earlier, Carathéodory had travelled to Holland, obviously to move Brouwer to accept the Berlin professorship. Brouwer's 1917–1918 papers concerning the foundations of set theory[396] and that of 1919 on the *Intuitionistische Mengenlehre* (Intuitionist Set Theory),[397] which introduced the notion of set, resulting from his search for a constructive procedure for the generation of elements of the continuum, probably marked the ensuing debate on the foundations of mathematics. So Brouwer was an attractive candidate for Berlin. From the Grand Hotel Central in La Haye, Carathéodory wrote to Klein on 1 September 1919,[398] that Brouwer had visited him, "which was pleasant as always." By the way, he recounted to Klein his impressions of Holland, which was so different from post-war Germany: "One lives here like in Cockaigne but the prices are fairy-like, if one calculates with German money: a simple sea bath costs about 11 marks and every drive on the electric tram 60–80 pfennigs; the cheapest Dutch paper costs 75 pfennigs and the German papers are sold on the street for 1–2 marks an issue!"

2.27
Carathéodory's Successor in Berlin

Hermann Weyl was proposed as the second candidate to become Carathéodory's successor. In 1918 Weyl was still firmly convinced that the Intuitionist criticism of formalist existence statements would gain acceptance in the future.[399] That year, in his booklet *Das Kontinuum* (The Continuum), he worked out some "long-cherished" thoughts about a new foundation of analysis[400] and in 1921 he gave many lectures on Brouwer's programme and his own approach to intuitionism. According to Weyl, "under the impact of undeniable antinomies in set theory, Dedekind and Frege had revoked their own work on the nature of numbers and arithmetical propositions, Bertrand Russell had pointed out the hierarchy of types which, unless one decides to "reduce" them by sheer force, undermine the arithmetical theory of the continuum; and finally L. E. J. Brouwer by his intuitionism had opened our eyes and made us see how far generally accepted mathematics goes beyond such statements as can claim real meaning and truth founded on evidence. I regret that in his opposition to Brouwer, Hilbert never openly acknowledged the profound debt which he, as well as all other mathematicians, owes Brouwer for this revelation."[401] Indeed, what had been only partly conjectured by Brouwer, was rigorously shown by Kurt Gödel in 1931 through his discovery of the undecidability of certain types of mathematical problems for which the completeness provided for by Hilbert's formalist school of mathematics cannot be achieved.

The basic argument of the appointment committee for listing Weyl in second place was that the great hopes which had been put on him then, had been in the meantime confirmed.[402] The third place in the list of candidates was filled by Gustav Herglotz, a full professor of mathematics at the University of Leipzig since 1909, whose versatility, sovereignty in mathematics and talent as a lecturer were praised together with the perfection of his works. Herglotz had contributed to partial differential equations, function theory and differential geometry and he would later succeed Runge in Göttingen in 1925.

In the event, none of the proposed candidates was prepared to accept the appointment. This was because of the precarious situation in Berlin. Even the subsequently proposed Erich Hecke rejected the appointment and chose Hamburg instead. In a new proposal, the faculty chose to look for a "geometer". Four mathematicians, who would guarantee a "worthy occupation of the chair", were proposed, but, as regards their scientific potential, none of them could be considered as being on a level with those proposed earlier. The candidates were, in order, W. Blaschke (Hamburg), L. Bieberbach (Frankfurt am Main), G. Hessenberg (Tübingen) and E. Steinitz (Kiel).

Finally, Ludwig Georg Elias Moses Bieberbach was appointed as Carathéodory's successor, as a full professor and the joint head of the Mathematical Seminar with effect from 1 April 1921. Together with Richard von Mises, he introduced a new era of mathematics at the University of Berlin. Bieberbach was known as an inspiring but rather disorganised lecturer. He had worked on the theory of functions. However, a racist theory he presented in the spring of 1934 at the annual meeting of the Society for the Promotion of Instruction in Mathematics and Natural Sciences (*Verein zur Förderung des Mathematischen und Naturwissenschaftlichen Unterrichts*) in Berlin, revealed him as a fanatical anti-Semite.[403]

2.28
The "Nelson Affair"

Carathéodory was involved, not only in the provision of a successor in Berlin, but also became involved in the appointment policies of the University of Göttingen. In this respect he was in contact with Hilbert and may even have acted on his instructions.

Hilbert's favourite candidate, Leonhard Nelson, was a philosophy lecturer who, following Fries, strove for a continuation of Kant's critical philosophy. He applied the critical method which, for him, possessed a programmatic character, to studies in the philosophy of mathematics and in the nature of mathematical axioms. He interpreted his own contributions as the philosophical foundation of Hilbert's axiomatic programme. His philosophical work concerned the scientific foundation of philosophy and the systematic development of philosophical ethics.

Nelson's habilitation of 1904 on *Die kritische Methode und das Verhältnis der Psychologie zur Philosophie – Ein Kapitel aus der Methodenlehre* (The Critical Method and the Relationship of Psychology to Philosophy – A Chapter from the Theory of Method) had been rejected in 1906 by the majority of the Göttingen philosophical faculty, obviously because of the polemical style of Nelson's texts, but formally on the basis of a separate vote by Georg Elias Müller, professor of experimental psychology, who found it incomplete, superficial and not independent. Hilbert and Klein, trying to secure the institutional establishment of research in the foundations of mathematics in Göttingen, had supported Nelson by a commonly signed appeal of 29 May 1906 to the commission in charge to decide orally on Nelson's request to habilitate. They stressed his mediating position in the philosophical (psychological, logical, epistemological) foundation of mathematics: against tendencies from left and right, Nelson advocated the Kantian view that, in the final analysis, mathematical judgements were neither of an empirical nor of a logical nature but were based on a specific activity of the human mind. In addition, Nelson had knowledge of recent mathematics, which young philosophers usually lacked. But Hilbert and Klein's request was declined by the dean on formal grounds and Nelson withdrew his own request on the dean's recommendation.

The sincere, clever and straightforward Nelson, whose "uncompromising liberalism and rationalism was offended by the reactionary tendency of the time", had incurred the dislike of the philosophy professor Husserl, whose "phenomenology" he despised and publicly ridiculed.[404] Nelson was later able to habilitate with his *Untersuchungen über die Entwicklungsgeschichte der Kantischen Erkenntnistheorie* (Studies in the Historical Development of Kant's Theory of Knowledge) in 1909, five years after he had gained his doctorate. He did not obtain an associate professorship for philosophy until 1919, after a failed attempt in 1917 to obtain the chair formerly occupied by Husserl. Husserl held a full professorship *ad personam* in Göttingen. After his appointment to Freiburg in 1916, the Philosophical Faculty of Göttingen University convened in spring 1917 to discuss a successor but was not able to come to a decision.

For the occupation of an associate professorship, the majority proposed Georg Misch in first place and Max Frischeisen-Köhler and Richard Hönigswald in joint second place. But in a minority report of 1 March 1917, the mathematicians Hilbert and Runge, the physicist Peter Debye, the physico-chemist Gustav Tammann and the historian Max Lehmann proposed Nelson instead. They established that the scientific treatment of epistemological problems by philosophers had made only limited progress since Kant, and the current epistemological problems emerging out of mathematical requirements demanded a direct inclusion of Kant's *Kritik der reinen Vernunft* (Critique of Pure Reason). They stressed that Nelson had dedicated his powers in that direction and that his co-operation with mathematicians was needed to set up this specific philosophical school in Göttingen so as to enable Germany to compete with parallel endeavours abroad, such as those of Federigo Enriques in Bologna, Léon Xavier in Paris and Bertrand Russell in Cambridge. Born describes Nelson's philosophy as a modified Critique in Kant's sense, namely a rationalism rejecting the dogmas of the older philosophical systems but accepting *a priori* principles, whereas Nelson's main interest was the application of these ideas to moral philosophy and politics. [405]

Neo-Kantianism, as a theory of knowledge and moral philosophy, had the intention of asserting itself against the almost total loss of meaning philosophy had suffered because of the triumphant progress of the natural sciences and positivism. Its success in this respect was not spectacular; nevertheless, it had been established as the predominant university philosophy. Neo-Kantianism did not, in fact, reject a positivist explanation of the world based on natural sciences; on the contrary, it confirmed such an explanation. Also neo-Kantianism, in the form of moral philosophy, attempted to make up for the loss of unity between *Sinn* (meaning) and *Sein* (being), by substituting for these notions *Sein* (being) and *Sollen* (to be). Between "being" and "to be", between facts and norms, moral values acquired their own existence which was neither a mere objectivity (understood in the sense of natural sciences) nor a mere subjectivity. [406]

An "opposing statement of the faculty majority" of 8 March 1917 was signed by, amongst others, Carathéodory and Landau. They distanced themselves from the negative evaluation of German contemporary philosophy by the minority. They reproached the minority for: (a) ignoring the philosophical literature of the past 50 years, especially the *Kantbewegung* (Kant movement) that had been spreading since the 1860s and contributing to the theory of knowledge; (b) identifying productive philosophy with that philosophical work aimed at revealing the axiomatic foundation of mathematical sciences, whereas mathematical logic, though significant, was only a small part of the extensive tasks which the philosophical theory of method had to solve; (c) the prejudice of attributing scientific value only to rational-aprioristic philosophies and ignoring the scientific value of the empirical direction. They found that Nelson's work as expressed mainly in his *Kritik der praktischen Vernunft* (Critique of Practical Reason) [407] was characterised by a "repulsive formalism" and that Nelson's attempt to found moral philosophy on unprovable axioms was doomed to fail, since moral philosophy ought not to be treated according to a set pattern borrowed from mathematics or natural sciences.

His opponents acknowledged Nelson's great impact on students but they attributed it to advertising for the purposes of agitation. Many of those who signed the statement found his influence on the students, especially his negative views on patriotism, to be hardly pleasant. Finally, they accused Nelson of attacks against contemporary philosophers of outstanding merit.

In the summer of 1918, the Philosophical Faculty submitted a list of candidates for the reoccupation of the full professorship of philosophy at the Department of History and Philology, which had been held by Heinrich Maier since the winter semester of 1911–1912. Maier had been appointed to Heidelberg and this department claimed the new professorship for themselves, since the two existing full professorships had to be equally shared between the two departments of the Philosophical Faculty, and Georg Elias Müller belonged to the Department of Mathematics and Natural Sciences. The faculty proposed Eduard Spranger as the first candidate followed by Ernst Cassirer and Georg Simmel as the second and third, respectively. They also requested that Georg Misch, who had been an associate professor since the winter semester 1917–1918, be made a full professor *ad personam*. Runge, Hilbert, Klein, Prandtl, the physicist Woldemar Voigt, the chemist Adolf Windhaus, the historian Max Lehmann, Landau and Carathéodory submitted a separate judgement on 29 July 1918 complaining that the proposals had been worked out by the Department of History and Philology with the aid of only Dean Conrad von Seehorst (from agrarian science) and the "psychophysicist" Georg Elias Müller,[408] whereas members of the Department of Mathematics and Natural Sciences had not been informed before the deciding session and therefore had not been able to include their wishes regarding the candidates in the proposal. To stress the claim for the participation of natural scientists in the appointment of philosophers, they referred to Poincaré and Russell to show the close connection between philosophy and the exact sciences in other countries and also drew attention to major changes in contemporary theoretical physics. They asked for a "post for systematic philosophy of the exact sciences", but if that were not possible, they would favour Misch's appointment to Maier's professorship and the faculty should present proposals for the associate professorship that would be freed through Misch's promotion. Finally, they drew attention to Cassirer's connection with the Marburg School, which was "known to our circles because of the superficial and amateur-like treatment of mathematical-physical problems". The majority replied with another report and the whole conflict ended with a divided faculty and the Department of History and Philology denying Hilbert the competence to judge philosophy altogether. They reproached the whole group around Hilbert for strong mistrust and for not considering different opinions.

Why had Klein, Landau and Carathéodory changed their opinion of March 1917 and sided with Hilbert in the summer of 1918? A possible explanation could be that they were united in their common aim to gain a philosophy professorship for mathematics. Also times had changed and the appointment of Nelson, who in one of his seminars in 1915 had treated the question of whether Belgium's invasion by German troops was morally justifiable, may not have appeared to them to be so outrageous now that the front was collapsing and Germany's defeat approaching.[409]

On 10 October 1918 Carathéodory informed Hilbert about his visit to the Speaker Councillor at the Prussian Ministry of Education:[410] "Yesterday, I was with Becker, who immediately referred to your letter. He is not in favour of a third full professorship. He said that G. E. Müller's professorship must be regarded for our discipline as an entirely valid philosophy professorship. He would have no objections against this if, in the case of a possible new appointment, another philosophical direction than just psychophysics would be considered. It seemed to me that he was prepared to do something for Nelson, provided that no great opposition would become apparent in Göttingen. Otherwise, he would rather like to find a place for N[elson] elsewhere."

Becker was of the opinion that his task was to get the right people into university positions, since the whole future of the university depended on the right choice. He stated that no professor would be appointed in Prussia without his order and that everyone appointed had to be intensively investigated beforehand.[411]

On 10 October 1918, one day after Carathéodory's visit to Becker, the ministry sent a decree to Göttingen University. The minister was expecting a new list of candidates from the humanities (Simmel had died in the meantime) and he was against a specific philosophical direction in the occupation of the associate professorship. However, he was considering the establishment of a special associate professorship for the philosophy of exact sciences, the planning of which would begin only after a peace agreement was reached.

The faculty proposals for Maier's successor put Misch in first place and Karl Groos second. Cassirer was also named without placement because, according to the faculty, the ministry had not given reasons for his rejection. On the basis of these proposals, Georg Misch obtained the full professorship formerly occupied by Maier at the beginning of the summer semester 1919.

On 22 December 1918, Hilbert sent a letter reminding the ministry of the prospect of a special associate professorship. In support of Nelson, he wrote that with Nelson's appointment Göttingen would probably become the first centre of scientific systematic philosophy in Germany. In the faculty's subsequent proposal of 9 January 1919, for an associate professorship of scientific systematic philosophy, Nelson was followed by Moritz Schlick. In February 1919, Carathéodory informed Hilbert that the ministry had proposed Nelson to the Berlin philosophical faculty for the occupation of a new associate professorship there that would combine exact sciences with political sciences. However, the Berlin faculty had rejected Nelson, because, as they argued, his mathematical works, as well as his works on moral philosophy and the philosophy of right, could not stand up to the faculty's critique, and, moreover, it would be noticed if the faculty ignored Ernst Cassirer. Hilbert responded immediately with a letter to the ministry in which he referred to Nelson as the only real philosopher in the old universal sense who would be suitable to help establish a philosophical school in Göttingen and it was only then that the resistance of the Göttingen faculty was overcome.

Nelson was finally appointed as an associate professor at the University of Göttingen with effect from 1 April 1919 and a schism in the philosophical faculty was avoided. In Hilbert's papers, there is a voluminous file entitled *Die Nelson Affäre* (The Nelson Affair) documenting his efforts to combine philosophy with natural sciences by promoting Nelson's academic career.

Becker's reform proposals of 1919 included the requirement that, alongside the recommendations of the faculty, there should also be reports by commissions of experts who would consult the government in the evaluation of appointment proposals so that the authority of the whole of German specialised science could be utilised.[412] The main purpose of his reform was to move towards a democratisation of the universities by strengthening the rights and privileges of younger lecturers, who would thus be prepared to work together with the Republic,[413] and to provide for the spiritual independence of the universities, interdisciplinarity and a broad general education. In his speech at the *Landtag* (state parliament) on 9 December 1919, Becker, then permanent secretary at the Prussian Ministry of Education, stressed that his aim was to maintain the aristocracy of the spirit.[414] Carathéodory, however sought an opportunity to leave. This was given to him by Venizelos who, as the Prime Minister of Greece for the second time, entrusted him with the foundation of a Greek University. Thus, after only two and a half semesters, Carathéodory departed from Berlin on 31 December 1919.

Residence of the Prince of Samos in Karlovasi, Samos (see 3.10).
Photograph: S. Georgiadis.

CHAPTER 3

The Asia-Minor Project

3.1

Preliminaries to the Greek National Adventure

The Prussian Minister of Science, Art and People's Education accepted Carathéodory's resignation to leave his office on 1 January 1920 and expressed his thanks and his "warm acknowledgement" for Carathéodory's "long and outstanding effectiveness in the service of the Prussian Universities and to German science". He wished Carathéodory success in the "great work" with which he had been entrusted on his "return to the homeland".[1]

Carathéodory had accepted Venizelos's offer because he considered it to be better than his professorship in Berlin. Greece belonged to the victors of the war and had gained new territories. Carathéodory, as any other Greek of the Diaspora, recognised in Venizelos the incarnation of the Great Idea. He also found in Smyrna the cosmopolitan character that suited him and the wealth that a city like Berlin lacked after the war. He believed he would be able to reunite himself with his family and secure a stable high income and be able to avoid the noisy bustle of Berlin.[2] Indeed, all this may well have been possible had it not been for the consequences of war in Asia Minor.

On 30 October 1918, ten days after the fighting stopped on the Western Front, the Ottoman Empire signed with Great Britain (representing the Allied powers) the Moudros Armistice, which put it at the mercy of the Allies. The armistice provided for the military occupation of the Straits, control by the Triple Entente (Great Britain, France and Russia) of all railways and telegraph lines, demobilisation and disarmament of the Ottoman troops, except for small contingents with police tasks, surrender of all Ottoman troops in the Arab provinces and the freeing of all Entente prisoners of war in Ottoman hands. The conditions of the armistice were put into effect immediately. The British occupied Mosul, the oil producing town of northern Iraq. Constantinople was formally placed under Allied occupation with military control mainly in British hands. B. Lewis describes the symbolism of conquest provided by the French General Franchet d'Espérey: on 8 February 1919, he entered the city, like Mehmed the Conqueror centuries before, on a white horse, a gift from the local Greeks. French troops advanced from Syria into Cilicia to the province of Adana. British troops occupied the Dardanelles and other strategic regions, as well as the entire length of the Anatolian railways. On 29 April 1919, Italian troops landed in Antalya (Adalia).

At the subsequent Peace Conference of representatives of the Allied and Associated Powers held in Versailles and opened on 18 January 1919, the Entente powers
were faced with difficult problems. A settlement in Anatolia had to include a homeland for the Armenians and a Jewish national home in Palestine – support for its
establishment had been promised by the British foreign secretary, Arthur Balfour,
to the leader of the Zionist movement in Britain, Lord Rothschild, in November
1917. There were also problems emerging from the secret war agreements of the Allies: the so-called Sykes–Picot agreement (French-Russian Agreement of 26 April
1916 and Anglo-French Agreement of 16 May 1916) provided for the annexation of
southern Mesopotamia (*vilayets* of Baghdad and Basrah) by Britain and of Lebanon
and a coastal strip of Syria, from a point between Akka (Acre) and Soûr (Tyr) up to
and including the province of Adana in Cilicia, by France. In the interior of Syria
and the *vilayet* of Mosul, Britain and France had agreed to recognise and protect
an independent Arab state, or a confederation of states, under the suzerainty of an
Arab chief. This state or confederation of states was in turn to be divided into French
and British zones of influence. In each zone of influence the respective power would
enjoy priority of the right of enterprise and local loans and would alone provide
advisers or foreign functionaries at the request of Arab states or confederation of
states. Russia would obtain the region of Erzurum, Bitlis, Van, Trabzon (Trapezunt)
and further territory in southern Kurdistan. Palestine was to be internationally administered, but Haifa and Akka were to be granted to Britain. Alexandretta (Iskenderun)
would be a free port as regards the trade of the British Empire and Haifa a free port
as regards the trade of France. The Sykes–Picot Agreement excluded Italy from the
share it had been promised by the Treaty of London of 26 April 1915. Greece and
Italy expected identical sections of the west coast of Anatolia including Smyrna. To
join the Entente, Italy had been promised, together with a recognition of its interests in the Adriatic and Africa, south-western Asia Minor under the London secret
war-agreement. For the same reason, offers of Smyrna and its hinterland had been
made by the British to the Greeks also in April 1915. The agreement of St. Jean de
Maurienne of August 1917, concluded between Italy, France and Britain from 19
April to 26 September 1917 and intended to remove Italy's objections to the Sykes–
Picot arrangement, redefined Italian claims by including Antalya and Smyrna with
its hinterland in the Italian zone. A vague zone of influence was to stretch north-
ward as far as Konya. The agreement was subject to the approval of Russia, but the
Bolsheviks who had come to power in November 1917 had repudiated all Tsarist
international commitments. In January 1919, two Greek divisions had strengthened
the French expeditionary force sent to help Denikin's White Russians against the
Bolsheviks in the Ukraine and the Greeks expected their reward from the French at
the Peace Conference. The strategic and political significance of Constantinople and
the Straits presented the Entente with another dilemma. The Constantinople Agreement (an exchange of notes in the period from 4 March to 10 April 1915) had called
for the incorporation of Constantinople and the Straits into Russia. Russia, in return,
had agreed to respect the special interests of Britain and France in the region of the
Straits and to view sympathetically the realisation of their possible plans for other
regions of the Ottoman Empire.

It took so long to resolve this situation that, when the Entente finally came to impose its terms on the Ottoman Empire in August 1920, the demobilisation of its forces following the end of World War I left it without the means to enforce them. The situation was exploited by the Greeks under Premier Venizelos, who offered to act in place of the Entente and to force the Turkish resistance movement in Anatolia to accept the peace terms. The result was a bloody war, which ended with a total Greek defeat in 1922.

Venizelos, who in 1917 was elected Prime Minister of Greece for the second time, was participating in the Peace Conference and was staying in Paris in September 1919. He invited Carathéodory, at that time in Berlin, to visit him there. He wanted contact with Carathéodory, as with other Greek scientists and intellectuals with European orientation and connections, such as, for example, Georgios Papandreou, Dimitrios Glinos, Alexandros Delmouzos, Manolis Triantafyllidis and also Nikos Kazantzakis (one of the most important Greek writers, poets and philosophers of the 20th century), in order to help design the educational programme for the Liberals and also secure their co-operation in national matters.

Since 1917 the Greek university system had been expanding. At the upper university level new practical directions were favoured instead of traditional theoretical studies. With law 2585/1917 "Concerning the Educational Reform" the School of Industrial Arts (*Σχολείον των Βιομηχάνων Τεχνών*) was reorganised and named the National Technical University of Athens (*Εθνικό Μετσόβιο Πολυτεχνείο*). Three new faculties were founded, for architecture, applied chemistry and topography, and the National Technical University was recognised as equal to the University of Athens. In the same year the Superior School for Forestry was set up. In 1919 the Chemistry Department of the Faculty of Natural Sciences and Mathematics and in 1920 the Superior Agricultural School and the Superior School of Commercial Studies were founded at the University of Athens. What Carathéodory and Venizelos had discussed in Paris is not known; however, in view of the later developments, one could assume with some degree of certainty that the main subject of their meeting had been the foundation of a new Greek university.

The idea of a new university outside Athens had initially been expressed in 1911 by Andreas Michalakopoulos,[3] one of Venizelos's close collaborators and a Liberal MP. Michalakopoulos had even proposed the appointment of foreign professors. The idea of a second Greek university in one of the "liberated" territories was not foreign to Venizelos. In early 1919, he secretly sent one of his close friends to Constantinople to look round some barracks that were destined to become the university building. Later, Venizelos spoke of a further university in Smyrna.

The story of how Carathéodory conceived the idea of establishing a second Greek university was told by Periklis Vizoukidis, professor for law at the University of Thessaloniki and Carathéodory's friend:

As soon as these territories[4] were liberated, Constantin Carathéodory and Georg [von] Streit, two sons of Greece who enjoyed a high scientific standing, spontaneously and full of enthusiasm proposed the foundation of a third[5] Greek university in the capital of Macedonia. [...] That bright plan, however, failed to materialise due to the world war [...] that broke out shortly afterwards. But after the end of the war and while Greece was still under arms, a new, more mature idea emerged, namely that the new Greek university ought to be founded in the source of Greek philosophy, that is on the soil of Ionia, Greek for three thousand years, on whose coasts the Greek blue and white flag was then proudly fluttering. [...] But unfortunately, soon arrived the tragic days of the national disaster and the lighthouse of science founded under great expectations was erased from the pages of the history book before being able to spread its light.[6]

Looking back at the plans for the foundation of the new university, Venizelos mentioned Carathéodory's name in an address to the Greek parliament in December 1929: "I remember that after the end of the great war the Government decided to proceed with the foundation of not only a second university, but also of a third. I even desired that the two universities were founded with a comprehensive programme; therefore, I invited Professor Carathéodory, who is not only a great mathematical expert but also an excellent organiser of universities, since the great Germany had entrusted him with the organisation of the Technical University in Breslau. [...] He went to Smyrna and devoted himself to laying the foundations of a university that would have attained a really honourable position in Greek science, had the Asia-Minor disaster not occurred."[7]

This was the second time Venizelos had chosen Carathéodory for the realisation of his plans. The first time, in 1911, he had asked Carathéodory to provide impartial advice concerning an appointment procedure at the University of Athens. Now, he was inviting him to help not only because of his university experience but also – as in the case of a university in Smyrna – because the Carathéodory family was very well known for their long-standing services to the Ottomans and Venizelos wanted to avert a probable Turkish reaction. As Venizelos's address to the parliament reveals, Carathéodory was invited to Greece because of his specific German know-how. This may have come as something of a surprise, since Venizelos's political position was just the opposite of that of the Germanophile royalists in the country. But it seems that Venizelos had complete confidence in Carathéodory and his abilities in pursuit of the national cause. By choosing Carathéodory he was also avoiding altogether the involvement of the Athenian cultural elite in his ambitious project.

3.2
The Greek Landing in Smyrna and the Peace Treaty of Sèvres

The foundation of Smyrna University was not only of great significance to science but most of all a highly explosive political issue with respect to the national aims of the Liberal government. During the peace negotiations at the end of World War I, Greece laid claim to Western Asia Minor, until then under Ottoman rule, but whose population, according to the Greek argument, had consisted of Greek inhabitants for the past three thousand years. In fact, two million Greeks were living in the Asia-Minor coastal areas at that time, but in only some of the cities did they comprise the majority of the population. However, the argument based on the descent of those people was, at least in the case of Western Asia Minor, of secondary importance to Venizelos's national politics. In a memorandum dated 30 December 1918 that he submitted to the Peace Conference on 19 January 1919,[8] he defined the will to belong to a nation as the exclusive nationality criterion. Britain supported the Greek claim both because of the anti-Muslim sentiment at home and because of its wish to have an allied state in control of the Aegean as a counter to any possible future Russian move.

Protesting against the refusal of the Allies to accept Italian demands on the Adriatic Sea, the Italian delegation left the Peace Conference on 24 April 1919, at the moment when the British and French Premiers David Lloyd George and Georges Clémenceau officially declared that they would confirm the Treaty of London. Although technically the agreement of St. Jean de Maurienne could not come into force, there was an endless controversy at the Peace Conference about the binding character of France and Great Britain's promise to Italy. Reports circulated during the absence of the Italian Premier Vittorio Orlando from Paris that Italian troops and ships were moving against Fiume and, on 30 April 1919, that Italy was sending warships to Smyrna. Initially, the American President Woodrow Wilson, Clémenceau and Lloyd George similarly ordered warships to Smyrna, but the Italians continued their disembarking. Without the consent of the Supreme Council, Italian troops landed in Antalya on 29 April 1919. On 6 May, when the Italian advance to Konya became known, in order to prevent the Italian deployment to Smyrna, the Supreme Council authorised Venizelos, through a mandate to restore "order in Anatolia", to send Greek troops to the coast of Ionia and to militarily occupy Smyrna and its hinterland. It was agreed to say nothing to the Italians or to the Turks before the Greek troops were embarked. As a consequence, an entire Greek division, accompanied by an armada of British, French and American warships, was brought into the harbour of Smyrna on 14 May 1919 and landed in the city the following day. Legal justification for the landing, especially towards the Italians, was found in article 7 of the Moudros Armistice (the supposed violation of which had been addressed to the Supreme Council by Venizelos on 12 April), allowing the Allies to occupy any strategic point in case their security was threatened. The landing itself contravened the letter of the armistice, for a report by an allied examination committee was unable to prove that either the security of the Allies or that of the local Greeks had ever been at risk in Smyrna. Moreover, the landing was accompanied by atrocities by some Greeks against Turkish civilians. The Greek army began moving into Anatolia and by the end of July 1919 they had advanced eastwards into the interior, to a greater extent than the Allies had originally intended, in the hope of being able to negotiate from a better position at the Peace Conference. The Greek invasion of Turkish Anatolia provoked a violent Turkish reaction, for the invaders were both a neighbouring people and a former subject people of the Empire. The Graeco-Turkish war began in Anatolia.

Fifteen months after the Greek landing in Smyrna, on 10 August 1920, the Peace Treaty of Sèvres[9] was imposed on a weak Ottoman government by the Entente powers, including Greece. The conditions of the treaty were much more severe than those that had been imposed on Germany and would have left the Empire a lacerated state living on the support of the powers who were annexing its wealthiest provinces. Syria and Cilicia were ceded to France; Iraq and Palestine were given to Britain, which also acquired the protectorate over Arabia. Armenia was recognised as a free and independent state. The contracting parties agreed to submit to the arbitration of the US President regarding the question of the frontier that was to be fixed between Turkey and Armenia in the *vilayets* of Erzerum, Trabzon, Van and Bitlis and to accept his decision thereupon. As an area of interest, the coast from Adramyti (Edremit in Turkish) to Antalya fell to Italy; Cyprus and Egypt were ceded to England. Predominantly Kurdish areas were granted local autonomy and the prospect of independence from Turkey. In favour of Italy, Turkey renounced all rights and title over Dodecanese and granted the Italians certain lucrative economical privileges in Southern Asia Minor as well as the right of occupation until the implementation of the Peace Treaty. The Treaty retained the Capitulations (the rules which exempted foreigners, and those Ottoman citizens on whom foreign consuls conferred protection, from the application of law and from the jurisdiction of courts in their state of residence, and which put them under the jurisdiction of the respective consuls), reduced the Turkish army to a symbolic force under allied control and put the Baghdad railway, the Dardanelles and the Russian harbour Batumi under international control. The navigation of the Straits had to be kept open, in both peacetime and wartime, to every vessel of commerce or of war and to military and commercial aircraft, without distinction of flag.

On the same day as the Treaty of Sèvres, a Tripartite Agreement on Anatolia between Britain, France and Italy was signed, recognising Cilicia as the area of French and southwestern Anatolia as the area of Italian special interests in return for French and Italian support to Britain elsewhere. Also on the same day, Greece and Italy signed an agreement by which Italy renounced in favour of Greece the Dodecanese islands with the exception of Rhodes, which would be given to Greece on the day Britain gave Cyprus to Greece, but in no case until fifteen years after the signing of the agreement.

On the basis of the Treaty of Sèvres, this document, fragile as the porcelain of Sèvres, as it was then said, "the Great Greece of the five seas and the two continents" was created and the Great Idea of the Greeks realised. While article 69 of the treaty provided that the city of Smyrna together with its hinterland, namely the entire *vilayet* of Aydin with the exemption of the county (*sancak* in Turkish) of Denizli that had almost exclusively a Turkish-speaking population, remained under Turkish sovereignty, Turkey would transfer the exercise of her rights of sovereignty over this territory to the Greek government. In witness of Turkish sovereignty the Turkish flag would remain permanently hoisted over an outer fort designated by the Principal Allied Powers in the city of Smyrna. According to article 70, the Greek Government would be responsible for the administration of the territory and would effect this administration by means of a body of officials which it would appoint especially for that purpose. The Greek government would be entrusted with the task of taking the necessary measures in order to set up a local parliament with an electoral system calculated to ensure proportional representation of all sections of the population, including racial, linguistic and religious minorities (article 72). The relations between the Greek administration and the local parliament would be determined by the Greek administration in accordance with the principles of the Greek Constitution (article 73). According to article 83, the local parliament could ask the Council of the League of Nations for incorporation of the city of Smyrna with its hinterland into the Kingdom of Greece only after a period of five years had elapsed after the Sèvres Treaty came into force.

Despite the fact that the landing of Greek troops in Smyrna was officially identified as taking place under the allied mandate, the Sèvres Treaty implied it as the prelude to a later annexation since, from the Greeks' perspective, the incorporation of the city of Smyrna with its hinterland appeared to be practically quaranteed; in fact, it would not be difficult for them to gain the necessary majority in the local parliament for the purpose of incorporation. Smyrna had a significant Greek population, which considerably influenced the character of the otherwise multinational city. And should that population prove insufficient to provide the requisite majority, Venizelos was confident that a majority could be secured through the immigration of Greeks from other regions of Asia Minor.

Events were to prove otherwise. By the time the Treaty of Sèvres was signed, the Constantinople government had lost credibility and were regarded as traitors by a militant Turkish nationalist movement. The agreements of the Peace Conference were not executed because the Treaty of Sèvres was never ratified and because the Turkish nationalists compelled the Allies to renegotiate the settlement of the Eastern Question.

3.3
Smyrna, a Cosmopolitan City

Because of its importance as a trading port, Smyrna was a cosmopolitan city in which "East and West met head-on".[10] Until World War I, more trade passed through Smyrna than through Constantinople or any other Mediterranean harbour of the Near East. The greatest part of Asia Minor's exports passed through Smyrna. Here were sold agricultural products, dyes, various spices, mohair from Angora (Ankara) and carpets. Meat and fish products, sugar, rice, coffee, herbs, fabrics, metals and wines and spirits were imported from France, Italy and Spain. Commerce was overwhelmingly controlled by Greeks. The figures for 1919 speak for themselves: Greeks 80.20%, Turks 10.47%, Armenians 2.02%, together with a handful of European traders. Smyrna was also the centre for ancillary economic activity. Some fifteen foreign insurance companies had their offices in the city as well as international companies dealing in tobacco, agricultural products and oil.

Smyrna also housed the overseas branches of many important European and American trading and industrial companies and some of these companies played a significant role in the economy of the region. The railway lines belonged to the British and French, electricity works and lighting installations were run by the British, the waterworks by the Belgians, the quay and the tram by the French. Alcoholic liquors and tobacco were largely the preserve of Americans, just like the petroleum depots at the Northern end of the coast line. British companies conducted the trade in carpets, grain, minerals and dried fruit.

Gaston Deschamps describes the international character of the city and its pulsating life in bright colours:

The Parisian-style fashion and clothing shops exhibit various products from all over Europe behind their big display windows. Apart from the legendary cotton products from Manchester flooding the Orient, there are pardessus, jackets, waistcoats and suits made of every kind of textile sent in huge parcels by a co-operative of Austrian tailors. Austria and Saxony are exclusive in almost all items of gentlemen's clothing: socks, underwear, flannel shirts and wool knitwear. But the pretty women of Smyrna, although they concern themselves little with difficult problems of political economy, render excellent service to French commerce; since they believe that to dress and adorn their beauty, nothing is more valuable than Lyon silk, Saint-Etienne ribbons, Saint-Quentin mousselines, woollen fabric from Roubaix and Reims, all these fine textiles which the Parisian dress-makers transform with their delicate hands into gracious and charming feminine clothing.

The feather hats with which the elegant ladies of the French part of the city endeavour to become irresistible are of French or English origin. The fez, which is considered the national headgear of the Turks and adopted by many travellers who intend to acquire the local colour, is produced in the Bohemian manufactures of Strakonitz. Dressed in products of the European industry, the inhabitants of Smyrna wear shoes made of leather finished in Toulon and Châteaurenault; they perfume themselves with fragrances of the famous Loubin, Pinaud and Botot; they cure their illnesses with castor oil from Milan and with quinine sulphate sold to them at reduced prices by Italians and Germans; they take their coffee with Austrian sugar cubes; they season their meat-balls with cinnamon, clove, cashew, ginger, nutmeg and spices sent from London, Marseille and Trieste; they go hunting with Belgian shotguns and bullets from Genoa; they write their letters and accounts on paper from Angoulême, Annonay or from Fiume, with French pens, German ink and Viennese pencils; they furnish their houses with mahogany from Antwerp and Paris; they check the time by their Swiss watches; they illuminate their houses and stores with oil from Baku; they knead their bread with grain from Odessa and Sevastopol; they drink cognac from Hamburg and they prepare their meals with Russian caviar, butter from Marseille, English codfish, French potatoes, Austrian smoked meat, Persian tea, Italian cheese and Egyptian onions. Having read some statistics, one cannot eat one's soup in a restaurant on the Rue Franque without discovering the whole world in miniature on one's plate. [11]

Carathéodory himself drew attention to Smyrna's French aura,[12] when years later during his lecture tour in the States he described his impressions of New Orleans to his cousin Penelope Delta: "The old part of New Orleans is still preserved and contains many houses of an age of 100 to 150 years, where French influence is visible. By the way, one can still hear some French in the streets and this part of the endless city reminds me in many ways of Smyrna, as I got to know it, since the French influence of the past century was also visible there." Smyrna's Levantines lived on the outskirts of the city and particularly in the fine parts of Cordelio and Burnabat. The Club Petrocochino, belonging to Carathéodory's family on his mother's side, was in Cordelio. The Muslim population was gathered at the foot of Mount Pagos, where the Jews had also settled. The Greek population occupied the entire lower city, the district at the sea and the trade area, except for some parts of the bazaar. But Smyrna's most impressive feature was the thorough mixing of various nationalities, which did not necessarily mean the fusion of their minds. The American Consul in Smyrna, George Horton, noticed that "in no city in the world did East and West mingle physically in so spectacular a manner as at Smyrna, while spiritually they always maintained the characteristics of oil and water".[13]

Industry and finances were in Greek hands to a considerable extent. Within Smyrna's urban agglomeration there were 391 factories: 344 Greek, 14 Turkish,

9 Jewish, 6 English, 5 French, 3 Italian, 3 Armenian, 3 Austrian, 2 German, 1 American and 1 Belgian. Significant industrial branches were engaged in leather finishing, the production of olive oil and confectionery, furniture, the production of wines, spirits and liquors, brick and tile production, soap and candle production, machine construction and food production. Out of a total of nine banks in the city, three that had been in Smyrna since the end of the 19th century were backed by Greek capital, namely the National Bank of Greece, the Bank of the Orient and the Bank of Athens. Furthermore, the Bank of Piraeus also had a representative in Smyrna and many of the management positions in all of the city's banks were held by Greeks who contributed significantly to Greek influence in the economy of the city and the region.

The educational system was also dominated by the Greeks. Sixty-five well-equipped Greek school units were in operation thanks to support from the Greek community, or the church, or private foundations. In addition, there were four schools for girls, as well as the Homer School for Girls (*Ομήρειον Παρθεναγωγείον*), established in 1881 as an institution for the training of future female teachers who were given lessons in foreign languages by British, French and German teachers. Among the other schools were the Graeco-German Lycée (*Ελληνογερμανικό Λύκειο*), founded in 1906 by a graduate of the Faculty of Mathematics and Natural Sciences of Athens University, with Dimitrios Glinos on its staff, the Graeco-French Lycée (*Ελληνογαλλικό Λύκειο*), which operated for three decades and, finally, the Evangelical School (*Ευαγγελική Σχολή*), the most important Greek educational institution aiming at the "education of our own children and of all poor foreigners"[14] in the spirit of the Gospel and according to the idea of "Greekness". This latter possessed an archaeological museum, a significant natural science collection and an excellent library which, shortly before its destruction in 1922, contained some 50 000 volumes and 180 manuscripts. Georgios Ioakimoglou was a graduate of that school.

In Smyrna, Greek was the language of trade. Fifteen Greek papers and journals were published in the city (*Αμάλθεια, Ιωνία, Σμύρνη, Πρόοδος* and *Αρμονία* were the most widely read papers; *Σμύρνη, Ερανιστής* and *Όμηρος* the scientific journals) and Turkish papers published their titles in both Turkish and Greek. The presence of a Greek-language theatre, designed by the Greek architect Ignatios Vafeiadis based on the Parisian *Théâtre du Chatêlet* and inaugurated in 1911, musical evenings, dances and teas given in luxurious *salons*, as well as sports clubs and collectors' societies all witness to the quality of life enjoyed by the Greek community. The Greek community was also engaged in major charitable work. It maintained the Greek orphanage, the Greek hospital of St. Charalambos, the biggest in the Orient open to all denominations and nationalities, and a hostel for the homeless. As Charles Vellay wrote in 1919: "If there is a city in the entire Asia Minor, whose Greek character it is impossible to question and which despite all changes over the years preserved its national traditions, then without doubt this city is Smyrna. As the centre of the crown of Greek cities from Kydonies to Ephesus it formed the first cradle of Hellenic civilisation in the Aegean."[15]

3·4
"Projet d'une nouvelle Université en Grèce"

When Carathéodory wrote to Felix Klein from Holland on 1 September 1919[16] that the project of the University in Smyrna seemed to be guaranteed, he probably had an agreement in mind that was to be signed by Venizelos and Stavros Palatzis in Paris two months later. Palatzis, a wealthy Athenian living in Paris, would undertake to pay the sum of 2 000 000 French francs in four half-year instalments towards the construction of the university buildings. A further 250 000 French francs per year would provide for the running expenses of the university, which Palatzis would pay in two instalments each year.

On 20 March 1920, a consultation among Venizelos, Leonidas Paraskevopoulos, the supreme commander of the Greek army in Ionia, Apostolos Psaltoff, a doctor and a respected member of the Greek community of Smyrna, and Carathéodory took place on the battle ship *Ιέραξ* (Falcon) anchored in Smyrna harbour. Three months later, in June 1920, when the Turks had been badly defeated and the Greeks had advanced in both Anatolia and Rumelia, Carathéodory took on the management of the Ionian University on the instructions of the Greek government. He was appointed Professor for Analytical and Higher Geometry to the Faculty of Natural and Mathematical Sciences of the National and Capodistrian University of Athens by the Ministry of Education and Religious Affairs on 2 [15] June 1920 and so acquired the status of a civil servant.[17] However, he had not sworn the oath and had not, indeed, even arrived in Athens.[18] Moreover, as Hondros informed the faculty, Carathéodory would delay taking up his duties, since he intended to travel in Europe for many months. Panagiotis Zervos, since 1917 professor of differential and integral calculus, was prepared to substitute for Carathéodory in the lecture programme that had been arranged by the faculty before Carathéodory's appointment.

A few weeks after his meeting with Venizelos in Paris, on 20 October 1919, Carathéodory had submitted his ideas regarding the foundation of a new university to the Greek government. In the memorandum, written in French and entitled *Pro-*

University of Athens. Photograph: S. Georgiadis.

jet d'une nouvelle Université en Grèce, présenté au Gouvernement Hellènique par C. Carathéodory (Project for a New University in Greece presented by C. Carathéodory to the Hellenic Government),[19] he opposed the concentration of scientific life in the capital of the Greek Kingdom. In his opinion, Athens clung too much to classical ideals which, while they had been admittedly justified in the 19th century as a counter-balance to the Byzantine heritage and the cultural after-effects of Ottoman rule, now prevented Greece from pursuing other significant interests. These latter demanded an openness to both the Orient and to Slav neighbours.

Carathéodory considered Greece and the Greek communities in other Mediterranean countries to be the connecting link between Europe, Asia and Africa. He, too, upheld the ideology of the Great Idea. His image of Greece included Slav, Turkish, Jewish, Armenian and Levantine minorities. He viewed the country as an outpost or prolonged arm of Europe in the Orient. For him this vision primarily meant cultural mediation "between the Slav and the Turco-Arabian world", as he put it in his memorandum, and the West. The Greeks who lived within the boundaries of the Ottoman Empire seemed to him to be predestined for such a task, firstly because they had maintained contact between their people and the culture of the Orient and secondly because they were able to preserve the Greek traditions, which had been carried on by the Patriarchate in a considerably hostile environment. This view questioned the role of Athens as an undisputed national centre, and thus also the role of Athens University as the core of Hellenic education. The financial wealth, as well the intellectual brilliance, of Carathéodory's ancestors in Constantinople had enabled them successfully to establish Constantinople as the centre of Hellenic education among the members of the semi-autonomous Greek *millet* (see 1.2); furthermore they emerged as the natural instruments of the "civilising mission of Hellenism", emanating from Constantinople, whether or not this "mission" was claimed by the University of Athens. This latter institution, embracing the call for the unity of all Greeks and the organisation of the Geek state according to the models of Western countries, saw itself as the agent for the introduction of the Enlightenment to the Orient.

Carathéodory's approach to the development of a national science programme was to recognise the essential Hellenic character of the University of Athens, but alongside that, to further the growth of new research and educational institutions, on a decentralised basis, through the restoration and strengthening of Greece's relations with all her neighbours without exception. The leitmotif in Carathéodory's conception of the new university was the Greeks' "civilising mission" in the Orient, an aspect of the Great Idea encompassing the ideas of "enlightenment" while at the same time absorbing modified national elements. Carathéodory chose *EX ORIENTE LUX* [20] as the motto of the new university. Since the foundation of the Greek state, numerous politicians, historians, scholars, writers, jurists and university professors propagated the idea that, because of her history and her geographic position, Greece was predestined to carry the European light and educational ideal into the Orient. Rhetoric about a "chosen people", hymns to "Greek superiority", "genius" and "extraordinary talent" assumed manic and at times grotesque proportions. The ideological construct of the "civilising mission" became one of the programmatic cornerstones of Greek foreign policy,[21] advocated indeed not only by the Greeks themselves, but, in exactly the same way, also by the Americans, French and British.[22] Together with the Greeks, the Allied forces had entered the Ottoman Empire absolutely convinced of the truth of their own propaganda intent on proving the superiority of Western civilisation over that of Islam. The "civilising mission", denoting the cultural part of a complex colonial ideology that was dominant in most of Western Europe, had a specific meaning for the Greeks arising from their national historiography. The "missionary" role which Carathéodory claimed for himself and for the Greek nation did not differ from that idea of a specific Hellenic destination that had been propagated by the proponents of the modern Greek Enlightenment and prophets of the Great Idea. Similarly, for Carathéodory the Greek "mission" was triply defined: by space, namely by Greece's geographic position, by time, namely by Greece's history, and transcendentally by the assertion that the Greeks were "inheritors of a sacred tradition that mingles with their history".[23]

Without at first examining the question of a particular location, Carathéodory expressed his intention to make the university the most modern one in the Orient. The foundation of such an educational institution would demand great efforts given the difficult task of combining local needs with national aims, let alone the fact that the required infrastructure was lacking in the particular locations under consideration. In his memorandum, Carathéodory expressed his opinion on the competitive and complementary nature of the Diaspora Greeks and those of the Greek Kingdom: "It is true that, until recent times, the civilisation of the Sultan's Greek subjects comprised the necessary complement to the intellectual culture which was developing in Greece and often contributed to the moderation of the one-sidedness of the latter."

The idea of an old, classically based, Athens University being balanced by a more scientifically oriented Ionian university was restated later when Thessaloniki was finally chosen as its site, following the defeat of the Greek army in 1922 and the expulsion of Greeks from Asia Minor. The driving spirit behind the idea of a new university corresponded entirely with the ideas Carathéodory had previously set out in his memorandum. According to the governmental declaration by the Greek Prime Minister Alexandros Papanastasiou in March 1924, the University of Thessaloniki had thus primarily "to include the practical sciences".[24] In July 1924, the Minister of Education and Religious Affairs, I. Liberopoulos, also emphasised that the foundation of the University of Thessaloniki was to serve the spiritual leadership of the Greek culture in the new territories and at the same time to be oriented practically in order to support the opening up and cultivation of the new territories.[25]

In his memorandum, Carathéodory saw the education of specialists for agriculture, engineering and trade as the immediate aim of the university and an important prerequisite for the economic development of the country. In addition, he proposed courses for Slavic, Oriental and Western European languages. Combined with these, he proposed classical studies and courses in humanities, with a slant towards Oriental culture, as well as law studies. Finally, Greek culture should be offered to the local minority groups. The foundation of a medical faculty and a faculty for Islamic law was planned for a later phase. Consequently, Carathéodory proposed the foundation of four faculties, namely engineering, agriculture, trade and oriental ethnology. He considered a plan for the distribution of the faculties in different locations within the Greek Kingdom as unfavourable. Both financially and with respect to academic exchange and concentrated research, a single location offered significant advantages. In order to substantiate the proposed combination of applied sciences and humanities within one institution, Carathéodory made a detailed excursion into the history of European universities. He reproached those institutions that were the most renowned and rich in tradition for a complete lack of flexibility during the industrial revolution, when they should have integrated practical sciences into their curricula. In France, this had resulted in the establishment of engineering schools independent of the old universities for which the "men of the French Revolution had little respect" – a model that was also adopted by Germany about fifty years later. The small-town universities predominant in Germany were hardly suitable for engineering, which was dependent on contacts with industry located in the large city areas; but the main reason that engineering schools in Germany were established outside the universities was that, under the influence of classicism, German universities viewed every "practical science" as incompatible with their dignity. In contrast, it was the English and American universities, characterised by the envisaged connections between the different educational fields, that, in Carathéodory's opinion, were suitable prototypes for the proposed new university. In this respect, Carathéodory appears to also be in agreement with Klein's plans for the integration of engineering sciences into the university curriculum.

Having set out his general thoughts, Carathéodory now made a statement about the location of the new university. Smyrna, Thessaloniki or one of the islands near the Asian coast were considered. Its Greek character and its proximity to Muslim Asia made Smyrna the ideal site for the faculty of Oriental ethnology; Smyrna also offered the advantage of being the centre of an extensive agricultural area and the greatest trade city of Asia Minor and also the prospect of developing into an industrial centre; finally, there were already numerous buildings to house the university. In favour of Thessaloniki were its proximity to Western Europe, and its link to the Balkan rail network, and its location within the economically significant wheat cultivation area of Macedonia. Chios, viewed by Carathéodory as one of the greatest islands near Asia Minor, would be appropriate as a peaceful site and, to a large extent, free from political fanaticism. However, traffic and infrastructure would present problems in this particular case. Parallel to these considerations, Carathéodory also drew attention to the necessity of establishing a Greek university in Constantinople as soon as possible. "The great capital, the polis, the dream and hope of all Greeks" (Colettis, 1843) was of course situated outside the Kingdom. However, from his statements, it cannot be ascertained under which political considerations Carathéodory saw Constantinople as the home of the Greek university. Admittedly, Venizelos had hoped for Constantinople as a trophy of the Asia Minor expedition, but he realised in time that he could not achieve his aim because of the balance of power in that area.[26]

In his memorandum, Carathéodory also elaborated on the balance of teaching staff and their means of recruitment. His suggestions at first provided a flexible system of several levels, comprising a small number of full professors, assistant professors, lecturers with commissions for specific subjects, and lecturers with fixed-term contracts. In addition, some personalities had to be invited in from abroad in order to build up laboratories over a period of two to three years. The university curricula, in Carathéodory's opinion, ought to parallel those of equivalent institutions abroad, with due consideration of local conditions. After acquiring a solid theoretical knowledge, every student had to specialise in a certain direction according to her or his individual inclinations; practical apprenticeships in workshops and laboratories were not to be neglected. The period of study had to vary according to the nature of the diplomas awarded. The opportunity to study should also be offered to those, such as farmers, who would not enter the university directly, but would be offered an introductory elementary course. Applicants must become proficient in the Greek language and they would be admitted only after an entrance examination. Unlike in Germany, intermediate examinations during compulsory and optional courses had to take place according to a strict programme.

Further, Carathéodory presented his ideas for establishing a comprehensive library and organising scientific laboratories for physics, chemistry, machine construction and electrical engineering, all of which should acquire equipment of the highest standards. A planned phased construction was supposed to allow the possibility of future extensions.

Thus, the concept of the new university was now set out. The only remaining issue to be clarified was that of the location. The final decision on that was a political matter and this is where Venizelos re-entered the stage.

3·5

Founding the Ionian University

Law 2251 of 14 July 1920, "Concerning the Foundation and Operation of a Greek University in Smyrna", was published in the Government Gazette 1920 A, II, p. 1347. The practical measures that were to lead to its realisation were discussed by Venizelos, Aristidis Stergiadis, Smyrna's High Commissioner, and Carathéodory himself on a battle ship in the harbour of Smyrna in August 1920. Venizelos did not disembark in Smyrna to avoid a possible Turkish or Allied reaction. Aristidis Stergiadis, a jurist from Iraklion, Crete, who had studied in Paris, had been summoned to Smyrna by Venizelos in 1919 and took up his office as soon as he arrived on 21 May. Until then, he had been Governor General of Epirus (1917–1919) and was reputed to be an honourable and conscientious man. In the past, he had co-operated with Venizelos on the preparation of various bills. He had also served the government of the Liberals as an expert in working out the articles of the Athens Treaty of 1913, which settled the affairs of the Muslims. Stergiadis was at first employed as a legal consultant to the general staff of the Greek occupying power, but was very soon appointed High Commissioner of Smyrna and thus vested with power in Ionia comparable to that of an absolute ruler. He possessed the authority to decide on every political or economical issue and, up to January 1920, also on every military issue. The High Commissioner's office also had further significant tasks, such as undercover surveillance of the Ottoman authorities in the area (police, law courts, etc.), until the Greeks took on the administration after the Peace Treaty had taken effect, and made all necessary preparations for the future annexation of Western Asia Minor to Greece. Although the

High Commissioner's office was renamed Greek Administration of Smyrna (*Ελλη-νική Διοίκησις Σμύρνης*) in August 1920, Stergiadis continued to be addressed or referred to as High Commissioner up until the Asia-Minor disaster (see 3.9). The Greek Administration of Smyrna took on the organisation of the administration in both the Sèvres Zone (*Ζώνη των Σεβρών*) and the Militarily Occupied Land (*Στρατι-ωτικώς Κατεχομένη Χώρα*) and the High Commissioner became a Minister without portfolio of the Greek Government and was entrusted with the task of representing Greece in Asia Minor.[27]

In Michael Llewellyn Smith's words, Stergiadis's "dearest project was the 'Ionian University' of Smyrna, which Professor [C. Carathéodory], an enlightened professor of mathematics at Göttingen University, came to Smyrna to direct. More than once he was heard to claim that Greece in Asia Minor could create the 'third or fourth Hellenic civilization'".[28] The thoroughness with which Stergiadis promoted the organisation of the university also attracted the attention of the King's son, Nikolaos, who made positive references to the enterprise in his diary.[29]

In mid-September 1920, Georgios Ioakimoglou, lecturer of pharmacology at Berlin University, was called to Smyrna by Stergiadis to assist Carathéodory in organising the university and to take on the full professorship for hygiene and microbiology. His appointment would follow on 1 January 1921 for a period of five years. Shortly before, Ioakimoglou met with Carathéodory and Venizelos in the latter's flat in Athens. Venizelos spoke to them with enthusiasm about the "civilising mission", which they were to accomplish in Asia Minor. Carathéodory and Ioakimoglou, accompanied by Alexandros Zachariou, who was in charge of the university buildings, embarked for Smyrna on 4 October 1920. On the evening of their arrival, they consulted with Stergiadis on measures to be taken regarding the foundation of the university. Stergiadis promised them his support and, indeed, the Greek Administration covered the expenses for the foundation, organisation and management of the new university and granted them a large half-erected building within an extended site on the hill Bahri Baba (now called Karataş) at the entrance of the city. The use of the site was not uncontested. In fact, it had formerly been used as a Jewish cemetery up until 1875 when the Jews relocated the cemetery to the outskirts of the city. The Ottoman government had used the site to provide temporary barracks for Turkish refugees from Eastern Rumelia and Bulgaria who had found refuge in Smyrna after the Russo-Turkish war of 1878. The Jews had protested against the violation of their sacred place and the land was restored to them in 1883, but on the condition that they woud not use it anymore as a cemetery and that they pay an annual rent. However, in 1914, the *Vali* of Smyrna, Rahmi Bey, went ahead with laying the foundations of a public library and with construction work for a Turkish higher school on the grounds that the Jewish community had not paid any rates for the use of the land. The Jewish community attempted to reassert their claim to ownership in November 1918, after the Moudros Armistice, but the Prefecture Council of Smyrna ruled that the site was public property. This was the situation in 1920, when the Greek Administration decided to use the site for the university. The Jewish community again protested and tried to obstruct the project, but the High Commissioner's Office rejected their complaint on the grounds that the site belonged to the Ottoman state.

Carathéodory rented a "simple" house for himself and his family in Boudja.[30] The High Commissioner's Office gave him a car with a driver; he himself did not drive. Boudja, a fashionable suburb, half an hour by rail from Smyrna, housed mainly descendants of British, French and Dutch settlers, who had arrived in the Near East about a century earlier. Many of the residences were surrounded by parks and rose gardens. Their proprietors had the opportunity of accumulating large fortunes, thanks mainly to the tax advantages the Capitulations bestowed on foreign residents, and were able to spend their lives in luxury and wealth. Whenever they were addressed as Levantines, they felt insulted, and instead they sought their identity in their original nationalities; their sons served the military in their home national armies during World War I with enthusiasm. The journey from Smyrna to Boudja led along the beautiful valley of Hagia Anna and the Meles river, at the banks of which Homer was said to have written his great epic poems,[31] a journey Carathéodory was to make every day.

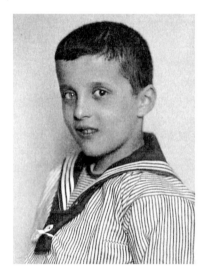

Carathéodory's son Stephanos in Smyrna, 1921.
Courtesy of Mrs. Rodopoulos-Carathéodory.

On 27 October [9 November] 1920, Carathéodory signed a contract of work with effect from 15 [28] July 1920; Stergiadis appointed him as organiser of the Ionian University and full professor of mathematics.[32] Shortly thereafter, a grave political change occurred in Greece. On 14 November 1920, Venizelos lost the parliamentary elections and the royalists came to power. The new government appointed their own people to numerous offices in the administration, justice, education, church and the military. The books written in the language of the compromise between reform supporters and reform opponents, Triantafyllidis's "human purified" language, were sent to be burned. Among them was *The Alphabet Book with the Sun* (*Το Αλφα-βητάρι με τον Ήλιο*), a product of co-operation between the three personalities who led the Educational Group (*Εκπαιδευτικός Όμιλος*) (see 2.8) and with illustrations

Maleas's painting in Carathéodory's house. In possession of
Mrs. Rodopoulos-Carathéodory.
Photograph: M. Georgiadou with permission of Mrs. Rodopoulos-Carathéodory.

by Konstantinos Maleas, an expressionist painter from Constantinople, a painting of whom hung on a wall in Carathéodory's house. The reform of 1919 was revoked, along with the use of demotic Greek in primary schools, and the triumvirate of reformers were characterised as nationally dangerous persons. Carathéodory himself did not escape censure; he was dismissed from his professorship at Athens University in December 1920 by one of the first decrees to be signed by the new Minister of Education.[33] Paradoxically, he remained in his position in Smyrna, as also did the High Commissioner, whose resignation was not accepted by the new government. Thus, the continuity of Greek policies in Asia Minor was assured and it is probably due to Stergiadis that many civil servants of Smyrna's Greek Administration were not dismissed. It was also thanks to Stergiadis that Carathéodory was able to continue his work in Smyrna until the disaster.

3.6
The High Commissioner's Decree

Thus, the High Commissioner's decree concerning the foundation of the Ionian University could be published in December 1920. It corresponded to the ideas that Carathéodory had presented in his memorandum and was a bold and detailed document.

The University was to comprise four schools, namely for agriculture and natural sciences, for oriental languages, for social and economic sciences and, finally, for trade and supervision of construction work. The first one would train future engineers, mechanical engineers, architects, electrical engineers, chemists, geologists, botanists, zoologists and farmers; the second would train future teachers of superior schools and would include in its curriculum courses in Turkish, Arabian, Farsi, Armenian and Modern and Ancient Hebrew; the third one would educate future civil servants in, among other things, administration law. The decree also provided for the creation of a superior Muslim institution for the education of future Islamic law experts (Muftis) and Islamic judges (Kadis), as well as an Institute of Hygiene, to fight contagious diseases. Finally, the university was to house a public library.

While Greek was the intended language of instruction, in some special cases Turkish or other languages would be allowed. The teaching staff would consist of full and associate professors and each subject to be lectured would be clearly identified as soon as the professors were employed. Academic freedom was explicitly guaranteed. The ranks and salaries corresponded to those of Athens University and any lecturing beyond what was compulsory would be rewarded with an additional payment. Also, university professors would be debarred from holding additional public offices. The number, qualifications and income of assistants and lecturers were to be determined by special decrees.

Admission to the university was to be open to all, irrespective of sex or nationality, provided the prerequisites determined through special regulations were met. The university would have the right to accept guest students and to award doctorates, diplomas and certificates. It would be the rector's task to build up the offices of the administrative authorities within the first five years after the foundation of the university. These authorities would be the rector, the senate, the deans of schools, the directors of the institutions attached to the university, as well as the heads of the councils and committees. For the treatment of legal issues, the position of a legal adviser, i.e. of an administrative or court's official, was provided. The central administration of the university would be accountable to the General Secretary.

3.7
The Development of the Ionian University

Several institutions of the university were set up quickly. The Institute of Hygiene was divided into a Department of Bacteriology and a Department of Hygiene. It was equipped with a library and extensive high quality technical apparatus. The aims of the Institute were to supervise the cleanliness of the streets, the water supply and the sewers and, in particular, to combat malaria by draining marshlands, to combat trachoma and tuberculosis and to produce vaccines. At the Institute, patients were examined without being charged. Midwives, nurses and medical doctors of the public health authority were to receive their training there. The Institute was responsible for the development of specific care programmes for babies, infants and adolescents. In addition, it was entrusted with the organisation of a registry office and also with the tasks of collecting medical statistics and working out health legislation for the various professions.

Parallel to the education of specialists in agriculture, the School of Agriculture had the task of organising seminars for landowners and farmers and advising them on more efficient cultivation and on how to fight plant and animal diseases, as well as conducting scientific research in its laboratories. It operated in conjunction with a big experimental farm at Tepeköy. A.Toynbee, who travelled the whole of Anatolia as a special correspondent of the *Manchester Guardian* in 1920–1921 in order to report on the military operations, was very impressed by his visit of the University farm: "In the director's mind (and in the mind of Mr. [Stergiadis] who appointed him) the real importance of [Tepeköy] is educational. By this example, the peasantry – Turks and Greeks alike – are to learn to exploit the agricultural riches of the Smyrna Zone. But the experiment of [Tepeköy] is of more than local significance. For all Anatolia it may mark the turn of the tide. For nine centuries now, the nomadism introduced by the Turkish conquerors from Central Asia has been divorcing Anatolia from agriculture, and now, perhaps, the plough (reinforced by the motor-tractor) is going at last to recover the ground it has lost."[34] So, Toynbee believed that Tepeköy could act as a catalyst for the whole of Anatolia. In turn, George Horton drew attention to the use of modern American equipment in Tepekoy. "A farm of thirty thousand acres situated at [Tepeköy], used by the Greek administration for the study of motor-culture, was bought and made exclusive use of American motor-plows. As a result, students completing the course recommended to the landowners the use of American motor-plows."[35] Indeed, "Oliver" ploughs, to a total value of 570 000 drachmas, were imported from the United States and given to the farmers independent of their nationality or religion. Twenty-seven fully equipped tractors were also imported.

On 14 December 1920, Carathéodory's former student of mathematics, Nikolaos Kritikos, then a recruit of the 31st infantry regiment of the Kydonies division, was appointed secretary to the Bureau for the Organisation of Smyrna University (*Γραφείο Οργανώσεως Πανεπιστημίου Σμύρνης*) on Stergiadis's order. This bureau was a special section within the Department of Education of the Greek Administration and was responsible for matters concerning organisational infrastructure. Similarly, the former assistant of the head of Athens Observatory, the meteorologist Christos Papanastasiou, then lance-corporal of the Greek army in Asia Minor, was detailed for development of the meteorological station of the university after Carathéodory had pressed for his employment. Kritikos was in charge of setting up the rules of the university management and the supervision of the completion and construction of the university building and the design of its environment. Following Carathéodory's suggestion, Stergiadis called upon the architect Aristotelis Zachos and invited him to come in person to Smyrna to supervise the project. Zachos was an architect of considerable renown. He had studied architecture in Munich, Stuttgart and Karlsruhe and worked in Baden for many years. In 1913, he took on the direction of the city planning of Thessaloniki and during 1915–1917, he was head of the Technical Services of the Athens City Administration. In Thessaloniki he supervised the restoration of the basilica of St. Demetrius, which had suffered significant damage from the great fire of 1917. He was interested in vernacular architecture and published studies on Byzantine monuments. In Smyrna Zachos redesigned and completed the half-finished building of the intended Turkish higher school, introducing the necessary adjustments to make it suitable for its new function as the Ionian University. The rearranged building consisted of seventy spacious lecture rooms flooded with light and an auditorium with a seating capacity of 320. Zachos added to the existing

Entrance of the Ionian University.
Courtesy of Mrs. Rodopoulos-Carathéodory.

complex new separate buildings for the Rector and the university professors. The whole project which was completed by the summer of 1922 cost 110 000 Turkish pounds.

Carathéodory described his activity as Rector of the Ionian University to Felix Klein in a letter dated 11 March 1921, which was exactly one day after the launch of the Greek offensive for the conquest of the Eski Shehir–Afyon Karahisar railway line. Despite its lack of equipment and its unprotected supply lines, the Greek army had been attacking the nationalist Turks since January 1921. Carathéodory wrote:

Many thanks for the lovely consignment of the first volume of your works, which I found on my table today as I returned from a short trip into the interior of Asia Minor. The book is absolutely wonderful. It is already a pleasure to take it into my hands, but when I turn the pages, I have to say immediately that the effort you made to survey the various works and to provide them with biographical notes was worth it. All collected works should actually look *like that*; only thus it is proved that the spiritual work of a man is something organic which is influenced by place and time. I am even more glad to have this book here with us, since I am not yet completely settled comfortably and the rest of my library still remains in Germany.

I would have written to you earlier but for many months now I have been about to travel to Europe in order to purchase some things for our university and also to appoint some scholars here, and I had hoped to have been able to tell you everything that had happened in person. However, on the one hand our work here has been delayed on account of recent political events and, on the other, I could not go away because our university building, which had been conceived as a boarding school by the last Turkish government and which I found half-built, required great alterations that could not be done in my absence. Now, I am ready and I hope to be able to travel in about three weeks. Who knows, maybe I will be in Göttingen for a few days, for your birthday.

The countryside over here is wonderful and we are lucky to have such a man at the head of the local government and, in fact, in my whole life I have not met anyone like him. He was able to restore public order within a few months to such a degree that could not be bettered in Germany. Last week, I was in Bergama [Pergamon] together with an English scholar and from there I rode over the mountains to Aivali [Ayvalik]; each time we arrived at our destination late at night, without thinking a single moment that the situation could become somehow uncomfortable. Nobody around here has seen such a thing for 1000 years. We were just coming through a purely Turkish area where the inhabitants were given arms this winter because the wild boars, which the Turks neither eat nor touch, had multiplied so significantly during the war years that they had destroyed the entire arable crop last summer. In the surroundings of Bergama alone, 3000 were shot in the last months. Mister Stergiadis's second great deed is his organisation of the repatriation of those Greeks whom the previous government had driven out of their houses since May 1914 (that is, before the war). The 11 000 Greeks of Bergama, for instance, had received the order to leave the city within two hours; about 8000 of them have returned. They were lucky to be able to embark for Greece immediately; most of them remained in Mytilene, others made it to Marseille. The inhabitants of Aivali were not as lucky, of 35 000 who left before the war, around 18 000 returned; the rest died of typhoid fever or other causes in the interior of Anatolia. 126 000 were brought back in total and the operation was organised so cleverly that it did not cost more than 5 million francs. In addition, the National Bank had lent the people 20 million, which they have, however, paid back. Everywhere, I found schools in full operation, similarly olive oil and soap factories, and within the population an optimism for the future that was absolutely worthy of admiration.

Unfortunately, our situation here is understood in Europe just as little as that of Germany and poses new difficulties for us again and again. But I am optimistic although I saw in the papers today that Düsseldorf would be occupied again. So, I hope to be able to tell you a lot more in person, soon. [...]

Many regards to Hilbert, Landau, Born and the rest of the Göttingen acquaintances.[36]

What Carathéodory received was Klein's *Gesammelte Mathematische Abhandlungen* (Collected Mathematical Works) that were published by Springer during 1921–1923. Just over two years before his letter to Klein, Carathéodory had expressed doubts about the prospect of a publication: "I am only today able to inform you of Springer's answer, which is not so pleasant, and, besides, it also contains a passage which made me very angry. I think the best thing now is to wait some months for – let's hope – quieter times; we could then try with Hirzel or Veit."[37] In his letter, Carathéodory's reference to the organic work of a man "influenced by place and time" is taken from Husserl, who refers to the empirical self which is bound to space, time and a certain person as juxtaposed to the transcendental self which precedes experience and determines its character, thus becoming the condition for the experience of the empirical self.

Carathéodory did not specify the destination of his "short trip into the interior of Asia Minor", but it could have been near the battlefields. It is also not improbable that the English scholar who accompanied him was the British historian Arnold Toynbee, who was visiting the region at that time. In his book *The Western Question in Greece and Turkey – A Study in the Contact of Civilisations*, Toynbee wrote: "During my journeys I acquired an affection not only for Smyrna (which had an indescribable charm of its own) but for Manysa [Manisa], Bergama, Aivali, and other smaller places in the hinterland, and I made friends with a number of people of almost every denomination and nationality."[38] Toynbee also described the organiser of the Smyrna University as his friend and expressed his admiration for him: "He was interested in everything – archaeology, hygiene, economics, languages – and constantly reminded me of what I had read about Ludwig Ross and the other German savants who came out to Greece in the thirties of the last century in the train of King Otto. In fact, professor [Carathéodory] was a Westerner abroad – constructive, broad-minded, humane, and out of water."[39]

The measures for the restoration of public order, to which Carathéodory referred in his letter, were taken in accordance with article 43 of the land warfare order agreed in The Hague during the Hague Convention II for the Pacific Settlement of International Disputes on 18 October 1907. They provided for the formation of a gendarmerie, i.e. of a local police force, and the foundation of a gendarmerie school in Smyrna, a militia and Greek military courts. Such a military court tried and severely sentenced those Greeks who were responsible for atrocities against the Turks during the Greek landing in Smyrna. Three of them were executed in public. Stergiadis was accused repeatedly by the Greeks of partiality towards the Turks in disputes between Turks and Greeks but this only proves his titanic attempt to be as impartial as that was possible for a man under such circumstances. The expulsion of Greeks, mentioned by Carathéodory, took place in two phases, one before World War I and then at the beginning of 1915. In order to force Greece to hand over the islands of the Eastern Aegean, which had come under Greek rule after the Balkan Wars, Turkey unleashed a systematic and violent expulsion of the Greek population from the areas of Eastern Thrace and the coast of Asia Minor.[40] Economic boycott, psychological terror, looting, murder and other atrocities prepared the way for their expulsion to Greece. Foreign trade companies were ordered to dismiss their Greek employees. The expulsions were planned as part of the preparations for war by German elite officers such as General Liman von Sanders and Admiral Usedom. By September 1914, altogether 80 176 Greeks are recorded as having to abandon their possessions and emigrate from Eastern Thrace and Asia Minor to Macedonia. A further 3000 wealthy refugees are not included in this number since they had not made use of the financial aid of the Care Commission. A second wave of expulsions affected the Greeks of the interior of Asia Minor. They were violently recruited by the Turkish army and coerced into work battalions (*amele taburu* in Turkish) in order to build roads in the Caucasus; almost all perished from cold and starvation. The action of the Turkish forces was carried out with the full knowledge and approval of the German government. The Greek King's adviser and convinced Germanophile, Georg von Streit, Carathéodory's brother-in-law, then Greek ambassador in Constantinople, accused Germany of a "plan to liquidate the Greeks" in an entry in his diary on 10 [23] March 1915 and quoted von Sanders, who described such a liquidation as "useful" in a statement that the latter denied.[41] The numbers for the repatriated refugees given by Carathéodory were confirmed by Michael Llewellyn Smith.[42] Like Carathéodory, Arnold Toynbee expressed his admiration for the reintegration of these people after the war, especially for the skilful handling by the High Commissioner's Office of the economic problem that had arisen, but he added that this could happen, at least partly, through expropriation of Turkish property.[43]

Carathéodory indeed applied for a twelve-week trip to Europe, which he had mentioned in his letter to Klein in May that year. He was going to organise the purchase of books for the university library and of apparatus and instruments for the physics laboratories, to receive offers for the furniture and equipment for the new institution and, above all, to contact renowned scientists whom he intended to call to Smyrna. He had in mind the Swiss physicist Paul Scherrer to organise the physics laboratory, the Greek physicist Phrixos Theodoridis for the Chair of Physics and the Greek chemist P. Kyropoulos to organise the future Faculty of Metallurgy. In 1921, Scherrer was in Zurich where he had been appointed professor for physics at the Swiss Federal Institute of Technology (ETH) since the summer semester of 1920. As Carathéodory's daughter Despina remembers, he was very fond of women and money. Theodoridis, who was living in Paris at that time, had studied physics at the ETH as a student of Pierre Weiss. In Carathéodory's opinion, he was the best available physicist among the Greeks. According to John Argyris, Carathéodory's judgement was based merely on the candidate's studies rather than his performance in teaching and research. Kyropoulos, whose experimental skill and easy lecturing was praised by Carathéodory, was living in Göttingen at that time. He had been an assistant in Gustav Tammann's laboratory for many years. Carathéodory must have known Tammann from his years in Göttingen.

Carathéodory succeeded in gaining Theodoridis as professor for physics, Kyropoulos as professor for chemistry and the German technical expert E. Paschkewitz, who was able to take the thread from living spiders for the cross-threads of optical instruments,[44] to build up the physics and chemistry laboratories. The employment contracts were signed during Carathéodory's trip. In Berlin, Carathéodory met Ioakimoglou to obtain his advice on the employment of staff for the microbiology laboratory. He was successful in appointing the Greek Theologos Kesisoglou, an internationally acclaimed personality in agriculture, for the setting up of an agricultural faculty. Kesisoglou came from Caesaria; he had finished his studies at the agricultural school of Gembloux (Belgium) and headed several agrarian missions to Asia (China) and Latin America (Uruguay and Columbia).

Carathéodory wished to equip the university with furniture and apparatus of the highest-quality. Typical, for instance, is his request to the famous Vienna furniture producer Thonet to design the seats of the auditorium. He ordered the seats for the lecture theatres and the furniture for the design studios from *Zelder & Platen* in Berlin. He contacted *Siemens & Halske* to invite bids for the supply of the central lighting installation of the university and the infrastructure for the physics and chemistry laboratories.[45] In Berlin, he was also able to purchase botanical models and teaching aids at favourable prices, with the help of Kesisoglou. From *Schott & Genossen* in Jena he bought glass instruments.

Carathéodory did his job in a non-bureaucratic, flexible, fast and extraordinarily skilful way. Of course, he enjoyed absolute freedom of action and Stergiadis's unconditional trust and unlimited financial support. Carathéodory sought approval for his plans and the High Commissioner was always willing to grant it by telegraphing his permission to the Greek embassies in the European capitals. In fact, Carathéodory was on excellent terms with Stergiadis. In Vienna and Berlin, Carathéodory was supported not only by his colleagues but also by the official authorities. He participated in various university celebrations; he visited the official celebration of the Prussian Academy of Sciences that always took place on the first Thursday of July, as well as the conference of the Swiss Naturalists' Society (*Naturforschende Gesellschaft*) in Schaffhausen on 24 August. Carathéodory used these gatherings to extend his contacts, which he considered indispensable for the foundation of the university.

During his trip to Europe, he endeavoured to obtain information on the events in Asia Minor and took note of public opinion regarding the Graeco-Turkish war in the countries he visited. On 12 July 1921, he reported to Stergiadis from Leipzig: "As far as the general situation in Greece is concerned, we have no news whatsoever. The papers bring telegrams from Paris, Rome and Ankara, but none from Athens. The Turkophilia of many people exceeds that of last year by far. Almost everyone has the impression that the Turks will expel us from Asia without a fight. I do not know whether the press office is responsible for this scandalous situation or whether it is the Turkish propaganda which makes the release of any news on the actual situation in Smyrna impossible. I thought it might be possible to use the telegraph to send some news directly from Smyrna to the Berlin embassy, which has ways of publishing in at least some of the major newspapers. Yesterday, for instance, I read the communication from the embassy on the condemnation of the chief culprits for the massacre of Jews in Aydin. But, is it not a pity that the papers print Kemal's communication daily, but never our own?" [46]

The fact was that military matters did not go well for the Greeks. Although repelled by Brigadier İsmet in April in an important battle fought at İnönü, the Greeks had renewed their attack in July, took Afyon Karahissar on 12 and Kütahya on 17 of that month and advanced beyond the Eski Shehir–Kütahya–Afyon Karahisar railway line towards Ankara. However, they were defeated by the Turks, commanded by the nationalist leader Mustafa Kemal, at the Sakarya river (24 August–16 September 1921). A grateful Turkish Grand National Assembly granted the victorious general the rank of field marshall (*müşür* in Turkish) and the title of a *Gazi*, a hero of the holy war. His victory was celebrated as the victory of Islamic Asia over Christian Europe. Kemal gained international recognition. Independent Turkish negotiations with France and Italy on 11 and 12 March 1921 respectively had provided for the withdrawal of French and Italian troops from Anatolia in return for economic concessions in south eastern Anatolia to France and in south western Anatolia to Italy. Friendly relations with the Soviets had been already established and the Turkish-Soviet frontier fixed through an agreement 'On Fraternity and Friendship' signed in Moscow on 16 March 1921. Kars and Ardahan were ceded to Turkey, and Nakijevan to Azerbaijan. The Supreme Council officially declared "strict neutrality" in the Graeco-Turkish conflict in August 1921. On 13 October 1921 Turkey signed the Kars

Agreement with the three Transcaucasian Soviet Republics of Georgia, Armenia and Azerbaijan, by which the Caucasian border between Turkey and the Soviet Union was recognised and the provisions of the earlier Moscow agreement, in which the three Tanscaucasusian republics did not participate, confirmed. Batumi was ceded to Georgia and served as a free port for Turkey. Soviet military and financial support for Kemal was thus secured for the following year. In order to avoid the costs and consequences of a large-scale campaign, fearing that a revival of Arabian nationalism in Syria would damage French economic interests in Turkey, and due to the unsettled Anglo-French relations in the Near East, France decided to end the fighting by signing an agreement with the Kemalist government. Thus, on 20 October 1921, almost simultaneously with the Kars Agreement, the French diplomat Henri Franklin Bouillon signed in Ankara the Franklin-Bouillon Agreement, by which France agreed to evacuate Cilicia and draw up a new frontier with Syria far more favourable for Turkey than the one laid down in the Treaty of Sèvres. France was also allowed, under the terms of this agreement, to keep its control over the region of Alexandretta on the condition that it would give it a special administrative regime and that the Turkish inhabitants of this region would enjoy every facility for their cultural development and the Turkish language would have official recognition. The Franklin-Bouillon Agreement was the first to be signed between the provisional Turkish nationalist government and a West-European power. The Italians also withdrew from their zone in southern Anatolia retaining only the Dodecanese islands. These agreements strengthened the military positions of the Turkish nationalist forces by enabling Kemal to withdraw his troops from the Syrian and Armenian front and put them into action against the Greek troops.

Carathéodory became a powerful personality in Smyrna and therefore received many requests for employment at the university and for his support in various matters. In June 1920, Ludwig Bürchner, an honorary citizen of Samos, a teacher for geography in Bavarian secondary schools and author of many works on the historical geography of Greece and Asia Minor, expressed to Carathéodory his interest in a professorship for general or historical geography at the Ionian University. Bürchner, who had good command of the Greek language, had published material on Samos in *Paulys Real-Encyclopädie der Classischen Altertumswissenschaften* (Pauly's Encyclopaedia of Classical Archaeology), vol. I, Stuttgart 1920. Another request concerned Eugenios Somaridis. In August 1920, the Greek Foreign Minister, Nikolaos Politis, asked Carathéodory to consider appointing Eugenios Somaridis, a philologist of modern Greek at the University of Vienna, to the Faculty of Oriental Languages. Somaridis's reduced post-war salary was insufficient to provide his needs in Vienna and he was looking for a better position. Again, in March 1921, the director of the country estate that belonged to the Greek school of Kydonies (Ayvalik), D. S. Liapis, asked Carathéodory for financial aid in order to revive the activities there which had ceased after the expulsion of Greeks by the Young Turks in 1914. In another case, the Metropolitan of Smyrna Chrysostomos asked Carathéodory, in April 1921, to employ Ioannis Philippidis, a doctor of law and the father of Chrysostomos' private secretary, as General Secretary of the Ionian University. In 1920, Chrysostomos, who had blessed the Greek troops on their landing in Smyrna, had sent his portrait to Carathéodory. He was photographed in the robes and mitre of his office, a pastoral staff in his right hand. The dedication reads: to the eminent professor of mathematics and wise organiser of the first Greek university in the liberated capital of Asia-Minor Greece, Mr. C. Carathéodory. In November 1921, we find A. Siniossoglou from Berlin telling Carathéodory of his intention of establishing a society largely funded by Greek capital that, in co-operation with the Berlin Society for Urban Development Works (*Gesellschaft für städtebauliche Arbeiten*), would carry out topographic measurements of Smyrna. The aim was to carry out a triangulation of the Gulf of Smyrna, at a cost of 6500 British pounds, and to draw up a city plan of Smyrna.

   One of Carathéodory's particular concerns was the university library, which he considered to be the backbone of the institution and which ought to be unequalled in the entire Orient. On 11 May 1921, the High Commissioner's Office agreed to his proposed expenditure of 18 000 Swiss francs for the purchase of the library of the Austrian Archaeological Institute in Smyrna, which was at that time closing down. The library had been set up during the excavations by Austrian archaeologists in Ephesus in the period from the turn of the century to the end of World War I. The library's collection contained many rare books concerning the history of Asia Minor. Carathéodory initiated the foundation of an informal department of the so-called Leipzig authority; this authority was in charge of replacing the stock of the Louvain University library in accordance with the Versailles Peace Treaty; the library at Louvain (Leuven) had been destroyed by fire by the Germans as a reprisal for alleged *franc-tireur* assaults during the war in August 1914. Carathéodory met Richard Oehler, Reich Commissioner entrusted with this task, who promised to support him in his work. The Leipzig authority would put its catalogues and staff at Carathéodory's disposal and would communicate favourable offers for book purchases to the Ionian University. Carathéodory justified his request with the argument that the procedure was politically non-problematic since the office also worked for the Belgian and French governments. For 20 000 to 25 000 British pounds, i.e. for one third or one fourth of the costs of rebuilding the Louvain library, Carathéodory could buy the basic stock of the Ionian University library. For a permanent supply of books from Germany and Austria, Carathéodory chose Carl Ausserer, an orientalist and employee at the Austrian National Library, who advised him in his negotiations in Leipzig. Carathéodory entrusted him with the task of listing books to be purchased on the subjects of history and Oriental languages. Carathéodory wrote about Ausserer[47] to Stergiadis from Leipzig on 12 July 1921: "he had lived in Constantinople and is well informed about the purchase of Turkish books, which is not that easy."[48] Professor Grohmann, a specialist in Arabian, offered to prepare a list of books on a payment of 900 Swiss francs. The Greek I. Kalitsounakis, professor at the Berlin Seminar for Oriental Languages, was responsible for the purchase of philological books, whereas Carathéodory himself and Phrixos Theodoridis were responsible for books on mathematics and physics and Th. Kesisoglou for books on agriculture. The books were bought and sent to Smyrna in thirty-six large crates. Carathéodory tried to gain financial support from private sponsors to cover the considerable expense. He also intended to encourage various publishers (local press, military newspapers, scientific institutions, etc.) to supply the library with their publications. The intention was for the library to expand to a public library and include books and journals covering subjects not taught at the university. Carathéodory did not abandon this plan until the beginning of August 1922, shortly before the Turkish occupation of Smyrna.

For the purchase of apparatus and equipment for the laboratory of microbiology and the purchase of specific material, the Greek Foreign Ministry granted 500 000 marks and the High Commissioner's Office in Smyrna an additional sum of 200 000 marks. Carathéodory acquired all this material, with help from Ioakimoglou, and made sure that they were packed into eighty-two crates and sent to Smyrna.

Thanks to Carathéodory's tireless efforts, by the middle of 1922, some one and a half years after the High Commissioner's decree setting up the Ionian University, considerable progress had been made in constructing and equipping the university. Some permanent appointments had been made, and others were anticipated. The payroll of the university for August 1922 lists the following appointed staff (with their monthly salaries):

- C. Carathéodory, 4 000 drachmas
- N. Kritikos, Secretary, 775 drachmas
- Th. Kesisoglou, organiser of the Faculty of Agriculture, 3000 drachmas
- Phr. Theodoridis, head of the Physics Institute, 3000 drachmas
- E. Paschkewitz, technician of precision machines, 1800 drachmas
- D. Dergalis, draughtsman, 900 drachmas
- N. Zographos, attendant, 562,50 drachmas
- S. Tokatoglou, typist, 300 drachmas

The names of three cleaners, a security guard, a porter and a scribe, who each earned 150 to 250 drachmas, completed the list. A comparison with the wages of other civil servants of the Greek Administration shows that the scientific and administrative staff of the university held financially privileged positions.

Although Carathéodory was very active in organising the Ionian University, he continued his studies in mathematics. After his famous initial work on the calculus of variations he produced, during his Smyrna period, significant contributions to problems involving multiple integrals. So, on 15 May 1921, he wrote a paper *Über eine der Legendreschen analoge Transformation* (On a Transformation Analogous to that of Legendre),[49] treating a new transformation for the case "of double integrals in which the necessary operations are entirely elementary"; on 4 August 1921 he generalised that work in his new paper *Über die kanonischen Veränderlichen in der Variationsrechnung der mehrfachen Integrale* (On the Canonical Variables in the Calculus of Variations of Multiple Integrals);[50] on 3 January 1922 in his paper *Über ein Reziprozitätsgesetz der verallgemeinerten Legendreschen Transformation* (On a Reciprocity Law of the Generalised Legendre Transformation)[51] he showed that the calculation of the generalised Legendre transformation could be made easier in many cases "by the remark that this transformation remains invariant under transposition of Greek and Latin indices".

3.8
"A Castle in the Air"

Despite Carathéodory's decisive commitment and achievements, the organisation
of the new university was not without its problems and complications: the support
promised by sponsors failed. For some unexplained reason, Palatzis did not contribute a penny. At times, doubts were expressed about the university's capability of
surviving. Arrangements for employing Ioakimoglou met with difficulties because
of his own hesitation and the delicate political situation. After the change of the
Greek government in November 1920, Ioakimoglou was seized with doubts and was
not willing to return to Smyrna. "I have written to Carathéodory that I would stay in
Berlin and await developments",[52] he said. Right up until June 1921, Carathéodory
still hoped he could be able to persuade Ioakimoglou to come to Smyrna; he was
even prepared to employ two female assistants from Switzerland, whom Ioakimoglou
considered essential for his work. On 13 September 1921, Carathéodory informed
Stergiadis[53] about the difficulties he faced and compared the case to the negotiations
he had had once before with Debye on behalf of the Philosophical Faculty of the
University of Göttingen in 1914. He attributed his success in winning Debye to the
free hand he had been given in the negotiations. In the event, Debye was appointed
professor of theoretical physics, replacing Born's former tutor Woldemar Voigt, who
was lowered to an associate professorship that had been created to attract Debye to
Göttingen. At the end of the war, Debye accepted an appointment in Zurich and his
post was offered to Born in 1920. "It could be said", Carathéodory wrote to Stergiadis, "that scientists are of the same kin as theatre tenors, namely whimsical. This
occurs because demand is higher than supply and because they tend to overestimate
their abilities. [...] On the one hand, it could be said that we would definitely have
to terminate every relationship with him, because of his unjustified or rather psy-
chologically justified attitude, but on the other hand, I believe that the common aim
of the organisation of the university should take precedence over personal matters.
It is inconceivable that the microbiological laboratory, which is being set up with
so much effort and expense, will be put into operation without his help." Carathéodory wanted to bring Ioakimoglou to Smyrna at all costs, even if it were only for a
few months. Ioakimoglou's hesitation in accepting the offer was not only because
he considered the political situation in Greece to be unpleasant, but also because of
various defamatory statements being made against the university and rumours from
Athens that it would not be able to survive. Ioakimoglou's age presented an additional
impediment. He had not reached the age of 40 at that time and he would run the risk
of being called up by the Greek army. Ioakimoglou had believed that the creation
of a genuine *universitas litterarum* would serve the balance of cultural differences
among ethnic minorities of the Smyrna region and that "Greece did not go to Asia
Minor to conquer alien populations, but to bring them her superior civilisation."[54]
Carathéodory's memorandum to Venizelos might have supported such a belief.

But the claim that the university was not founded with the aim to put foreign
groups in Asia Minor under Greek predominance was not, in fact, in accordance with
Greek political intentions.[55] As Venizelos had stated, the university was founded to

meet an acute necessity in Greek Asia Minor and not to contribute to reconciliation of differences. Was, then, Carathéodory's memorandum mere rhetoric and his project in fact in absolute accordance with Greek political intentions? To a certain degree, the whole character of his project for a new university becomes apparent when the question of language is considered. Only very few Turks had enough Greek language skill to meet the prerequisites for enrolment at the Ionian University, despite the aspirations of the founders that it would serve as a means of also educating the Turkish population through the medium of Western knowledge. *Ex Oriente Lux* was the wrong motto. All appointed professors were Greeks who had studied and worked in Europe and were supposed to lecture in Greek. Only through the filter of Greek culture could the Turks have access to Western civilisation. In Toynbee's opinion, this was a practically insurmountable obstacle: "While Greek propagandists in Europe and America, and their Western sympathisers, were representing Greek aspirations in Anatolia as the cause of civilisation, the Turkish population under Greek rule was actually being cut off by it from access to Western culture."[56] He estimated that "the new Greek University was a castle in the air, without local foundations – a doubtful experiment even if made at the cost of nothing else."[57]

According to Toynbee, the Greek administration policy, in the process of re-structuring the educational system, aimed at encouraging Turkish primary education and yet, at the same time, at sabotaging higher Turkish education. Stergiadis was indeed generous as regards primary education. A Turkish educational commission was employed and even granted wide autonomy in the administration of provincial endowments for primary education and also of the proceeds of the special tax previously collected for educational purposes by the Ottoman government. On the other hand, Stergiadis expropriated the Sultaniyyah School, which prepared students for entry to higher education in Constantinople. The Greeks argued that if the Turks of the Smyrna Zone desired higher education, they could go either to the new Greek university in Smyrna or to Athens.[58] However, this would only mean the recognition of the Greek cultural predominance. Therefore, Turkish resistance to the project of the university can be well understood. The views expressed in Toynbee's book evoked a series of protests among the Greeks, especially since its publication coincided with the defeat of the Greek army in Asia Minor. Renowned members of the Greek community in London, who sponsored the Korais Chair for Modern Greek and Byzantine History, Language and Philology ('Εδρα Κοραή Σύγχρονης Ελλη-νικής και Βυζαντινής Ιστορίας, Γλώσσας και Φιλολογίας) which had been founded in 1919 at King's College, London, then held by Toynbee, had attempted to force his dismissal. Finally, in 1924, Toynbee was obliged to resign.[59]

3.9
The Asia-Minor Disaster and the End of the Ionian University

In the two years after the decision to found the university, Carathéodory worked unwaveringly to build up this institution systematically. With respect to the internal organisation, nothing seemed to block his way, but things were different as regards the external political conditions. The situation deteriorated rapidly in the months after the summer of 1921. On 2 February 1922, a dramatic memorandum was sent by the Greek government to the British Foreign Minister Lord Curzon, one of the basic share holders of Turkish Petroleum. It was a last appeal for help and a threat, that if no help was forthcoming, Greece would evacuate Asia Minor leaving Great Britain to settle her own affairs with Kemal. The memo was ignored and both the British and the French governments refused to provide guarantees for a loan to Greece. The British had promised the Greeks that they would strive for a diplomatic solution, but still let them take the decision to withdraw from Anatolia and Stergiadis agreed to the withdrawal of the Greek army from Asia Minor. It appeared that a bitter end to the Greek adventure was inevitable. The various plans to set up an autonomous state of one or the other kind in Asia Minor, following the withdrawal of Greek troops and the termination of the immediate presence there of official Greece, can only be seen as the swan song of Great Greece.

Carathéodory was very well informed about the diplomatic and military developments and he did not seem to have any illusions about the political prospects. This can also be seen from a short letter to Felix Klein on 4 February 1922, seven months before the disaster: "Your second volume has just arrived at the eleventh hour, because we are leaving the country, as you already know. Many thanks and I hope to have more time soon to write to you in more detail."[60] On the other hand, Carathéodory held on to his project with absolute determination until the last moment. Only twenty-five days before the Turks entered Smyrna, everything looked as if nothing dramatic was to be expected. Despite war raging in the interior of Anatolia, Smyrna remained unaffected. University business still carried on, and also naturally social life. A letter that Carathéodory received from the General Secretary of the YMCA on 15 August 1922 was full of euphoria.

Dear Dr. Caratheodore,

I wonder if you could care to have a little reminder of our camp in the shape of the enclosed photographs? I am sorry that they did not all succeed as well as these. Last Wednesday Mr. Harlow, Mr. Birge and I went to Ephesus and had a splendid morning with Professor Soteriou. We were deeply impressed with the efficiency of his work. He also put himself most unreservedly at our disposal and made the old church[61] live again for us.

Next Saturday I expect to join my family at Phokia. Surely in this kind of weather one may be pardoned for wanting to get away from Smyrna.

With hearty greetings, I remain,
Yours cordially,

E. O. Jacob
General Secretary.[62]

Only on 1 September, when the news reached Smyrna that the Greeks had deserted Ushak, a place 250 km east of Smyrna, and the first Greek injured soldiers arrived in the city, did the British residing in Boudja and Burnabat start thinking about whether it would be better to leave than stay. Rich Armenians and Greeks decided quickly to take a vacation abroad. Carathéodory's daughter, Despina, a ten-year-old girl at that time, still remembers the panic that had broken out as soon as the news spread that "the Turks are coming".

On the eve of the Asia-Minor disaster, all the measures necessary for managing the university had been taken. The Institute of Hygiene[63] and the library were scheduled for inauguration 10 October 1922. All was in vain: the university would not be able to open. Two days before Turkish troops marched into Smyrna, the lower civil servants of the High Commissioner's Office left the city. On 8 September, all archives were cleared away and the higher officials and civil servants of the High Commissioner's Office departed as well. They embarked the Greek battle ship *Naxos*, calling first at Syros. On the evening of that day, the High Commissioner himself boarded HMS *Iron Duke*, a British flagship and his departure officially marked the end of Asia Minor's occupation by the Greeks. Stergiadis travelled by way of Constantinople and Rumania to arrive in Paris, where he stayed for a short while, and at the beginning of 1923 he moved to the French city of Nice, where he spent the remaining twenty-seven years of his life in self-imposed exile and great poverty.[64] Carathéodory, accompanied by the university treasurer Revelis, was among the last of the prominent passengers of *Naxos*; Theodosios Danielidis, a journalist from Smyrna, had taken him there in his personal boat. "Inexhaustible, fearless and as a good captain watching his ship going down, Carathéodory is at first concerned to save all passengers one by one and then anything he can from the load that he has been entrusted with. Thus, he himself departs on the last Greek ship from Smyrna, only after succeeding in loading the books and some of the instruments of the university for Athens", recounted Nikolaos Kritikos.[65] "One of the last Greeks I saw on the streets of Smyrna before the entry of the Turks, was Professor [Carathéodory], president of the doomed university. With him departed the incarnation of Greek genius of culture and civilization in the Orient", wrote George Horton in his *Report on Turkey*.[66]

After Kemal's breakthrough near Dumlupinar, the second division of the Turkish cavalry forced their way into Smyrna and occupied it around noon on 9 September 1922, thus completing the reconquest of Anatolia. Metropolitan Chrysostomos was murdered, the non-Turkish population were subject to widespread rape and murder and their possessions looted; a great fire lasting three days destroyed the city; those who could fled the city in panic. "In these days and later", Schramm-von Thadden wrote, "1.3 million Greeks of Asia Minor left their homeland. Anatolia's entire Hellenism was suddenly destroyed, a culture of at least three thousand years completely dissolved."[67]

The leader of the Turks, Mustafa Kemal, born in Thessaloniki in 1881, was a graduate of the War College in Istanbul and had been the successful defender of the Straits against the great British assault in 1915. He had emerged from the circles of Young Turks, who actively promoted the ideas of modernisation and westernisation of the Ottoman Empire, and these Young Turks had sprung out of the Young Ottomans. These latter had resisted autocracy in the Empire and had adopted specific political demands for a constitutional and parliamentary government in 1878, 1889 and 1905. It is interesting to note that among those who shared the same ideological background were, among others, Carathéodory's grandfather Constantin and Carathéodory's father-in-law, Alexander Carathéodory Pasha. European influence in the Ottoman Empire was not only expressed in the increasing incorporation of the Ottoman economy into the capitalist system and in European attempts to control the

Ottoman state, but also in the impact of the European revolutionary ideologies of nationalism and liberalism and of European ideas such as secularism, which reached the Orient with the expanded involvement of the Empire in European politics and diplomacy after the Napoleonic Wars. The reform period (1839–1876) of the Ottoman Empire was marked by the beginning of an Ottoman constitutional movement. Alexander St. Carathéodory had participated in the 28-membered constitutional committee under the pro-British liberal politician Midhat Pasha (1822–1884), which drew up the liberal, bicameral constitution for the Empire. This constitution proclaimed in December 1876 was modelled on the French constitution of 1814 and the Belgian one of 1831 and included human rights and institutions that had been developed through the reform process. The constitution was also adopted by Ottoman Greeks who were persuaded that a strong and endurable Empire would destroy Slavic hopes for rule over the Balkans and Constantinople. However, only two years after its introduction, the constitution was abolished by Sultan Abdülhamid II and the parliament did not convene for the following thirty years. Carathéodory's grandfather had been professor at the Imperial School of Medicine and it was in this school that the first nucleus of constitutionalists was formed in 1887. The group's goals were freedom and Ottomanism (*Osmanlılık* in Turkish), the Ottoman system of co-existence of various national groups of different denominations under a uniform political system.

3.10
Fleeing from Smyrna to Athens

Carathéodory himself reached Athens, where he was reappointed as a full professor of Higher Mathematical Analysis at the National University on 2 [15] September 1922[68] and, in the following year, as a professor of mechanics at the National Technical University of Athens. From Athens, he sent the following message to von Kármán on 11 December 1922: "We departed from Smyrna just in time and I have saved my books and manuscripts – also the library and the biggest part of the apparatus of the university."[69]

Typical of the feelings evoked in Greece by the disaster are the those expressed in a letter that the royalist physics professor Dimitrios Hondros addressed to Sommerfeld on 14 September 1922:

Our unexpected misfortune in Asia Minor has stricken us like lightning. [...] I expect Herr Carathéodory from Smyrna any day. He has sent his family on a small sailing boat to Samos. The tragedy of Asia Minor begins anew, a slaughter without end. Today, we heard that the Archbishop of Smyrna was murdered. In these ten days, I have aged ten years.[70]

Carathéodory had chosen Samos as the first refuge for his family on their way to Greece, obviously because there were still relatives on the island who could receive them in their homes. Between 1832, when the island was acknowledged by the Sublime Porte as an autonomous principality with a Christian prince appointed by the Sultan, and 1912, when the liberation war under Themistoklis Sophoulis[71] broke out, the Princes of Samos were almost exclusively Greeks from Constantinople. Ten

Technical University of Athens. Photograph: S. Georgiadis.

of Carathéodory's relatives were among them, such as his father-in-law, Alexander Carathéodory Pasha, and the latter's brother Konstantin Carathéodory, who had introduced a series of measures promoting economic growth and modernisation of life on the island.

Not until several weeks after his arrival in Greece, Carathéodory was able to get in touch with his family. From Athens, he described their flight to relatives in a letter to Hilbert on 25 December 1922:

Dear Hilbert,

I had planned to write to you for Christmas, but Christmas day is suddenly here and I do not know exactly how it came. But here is also the winter as I have never seen it – not even in Smyrna: only now do the leaves start to fall from the trees and the heath is blooming – at the same time, we have the most beautiful spring weather that we could imagine. Despite all the difficulties that we have experienced, we have still had a lot of luck: just two days before the catastrophe, I was able to put my family on a sailing ship that brought them to Samos. My wife managed to take all her possessions and my books with her – it was like a pleasure trip and she did not need to notice that it was a flight. Of course, I had stayed up to the last hour, but since I took care, nothing happened to me – only that I had to sail from one island to the other for ten days within all this chaos until I reached Athens. Now we are living in the countryside and I am giving some lectures at the university here. Berlin proposed that I give a lecture as a guest professor there in the summer – if it can be arranged, and I travel to Germany, I would be very happy to visit you in Göttingen then. Your picture that you sent me in Smyrna was a great joy; it is an excellent picture. Unfortunately, it arrived rather damaged and I would be more than pleased if you still had another copy. We recently heard from Miss Burger that your wife has been ill during the summer. We hope that she has fully recovered. Please, give her our hearty regards from both of us – we speak so often about you and Göttingen and it is only due to the circumstances that we haven't written before. Please, convey our regards to all our friends, colleagues and acquaintances.

Yours, Carathéodory. [72]

3.11
Professor in Athens

Carathéodory must have spent the summer of 1923 in Germany, for he notes that he started his work on *Die Methode der geodätischen Äquidistanten und das Problem von Lagrange* (The Method of Geodesic Equidistant [Surfaces] and the Problem of Lagrange)[73] in Berlin in August 1923, whereas his paper *Über die Hencky-Prandtlschen Kurven* (On the Hencky-Prandtl Curves)[74] was written in September 1923 at the Marburg assembly and was edited by Erhard Schmidt.

Twenty-one years after the Asia-Minor disaster, in the middle of World War II, Carathéodory confessed to van der Waerden's wife: "The most significant thing is for your husband to return to mathematical work as soon as possible. After the catastrophe of Smyrna that was the only thing that kept my head above water."[75] And we find a number of works, mainly in the calculus of variations, confirming this. Carathéodory began one of them in August 1922, when he was still in Smyrna, and completed it during his flight to Syros in September 1922.[76] He continued his production in the years 1923 and 1924, which he spent in his sister's house at Kiphisia near Athens.[77]

In Athens, Carathéodory was obliged to instruct first-year students at the Department of Chemistry at the National Technical University of Athens in the introductory course on Elements of Mathematics. This was an evident sign of his mathematical colleagues' pettiness, as one of his students of that department and later professor at the University of Athens, Michalis Anastasiadis, remarked.[78] The mild and loveable lecturer, as described by Anastasiadis, still performed this ungrateful task. His audience behaved without interest, enthusiasm and at times without respect towards him. Once a student interrupted his lecture, which he was delivering in Greek, with the German phrase *"noch einmal"* (once more). Thereupon, Carathéodory left the amphitheatre screaming in rage: "I am a Greek, I am a Greek". Later, some of his students said that his Greek was very poor and they could not understand him. He wrote to von Kármán that he held "a lecture for beginners and another one on conformal representations, which was visited by a few sensible people."[79] Most of his teaching was at a fairly elementary level, including a seminar for second-year students of physical sciences. He took part in the session of the Faculty of Natural and Mathematical Sciences of Athens University on 6 [19] October 1922, at which the dean, Georgios Remoundos, expressed his joy for the re-appointment of professors K. Maltezos (mechanics and theoretical physics), I. Politis, and C. Carathéodory (mathematical analysis) and his wish to see D. Aeginitis and P. Zervos also being reappointed. Konstantinos Zeghelis, professor of inorganic chemistry since 1906, said that the scientists of the only national university ought not to content themselves with the passive expression of joy or sorrow every time the state sought to raise or insult the institution according to the political barometer. They ought to feel obliged to actively defend its rights, its honour and its dignity whenever these were put at risk.[80] In the university yearbook of 1923–1924 Aeginitis is mentioned as professor of astronomy and Zervos as professor of geometry.

Carathéodory seems to have been well acquainted with Panagiotis Zervos, Nikolaos Hatzidakis and Georgios Remoundos, professors of Athens University, who comprised the so-called second mathematical school of Athens. All three represented Greece at international congresses and shaped mathematical education in the country in the first half of the 20th century. Zervos researched and contributed to set theory, Lie groups, partial differential equations and the philosophy of mathematics. The use of Greek terminology in set theory is owed to him. In co-operation with his other two colleagues, he introduced seminars with free admission in order to stimulate research. Remoundos, who had been appointed professor of Higher Mathematical Analysis at the University of Athens with Carathéodory's help in 1912 and was also appointed to the Technical University in 1916, was engaged in research in function theory. He had published about twenty scientific works in European mathematical journals and the *Bulletin of the Greek Mathematical Society*. He was one of the founders of the Society in 1918, a member of the editorial board of the *Bulletin*,[81] and author of text books on the theory of differential equations, analytic functions, higher algebra and analytical geometry. A delegate of Greece to the League of Nations and an academician since the foundation of the Academy of Athens, Remoundos died at the age of 50 in 1928. Hatzidakis had studied in Paris under Poincaré, Picard and Darboux, in Göttingen under Hilbert, Klein and Schönflies, in Berlin under Schwarz, Fuchs and Knoblauch. In 1896, he abandoned his studies in Paris to return to Crete to join the rebels together with his uncle Georgios Hatzidakis, who had established the science of linguistics in Greece. Hatzidakis researched in differential geometry and generalised Darboux's theories in the n-dimensional space (1900). Appointed professor at the University of Athens in 1901, the year in which the Faculty of Natural and Mathematical Sciences was created, he held lectures on the theory of functions in general, on the theory of elliptic functions specifically, and on the kinetic theory of curves.[82]

In May 1924 Carathéodory wrote the foreword to Sakellariou's book on analytical geometry. Nilos Sakellariou was professor of analytical geometry at the University of Athens. In his text, *Foreword to the Analytical Geometry of N. Sakellariou (Πρό-λογος Αναλυτικής Γεωμετρίας Ν. Σακελλαρίου)*,[83] Carathéodory identifies the point when modern mathematics departed from the science of antiquity with the invention of analytical geometry by Descartes (1596–1650), whose initial idea "aimed at combining algebraic methods with the science of space". As Carathéodory wrote in the introduction, N. Sakellariou, was constrained to follow the teaching philosophy of Athens University and had perforce to adhere to classical methods. "At the same time however, this opus is a very good introduction to analytical geometry and thanks to its great number of well chosen exercises, it offers the lecturer the means to dedicate himself to the study of higher areas of analytical geometry." On 19 May 1924, Carathéodory gave a lecture "About Mathematics in Secondary Education" *(Περί των μαθηματικών εν τη μέση εκπαιδεύσει)*[84] to the Greek Mathematical Society. He established that the mathematical curriculum in Greece was overloaded,

Building of the Ionian University.
Photograph: Ali Onur.

whereas arithmetic exercises and mental arithmetic were neglected. In his opinion, it was necessary for the secondary school to combine the teaching of mathematics with practical applications. This direction could be attained by the drawing of graphs, which would contribute to a more profound notion of functions, by exercises on the third dimension through drawings and models and finally by many applications of the geometric theorem of similarity. Further, it would be useful to persistently draw attention to facts with a certain general character and to teach the students the way to see things in their context. The subject matter should be presented without cumbersome formulas and tedious theory in such a way that the essence of things would not be lost.

Carathéodory's future career would lie in Germany and the United States, but he was not forgotten in Greece. In 1994, the Greek Post Office issued a series of stamps depicting Carathéodory and Thales of Miletus. In 1999, the Greek Ministry of Education (Centre for Educational Research) decorated the cover page of its publication *Assessment of the 3rd-Grade Lyceum Students in Mathematics* (Αξιολόγηση των μαθητών της Γ'Λυκείου στα μαθηματικά) with Carathéodory's portrait, photographed in Smyrna, and published a short version of his curriculum vitae. In the year 2000, celebrations in his honour were organised in various Greek places. But at that time, the Greek capital showed its most inhospitable side towards this Smyrna refugee with a German background. Newspapers spread the quite unsubstantiated rumour that Carathéodory had suffered from syphilis and had as a consequence brought a disabled child into the world.[85]

Building of the Ionian University.
Photograph: Ali Onur.

To conclude the unhappy affair of the Ionian University, the books and scientific apparatus rescued by Carathéodory were distributed to various faculties of Athens University and other scientific institutions. Volumes bound in leather with works by Einstein, Curie, Heisenberg and others, are now stored in the stock-room of the former chemistry building at Solonos Street 104 in Athens. The same building housed the book-shelves of the Ionian University library, rerouted directly to Athens instead of Smyrna because of the swift development at the war front in Asia Minor. Carathéodory passed the archive of the Ionian University on to Hondros, who was responsible for the university library in Athens. As late as 1962, Caesar Alexopoulos, the physics professor of Athens University, whom Carathéodory held in high esteem, was the archive curator. The archive, unfortunately, no longer exists. The university building in Smyrna was burnt out during the great fire but later repaired and used as a school for girls (*Kız Lisesi* in Turkish).[86]

3.12
The Lausanne Treaty: Defeat of the Great Idea

In Athens, the government's reaction was hopelessly inappropriate. Following his unsuccessful negotiations with the Allies, Premier Dimitrios Gounaris resigned on 16 May. Petros Protopapadakis, the Minister of Finance in Gounaris' government, became prime minister of a Royalist Coalition Government on 22 May after a short interlude of rule by the Cabinet under Nikolaos Stratos. However, Gounaris and Stratos remained in the government becoming respectively Minister of Justice and Foreign Minister, and Minister of the Interior. Immediately following the fall of Smyrna, the Protopapadakis Government resigned to be replaced by the Cabinet of Nikolaos Triantafyllakos. In other words, the Athens politicians faced the events from a party political point of view, being interested only in retaining power.

The victorious Kemal was now intent on following up his successes in Anatolia by also driving the Greeks out of Thrace. A radical change of Greek politics was demanded to meet the new danger. This was the opinion of a group of Greek officers evacuated from the front to Chios and Mytilene, who, under Colonel Nikolaos Plastiras, organised a successful military coup against the monarchy, the so-called Revolution of the Army and the Fleet. Given no alternative, the Triantafyllakos government resigned on 27 September. The "revolutionaries" sent a telegram to Venizelos in Paris asking him to represent them abroad with the Entente powers and he accepted. King Konstantin abdicated, to be briefly replaced by his eldest son, and left Greece. At the beginning of 1923 Konstantin died in Palermo. In the meantime, in Athens, a new government had been formed under Sotirios Krokidas. The character of the coup was determined by the myth that the Greek army had not been defeated but betrayed. The consequences were brutal. Six of the accused politicians and high-ranking military officers, the scapegoats, were sentenced to death by a court martial on charges of high treason on 28 November 1922 and executed on the same day by firing squad. As Pangalos later admitted, the sentence was an act of "national expediency", necessary to put an end to the drama of the disaster. It was clear that not only had there been no high treason but, as Venizelos revealed in his address to the Greek parliament eight years later, on 25 June 1930, there had been no evidence against the former Prime Minister, Gounaris, who had stated in March 1922 that he was not prepared to retreat from Asia Minor. This did not prevent him from being among those executed following the Trial of the Six. The Krokidas Cabinet resigned a day before the executions and a new government under Colonel Stylianos Gonatas was formed. The organiser of the coup, Plastiras, remained a major player on the political scene until the 1950s. The year following the coup, he sent his portrait in military uniform to Carathéodory, in November 1923, with the dedication: to the honourable and distinguished professor Mr. C. Carathéodory.

Among Carathéodory's papers in the archive of the Bavarian Academy of Sciences there is the cover of Edward Hale Bierstadt's book *The Great Betrayal*.[87] The photograph under the title depicts the fleeing population pushing their way to the ships anchored at the quay of Smyrna. The publisher's blurb reads: "A candid and impartial presentation of the facts in the Near East situation of to-day, with a study of the historical backgrounds. The dramatic story of the betrayal of the Christian minorities, the wanton sacrifice of the American philanthropic investment, the desertion of Greece by the Allied Governments, and the struggle for economic control in the Near East." Carathéodory's opinion about the French allies is registered by Penelope Delta in her diary on 21 April 1923: "All the young French officers who fought with us in the great war and who acknowledge that we fought well have returned full of friendship for the Turks and hatred for the Greeks. Kostias Carathéodory claims that after the war the Turks made sure of winning them over by pumping their pockets full of gold coins and putting their harem at their disposal."[88]

Conflicts between France and Britain, between Britain and the USA, between France and Italy, and between Britain and Italy had been expressed in various forms and combinations in the oil producing region of the Near East from 1919 to 1922. Scott Nearing saw the causes of war in Asia Minor in the international struggle for petroleum sources, markets and investment opportunities.[89] Conflicts between the Balkan states and between Greece and Italy, although of secondary significance in themselves, extremely complicated the attempt for a new political realignment after World War I. Furthermore, Greek and many Western historians have been apologetic as regards Greek expansion in Asia Minor and, in fact, sympathetic to those policies that led to the dismemberment of the Ottoman Empire and even the subjugation the Turkish nation. The Greeks acted within the Ottoman Empire mainly by the authority of the Entente's mandate, but they also pursued their own goals and, in doing so, they endangered the local Greek minority who, in the end, became the tragic victims of the head-on collision between Greek and Turkish nationalism.

It seems that Carathéodory never questioned the character of the Greek involvement in Asia Minor. Although Penelope Delta too, advocated the Great Idea, she considered that the Greek defeat was determined by the so-called National Schism (*Εθνικός Διχασμός*). Ten years later, she took stock of the disaster: "Other wars followed which cleared the disgrace of 1897. There were sacrifices; blood was shed; we held our heads high; we took heart. We felt like a nation. We became proud. However, deep in our hearts we remained unchanged. We continue to think first of ourselves; we cannot forget ourselves, nor put ourselves aside; we are not able to fall into line with a common Greek interest. The fanaticism of the two great parties was enough to split the country, to strangle patriotism, to knock the Great Idea over, to destroy the achievements of the national movement, to push the general interest aside, to even extinguish humanism. The individual small interests gained the upper hand through stubbornness and pseudo-dignity. Everyone followed their own way until the country fell apart and the disaster of 1922 arrived."[90]

In order to reach Thrace, Kemal would have to cross the Dardanelles, still then occupied by an inter-Allied force. French and Italian contingents withdrew first but the British remained, only to give way to Kemal's demands at the eleventh hour to avoid an armed confrontation. Under the armistice, signed at Mudanya on 11 October 1922, the Allies agreed to a restoration of Turkish sovereignty in Constantinople, the Straits and eastern Thrace. The Greeks acceded to the armistice on 14 October and five days later the Turks entered Constantinople.

The Lausanne Treaty, which put an end to the Graeco-Turkish war and fixed the conditions for peace, was concluded on 24 July 1923.[91] For the Greeks, it meant the failure of the Great Idea. It obliged Greece to return Eastern Thrace and the islands of Imbros and Tenedos to Turkey, as well as to give up her claim to Smyrna. For the Turks, it meant the re-establishment of Turkish sovereignty in almost all the territory included in today's Republic of Turkey. The republic was declared on 29 October 1923, and Kemal Atatürk was elected its first president. The Lausanne Treaty also confirmed the special agreement between Venizelos and Mustafa İsmet Pasha (İnönü), signed on 30 January of that year, on the exchange of Greek and Turkish parts of the population – around 1.3 million Greeks for 0.5 million Turks – which, however, only put the formal seal of approval on what had already been "accomplished" by the demographics of war.

3.13
The Refugees

The Greek national disaster of 1922 also affected other nationalities. In all some 45 000 Armenians fled to Greece and 92 000 Bulgarians to Bulgaria and thousands of Christians migrated to Europe or America. Morally, politically, socially, demographically and financially weakened after ten years of war, Greece was now forced to receive hundreds of thousands of refugees, without having any suitable plans. The majority of the refugees (53%) settled in Macedonia, at first in the areas formerly inhabited by the Muslims who had migrated to Turkey in implementation of the agrement on the exchange of population. Up to June 1923, the refugees received support from the Greek state, some private organisations and the American Red Cross. The main problems facing the Greek state in the initial stage were the restoration of the agricultural estates deserted by the Turks and Bulgarians, the building of new houses and settlements, road construction, the distribution of land to the farmers among the refugees and the purchase of farming tools and machines. In the following years, the central task in Greek domestic policy was the integration of the refugees, which changed Greece's political and economic character.

The Greek state was also in desperate need of financial support. Help came from the League of Nations, which contributed to the International Refugee Settlement Commission. The Commission was an independent organisation founded according to the Geneva protocol on 29 September 1923. It administered foreign credit to the rural resettlement of the refugees, the so-called 1924 League of Nations Refugee Loan. Foreign capital began to enter Greece after 1924 with the flotation of the Refugee Loan. The Commission was required to work independently of the Greek state and report to the League of Nations every three months. The Central Council of the Commission was to consist of four nominees, two Greeks employed by the Greek government and approved by the League of Nations and two foreigners employed by the League of Nations. One of the latter, exclusively an American, would become the Chairman of the Commission. The Greeks were (a) the "sagacious, benevolent and most loveable philanthropist and patriot"[92] Stephanos Deltas, President of the Greek Red Cross, First President of the Agrarian Bank of Greece and husband of Penelope Delta and (b) Periklis Argyropoulos, Governor General of Macedonia in the years 1917–1918 and one of the founders of the Venizelian National Defence ($E\theta\nu\iota\kappa\dot{\eta}$ $\text{'}A\mu\upsilon\nu\alpha$) movement in Constantinople in the years 1921–1922. Henry Morgenthau, Sr., chairman of the Democratic Party Finance Committee in Woodrow Wilson's 1912 presidential campaign, and American Ambassador to the Ottoman Empire in the years 1913–1916, became the first Chairman of the Commission. Known for his humanitarian aid during the Armenian genocide and the expulsion of Greeks in 1915, Morgenthau was honoured in Greece several times. In the Versailles Peace Conference, he had participated in a half-official mission. He was sent by Wilson to Poland to lead the delegation investigating anti-Semitic acts of violence there. What he saw in Poland made him understand "what the collective pain of a hunted people was".[93] Morgenthau described the story of the Greeks in Asia Minor, of the national disaster and also of the refugee settlement in his book *I was sent to Athens*. Carathéodory was himself acquainted with Morgenthau. The American was also mentioned in a letter from Carathéodory to Einstein written in Thessaloniki on 12 May 1930:

Dear Einstein,
Herr Morgenthau, who has been the American ambassador in Constantinople for a long while and is one of the most influential American Jews, told me that he would travel to Berlin to visit you, if you are there.

Since he does not know you personally, he wanted to write to you through Freundlich, if I did not wish to do it. I could not deny him this request since he has achieved exceptional things here in Greece for the 1 500 000 refugees in 1922 and we are all very indebted to him.[94]

The Greek state put 500 000 hectares of land at the Commission's disposal for the resettlement of the refugees. This land was either owned by the state or expropriated private land, or land vacated by Muslims who had migrated to Turkey under the population exchange agreement. Later, by the end of 1924, the Greek state made more land available by accelerating the land reform in the north of the country: the expropriated land was distributed to both the refugees and the farmers of Macedonia, Thrace, Thessaly, Epirus and Central Greece who had possessed no land until then, in accordance with a decree of the "revolutionary government" of Plastiras (14 February 1923, no. 3473). The refugees themselves contributed to the success of the programme and played a major part in the revival, progress and wealth of the country within only a few years. They greatly increased the production of tobacco and the production of cotton and silk enabled a growth of the textile industry and the export of silk. As Morgenthau put it, the refugees proved that they had been a "blessing" for Greece.

3.14
Carathéodory's Report to Henry Morgenthau

In 1926, when the population exchange was almost over, Carathéodory travelled to Thrace and Macedonia and at the end of his trip wrote a report[95] to Henry Morgenthau, Sr. He suggested that these areas needed to be put under the international guarantee of the League of Nations:

If there exists one reason for which I congratulate myself for having undertaken this voyage, it is that it allowed me to see for myself, how far the work which is carried on in the Northern part of Greece is in accord with the conclusions to which the hard logic of facts led me. Were you to visit Macedonia and Thrace, you will encounter in the Colonisation bureaus and the refugees surrounding them, but one and only care: how to render the country responsive to its new destination, that is, how to ensure the existence and perhaps even the prosperity of this million of new residents of Greece who have come to increase the number of the old inhabitants of the two provinces. From a distance an estimate can be made of the sums that need still to be spent in order to support the agrarians, to dry the swamps, to regulate the flow of waters; it is only on going near, on the spot, that one learns to appreciate the enthusiasm and the faith which predominate in this work and the confidence, I would even say the blind faith, in a long future of tranquillity and peace, which is taken for granted by those who collaborate in this great task.

But it is not only this arduous task which they have before them, nor yet the idea that they have no longer any claims to defend beyond their national boundaries that inspire these people with an immense desire for peace. United, all of a sudden, on a free soil, in a compact homogeneous mass, after having lived during many ages in dispersion, they know that they have much to forget and much to learn, that they must undergo a complete re-education which alone will permit them to adapt themselves to the new conditions of their existence.

Many moral and material forces have been abolished and there are others that must be developed. But every disturbance would be fatal to such a development.

These things are easily forgotten when one is found in the turmoil of the city of Athens and they would hardly furnish enough subject-matter for discusion during a drive from Kifissia to Phaleron; and yet the same questions become realities when one finds himself transported to the New Provinces of Greece. One feels then a bitter regret at the thought that one half of the nation ignores or only partially knows this immense effort that the other half of the nation is putting forth, especially as the beneficial reaction of this effort will, sooner or later, be felt by the first as well as the second.

It would be surprisingly interesting to make known to the foreign world the results obtained by the Greek Colonisation, results due entirely to the protection which the League of Nations has accorded to this country at one of the most critical stages of its history. Public opinion would judge the work accomplished as being worthy of the patronage of the League and the sympathies of this Assembly of the Nations would be deepened towards a people who, through the accomplishment of this task, undertook a veritable pacifist campaign. This Assembly would perhaps wish to make its voice heard, should certain neighbours of Greece, inspired by the same greedy appetites which caused the world to rise up against them in 1914, oblige the refugees to ask themselves whether they will not be constrained to defend with their firearms this corner of land, where at the risk of so many sacrifices they have been allowed to rest their tired heads. To such a Work and such workers there should be given *special guarantees of an international order.*

Freed from this anxiety, the Greek Government could follow up the accomplishment of its program. Greece would then be the first country to furnish a striking proof of the efficacy of the principles which have inspired the constitution of the League of Nations, principles admirably summed up by an inscription found on a monument in one of the ancient Hanseatic cities: "Concordia domi pax foris".

July 1926.

In his report Carathéodory hinted at the end of the Great Idea with the phrase: "Smyrna, Philadelphia, and the Seven Churches[96] will be Greek only in memory". A peaceful future, he added, was the way which Greece had to follow and he reached the conclusion:

I had not revisited Thrace and Macedonia since 1920 and I have undertaken this study without any preconceived idea. As I advanced in my work certain conclusions imposed themselves on my mind with the force and the reality which are infallibly characteristic of truth. This study appeared to me to be without any meaning, useless and mutilated, if I did not now that I am at the end, condense the convictions to which it led me, in two or three pages.

The Greeks were of old a very numerous people who were decimated by the wars they were obliged to engage in because of their geographical position. The result of this history was that they retained their hold, in tight communities, along the coasts of the European and Asiatic territory which they had occupied since very ancient times, and in the cities which they had built and their environs. But the question soon arose of the cohesiveness of the different Greek communities and whether they would break apart to a greater or lesser extent.

The Greek patriots deplored the fact that Hellenism found itself in the difficult situation of being unable to unite the different broken trunks to the mother-country – as they had the right to aspire to such a unity, being large enough to constitute of themselves a part of a Great Greece – without incorporating foreign elements.

However, the genius of a great man, seconded by the impetus of the Nation, came very near to the fulfilment of this great task. But at the decisive moment, the energy of the people failed and the Greeks missed their faith in their fortune and their destiny.

The catastrophe which followed this event – more important according to Lloyd George, than the taking of Constantinople in 1453 – has, as its consequence thrown the entire Greek race within the Greek boundaries and led to the unity of Hellenism through its extermination. It is permitted here to look for a consolation in the midst of so many misfortunes; we shall find it in the thought that, henceforth, the minorities in Greece will constitute so very slight a portion of the Greek population that they can no longer serve as a pretext to whomsoever would attempt to call into doubt the Greek character of the smallest bit of territory.

Greece has no longer any children, that she must free, living outside her territory. Smyrna, Philadelphia, and the Seven Churches will be Greek only in memory, just as the city of Syracuse and Agrigente were of old. The small islet formed by the Greeks of Constantinople, separated on all sides from our frontiers by foreign populations, cannot of course justify our claims to a city.

These truths are cruel; more cruel perhaps for those who utter them than for those who hear them. But they are only the expression of the necessity imposed by the events, and whose consequences must be submitted to; they show to the Greek people and to its Government the path in which they will, from now on, have to walk and from which they cannot deviate, for a long time to come, without facing mortal danger. This path is the path of peace, of work and of recovery.

3.15
In the Hope of Venizelos's Return

Two years after his report to Henry Morgenthau, Sr., Carathéodory wrote a paper[97] on the influence of the refugees on Greek life and on the work so far completed for their integration, which he wanted to have published in Greece. Carathéodory judged that the country had recovered from the disaster to a great extent and he even implied a revival of the Great Idea. His conclusions were as follows:

Eleutherios Venizelos ca. 1923. Courtesy of the Venizelos Archives, Historical Archives, Benakis Museum.

– The concentration of the bulk of the nation into a single country would, in a few decades, outweigh the human and material losses. This reinforced nation would undoubtedly expand in the future in the same direction and under more favourable circumstances.
– The thorough Hellenisation of provinces as a result of the Asia-Minor catastrophe led to a national homogeneity and thus reduced the other painful sides of the national tragedy to a certain degree by making these territories incontestable.
– The settlement of a million and a half refugees in agricultural and urban areas required a steady amount of work, study and experimentation. Thus, competent and capable personnel were trained for government posts or for the private economy, who dedicated themselves especially to agriculture and the building of dwellings and contributed in this way to national wealth and prosperity.

– The charitable work of the Greek Red Cross, the Patriotic Institution of Assistance and secondary organisations had a significant civilising effect.

The reason for Carathéodory's absolute euphoric mood might have been the political advancement of Venizelos, who came to power again in August 1928. Almost exultant, Carathéodory gave his views on this event in a letter from America to Penelope Delta. He made no secret of his joy for the "final crush of the royalists" and added: "Venizelos has now at his disposal all the means that would allow him to consolidate the democratic beliefs once and for all and to convert the ephemeral vote of a wavering people tired of the opponent's insults and abuses to an unshakeable symbol of self-confidence. I am convinced that once again we will experience that wonder of internal reorganisation to which he has accustomed us every time when the people granted him carte blanche, which he apparently needs in order to work effectively. At last, that miserable chapter of our country's history that has not stopped bothering us since 1915 is over. At last, we see light again! The brilliant and strange raising of the Greek nation which has taken place in the last years will continue, will even continue rapidly."[98]

CHAPTER 4

A Scholar of World Reputation

4.1
Appointment to Munich University

On 22 November 1923, the dean's office of the Philosophical Faculty, Section II of Ludwig-Maximilian University, Munich presented a list of candidates for the post of a full professorship for mathematics that had just become vacant. Carathéodory, "from the mathematical seminar of Berlin University", was proposed as the first candidate on the list, followed by Paul Koebe, professor at Jena University. The professorship had become vacant on the retirement of Ferdinand Lindemann. Lindemann, originally from Hannover, who had been the supervisor of Hilbert's thesis in Königsberg and full professor in Munich since 1893, had become famous when, with his 1882 work *Über die Zahl* π (On the Number π) published in the *Mathematische Annalen*, he succeeded in proving that π is transcendental, after Hermite had established the same, some years earlier, for the number e.

The relevant document of the philosophical faculty concerning Carathéodory's career reads: "Since 1900 Carathéodory has lived [...] almost continuously in Berlin, which he considers his actual homeland, despite the fact that he is not of German nationality. [...] In 1920, he accepted an appointment to his Greek homeland, at first in Athens, then in Smyrna, where after the victory of the Greeks over the Turks he was to build a new university. But since this plan could not be executed because of the Greek defeat, [...] he returned at once to Berlin, where, as a member of the Academy, he has the right to give lectures at the university. Carathéodory is exceptionally versatile as a mathematician. But a complete picture of him cannot be gained from his publications alone, since he developed many of his ideas, even significant ones, simply through discussion, and therefore in the works of other mathematicians is found the not infrequent remark that they owe the one or the other idea to Carathéodory."[1]

The idea of bringing Carathéodory to Munich was not accepted favourably by everyone. Even the departing Lindemann attempted to influence the decision on 13 December 1923: "My differing opinion with respect to the election of my successor", he noted in a text, "is due to the fact that the faculty has proposed a foreigner (the Greek Karatheodory). As if there were no competent mathematicians in Germany who could be named next to Karatheodory! Apart from objective considerations, it seems here that personal considerations must have played a role, for he (certainly a distinguished scholar) is now to find a post again through his friends of the Göttingen school. Anyway, now that the number civil servants will be reduced in Germany, I think that the reduction in the numbers of positions open to younger German scholars through such an appointment would not be a responsible decision, and of course it is true that in Heidelberg the earlier professorship of Perron has fallen victim to the reductions."[2] Lindemann's comments show that he was in no way immune to hostility towards foreigners; he even brought his opposition to Carathéodory to public attention. An article on this subject published anonymously in the right-wing *München-Augsburger Abendzeitung* of 6 January 1924 leaves no doubt about the identity of its author, especially if one compares the text with Lindemann's earlier comments: "It seems strange that a foreigner is appointed at a German university at this time of a reduction in the numbers of civil servants. [...] However, the appointed scholar has grown up in Germany, has studied in Germany and has held professorships for mathematics in Göttingen and then in Berlin, but he had given up his office there in order to be active in his homeland at the new University of Smyrna. After these plans fell through he returned to Berlin where, contrary to expectations, his position had been occupied; but with his appointment in Munich, no position for a younger scholar would become available, and the prospect of advancement for such a scholar again would be impeded. Naturally, this consideration does not say anything against Karatheodory's scientific performance; but wouldn't it be possible to find a competent German scholar for the position to be occupied in Munich?"[3]

Both in Lindemann's text and the article in the *Abendzeitung*, Carathéodory's origin was the exclusive reason for his rejection. The praise expressed in both cases for the "scientist" strengthened the impression of hostility against the foreigner, since the plea for a native scientist was neither specified anywhere (for instance by mentioning a name) nor justified by a comparison of scientific qualifications. Here, nationalism fused in a peculiar way with provincial bigotry and, even then, heralded the cataclysm that was to affect the University of Munich a decade later. After World War I, there were signs of increasing provincialism and traditionalist cultural attitudes in Munich.[4] A militant anti-Semitism, which was also apparent in the universities, had emerged after the end of the *Räterepublik* (Republic of Councils) in May 1919. Nationalist resentment had never been absent from the University of Munich since the early 1920s. This once great institution, famous for its humanities, gradually developed into a breeding ground for *völkisch* ideology among professors and students alike. Here Hitler's "movement" found fertile soil.[5]

The provincialism emerging from Lindemann's text was indeed not untypical of Munich. In a previous official letter dated 20 February 1923 to the Dean's office of the Philosophical Faculty, Section II, signed not only by Lindemann but also by Prings-heim, Voss and Perron, the authors argued against Herglotz's appointment as Voss's successor. In addition to a series of sound reasons, they presented their rejection as a resistance to E. Schmidt's recommendation of Herglotz: "In any case, the signing mathematicians refuse to be given instructions on their duties from abroad, especially from Berlin". A further recommendation of Herglotz was made by D. Hilbert.[6] Berlin, being a symbol of liberalism, socialism and cosmopolitanism within the Weimar Republic, attracted the hatred of conservatives. Obviously, Carathéodory had expressed to his friends the view that Munich was a cosmopolitan city and it was probably this opinion of his which Kármán had in mind when he commented ironically in 1927: "I am already looking forward to imagining you going from one bank to the other in Munich in order to cash [21 yen]. We will thus see whether Munich is really a cosmopolitan city." [7] (See 2.10).

Not everyone was able to cope with the atmosphere in Munich. The 1915 Nobel Prize winner for chemistry, Richard Willstätter, resigned from his Munich professorship in 1925.[8] He took this step because of the "compulsive" anti-Semitism of the "narrow-minded, intolerant" student majority who opposed the constitution and also because he noticed that similar points of view, combined with opportunism, started to play a veiled role in the faculties' consultations for appointments.[9]

Anti-Semitism had existed in Germany latently since pre-modern times, but was partly cloaked by the economic liberalisation of the 19th century and the related improved social standing of sections of the Jewish community. It emerged with power again by the end of the century and gained an explosive dimension, especially in Bavaria, after the liquidation of the *Räterepublik*. Anti-Semitic tendencies in the bourgeoisie were cultivated in nationalist circles such as the Thule Society, which was founded in Munich at the beginning of 1918. But there was also a plethora of other nationalist organisations and militant groups in which hostility against the Jews found fertile ground. The Jews were held responsible for the military defeat of Germany in World War I, the revolution of 1918 and its aftermath. Of all radical anti-Semitic groups, the National Socialist German Workers' Party (NSDAP = *Nationalsozialistische Deutsche Arbeiterpartei*) was the most vociferous in Bavarian politics. The Bavarian government, especially under Prime Minister Kahr, tolerated the unbridled anti-Semitic propaganda. The newspapers *Miesbacher Anzeiger* and *Völkischer Beobachter*, which were widely read in Upper Bavaria, were exceeded in their significance as anti-Semitic agitation forums, by the magazine *Der Stürmer* published from 1923. Despite an obvious moderation of Bavarian government policies towards the Jews after the failure of the Hitler-Putsch in November 1923, numerous attacks on Jews and the desecration of Jewish cemeteries and synagogues took place, mostly in Franken, up to 1933.[10]

With effect from 1 May 1924, Carathéodory was "appointed budgeted full professor for mathematics at the Philosophical Faculty (Section II) of Munich University". On 11 April 1924, the Bavarian State Ministry of Finances approved a final salary of a full professor plus a 10% supplementary salary to be granted to Carathéodory.[11] The total sum amounted to 6171 goldmarks. Carathéodory felt that the relatively low supplementary salary was a discriminating measure due to his foreign descent.[12] His appointment as co-director of the mathematical seminar of the University of Munich did not have any special remuneration connected to it.[13]

One of the reasons Carathéodory chose Munich was, according to his daughter Despina, his belief that the philhellenic tradition derived from Ludwig I was still alive there. As he wrote to Kalitsounakis on 24 February 1924, he would have stayed in Greece, if he were convinced that he could serve the country in any way. "But as things in our faculty are, and still will be for many years to come, this seems to be very, very difficult. Therefore, I intend to emigrate. Against that, I think that you and

Ioakimoglou should come here." Kalitsounakis attributed Carathéodory's decision to accept the Munich professorship to the collapse of his Smyrna project, to the Asia-Minor disaster and its political implications and to the situation in Athens, which was restrictive for any scientific activity.[14] This was the first time that Carathéodory asserted that he was "useful" to Greece when he was outside the country but urged others to leave their posts abroad and return. L.Young gives the reason for Cara-théodory's decision: "At Smyrna he lost his belongings in the flames of the Turkish invasion: he did not speak of such things, but once, when I admired his beautiful books, he called them his own extravagance, and continued, 'And now they are all I was able to save.' He had no cause to love the Turks: his mother was from Chios – this speaks volumes. He gladly returned to Germany to join Perron and Sommerfeld in Munich."[15]

That Carathéodory had lost his patience in Athens, and left behind a situation that he perceived as an insult to his person, is revealed by both the incident told by Michalis Anastasiadis, when a student had infuriated Carathéodory by using the German ex-pression "noch einmal" in one of his lectures (see 3.11), and his outburst during a faculty discussion. The 1924–1925 minutes of the Faculty of Natural Sciences and Mathematics of Athens University mention a debate between, among others, Carathéodory, Skoufos, Maltezos, Zeghelis, Hatzidakis, Remoundos and Sakella-riou, on 23 February 1924, concerning the post of an associate professorship for the history of natural sciences that had been founded two years earlier. Carathéodory left in disgust in the middle of the debate. The minority attempted to revoke the decision for the foundation with the ridiculous argument that the candidate, M. Stephanidis, who lectured on the subject as deputy professor ($\upsilon\varphi\eta\gamma\eta\tau\acute{\eta}\varsigma$), was not in command of the whole history of natural sciences. Of the arguments of the majority, the most remarkable are those from Zeghelis, who supported the claim that the history of nat-ural sciences was necessary to restore the connection between philosophical thought and scientific research, without which research would be degraded to practical ex-perience, and from Carathéodory, who said that the candidate did not need to know the whole subject and that the faculty ought to be considered fortunate to have such a candidate and ought not to lose him. Carathéodory said that when Hilbert had been appointed professor, he had not engaged himself in any other branch of mathematics than the invariants,[16] but now he was perhaps the most knowledgeable person in mathematics and mathematical physics.

The same minutes[17] confirm that the "text of the Rector's office no 16539/4-7-24" announced that Carathéodory's resignation from the chair of analysis had been accepted. Maltezos proposed the election of Carathéodory to honorary professor of the university.[18] That Carathéodory had developed a concrete plan to return to Germany is evident from his visit to Berlin in August 1923, which he noted in the chronological listing of his works. He must have also been in contact with Hamburg, for Blaschke mentioned in 1945 that "when the outstanding Greek mathematician had to flee from Smyrna during the Greek-Turkish War, my attempts to bring him back to Germany were successful."[19] Hondros also mentions an incident at the *Arkaden-Café* in Munich confirming that Carathéodory might have been considered for Hamburg. Hondros was recounting to a group of physicists and mathematicians that when, after the Asia-Minor disaster, Carathéodory had no academic position, he had received a letter from Hamburg assuring him that he would be welcome at their university. And someone commented: "how conceited they are in Hamburg"! On the other hand, reacting to Hondros's view that Carathéodory had not been a successful teacher of elementary mathematics to first-year students of chemistry at Athens, Sommerfeld commented that the University of Munich was big enough to allow itself the luxury of someone like Carathéodory.[20]

In Munich, Carathéodory bought a three-family house in Bogenhausen, Rauchstraße 8, from the engineer Matte. It was situated on a 900-m^2 site[21] and he was to remain there for the rest of his life.

With Carathéodory's appointment the three posts of full professorships of mathematics at the University were excellently filled. With the Carathéodory–Tietze–Perron trio, mathematics could flourish in Munich.[22] Perron was seven years younger than Carathéodory and came from Bavaria. Thanks to his versatile and excellent research in the theory of continued fractions and of differential and integral equations, his career developed rapidly. He had been appointed in 1914 as full professor in Heidelberg and in 1923 as the successor of his tutor Pringsheim in Munich. Heinrich Tietze, who was the same age as Perron and close friends with Paul Ehrenfest, Hans Hahn and Gustav Herglotz, came from Austria. He contributed to the development of topology in research and teaching and also to the field that was to be called bio-mathematics. Appointed to Munich a year after Carathéodory, he occupied the Chair of Mathematics as the successor to the "sensitive and subtle scholar" Aurel Voss.[23] His appointment on 1 April 1925 was unanimously approved by the faculty on the basis of a report read by Perron and written with Carathéodory's consent.[24] Carathéodory, Tietze and Perron comprised a harmonious trio in Munich. Characteristic of this co-operation is Tietze's dedication of his two-volume work of 1949 *Gelöste und ungelöste mathematische Probleme aus alter und neuer Zeit* (Solved and Unsolved Mathematical Problems of the Past and Present) to both of his colleagues.

During his time in Munich, Carathéodory supervised ten dissertations as the first doctoral supervisor, four of them together with Perron and three of them with Tietze; he also supervised seven dissertations as the second doctoral supervisor, four of them with Perron and two of them with Tietze.

4.2

Life in Munich

As soon as Carathéodory settled down in Munich, he started working with great zeal. He went daily to the university on foot through the English Garden or he took the tram, but never went by car. He used this time to prepare his lecture. The way he lectured was "only productive and never recapitulating" and corresponded to a view that he shared with other Hilbert students, namely that the lecture was made more stimulating for the students in this way. He opened the lecture by telling the students: "On my way through the English Garden I thought that one could derive the proof in another way and this is what we are going to do now."[25] Kritikos describes Carathéodory's method of lecturing as the product of a certain didactical concept that aimed at demonstrating the wavering of thought in the process of becoming. Even when he was lecturing on old, well-known topics, he was keen to present the ideas as if they were just conceived and to elaborate on them gradually, sometimes putting in great effort in front of the audience, until he could bring them to a final, satisfactory form. This way of lecturing was also a product of his own mentality; he disliked imitating others and preferred to use his own methods, which suited his profound particular mathematical nature. He perceived every invention, however small, with the highest satisfaction. Beside this, he refrained from any pretentiousness, affectation and empty rhetoric. Anyone with a mathematical background who listened to him and desired to enter deeply into the subject and at the same time to learn the method of thinking and researching not only profited from Caratheodory's lectures but was fascinated by him.[26] He expressed every thought clearly and slowly just once, in best German, recounts professor Romberg; Perron, conversely, spoke quickly and said everything three times, every time in different words. To listen to Perron and take notes after Carathéodory's lecture was very tiring, for everyone tried to pay attention to every sentence; but a lecture by Carathéodory after Perron demanded specific attention.[27]

According to Tietze, Carathéodory strove to lead his students to boundaries of knowledge that presented new tasks for research. He often chose the subject for the seminar for advanced students himself. He wished to promote the most talented of emerging mathematicians, although the inflexible nature of German institutions frustrated such initiatives. To overcome this inflexibility, he regularly invited a chosen circle of students to his house in the evening and discussed with them mathematical questions, while offering them tea and sandwiches prepared by his wife. He had acquaintances, friends and students throughout the mathematical world. Many came from abroad to listen to him and study further under his supervision. Several of his Munich students became successful university lecturers and researchers.[28] The undergraduate L. Young describes his impressions of Carathéodory as a teacher and a host in Munich:

Carathéodory and Perron had me as virtually their only 'graduate type' student, until I was joined, in my third semester there [...] by J. L. Walsh [...], later professor at Harvard. In Munich, in addition to my lectures and seminars, which were inspiring by any standard, I had above all the inestimable advantage of almost weekly visits to Carathéodory's home: I did not perhaps grasp fully more than half he told me, but half was a great deal for someone with little preparation. I asked few other than obvious questions: it was up to me to find out what he meant by reading in the library later. I was, of course, dreadfully conscious of imposing on his time, but in retrospect I believe he enjoyed my coming. Once I had been all the way up to his apartment and had decided to slip down again quietly: but Mrs Carathéodory had seen me coming, and I explained myself as best I could.

Carathéodory later told my father that I was afraid of my shadow: this shows, I think, the remarkable understanding Carathéodory had of young people. [...] Carathéodory was alluding to the horse, Bucephalus, that Alexander alone had understood: in my student days the stories from Plutarch had not been relegated to oblivion. Plutarch does not say *why* Alexander understood the supposedly untamable horse: it could only be because he had been through a similar experience himself, and had, doubtless, been understood by the great Aristotle, his teacher ... This is how, after being with Carathéodory, I was more than ready for the strange ways of Cambridge, and for the tough competition of my fellow students at Trinity.[29]

Analytische Mechanik.

Einleitung.

1. Die Analytische Mechanik ist die Wissenschaft in der man sich zum Ziele setzt die Bewegungen der wägbaren Materie, die man in der Natur beobachtet /und misst

/oder hervorruft .

mit Hilfe ~~durch~~ einer mathematischen Theorie zu beschreiben ~~und~~ vorauszusagen und zu berechnen.

Die Grundbegriffe und Prinzipien, die wir zum Aufbau dieser Disziplin benötigen, hängen — wie übrigens dies für jedes andere Kapitel der Physik der Fall ist — in erster Linie von der Deutung^{ab}, die wir unseren Beobachtungen und Messungen zukommen lassen.

Diese Deutung der beobachteten Erscheinungen wird nun teils durch die besondere Beschaffenheit unserer Sinnesorgane, teils durch die unveränderlichen Gesetze unseres Denkvermögens bestimmt, drittens aber auch durch die allgemeine philosophische Einstellung, die jeder von uns besitzt, und die nicht nur, wie die Erfahrung lehrt, sich im Laufe der Geschichte verändert hat, sondern auch - allerdings in geringerem Masse — von Individuum zu Individuum variiert.

Aus allen diesen Gründen ist ~~also~~ das Bild, das wir schliesslich erhalten ~~bis zu einem gewissen Grade~~ in einer nicht restlos aufzuklärenden Weise subjektiv beeinflusst. Sind wir schon allein dadurch, dass wir unsere Aufmerksamkeit jederzeit nur auf eine einzige Tatsache konzentrieren können, gezwungen, das allgemeine Geschehen der Welt ~~in eine Reihe zu zerglieder, die der Kritik der Philosophen niemals wird standhalten können~~ in Einzelerfahrungen zu zergliedern, und ~~dann~~ wiederum

Carathéodory's notes for his lecture on "Analytical Mechanics" in the winter semester of 1925–1926. Courtesy of the Bavarian Academy of Sciences, Archive.

Carathéodory used to distinguish between graduates of the *Gymnasien*, i.e. humanistic schools offering Ancient Greek and Latin, and those from either the *Realgymnasien*, i.e. schools offering Latin but no Ancient Greek, or the *Oberrealschulen*,
i.e. non-classical schools offering neither Ancient Greek nor Latin. In Carathéodory's
opinion, the *Gymnasien* graduates, having studied Latin and Ancient Greek, had acquired the logical structure of thought necessary for them to attend mathematical
university lectures.[30] The distinction he made was not in accordance with the official
programme of the German state, which, in its early years, had already granted the
Realgymnasien and the *Oberrealschulen* the right to send their students to the universities to study natural sciences and modern languages, although many government
positions and most university programmes remained closed to graduates of these
schools until 1900.

At the university, Carathéodory shared an office with Perron and Tietze. It was
furnished with three tables and was used for oral semester examinations, as well.
The assistants had their desks in the adjacent room of the teaching staff's reference
library.[31]

At home, he had reserved two rooms for his exclusive use. He locked himself in
there to work undisturbed or to write love poems in French in his spare hours. He
moved from one to the other when the air became too thick from his cigar smoke. He
smoked continuously, and his cigar became legendary, something like a trade mark:
his students considered it a hobby; Born had spoken of Carathéodory with "a cigar
in his hand"; in his *Gesammelte Mathematische Schriften* (Collected Mathematical
Writings) Carathéodory appears in one of his official portraits again with the cigar;
his daughter still believes that cigar smoking caused his death.

Euphrosyne, his wife, rarely left the house. She was busy receiving relatives,
family friends and Carathéodory's colleagues. She liked reading French sentimental
novels and playing the piano. Despina, their daughter, also entertained their guests
by playing the piano and often accompanied her father to the university, to various
cultural events and on journeys. Stephanos, who was reluctant to follow his father's
profession, despite his interest in number theory, engaged in philological studies but
led a very limited life because of his restricted mobility. Frau Köhler took care of the
household and the children, who called her Elilein. She was very close to Despina and
they went to the pictures together secretly to avoid Carathéodory's comments, whose
views about the sort of entertainment appropriate for his daughter were rather different from Despina's. With Despina, Carathéodory frequently attended performances
of Wagner's *Flying Dutchman*.[32] Composed in 1841, in the tradition of German romantic opera, it centred on the clash between the human world and the world of the
spirits and was combined with the theme of *Weltschmerz* (world-weariness).[33]

After the failed revolution of 1848 Wagner, a political emigrant, had become a
convinced anti-Semite as evidenced by his sarcastic text *Das Judentum in der Musik*
(Jewry in Music) of 1850.[34] In 1922 Hitler was received among the social circles of
prosperous Wagnerians. The *Sturmabteilung* (SA = storm unit, a paramilitary unit
of the NSDAP, also known as the "Brown Shirts") were required to attend Wagner's
operas.[35] Carathéodory admired Pringsheim's friendship with Richard Wagner and
mentioned that Pringsheim himself had produced the first piano scores of the *Ring*

for his personal use and that, with the same great pleasure, he went to Bayreuth in the summer of 1930 just as he had done fifty-four years earlier.[36] Pringsheim had even become entangled in a quarrel with a Wagner critic in Bayreuth that led to a duel with him.[37] The effect of Wagner's operas reflected the pure Wilhelminian spirit in Bayreuth before World War I. In 1913, the *Münchner Neuesten Nachrichten* published a text about Wagner that reflected contemporary public opinion: "it is as if, under his hands, everything becomes German." The person in charge of the Bayreuth Festival described Wagner in 1924 as a "leader in German nature".[38] Carathéodory was clearly comfortable with this sort of entertainment that belonged to the culture of the broadly educated bourgeois.

But also, in contrast, Carathéodory's taste in arts leant towards academicism. He seemed to show a preference for Greek painters who had studied in Munich, such as Georgios Iakovidis and Nikolaos Gyzis. Iakovidis, who came from Smyrna, had been a student of the Academy of Fine Arts in Munich and had won many prizes in Germany (Bremen 1890, Berlin 1891, Munich 1893). In 1900, he went to Athens to become director of the National Gallery. His works include nudes, scenes of everyday life and portraits of children and they exhibit a touch of impressionism. Gyzis had studied at the School of Arts in Athens and at the Academy of Fine Arts in Munich and had worked in Greece and Asia Minor until 1874, when he settled down in Munich. In 1882 he became an associate professor, and in 1888 a full professor, of the Academy of Fine Arts in Munich. He received many international honours for his portrait painting, history painting, and allegories and scenes of everyday life, and in 1898 he was honoured with the *Ritterkreuz* of Bavaria. The *Pinakothek* in Munich bought some of his works; some of his frescoes appear in the museums of Kaiserslautern and Nuremberg. In 1894, Gyzis presented Prince-Regent Luitpold with a painting of the festival service in the *Salvatorkirche* (Church of the Redeemer) on the occasion of its 400th jubilee, as if viewed through the west rose-window into the interior of the church. Iakovidis and other members of the parochial church council were present in the *Residenz* and attended the event. Carathéodory later belonged to this church council which, in 1927, was presided over by Greece's Consul General in Munich, Professor Dr. von Bassermann-Jordan. Carathéodory himself, wishing to play a role in the Greek Orthodox community of Munich, became a donor to the church, which had been given to the Greek Orthodox community by King Ludwig I in 1829. Also in 1927, Carathéodory was instrumental in blocking a decision by the Bavarian Ministry of Education and Cultural Affairs that would have allowed Old Catholics to use the church.[39]

4.3
Planning an Institute of Physics at Athens University with Millikan

On 26 July 1924, Carathéodory wrote a letter to the American physicist Robert Andrews Millikan, who was at that time in Geneva. Professor Millikan had measured the charge of the electron (elementary charge) in 1909, establishing the fact that both charge and energy are quantised; he had proved experimentally Einstein's law for the photoelectric effect in 1916,[40] confirming the corpuscular structure of electricity, and was awarded the Nobel Prize in physics in 1923 "for his work on the elementary charge of electricity and on the photoelectric effect". Millikan was president of the California Institute of Technology in Pasadena from 1921 to 1945 and was very active in promoting the teaching of physics throughout the USA and the rest of the world.[41]

Millikan must have heard of Carathéodory or, perhaps, had met him in Athens during his trip to Europe in the summer of 1924. On behalf of the trustees of the Charles M. Hall Education Fund (Cleveland, Ohio), Millikan was exploring the possibilities of establishing in Greece an American school, such as Robert College in Constantinople, and presumably would have asked Carathéodory and Dimitrios Hondros for their opinion. It is probable that Carathéodory and Millikan became acquainted through Henry Morgenthau, Sr., who was a member of the executive committee and the national board of trustees of the Near East Relief. The philanthropic work of this organisation in Athens was aimed at supporting and educating orphans from Asia Minor who had reached Greece after the disaster. While in Athens, Millikan had visited orphanages under the care of the Near East Relief and was asked by this organisation for his observations about their work, and for reports, photographs, materials and every help he could give them to enable the project to achieve its goal "to the lasting benefit of these countries [blasted and blighted by war] in their days of reconstruction". In a letter asking for Millikan's help, the Near East Relief representative Frank W. Ober referred to Mr. Homer Johnson, who highly appreciated the opinion of his friend Millikan on this enterprise.[42] Johnson happened to be a trustee of the Hall Fund and Carathéodory sent him a copy of his letter to Millikan.

In this letter, Carathéodory resumed his remarks on the project of the Hall foundation. He rejected the idea of a Robert College for Greece and pointed to the inherent dangers: A "true American College transplanted in the East" would educate people without identity, people who would be "neither Greeks nor Americans or English, but, if you take the best of them, a type of well educated Levantines." He went on to say that such an institution might only be justified in an international city such as Constantinople or Smyrna before its fall to the Turks. With respect to Greece, such a school would not be a sort of universal remedy for the educational crisis in Greece, as the Greek supporters of the project were arguing. The educational crisis was in Carathéodory's eyes a symptom of a more general crisis, which had been transforming the Greek society for the past thirty years. What Greece needed, now that Greeks were united in their own homeland, was "Greek" education, that fostered a national spirit and a national identity, not an imported educational programme, no matter how liberal.

On the other hand, Carathéodory was in favour of Millikan's suggestion of creating a modern scientific institution in association with the University of Athens, especially if Millikan's idea of a mix of Greek and foreign staff could be realised. So, he desired American intervention but only under Greek control, that is under the control of the Greek state university. "Isolation is the greatest draw-back for the development of science in my country", he wrote.

To Carathéodory, the question of which particular field of science should be promoted at the institute was not important. Only the "impulsion" that such an institute would give to the university as a whole was significant, provided that its programme was approved by the majority of Greek educated people. There were a number of choices: for an institute of pure and applied physics, Carathéodory recommended that the details of its organisation ought to be elaborated and that the project be presented in an appealing way to the public and possibly be partly linked to Athens Polytechnic; alternatives could either be an institute of pure and applied chemistry or, even better, an institute of geology, both of practical use for Greece and of interest to pure science or, finally, an institute of Oriental and Slavic languages. With respect to the latter, Carathéodory informed Millikan that the library of the Ionian University, which included a collection on related topics, was stored in Athens.

The success of the enterprise would presuppose harmonious co-operation between Americans and Greeks. The previous experience with the *Institut Pasteur* in Athens, which had been created with Greek private money but whose only effect had been to provide large salaries for French scientists, ought not to be repeated. Carathéodory suggested that 10 percent of the revenues administered by the board of trustees should be spent on publishing books or even translating good textbooks into Greek, which, in the case of a physics institute, had to also include the main mathematical subjects. A year's leave ought to be granted every six or seven years mainly to the Greek staff of the institute for research and travel, in order to give them the opportunity to keep up with the advancement of science in foreign countries. Finally, Carathéodory asked Millikan to give a short account of his letter to Venizelos.[43]

Millikan reported to the Hall Fund that Greece was the most influential of the Balkan states and that the Greek people were capable of profiting by anything done for them. Millikan assured the Fund that the enterprise would be justified by the fact that through it the Greeks would again become capable of making great contributions to their neighbours and to the world. An institute of pure and applied physics, in affiliation with the University of Athens, that would concentrate on scientific development, then lacking in Greece, could be the most that could possibly be done in introducing American standards of religious and political freedom and by American example promote character building in the country. The trustees Homer Johnson and Arthur Davis appreciated the importance of Millikan's survey of the situation fully and promised that Greece would attract their attention "quite as much as the beneficiary of the payment of some part of the debt which the west owes to her for her contribution to science, philosophy and art in past ages."[44]

On 27 September 1924 Millikan sent Carathéodory his views. He sided with him as regards the establishment of an institute of science in Athens. "It must be said, however, that some of your compatriots who are experienced men of affairs, as, for example, Mr. Politis, Greek ambassador to France, would say that the views which you and I and our scientific friends at Athens hold are those of specialists in science, and represent the bias which all scientists might be expected to have in their estimates of relative values."[45] Politis seems to have favoured intermediate educational facilities for Greece, such as a Robert College. Venizelos on the other hand, with whom Millikan had a two-hour discussion, had shown great interest in both types of plans, a Robert College and an institute. Millikan ruled out the possibility of establishing a Robert College on the grounds of its high cost. The half a million dollars that could be raised in Greece – according to the estimate of Politis and Venizelos – and another million at most from the Hall trustees could not make more than a third or a quarter of the sum needed for a Robert College. So Millikan, *de facto*, ended up by supporting Carathéodory's proposal in favour of a scientific institute. Professor Hondros and Rector Zeghelis of Athens University had promised him they would study the matter further, consult Carathéodory and submit their own suggestions. "The best procedure would now be for some of those among the Greeks who are most ably and intensively interested in Greek education to formulate specific plans", Millikan commented. Whether the trustees would be well disposed to the proposals would depend in Millikan's opinion on whether such a plan would be really desired by the Greek people and whether it would meet with their enthusiastic support and co-operation. Millikan also believed that the trustees would attempt to check up on his judgement by independent opinions and would only make definite proposals a year later.[46]

On 6 October 1924, Millikan wrote to A. Andreadis, professor of international law at the University of Athens and a member of the Greek Legation in Paris, to inform him of the developments. He made a specific proposal: "If the Greek State or private Greek philanthropists could find a way to build a building for the University of Athens, which would be called Institute of Pure and Applied Physics, the Hall trustees might furnish the income from $1 000 000 toward the maintenance and upkeep of that Institute, this income to be expended in accordance with the terms of the testament of Mr. Hall under the authority of an Anglo-Saxon board." As the next step to be taken, Millikan requested the sharpest possible formulation of the views of Andreadis, Hondros, Zeghelis and Carathéodory "as to the way in which such assistance as the Hall trustees may be able to give can be most effectively furnished".[47]

On 11 November 1924, Andreadis replied to Millikan from Paris favouring the foundation of a Robert College. He could not imagine, he wrote, how more than a million dollars would not be enough to start the institution. As for the building, he thought that the Greek government could provide them with one, or at least a piece of land to build on. Half a million dollars could cover building or extended repair costs. The income of the remaining half million would then be "more than sufficient to keep up a certain number of poor children; the wealthy ones, and they will flock from every corner of hellenism, will pay their expenses. Fees can be kept at a low level."[48]

On 25 December 1924, Hondros wrote to Millikan that, after having conferred with the professors of physics and chemistry, Rector Zeghelis and himself found "an area of agreement" between his views and theirs in which "a collaboration might be established with good hope of success". He sent him the letter through Carathéodory. Hondros proposed the use of the financial contribution of the Hall Fund for enlarging the existing university building housing the laboratories of physics and chemistry. A definite proposal on Millikan's part would enable Hondros to obtain funds from the university, and maybe also from the state, and very probably an additional half a million goldmarks, or $ 125 000, for equipment and apparatus, which could be set against German reparations. "We picture the new enterprise as a general institute of the natural sciences belonging to the faculty of science of the university, an institute in which, in addition to scientific research in the fields of physics and chemistry, mineralogy and perhaps physiology, studies in telephony and telegraphy and allied fields would be pursued. The university would have to provide for the appointment of the Greek professors; the Hall Foundation could be used for paying the salaries of the foreign professors and of secondary personnel and for contributing to the renewal of equipment and general laboratory maintenance". Hondros asked for Millikan's opinion on his views and promised to send him another proposal by professor Maltezos that would "open up another point of view".[49]

In the minutes of the Faculty of Natural Sciences and Mathematics covering the period 1924–1925 there is no debate concerning the foundation of an institute for theoretical physics at the University of Athens. However, it is mentioned that the faculty approved the requests for new buildings and auditoriums made by professors at the session on 23 February 1924. The directors of the laboratories for physics and chemistry needed a space of 9940 m^2 that could be acquired by rearranging their laboratories and adding a third storey to the existing chemistry building ($X\eta\mu\varepsilon\acute{\iota}o$), as well as by building a new physics laboratory on the site of the university hall for physical education. The directors of the museums of natural sciences and of their attached laboratories requested an area of 2400 m^2 plus a new lecture hall big enough to house an audience of five hundred. Finally, the professors of the mathematics department, N. Hatzidakis, G. Remoundos, N. Sakellariou and C. Carathéodory asked for three lecture halls, a seminar hall and a special room to be used for faculty meetings and examinations.[50]

Scholars with experience in the Orient were ready to offer their services to the establishment of an educational institution in Greece. L. A. Kenoyer, for example, with a PhD in botany from the University of Chicago and the Iowa State College, who was previously engaged as a professor in the educational work of the American Presbyterian Mission in North India and was Officiating Principal of the Allahabad Agricultural Institute, was strongly attracted by the idea of helping the development of an institution such as an American College in Greece.[51] L. P. Chambers, having lived in Turkey for many years and having taught at the Constantinople Woman's College and at Robert College, was greatly interested in the scheme and asked Millikan to put him in touch with the trustees of the Hall Foundation. As a reference for himself he mentioned a full statement of his academic career and experience that was on file at the appointment office of Harvard University.[52] He wrote to Millikan: "I am greatly interested in the Armenians and Greeks and believe that the traditional 'forwardness' of the one and 'fickleness' of the other are superficial vices which blind the casual observer to certain sterling qualities." Chambers said that although he preferred to teach philosophy, he was prepared to teach history, English, politics and economics, Biblical literature, history of religions or similar subjects were he given the opportunity to do so in some Near East educational institution. The "lure of the Near East" was stronger to him than "the lure of the subject".[53]

Parallel to gaining information on the possibility of educational investment by the Hall Fund in Greece, Millikan was pursuing similar explorations about the possibility of setting up scientific institutions in Turkey and also in Eastern Europe. For this purpose, Professor H. von Euler of the University of Stockholm provided him with the names of two persons interested in the scientific development of Turkey who could give the best advice and information concerning his plans: Dr. Adnan Bey, representative of the Ankara Foreign Office in Constantinople, and Dr. Emin Bey, Governor of Constantinople. Von Euler informed Millikan that the Turkish people distrusted the political purpose of Robert College and that only a new scientific institution free of political influence could persuade them and could create the atmosphere of a closer connection with the Turks than had been the case with Robert College or the German School of Constantinople which had to close following the Treaty of Sèvres. The main purpose was to "destroy the Turks' suspicions against the motives of foreign help". Von Euler also put Millikan in touch with the head of the institute of plant physiology at Masaryk University in Brno, professor Vladimir Úlehla, to provide him with information concerning scientific organisations in Czechoslovakia. For Hungary, he added the address of the biology professor Alexander Gorka, General Secretary of the Hungarian Society of Sciences.[54] At the end of 1924, Millikan wrote to von Euler that he had contacted most of the sources of information but was yet undetermined whether anything useful would come from this effort.[55] At the same time he stated to L. A. Kenoyer that it was not yet certain whether there would be any opportunities in Greece.[56]

In 1925 an American-style, elite private boys' high-school, the Athens College, was founded by the Hellenic-American Educational Foundation, which had been created the same year in Greece. Athens College was largely funded by Stephanos Deltas and Emmanuel Benakis. Most of its founding faculty were Robert College graduates and many of its graduates chose North America over Europe for their higher education.[57]

4·4

Reichenbach and the Berlin Circle

Shortly after his appointment in Munich, Carathéodory sent "a small note [...], which appeared some months ago"[58] to the philosopher and assistant professor at the Technical University of Stuttgart, Hans Reichenbach. This was obviously a referrence to his paper *Zur Axiomatik der speziellen Relativitätstheorie* (On the Axiomatics of the Theory of Special Relativity). He remarked that the subject matter of his paper lay "entirely within the compass of your [Reichenbach's] nice book" and added: "I regret not to have known of your studies earlier, of which I learnt through your book only yesterday. There are some differences in the two conceptions that might interest you." He apparently referred to Reichenbach's *Axiomatik der relativistischen Raum–Zeit-Lehre* (The Axiomatics of Relativistic Space–Time Theory).[59]

In Carathéodory's axiomatics, events satisfied three groups of axioms: those of time order, those of light propagation and those containing assumptions about the space surrounding us. His three axioms of time order corresponded to Reichenbach's axioms of power, order and coherence, while Carathéodory's axioms of light prop-

SITZUNGSBERICHTE

1924.
V.

DER PREUSSISCHEN

AKADEMIE DER WISSENSCHAFTEN.

Sitzung der physikalisch-mathematischen Klasse vom 14. Februar.

Zur Axiomatik der speziellen Relativitätstheorie.

Von C. CARATHÉODORY.

Separate print of Carathéodory's paper On the Axiomatics of the Theory of Special Relativity. *Courtesy of the Bavarian Academy of Sciences, Archive.*

agation corresponded to Reichenbach's axioms of time comparison. Reichenbach characterised his own two groups of axioms as "topological axioms of time order". Finally, Carathéodory's third group of axioms corresponded to Reichenbach's metric light axioms.

Carathéodory's axiomatics of the special theory of relativity was reviewed by Einstein in an undated one-page summary submitted to the Prussian Academy of Sciences. Einstein remarked that "an axiomatic presentation of space–time theory of the theory of special relativity is given, which is based only on premises about the behaviour of light but does not make direct use of the idea of the rigid body". The name "Carathéodory" appears on the top margin of this document, presumably in Planck's handwriting. In his paper, Carathéodory indeed showed "that this axiomatics can be simplified in an unexpected way, if one bases the whole theory only on time observations."[60] In his book, Reichenbach made no use of the rigid body either: "Space geometry", he wrote, "is not arbitrary in our order of definitions but an empirical fact, since the equality of distances is defined independently of geometry. At the same time, this is an example where Euclidean geometry is not a requirement in every comparison of quantities; we have defined congruency in a different way and we did not even use the rigid body, but only the motion of light" (p. 41). Reichenbach had started research on his book in autumn 1920 and had almost completed it in March 1923. He reported on his results during the physics conferences in Jena in 1921 and in Leipzig in 1922. Carathéodory, however, did not seem to have been informed about it.

Hans Reichenbach got in touch with Carathéodory again in Munich on 20 March 1933 before emigrating to Turkey. Reichenbach had obtained his doctorate in Erlangen in 1915 with a dissertation on *Der Begriff der Wahrscheinlichkeit für die mathematische Darstellung der Wirklichkeit* (The Concept of Probability for the

Mathematical Representation of Reality), an epistemological work dealing with the application of mathematical probability theory to physical reality. In 1920, he habilitated in physics at the Technical University of Stuttgart but in the following years he distanced himself increasingly from "pure" physics in order to turn to philosophical issues of the natural sciences. He was already well-known in scientific and philosophical circles at that time, but his work in both the spheres of natural sciences and philosophy resulted in philosophers regarding him with suspicion and his ambition to obtain a chair of philosophy was thwarted. At the same time, scientists were also reluctant to accept him as one of them, because of his socialist political attitude and also because of his understanding of philosophy. Finally, after a successful trial lecture on *Zeit und Kausalität* (Time and Causality), he obtained the authorisation to lecture as an associate professor at the physics department of Berlin University in August 1926.

On 13 July 1925 he wrote to Carathéodory: "Thank you very much for wishing to submit my manuscript to the Bavarian Academy, as I have just heard from Herr Sommerfeld. After all, it did not seem to you appropriate for the *Annalen*. I had actually thought of the *Berliner Akademieberichte* [Berlin Academy Reports] but, of course, Munich is also right to me. I would be very interested, if you could send me your view on the matter some time."[61]

Reichenbach obviously referred to his paper *Die Kausalstruktur der Welt und der Unterschied von Vergangenheit und Zukunft* (The Causal Structure of the World and the Difference between Past and Future), which Carathéodory submitted to the Bavarian Academy at the session of 7 November 1925. Reichenbach showed in this work "that a quantitative description of natural occurrences is also possible without the hypothesis of strict causality, since it just performs everything which physics could ever perform, and which, in addition, is appropriate for solving the question of the difference between past and future to which the strict hypothesis of causality has no answer."[62]

Reichenbach substituted the "implication form" of the causality hypothesis (if *A*, then *B*) and the "determination form" of the causality hypothesis (namely that the course of the world as a whole is fixed and remains unchanged, so that with a cross-section of the four-dimensional world, past and future are absolutely determined) by the assumption that a probabilistic connection exists between cause and effect: "if *A*, then it determines a *B* according to the laws of probability". The difference between past and future corresponding to the human perception emerges in the theory of probabilistic connection. With the aid of this theory Reichenbach showed in which sense the past can be called "objectively determined" and the future "objectively undetermined". He founded the notion of world coherence on topological properties of a net structure of observed individual events which are connected to each other through chains of conclusions.

Reichenbach had submitted *Die Kausalstruktur der Welt und der Unterschied von Vergangenheit und Zukunft* to the philosophical faculty of Berlin University to qualify for a habilitation, but he took it back on 5 June 1925 saying that he would like to add some further remarks. He presented instead the text *Der gegenwärtige Stand der Relativitätsdiskussion* (The Present State of the Discussion on Relativity).[63]

Reichenbach had close contact with the Vienna Circle of Logical Positivism and was co-editor of its journal *Erkenntnis* (Knowledge). The aim of the Vienna Circle was to determine the contents of philosophy, in a scientific way with the help of modern logic, on the basis of everyday experience and the experience of individual sciences. The leader of the Vienna Circle between 1924 and 1936 was Moritz Schlick, who in 1922 succeeded to the chair of philosophy of the inductive sciences at the University of Vienna, previously held by both Mach and Boltzmann. By 1924 an evening discussion group had been formed with the active participation of Schlick, Hans Hahn, Otto Neurath, Victor Kraft, Kurt Reidemeister and Felix Kaufmann. Rudolf Carnap joined them in 1926. One of the early activities was the study and critical discussion of the *Tractatus Logico-Philosophicus* (1921, *Annalen der Naturphilosophie*) of Ludwig Wittgenstein, an Austrian-British thinker who worked on analytical and linguistic philosophy. At that time the views of Carnap and Wittgenstein were congruent, although they had been formulated and elaborated quite differently. There were parallel developments in Berlin, by a group of which Hans Reichenbach, Richard von Mises, Kurt Grelling and Walter Dubislav were the leading spirits.

The Berlin group developed an institutional framework for their deliberations, namely the Berlin Society for Empirical Philosophy (*Gesellschaft für empirische Philosophie*). Reichenbach put particular effort into the building up of the Berlin Circle, which was prepared to follow his own specific philosophical views, which deviated from those of the Vienna Circle in Reichenbach's conception of probability as a limit of relative frequency. Driven by the need for an interpretation of the concept of probability that was thoroughly empirical, Reichenbach argued that a theorem can only have a cognitive meaning if observations enable one to attribute a probability to it.[64] Carnap acknowledged the importance of this concept, especially in modern physical theories, but attempted to define a concept of the "degree of confirmation" which was purely logical. Reichenbach, on the other hand, in his *Experience and Prediction* (1938), asserted that if hypotheses, generalisations and theories can be made more or less probable by available evidence, then they are factually meaningful. Richard von Mises, following Schlick, deviated from Reichenbach's probabilistic definition of factually meaningful theories. He advocated the view that theories are factually meaningful only if the concepts, in terms of which they are formulated, can be related, through chains of definitions, to concepts that are definable in an obvious way, namely by pointing to aspects of direct experience. Carathéodory's reasoning in his paper *Über die Bestimmung der Energie und der absoluten Temperatur mit Hilfe von reversiblen Prozessen* (On the Definition of Energy and Absolute Temperature with the Aid of Reversible Processes), which he submitted to the Bavarian Academy of Sciences at the time when he was corresponding with Reichenbach, was in tune with the logical arguments of Schlick and von Mises. In his *Positivism*,[65] von Mises argued that theorems of logic or pure mathematics are said to be "correct" if they are in agreement with a system of accepted definitions and rules, but the application of mathematical methods can never guarantee the correctness of a non-mathematical proposition regarding, for instance, connections between observable phenomena. If this is the case, the scientist will be subject to control by future experiences. Therefore, attempts to arrive at results of absolute certainty in areas other than mathematics, a problem to which Kant's epistemology is devoted, are illusory.

Reichenbach's position was extremely anti-Kantian and denied Kant's *a priori* concepts of space and time, because measurements had shown that real space was curved and Euclidean space could only be regarded as a special case. Reichenbach did further work on the philosophical foundations of the theory of relativity and quanta.

4.5
Suggestions to Hilbert on Quantum Mechanics

During the first period of Hilbert's engagement with physics (1912–1914), he established a new basis for the results of classical physics in a strict way through methods of integral equations. Hilbert returned to the study of physics a decade later when he looked for a theory of quantum mechanics based on the fewest possible theorems free of contradiction, with the requirement that the theory should be derived purely mathematically. Carathéodory contributed to Hilbert's work in this later phase, as he had done in the earlier research. Hilbert announced his first lecture on quantum mechanics for the summer semester of 1926.[66] In the winter semester 1926–1927 he held a lecture on the recent developments of quantum mechanics in the preparation of which he was considerably assisted by Lothar Nordheim, Born's assistant.

That lecture was the basis for Hilbert's final work in physics *Über die Grundlagen der Quantenmechanik* (On the Foundations of Quantum Mechanics),[67] which was submitted to the *Mathematische Annalen* in April 1927. The paper explained the logical structure of the theory and presented its analytical formalism. In a letter to Hilbert of 14 April 1926, Carathéodory told him that he "had read almost all recent works on quantum mechanics" and that great progress had been reported in this area. He recommended Hilbert to lecture "especially on Dirac's second work", by which he meant *The Elimination of the Nodes in Quantum Mechanics*.[68] "But the direction of the Schrödinger works is much more interesting. By the way, Schrödinger is supposed to have recently proved that his own method and that of Heisenberg will necessarily lead to the same results", Carathéodory commented. In his paper *Über die Grundlagen der Quantenmechanik* Hilbert wrote: "The relationship between Schrödinger's theory and Heisenberg's matrix mechanics is, as is well known, delivered by a calculus of operators". Most of the results obtained by Dirac in *The Elimination of the Nodes in Quantum Mechanics* had been found earlier by German theorists using matrix mechanics, but Dirac was able to derive them from his own quantum-mechanical system and improve on some of them. Dirac wanted his algebra for quantum variables to be a general and purely mathematical theory that could then be applied to problems of physics. Although it soon turned out that his theory was equivalent to matrix mechanics, in 1926 it was developed as an original alternative to both wave mechanics as developed by Schrödinger[69] in Zurich (drawing on ideas previously suggested by Louis de Broglie in Paris) and matrix mechanics as developed by Heisenberg, Born and Jordan.[70] As Carathéodory had written to Hilbert, Schrödinger was indeed the first to prove the formal equivalence between the two theories.[71] Schrödinger became Planck's successor, when the latter retired in 1927.

Through Pauli's assumption of an additional degree of freedom for the electrons, "a classical, non-describable ambiguity", and through his exclusion principle, namely the demand that every quantum state is allowed to be occupied by only one electron, the Bohr–Sommerfeld atomic theory was put to rest in 1925.

"The young Pauli is very stimulating; I will never have such a good assistant again", Born had written to Einstein on 29 November 1921[72] and a month later Einstein found that Pauli could be proud of his *Encyklopädie* article,[73] which Born considered to be a fundamental work exceeding all other presentations of relativity theory in depth and thoroughness, but also such famous works as that by Sir Arthur Eddington.[74] Already in 1926, after Heisenberg had formulated quantum mechanics in terms of matrices, but before Schrödinger developed the wave picture, Pauli was successful in obtaining a derivation of the hydrogen spectrum with the help of matrix mechanics. The following year, he showed how electron spin could be accounted for through the extension of Schrödinger's equation. In his important quantum-mechanical theory of spin,[75] he developed the idea of representing spin by three spin variables, the so-called 2×2 Pauli matrices, which yielded a two-component wave function for the electron. The extended equation itself was proved to be a limiting case of the relativistic wave equation set up by Dirac. That was Pauli's greatest accomplishment in the period subsequent to the Heisenberg–Schrodinger revolution.

Although in the Schrödinger and the early Heisenberg–Born–Jordan treatment of quantum mechanics the spin effect was introduced as an *ad hoc* assumption, it arose in a natural way in Dirac's relativistic quantum mechanics.[76] Heisenberg[77] and Dirac[78] showed that the phenomenological vector-coupling model provides a very simple spin Hamiltonian which gives the correct energy splitting between the singlet and the triplet states and denotes that the lower energy state is achieved by parallel spins when the exchange Coulomb integral is positive. Pauli, professor of theoretical physics at the ETH Zurich since 1928, visited Carathéodory in Munich on 23 July 1929, the year in which he and Heisenberg co-operated in one of the earliest subjects of quantum field theory, the interaction of radiation and matter.[79] With their investigations on the quantisation of the field equations, wave mechanics passed from the theory of a single particle to that of the interaction of an indefinite number of particles, a step for which Dirac's quantum theory of radiation had paved the way.

Carathéodory, who had "the impression that the mathematicians will also get a number of things out of it [quantum mechanics]",[80] had obviously foreseen these developments and shared the opinion of all those who worked on the mathematical aspects of quantum physics. The mathematics of Dirac's equation was indeed explored by mathematicians like Weyl, van der Waerden, von Neumann, Fock and others. One of the most important results of this work was spinor analysis that was built up on a generalisation of the properties of Dirac matrices. Van der Waerden's book *Sources of Quantum Mechanics*[81] contained all significant treatises on the emergence of quantum mechanics and a detailed introduction to their relationships. John von Neumann's book on the *Mathematische Grundlagen der Quantenmechanik* (Mathematical Foundations of Quantum Mechanics)[82] contained the strict foundation of the mathematical concepts and methods used by Heisenberg, Jordan and Born, especially of the interpretation of matrices as operators in Hilbert space introduced by Born.

4.6

Calculus of Variations

Among the many areas of contribution to mathematics made by Carathéodory, his work on the calculus of variations is of major importance and is the work for which he is best known. He worked on this for many years and, building on the work he had done earlier, his contributions include a comprehensive theory of discontinuous solutions, in which previously there had been only limited findings. Except for discontinuous solutions, Carathéodory introduced conformal mappings and multiple integrals in the calculus of variations. This allowed him to investigate problems in mechanics and in multi-dimensional space which exhibited parallels with problems in differential geometry and physical applications as had appeared in Euler's work.

4.6.1

General Theory

Perron notes that "the beginnings of a new treatment of positive definite variational problems, which are connected with Johann Bernoulli's method for the solution of the brachistochrone problem," appeared in Carathéodory's doctoral thesis, "whereas far less demands for differentiability were set on the function under the integral than was usual earlier." This method was extensively presented in the paper *Sur une Méthode directe du Calcul des Variations* (On a Direct Method of the Calculus of Variations)[83] and consisted in the following:

In the case of the integral

$$\int_{t_1}^{t_2} F(x, y, x', y') \, dt \qquad \left(x' = \frac{dx}{dt}, \quad y' = \frac{dy}{dt}, \quad t_1 < t_2 \right)$$

the function F should be of the first order, positively homogeneous in x, y and in the case of constant x, y the curve $F(x, y, \xi, \eta) = 1$ should be a strictly convex oval, including the zero point, on the (ξ, η) plane. Thus, F, as a function of both the last variables, is positive definite. Further, the partial derivatives $F_{x'}$, $F_{y'}$ should exist and be continuous, whereas the existence of other derivatives is not demanded, so that the Euler differential equations of the problem do not exist at all. Now a family of curves $\varphi(x, y) = t$ will be called a geodesic parallel, if both equations

$$\frac{\partial F}{\partial x'} = \frac{\partial \varphi}{\partial x}, \qquad \frac{\partial F}{\partial y'} = \frac{\partial \varphi}{\partial y},$$

whose left sides are homogeneous of zero order as regards x', y', that is, depend only on the angle θ defined through the equations

$$x' = \sqrt{x'^2 + y'^2} \cos\theta, \qquad y' = \sqrt{x'^2 + y'^2} \sin\theta,$$

produce the same value $\theta = \theta(x, y)$. Then, along a curve C, which infiltrates the geodesic parallel family in such a way as to form the angle $\theta(x, y)$ with the positive X-axis in every point x, y

$$F(x, y, x', y') = x' \frac{\partial F}{\partial x'} + y' \frac{\partial F}{\partial y'} = x' \frac{\partial \varphi}{\partial x} + y' \frac{\partial \varphi}{\partial y} = \frac{d\varphi}{dt} = 1$$

is true, whereas it can be proved that along every other curve

$$F(x, y, x', y') > 1 \quad \text{is always true.}$$

Thus, the curve C is certainly a minimal for the integral.[84]

Carathéodory himself remarked in his paper that his method presented considerable analogies with the classical method of Weierstrass, but had the advantage that it was more general. A field in Weierstrass' theory can always be derived from a suitable family of surfaces by the method of geodesic descent, whereas the opposite is not true, as can be seen in the case of the existence of discontinuous solutions. Then, the method of geodesic descent provides a field of solutions although Weierstrass' theory does not apply. Indeed, Euler's equation may not even exist.

Although Carathéodory used the term "direct method", in fact, he applied the method of variations, i.e. the study of necessary and sufficient conditions to be satisfied by the function which realises the extremum of the functional. Fleming notes, that "after Hilbert, 'direct methods' were widely used in the calculus of variations, although the term was not used consistently. What Carathéodory meant by 'direct method' in his 1908 *Rendiconti Circolo Mat. Palermo* paper [...] is not the same as the 'direct methods' of Tonelli and others afterward. In the Tonelli approach (which also is derived from Hilbert), the first step is to prove that there exists a minimum in the calculus of variations problem, using semicontinuity and compactness properties for the functional to be minimised. The minimum may not at first be known to have enough smoothness for the classical necessary conditions of Euler, Weierstrass, etc. to hold. A second step (the 'regularity' problem) is to prove that the minimiser is indeed smooth, under assumptions which guarantee that smoothness in fact holds. Such direct methods have been widely used for multiple integral problems in the calculus of variations, including total energy minimisation and the solution of the Plateau problem independently by Radó and Douglas."[85]

Carathéodory sent this "small note on the calculus of variations" to Königsberger[86] briefly commenting on an example from Maxwell, since Maxwell "gives no proof and expresses the result in a much too complicated form." It reads: "A street has to be constructed on a hemispheric hill in such a way that a car with velocity proportional to the third power of the cosine of the inclination needs the least time in order to reach the peak from a point at the hill's base."[87] He had included this example in his paper and repeated it in his 1935 book *Variationsrechnung und Partielle Differentialgleichungen Erster Ordnung* (Calculus of Variations and Partial Differential Equations of the First Order)[88] as a case of an analytical variational problem, in which the strong extremals that contain a singular element of length present no corners and are, despite that, not analytic.

Carathéodory's method in the calculus of variations was later to be concisely presented in his contribution, *Variationsrechnung* (Calculus of Variations), to Frank–Mises' textbook *Die Differential- und Integralgleichungen der Mechanik und Physik* (The Differential and Integral Equations of Mechanics and Physics).[89] There Carathéodory attempted to construct the theoretical building blocks of mechanics and physics on the foundation of concepts of the calculus of variations, which he related to the theory of differential equations with emphasis on the applications of analytical mechanics.

Carathéodory developed his theory as follows. The task set in the calculus of variations is to examine whether there is any curve represented by arbitrarily often differentiable n functions $x_i = x_i(t)$ $(i = 1, 2, \ldots, n)$ which connects two points P_1 and P_2 of an $(n+1)$-dimensional space R_{n+1} for which the value of the line integral

$$I = \int_{t_1}^{t_2} L(x_i, \dot{x}_i, t)\, dt$$

taken along the curve over the positive analytic function $L(x_i, \dot{x}_i, t)$ of $(2n + 1)$ variables $x_i, \dot{x}_i, t$ is smaller than that for any other curve which connects the same two points P_1 and P_2 and which lies within a certain neighbourhood of the initial curve and, if necessary, to determine the curves for which this minimal value is reached and which are called the extremals.

Carathéodory starts by considering a family of surfaces

$$S(x_1, \ldots, x_n, t) = \lambda = \text{const.}$$

He generates from it paths cutting these surfaces by geodesic descent. These paths, which must be solutions of Euler differential equations

$$(dL_{\dot{x}_i}/dt) - L_{x_i} = 0\,,$$

are the solution of the variational problem, whereas the family of surfaces is determined as the solution of Hamilton–Jacobi partial differential equation

$$S_t + H(x_1, \ldots, x_n, S_{x_1}, \ldots, S_{x_n}, t) = 0\,.$$

The family of surfaces together with the paths of geodesic descent constitute Carathéodory's "complete figure" for the variational problem. This figure is the solution of the variational problem presented in clear geometric form.

Surface elements of S are transversal to line elements of the paths. The integral I over the piece of every path cut by two of the surfaces, for example $S = s_1$ and $S = s_2$, has the value $s_2 - s_1$. Along every other curve from $S = s_1$ to $S = s_2$ it acquires a greater value. Thus, these surfaces can be called "geodesic equidistant surfaces" when one takes the minimum value of the integral I in considering the paths.

Carathéodory's sufficiency proof for the minimising property of paths of geodesic descent is based on the inequality

$$S(x_2, t_2) - S(x_1, t_1) \leq \int_{t_1}^{t_2} L(x, \dot{x}, t)\, dt\,,$$

whereas, with the introduction of the Weierstrass excess function

$$E(x_i, x_i', \dot{x}_i, t) = L' - L - \sum L_{x_i}(x_i' - \dot{x}_i),$$

his fundamental formula becomes

$$\int_{t_1}^{t_2} L(x, \dot{x}, t) \, dt = [S(x_2, t_2) - S(x_1, t_1)] + \int_{t_1}^{t_2} E(x_i, x_i', \dot{x}_i, t) \, dt.$$

Along a path of geodesic descent E vanishes.

It has to be remarked that the direction of geodesic descent is not necessarily unique and this fact may result in the discontinuous solutions investigated by Carathéodory in his doctoral thesis and his habilitation. A sufficient condition for uniqueness is the strict convexity of L as a function of $\dot{x}$.

Carathéodory's method in the calculus of variations can be most easily understood in connection with the application of variational calculus to dynamics and optics.[90] Here, his method is closely related to Hamilton's ideas. In the dynamical analogy, with $L(x_i, \dot{x}_i, t)$ representing the Lagrangian of a dynamical system, $S(x_1, \dots, x_n, t)$ becomes the Hamiltonian representing a family of surfaces of constant (least) action. These surfaces are crossed in steepest descent by the trajectories of the point representing the dynamical system. In the optical analogy, the surfaces S are wavefronts, the new wavefront is determined by Huygens' principle and the direction of geodesic descent is the direction of ray propagation. Thus, the direction of ray propagation can be defined as that which maximises the normal wavefront velocity.[91]

The fruit of Carathéodory's efforts and thoughts about how he could make the mechanism of the field method more transparent was his work on *Die Methode der geodätischen Äquidistanten und das Problem von Lagrange* (The Method of Geodesic Equidistant [Surfaces] and the Lagrange Problem).[92] Together with problems of multiple integrals it comprised Carathéodory's second great contribution to the calculus of variations, after his famous works at the beginning of the century. His love for the method of fields of extremals, which is not modern, is attributed by his student Boerner to Carathéodory's proclivity for history. In the case of the so-called Lagrange problem, it offers "except for greater geometric transparency various other advantages. For example, the establishment of the Weierstrass E function, which in the usual treatment would lead here to a difficult construction, is almost direct. Further, one obtains the Hilbert independence integral almost without calculation. Finally, one is able to work out all the relations comprising the Hamilton–Jacobi theory without having to construct a field of extremals from the beginning. On the contrary, one can put aside the rather complicated field construction until the introduction of canonical co-ordinates, and all difficulties that arise from the side conditions are then eliminated by themselves." [93]

In the introduction to his paper *Über die Existenz der absoluten Minima bei regulären Variationsproblemen auf der Kugel* (On the Existence of Absolute Minima in the Case of Regular Variational Problems on the Sphere),[94] Carathéodory wrote that the result that any two points on a regular surface can always be connected by an absolute shortest geodesic line, had been transferred by various authors to general positive definite variational problems. However, if a problem is not definite, it can be shown that the existence of an absolute minimum locally is absolutely not sufficient to secure the existence of such a minimum globally. This is related to the fact that Hilbert's method employed to prove the result, with the help of Cantor's diagonal process, presupposes the uniform boundedness of all comparison curves. Tonelli had remarked that Hilbert's approach can be used as soon as one can derive an upper bound for the length of the curve C from the value of the curvilinear integral I_C of the variational problem along C following any arbitrary path. He tested this general theorem for many special cases and obtained results particularly concerning semi-definite variational problems, which Hahn had summarised in one single theorem. Carathéodory found that Tonelli's theorem could not be substantially generalised, since Tonelli's conditions made use of assumptions rooted in the nature of Hilbert's method. However, it appeared appropriate to Carathéodory to try to substitute Tonelli's condition by another, in which the lengths of the curves over which the integration is carried out do not appear at all, and which depends only on the curvilinear integral under discussion.

Carathéodory described similar considerations to Heinz Hopf, known for his work on algebraic topology, in a letter on 14 January 1931:

Dear Herr Hopf,

I was very pleased that you have thought of me again. The theorems of Herr Rinow have interested me extraordinarily, not at least because of their nice, smooth formulation. Surely, they are new, i.e. they have never been published. However, I have known theorem I for about the past 15 years; it is true for regular variation problems on the sphere if the function f under the integral has a positive non-vanishing lower bound. For geodesic lines on convex surfaces Blaschke has included my result (four-peaks theorem) in his book (Differential Geometry I).[95] In my proof *two* separate rays with conjugate points emerge in fact from a.

I conclude roughly like this: On every geodesic ray from a there is a point P for which the path aP is the shortest and another aQ for which this is not the case. Let b be the point which divides the two classes. Then, either the path ab is the only one out of the shortest paths that connects ab and then conjugates b to a, or there are two equally long geodesics g_1 and g_2 through a and b. If one constructs in the same way a ray g_3, whose tangent at the origin lies in the interior of the lune $g_1 g_2$, one obtains a figure which *has to* lie entirely within the first and one can demand that the angle in a between g_3 and g_4 be at most $a/2$. By iteration one obtains a point w that is conjugate to a. On the sphere, two domains are defined by $g_1 g_2$, each of which must contain a point like w in its interior. Naturally, $b' \ldots$ and therefore also w could coincide with b; but then b is conjugate to a on infinitely many rays through a.

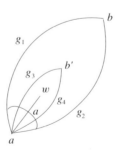

I have not thought the matter over for the projective plane,[96] but one should all the same come through with similar methods, which actually go back to Darboux and Hilbert.[97] Has Herr R[inow] had similar thoughts or are his proofs built up entirely differently?

Theorems II and III were completely new to me; but they can be derived also very easily, if I am not wrong, through increasing numbers of shortest [geodesics].

For me the main thing is the perfect posing of the problem and I congratulate you and your student Rinow for these extremely interesting results and especially for the 'analytic continuation'.

For the case that the completeness requirement is not satisfied, I have found, as you may know, a very instructive example (*Bullet. Un. Matem. Italiana*, vol. II).[98]

For the case that the sign of F in $\int F \, dt$ is arbitrary, the difficulties are of course much greater. In limited domains or on closed surfaces one can derive also here the existence of the shortest if $\int F \, dt > 0$ on all closed curves. But I have not published it, either. [...]

"It is self-evident", Carathéodory added in a remark at the end of his letter, "that if for a closed curve γ, $\int_\gamma F \, dt < 0$, no minimum can exist." [99]

The question of the effect of the curvature on the extension of the arc between conjugate points on which the geodesic line remains shortest is of importance in the study of the connection between the curvature and the topological structure of the space.

Carathéodory writes that "Hilbert's process [...] is an application of the 'principle of choice in set theory' explicitly formulated by *E. Zermelo*;[100] starting from the existence of a lower bound of the curvilinear integral for curves connecting two points P_1 and P_2, it is assumed that there is a countable set of curves, for which this curvilinear integral converges to its lower bound. This way of concluding would not be thinkable without the use of the afore-mentioned principle. It is similar when, out of an infinite point set, we pick out a subset which possesses a given limit point of the initial set and otherwise does not have any other limit point. *Hilbert's* original proof is based on certain general properties of families of curves, which are independent of the calculus of variations." [101]

4.6.2

Multiple Integrals

Using his notes from Smyrna, Carathéodory wrote a paper *Über die Variationsrechnung bei mehrfachen Integralen* (On the Calculus of Variations with Multiple Integrals)[102] in 1929 and produced with it a more lucid presentation of his own earlier results in this area. It is a detailed treatment of the general theory of m-fold integral problems, whose Lagrangian depends on n dependent functions. Through a slight modification he was able to present his old calculations in a much more symmetrical way using a technique which provides an immediate access to the Hamilton–Jacobi theory. His formulation is based on the extension of the classical notion of Legendre transformation, which Carathéodory had developed in a series of earlier papers when he was in Smyrna.[103] As he admitted, he was led to do so after studying Alfréd Haar's paper *Über adjungierte Variationsprobleme und adjungierte Extremalflächen* (On Adjunct Variation Problems and Adjunct Extremal Surfaces)[104] and "Tibor's hints" also helped him. Also, in that paper, Carathéodory set up the E function belonging to the problem

$$\int \ldots \int f\left(x_1, \ldots, x_n; \ t_1, \ldots t_\mu; \ \frac{\partial x_1}{\partial t_1}, \ldots \frac{\partial x_n}{\partial t_\mu}\right) dt_1 \ldots dt_\mu$$

Separatum.

ACTA
LITTERARUM AC SCIENTIARUM

REGIAE UNIVERSITATIS HUNGARICAE FRANCISCO-JOSEPHINAE

SECTIO

SCIENTIARUM MATHEMATICARUM.

TOM. IV. FASC. IV.

C. Carathéodory

Über die Variationsrechnung bei mehrfachen Integralen.

A M. KIR. FERENCZ JÓZSEF-TUDOMÁNYEGYETEM

TUDOMÁNYOS KÖZLEMÉNYEI

MATHEMATIKAI TUDOMÁNYOK.

IV. KÖT. IV. FÜZ.

1929. XI. 5.

SZEGED.
A M. KIR. FERENCZ JÓZSEF-TUDOMÁNYEGYETEM BARÁTAI EGYESÜLETÉNEK
KIADÁSA,

Separate print of Carathéodory's paper
On the Calculus of Variations with Multiple Integrals.
Courtesy of the Bavarian Academy of Sciences, Archive.

in both usual and canonical co-ordinates for the first time. Finally he showed that, if a geodesic field cuts a surface transversely, this surface should necessarily be a solution of the Euler–Lagrange equation.

Fleming remarks that Carathéodory's methods do not seem to have enjoyed the same success for multiple integral problems as for single integral problems for which there is an associated Hamilton–Jacobi equation.[105] H. Rund explains that it is probably on account of Carathéodory's canonical form, which does not reduce to the standard form when $m = 1$, nor does it possess a Hamiltonian function in the usual sense of the term, that it has never been applied to any of the physical field theories based on multiple integral variational principles.[106]

Carathéodory himself found that his system presented in his 1929 paper was "formally the most elegant" and "could be useful to physics", as he wrote to Born on 24 January 1935.

Dear Born,

through a work of Weyl, which I have received today, I see that you have used Hilbert's independent integral for multiple integrals in a work that I do not know. Judging from the work of Weyl, I assume that, for this case, you are not acquainted with the canonical variables, which I found with great effort 12 years ago. Unfortunately, my works are not very well-written and are based on a very complicated transformation. But it is the *right* substitute for Legendre transformation, also used by Weyl, which is valueless here. So it would be worth to take a look at my note, which I am sending you right away. It is possible that the one or the other of my two formulary systems (that in the *Acta Szeged* is formally the most elegant) could be useful to physics.

On this occasion, I would like to ask you to also send me your works on the modified system of Maxwell equations; I would be very interested in it.

The Calculus of Variations, on which I have been working for many years, will appear in a few weeks. I have tried to stress the connection with mechanics. The English mathematicians will have no particularly positive feelings for my book. Maybe, it would be possible to let the book be reviewed somewhere. In any case, I will send you the book and see to it that Teubner sends a review copy to *Nature*. [...][107]

Born's "works on the modified system of Maxwell equations" have their origins in his publication *On the Quantum Theory of the Electromagnetic Field*.[108] There, Born studied the properties of a field described by a set of equations which were a special form of Mie's general field theory. His main result was the establishment of the existence of an electron with finite radius and finite energy, whose potential agreed with Coulomb's law for large distances. The electron behaved in external fields like a point-charge, provided the fields were almost constant in a region that was large compared with the dimensions of an electron, but behaved quite differently for short waves. Born suspected that the new field laws might have the properties necessary to explain the inconsistencies of the older theories. In a proposed quantisation of the field equations, which was based on Hilbert's independence theorem of the calculus of variations, he indicated a method by which the non-linear field equations could again be linearised with the help of Dirac matrices. Linearisation of the wave equation was necessary for the principle of superposition to be satisfied.

The new Lagrangian which ensured relativistic invariance and offered the possibility of a systematic quantisation of the field was presented by Born in his *Modified*

Field Equations with a finite Radius of the Electron.[109] In *The Absolute Field Constant in the New Field Theory,*[110] by a rough estimate of the "absolute field", Born and Schrödinger discovered that, taking into account the electron spin, the radius of the electron increased by a factor of about 10. In a subsequent paper entitled *Quantised Field Theory and the Mass of the Proton,*[111] Born applied the same method to estimate the mass of the proton, which he found to be sufficiently in agreement with the experimental value.

Intent on exploring the possibility of applying his mathematical formal system to physics, with the aid of Born's work, Carathéodory wrote to Born on 26 February 1935:

Dear Born,

[...] I would actually like to study the physical significance of my canonical co-ordinates for your general field equations (that the Legendre condition is not satisfied here is of course irrelevant) before I write to you; but I see that it would take some time before I come to that, since I must first of all finish my book.

Certainly, the whole book has already been announced to be ready for press, but I am still working on the list of references plus an index. Actually Herglotz, who was in Munich at the end of October, drew my attention to your field equations. Herglotz presented the matter in such a way that also general electrodynamic equations could be directly described as three-dimensional minimal surfaces in a four-dimensional space, if one takes the line element $ds^2 = dt^2 - dx^2 - dy^2 - dz^2$ (maybe I have misunderstood him and it only concerned the electrostatic problem). Anyway, this solution for the electron corresponds to a minimal surface, symmetric with respect to rotations, similar for instance to the rotation surface with the chain line [catenary] as meridian which is the only minimal surface symmetric with respect to rotations.

Now, in 4-dimensional space there are still other two-parametric Lie motion groups and Herglotz has calculated the corresponding solutions of your field equations for all these groups. But he told me that, if he had not made a mistake, these solutions would be useless to physics. [...][112]

Weyl's work, which Carathéodory had mentioned in his letter to Born on 24 January, was obviously the *Observations on Hilbert's Independence Theorem and Born's Quantization of Field Equations.*[113] There, Weyl gave a simple and explicit formulation of the independence theorem and suggested a modification of Born's proposal of a quantisation of electromagnetic field equations, which allowed a comparison with Heisenberg–Pauli's quantisation under the simplest circumstances. In a letter to Weyl on 5 March 1935, Carathéodory resumed his own method which is a clear, concise description of his calculus of variations techniques in the context of mathematical physics:

Dear Herr Weyl,

I have read your work about the invariant integral in the *Phys. Rev.* with great interest. But since it seems to me that the integral you use is wholly different from the one I have studied earlier than you, and since I have very little time at the moment to confirm this, I send you my old works. You will surely have someone at Princeton, who could look into the problem. Because it is not easy to find the basic idea in those works that are not very well written, I resume the contents of my method.

1. Every functional determinant Δ can be written as a divergence; consequently, the integral over Δ depends only on the boundary. If one considers the functions $S_\alpha(t_\beta, x_i)$ ($\alpha_1\beta = 1$,

$...m_i \ i = 1, ...n)$ and puts in these the x_i as functions of t_β, then with the notations

$$P_{i\alpha} = \partial x_i / \partial t_\alpha, \quad c_{\alpha,\beta} = (\partial s_\alpha / \partial t_\beta + \partial s_\alpha / \partial x_i) p_{i\beta}, \quad \Delta = \begin{vmatrix} c_{11}, & \cdots \\ \cdots, & c_{mm} \end{vmatrix},$$

the integral over Δ is an invariant integral.

2. Consequently, two variational problems with the basic functions $L(t_\alpha, x_i, p_{i\alpha})$ and $L^* = L - \Delta$ have always the same extremal surfaces and are equivalent.

3. If there are place functions $\Psi_{i\alpha}(t_\beta, x_j)$ for which

(1) $L^*(t_\alpha, x_i, \Psi_{i\alpha}) \equiv 0$

and

(2) $L^*(t_\alpha, x_i, p_{i\alpha}) > 0$,

in the case where not all $p_{i\alpha} = \Psi_{i\alpha}$, then the surfaces defined through the equations $\partial x_i / \partial t_\alpha = \Psi_{i\alpha}$, if these exist, are extremal surfaces for the problem over L^*.

4. Thus, I seek to determine the functions $S_\alpha(t_\beta, x_i)$ in such a way that, with $L^* = L - \Delta$, relations (1) and (2) exist. In addition, one must necessarily have

(3) $L(t, x_i, \Psi_{i\alpha}) = \Delta(t, x_i, \Psi_{i\alpha})$,

(4) $L_{p_{i\alpha}}(t, x_j, \Psi_{j\alpha}) = \Delta_{p_{i\alpha}}(t, x_j, \Psi_{j\alpha})$.

From (3) and (4) follows (2), if the Legendre condition is satisfied, i.e. if a certain quadratic form is positively defined.

Through elimination of $\Psi_{i\alpha}$ between (3) and (4), one obtains one partial differential equation for the functions S_α, so that one can still choose $(m - 1)$ of these functions. Thus, through integration of these equations, one obtains geodesic fields and calculates that every surface which cuts such a geodesic field transversally has to satisfy the Euler equations. A short time ago, Boerner proved that, inversely, one can embed every solution of the Euler equations in a geodesic field. The equations $p_{i\alpha} = \Psi_{i\alpha}$ are generally integrable only along this *one* extremal surface, which however does not matter at all.

The elimination of $\Psi_{i\alpha}$ becomes very symmetric, if one uses the canonical variables appearing in my works; but this is only a detail.

Many regards,

Yours obediently,

C. Carathéodory

P.S. Consequently, it should be examined whether the invariant integral you have put forward could be retained through specialisation of the geodesic fields. Obviously, there are variational problems for which this is the case. But I assume that there are variational problems for which the two constructs are completely different.[114]

Weyl replied on 23 March 1935:

Dear Carathéodory:

I am sure you won't mind my addressing you in English; I can avail myself of secretarial help if I do so.

Thank you so much for sending me your reprints, I did not know these papers before. You know that my connection with calculus of variations is rather spurious. Only when I was called upon to report on Born's Ansatz [approach] for the quantization of field equations in our Physical colloquium, it seemed useful to me to arrange the formal foundations of calculus of variations for several variables in that concise form which I published in the *Physical Review*. I found the distinction of "three stages of independent variables" particularly convenient for the sake of exposition; but I did not claim to give anything new.[115]

Weyl proceeded to explain how he arrived at his own theory. In his paper concerning *Geodesic Fields in the Calculus of Variation for Multiple Integrals*,[116] he remarked that his note was drafted to meet Carathéodory's question and, in order to facilitate comparison, he presented the work according to Carathéodory's style and then expounded the essentials of Carathéodory's theory. Finally, he solved "the problem of embedding a given extremal in a geodesic slope field – this notion taken in the sense of trace theory," something that, as Carathéodory told him, "Boerner did [...] for his more sophisticated theory." In fact, Weyl obtained a simplified form of the theory by linearising Caratheodory's construction. Both Caratheodory and Weyl's fields are non-integrable and the equations defining them constitute undetermined systems.[117]

4.6.3

Carathéodory's Book on the Calculus of Variations and Partial Differential Equations

Carathéodory's book *Variationsrechnung und partielle Differentialgleichungen erster Ordnung* (Calculus of Variations and Partial Differential Equations of the First Order) was published in 1935 by B. G. Teubner in Leipzig and Berlin. As Carathéodory had wished, the book was immediately reviewed in *Nature*. The reviewer (E. L. I.) reported that, in the first ten chapters, Carathéodory had lucidly and completely established the basic theory of characteristics, canonical transformations of systems of equations, contact transformations and the Pfaff problem. In the second part of the book, he had founded the theory of the calculus of variations weaving new matter into the older fabric of Jacobi and his pupils' works on the connection between the calculus of variations and partial differential equations and stating almost everything in terms of a multi-dimensional space. The author had treated the simple and the parametric variational problems, the second variation, the boundary problem, closed extremals and the Lagrangian problem. E. L. I. ended his review by the comment that the book was "indispensable".[118]

Carathéodory's motive for writing the book, as he reveals in the foreword, was to shed some light on the ideas regarding the connection of the calculus of variations with the theory of partial differential equations, which had remained unnoticed so long. In the first part of his book, he attempted to simplify the representation of the theory of the first-order partial differential equations as to make it accessible to the average student of mathematics. The second part contained the calculus of variations and, except for those chapters regarding the discussion on the foundations of the calculus, it was not written to be complete.

Carathéodory presented the calculus of variations as primarily of service to mechanics and intended to convince the knowledgeable reader that, at that time, there were three main directions of the calculus of variations: that of Lagrange, which had become part of tensor calculus, Tonelli's theory, in which the more elegant relations of the minimum problem to the theory of sets were developed and, finally, the direction followed in his own book, which was oriented to the theory of differential equations, to differential geometry and to applications in physics. Carathéodory had shown that the Weierstrass theory of the calculus of variations belonged to this last direction.

W. Fleming remarks that Carathéodory's statement regarding Tonelli's theory is quite misleading, since set theory underlies all of mathematical analysis. (See Fleming's comments on Tonelli and "direct methods" in 4.6.1).[119]

Perron's entry on Carathéodory in the *Jahrbuch* remarks that in the foreword to the book

the author expresses his astonishment at the fact that the close connection between partial differential equations and the calculus of variations had not become quite clear to mathematicians until now. He writes: 'While fostering the actual calculus of variations neither Jacobi, nor his disciples, nor the many other excellent men, who have so brilliantly represented and promoted this discipline during the 19th century, have thought in some way of the relationship connecting the calculus of variations with partial differential equations. This is all the more obvious, since most of these great mathematicians have also engaged especially with partial differential equations of the first order.' On the contrary, Carathéodory endeavours now to bring out this connection sharply and, therefore, in the first part of the book, what is essential about partial differential equations of the first order, namely, the theory of characteristics, the Poisson brackets, the canonical transformation, the integration theories of Lagrange, Jacobi, Adolf Mayer and Lie and some other things, is presented with impeccable clarity.

In the second part, the actual calculus of variations, the author's achievements until then in no way come to the fore, on the contrary, they play a very humble role; the discontinuous solutions, for example, are only mentioned briefly in a small footnote. On the other hand, what is basic in the calculus of variations is completely dressed up as new, which differs considerably from what had been usual until then. From the beginning, the theory is carried through in an $(n+1)$-dimensional $(t, x_1, \ldots, x_n)$-space. By simply writing (t, x_i) and omitting the summation sign as usual, the formulae do not appear to be longer than in the case $n = 1$. The extremals (minimals, maximals) are not defined as the integral curves of a certain system of differential equations, as usual, but as curves, along which the given integral $\int_{t_1}^{t_2} L(t, x_i, x_i') dt$, where $t_1 < t_2$, has at least one weak local extremum. At first, it is required to assume only continuous first and second derivatives of the function $L(t, x_i, x_i')$ with respect to x_i'. It is then shown that a minimal can go through an element of length $t^0, x_i^0, x_i^{0'}$ only if the quadratic form $L_{x_i'x_j'}(t^0, x_i^0, x_i^{0'})\xi_i\xi_j$ is positive definite or at least semi-definite, in which case the indices of L denote partial derivatives. If it is positive definite everywhere in a certain domain B of the (t, x_i, x_i')-space, then a minimal goes really through every point of B, as well. This is proved in the following way: if an integrand $L^*(t, x_i, x_i')$ has the property that there are n functions $\psi_i(t, x_k)$, for which $L^*(t, x_i, \psi_i(t, x_k)) = 0$, whereas $L^*(t, x_i, x_i') > 0$, it is clear that, as long as only the n differences $x_i' - \psi_i(t, x_k)$ are absolutely sufficiently small and all of them do not vanish, the integral curves of the system of differential equations $x_i' - \psi_i(t, x_k)$ are minimals, since along them $\int_{t_1}^{t_2} L^*(t, x_i, x_i') dt = 0$ because the integrand vanishes; on the contrary, along every neighbouring curve (close neighbourhood) the integral is > 0, because the integrand is positive. Now, if it is a matter of the minimum of a given integral $\int_{t_1}^{t_2} L^*(t, x_i, x_i') dt$, one cannot demand from the function L to have the just mentioned required property of L^*. To overcome this evil, let $S(t, x_i)$ be any function with continuous partial derivatives and let us put:

$$L(t, x_i, x_i') - \frac{\partial S}{\partial t} - \frac{\partial S}{\partial x_j} x_j' = L^*(t, x_i, x_i').$$

Then, exactly the same minimals belong to the integrands L and L^*, because the difference of the integrals is

$$\int_{t_1}^{t_2} \left(\frac{\partial S}{\partial t} + x_j' \frac{\partial S}{\partial x_j} \right) dt = \int_{t_1}^{t_2} \frac{dS}{dt} dt = S(t_2, x_{i2}) - S(t_1, x_{i1}),$$

and so it is independent of the path. Consequently, one returns to the previous case, in which one can find a function $S(t, x_i)$ and n functions $\psi_i(t, x_k)$ such that L^* as a function of x_i' has a strict minimum for $x_i' = \psi_i(t, x_k)$, which is zero. This easily produces the conditions

$$(A) \begin{cases} \frac{\partial S}{\partial x_j} = L_{x_j}(t, x_i, y_i) \\ \frac{\partial S}{\partial t} = L(t, x_i, y_i) - \psi_i L_{x_j'}(t, x_i, y_i), \end{cases}$$

whereas the sufficient condition for the strict minimum

$$L^*_{x_i, x_j'} \xi_i \xi_j \quad \text{is positive definite}$$

is satisfied by itself, because $L^*_{x_i, x_j'} = L_{x_i, x_j'}$. Carathéodory calls the formulae (A) the fundamental equations of the calculus of variations. From here one arrives almost necessarily at the Weierstrass E function and the condition for a strong minimum. Thus far, anything goes, without the necessity of assuming the existence of further derivatives, as, for example, L_{x_i}. The functions S and ψ_i, which satisfy the requirements (A), can be found in many special problems. However, for a general valid existence proof, the existence and continuity of further derivatives must be assumed. Then, the reasoning for the proof consists in gaining a partial differential equation of the first order for the function $S(t, x_i)$ (Hamilton–Jacobi equation) through elimination of the ψ_i, whereupon, for the extremals, also the Euler differential equations result at once.

Thus, Carathéodory throughout proceeds in the opposite direction from that, which one followed up to then. Initially he does not start from the necessary but from the sufficient conditions and only at the end obtains the Euler differential equations, which otherwise came first, and he obtains them of course in a very simple manner, and not as earlier through a very tricky transformation of the first variation. Then, it is only later also shown that all extremals are obtained in such a way and that the conditions are thus also necessary. That's all about what is fundamentally new in the book which, by the way, also offers many new details and, in addition, an abundance of thoroughly checked examples. [120]

Thus, with this method later called by Hermann Boerner the "royal way in the calculus of variations", [121] Carathéodory was able to derive all the classical results of the calculus of variations, from Hilbert's integral up to the Euler equations; he also endeavoured to achieve sufficient conditions for weak minima of variational problems, and made significant progress in solving the so-called problem of Lagrange, with regard to which the words "geodesic field" or "geodesic equidistant surface", used in his famous *Acta matematica* paper of 1926, no longer appeared. Boerner mentions that with this tool Carathéodory was able to treat the famous Zermelo navigation problem "in a very elegant way." [122]

In an attempt to legitimise his use of a classical method in the calculus of variations, Carathéodory closed his speech [123] at the tercentenary celebration of Harvard University with the following words: "I will be glad if I have succeeded in impressing the idea that it is not only pleasant and entertaining to read at times the works of the old mathematical authors, but that this may occasionally be of use for the actual advancement of science. Besides this there is a great lesson we can derive from the facts which I have just referred to. We have seen that even under conditions which seem most favorable very important results can be discarded for a long time and whirled away from the main stream which is carrying the vessel of science. Sometimes it is of no use if such results are published in very conspicuous places. It may happen that the work of most celebrated men may be overlooked. If their ideas are too far in advance of their time, and if the general public is not prepared to accept them, these ideas may sleep for centuries on the shelves of our libraries. Occasionally, as we have tried to do today, some of them may be awakened to life. But I can imagine that the greater part of them is still sleeping and is awaiting the arrival of the charming prince who will take them home."[124]

In his *Variationsrechnung*, Carathéodory included an existence theorem for the solution of ordinary differential equations $y' = f(x, y)$ and its proof. This theorem, an application of modern integration depending on unsymmetrical assumptions, also appeared in his 1918 book on real functions.[125] There, the assumptions for f were that f is bounded by x, y, continuous for constant x and measurable in x for constant y. According to L. C. Young, continuity is a natural restriction, but measurability is so general that nobody has been able to name a single non-measurable function unambiguously. In 1935, Carathéodory removed the question about the symmetry of the variables by replacing the word measurable by continuous.[126] Young commented:

Since then, however, the original lopsided theorem has turned out to be essential from the most practical standpoint. [...] Engineers who send rockets to the Moon and elsewhere are no longer concerned with classical differential equations, in which everything is determined by initial conditions: even Jules Verne knew very well that it is virtually impossible to reach the Moon in this manner. We have to *control* the projectile during flight.

Basically we have then not a differential *equation* but an *inclusion*: the velocity vector x is now a member of a certain set F, depending on the position x and the time t. To express this by a differential equation, we introduce a control parameter u, allowed to vary at our discretion. The equation is $\dot{x} = g(t, x, u)$, where u is a function of t entirely at our disposal. The equation exhibits the natural symmetry of the variables: the function g is continuous in all three. However, we have to substitute for u an absolutely arbitrary function of t! That makes quite a difference: g is turned into a function of t, x, continuous in x, but possibly quite wild in t. Even the slight Caratheodory restriction of measurability in t for constant x is only enforcable thanks to a remarkable theorem of Filippov, whose general form requires the famous *continuum hypothesis*, and which thus illustrates the extent to which Set Theory now permeats analysis. Only in this very general manner, thanks to Caratheodory, Filippov, and Cantor, has analysis been able to attain the essential freedom which matches the radio and engineering techniques for controlling a Moon shot. Even so, this freedom, now finally reached, goes further than the most general notion of function: it requires the weak solutions of which I speak at various times.[127]

4.6.4
Control Theory, Dynamic Programming and Pontryagin's Principle

H. J. Pesch and R. Bulirsch "show that the most prominent results in optimal control theory, the maximum principle and the Bellmann equation, are consequences of Cara-théodory's results published more than a decade before optimal control theory had begun to develop from the calculus of variations."[128] The so-called Hamilton–Jacobi–Bellmann equation $S_t = \min_{x'}\{L(t, x, x') - S_x x'\}$ appearing in Carathéodory's 1935 book is, according to Pesch, wrongly named after Bellman, for Bellman's results regarding equations of this type go back to 1954. Pesch asserts that Bellman knew Carathéodory's works, but has never cited them.[129]

Fleming explains that "Bellman's dynamic programming is a widely used technique for optimization problems, in which decisions are made at each time instant using the current state of the system under consideration. This is the idea of feedback control. The optimization problem can be deterministic or stochastic and formulated in continuous or discrete variables. During the 1950s, Bellman wrote several papers on dynamic programming and the calculus of variations, in which he failed to give credit to Carathéodory. For calculus of variations problems, Bellman's dynamic programming principle leads to the Hamilton–Jacobi PDE. As Pesch and Bulirsch observe, it is unjust to call this the Hamilton–Jacobi–Bellman PDE. Unfortunately, the Hamilton–Jacobi–Bellman terminology has persisted in recent times."[130]

Pesch also mentions that the maximum principle in optimal control theory, which is usually attributed to a Russian research group around Pontryagin, should be called the Carathéodory–Hestenes–Isaac–Pontryagin principle, for Carathéodory's book contains a formula from which one can easily prove the principle for a special case.[131] Namely, the only small step required to obtain a special case of the maximum principle is to substitute the variational form of an optimal control problem into equation $H(t, x, y) - H(t, x, y') - H_y(t, x, y')(y - y') \geq 0$.[132] Fleming notes that this last sentence understates the importance of Carathéodory's work to control theory. Also, by using the name Carathéodory–Hestenes–Isaac–Pontryagin principle, two different approaches to control theory via Bellman's dynamic programming become confused.[133]

The maximum principle plays a central role in the numerical calculation of optimal trajectories, optimal control of robots, or optimal control of process-engineering processes. In their paper, Bulirsch and Pesch put the question of how a plane landing under a windshear has to be piloted so that the landing approach stops and the plane begins climbing again while having the possible greatest distance from earth. This is an optimal control problem of the minimax type, i.e. the minimum altitude has to be maximised. It is an application showing, according to Pesch, how mathematics can save human life.[134]

Again, to quote Fleming,

the Pontryagin approach to optimal control began in the late 1950s. It became very popular in the 1960s and afterward. Pontryagin considered problems in which the states evolve in time according to differential equations, which depend on control variables subject to inequality constraints as well as initial and possibly final conditions. Pontryagin's principle gives a necessary condition for a minimum. The "costate" variables in Pontryagin's principle have a role similar to variables introduced by Carathéodory via canonical transformations. Shortly before Pontryagin, M. Hestenes published in a RAND Corporation report (written for engineers) results quite similar to Pontryagin's.

R. Isaacs invented the subject of two-player, zero sum differential games. These are control problems, in which two controllers (maximizing and minimizing players) oppose each other. Isaacs' technique depends on solving a first-order PDE of Hamilton–Jacobi type. This PDE is now often called a Hamilton–Jacobi–Isaacs PDE.[135]

4.6.5
Viscosity Solutions to Hamilton–Jacobi PDEs

The Carathéodory theory is also helpful in situations leading to viscosity solutions which have only recently been explored. Fleming comments:

The Carathéodory theory applies in regions in which solutions to the corresponding Hamilton–Jacobi equation remain smooth. Unfortunately, discontinuities typically appear at certain lower dimensional "singular sets" which are not known in advance. There is the serious issue of choosing the "physically correct" solution among many possible solutions to the Hamilton–Jacobi equation with given boundary conditions. This was resolved in the 1980's through the M. Crandall and P.-L. Lions theory of viscosity solutions. Viscosity solutions also provide a powerful method to study a variety of asymptotic problems in analysis, including "vanishing viscosity" limits from second-order PDEs of parabolic type to first-order Hamilton–Jacobi PDEs. P.-L. Lions received a Fields Medal at the 1994 ICM in Zurich, in part for his seminal work on viscosity solutions.

Recently, Hamilton–Jacobi PDEs have become important in applications to modern technology in addition to their traditional importance in mechanics, wave propagation, and other areas of mathematical physics.[136]

4.7
Member of the Academy of Athens

Under the presidency of Phokion Negris,[137] the Greek academy of sciences, called the Academy of Athens (*Αχαδημία Αθηνών*), elected Carathéodory to a full member on 13 December 1926. He had been proposed during the academy session of 26 November that year, only a few months after the foundation of the institution. Carathéodory was one of the first three elected academicians to join the initial thirty-eight nominated members. He was elected to the Class of Positive Sciences; Dimitrios Kambouroglou was elected to the Class of Letters and Georg von Streit to the Class of Moral and Political Sciences. The fact that Carathéodory attained the status of a full member was particularly unusual, since he had his permanent residence abroad and he was in the employ of another state. This honour was not bestowed again on any other academician until 1997, when the theoretical physicist Nanopoulos, who was living in Texas, was also nominated to a full academy member.

The Academy of Athens was founded by a decree of Pangalos' dictatorial government[138] on 18 March 1926 "to assist the state authorities, to explore the history of the Greek language and literature, to write an official grammar and a dictionary of the language, to study the ancient Greek authors and publish their works and to strengthen public education."[139] Further, the academy was required to explore flora, fauna and mineral resources in Greece and utilise them to the advantage of state interest and to supervise history, jurisprudence, customs and traditions and folklore. The opening ceremony of the academy on 25 March 1926 coincided with the Greek national celebration of the War of Independence. The Minister of Education and Religious Affairs, Professor Dimitrios Aeginitis,[140] gave the inaugural speech. A project with a history of several decades was thus completed. The foundation stone of the academy building had been laid already by King Otto over seventy years earlier, on 2 August 1850. The Danish architect Christian Hansen had designed the neo-classical building, which was executed by the German architect Ernst Ziller who was working in Greece. Funding was provided by the Greek patron, Baron Sinas, then residing in Vienna. Alexander Rangavis, the former envoy of Greece in Berlin (1875–1877) and one of the most versatile scholars of the young Greek state, was the first to urge for the enactment of a Greek statute for an academy. However, the foundation statutes of 1926 were based to a large degree on those of the Belgian Royal Academy.[141]

The Academy of Athens. Photograph: S. Georgiadis.

In a letter to Kalitsounakis on 21 November 1926,[142] Carathéodory mentioned that Angelos Ginis and Nikolaos Genimatas, both professors at the Technical University of Athens, had congratulated him on his election to the academy. He expressed his joy at being honoured but he added that the subject domains of the academy had very little to do with his own scientific discipline. In his opinion, the academy had to avoid dissipating its energies and should concentrate its work on those areas which would promote the country, such as geology, botany and zoology in the natural sciences, and philology, ethnology, music and archaeology in the humanities. Carathéodory detected inconsistencies with regard to the criteria of the academicians' nominations, which he partly attributed to the fact that too little attention had been paid to the Greek character of the institution. So he expressed his unease that Nikolaos Balanos,[143] the Acropolis restorer and "an excellent connoisseur of the topography of Athens", had not been included in the academy. Despite all his criticism, Carathéodory stressed that the foundation of the institution was a great effort. "Maybe", he added, "it will survive as the only trace of the past dictatorial regime [of Pangalos], just as the statutes of the *Comédie Française* survived as the only trace of Napoleon's great campaign to Moscow."

In the summer of 1927, in the Athens National Theatre, Carathéodory represented the academy at the celebration of the 100th anniversary of Wilhelm Müller's death. The poet, called *Griechen-Müller* (Müller the Greek) because of his love for the Greeks, was the publisher (1821–1826) of the *Griechenlieder*,[144] whose proceeds went to Greek organisations, and a passionate advocate of the Greek War of

Independence. In his speech, Carathéodory defended the cause of Greece with great enthusiasm and made critical remarks against the contemporary Turkophile politics of Western powers.

4.8
Caring for Munich's Scientific Life

Towards the end of 1924, Carathéodory tried to obtain Gustav Herglotz for the chair of astronomy at Munich University as the successor to Hugo Ritter von Seeliger,[145] the director of the Munich observatory, who had died. However, Herglotz refused the position, as Carathéodory informed Runge,[146] to whom he had turned for advice on who should be appointed, given that it was intended to change the post into a chair for theoretical astronomy. As mentioned earlier, Herglotz later succeeded Runge in Göttingen in 1925.

Carathéodory's own recognition in Bavaria came very soon after his appointment, when Aurel Voss, Oskar Perron, Sebastian Finsterwalder, Georg Faber, Alfred Pringsheim and Walther von Dyck proposed his election to the Bavarian Academy of Sciences on 7 February 1925. Two weeks later he became a full member.[147]

On 17 January 1927, Carathéodory, Perron and Tietze, referring to Hartogs' research and teaching activities, asked for him to be promoted to a full professorship. On 5 February 1927 the seven mathematicians[148] of the Bavarian Academy of Sciences proposed Hartogs' election to a full member of their class, referring not only to his research results but also to his new methods of proof in the complex analysis of several variables. But their proposal did not gain the support of the three-quarters majority required, so Hartogs' election was rejected.[149]

On 4 February 1927, Carathéodory and other academicians[150] proposed Einstein's election as a corresponding member of the Bavarian Academy of Sciences. They honoured Einstein's theory of relativity but they underlined that the name of the theory was "not especially appropriately chosen, since it suggests ethical analogies having nothing to do with the Einstein theory".[151] Einstein was admitted shortly thereafter.

On 28 March 1927, Carathéodory congratulated Sommerfeld, professor for theoretical physics at the University of Munich since 1906, on his appointment to Berlin. He said he had already been to the Ministry in order to make Sommerfeld's significance for Munich clear to them[152] and thus prevent a great loss for Munich University. In the end, Sommerfeld remained in Munich and kept family contacts with Carathéodory.[153] On 23 February 1929, Carathéodory, jointly with others, congratulated Sommerfeld on the rejection of Stark's appointment as Wien's successor.[154] Stark had written the polemical book *Die gegenwärtige Krisis in der deutschen Physik* (The Present Crisis in German Physics) in 1922, in which he criticised both the theory of relativity and the Bohr–Sommerfeld quantum theory as harmful for German experimental research. In 1929, he directed a detailed critique against Schrödinger's quantum-mechanical theories, although they offered a good interpretation of the Stark effect, as well as against Sommerfeld's support of these theories, and he thus isolated himself so much that nobody engaged in atomic physics took him seriously.[155]

4.9

First Visiting Lecturer
of the American Mathematical Society

On 9 May 1927, Carathéodory asked the Bavarian Ministry of Education and Cultural Affairs for paid leave of six months. Harvard University had invited him to deputise in spring and summer 1928 for George Birkhoff, who was lecturing there and was generally regarded as the "best American mathematician".[156] Birkhoff had become known to Göttingen mathematicians when he proved[157] Poincaré's Geometric Theorem,[158] from which the existence of periodic solutions of the three-body problem could be deduced.[159] In his 1917 paper *Dynamical Systems with Two Degrees of Freedom*[160] and in his book *Dynamical Systems*, published in 1927, Birkhoff used ideas developed by Poincaré to lay the foundations for the topological theory of dynamical systems by "qualitative determination of all possible types of motion and the interrelation of these motions", as he wrote.[161]

Carathéodory's invitation originated from a unanimous decision taken by the Third Division of Mathematics of Harvard University on 18 March 1927. The division consisted of professors Osgood, Huntington, Kellogg, Graustein, Morse and Walsh and the Chairman, Dr. Brinkmann, who was the one to communicate the invitation to Carathéodory.[162] Carathéodory had also been invited by the American Mathematical Society (AMS) to hold lectures at about twenty universities, as he told Penelope Delta[163] and was, in fact, the first to receive a Visiting Lectureship of the American Mathematical Society. The previous year, on President Birkhoff's suggestion, the Council of the AMS had approved the plan to create this lectureship with the primary aim of attracting distinguished mathematicians of interested mathematical centres to offer lectures in the USA. The committee appointed to nominate the lecturer to the Council and to issue the invitation to the Visiting Lecturer and supervise the general arrangements, consisted of Birkhoff (chairman), G. A. Bliss and E. R. Hedrick. The financial responsibility was taken on by the colleges and universities concerned and was binding for the year of the invitation. Apart from Carathéodory, the European mathematicians H.Weyl, E. Bompiani and W. Blaschke were also invited in this way to America.[164]

4.10

Hindered by the Bavarian Ministry of Finances

To support his application to the Ministry, Carathéodory argued that the University of Munich would benefit from the experiences he would gain from a one-semester lecture course at one of the best American universities. He then made proposals for how the whole enterprise was to be staged and for the summer semester in Munich. Since the spring term of the American universities began in early February, Carathéodory planned to finish his lectures in Munich in January. Therefore, he intended to condense his normal programme there and hold three-hour lectures on the calculus of variations. He proposed the assistant professor Dr. Fritz Lettenmeyer[165] as his deputy in analytical geometry. Lettenmeyer had been released from the army during World War I after a bad leg injury. In 1918 he gained his doctorate under Ferdinand Lindemann. He worked as an assistant at the Technical University of Munich for a short while, then in 1920–1921 he went to Göttingen for further studies and became Konrad Knopp's assistant in Königsberg. During the period 1923–1933 he held an assistant position at the University of Munich, where he habilitated in 1927. He belonged to the SA and was not appointed as associate professor until 1933.[166]

The rector and the academic senate of the University of Munich approved Carathéodory's application immediately, but the Ministry of Finances adopted a negative attitude. On 11 July the Bavarian Ministry of Education and Cultural Affairs reacted with remarkable sharpness.[167] They considered the rejecting statement of the Ministry of Finances against their draft of a resolution for Carathéodory's leave to be in regrettable contradiction to the significance which "the German education administrations, the Foreign Ministry and the Reich Ministry of the Interior, in harmony with the broadest scientific circles of Germany, attach to the redevelopment and strengthening of the scientific communication of German scholars and scientific organisations with foreign countries." They felt that the scientific and personal damage suffered by representatives of German science after the outbreak of the war was ignored by their foreign colleagues. After the war, the "financially much stronger foreign countries" had used considerable means to promote science, "whereas in the economically and politically heavily damaged Germany the preservation, care and building of the scientific institutes" in both the humanities and sciences "from which one expected direct promotion of science" were jeopardised. In addition, after the war, Germany's enemies had established new scientific unions, namely the Academic Union (*Union Académique*) for the humanities and the Research Council (*Conseil des Recherches*) for natural sciences, from which Germany remained excluded. Germany was highly interested in winning the sympathy of these organisations and promoting personal relationships between German and foreign scholars. In a review of the situation at the Foreign Ministry in Berlin in April that same year it was affirmed that an opportunity to curb the cultural propaganda against Germany, which especially France was pursuing with "emphasis and success", would be given by the personal appearance of respected German scholars abroad. "Consequently, the Reich government, the Emergency Action Organisation for German Science [see 4.24] and the German education administrations, should, if possible, support scholars who are invited abroad or make study trips abroad on their own initiative. It is of course clear that it is not a matter of a pleasure trip, even if the inviting institutes offer favourable conditions, but that such enterprises demand a high expenditure of work and energy." Finally, the Ministry of Finances was admonished not to hinder the mobility of such scholars either through doubts or through other considerations, since cultural-political intentions were given priority in such cases.

The attitude of the Ministry of Finances demonstrates the narrow-mindedness of which Karl Vossler, the classical scholar and rector of Munich University, had warned when he spoke of "provincialism as a danger for the spirit and as a restrictive conviction".[168] On the other hand, the reaction of the Ministry of Education and Cultural Affairs clearly shows that Carathéodory was not only considered as a bearer of "German science" but also its ambassador abroad, entrusted with the difficult task of contributing to the restoration of Germany's destroyed image.

On 14 August 1927, Carathéodory wrote to Mrs.Young that he would spend the summer term of 1928 at Harvard and that he was also invited to Berkeley, California. He thought that he would leave Munich directly after Christmas, since he was asked to be in America a bit earlier than the beginning of the second half-year. [169] He was finally granted leave to do so.

4.11
At the University of Pennsylvania

Accompanied by his wife, Carathéodory travelled to America on board the ocean-going steamer *Aquitania*. His first lecture in America was at the University of Pennsylvania on 10 January 1928. In that day's edition of the student newspaper *The Pennsylvanian* there appeared a small article with the heading *Prominent Scientists Present Five Free Lectures This Week*, which told its readers that Carathéodory would "deliver the first of a series of three lectures on the 'Application of the Lebesgue Theory to different problems of analysis, geometry, and mechanics' today at 4 p.m. in the Houston Hall Auditorium. Dr. Caratheodory will complete this group of talks, given under the auspices of the Mathematics Department of the College, on Wednesday and Thursday at the same hour in the Houston Hall Auditorium." [170]

4.12
At Harvard

At Harvard, in the second half-year 1927–1928, Carathéodory took up his position as a visiting lecturer of mathematics for undergraduates and graduates at the Faculty of Arts and Sciences. Three times a week, on Monday, Wednesday and Friday at noon, he held lectures on "space, time and relativity". His course was open to students who had either taken the course on differential and integral calculus and analytic geometry held by professors Osgood and Huntington and by Dr. M. S. Demos, or had an equivalent preparation in differential and integral calculus. [171] The textbooks used at Harvard University for freshmen were Graustein's Plain and Analytic Geometry and Osgood's Introduction to Calculus.

On 28 April 1928, Carathéodory wrote to Sommerfeld: "I am quite satisfied with my audience here; I have more than 20 people in every lecture, which is very much for Harvard. In June, I will go at first to Stanford for 8 days and then to Berkeley for 6 weeks. Unfortunately, the summer course is somewhat shifted and lasts until the middle of August; consequently, I will not be in Munich before 10–15 September. *Will I still meet you there?* [...] We recently visited President Lowell, who has a wonderful house, and also his sister, Mrs. Putnam, whose husband is said to have shown interest in the development of mathematical instruction. I had a very interesting almost 2-hour discussion with Millikan, who was here and held almost a whole lecture (very popular) about your theory of fine structure. I see quite a lot from the physicists. From the younger I like Slater the best!" [172]

A. Lawrence Lowell was the president of Harvard University; his sister, Elisabeth Lowell, was married to William Putnam, a member of the Harvard class of 1882. Putnam, who, according to Carathéodory, "was said to have shown interest in the development of mathematical instruction", described the merits of academic competition between teams of undergraduate students of different schools in regular college studies in an article for the December 1921 issue of the Harvard graduates' magazine. His views were shared by his wife and her brother. After Putnam's death, his widow created the William Lowell Putnam intercollegiate memorial fund in 1927. Her brother retired from his position as the president of Harvard in 1933; two years later, Elisabeth Putnam died. After her death, her sons consulted George Birkhoff and introduced the annual William Lowell Putnam mathematics competition in 1938, which still enjoys a high standing. Carathéodory gave the William Lowell Putnam speech on 21 March 1928.[173]

Millikan's popular lecture on Sommerfeld's theory of fine structure was probably based on Sommerfeld's book *Atombau und Spektrallinien* (Atomic Structure and Spectral Lines), which had been published in 1919. It is not sure whether Sommerfeld knew of the earlier contact between Carathéodory and Millikan in 1924.

The American physicist John Clarke Slater, who made such a positive impression on Carathéodory, was then 28 years old. He had gained his doctorate from Harvard in 1923, visited Cambridge University on a travelling fellowship and stayed for six months in Copenhagen near Niels Bohr. In 1924, Slater worked with Bohr and Hendrik Anthony Kramers on dispersion theory, which played an important role in the creation of matrix mechanics. In a letter to Bohr in 1926, he mentioned that he had independently duplicated most of Dirac's results.[174] In 1927 he published a wave mechanical calculation of the Einstein B coefficients.[175] Carathéodory's evaluation of Slater proved to be particularly accurate. Slater came back to Europe in 1929 to work with Bohr on quantum theory. That year he published his important work on the method of wave-function determinants,[176] which was immediately adopted by atomic physicists. The determinant that bears his name can be used for the construction of the anti-symmetric total wave function for any system with n electrons. In 1930 Slater became head of the physics department at the Massachusetts Institute of Technology (1930–1951), where he strengthened and expanded its research and scientific programme. In 1951 he was appointed to the Harry B. Higgins Professorship and became Institute Professor emeritus in 1966.[177]

4.13

At Princeton

Carathéodory was delighted with his experiences in America. His euphoria about his life and academic contacts is apparent in a letter to Weyl on 17 April 1928: "I was recently in Princeton and was very pleased to hear that you were going to go there on a 'trial year'. I think that you would like it excellently there. I myself found life here much more pleasant than I had expected and my wife is very satisfied with her stay. I am curious about California but I do not know whether we will return via Canada or the Panama Canal."[178]

Carathéodory might have been to Princeton to attend either the lecture on Delphic festival music on Monday 30 January 1928 by Eva Sikelianos, who was professor of Byzantine music at the Conservatory of Greek Music in Athens, or the lecture by Professor Edward Capps on the excavation of the ancient market place of Athens on 29 February 1928. Of course, he also went to talk with the Princeton mathematicians.

The Princeton Herald of 13 April 1928 announced that "Professor Weyl, who is a leading member of various European learned societies and who is generally recognized as one of the outstanding mathematical physicists of the world will join the Princeton Faculty next autumn. In 1927 he was awarded the [Lobachevsky] Prize of the Physico-Mathematical Society of Kazan."

4.14
An "Excellent Man" but not to be Appointed

Carathéodory's satisfaction in Harvard was not at all exaggerated. The positive impressions that his activity left led to considerations about a possible offer of appointment. On 7 May 1928, at the tenth division meeting attended by professors Osgood, Huntington, Graustein, Morse, Walsh and Brinkmann and Dr. Stone, chairman Coolidge reported that "no answer had been received from Professor Kellogg in reply to the cable asking his opinion concerning the plan to secure a permanent appointment for Professor Carathéodory and that Professor Birkhoff had replied as follows: Excellent man, excellent appointment if not preferring Germany and not delaying young men." The division voted to approve of the plan to secure a permanent appointment for Carathéodory.[179] Birkhoff expressed the same attitude in another instance. He recommended to Warren Weaver, head of the mathematics department at the University of Wisconsin in Madison, who was also considering Carathéodory's appointment, to prefer a young American mathematician instead.[180] In Birkhoff's opinion, later expressed in his 1938 address *Fifty Years of American Mathematics*, the main focus of mathematical research shifted from France and Germany to the USA at the beginning of the century. Young American scientists who secured this development had an educational background in Germany but they were subsequently working independently and strictly within the borders of the United States. At about the same time, when the University of Chicago came into existence in 1892, with its mathematical department consisting, among others, of Eliakim Hastings Moore, Oskar Bolza and Heinrich Maschke from Germany, the mathematical tradition at Harvard was strengthened by Osgood and Bocher, who were inspired by their stay in Germany and in particular by Felix Klein in Göttingen. "A few years later, under the guidance of Dean Henry Burchard Fine, who had been strongly influenced by his studies under Leopold Kronecker, promising young men were called to the mathematical staff at Princeton, in particular L. P. Eisenhart, Oswald Veblen, and J. H. M. Wedderburn. From that day forth there has always been an important mathematical group at Princeton."[181]

Exactly one year after the session of the mathematics division at Harvard mentioned above, on 9 May 1929, Coolidge expressed his regret to the board of overseers of Harvard College about the fact that Harvard had missed the opportunity to appoint

OSIRIS

Studies on the History and Philosophy of Science, and on the History of Learning and Culture

EDITED BY

GEORGE SARTON, S.D., L.H.D., LL.D.
Associate of the Carnegie Institution of Washington

———

VOLUME III

PART 1

1937

C. CARATHÉODORY

The beginning of research in the Calculus of Variations

———

THE SAINT CATHERINE PRESS LTD.
51, Tempelhof, BRUGES (BELGIUM)

Printed in Belgium.

Carathéodory's historical paper
The beginning of research in the Calculus of Variations.
Courtesy of the Bavarian Academy of Sciences, Archive.

Carathéodory: "not only was his scientific standing of the highest but he showed a remarkable aptitude to inspire every student who came near to him. We never made him an offer for we never saw the money available .... Recently an American University of high standing has offered him $10 000." [182] Of course, Coolidge avoided talking about the resistance within the mathematical faculty against an offer of appointment to Carathéodory that, in retrospect, appears to have been the biggest hindrance. However, his regret was genuine. He would have been only too pleased to have the mathematician from Germany at Harvard.

On 25 July 1928, Carathéodory conveyed his thanks to Coolidge for his hospitality and for the chance "to spend some time at that great old University". [183] Three years later, on 28 July 1931, Julian and Theresa Coolidge visited the Carathéodory family in Munich. Next to his signature in the guest-book, in parentheses, Julian Coolidge wrote the word "*Philister*" (Philistine), an attribute used by students of German universities to describe the ordinary citizens of their university town, whom they considered inferior and treated with contempt and at times violently. There is also an entry by Theresa and Rachel Coolidge in Carathéodory's guest-book dated 29 July 1934.

Carathéodory remained loyal and devoted to Harvard. On the occasion of its tercentenary, he gave a speech on "The beginning of research in the Calculus of Variations" [184] at the meeting of the American Mathematical Society on 31 August 1936.

4.15
The "Bochner Case"

As can be deduced from a letter by Perron to Carathéodory dated 6 July 1928, [185] Salomon Bochner was to be appointed assistant professor to Munich University, but he had been blocked firstly by the university senate and then by the local government although the Minister of Education and Cultural Affairs had himself supported him. Because of Bochner's excellent scientific performance, the faculty had asked for his appointment although he was a foreigner; "formerly an Austrian, he became Polish through a peace treaty". The Senate were prepared to accept Bochner's appointment only if he became "a member of the Reich". This led to his official employment as a "commissioned lecturer", but not as an assistant professor. [186] Perron referred to the "Bochner case" that could damage the university's reputation. He and Carathéodory had thought about the possibility of an appointment to Harvard for the young mathematician whom they both held in high esteem. Perron asked his colleague to accelerate Bochner's invitation to America, "since we could thus have a new means in our hand to make even clearer to the Ministry that Bochner is quite a character and that we make a fool of ourselves in front of the scientific world if we put him off life here." Carathéodory strongly recommended Salomon Bochner to Harvard. The question of substitute appointments, in case one or more members of the division received grants, was discussed at the twelfth division meeting of mathematics at Harvard University on 2 June 1928. Bochner's qualifications were carefully considered and the chairman was requested to confer with Carathéodory with regard to that matter. "After the conference, the Chairman appointed Professors Birkhoff, Kellogg, and Graustein as a committee with power, to investigate Dr. Bochner's qualifications at the Bologna Congress and, if it appears advisable, to offer him a position for the second half of the academic year." [187]

Carathéodory himself was convinced that the Munich deviousness and manipulation stemmed from a morbid political ideology. With respect to Bochner affair he wrote to Julian Coolidge on 25 July 1928 that "these silly people mix their political passions with the scientific development of our university." [188] Obviously he alluded to an anti-Semitism that was becoming more and more perceptible and was accompanied at the universities by the openly expressed demand for the exclusion of Jews such as Bochner. [189] Carathéodory, who wrote from Berkeley, asked Coolidge to accelerate the procedure for Bochner's invitation to Harvard and to address a private letter to Bochner that would suffice to strengthen the latter's position with respect to Munich. So Carathéodory pressed for Bochner's promotion in Munich by Bochner's employment in Harvard.

Four days later, from his summer residence in North Haven, Maine, Coolidge informed Carathéodory that a new element had significantly worsened the situation for Bochner: the distribution of research grants had been delayed and it was not sure whether there would be any available until February or even September 1929. Therefore, Coolidge was going to turn the whole matter over to Birkhoff to handle. [190] What Carathéodory did not know was that chairman Coolidge had in the meantime received a letter from Birkhoff, who was then in Paris. It would be more appropriate as university policy, Birkhoff wrote on 7 July 1928, "to get some promising young American, of about equal standing and achievement, to come to us for half a year at a similar salary. [...] [Bochner's] conversation gives the impression of genuine devotion to his science. He says he is interested in the whole field of analysis. Personally I have heard nothing much of Bochner in Vienna, Budapest, Szeged, Göttingen, or Berlin, which seems to me to indicate that he is not the man of outstanding promise. It is easy to understand that C[arathéodory]'s recommendations are slightly tinged by his personal relations to B[ochner], and also perhaps by a feeling (not rare here) that any second rate European youngster is good enough for us." [191] Birkhoff had been gathering this information on Bochner as an adviser of the International Education Board, which shortly after being founded by John D. Rockefeller, Jr., announced in 1923 a series of fellowships to help young scientists study in other countries what they were not able to study at home with equal advantage. [192] This "second rate European youngster" had gained his doctorate under Schmidt and Schur with a work *Über orthogonale Systeme analytischer Funktionen* (On Orthogonal Systems of Analytic Functions) in Berlin on 13 June 1921, and had worked together with Harald Bohr in Copenhagen, G. H. Hardy in Oxford and J. E. Littlewood in Cambridge in the years 1924–1926. He had been lecturing as an assistant at Munich University since 1926 and habilitated there with a work on the theory of almost-periodic functions in 1927. [193] Birkhoff's views were clearly tendentious and this is supported by the fact that, some years later, in 1933, Bochner was appointed associate professor to Princeton University. "It seems likely that Birkhoff shared the somewhat diffuse and varied versions of anti-Semitism held by many of his contemporaries." [194]

4.16
At Austin and San Antonio

As Carathéodory had told Sommerfeld, he would lecture at the University of Berkeley in June. On their way to California, Carathéodory and his wife stopped off in Washington, DC and directly afterwards in "the old slave city New Orleans", which he described as "full of negroes",[195] and old town reminding him of Smyrna.[196] His next stop was Austin, Texas, where, as a guest of the university, he gave a lecture on *Conformal Mapping* in front of a densely packed audience in Garrison Hall on 12 June 1928. Presenting the theory of functions of complex variables, he addressed a public of advanced students of mathematics. He received an enthusiastic welcome in Austin. He was introduced by Dr. E. L. Dodd and Dr. D. L. Clark to the university community as the greatest living mathematician and editor of the well-known mathematical journal *Mathematische Annalen*. His teaching activity in Germany and Harvard was praised and his political interests were stressed; his career was even compared with that of the piano virtuoso and Poland's former Premier Paderewski.[197] With a certain degree of exaggeration he was called an authority in Near East politics and he was mentioned as "the founder of the famous Smyrna University which was destroyed by fire by the Turks". The president of the university, H.Y. Benedict, invited him and his wife for lunch at his home, while the student paper *The Summer Texan* in the issue of 14 June drew attention to the "many hospitalities"[198] offered to Carathéodory. The host's zeal was really exceptional. Those days in Austin started with the obligatory common breakfast, continued with various invitations and car trips into the countryside and ended with an evening visit to the theatre. This daily routine was quite strenuous for the couple, especially for Constantin, who had to keep an eye on his academic engagements.

Carathéodory held the University in Austin in high esteem and was surprised by the mathematical faculty because of the large numbers of good mathematicians. He described it as first class, a remark accepted by its members with gratitude. As an organiser of science, he was impressed especially by the way in which the university had made money. An investment of its fortune in infertile land, enforced by politicians half a century earlier, changed into an exceptional stroke of luck when petroleum was discovered under that soil. Carathéodory believed that the university, being in possession of such means, could soon develop into one of the best in the world.[199]

Among the towns of Texas, San Antonio was the most attractive in his eyes. He described it as a fairy town in the middle of the hot desert, with tropical gardens, millionaires' palaces and old Spanish churches,[200] and he compared it with Nice, although a Nice of immense dimensions. From San Antonio he travelled directly to California. The drive lasted forty hours, more than half the time across the desert under temperatures of 45° C.[201]

4.17
Impressions of America

The first stop the Carathéodory couple made in California was Los Angeles. From 2 to 20 July, Carathéodory delivered fourteen lectures on the calculus of variations and selected subjects from analysis at the University of Berkeley, for which he was granted a honorarium of $1000.[202] He and his wife remained in Berkeley twenty more days after this. Shortly after their arrival he wrote to Käthe Hilbert full of enthusiasm for the Californian landscape: "Gardens, flowers, fruit trees and an everlasting spring. The plants do not know when to bloom; only the fig tree is wiser and one knows that it is winter when it has no leaves."[203] Carathéodory's wife is likely to have shared this euphoria, since he wrote to Penelope Delta on 17 June that Euphrosyne was very fond of the people and landscape and made remarkable progress in the English language.[204]

The couple used their stay in the Far West to visit the observatories of Mt. Wilson and Mt. Hamilton.[205] They found the landscape familiar, at least in Southern California, and Carathéodory even spoke of striking similarities to Greece as regards colours and vegetation,[206] but they also met quite a lot of curiosities. One of them was the American camping system, which impressed them because of its conveniences. They encountered it in Yosemite National Park, a deeply carved valley in Sierra Nevada, which, in Carathéodory's words, was more beautiful than the valley of the Swiss Engadin. On 20 August the couple was in Portland, Oregon from where they started out on an excursion along the banks of the Ohio. A day later they arrived in Seattle, the following day in Vancouver. They returned to New York on the *Canadian Pacific* railway and from there they embarked for Europe on 7 September. Carathéodory needed the peace of his own domicile again in order to realise his plans for several publications.[207] He had used the time during his long drives through America to prepare them. According to his own information, his paper *Stetige Konvergenz und normale Familien von Funktionen* (Continuous Convergence and Normal Families of Functions) originated "in the Rocky Mountains of Canada, mid-August 1928".[208]

From Boston to San Francisco, Carathéodory met several Greeks and many of them, the best one could imagine, as he wrote, were among the audiences of his lectures. Although he was appearing everywhere as the professor from Germany, he used every opportunity to propagate Greece's cause. "I assure you", he wrote to Penelope Delta from Los Angeles, "that in the few months of my stay here I have achieved much more than during my two years in Athens." He had not been able to carry through his ideas in Greece, but that was different in Europe and, most of all, in America.[209] It is the second time that Carathéodory assured another person that he was more useful to Greece when abroad, whereas he encouraged Penelope to stay in the country.

4.18

"A Great Catch": Appointment to a Full Professorship of Mathematics at Stanford University

On 1 November 1928 Hans Frederick Blichfeldt, a professor of mathematics at Stanford University, wrote to the president of the University, Dr. R. L. Wilbur. Blichfeldt was spending some days in the White Mountains at that time, together with Professor

Roland G. D. Richardson, dean of the graduate division of Brown University, secretary of the American Mathematical Society and a close friend of George Birkhoff. He informed Wilbur that Richardson was a man of considerable influence in mathematical circles in the United States and had acted as a confidential mathematical advisor to Yale, Columbia and other universities. In the previous year, Richardson thought that the prospect of real mathematical development on the Pacific coast would concentrate in Pasadena and Stanford in the near future. In the summer he had excluded the possibility of obtaining Bliss or Veblen[210] for the Pacific coast, unless they were offered $12000 to $15000. Richardson argued for a bigger salary for young talented men in order to attract them to the West Coast, since competition with other universities was severe and, according to common opinion, the West Coast climate did not stimulate productivity. Blichfeldt himself had asked Richardson if it were advisable and possible to bring Carathéodory to Stanford. Richardson replied that Carathéodory would be "a great catch"; several universities would like to get him, but they did not stand the slightest chance. Both Carathéodory and his wife were fond of California and it might not be that difficult to get him to come to Stanford. However, he would not be the appropriate person to teach undergraduates. Richardson countered Blichfeldt's argument by saying that Stanford ought to go after Carathéodory by all means and that it did not matter much in which way Carathéodory would be used in lecturing. After all, any one of the young talented analysts could take on this task as an instructor or assistant professor and would be glad to profit from contact with Carathéodory. Richardson strongly urged Blichfeldt to find out from Carathéodory under what conditions he would be prepared to go to California. Blichfeldt indeed contacted Carathéodory in New York on the day of the latter's departure and he indicated his requirements: a satisfactory salary and expert medical services for his son, who was suffering from the effects of an infantile paralysis. Better medical care for his son was after all the reason that he had given up his professorship in Göttingen to accept the one in Munich. In addition, Carathéodory had stated that his pension in Munich would equal his salary. Blichfeldt informed him of Stanford's building fund, from which professors could borrow money without interest, and Carathéodory replied that he was not interested in building a house that would cost more than $10000. Blichfeldt also informed Wilbur that, although he was reputed to be wealthy, Carathéodory was living very simply. From this point of view the salary question would not be very important to him unless he thought it ought to measure up quite well with the salaries of prominent mathematicians in the States, who were getting $9000 to $10000. Blichfeldt concluded that Carathéodory ought to be offered a salary of $8000, a loan of $10000 for building purposes from Stanford's building fund and finally that, regarding the climate, Stanford would be better than Munich for Stephanos and that Stephanos would receive at least equally good medical treatment in Stanford as in Munich. If Carathéodory was prepared to go to Stanford, then they would have to employ a young man to lecture analysis to the undergraduates. That young man would guarantee continuity upon Carathéodory's retirement. At the end of his letter to Wilbur, Blichfeldt added the remark that the few men in the Stanford faculty who had come in close contact with Carathéodory during the latter's stay in Stanford were very impressed by him.[211]

This letter reveals bargaining on both sides: Carathéodory's two untrue statements, namely that Stephanos' health was the reason he had left Göttingen for Munich and that his pension would equal his salary in Munich, and the Americans' attempt to cut Carathéodory's salary despite the fact that they acknowledged him as a great catch. That bargaining was to continue, when two weeks later, on 14 November 1928, Blichfeldt communicated Wilbur's telegraphed response to Carathéodory: a salary of $ 7500, the possibility of getting a loan free of interest within the following year or two to build a house on university ground, excellent facilities in California for his son's treatment. Wilbur himself was one of the best doctors in the country and for several years he was dean of the Stanford medical school in San Francisco.

Blichfeldt also informed Carathéodory that staff members of the university retired at the age of 65 and the amount of their pension depended on the amount of the yearly payments into the Teachers Insurance and Annuity Association, each payment being equal to one tenth of each professor's salary. Finally, Blichfeldt informed Carathéodory about the taxes he would have to pay, which were state and county taxes and income tax, all of them in fact being very low.[212] On 10 December 1928, Carathéodory announced to the dean of his Munich faculty that he had received an offer for an appointment at Stanford.[213] He replied to Blichfeldt on 26 December 1928 in a very reserved manner.[214] A new offer, this time from Wilbur himself, reached him on 23 January 1929. The salary offered had been raised to $ 8000 a year and Wilbur promised to expand the library facilities. He concluded his offer by saying that he would be very glad to recommend to the board of trustees that Carathéodory should be appointed a professor of mathematics at Stanford University commencing on 1 September 1929.[215] On 14 February 1929 the board of trustees did in fact appoint Carathéodory to a full professorship in mathematics and fixed his salary at $ 8000 for the academic year 1929–1930.[216]

4.19
Carathéodory Negotiates to Remain in Munich

Now armed with a definite offer from Stanford, Carathéodory began to use it to negotiate better conditions for remaining in Munich. He explained that the conditions offered to him by Stanford were more advantageous than those he had in Munich and he demanded a supplementary salary of RM 3000 that would count towards his pension. Until then he was receiving an additional salary of RM 1960 which amounted to 10% of his final salary as a full professor. Munich University argued for the retention of Carathéodory since "regarding scientific reputation, he stands at the top of the mathematics representatives here and [...] he has proved to be a very good lecturer." The university asked the Ministry of Finances to meet Carathéodory's demand with the following suggestion: "The increase in salary desired by him is significantly smaller than the one recently granted to Professor Dr. Tietze in order to avert his appointment at Leipzig. The new supplementary salary would not even amount to 24% of the final basic salary of the full professors."[217]

The Bavarian Ministry of Education and Cultural Affairs also asked the Ministry of Finances to consent, at least partly, to Carathéodory's demand. On 3 April 1929 they drew attention to the fact that Carathéodory had been treated unfairly already during the negotiations for his appointment to Munich, since his supplementary salary did not count towards his pension. Carathéodory had swallowed the insult, at first without protest, "maybe because he was then in Athens and there was hardly any time for the resumption of negotiations for the acceptance of a professorship that had to be occupied already on 1 May 1924. [...] But as professor Dr. Carathéodory later became aware that he was almost the only university professor with an additional salary not counting towards his pension, he felt that the settlement made for him was more and more a striking and objectively illegitimate exception." That treatment was inappropriate. "Since professor Dr. Carathéodory is by descent a foreigner, it is likely that he assumes that he has to be left behind because of this property, whereas by his scientific importance he surely has to be placed to the fore no less than the native professors. In fact, as a scholar he enjoys an international reputation."[218] Finally, the Ministry of Education argued for an increase in Carathéodory's supplementary salary to RM 2400 and its incorporation into the pension.[219] That request was less than Carathéodory's claim. The fact that Carathéodory had possessed German nationality since 1924 seems to have been no guarantee of equal treatment to that of his German or even his Austrian colleagues, such as Tietze. The Ministry's argument, namely that Carathéodory should not be left with the impression that his unjust treatment had to do with his non-German descent, is typical of this lack of guarantee.

Why was Carathéodory not prepared to leave his chair in Munich for another one in Stanford at that time? Leon Warren Cohen's account of his meeting with Carathéodory may answer the question: "The summer of '29 [...] we went to Munich. I visited the university and was introduced to this young man, Salomon Bochner. I had another interesting meeting there. Caratheodory had lectured in the United States and come to Ann Arbor while I was there. A very impressive fellow. One of the few books I ever bought was his *Vorlesungen über Reelle Funktionen*. So in Munich I reminded him that we had met at Ann Arbor. He insisted that I come to visit him at his home, and for the first time I saw what a well-endowed chair could do for you. He lived in great comfort. He was a very pleasant host; he put me at my ease, and we had a pleasant discussion for about half an hour."[220]

On 17 May 1929 the Ministry of Finances granted Carathéodory the increase in his supplementary salary, which was to count towards his pension, proposed by the Ministry of Education.[221] With this in hand Carathéodory informed Blichfeldt on 12 May 1929 that, to his great regret, he and his wife had decided not to leave Munich. The main reason for the rejection of the offer was the difficulty with his pension. By leaving Germany, he would lose all the money he had saved during the previous twenty-four years of work in that country. However, Carathéodory gave the impression that he would be pleased if he were invited as a guest professor to America for a year or even longer. But he added that it would be too risky for all his family to move "to the other side of the world".[222]

As Carathéodory's letter was on its way to America and on the same day (17 May) on which the Bavarian Ministry of Finances approved the increase in Carathéodory's supplementary salary, Blichfeldt wrote to Wilbur, who had just become Secretary of the Interior in Washington, DC, that the pension plan for Carathéodory was advancing due to his efforts and that he could gain the co-operation of Dr. Coryllos, a former colleague of Carathéodory in Athens, of Mr. Alfange, a young American banker and a Philhellene, and of a wealthy, liberal and educationally minded Greek.[223] However, Blichfeldt's efforts were in vain.

4.20
Carathéodory and Radó

On 3 July 1928, from Berkeley, Carathéodory had informed Radó that Radó's application for a post at Munich was excellently well-founded and adequately supported. He promised to write directly to Trowbridge and send a copy of his letter to Radó. He was happy that Radó had decided to go to Munich and was sure that he would have a good opportunity there of proceeding with solutions of his problems, which were all of great interest to Carathéodory.[224] In Kérékjartó's opinion, Radó was one of the most diligent and most talented mathematicians of South Eastern Europe.[225]

Tibor Radó was born in Budapest in 1895 and was able to study for only two years at Budapest Technical University before being recruited by the Austro-Hungarian Army in the First World War. He was taken prisoner by the Russians in 1916 and spent most of the following four years in a prisoner-of-war camp in Siberia. "He gradually worked his way from Vladivostok back to his native land, doing all sorts of odd jobs, living in the Arctic, and studying science from a textbook written by an American"; he was "full of good wit and humor."[226] He continued his studies with professors Alfréd Haar and Frigyes Riesz at Szeged University, where he wrote his doctoral dissertation under Riesz in 1922. During the period 1923–1929 he spent three years in Germany with the help of a fellowship of the Rockefeller Foundation, which allowed him to work together with Carathéodory in Munich (1928–1929) and with Paul Koebe and Leon Lichtenstein in Leipzig. While with Carathéodory, Radó researched conformal functions and the calculus of variations. His main result connected to this research was the solution of the Plateau problem.[227] In 1929 he went to the United States as a visiting lecturer, at first to Harvard and subsequently to the Rice Institute.[228] In the following year he was appointed to Ohio State University in connection with the foundation of a graduate programme for mathematics. He remained there until his retirement in 1965. Radó's name figures in Carathéodory's guest-book together with the signatures of H. F. Blichfeldt, J. Lense, S. Bochner and F. Hartogs under the date 20 June 1929. A year later, on 12 June 1930, Radó visited Carathéodory again. His name in the guest-book is accompanied by the words "Houston–Szeged–Columbus". When professor at Columbus, Ohio, he and Erhard Schmidt had contributed to the mathematical proofs in Carathéodory's book *Conformal Representation*. It was published in July 1932 as the 28th volume of the Cambridge Tracts in Mathematics and Mathematical Physics and had resulted from lectures Carathéodory had given at Göttingen, Berlin, Athens, Munich and Harvard University. It contained the theory of conformal representation as developed during the previous two decades. His manuscript, written in German, was translated by B. M. Wilson of the University of Liverpool and by Margaret Kennedy of Newnham

College. The latter had participated in the mathematical colloquia of the previous year held in Carathéodory's house on March 6 and July 31 and suggested ways to simplify his text.

For the second edition of his *Conformal Representation* after the war,[229] Carathéodory appended a new chapter (Chapter VIII) to his original work, on the general theorem of uniformisation. As he wrote in the preface, he had succeeded in making this chapter rather short thanks to the beautiful proof from van der Waerden,[230] which enabled him to expound the topological side of the problem in a few pages. In this chapter he also referred to Radó's paper *Über den Begriff der Riemannschen Fläche* (On the Notion of the Riemann Surface)[231] and to his own *Bemerkung über die Definition der Riemannschen Flächen* (Remark on the Definition of Riemann Surfaces),[232] a paper dedicated to Perron on his 70th birthday. Carathéodory had asked Radó on 17 November 1949 to provide the relevant reference as soon as possible, since the new editor of the Cambridge tracts, Hodge, was leaving for the States and Carathéodory wished to send him the complete manuscript before his departure.[233] But just the following day, Carathéodory found Radó's reprints, which he had mislaid and lost for many years and could study again Radó's "*beautiful paper*".[234] Radó replied ten days later.

Dear Professor Caratheodory:

Thanks for your letters of November 17 and 18 which I received simultaneously. I shall, of course, be very happy if my remark on the concept of a Riemann surface will be referred to in the new edition of your book. In this connection, you may be interested in a recent paper by M. Heins, *Annals of Mathematics*, vol. 50, 1949, pp. 686–690, in which he shows that the basic existence theorem can be established by using subharmonic functions, without triangulation, if the definition of a Riemann surface is used in the modified sense I proposed. Of course, I became aware of this possibility myself when I worked on subharmonic functions, but somehow I could not make up my mind about publishing the thing. I am quite happy about the Heins paper, since the method is quite nice.

During the last year or two, I started on the difficult problem of studying the topological foundations for [the] Calculus of Variations for n-tiple integrals. Thus I spent most of my time studying Algebraic Topology. It is an enormous and extremely beautiful field, but it begins to look as though most of the things I seem to need are quite remote as yet. However, with one of my very able younger friends here, we obtained an extremely general theory of the transformation of the independent variables in multiple integrals, based on certain generalized Jacobians. In addition to practically everything in the theory of functions of real variables, we have to use all of the so-called Cech homology theory of compact metric spaces. We are working now on a monograph containing this theory, and since the results are primarily meant for use by Analysts, the matter of presenting the topological portions is quite a problem.[235]

As Heins wrote, his approach was based upon (1) the monodromy principle for simply connected surfaces, (2) the modern treatment of the Dirichlet problem with the aid of subharmonic functions (employed by Perron and Carathéodory) which was directly available for regions on a Riemann surface with compact adherence, and (3) the relationship between his proposed conformal maps and the groups of conformal transformations of the canonical models onto themselves, which was more characteristic of his approach.[236]

4.21
A *"Pack of Wolves"*

On 30 October 1928, Carathéodory travelled to Laren to meet Brouwer and clear up matters raised by the contents of two letters addressed to Brouwer from Göttingen, one written by Hilbert, the other by Carathéodory himself. He had already discussed the problem with Blumenthal in Göttingen and Schmidt in Berlin[237] but not with Hilbert. In his letter, Hilbert had announced the impossibility of his further co-operation with Brouwer because of their incompatible views on fundamental matters. Through this step Hilbert expressed his basic disagreement with both Brouwer's hostile position towards foreign mathematicians that had been expressed in the editorial work of the *Annalen* and with Brouwer's opposition to the International Congress of Mathematicians in Bologna in September.[238] Having secured the consent of the Academy in Göttingen and the Prussian Academy, and believing that it was "a command of rectitude and the most elementary courtesy to take a friendly attitude towards the Congress",[239] Hilbert had led more than sixty German mathematicians there, although he was suffering from a recurrence of the illness that had almost caused his death some years earlier. However, a group of "nationally" minded mathematicians objecting to the participation had called for a boycott of the Bologna congress, arguing that the congress was connected with the International Mathematical Union and the International Research Council, which were still hostile to German science. Three Berlin mathematicians, namely Bieberbach, von Mises and Schmidt, belonged to the initiators of this action.

Carathéodory had not participated in the congress, he was then on his way from the United States back to Europe. Of course, he could have arranged his return in such a way as to be able to participate, but he probably preferred avoiding getting caught up in the conflict. However, he could not avoid his entanglement in the *Annalenstreit* (see 2.10), that was to follow in a very unpleasant way.

In his letter, Hilbert also informed Brouwer that he was authorised by Blumenthal and Carathéodory, members of the board of chief editors, to remove his name from the editorial board of the *Annalen*.[240] To substantiate his request for authorisation, which he also addressed to Einstein, Hilbert explained that Brouwer had offended him and the overwhelming majority of the German mathematicians through his last circular letter to the German mathematicians at the international congress in Bologna; that, especially through Brouwer's pronounced hostile position against foreign mathematicians who were well-disposed towards the Germans, his participation in the editorial board of the *Mathematische Annalen* was not appropriate; that Hilbert would, after all, like to keep Göttingen as the main base of the *Mathematische Annalen*. His decision was firm and irrevocable.[241]

Hilbert's accusations were not entirely unjustified. It is true, for instance, that Brouwer had expressed his disagreement with the invitation of the French mathematician Painlevé to co-operate in the special volume of the *Annalen* in honour of Riemann "because in the public session of the French Academy Painlevé had characterised the German scholars of the 19th century as 'slave-like builders of the most colossal murder machines that mankind had known'".[242] Brouwer's attitude against the international congress in Bologna is also a fact. As Hilbert said: "it was awful, but the worst is that although most of the German mathematicians and all foreigners, especially his close colleagues in Holland, have recognised his wheelings and dealings, some German mathematicians still stick by him".[243] Hilbert obviously meant the three Berlin colleagues. In a letter to Einstein, Born tried to explain Hilbert's decision: "In terms of politics, Hilbert has not at all a proclivity to the left, I should say quite the reverse; for my taste and even more for yours he is rather reactionary. But he has quite a sharp awareness of what is necessary in the communication of scientists of various countries for the benefit of the whole. Hilbert, as we all did, considered Brouwer's behaviour in the matter to be a folly, since he appeared more nationalist than the Germans themselves, but the worst was just that the Berlin mathematicians were taken in by Brouwer's tomfoolery. I would like to add that the Bologna matter was not crucial in Hilbert's decision to remove Brouwer, but only an opportunity. In the case of Erhard Schmidt, I can understand this; he was politically always right-wing and, to be precise, really out of genuine feelings. But in the case of Bieberbach and Mises it is a rather lamentable symptom."[244]

Despite the fact that Einstein regarded Brouwer "as a psychopath" and an "involuntary advocate of Lombroso's theory of the close connection between genius and madness," he had great respect for his spirit and considered it "neither objectively justified nor appropriate to undertake anything against him". With a touch of irony, he wrote to Hilbert: "Sire, give him the freedom of a court jester!" Einstein also sent Carathéodory a copy of his letter to Hilbert.[245] Carathéodory answered that the conflict in Bologna was only the "pretext of Hilbert's action", whereas the real grounds were deeper and partly went back to a decade ago. Besides, Hilbert feared that after his death the further existence of the *Annalen* would be at stake; his stubbornness due to his disease was confronted with Brouwer's unpredictability. Finally, Carathéodory held that Einstein's picture of Brouwer was drawn wrongly.[246]

Carathéodory's view at that time seems to be partly confirmed by what Hilbert said to Born about Brouwer: he was "an eccentric and unbalanced man, to whom he would not like to entrust the heritage of the management of the *Mathematische Annalen*".[247]

Einstein responded to Carathéodory by asking him for a genuine answer to the question of who and what was behind that "systematic action" and he once again expressed his opinion about Brouwer. Despite Brouwer's weaknesses, Einstein admired him not only as an exceptionally perceptive spirit but also as a direct and steadfast man. Einstein assured Carathéodory that he would never tell anyone that Carathéodory was the one to have informed him.[248] As expected, Hilbert did not follow Einstein's recommendation. On the contrary, he told Einstein that he considered it "a great fortune for the *Annalen* that Brouwer's expulsion from the editorial board could now occur."[249] During his visit to Holland, Carathéodory apologised for the contents of his own letter (which, however, was returned unopened to him by Brouwer), but begged Brouwer to accept Hilbert's decision without resistance. He appealed for compassion for Hilbert because of the latter's disease.[250] When parting, Brouwer told Carathéodory that if Hilbert accepted Einstein's argument, he would definitely be able to revoke his decision, as well.

In a circular letter to the publisher and editors of the *Annalen*, Brouwer asserted that Carathéodory was not able to give him a "reasonable answer", but only exclaimed: "'What should one do' [...] 'I do not want to kill a man' [...] 'I do not understand you anymore', 'I consider this visit as a parting visit' and 'I feel sorry for you'." Indeed, a desperate and embittered Carathéodory had left his former close friend. Brouwer was insulted, Carathéodory's attitude had caused him "astonishment, uncertainty and anger".[251]

Three days after their meeting Brouwer sent Carathéodory a rather insensitive letter, which he also addressed to Blumenthal. He wrote that he would be prepared to behave towards Hilbert in the way Carathéodory had demanded, i.e. as if Hilbert was "not responsible for his actions", only if Frau Hilbert and Hilbert's physician requested him to do so.[252] Blumenthal reacted with a letter to the publishers and editors of the *Annalen* on 16 November: "For this alarming and repulsive letter, of which Brouwer informed me too by copy, I have only one explanation, that of all things out of Cara's expressions and requests, Br., according to his views, has (intentionally or spontaneously) constructed the most hateful."[253] Thus, Brouwer also lost Blumenthal's potential support. Carathéodory considered Brouwer's letter to be crazy and believed that his own initiative had fallen through. He saw the only positive outcome in the withdrawal of the Berlin mathematicians' unconditional support for Brouwer and consequently in keeping the conflict away from publicity. Since his mediation had failed "so miserably", he felt that he had to resign from the editorial board as soon as possible.[254]

On 5 November, Brouwer fuelled the flames by appealing directly to the sense of chivalry of the publisher and editors of the *Annalen* and to their respect for Klein's memory. Brouwer argued that the moral prestige and scientific contents of the *Annalen* were going to be sacrificed through the handling of the situation by Blumenthal and Carathéodory, who "regarded the expected advantages for Hilbert's state of health" more highly than his own "liberties and chances to be active". The chief editors should either give up their plan or the other editors should continue Klein's tradition and manage the journal themselves. Brouwer reminded the receivers of his circular letter of the obituary to Klein, in which Carathéodory, in the name of the editorial board, had written:

He made sure that the various mathematical trends were represented in the editorial board and that the members of the board worked alongside him with equal rights. As always in his life, he never paid any attention to himself, but only kept his eye on the aim to be achieved.

Brouwer believed he recognised Hilbert's wish "to harm him and to insult him in any possible way" in the latter's intention to remove him from the editorial board of the *Annalen*.[255]

Carathéodory handed in his resignation but permitted Blumenthal to postpone its announcement so as not to feed any rumours that he had turned against Hilbert. On 13 November, Brouwer and Bieberbach unexpectedly visited the publisher Ferdinand Springer, who told them that he would not like to take sides, but if forced to do so, he would choose Hilbert's. As they were leaving, they threatened Springer with damage to his publishing company if he acquired a bad reputation among German mathematicians because of his behaviour in the *Annalenstreit*. Springer had the impression that Brouwer was determined to carry the fight to its bitter end.[256]

Blumenthal's circular letter to the editors, written after consultations with R. Courant, Carathéodory and H. Bohr, reviewed, in its final version of 16 November, the story of the conflict thus far and drew the conclusion that

> [Hilbert] had gained the firm conviction that Br's activity was harmful for the *Annalen*, and that he therefore could not take the responsibility of appearing as a chief editor in an editorial board to which Br. belongs. [...] Hi. had recognised in Br. an obstinate, insane and domineering character. He feared that once he had parted from the editorial board, Br. would force the editorial board, according to his own will and he considered it to be such a great danger for the *Annalen* that he wanted to confront it as long as he was still able to do so. Probably under the impression of his recent illness he considered himself, to the interest of the *Annalen*, obliged to cause Br's resignation from the editorial board and to take this measure immediately and with all his power.

Blumenthal's letter ended with the request "for either a quick statement or a silent approval" and for permission to remove Brouwer's name from the cover of the following issue of the *Annalen* and to refrain from sending further notifications regarding the *Annalen* to Brouwer.[257] Four days after his circular letter, Blumenthal contacted Carathéodory for a meeting with Springer, saying that it would be necessary to try their luck with Bieberbach by influencing him personally, so that they would be able to come out of this affair without getting their fingers burned.[258] Born, who had been let in on the details by Bohr and Courant, informed Einstein on that same day that in the impending meeting at Springer, everything depended on a concerted action from the closer editorial board and asked Einstein to insist on his neutral attitude and not to undertake anything against "Hilbert and his friends". Born attributed his interest in this affair to his concern for Hilbert and sympathy for his tutor and friend. In his opinion, Hilbert's spirit was clearer than ever and Brouwer's allegation about the unsoundness of Hilbert's mind was extremely heartless. Born sent Einstein a text from Springer to Bohr and Courant, proving that Brouwer and Bieberbach would suspect Springer of not being national and harm him if he stuck by Hilbert.[259] Bieberbach turned absolutely decisively against the intended exclusion of Brouwer, which he described as "objectively ungrounded". Brouwer was for him a man of best intentions and in every way worth of appreciation.[260] Springer abstained from voting on Blumenthal's proposal; Bieberbach sided with Brouwer, but without attacking Hilbert; von Dyck justified neither Brouwer's views nor Hilbert's action; Hölder disapproved of Brouwer's removal by force.[261] Finally, Einstein distanced himself from the dispute of that "mathematical pack of wolves", in which he was caught as an "innocent lamb". Ironically, he expressed the wish for "an extensive continuation of this just as noble as significant struggle".[262]

Blumenthal wrote to von Kármán on 22 November: "In the *Annalen* everything is topsy-turvy. You have received a document from Brouwer, of which you probably do not quite know what to make. My reply to this document, which I hope is clear, is enclosed. Brouwer's letter to Carathéodory is after all an amazing performance."[263] In order to obtain information on the legal question[264] and have the conflict resolved in such a way as to prevent any injustice being done to Brouwer, on one hand, and spare Hilbert from getting irritated, which would endanger his life, on the other, Carathéodory asked his colleague and friend Müller-Erzbach, from the law faculty of Munich University, for legal advice. Taking account of all the possibilities that Müller-Erzbach had put before him, Carathéodory came to the conclusion he should recommend to Blumenthal the dissolution of the entire editorial board and the creation of a new one.[265] Courant basically agreed to this solution but thought that it was necessary to both parties (publisher and editors) to cancel the contract voluntarily and to order the new relations by contract directly afterwards. He and Bohr intervened in the whole matter more than they would have liked, because they were representing Hilbert officially and they were trying to keep him away from unpleasant things. Courant believed that a quick handling of the situation in the direction that Carathéodory had proposed would not be possible and, therefore, the meeting in Berlin proposed by Springer and Blumenthal would be appropriate.[266]

On 1 December Carathéodory wrote to him that he had doubts of principle about his presence at the "Berlin conference". He felt obliged to play a role only in those matters serving the "liquidation of the affair", most of all because of his devotion to Hilbert, but he would not like to intervene in matters of the future. In case of an oral review it would not be possible to distinguish between these two blocks of problems.[267] The heralded meeting between Carathéodory, Courant, Blumenthal and Springer had been repeatedly postponed and was finally cancelled. On 3 December, Carathéodory wrote to Brouwer that he did "everything possible to find an acceptable solution to this matter" and asked Brouwer urgently "to be patient for some more weeks". Brouwer hoped and trusted that Carathéodory's efforts would succeed.[268]

The same day that he wrote to Brouwer, Carathéodory announced to Einstein that he had asked Blumenthal to keep his resignation concealed from the editorial board, so as not to give the impression that he had turned against Hilbert. He also tried to find a way out of this situation that would be bearable for all. That day, Carathéodory had received a letter from Bohr containing "a very remarkable" proposal. It was based on the idea, which, as Carathéodory said, was also the foundation of his own attempt for solution, that the change in the editorial board should not be a legal trick aimed at breaking Brouwer's resistance, but should instead neutralise the offence caused by Hilbert's initial action as much as possible. The method that Bohr would like to apply was to reduce the whole editorial board to four persons and to dispose of the so-called secondary editorial board. In case Einstein, just like Carathéodory, did not want to be a member of the new editorial board, Weyl could be proposed as a suitable substitute. Carathéodory had already proposed Erich Hecke to replace him. Only Hilbert and Blumenthal would remain from the old editorial board, namely the initiator and the main supporter of the campaign against Brouwer. Carathéodory thought that the dissolution of the present editorial board ought to be prompted by the publisher, so that neither he nor Einstein need worry about what would happen next. He communicated this idea to Einstein and asked for his opinion confidentially. Einstein agreed.[269]

The matter was solved when Courant and Springer agreed on the following compromise. Springer would sign a contract only with Hilbert, something that would not hinder Blumenthal being accepted in the new editorial board at a later time. On

12 December, Carathéodory congratulated Courant on his skilfulness in handling the situation and maintained that the solution suited his own intentions better than the plan he had proposed himself. He confessed that he had been deeply worried during the previous weeks and that he feared he would lose all his friends after having lost Brouwer. He had even played with the idea of turning his back on Europe once and for all, after having received an offer for an appointment at Stanford. His last letter, he added, seemed to have angered Bohr, but Bohr's reaction was due to a lack of knowledge of the details. Otherwise, Bohr would have probably approved of his reasons. Also Einstein had agreed to the solution and he had authorised him to inform Courant accordingly.[270]

The plan was accepted by Blumenthal,[271] whereas Einstein asked Courant not to include him in the new editorial board after the end of the "frogs-mice battle".[272] Even after the resolution of the *Annalenstreit* and having refused to participate in any future organisation of the *Annalen*, Carathéodory stuck to his opinion as regards Hilbert's motives, which he considered to be partly personal, despite Bohr and Courant's attempt to persuade him of the opposite.

In a letter dated 17 December, Courant announced to Carathéodory that Hecke was very pleased to join the editorial board of the *Annalen*. He further assured Carathéodory of Hilbert's appreciation of the role he had played during the quarrel. In fact, Carathéodory had done everything possible for Hilbert, who was absolutely satisfied with the outcome. Courant added his conviction that no personal motives were involved in Hilbert's initial moves, but only his concern to protect the journal from unpredictable and maybe fatal influence.[273]

Two days later, Carathéodory reminded Courant that Hilbert now presented Brouwer's insult against him as the only reason for his decision then, and therefore it would be humiliating for Hilbert if someone belatedly constructed a scenario excluding personal motives from Hilbert's behaviour. With respect to the new contract, Carathéodory did not agree with Hecke, who believed a tighter organisation of the editorial board was necessary. He reminded Courant that Klein had organised the editorial board in such a way as to create a kind of academy with equal rights for everyone; this was exactly the main reason why the *Annalen* could lay claim to being the first mathematical journal in the world. But now the *Annalen* would become a journal like any other and this would happen because there would be no other choice. However, Carathéodory saw Hecke's acceptance to join the editorial board as a guarantee that the quality of the *Annalen* would not deteriorate.[274]

On 23 December, Courant tried once more to convince Carathéodory of Hilbert's motives. He said that, now that it was again possible to talk with Hilbert peacefully and extensively, he knew for sure that Hilbert's motives were objective and rooted in his sense of responsibility for the *Annalen*, as well as in his conviction that when he was no longer able to counterbalance Brouwer, the latter's personality could turn out harmful and dangerous for the *Annalen*. Courant added that Hilbert had stressed to him and to Bohr that he had no feelings of hate, wrath or offence towards Brouwer.[275] Bohr suggested that if Carathéodory was not convinced by him and Courant, he should openly ask Hilbert about his motives. Bohr did not accept that Hilbert was unpredictable and subjective, particularly since Hilbert was not informed

MATHEMATISCHE ANNALEN

BEGRÜNDET 1868 DURCH

ALFRED CLEBSCH UND CARL NEUMANN

FORTGEFÜHRT DURCH

FELIX KLEIN

UNTER MITWIRKUNG

VON

LUDWIG BIEBERBACH, HARALD BOHR, L. E. J. BROUWER,
RICHARD COURANT, WALTHER V. DYCK, OTTO HÖLDER,
THEODOR V. KÁRMÁN, ARNOLD SOMMERFELD

GEGENWÄRTIG HERAUSGEGEBEN

VON

DAVID HILBERT   ALBERT EINSTEIN
IN GÖTTINGEN   IN BERLIN

OTTO BLUMENTHAL   CONSTANTIN CARATHÉODORY
IN AACHEN   IN MÜNCHEN

100. BAND

BERLIN
VERLAG VON JULIUS SPRINGER
1928

MATHEMATISCHE ANNALEN

BEGRÜNDET 1868 DURCH

ALFRED CLEBSCH UND CARL NEUMANN

FORTGEFÜHRT DURCH

FELIX KLEIN

GEGENWÄRTIG HERAUSGEGEBEN

VON

DAVID HILBERT
IN GÖTTINGEN

UNTER MITWIRKUNG VON

OTTO BLUMENTHAL   ERICH HECKE
IN AACHEN   IN HAMBURG

101. BAND

BERLIN
VERLAG VON JULIUS SPRINGER
1929

The change in the editorial board of the Mathematische Annalen *as seen
by comparison of the cover pages of volumes 100 and 101.*

about the earlier procedures and was thus unable to defend himself.[276] One day later, Carathéodory replied to Courant that it was a very complicated business to judge Hilbert's motives. He believed that he was able to see through them, since he had been used to his way of thinking for more than twenty-five years. He cited from Hilbert's letter of 15 October to the co-editors to prove that the only motive for Hilbert's initial action against Brouwer was that Hilbert felt he and the majority of German mathematicians had been offended. Carathéodory hoped that he had convinced Courant. However, he thought that this debate was obsolete; at that moment the only important thing was the agreement over the final aim, and that had been achieved. Otherwise, the unimportant difference in their opinion referred only to the assessment of what could tactically be more advantageous.[277]

Despite all the efforts of the members of the editorial board to discover objective motives behind Hilbert's behaviour, Carathéodory finally came to the conclusion that personal reasons had driven Hilbert to decide on Brouwer's expulsion from the editorial board of the *Annalen*. Thus, Carathéodory's view changed from his initial impression as expressed to Einstein, namely that Bologna was only the pretext for Hilbert's actions. In the end he rather favoured Einstein's view that the quarrel could have been simply a clash between a "Sire" and a mad genius. He recognised that his partiality for Hilbert had forced him to do injustice to his friend Brouwer, and in the course of the quarrel he did not hesitate to change his mind and face the consequences by resigning from the *Annalen*.

On 23 January 1929, after the arrangement between Hilbert and Springer had been sealed, Brouwer turned to the chief and associate editors and called upon them to revolt against Hilbert, Blumenthal and Springer: "Consequently, to the remaining editors falls the task to further run Felix Klein's edifice and to continue Klein's tradition in the management of the mathematical journal."[278] Of course, Brouwer refrained from attacking Carathéodory and Einstein, who were no longer on the board of editors.

The changes in the *Annalen* board happened from volume 100 (1928) to volume 101 (1929). Only Hilbert remained a chief editor; Carathéodory, Einstein and Blumenthal were absent. Also Bieberbach, Bohr, Brouwer, Courant, von Dyck, Hölder, Kármán and Sommerfeld no longer appeared as associate editors and were replaced by Blumenthal and Erich Hecke.

Thus, the *Annalenstreit* put an end to the Brouwer–Hilbert or intuitionist-formalist discourse on the foundation of mathematics, which had grown increasingly subjective throughout the 1920s, to the point that it turned into a personal dispute, with consequences for Carathéodory and Einstein, as well as for the *Annalen*. The *Annalen* then came under Hilbert's sole supervision.

4.22
Carathéodory's View of Rosenthal

While Carathéodory was concerned with the future direction of the *Annalen*, and having to absorb Brouwer's criticism of him, there were also other matters for him to attend to. In September 1928 his old friend Steinitz, a professor at Kiel, had died and the question of his replacement arose. Fraenkel had been appointed that year to Kiel, from his post of assistant professor at Marburg, and he wrote to Carathéodory for advice on finding a suitable replacement. Carathéodory's reply of 5 November, the same date as Brouwer's circular letter to the publisher and editors of the *Annalen*, made a clear recommendation of Rosenthal for the post. Carathéodory wrote:

I do not want to express my views on Herr Hasse, whom I know only very superficially but he has made a very good impression on me. He is anyway considered for the 1st place by your commission, isn't he? From all other gentlemen, I would by far prefer A. Rosenthal for Kiel.

Rosenthal is not only very well versed in most areas of mathematics as only *a few* could be, he knows also physics, he is an outstanding lecturer, excellent dissertations have been suggested by him, his scientific works are exemplary. His summarised accounts, for instance, in the *Enzyklopädie* are so scholarly that surely many people who do not know him, believe that he is somewhat boring. Still, this account is a first-class performance; I use it all the time and I admire the clarity of style as well as the completeness and exactness of the information. But I can assure you that no judgement about him could be more wrong. On the contrary, Rosenthal is a man who shows great taste in everything he does. Apart from this, he is loveable, modest and one of the most pleasant colleagues one could wish to have.

In second line, in any case after Rosenthal, I would name Haupt in Erlangen. Haupt is a good mathematician, a quiet, kind, pleasant man and, as I judge from his entire personality, quite a good teacher.

Thirdly, one could think of Brandt in Aachen, who writes excellent works and all who know him intimately praise.

Naturally, someone like Rademacher would also be appropriate if you think that there is the prospect that he would exchange Breslau for Kiel. From the younger people you name, v. Neumann is surely the most versatile. He is fantastically approved of, but still very young.[279]

Also on the same day, Landau officially gave his "vote" to Fraenkel, who had asked him for his view in the name of the appointment committee. In Landau's assessment, only three of the named had the outstanding figure of Steinitz. These were Hasse, in the first place, and both Rademacher and Rosenthal in alphabetical order, in the second.[280]

In an earlier discussion, Landau had found Haupt to be insignificant, v. Neumann an excellent lecturer, but without much experience, Rogosinski good and Rosenthal very good, while the name of F. K. Schmidt was new to him. In a matter relevant to reviews, Reidemeister had behaved indecently towards Landau's colleague Geiger and Landau was able to produce the evidence to the appointment committee, if asked to, but he hoped to be spared that ordeal.[281]

4.23
Works of Art for Delta

At the end of September 1928, Carathéodory spent a few days in Berlin, probably in order to consult H. A. Schwarz's papers at the Prussian Academy of Sciences so as to be able to include a history of the Delaunay problem in his paper on the *Untersuchungen über das Delaunaysche Problem der Variationsrechnung* (Studies on the Delaunay Problem of the Calculus of Variations).[282] On his return he found a telegram from his cousin Delta, who asked him to purchase some paintings in Munich. These were obviously intended for the Benakis collection, which was to open its doors to the Athens public two years later.[283] From January 1929 onwards, Carathéodory conducted searches and carried out negotiations in Munich in his attempt to fulfil his cousin's wish. Two months later, he was able to send the paintings to Trieste, to be forwarded on from there by ship to Piraeus. The chosen paintings were a portrait of General Chatzichristos by Helmi, a portrait of Neophytos Vamvas by D. Tsokos, which Carathéodory wrongly considered to be the portrait of Oeconomos ex Oeconomon by Ricquet,[284] and a view of Athens with the palace by Lanza.[285] Included in the consignment were drawings, some of them of historic significance, and also twenty lithographs and copperplate engravings, some of them in several copies, such as, for example, a picture of Otto. In total, Carathéodory sent fifty prints worth a total of one hundred marks. The consignment also contained an oval painting (25 cm × 34 cm), with a value of RM 350, that depicted Greek peasants by Peter

Hess,[286] who was well-known for his Greek-revolution pictures of *Hofgarten*, and a water-colour from the *Residenz* that showed the bay of Avlis by Karl Rottmann,[287] painted some ninety years earlier. Carathéodory was able to get the price of this latter painting down from RM 1400 to RM 1200. He insured the consignment for a sum of RM 4000. It included also some drawings from the *Residenz*, which he had been given for free. These were works by Greek painters who had received grants from Ludwig I of Bavaria and which, Carathéodory assumed, could be of significance to art history in Greece. Carathéodory also mentioned the existence of a painting by Heydeck (*Paysage un peau de fantaisie*),[288] for which he was asked to pay RM 600.[289] The paintings sent to Greece are now in the possession of either Benakis's heirs or of the Benakis Museum.

4.24
Honour to Schmidt-Ott

In February 1930 the chairman of the Kaiser Wilhelm Society (*Kaiser-Wilhelm-Gesellschaft*), Adolf von Harnack,[290] asked for contributions to honour Minister Friedrich Schmidt-Ott,[291] the president of the Emergency Action Organisation for German Science (*Notgemeinschaft der Deutschen Wissenschaft*), on the occasion of his 70th birthday. A commission consisting of Walter von Dyck (Munich), von Müller (Munich), Tillmann (Bonn) and Harnack took on the preparation of a Festschrift entitled *Aus fünfzig Jahren deutscher Wissenschaft* (From Fifty Years of German Science) that was to illustrate "the development of individual areas of science in the course of the last 50 years in summary overviews". Dr. Abb, department director at the *Staatsbibliothek* of Berlin, was responsible for the edition.[292]

The history of the *Notgemeinschaft* goes back to 13 March 1920, when the participants of a conference initiated by the Reich Association of German Technology (*Reichsverband Deutscher Technik*) asked the former Minister of Education, Schmidt-Ott, who had resigned in November 1918, to develop and realise the idea of a union of all scientific organisations. The foundation of the *Notgemeinschaft* followed on 30 October with the participation of all university rectors and academy presidents and the representatives of the most significant scientific umbrella organisations. Schmidt-Ott was unanimously elected to be the president, with the Munich mathematicians Walter von Dyck and the director of the Kaiser Wilhelm Institute for Chemistry, Fritz Haber, as vice-presidents. Adolf von Harnack became chairman of the main committee. The main idea of the *Notgemeinschaft*, Schmidt-Ott wrote in his autobiographical memoirs, and its purpose was to help science recover from the severe damage resulting from the war. In addition to the terrible loss of human life, institutions had to cope with the consequences of severe economic inflation; the sense of desolation and despair that permeated scientific institutions was palpable. The *Notgemeinschaft* was set up with the aim of providing support directly to German scientists, rather than to their institutes or to official authorities. Grants were therefore made available to budding researchers, but not their academic teachers, and the organisation was especially responsible for the development of a numerous and capable young academic generation.[293] The *Notgemeinschaft* became, in effect, an autonomous organisation, relating to all universities, academies and societies with the aim of supporting the academic activities of the impoverished state.

As was the case of the *Kaiser-Wilhelm-Gesellschaft*, the funding of the *Notgemeinschaft* became the responsibility of the Reich.[294] Within a short period of time, Schmidt-Ott was able to promote several small research projects, encourage new talent, patronise scientific publications, support libraries, begin excavations and expeditions, and further co-operation with foreign countries through the organisation of visits and lectures. However, in 1928 the Minister of Education and Cultural Affairs, C. H. Becker, considered the *Notgemeinschaft* to be an absolutely autocratic institution directed by Schmidt-Ott, who, he believed, obstructed investigation into its management and did not permit access to it by younger persons. The various contributions to the Festschrift in honour of Schmidt-Ott, and the associated correspondence between the interested parties, were written at a time when Becker was trying to exert influence over the *Notgemeinschaft*.

Carathéodory's contribution to the Festschrift was completed before he left for Athens, where he stayed at the Hotel Palace, taking up leave granted to him by the Bavarian state government to start on 15 March 1930 (see below). He had shown his contribution to von Dyck, who suggested a few changes.[295] In their common contribution, called *Mathematik – Aus fünfzig Jahren deutscher Wissenschaft* (Mathematics – Fifty Years of German Science),[296] Carathéodory contented himself with "demonstrating the great lines to which research had given preferential treatment" and once more he proved to be a historian of mathematics. At the same time he established that further development of mathematics would follow through "shared, meshed work" by an ever extending group of scholars and that "exactly the meshing together of various areas, their mutual fertilisation, the reduction of their foundations to the simplest common premises" was what tied mathematicians together.

4.25
Expecting a New Mission in Greece

Carathéodory wrote to Penelope Delta from Seattle on 21 August 1928[297] and described how he and his wife had reacted to the results of the recent elections in Greece, which had brought Venizelos to power again. In Carathéodory's opinion, "the miserable chapter of Greek history, which since 1915 had not left the country in peace, was now over". Carathéodory was referring to the so-called National Schism. For many years an abysmal rift had indeed divided both the Greek political leadership and the Greek people into two camps. On one side stood the Venizelists, on the other the royalists. Their conflict reflected the antagonism of the powers since World War I, when Venizelos had sided with the Allies, whereas King Konstantin had tried to fulfil the wish of his brother-in-law, the German Kaiser Wilhelm II, for a neutral Greece. It also reflected the conflict between modernisers and conservatives, between the populations of the new territories gained in the Balkan Wars and the established population of Greece. The leaders of the royalist People's Party (*Λαϊκό Κόμμα*) accused Venizelos of unbridled ambitions and war intentions. Their propaganda, concealed

The map has been published in The Observer *on 21 November 1920.*

under the mask of pacifism and anti-militarism, even persuaded Venizelists from time to time. Penelope Delta for instance, a fanatic supporter of Venizelos, defended the politics of neutrality even until spring 1916. In fact, peace had initially also been the basic doctrine of Venizelos's foreign policy, since he considered it necessary for the stabilisation of the country, which had doubled its territories after the Balkan Wars. But for Venizelos the fate of Greece was bound to the interests of the Allies in the long term.

If the year 1915 was characterised by the National Schism, the year 1916 was marked by limitations of Greece's sovereignty, implemented by both the Central Powers and the Allies. Political power in the country was exercised by two governments, one in the north, the other in the south. In Thessaloniki the Provisional Government (*Προσωρινή Κυβέρνηση*) under Venizelos took over the leadership of the officers' movement, National Defence, in October and declared war against Germany and Bulgaria the following month; the Athens government, under the King, started to persecute the Venizelists by mobilising private terror groups.

Penelope's father, Emmanuel Benakis, Mayor of Athens from 1914, was arrested in November. The Archbishop of Athens publicly accused Venizelos of betraying his fatherland and uttered a curse against him in December. In February 1917, King Konstantin lost the support of his relative, the overthrown Russian tsar, and the war was decisively moving in favour of the Allies. In June 1917, France enforced Konstantin's abdication. The King left the country accompanied by Carathéodory's brother-in-law, Georg von Streit. Crown Prince Alexandros was sworn in as king. Some days later, Venizelos set up a new government and this time, as Prime Minister of the whole of Greece, declared war anew against Germany and its allies. Thus, shortly before the Versailles Peace Conference, Greece had joined the victorious powers.

The National Schism did not leave the Carathéodory family untouched. Constantin's sister Ioulia and her husband, von Streit, sided with the royalists; the rest of the family with Venizelos. Their support for Venizelos was not motivated by an anti-monarchist attitude, but was directly related to their esteem for the protagonists of the conflict and, of course, to their perception of the national aspirations of Greece. What Georg von Streit's diary entry for 16 [29] July 1915 reports is typical in this respect. After receiving a visit by Konstantin Carathéodory, a great-uncle of Constantin and a brother of his father-in-law, he wrote: "In the evening we had a visit from the Carathéodorys. To K. Carathéodory's wailing about the alleged Turkish-Bulgarian co-operation, which Germany would not have allowed, if it were a friend of Greece, to his attacks against Germany, whose defeat he wishes, I reply that such unthinking interminable speeches have damaged and continue to damage our own policies very much by inciting Germans and Turks and causing persecutions of the Greek element without reason, particularly if they are expressed by persons like him, who, on the other hand, admit the existence of a far greater danger threatening us if Russia or Italy gain predominance. At present, Greece is in dire straits, much caution and great reserve is an imperative on our side."[298]

Both Constantin Carathéodory and Penelope Delta not only kept their friendships alive during the National Schism, but they did so independently of the political attitude of the persons involved. As regards Carathéodory, Georg von Streit was the most remarkable case in this respect. Delta describes him as a man of intrigues, who, hidden behind a folding screen, listened attentively to the King's discussions. Venizelos, who had appointed Georg von Streit as foreign minister, is said to have admitted: "I could not have imagined that he would deceive me, that behind my back he would use everything I have entrusted him with because of his office in order to pursue hostile policies. I had to see the evidence first in order to believe it."[299]

In his book *Venizelos and the War*, Crawfurd Price presents a gloomy portrait of Georg von Streit.

The inner cabal which assisted the King to direct the ship of state was composed of three men – Dr. Georges Streit, General Dousmanis and Colonel Metaxas. [...] With the assistance of this triumvirate, King Constantine 'ran the show' [...].

The *rôle* of Dr. Georges Streit was that of his Majesty's adviser on foreign affairs. A perfect gentleman and a charming companion, he remained a popular and respected private citizen until M. Venizelos dug him out of his professorship at the University of Athens and sent him

Georg von Streit. Courtesy of Mrs. Rodopoulos-Carathéodory.
The photograph is published in vol. 1 of the book Η Ελλάς του 1910–1920:
ιστορική μελέτη Γ. Βεντήρη *(Hellas in the Period 1910–1920: A Historical Study*
by G.Ventiris) (2 vols., 2nd edn. Athens: Ίκαρος *1970).*

as ambassador to the Court of Francis Joseph. His success in the congenial atmosphere of
Vienna was such that after the second Balkan war (1913) M. Venizelos appointed him Foreign
Minister in his Cabinet. He remained in this capacity until August 1914, when a disagreement
with the Cretan Premier on the subject of the unconditional participation of Greece in the
European conflict led to his resignation. From that day he has been the King's guide and
philosopher, and has had much to do with the general direction of Greek foreign policy.

Dr. Streit has, as his name implies, German blood in his veins. His grandfather, a Bavarian,
accompanied King Otto to Greece, and his father ended an honourable career as governor of the
National Bank. Though he undoubtedly tries to discuss international politics from a Hellenic
standpoint, he has never succeeded in ridding himself of a certain leaning towards Teutonism.
[…]

Such, then, were the men around King Constantine. They were aided and abetted by a
gang of court flunkeys and flatterers, who sought little save to fawn in the grace of their
Royal master and mistress; but to Streit, Dousmanis and Metaxas must be attributed the chief
responsibility.[300]

Prince Nikolaos writes in his diary that

an insinuation tending to represent Greece's King as an unavowed but certain friend of Germany, was adroitly calculated to discredit him forever in the eyes of the Entente. The fact that
Queen Sophie was the German Emperor's sister; that King Constantine had a German Field
Marshal's bâton; that Mr. Streit was of German origin; that several members of the General
Staff had finished their military training at the Berlin *Kriegsakademie* added weight to the
rumour that had been set afloat. It was impossible for King Constantine to fight against such
odds, and one may say that from the moment the Press got hold of this formidable weapon

against him, his doom was sealed; from that day onward, King Constantine and all his followers were irrevocably classed as pro-Germans and nothing, even to the present day has shaken this conviction.[301]

Carathéodory's friendship to Streit remained unbroken. But at the time when Carathéodory considered the "miserable chapter of Greek history" to be over, even anti-Venizelists were beginning to change their views and become more subdued in their criticisms. "'Every Greek in his right mind', remarked the royalist George Streit early in 1930, 'must hope that [Venizelos] would be able to carry through his plan of building up the country economically.'"[302]

On 21 October 1928, about a month after his return to Germany and two weeks before the start of the *Annalenstreit* affair, Carathéodory contacted Delta[303] to express his sympathy upon her mother's death. Mrs. Benaki, he wrote, was kind and obliging; she personified the Greek woman in an exemplary way. Carathéodory and his wife, so he continued in his letter, were enjoying being together with their children again, who were good children and most of all the best Greeks. In 1925 Penelope's book *The Life of Christ* (*Η ζωή του Χριστού*) was published. She dedicated it to her mother, who never read it because it was written in demotic Greek (*δημοτική*), the spoken language. Some months after Mrs. Benaki's death, Delta's father also died. Benakis's death, Carathéodory wrote to her on 24 March 1929,[304] filled every Greek with sorrow; he had been an irreplaceable symbol of Hellenism; his name was bound to the healing of the wounds of Hellenism, despite all the embitterment[305] he had had to swallow during part of his life.

The intentionally nationalist rhetoric is conspicuous in Carathéodory's stated views. Undoubtedly, it had something to do with Venizelos's return to political power, from which Carathéodory surely expected a new national assignment comparable to his mission in Asia Minor in the years 1920–1922. That was the moment in which Venizelos tackled the re-establishment of Graeco-German relations. His statement to the press during his trip to Berlin in autumn 1929 reads: "Great financial interests connect Greece and Germany; primarily, the exchange of goods between the two countries. Therefore, mutual relations should be especially friendly. I have always admired the civilisation and agility of the great German nation. We have never had anything to share with Germany. We were found in enemy camps during the war only because of historic reasons."[306] At the celebration to commemorate Georges Clémenceau, organised in Athens by the Association of Reserve Officers on 10 December 1929, Venizelos talked on the same subject. "We Greeks have never had anything against Germans. Our aversion concerns only German militarism and imperialism. If Germany had won the war, then monarchy would have won over democracy."[307]

4.26
Venizelos Calls Carathéodory to Rescue the Greek Universities

The educational reform of Venizelos's second government (1917–1920) was interrupted by his political defeat in November 1920. Re-elected in 1928, he attempted once more to achieve the objectives of his educational policy. The reform was introduced by Konstantinos Gondicas and Georgios Papandreou,[308] who served him as ministers of education. Papandreou was a minister from January 1930 to May 1932, the period in which 770 official documents regarding the educational reform were published in the Government Gazette.

Venizelos's election speech in Thessaloniki on 22 July 1928[309] reveals that, in his eyes, the educational system had the task of serving the stabilisation of a certain social status quo that was threatened by increasing social movement. To accomplish this task, knowledge acquired through school instruction must serve economic growth; against that, classical education ought to be restricted to a small elite of particularly gifted students. Indeed, quality was emphatically put before quantity. Awarding grants, and steering students to directions favoured by the state, were new elements of the government's educational philosophy. During Venizelos's four-year government, compulsory school instruction was extended from four to six years and for the first time the spoken language was taught in school. Several technical, vocational and agricultural schools were founded to promote secondary education. A total of 3167 new schools were built with the help of a credit of £ 1 000 000.[310]

The reform proposals also included the universities of Athens and Thessaloniki. They were based on a concept developed by Carathéodory, invited to Greece for the third time by Premier Venizelos in January 1930, and entrusted with their reorganisation. Carathéodory himself seems to have recognised the potential for a fundamental change in Greece's fate, although he was rather pessimistic about his chances of successful intervention. He sensed that the end of a great politician was approaching and a struggle was beginning for the successors. He felt, as it was then being said, that Venizelos refuted Venizelism and Venizelism questioned Venizelos.

In explanation of the reasons behind Carathéodory's acceptance of Venizelos's invitation, he wrote to Kalitsounakis on 3 February 1930:

When the British went to Egypt in 1883, the foundations of the dam at the head of the delta were so clapped out that the construction was absolutely useless. With great expenditure and 4 years' work they achieved a barrage that could withstand 10 cm of water. All people said that 10 cm of water was not worth £ 40 000. But still, it was the British work that caused the start of change in Egypt. I believe that the University of Athens needs something similar, but if I succeed in achieving anything there, this would be hardly visible, and all people will shout that I have failed, most of all because they would have expected something different. Therefore, I reckon on no success and neither do I aim at achieving it. And the only reason why I have accepted the personal invitation by Venizelos, who wrote me a letter by his own hand, is my belief that, since I am out of the local circle, my intervention would be easier to tolerate and my hope that I can persuade the government to take mutually agreed measures. For this purpose, I have seen various types of universities and in America I got acquainted with persons who could be useful to us. I intend to depart during the first days of March.[311]

Carathéodory considered that his relative independence from local centres of decision would facilitate his initiatives. Thus for the third time he clarified how he understood the value of his service to Greece: as help from abroad.

At that time, D. Kalitsounakis, one of Kalitsounakis's relatives, was professor at the Superior School of Economic and Commercial Sciences in Athens and later a member of the Superior Educational Council (*Ανώτατο Εκπαιδευτικό Συμβούλιο*) established by law 4653/1930 "Concerning the Administration of Education".[312] Composed of representatives drawn from science, education, administration, trade-unions, the Chamber of Industry, the Chamber of Commerce and the Technical Chamber, and presided over by the Minister of Education, the council aimed at an organisational and administrative restructuring of Greek education.[313]

Carathéodory asked the Bavarian state government for leave of absence to carry out his work for the Greek government. The reasons he used to support his request for leave can be gleaned from a communication from the Bavarian state government (Decker) to Dr. Terdenge, the Speaker Councillor at the Foreign Ministry in Berlin, on 10 February 1930:

> About a month ago, professor Carathéodory informed me that Premier Venizelos, who is personally on close terms with him, had asked him to go to Athens 'on a university matter' for about a month and that they had submitted the relevant request through diplomatic channels. The Bavarian state government would also be asked for its approval and for leave of absence for professor Carathéodory through this way. Professor Carathéodory intends to accept the offer. His absence would fall mainly in the Easter vacation. In his statement to me he was not yet absolutely clear about the reason for the offer of appointment. He assumed, as I may remark confidentially, that he would have to settle differences of opinion at the university and give advice, since this university, which had been excellent up to the turn of the century, has not really moved forwards since then. Some appointments do not seem to have been fortunate; the university also suffers from student inflation, just like German universities do. Secondary schools have somewhat declined in quality through the long war years. University professors are culturally split into two parties, one of which is inclined to France. Almost all of the medical faculty are students of German teachers.
>
> The Bavarian education administration would surely not object to professor Carathéodory's request for leave of absence. [...] Professor C. informed me in the above-mentioned discussion on 8 January that, if necessary, the Greek government would continue to pay him his salary during his leave."[314]

Carathéodory was indeed granted leave from 15 March to 15 May 1930. He received his Bavarian salary without cuts for March and April, while for May it was halved. Bochner replaced him with an unpaid stand-in, as Carathéodory had suggested on 25 February 1930.[315]

The use of the diplomatic channel shows that Venizelos had made Carathéodory's appointment a state affair and a matter of bilateral relations between Greece and Germany. On his side, Carathéodory granted German authorities insight into the internal affairs of a Greek institution, the University of Athens. Therefore, it is perhaps not surprising that, after his arrival in the Greek capital, he came under close observation by the German legation, who informed the Berlin Foreign Ministry about him on 15 April 1930:

> Professor Carathéodory has been staying in Athens since a few weeks ago. He has immediately got in touch with the Legation and also contacted German circles such as the German School and the Archaeological Institute. As I hear from him and from another side, the purpose of his stay, co-operation in the reorganisation of the University of Athens, presents him with difficult tasks. Since he is expected to act, on one hand, as a consultant to the government and, on the other, to mediate between the government and the faculties, it is natural that individual faculties would attempt to use him to satisfy special requests and oppose him should he ignore them.

Dimitrios Hondros's portrait of 1930 published in a dedication
on the 100th anniversary of his birth (note 44, Chapter 3).
Courtesy of Professor M. Roilos.

An article published in the *Messager d'Athènes* on 7th of this month, which I have the honour to enclose, provides information about the details of the university problem. The issue of the creation of a big university clinic stands in the foreground of interest. Further, the demand from the philosophical faculty for the introduction of modern language teaching and the employment of French, English, German and Italian lecturers deserves consideration.[316]

During his stay in Athens, Carathéodory was frequently asked for advice on a variety of matters, related or not to his main task. Thus, on 25 April the Minister of Education appointed him as the president of a committee consisting of Hondros, Genimatas, a professor of the Technical University, and Gryparis, the director of the Ministry of Education, whose task was to examine the training of secondary-school teachers specialised in natural sciences, and to propose the most appropriate methods to the government. On 28 April, the general director of the Greek Tourism Organisation (*Ελληνικός Οργανισμός Τουρισμού*) asked Carathéodory to propose two architects and two civil engineers specialised in hotel building, who could establish a Greek hotel, of a type combining international demands with local conditions. Carathéodory thought of the French architect Hébrard[317] for the buildings and the Greek architect Pikionis[318] for their interior design.

Carathéodory also received various honours during his stay in Greece. The National Technical University of Athens made him an honorary member on 7 April 1930 and in summer 1930 he was awarded the title of Honorary President of the Greek Mathematical Society.

4.27
Carathéodory's Report

Only a few months after his arrival in Athens, Carathéodory produced a detailed expert report *The Reorganisation of the University of Athens* (Η Αναδιοργάνωσις του Πανεπιστημίου Αθηνών).[319] In this remarkable text Carathéodory translated his assignment into a programme for a thorough modernisation of the Greek university. It was an attempt to combine the solution of current problems with long-term considerations about the production and implementation of knowledge at the service of society. Thereby he courageously defied organised interests and avoided short-sighted compromise.

He began his report by establishing that during the previous years the University of Athens had been unable successfully to fulfil its "mission" because of continuous political turbulence, excessive student numbers, its limited assets and a restrictive financial situation that was no longer supported by private donations but solely by state subsidies. He believed a simple renewal of the university law would not be sufficient to rescue the university, since universities were sensitive living organisms, sometimes reacting in an unexpected way to interventions from outside. Consequently, he aimed at the establishment of conditions that would allow the creation of an independent and lively institution with structural stability and administrative continuity in which the most competent students, the future Greek intellectual elite, would be educated. He opposed the imitation of Western prototypes, both in academic debate and in practice, pointing out that specific conditions would require specific and contextually adjusted solutions not, as he wrote, in a Greece that was imagined or desired, but in a Greece as she really existed at that time.

Carathéodory's concrete proposals dealt with all aspects of the university, from the buildings and the technical equipment, up to admission regulations and financial questions. He suggested the erection of a clinic for surgery, internist medicine and gynaecology with at least 600 beds, a great building complex with auditoriums, rooms for seminars and the university library, and finally, a building for all biological institutes and museums for geology, palaeontology and mineralogy. The medical institutes then under construction needed to be completed. In addition, Carathéodory demanded annual credits to purchase equipment and material to meet the needs of the institutes. He advised that the university library should be separated from the state library and transformed from a book archive into an instrument equipped with the entire Greek and foreign scientific literature, easy for the scientist to use. The seminar, institute and reading-room libraries had to be reorganised as well. Two or three young persons had to be sent to America to gather information on the library system. The bookshelves of the former Ionian University, which had been designed for around 60 000 volumes, and were stored in the basement of the Institute of Chemistry, ought to be refitted in the university library, which would provide an immediate solution. Carathéodory pointed out the need for centralising the production, printing, publication and distribution of the most significant books required for teaching and professors would then be relieved of this burden. The books ought to be sold to students without profit and legal arrangements could be made

Η ΑΝΑΔΙΟΡΓΑΝΩΣΙΣ

ΤΟΥ

ΠΑΝΕΠΙΣΤΗΜΙΟΥ ΑΘΗΝΩΝ

ΥΠΟ

Κ. ΚΑΡΑΘΕΟΔΩΡΗ

ΚΑΘΗΓΗΤΟΥ ΤΟΥ ΠΑΝΕΠΙΣΤΗΜΙΟΥ ΜΟΝΑΧΟΥ

ΕΝ ΑΘΗΝΑΙΣ

ΕΚ ΤΟΥ ΕΘΝΙΚΟΥ ΤΥΠΟΓΡΑΦΕΙΟΥ

1930

Carathéodory's report
The Reorganisation of the University of Athens.
Courtesy of the Bavarian Academy of Sciences, Archive.

for royalty payments to authors. Professors or lecturers would be commissioned by a small committee to write books or translate foreign scientific literature. A control authority would then be set up to supervise these procedures.

Carathéodory favoured an increase in the number of seminars and exercises and a reduction of *ex cathedra* teaching hours, which would help consolidate scientific knowledge and simplify examination procedures. By making the examinations for admission more difficult, the number of students could be reduced from 6400 to 2850. In addition, elementary knowledge of at least one foreign language would be obligatory as a condition for entrance. Tuition and examination fees would be increased; poor students would be exempted from part or all of the tuition fees.

Carathéodory proposed a major change in the procedures for the appointment of new professors. Before a university job could be offered to anyone, foreign scholars had to be asked for their opinion. A summary of the references had to be published. The occupation of new or vacant chairs had to be assigned to the administration or the university senate and not to the faculties. Second-job activities of professors had to be strictly limited and the salary of those who would dedicate themselves entirely to the university had to be doubled. Carathéodory's plan required an increase in the number of assistant and associate professors and also of assistants who would engage themselves in both lecturing and research, as well as a reduction in the number of full professors. These measures aimed at promoting a new academic generation. The plan also aspired to relieve the academic staff from administrative tasks and, through delegation of simple jobs to small committees, to release the senate from matters of minor importance.

Carathéodory completed his proposal with an estimation of the costs for a period of ten years. He asserted that the maximal yearly expenditure needed to secure a first-class education for 3000 students would be smaller than the sum spent by the state to train the same number of soldiers. He thus demonstrated the value that education ought to enjoy in comparison with other state priorities.

On 6 August 1930 the German Legation in Athens reported to the Foreign Ministry in Berlin: "As expected, Professor Carathéodory's reform proposals, namely to limit the number of students by introducing a numerus clausus, face a lively critique, not least from the circle of university professors."[320]

4.28

In Thessaloniki

In 1930 Carathéodory travelled twice to Thessaloniki: in May and again from the end of September until 3 October.[321] The task still to be accomplished there was the organisation of the young university not, as in Athens, a reorganisation of an older institution.

Thessaloniki was the largest harbour city of the northern Aegean. Its development had differed from that of Athens. Its character was more Balkan than Mediterranean; domestic and foreign trade was always more significant than administration. It was more multinational than national and had its own network of connections with the rest of the world. In the 20th century Thessaloniki had been both a national outpost and a second Jerusalem and then, after 1922, a refugee capital and the "big mother of the poor".[322] In the first quarter of the century it became the scene of significant political events, crucial for the fate of Greece.

The University of Thessaloniki was founded following a promise given by Alexandros Papanastasiou on 25 March 1924, that is on the same day he proclaimed the first Greek Republic, abolishing the monarchy. Law 3341 "Concerning the Foundation of a University in Thessaloniki" was passed by the Michalakopoulos government on 14 June 1925.

Papanastasiou had expected the new institution to stabilise democracy in the newly gained territories to the north of Olympus and to contribute to their organisation. He also hoped that the new university would stimulate its Athens counterpart to improve its performance through competition. Two short-lived democratic governments under Alexandros Papanastasiou and Andreas Michalakopoulos, a short-lived dictatorship under Theodoros Pangalos and its overthrow, and the short-lived regime under Georgios Kondylis, which was anything but democratic, marked the first three years of the university's history. Only a year after law 3341 was passed, the Minister of Education and Religious Affairs of the Pangalos government, D. Aeginitis, abolished it by decree and put the University of Thessaloniki under the sovereignty of the Academy of Athens. The Academy obtained the right to decide on the number and organisation of professorial chairs and university institutes, as well as the award of diplomas and certificates. Shortly thereafter, the new Kondylis government dismissed all the professors who had been appointed. Law 3341 was reintroduced on 21 September 1926 and provided for the foundation of five faculties, namely theology, philosophy, law and economics, natural sciences and mathematics, and medicine.

Just like the University of Athens, the University of Thessaloniki was a public corporation under the supervision and control of the Minister of Education and Religious Affairs. It was administered by the rector and the senate and its budget was subsidised by the state. The university buildings were erected on the site of the earlier Jewish cemetery of the city, coincidentally just as those of the Ionian University had been. The number of seminars at the new university exceeded those in Athens. Scientific work was produced more intensively here than in Athens. Unlike the conservative Athens institution, the University of Thessaloniki represented more progressive tendencies.[323]

During his visit, Carathéodory gave directions to the rector's council on study programmes, the appointments of professors, the work of various councils, and on ways to combat the spirit of factionalism current among the students. He expressed his conviction that the two Greek universities should have compatible programmes to enable students to change their place of study. Professors should be encouraged to remain at their posts for a minimum period of ten years by imposing a loss of financial privileges.[324]

The most important subject of his discussions in Thessaloniki was a new organic law for the university. On 27 September 1930, in a session of the rector's council, Carathéodory expressed the opinion that the organic law ought to contain only general regulations. He criticised the internal law of Athens University, because it regulated every single detail, as well as law 3341, because it omitted basic regulations and included unnecessary ones instead. He advocated the substitution of law 3341. Referring to the German example, he pleaded for a new law that would provide a general framework of function but otherwise entrust decisions on all specific issues to university organs. In the internal procedures of the university an attitude of bureaucracy ought to be avoided.

The senate of the University of Thessaloniki had appointed a committee, with the participation of Kritikos, which worked out a draft for the new organic law of the university. This draft was submitted by Carathéodory to the Minister of Education and Religious Affairs. More than a year later, on 22 February 1932, Carathéodory brought a law proposal to the rector's council of the University of Thessaloniki. He had prepared it in co-operation with Professor Papoulias of the law faculty of Athens University and Bertos, the General Secretary of the Ministry of Education and Religious Affairs. It enjoyed the approval of Georgios Papandreou and was to replace law 3341.

4.29
"The Crown of Thorns"

Carathéodory spent the remaining 27 days of October 1930 in Athens. On 31 October, he addressed a letter to Venizelos including an account of his work for the two Greek universities. He also submitted to the Premier new proposals concerning the location of the new Athens university buildings, presenting his arguments for effective use and financial advantages. He expressed the wish to have the architect of the buildings for two to three weeks in Munich, where he could introduce him to the director of the architectural office of the Bavarian Ministry of Education. In Carathéodory's view, this man had considerable experience with buildings of this type. Finally, he asked for an increase in the state subsidy for the university in the budget of the following year by at least 20 million drachmas.

In the winter semester 1930–1931, Carathéodory lectured in Munich for four hours each week on mechanics and another four hours on function theory. He also held two-hour exercise classes in mechanics and, together with Perron, Tietze and Hartogs, supervised a two-hour mathematical seminar. From November 1930 he worked with de Possel. Because he was going to stay in Greece in the winter of 1931, Carathéodory planned to send the young Frenchman, who was working at that time "on a rather difficult subject of conformal mappings of certain star regions related to complicated studies on Stieltjes's integrals", to Göttingen where Courant could help him. Carathéodory had advocated an eight-month stay, whereas Jones, with whom Carathéodory had spoken in Paris, had in mind a three-month stay for de Possel in Göttingen and therefore desired Courant's permission for this "change of programme".[325] Courant reacted promptly to Carathéodory's request to try to do something about the matter and wrote to Jones.[326]

On 13 January 1931, the Greek Minister of Education and Religious Affairs, Georgios Papandreou, introduced the bill on university education to the parliament, explaining that the "mission" of the university was to guide the endeavour for national rebirth. The government's philosophy regarding its structure, organisation and orientation was almost a full adoption of Carathéodory's memorandum on the reorganisation of the Greek universities.

On 4 February 1931, Carathéodory informed the philosophical faculty of Munich University that the Greek Premier had asked him by letter to take on the office of rector of Athens University for ten years in order to carry out the reorganisation of the institution. Although there had been no negotiations of any kind about the offer, he considered it his duty to inform his superior authority immediately about the proposal. Carathéodory was probably obliged to offer this explanation because the news that he had already accepted the appointment was already becoming known in Athens. Indeed, this must have been why the German Legation in Athens was prompted to issue an unattributed, and unofficial, denial that Carathéodory had accepted the offer, published on 19 January in three national newspapers.[327] One day after receiving Carathéodory's letter, the Munich Philosophical Faculty, Section II circulated his communication among core members of the faculty for their attention and comment. In Dean Tietze's opinion, the loss of Carathéodory would be "deeply regrettable". Carathéodory was internationally recognised and highly regarded as a scholar and a personality; scientific life in Munich would suffer if he left. Oskar Perron believed that the faculty ought to contact the Minister of Education at the soonest opportunity in order to prevent Carathéodory's departure "by all means". Everyone agreed with Perron's proposal and Sommerfeld expressed the hope "that colleague Carathéodory would not put on the crown of thorns for 10 years!" Tietze announced finally that their visit to the ministry would follow on 11 February.[328]

Carathéodory travelled to Athens again on 25 March 1931 on behalf of the Greek government and returned to Munich at the end of April. This time his teaching duties were taken over by Bochner, Lettenmeyer and an assistant. Carathéodory relinquished his salary, which was used to pay the three mathematicians replacing him. In fact, Munich University demanded that the Greek state should pay the salaries of the mathematicians taking over Carathéodory's work, which was objected to by Dean Tietze, as being unreasonable.[329]

On 18 May 1931, Carathéodory wrote to Courant to report that the institutes he had proposed in Athens were being built and, therefore, he would like to meet him.[330] It is probable that Carathéodory wished to ask his opinion about matters of organisation and funding, drawing on Courant's earlier experiences with the Göttingen institutes.

Three days earlier, the Greek Legation in Berlin had asked the German Foreign Ministry to support Carathéodory's leave of absence, which was to be granted by the Bavarian state government. Carathéodory was going to "hold his office in Greece as Commissioner of the Greek government for the reorganisation of the Greek universities in the winter semester 1931–1932", a responsibility that would possibly extend to the winter semester 1932–1933. To support their request, the Greeks alluded to the strong historical bonds between Bavaria and Greece: "The Greek government is convinced that if the Bavarian government take into consideration that Athens University was founded at the time of King Otto, following the example of Bavarian universities, they will not deny their approval to a work striving for the further development of this university."[331]

4.30
Commissioner of the Greek Government

It seems that during his visit to Greece in early spring 1931, Carathéodory had managed to change the terms of his employment by the Greek government and, instead of taking the rector's office of Athens University for a period of ten years, he had accepted the post of governmental commissioner for six months per year. The most plausible reason for this decision seems to be that by accepting the office of rector Carathéodory would have lost his professorship in Munich, with all the consequences that such a loss might have entailed, with respect to his pension for instance. Unlike what had happened in the case of Smyrna, this time there would not be an opportunity to return again to Germany and obtain a professorship. After ten years of activity in Greece he would have even exceeded the age of retirement. It was the second time within three years that Carathéodory failed to avail himself of the opportunity to leave Germany. From then on, his own fate would be inextricably linked with that of Germany. In any assessment of Carathéodory's decision, one must also not underestimate his lack of confidence in the success of his own intervention in Greek educational politics, as he had expressed to Kalitsounakis on 3 February 1930.

The Berlin Foreign Ministry granted their approval to the request from the Greek Legation and, in a letter to the Bavarian Ministry of Education on 23 May 1931, they affirmed that Carathéodory's activity served the interests of German foreign policy. In this way, the request to the Bavarian government acquired an almost compelling character and its rejection became practically impossible. "The Foreign Ministry", reads the relevant text, "attaches particular cultural-political importance to the fact that Professor Carathéodory is entrusted with the reorganisation of the Greek universities. It is known here that Professor Carathéodory enjoys a high standing in Greece. Since he is associated with Germany in an equal way, it can be hoped that the task given to him by the Greek government will bear valuable fruit for the revival of scientific relations between Greece and Germany. The Foreign Ministry would therefore be particularly obliged if the request of the Greek government would be fulfilled."[332]

On 30 May 1931, the Bavarian Ministry of Education asked the senate of Munich University to report on whether there were any doubts concerning Carathéodory's leave of absence that had been approved by the Foreign Ministry and, if that were not the case, to submit proposals for his replacement. However, the most significant announcement made to the senate by the Ministry was that Carathéodory did not wish to accept a ten-year appointment as rector of Athens University, but preferred instead to travel once or twice a year to Athens in order to give advice and to supervise the reorganisation. This means that the Greek government's request corresponded to Carathéodory's plans. Carathéodory had asked orally for a discussion about the possibility of becoming a civil servant in Athens without losing his civil-servant status in Bavaria.[333] In this way he was seeking economic and social security by trying to achieve the same status in both Greece and Germany.

As seen from the approach of the Greek Legation in Berlin to the German Foreign Ministry on 15 May 1931, the Greek government had already decided by that time, not only to create the office for a commissioner at Athens University, but also that Carathéodory should occupy it. However, the bill "Concerning the Establishment of the Office of a Governmental Commissioner at the Universities of Athens and Thessaloniki" was submitted to the Greek parliament by Georgios Papandreou only on 3 June 1931. The discussion on it opened eight days later. A member of parliament argued that if the engagement of a commissioner was necessary, then the holder of that office had to stay in Athens in order to be able to fulfil his duties. The bill, he said, prescribed the qualifications of the future commissioner, which corresponded to a specific person. Papandreou took on the political responsibility for the bill and confirmed that the office was indeed destined for Carathéodory, a scientist who honoured Greece's name abroad and showed himself willing to offer his knowledge to his homeland. Further, Papandreou explained that the office of the commissioner was not an administrative position that would require Carathéodory's presence for more than six months a year, but a function of the government's consultant on the universities. Premier Venizelos also supported Carathéodory decisively in that discussion. The parliament passed the law as it had been proposed on 17 July 1931, as law 5143 "Concerning the Governmental Commissioner at the Universities of Athens and Thessaloniki". It was published in the Government Gazette vol. I, p. 213. According to that law, the commissioner's income and rank corresponded to those of rector of Athens University. However, with regard to the salary, there was the possibility of a special arrangement between government and commissioner if the latter was a full professor at a foreign university. His salary was to be set against the budgets of the universities of Athens and Thessaloniki equally. He had the task of controlling the implementation of law at the universities, to participate, if he considered it necessary, in all committees without a vote, to supervise all university authorities and their branches, to report to the Minister of Education and to consult him on matters concerning the university. The minister was responsible for granting leave to the governmental commissioner. If the latter was a professor of a foreign university, he had to accept the obligation to spend at least one semester per year in Greece. All rules applying to professors at Athens University with respect to age limit, disciplinary proceedings and so on applied also to the governmental commissioner. He was allowed to lecture on his subject at both universities if he was, or had been, a professor. He had his seat at the University of Athens but he ought to maintain his own bureau at both universities. The personnel at his service, namely a secretary, a typist and a porter, would be chosen from the university's personnel. Both universities had to cover the expenditures of his office.

By a decree of 3 October 1931, Carathéodory was appointed as the Commissioner of the Greek Government according to Papandreou's proposal.[334] In less than a week, on 9 October 1931, the Greek legation in Berlin asked the Foreign Ministry to contact the Bavarian Ministry of Education and Cultural Affairs so that Carathéodory would be allowed to have leave also for the winter semester 1932–1933.[335]

Although the decision to create the office of a governmental commissioner can be explained by the spirit of the time and the attitude of a small number of academics, the decision itself can only be interpreted as the will of the government to gain complete control over education, as the expression of an absolute centralism characteristic of Greek education that had become established since Capodistria's time. The 1836–1837 legislation regarding the universities already provided for the office of a governmental commissioner. This time, however, the introduction of the office was expected to meet strong opposition.[336]

Papandreou's speech on 19 March 1932 during the debate on law 5343/1932 "Concerning the Organic Law of the University of Athens" was mainly a reply to criticism from both the left and the right, expressing the opposition of social, political and educational groups, as well as that of individuals. Objections to the reform had already surfaced in 1929. Some days after Papandreou took on the Ministry of Education and Religious Affairs on 10 January 1930, opposition led to student protests, occupation of university buildings and clashes between students and professors and provoked police intervention and numerous arrests. In March 1931, the People's Party (*Λαϊκό Κόμμα*) and the Progressive Union (*Προοδευτική Ένωσις*) argued in parliament that the reform was a luxury rather than a primary necessity and that its implementation was being attempted despite the lack of the necessary financial commitment. Also conservative personalities from the philosophy faculty of Athens University declared their resistance against the reform.[337] Criticism from the left was articulated mainly by Glinos who, after the splitting of the Educational Group in 1927, aligned with the Marxist left. In Papandreou's statement, that academics had the right to preserve their convictions but not to express them, Glinos detected the unconcealed admission of "class tyranny".[338] Also democrats criticised the abolition of the university's independence. The writer and critic Spyros Melas satirised Papandreou's policies as "deciding before thinking".[339]

So, three groups emerged in the field of education which more or less corresponded to the existing political parties: on one side the "conservatives" were aiming at the preservation of the traditional educational scope and accepted only the need for change in the university's organisation and operation; in the middle, the "progressive" bourgeois believed in the necessity of educational reform which, through the newly emerged social forces, would contribute to the preservation of balance and improve the quality of life for the whole population; at the other end, the "left" framed the educational endeavour as a class struggle and believed that it was defined by the needs and expectations of the oppressed classes.

Papandreou himself maintained that the government had nothing to fear from discussions concerning the new organic law of the university, and he himself desired to stimulate debate; the government intended to produce constructive proposals and was not animated by perfectionism. The work that had been done so far was the result of a two-year toil. The government had called on the prominent Greek professor, Carathéodory, and entrusted a finance commission with the study of finances. The chairmen of that commission were D. Maximos, former governor of the National Bank and economic adviser to the People's Party, and E. Tsouderos, governor of the Bank of Greece. The members of the commission were the university professors Petimezas and K. Varvaressos, the latter being an adviser to the Bank of Greece, one of the outstanding economists and administrators of the inter-war period and a future representative of Greece in the International Monetary Fund. The report on the finances made up a whole volume. Another commission under Papoulias and Carathéodory, acting with the participation of the general secretaries of the Ministry of Education and the university, had worked out and published a rough plan of the organic law.

As to the criticism regarding the abolition of the university's independence, Papandreou argued that it implied the university ought to be administered by its professors and the state ought not to intervene, as if the university were a private corporation pursuing private aims and not the highest national institution on whose performance the fate of the nation depended.[340] However, this criticism of the opposition can be well understood. Despite Venizelos's unreserved trust in Carathéodory's capacities, Carathéodory had been posted by the Premier to implement the reform. He had not been elected by any university and therefore his authority was not legitimate. That his power had not been the result of a democratic procedure can be clearly seen in another instance: with article 17 of the so-called "Constitution of 1968", the Greek military dictatorship of 1967 reintroduced the post of the governmental commissioner with the task of "supervising the highest educational institutions". The decree 180 "Concerning the Governmental Commissioner of the Highest Educational Institutions" was issued by the military government on 30 April 1969. The main difference from Venizelos's law was that this decree provided for more than one commissioner, namely one for each university, and that the post be occupied by either a former full professor of one of the universities, or an acting or retired senior civil servant or military person, or a senior judge, or a renowned scientist.

With only a few modifications and additions, Carathéodory's text *The Reorganisation of the University of Athens* comprised the basis for university law 5343 of 23 March 1932. It contained 357 articles and was to last for half a century before Georgios Papandreou's son, Premier Andreas Papandreou, introduced a new reform for the universities with law 1268/1982.

4.31
Undesirable Reform

Carathéodory complained later about the use of spurious arguments to impede the implementation of his initiatives. His name appears in the minutes of the senate sessions for the last time on 9 April 1932. In the next two sessions, on 19 and 26 April, the minutes state that the governmental commissioner C. Carathéodory was absent. He is not mentioned at all in the senate's minutes after that date. G. Papandreou granted him a five-month leave of absence with effect from 1 May 1932.[341] Carathéodory's daughter says that his dismissal was typed on half a page and was signed by a civil servant of the Ministry of National Education and Religious Affairs on 26 July 1932, during the Papanastasiou government. This is obviously wrong. With grave reservations, Papanastasiou had taken on responsibility for the government on 26 May to resign only eight days later. As the reason for his resignation he named restrictions concerning social security, land tax and workers' claims imposed on his government by Venizelos. On 5 June Venizelos had formed a new government. The parliament ceased to convene on 19 August and elections were proclaimed for the 25th of September. During the summer, Venizelist military personnel were preparing to intervene if Venizelos should lose the election.[342] This year Carathéodory also stopped corresponding with Penelope Delta.

The educational reform of 1930–1932 dealt with the organisation and administration of Greek higher education rather than a programme of studies. It experienced the same fate of the previous two reforms, namely that its implementation was postponed. The conservative faction of the bourgeoisie that came to power blocked the reform. The constitution itself was at the centre of the parliament's concerns; the acute social and economic crisis that seized the country was ignored. The period from now until the beginning of the Second World War was marked by the emasculation or abolition of solutions that had been introduced during the reform, and by tentative and not co-ordinated actions, which suggested the impression of progress whereas in reality they contributed to the perpetuation of those educational characteristics that conformed to the old system.

4.32
Academic Contacts in Greece

The hostile attitude of several personalities at Athens University did not deprive Carathéodory of contact with individuals belonging to both the university and the technical university, including also the Superior School of Fine Arts. Professors of both Athens universities visited Carathéodory in Munich, as did the jurists Georg

Carathéodory's portrait sketched by S. Vikatos.
Courtesy of Mrs. Rodopoulos-Carathéodory.

The Faculty of Philosophy of the Aristotle University of Thessaloniki.
Photograph: S. Georgiadis.

von Streit, Ieronymos Pintos and Konstantinos Petropoulos, the physicists Dimitrios Hondros, Achilleas Papapetrou and Theodoros Kujumtzelis (who had served Carathéodory as secretary to the governmental commissioner), the mathematicians Nilos Sakellariou, Aristotelis Oeconomou,[343] the "inspired and punctilious" Nikolaos Kritikos and Dimitrios Kappos,[344] the pharmacologist Georgios Ioakimoglou, the surgeon Marinos Geroulanos, the economist Ioannis Kalitsounakis, the civil engineer Nikolaos Kitsikis, the painter Spyros Vikatos and the librarian of the student reading room at Athens University, Anargyros G. Giambouranis. The Athens University mathematician N. Hatzidakis[345] does not, however, seem to have visited Carathéodory in Munich.

Carathéodory's relationships with personalities of the second Greek university in Thessaloniki were just as strong. The signatures of many of its professors are to be found in his Munich guest-book. Among them are professors of the Faculty of Philosophy, the first faculty founded at Thessaloniki, such as Linos Politis (later director of the National Library and mostly known for his works on the Greek "national poet" Dionysios Solomos) and the linguists Manolis Triantafyllidis and Ioannis Kakridis. The latter two are typical of the intellectuals rejected by the University of Athens.

Triantafyllidis, after concluding his basic studies in Greece, studied linguistics in Munich and gained his doctoral degree under the renowned scholar of Byzantine studies Karl Krumbacher. While in Munich, he had his first study *Expulsion of Foreigners or Equality of Purpose. Study Concerning Foreign Words in the Modern Greek Language* (Ξενηλασία ή ισοτέλεια. Μελέτη περί των ξένων λέξεων της Νέας Ελληνικής) published in 1905. The study referred to the problem of the foreign origin of some words in the Greek language and attacked established views in linguistics, and conceptions that were deeply rooted in the ideological foundations of Greek society. Publication of about 175 of his works upset established balances, for he was one of the most significant representatives of the educational reform and the establishment of demotic Greek. Triantafyllidis had been a candidate for a chair at Athens University in 1924, in 1938 and for the third time in 1948, but he was never appointed. Also the Athens Academy, fearing his candidacy, refused to advertise a chair for linguistics.[346]

The other linguist, Ioannis Kakridis, the most significant classical philologist of post-war Greece was widely recognised abroad. He was known for his avant-garde contribution to Ancient Greek studies, his continuous concern about educational practice, his admirable frankness in the defence of his ideas and his talented lecturing.[347]

Members of the Faculty of Natural and Mathematical Sciences, which was founded in 1927–1928, also belonged to Carathéodory's circle of friends: these included Philon Vasileiou, Othon Pylarinos,[348] Rector of the University of Thessaloniki in 1944–1945, and Ioannis Gratsiatos, a Sommerfeld doctoral student.[349] Carathéodory's contacts in the Department of Law/Faculty of Law and Economy, which started its activities in 1930, were professor Periklis Vizoukidis, Rector of the University in 1931–1932, Ioannis Spiropoulos, a professor of international law and elected member of the International Tribunal in The Hague by the UN General Assembly in 1957, and finally Vasileios Vojatzis, the director of the Social Security Foundation (ΙΚΑ = Ίδρυμα Κοινωνικών Ασφαλίσεων) in the years 1937–1946. The future member of the Faculty of Forestry and Agriculture, which was founded in 1937 to compensate the abolition of the Agricultural Superior School in Athens, Manthos Kotsiopoulos, also visited Carathéodory in Munich in the summer of 1935.

Among Carathéodory's other friends were Dimitrios Syllaidopoulos and the gynaecologist Georgios Tsoutsoulopoulos, who were appointed professors to the Faculty of Medicine. Archimandrite Averkios Papadopoulos, chairman of the Greek community of the Church of the Redeemer (*zum Erlöser*) in Munich in the years 1933–1939 and professor at the Faculty of Theology in Thessaloniki, was also Carathéodory's guest for the new year's-day celebration of 1935.[350]

During the years after he returned to Munich, Carathéodory continued to keep up his contacts with Greek intellectuals, the overwhelming majority of whom had either studied in Germany or maintained connections with Germany. Many of them were active in politics and participated in international expert commissions and organisations. These had often made brilliant careers abroad but had returned to continue their work in Greece. Kitsikis, Kritikos and Papapetrou belonged to the left-leaning intelligentsia. At the University of Thessaloniki, Carathéodory's contacts were personalities who, finding themselves in confrontation with both the conservative segment of the social class from which they originated and with the growing workers' movement, were able to cultivate their political and intellectual liberalism within the second university of the country. It may be noticed that Carathéodory showed respect for any personality with a solid intellectual performance, independently of their political convictions or their attitude towards the language problem, in a period where these questions were dividing the country.

4.33
Goethe: A Graeco-German Bridge

On 19 January 1932, some time before the end of the third Greek episode in his career, Carathéodory wrote to Hilbert to congratulate him on his 70th birthday.[351] It is obvious that Carathéodory was enjoying his stay in Greece. Venizelos was still in power and all that interested Carathéodory, painting, sculpture, music, literature, archaeology, were blossoming. The Superior School of Fine Arts (*Ανωτάτη Σχολή Καλών Τεχνών*), the National Theatre, the General Council of Greece's Libraries were established by law. Measures were taken to purchase works by Greek artists and to display them in the National Gallery, to protect works of art and literature by copyright, to build archaeological museums and to preserve ancient monuments and to present Greek art abroad. All this happened in the period 1930–1932. Full of enthusiasm, Carathéodory described the countryside he saw while flying from Thessaloniki to Athens:

Some weeks ago, I flew in less than two hours from Salonica to Athens, a trip which in my youth had taken several days. It is the nicest trip that I have made in my life, also with respect to the landscape. Every five minutes the picture changes: snowy mountains such as Olympus, the blue sea, the green forests, then again flat corn-land. One cannot even imagine how varied the picture is.

Carathéodory must have also visited Mistras, the Peloponnesian castle, once the centre of Byzantine spiritual life and the crowning place of the last Emperor of Byzantium, with its magnificent palace, the fresco-painted churches and the monastery. He sent Hilbert a picture of a church in Mistras and did not omit to remind him of the presence of the site in a classic work of German literature: "Mistras is the Sparta of the Middle Ages, which appears in *Faust*, part II."[352]

Naturally, Carathéodory was familiar with *Faust* and its author Johann Wolfgang von Goethe. Carathéodory's library included Goethe's collected works. On 1 September 1931, Werner Deetjen, a member of the Goethe Society and a Goethe scholar, visited Carathéodory in Munich; he left "highly pleased about the meeting after such a long time". Deetjen, a historian of German literature, had been appointed to the Technical University of Hannover in 1909, just like Carathéodory. In 1916 he became director of the Library of the Grand Duke (*Großherzogliche Bibliothek*) in Weimar and he also lectured at the University of Jena. His visit to Munich was probably connected with the preparations of the Goethe Year that was to be celebrated in March 1932 on the occasion of the 100th anniversary of the poet's death. As Large remarks,[353] in the heavy atmosphere of ideological radicalism and in the aftermath of the financial crash, Goethe was raised up as a symbol, opium for the soul. Hardly anyone read him, yet still he was cited everywhere.

Since the beginning of the 19th century, Goethe had been a significant component of standard educated bourgeois, *bildungsbürgerlich*, ideology. His proclivity to Ancient Greece was considered a prerequisite for a genuine humanisation of civilisation. "The clarity of the view, the cheerfulness of perception, the lightness of communication" Goethe wrote, "that is what delights us; and if we now claim that we find all this in the authentic Greek works performed on the noblest material, on the most dignified content in a sure and perfect execution, one will understand why they are always our point of departure and always our point of reference. Let everyone be a Greek in his own way but let him really be a Greek." [354] This Graecomania, a passion for all things Greek, exhibited by Goethe and his followers was nothing but a symptom of what Eliza Butler called the tyranny of Greece over Germany.[355] Carathéodory obviously classified Hilbert as belonging to the *bildungsbürgerlich* ideology.

4.34
A Timely Overview of Mathematics

Carathéodory wrote the entry for mathematics for the *Great Greek Encyclopaedia* of 1931.[356] He provides a concise overview of the development of the subject, claiming that the Greeks were the first to think of geometry and arithmetic as independent and abstract disciplines. He then suggests that geometry had mostly influenced the general progress of European peoples and that mathematics owed its progress over recent centuries to the spread of Arabic numerals as well as to the development of the algebraic language of symbols. This led to early research in algebra carried out in Italy during the Renaissance.

The brief survey of the subject, following the discovery of the infinitesimal calculus in the 17th century by Newton and Leibniz, takes in the Bernoulli family and their contributions to the promotion of analysis, as well as to the calculus of variations, followed by Euler and Lagarange in the 18th century. Carathéodory honours Carl Friedrich Gauss as the most brilliant spirit of the 19th century, when mathematics experienced a growth in a variety of different directions initiated by both theoretical formulations and demands set by natural sciences. Carathéodory mentions Dirichlet, Riemann, Dedekind, Kronecker and Hilbert as famous researchers in number theory and Cauchy, Abel, Jacobi, Riemann and Weierstrass in the theory of analytic functions. Projective geometry discovered by the French mathematician Poncelet was later extended by Möbius, Steiner, von Staudt, Plücker and Kummer in Germany. The geometries of Riemann, Lobachevsky and Bolyai were later related to each other by Klein and used by Einstein in the construction of the theory of general relativity and gravitation. Carathéodory believed that the most significant event of the previous sixty years had been the re-establishment of that exactness in mathematical trains of thought that had marked the works of ancient Greek mathematicians. This period of time included Cantor's set theory, the systematic exploration of geometric axioms and axiomatics. Mathematical science, according to Carathéodory, exhibited two tendencies: on one hand, to continuously find new logical arguments and theories and, on the other, to summarise and generalise already known theories.

Carathéodory's own mathematical work in the years 1931–1933 lay in the second of the two tendencies he identified in his article, namely summarising and generalising known theories. In his studies into the calculus of variations, function theory and mechanics he was able to make original contributions.

4.35

Neugebauer, Courant, Springer

In the spring and summer of 1931 Carathéodory had frequent contact with Courant. On Pringsheim's request he asked Courant to send Pringsheim a critical review of Lambert's and Legendre's works, which was kept in Gauss's papers in Göttingen. He explained that in the 8th volume of Gauss's works, pp. 27–29, there was a proof of irrationality of π and also in a note on page 29. Robert Fricke had attributed the first proof of that irrationality to Legendre, despite the fact that as early as the 1890s Pringsheim had shown that Lambert had been successful in the proof and Legendre was the one to have destroyed Lambert's proof. Fricke mentioned a critical review of the works of Lambert and Legendre recorded by Gauss, but Fricke did not publish it, although he believed that it was much more interesting than the calculations themselves.[357]

Carathéodory planned to visit Göttingen on the 16th of June, on his return from the centenary of the Technical University of Hannover, stay there for the whole day and even give a small speech to the Mathematical Society.[358] Courant wanted to consult him on the edition of Hilbert's collected works[359] and he consequently arranged their meeting with Blumenthal and Hecke.[360] Carathéodory, who had to remain for a few hours in Berlin en route to Hannover or on his return, saw no reason why Courant and Blumenthal could not confer earlier, since his own role would be limited to encouraging Hilbert to tackle the thing as soon as possible.[361] Courant accepted Carathéodory's suggestion; he would pick him up from the railway station and invited him to lunch at home.[362]

Courant asked for Carathéodory's suggestions concerning the edition of Hilbert's works and informed him that Neugebauer had developed all sorts of concrete plans on the Neugebauer–Springer idea of publishing small monographs. Courant assumed that Neugebauer would also turn to Carathéodory in this respect.[363] In 1932, Neugebauer turned down an offer for appointment as a full professor at the Darmstadt Technical University to remain in Göttingen. He was freed from financial worries because the Springer publishing house had entrusted him with the editorship of two mathematical periodicals and a monograph series.[364]

4.36
At the International Congress of Mathematicians in Zurich

In 1932, the International Congress of Mathematicians was held in Zurich. Its president was Karl Rudolf Fueter, a full professor at Zurich University, who, together with the ETH, sponsored the meeting. Organising the congress had been a difficult task, since the statutes of the International Mathematical Union had been allowed to expire in 1931. During the congress, many mathematicians, including Veblen, Wiener, Bohr, Watson and von Schouten, described the union as a "useless" organisation. On their insistence, it was agreed to disband the IMU and to ask the congress to appoint a committee to investigate the need for a continuing body of some sort. The organising committee invited Ludwig Bieberbach, who had opposed the Bologna congress of 1928, to deliver one of the plenary addresses. Carathéodory held the first of the "great speeches" *Über die analytischen Abbildungen durch Funktionen mehrerer Veränderlicher* (On Analytic Mappings Through Functions of Several Variables)[365] after Fueter's opening speech.

Behnke mentions that he had sent a work by one of his students to Carathéodory shortly before the congress and discussed it in an accidental meeting with Carathéodory, who decided to change his speech. Behnke remarks that this incident

*Detail section of the following photograph with Carathéodory
at the International Congress of Mathematicians, Zurich 1932.
Courtesy of Dr. Antal Varga, Bolyai Institute, University of Szeged.*

Participants of the International Congress of Mathematicians, Zurich 1932

Mrs. Adams	Mr. Feinler	Prof. Kuratowski	Prof. Simon
Prof. Adams	Prof. Fejér	Prof. Kyrtsis	Ms. Simons
Prof. Ahlfors	Prof. Fekete	Dr. Lehmann	Mrs. Skewes
Dr. Alt	Prof. Kampé de Fériet	Mrs. Lense	Prof. Skewes
Dr. Amira	Prof. Finsler	Prof. Lense	Prof. David Smith
Prof. Archibald	Dr. Anna Fischer	Dr. Liebl	Prof. Paul Smith
Rector Buchner	Dr. Fjeldstad	Lietzmann	Prof. Snyder
Dr. Barnett	Prof. Forsyth	Mr. Linder	Prof. Speiser
Prof. Bays	Prof. de Franchis	Dr. Locher	Prof. Spiess
Dr. Bergman	Prof. Franel	Mrs. Lövenskiold	Dr. Staehelin
Prof. Bernays	Mr. Fueter	Prof. Lorey	Prof. Stark
Prof. Berwald	Prof. Fueter	Dr. Mahler	Prof. Steggall
Prof. Bieberbach	Dr. Gangulliet	Maillet	Mrs. Störmer
Prof. Blichfeldt	Prof. Giambelli	Prof. Marchand	Prof. Störmer
Mrs. Blumenthal	Dr. Gölz	Dr. Maspoli	Dr. Stohler
Prof. Blumenthal	Prof. Goldziher	Prof. Mentré	Prof. Stone
Dr. Bochner	Prof. Gonseth	Prof. Merriman	Mrs. Stouffer
Prof. Bohr	Mrs. Graustein	Mr. Ping-Ling Miong	Prof. Stouffer
Prof. Bompiani	Prof. Graustein	Prof. Mitchell	Prof. Straszewicz
Prof. Bortolotti	Mr. Guldberg	Prof. Mohrmann	Dr. Sudan
Prof. Brusotti	Dr. Gut	Prof. Moore	Prof. Synge
Prof. Brink	Prof. Hadamard	D. L. Mordell	Prof. Szász
Prof. Brouwer	Prof. Hamel	Ms. Mordell	Prof. Takagi
Brunner, dipl. math.	Prof. Harshbarger	Prof. Mordell	Prof. Tamarkin
Prof. Buhl	Prof. Hasse	Prof. David Morse	Prof. Tiercy
Mr. Büsser	Prof. Heffter	Prof. Marston Morse	Prof. Togliatti
Prof. Cairns	Mr. Herter	Prof. Müller	Prof. Tonelli
Dr. Candido	Mrs. Hille	Prof. Nielsen	Prof. Tonolo
Prof. Carathéodory	Prof. Hille	Prof. Papaioannou	Prof. Tortorici
Prof. Élie Cartan	Mr. Kiu-Lai Hiong	Prof. Du Pasquier	Mr. Trost
Prof. Henri Cartan	Mr. Ping-Ling Hiong	Mr. Perna	Mrs. Tricomi
Dr. Cartwright	Dr. Hofreiter	Prof. Picone	Prof. Tricomi
Prof. Cèch	Hollcroft, jr.	Rector Plancherel	Prof. Turnbull
Dr. Marie Charpentier	Mrs. Hollcroft	Dr. Politzer	Prof. Tschebotarow
Prof. Jules Chuard	Prof. Hollcroft	Dr. Pollaczek	Prof. Tyler
Dr. Silvio Cinquini	Mr. Honegger	Prof. Pólya	Prof. Tzitzéica
Ms. Maria Camelia	Prof. Hurwitz	Ms. Rainich	Dr. Ullrich
Comesatti	Prof. Husson	Dr. Rellich	Prof. de la Vallee
Mrs. Comesatti	Dr. Jacob	Prof. Reymond	Poussin
Prof. Annibale	Prof. Janet	Prof. Riabouchinski	Prof. Vetter
Comesatti	Prof. Jarnik	Mrs. Riabouchinski	Prof. Vincensini
Prof. A.W. Conway	Dr. Jessen	Prof. Richardson	Dr. Völlm
Prof. L. Crelier	Dr. Johansson	Prof. Riesz	Mr. Waldmeier
Prof. N. Criticos	Prof. Julla	Prof. Roever	Prof. Walsh
Prof. Louise Cummings	Prof. Juvet	Prof. Rosenblatt	Prof. Wavre
Dr. K. Dändliker	Dr. Kálmár	Prof. Rosenthal	Prof. Weeks
Prof. Daniell	Mr. Kemmer	Prof. Rowe	Prof. Weiß
Dr. Dantzig	Prof. Kiepert	Mrs. Saks	Prof. Weyl
Mr. Dell'Agnola	Ms. Kenedy	Dr. Schaertlin	Mrs. Wiener
Prof. Dell'Agnola	Dr. Kistler	Mrs. Schieldrop	Prof. Wiener
Prof. Dines	Prof. Knaster	Prof. Schieldrop	Prof. Wilkosz
Prof. Dive	Prof. Koebe	Dr. Schmidt	Prof. Wirtinger
Prof. Doetsch	Prof. Kogbetliantz	Prof. Schur	Prof. Wolff
Prof. Dumas	Prof. Kollros	Prof. Scorza	Dr. Wrinch
Prof. Fantappié	Prof. Koopmann	Prof. Severi	Mr. Young
Prof. Favard	Prof. Kraitchik	Mr. Silverman	Ms. Young
Prof. Fehr	Prof. Kueser	Prof. Silverman	Prof. Zöllig

At the International Congress of Mathematicians, Zurich 1932. Courtesy of Dr. Antal Varga, Bolyai Institute, University of Szeged.

VERHANDLUNGEN DES INTERNATIONALEN
MATHEMATIKER-KONGRESSES ZÜRICH 1932

ERSTER BAND

BERICHT UND ALLGEMEINE VORTRÄGE

SEPARATABDRUCK

C. CARATHÉODORY
Über die analytischen Abbildungen von
mehrdimensionalen Räumen

ORELL FÜSSLI VERLAG ZÜRICH UND LEIPZIG

*Separate print of Carathéodory's paper
presented at the International Congress of Mathematicians, Zurich 1932.
Courtesy of the Bavarian Academy of Sciences, Archive.*

demonstrates Carathéodory's great agility, which, next to his broad knowledge and his astonishing linguistic fluency, made him one of the most important personalities in international congresses.[366]

Carathéodory's former Göttingen colleague, Emmy Noether, also participated. A year later, referring to her speech on *Hyperkomplexe Systeme* (Hypercomplex Systems), Birkhoff accused her of intentional ignorance of Anglo-American scientists.[367] He could have done the same for Carathéodory, whose own list of references included, along with his own name, the names of Wirtinger, Weyl, Radó, Thullen, Horstmann and Cartan. None of these were British or American, but Birkhoff did not make the same criticism of Carathéodory.

Such international conferences also offered parallel entertainments and visits and after the Congress participants were able to visit the scientific station on the Jungfraujoch.[368] Carathéodory's daughter took advantage of the programme arranged for ladies and Carathéodory himself took a trip to the top of the Rigi on the famous cogwheel railway.[369]

4.37
Mechanics

In addition to his axiomatics of the theory of special relativity and his article about the notion of space–time, Carathéodory produced a series of papers on mechanics starting in 1933.

In his major contribution *Über die strengen Lösungen des Dreikörperproblems* (On the Strict Solutions of the Three-Body Problem),[370] he presented a new derivation of the Lagrange solutions of the three-body problem. This problem can also be considered for four or more bodies that are on a plane or on a straight line and Carathéodory showed that apart from certain trivial cases there exists no spatial configuration of planets for which the Lagrange motion is possible.

In the paper *Sur les Équations de la Mécanique* (On the Equations of Mechanics)[371] presented to the Inter-Balkan Congress of Mathematicians in 1934, he generalised the equations of motions of mechanical systems for spaces of more than three dimensions.

On 2 September 1940[372] he showed Herglotz the way in which the integration of the two-body problem could be performed "most elegantly". He had leafed through Herglotz's old calculations for the determination of planet orbits and presented to Herglotz his own calculations *Über die Integration der Differentialgleichungen der Keplerschen Planetenbewegungen* (On the Integration of the Differential Equations of the Kepler Motion of Planets).[373] He ended this work with two equations:

$$r = C^2 (f + F \cos v)^{-1}, \quad \dot{r} = C^{-1} F \sin v.$$

The first equation says that the orbit lies on a conic section and the second gives the radial velocity of the planet. The differential equations of the two-body problem can be treated most easily if all auxiliary quantities in question that remain invariant under an orthogonal transformation of the space co-ordinates are systematically introduced. An advantage of this method is that all formulae for the space problem can be at the beginning and all data of the trajectory can be calculated from the most general initial conditions, without having to assume the results of the theory of the two-body problem in the plane. Five years later, and under the same title, Carathéodory submitted to the Bavarian Academy of Sciences a paper simplifying and completing his former presentation.[374]

CHAPTER 5

National Socialism and War

5.1
"Gleichschaltung"

The rise of national socialism as a political force in the 1930s did not only impact on German universities from outside, in the sense of politically violent measures to enforce conformity. Antidemocratic traditions, economic problems of the Weimar Republic and the crisis of the parliamentary system had paved the way and enabled national socialism to grow on an already fertile soil of pre-existing political, ideological, social and scientific views and dispositions and to merge with them. Since 1918, university intellectuals and those belonging to the wider *Bildungsbürgertum* had adopted an attitude of "wait and see". The Republic had not been loved by university professors and the "apolitical" ones had behaved reservedly. Once nazism became an established political force, the Association of German Universities (*Verband der Deutschen Hochschulen*) stated their belief in national socialism, because they realised that it best represented nationalistic aims. The *Verband* had been founded in 1920 and was a kind of an umbrella organisation. It was characterised by a nostalgia for the *Kaiserreich*, was consciously conservative in outlook and represented and safeguarded the scientific, but most of all, the financial interests of both the universities and their academics. The German University Day, organised annually by the *Verband*, was a major event in its calendar. Prior to the elections of 5 March 1933, just over three hundred professors put their names to The German Intellectual World for List 1 (*Die deutsche Geisteswelt für Liste 1*) in support of national socialists and Adolf Hitler. On 3 April, the association published a statement "full of indignation and with the sharpest protest against the horror propaganda from abroad lacking every basis", referring to reports on the Nazi regime. On 12 April, the German Rectors' Conference set up a committee to prepare the firm integration of universities into society, in other words, their nazification. On 13 April, the German Students' Corporation (DSt = *Deutsche Studentenschaft*) started an Action against Non-German Spirit (*Aktion wider den undeutschen Geist*), which culminated on 10 May in the public burning of books. Also on 13 April, a law regarding students, issued by the Prussian Minister of Science, Art and People's Culture, introduced the "student self-administration" that had been abolished by the ministry during the time of the Weimar Republic, and limited its membership exclusively to "Aryans". There soon followed regulations prescribing quotas for "non-Aryans" at German schools. On 21 April, the Association of German Universities stated that the rebirth of the German people, and the rise of the new German Reich, would provide universities with the fulfilment of their longing and the confirmation of their passionate hopes. On 28 April, an executive committee consisting exclusively of National Socialists was appointed to the German Association of Academic Assistants (*Deutscher Akademischer Assistentenverband*). At the beginning of May 1933, all executive-committee members of the Bavarian Association of University Lecturers (*Bayerischer Hochschullehrerbund*) became members of the National Socialist Teachers' Association (NSLB = *NS-Lehrerbund*) at the request of Hanns Dorn, professor for social sciences at the Technical University of Munich and first president of the Bavarian Association of University Lecturers. Albert Rehm from the University and Sebastian Finsterwalder from the Technical University of Munich were among them. On 21 May, without external pressure, the executive committee of the Association of German Universities formulated a declaration introducing the *Gleichschaltung*, or ideological levelling down, of the universities and their integration into the National Socialist power and state machinery. The declaration introduced the *Führerprinzip* (leader principle) as the fundamental principle for a university and was adopted unanimously. University autonomy was thus abolished. Such hastily established connections with national socialism show that a greater part of the academics were familiar with, or at least sympathetic towards, national socialist ideas. In June 1933, the newly elected executive committee of the Association of German Universities stated its "belief in the national socialist world view without reservation" and signed its statement with *Heil Hitler*. On 11 November 1933, 700 out of 2000 professors supported a Declaration of Belief in Adolf Hitler and the National Socialist State (*Bekenntnis der Professoren an den deutschen Universitäten zu Adolf Hitler und dem nationalsozialistischen Staat*).[1] On that same day, local organisations of academics with temporary positions at the universities were required to disband. In the winter term of 1934–1935, with fourteen percent of university professors already dismissed, a declaration of the Association of German Universities paved the way to the deliberate *Gleichschaltung* of the universities. It referred to the selection of lecturers according to the ideal of discipline and ability.

National socialism was also particularly attractive to students. They were one of the first social groups in which National Socialists were able to achieve political hegemony through democratic means at the beginning of the 1930s. Students attracted to nazism were not only the children of the old elites who had the most to lose through the rise of modernism, but also the children, increasingly represented at the universities, of the lower and middle classes. Already in 1931, the DSt had elected a party member as their first chairman. Up to 10 January 1933, about half of all students organised in societies were members of National Socialist organisations, such as the National Socialist German Students' Association (NSDStB = *Nationalsozialistischer Deutscher Studentenbund*), the National Socialist German Workers' Party (NSDAP = *Nationalsozialistische Deutsche Arbeiterpartei*), the Storm Troopers (*Sturmabteilung*), and the SS (*Schutzstaffel*, a paramilitary combat unit of the NSDAP). In 1933–1934 student functionaries became the real centre of power in the universities and about twenty-nine percent of all students belonged to the NSDStB, which, founded in 1926 and led by Baldur von Schirach, had evolved into the strongest political force among German students. Since June 1933, the Nazi Party and this association had exercised a policy of enforced gradual nazification of the student societies aiming at undermining their independence. At the end of 1934, the Work Community of National Socialist Female Students (ANSSt = *Arbeitsgemeinschaft NS Studentinnen*) was disolved as an independent organisation and almost entirely incorporated into the NSDStB hierarchy. The student contribution to the destruction of democracy is attributed by Craig[2] to the "systematic discouragement of student reform movements" throughout a whole century and to the "deliberate fostering of political indifference by regional governments". German historians attribute the strong Nazi influence on students, not only to the decline of the *bildungsbürger- lich* value system after World War I, but also to the concern of young intellectuals about their social degradation as a consequence of the economic crisis of 1929.[3] Lack of prospects in the Weimar economy for university graduates had made them susceptible to Nazi rhetoric.

Furthermore, among students a radical anti-Semitism became dominant when most of the traditional student associations had come together in 1921 to create the German University Ring (*Deutscher Hochschulring*). In the following year, the *Hoch- schulring* gained two thirds of the seats in student parliaments. With their major participation in a ballot in 1927, seventy-seven percent of Prussian students voted for the retention of a membership formula which excluded Jews. Student anti-Semitism was not only racist but also extremely elitist and strictly directed against "rowdies": the "Jewish problem" had to be "solved" quickly, radically and "in a functional man- ner" through laws for foreigners and through the expulsion of all Jews from Germany by means of measures taken by the state and not through pogroms and violence.

As disturbing as the opinions of the students might have been, it was the nazism of the professors that evokes concern. Mathematicians were confronted with both the government's attempt to force academics into line, a move which resulted in the dismissal of Jews from any state service, including the universities, and with the un- successful attempt of a small group of mathematicians to introduce a racial agenda into mathematics.[4] Up to the end of 1934, the German Union of Mathematicians (DMV) was a battle ground between those who attempted to enforce Nazi policies and those who resisted them, arguing for the necessity of maintaining an interna- tionally open and purely scientific way of proceeding in the DMV, of which many members were foreigners. A strong dispute broke out among representatives of the two tendencies in the annual assembly of the DMV in Bad Pyrmont on 13 September 1934. An attempt of a small group around Bieberbach to gain control of the organ- isation failed. Bieberbach and Doetsch were designing plans for various changes in the union. In Doetsch's opinion Perron, who had been elected president in Septem- ber 1933, could no longer be tolerated, Hasse and Knopp had to be thrown out of the *Jahresbericht* (Annual Report) and the only timely solution would be a perma- nent presidency of Bieberbach with the characteristic status of a *Führer*.[5] Blaschke and Hecke belonged to the "internationalists" and Blaschke was even elected chair- man for the following year. Hecke proposed a weak "leader principle", according to which the chairman would continue to be elected by the assembly of DMV mem- bers but would himself appoint and dismiss the other members of the committee. Although Bieberbach was blocked in his attempt to introduce a strong "leader prin-

ciple", he was able to sabotage the proposed change of statutes. On the intervention
of the Reich Ministry of Education provoked by Bieberbach, both Bieberbach and
Blaschke had to resign their offices. The DMV remained formally independent but,
in fact, was obliged to conform with ministry demands. Bieberbach was succeeded
by Emanuel Sperner in Königsberg, who became the new secretary and Blaschke
was suceeded by Georg Hamel from the Technical University of Berlin, who became the chairman of the union. The result of these changes was that, after 1934,
the DMV was in the hands of mathematicians who, though receptive to compromise
with the Nazi regime, were more moderate than their predecessors and provided the
opportunity for an association of mathematicians based on scholarship. The president of the DMV declared for example in 1941 that he was pursuing "the imperialistic aim" of concentrating all rights and duties concerning mathematics in the
hands of the union alone, while an instruction or training committee of the DMV,
under the direction of an experienced and active man, should resume the duties
of the Reich Mathematics Association (*Reichsverband Mathematik*). Officially, the
Reichsverband Mathematik might have been the umbrella organisation of the DMV
but, in fact, it was a branch institution.[6] Since its foundation in 1921, it had been led
by Georg Hamel and, although it was an idependent organisation for the political
pursuit of the interests of mathematicians, it ensured the mathematicians' connection
to the government. In 1933 Hamel spoke of a spiritual bond between mathematics
and the "Third Reich".[7]

5.2

Carathéodory's Friends:
Victims of the 1933 Racial Laws

According to the "Law for the Restoration of the Civil Service" (*Gesetz zur Wieder-
herstellung des Berufsbeamtentums*) of 7 April 1933, aimed at re-establishing the
body of German civil servants which had been dissolved after 1918, civil servants
could be dismissed or sent to retirement on political or "racial" grounds or simply
for the "simplification of the administration". Those affected by the law were persons who had entered the civil service after 9 November 1918, civil servants whose
political activity did not guarantee their unconditional support of the national state
and civil servants of "non-Aryan" descent. The "Law against Overcrowding of German Schools and Universities" (*Gesetz gegen die Überfüllung deutscher Schulen
und Hochschulen*) was made on 25 April 1933 to impede registration, study and
graduation of "non-Aryan" students. Decrees of the Prussian Ministry of Education
ensuring the implementation of that law provided for a proportion of "non-Aryan"
students not exceeding five percent of those already registered at any university or
faculty. Jewish students could only register if the proportion of "non-Aryans" at their
faculty would remain less than 1.5 percent. Exemptions were similar to those provided for by the Law for the Restoration of the Civil Service. The latter was extended
by a third decree on 6 May 1933 to include lecturers, even if they were not civil servants and did not obtain a fixed salary from the universities. Thus, all lecturers had to
become civil servants. Also by this third decree the category "politically unreliable"

was extended to include persons who had been dismissed because they were involved in opposition activities. In many cases, the exception made to old civil servants and front-line soldiers of World War I was cancelled through pressure, harassment and denunciation.

Almost all ministries of education were taken over by long-serving Nazis: Bernhard Rust in Prussia, Christian Mergenthaler in Württemberg, Otto Wacker in Baden, Fritz Wächtler in Thüringen, Hans Schemm in Bavaria. Schemm was a former primary-school teacher who, as a member of Epp's volunteer corps (the *Freikorps Epp*), had taken part in the suppression of the Munich *Räterepublik* in 1919. A member of the NSDAP since 1923, he had been elected to the Bavarian *Landtag* (state parliament) in 1928 and to the *Reichstag* two years later. From 1932 he held the post of *Gauleiter* of the Bavarian Ostmark. In 1929, he founded the National Socialist Teachers' Association, which he led until his death in 1935. Schemm wished to impose members of that association on every university as their heads (rectors). They had to be spokesmen of the NSDAP, be proved Nazis, indifferent towards material things, acknowledged as specialists in their subject, and not too unworldly. The NS Teachers' Association claimed for itself the task of evaluating the political reliability of persons considered for university posts. But in practice it lacked authority and financial means and was ineffective under Schemm.

On 18 July 1933, Schemm asked the senate of Munich University to implement the first decree (11 April 1933) of the Law for the Restoration of the Civil Service on the basis of certificates submitted from, and questionnaires completed by, university members. Schemm attached a list of names of those academics who had proved their "Aryan" descent through certificates. In fact, not all were required to provide certificates. This obligation was not required of those who had held a budgeted position in the civil service before 1 August 1914, or had been a front-line soldier in World War I or who, in 1914 had lost a father or a son in the war. Carathéodory was listed among the professors of "Aryan" descent, whereas his close friends and colleagues Friedrich Hartogs and Alfred Pringsheim, both civil servants and full professors, were listed among the "non-Aryans" and had already been dismissed on 27 June 1933.[8] The recipients of the questionnaires had to give detailed information concerning the descent of their parents and grandparents and, by 14 June 1933 at the latest, to send them in duplicate to the rector of Munich university, who would in turn forward them together with the certificates to the ministry. This obligation marked the start of a series of discriminatory acts against Jews: according to the first decree of 11 April, a civil servant was considered a "non-Aryan" if one of his parents or grandparents was a "non-Aryan" and these persons were in turn considered Jews if they were of Jewish religion.

Carathéodory completed his questionnaire just on the deadline and noted that he would submit his own birth certificate and his parents' wedding certificate later. He stated that his citizenship was German at that time, but had been Turkish at his birth, and his confession, as well as that of his parents and grandparents, was Greek Catholic.[9] His parents' wedding certificate was issued on 15 June 1933 by the head of the Greek Orthodox church of Marseille where the matrimonial ceremony had taken place.[10] His own baptismal certificate was issued according to the register of the

Russian Orthodox church in Berlin a month later.[11] In all official documents until his Bavarian period, Carathéodory's confession had been declared as Greek Orthodox.

In an addendum to the questionnaire, Carathéodory answered the question of whether he belonged to or was active in the communist party or the national-communist movement Black Front (*Schwarze Front*) with "no". Next to the questions "which political party have you belonged to up to now, since when?" and "have you been a member of the Black-Red-Gold Reich Banner [*Reichsbanner Schwarz-Rot-Gold*], of the Republican Association of Judges or Civil Servants [*Republikanischer Richter- oder Beamtenbund*], of the Iron Front [*Eiserne Front*], or of the League for Human Rights [*Liga für Menschenrechte*]?" Carathéodory replied *"Fehlanzeige"* (no chance).[12]

In Munich, a total of 333 civil servants were dismissed in implementation of an ideological construct with devastating consequences, especially with regard to the universities. The dismissals in Germany immediately provoked a reaction abroad. The Emergency Committee in Aid of Displaced German Scholars was created in New York under the liberal political scientist Stephen Duggan and sent their first circular letter to the presidents of American colleges and universities as early as 27 May 1933. Funding of emigrant scientists was shared by both universities and the Rockefeller Foundation, which primarily promoted advanced research and had set up a special fund for the exiled scholars from Germany on 12 May 1933. For this purpose, Duggan co-operated closely with Oswald Veblen and Hermann Weyl. In many cases, international research grants from the International Education Board of the Rockefeller Foundation had in the past – as was to happen now – served the purpose of attracting mathematicians from German-speaking Europe. Such examples are Eberhard Hopf, Carathéodory's future successor in Munich who, with the help of such a grant, had gone to Cambridge, Massachussetts, in 1930, Carathéodory's friend Isaak Schoenberg, a Rumanian by birth, who had gained his diploma and doctorate at the University of Jassy and studied further in Berlin and Göttingen and, finally, Tibor Radó.

Besides setting up contacts between exiled scholars and scientific institutions interested in having them, all aid organisations also endeavoured to obtain additional money to create new positions especially for emigrants, and so counter the suspicion that scientists at home were underprivileged compared to the emigrants. With such additional funding, some universities acquired outstandingly good scientists, without having to pay their salaries at least for a period of time. Very soon, lists with names of prominent scientists began to circulate. There were enough offers for famous physicists, such as Born or Franck, but things were difficult for less well-known scientists and for those over 40 years of age or for the very young ones, who had not as yet had the opportunity of establishing themselves. The dilemma of choosing between helping refugees or supporting young Americans was resolved fairly and the differing views of Veblen, who was active in finding positions for European emigrants in the United States and Birkhoff, who pressed the cause of promising young Americans, served to maintain a balance.[13]

In England, the Manchester Guardian of 19 May 1933 printed a list of 196 academics of all faculties who had been dismissed in Germany between 13 April and 4 May. The London Times of 24 May released an appeal for the foundation of an Academic Assistance Council. The council aimed to support researchers or university teachers who, "on grounds of religion, political opinion, or race", were unable to carry on their work in their own country. Aid was not limited to scientists from Germany and its purpose was restricted to "the relief of suffering and the defence of learning".[14]

A great many of Carathéodory's friends and colleagues were affected by the racial laws of 1933. The list is a long one but those we mention here will provide an illustration of how deeply the application of these laws in 1933 cut into the fabric of German scientific institutions.

The 55-year-old Felix Bernstein, for example, director of the Institute of Mathematical Statistics in Göttingen, was dismissed from his office in Germany while he was in the United States. He remained in the United States as a guest professor at Columbia University in New York.

The 57-year-old Otto Blumenthal, of Jewish origin but a convert to Protestantism, and also a front-line soldier in World War I, was denounced by the Students' Union (AStA = *Allgemeiner Studentenausschuß*) and was held in "preventive detention" for fifteen days. He was compelled to give up his professorship at the Technical University of Aachen and lost the editorship of the *Jahresbericht*, in which he was succeeded by Konrad Knopp. Two years later, still in Aachen and fully demoralised, he wrote to von Kármán, who had been director of the Aeronautical Institute in Aachen up to 1930: "I am not suitable for a high position such as Princeton for, on the one hand, I am not productive enough so as to occupy it to my complete satisfaction and on the other, I am not a good actor so as to play at being the productive spirit without being one. On the contrary, I feel qualified for a post with average demands. I can understand new things, I can imbue them with something of my own and lecture in such a way that the audience will benefit therefrom. [...] In the United States, especially in the Middle, the South and the West, there are surely universities that could avail themselves of such lecturers."[15] Blumenthal fell exactly within the category described by Born: "What can one do with a 55-year-old dentist? The secret state police will put him in a concentration camp if he does not emigrate soon. But his American register number is 60 000!"[16]

The 34-year-old Salomon Bochner, for whom Carathéodory had tried to find a job at Harvard five years earlier, was dismissed from the Munich Mathematical Seminar by a rector's decision on 18 May 1933. Carathéodory's daughter remembers that he was a quiet young man, who resisted the idea of emigrating, a thing that made her father become frantic and yell at Bochner behind the closed door of his office. In the end, Bochner emigrated to the United States and spent the next thirty-five years at Princeton University where, from 1950 until his retirement, he worked as a Henry Burchard Fine Professor of Mathematics. He was described by his associates as an "analyst of exceptional power and originality." He made significant contributions to a wide range of fields of mathematics, including probability, complex analysis and differential geometry.[17]

Max Born, a convinced democrat,[18] was forced at the age of 51 to take "leave" of office and left Göttingen accompanied by his wife on 10 May. He had "never felt particularly as a Jew" despite the Jewish descent of his parents;[19] he had found a spiritual home in German philosophy, literature and music and shared with German Christians an educational blend of classical knowledge and humanist culture with a Christian ethical tone.[20] He was informed unofficially that an offer of appointment was available for him in Belgrade where everything depended on personal relations and where it was considered much more important to tell amusing stories for the entertainment of some minister while drinking wine than to carry out research in a scientific way. Of course, such a description of scientific relations was a deterrent to Born.[21] Soon he accepted an invitation for a temporary post in Cambridge, England, where he had studied and had friends. In 1936 he accepted a professorship of theoretical physics in Edinburgh. Meanwhile, Nordheim, Born's year-long assistant in Göttingen, left for California.

The 45-year-old Richard Courant, to whom Göttingen owed the foundation of the prestigious mathematical institute by American funding after World War I, resigned as the director and at the end of August 1933 accepted a position as a visiting lecturer in Cambridge, England. Later, he moved to New York University and became recognised as the *éminence grise* among German exiles, whom he helped to integrate into American society. With Courant, Born and James Franck, the heads of three out of the four institutes for mathematics and physics in Göttingen were lost because they were Jews. Together with twenty-seven other scientists, Sommerfeld among them, Carathéodory signed a petition in support of Courant, which had been initiated by Kurt O. Friedrichs from the Technical University of Braunschweig and Otto Neugebauer from the University of Göttingen. The petition underlined the financial advantages which Courant had brought to Göttingen and Germany, his reputation as a scholar, his exceptional success as an academic teacher and, finally, his behaviour according "to the old law of Prussian civil service" prescribing defence of oneself against circulating rumours by actions and not by words. The petition was submitted to the Ministry of Education around mid-June 1933 but received no response. In the summer of 1933, while on leave, Courant was considering the possibility of accepting a professorship at the newly organised state university in Istanbul. Therefore, he needed the opinion and advice of a European expert in matters concerning Turkey and turned to Carathéodory. On 19 July he wrote:

Dear Carathéodory,

after all, I have returned rather quickly without going the long way round via Munich, since I considered it necessary to inform the university *Kurator* [university officer dealing with financial and legal matters] of the somewhat fantastic Istanbul matter as soon as possible, before distorted and misleading news spread by another route. It seems to me that the matter itself is still substantially not clarified. Anyhow, one in my present position has to take all possibilities seriously and, therefore, it means a lot to me to really get some information from you or Kritikos. The matter seems to me to be the following:

It has been suggested to the Turkish government to use the favourable opportunity for them in order to appoint a large number (over 30) of good European scientists with comparably small salaries to the university of Istanbul, which is to be reorganised. For good reasons it seems that it means a lot to the Turks to settle the matter quickly. They have a gentleman who opened a mediation centre for German scientists and apparently tries hard, having been given this task, to mediate these offers of appointment. When I met this gentleman in Zurich on Saturday, I was surprised by the whole matter, especially by the fact, that to a certain extent it had to do with a mass action. My first reaction was that the matter should not be bound to any kind of secrecy or conflict against the German authorities under any circumstances. On these grounds, I have also prevented the gentlemen in question gathering together for a general discussion of the plan; likewise, I have refused to travel straight away to Constantinople together with Franck, in order to examine the situation closely. It is now probable that some Zurich colleagues will make this journey next week in order to gather information and maybe a clearer situation will result then. What I would like to know from you is the following:

1. Your general impression, if and to what extent such an enterprise should be taken seriously and to what extent one could rely on the firmness of Turkish intentions, should the occasion arise.
2. Whether a salary of 500 Turkish pounds per month can be considered sufficient or reasonable for such a post.
3. What the living and health conditions in Constantinople are like.
4. What the school system conditions are like.
5. How one carries out individual negotiations expediently with a person authorised by the Turkish government, should that be the case.

I could imagine that, if indeed a series of good people come together and the living and working conditions are not unfavourable, such an enterprise could be quite appealing. One of the essential prerequisites, however, would always remain that the taking on of this task should happen in accordance with the German authorities and be recognised as being in Germany's interests.

Beside all this, my personal situation here is still not clear. In any case however, the information I have asked for would be of decisive importance also for other colleagues. All the same, I would perhaps seek to get into contact with you, that is in any case certainly not before the end of the first week of August, if the matter has progressed further and the report of the Zurich colleagues is available. What are your travel plans?

Many kindest regards

Yours [22]

Courant was obviously referring to the Emergency Action Organisation for German Scientists Abroad (*Notgemeinschaft deutscher Wissenschaftler im Ausland*) the successor to the Advice Centre for German Scientists (*Beratungsstelle für Deutsche Wissenschaftler*), which had emerged in Zurich almost simultaneously with the Law for the Restoration of the Civil Service and was directed by Philipp Schwartz, an associate professor of General Pathology and Pathological Anatomy (*Allgemeine Pathologie und Pathologische Anatomie*) at the University of Frankfurt am Main who had fled to Switzerland. The existence of this centre was made known through a small note in the *Neue Zürcher Zeitung* in mid-April. Soon able to expand its finances and establish an office with the help of wealthy Swiss friends, it acquired the name Emergency Action Organisation for German Scientists Abroad. [23]

In the summer of 1933 they managed to find jobs for about thirty scientists in Istanbul. Born belonged to the executive committee of the organisation and was extremely active. Courant and Franck travelled to Istanbul later in the summer, but Courant did not find nationalist Turkey to his taste. Judging by his letter to Carathéodory, he may still appear to be incapable of recognising the dramatic political change in Germany, but maybe he was only trying to avoid the *Reichsfluchtsteuer*, namely the tax imposed on those fleeing the Reich, by attempting to be characterised as a man working "for the German interest" during his leave abroad,[24] or maybe he simply believed that the Nazis would not stay in power for long. Courant would retire formally with effect from 31 December 1935. It should be noted here that in Born's opinion, Courant, just like Franck, was inwardly a German.[25] Most of the intellectual German Jews affected by Nazi persecution were assimilated and only a few could identify with their Jewish heritage.

Interestingly, a letter from Blichfeldt to Wilbur about Courant, shows clearly that Carathéodory certainly had the opportunity to leave Germany at that time for a post in the United States. He had been at Stanford in 1928 (see 4.18) and was still welcome to go there if he desired. As Blichfeldt wrote, "Professor Courant of Göttingen is a very active man, age 45, of fine business ability, an excellent teacher, and a mathematician of prominence, though not of the rank of Carathéodory, Harald Bohr, Schur, etc.; nor of the rank of quite a number of mathematicians in this country. He is not quite an analyst of the attainments and originality that we had hoped to get for Stanford sometime".[26] At Bohr's request, G. H. Hardy tried to get a permanent position for Courant in Cambridge, England, but without success.

Courant's successor in Göttingen, the 34-year-old lecturer Otto Neugebauer, refused to take the oath of allegiance to Hitler and had to flee Göttingen as a political opponent. He served only one day as the head of the mathematical institute there. Neugebauer transferred the editorial board of the *Zentralblatt für Mathematik* to Copenhagen. But even by 1936 "certain circles of German specialists" complained against Neugebauer's activity in the *Zentralblatt*. The claim was that the racial origins of authors and referees ought to be taken into account. In November 1938 the crisis reached its climax and in 1939 Neugebauer emigrated to the States where he founded the *Mathematical Reviews*.

Einstein stated his views of events in Germany openly, called for a world-wide protest against Hitler during a lecture tour across the USA and announced his decision not to return any more to Berlin. On 28 March 1933, he resigned from the Prussian Academy of Sciences and a few days later he applied in Brussels for his release from German nationality. With this action, he anticipated the disciplinary proceedings that the Prussian Academy of Sciences would open against him. To the question by the Bavarian Academy of Sciences, of how he viewed his relation to it, Einstein answered on 21 April that he wished to be crossed off the list of its members because, although academies had as their prime task to promote and protect the scientific life of a country, German academic societies had silently put up with the fact that a considerable proportion of German scholars, students and professionals with an academic training were robbed of their chance to work and earn their living.[27]

Exactly three years earlier, Carathéodory had been ready to arrange a meeting between Einstein and Morgenthau (see 3.13). But now he did not raise a word of protest against Einstein's treatment by the two academies. Carathéodory was not opposed to the idea of protest through a petition to the authorities, provided the scientist under persecution had not come to the point where the authorities considered there was a conflict with them. Four years earlier, in his paper *Deutsches Wissen und seine Geltung* (German Science and its Importance) he had included Einstein among German scientists (see 2.14). On the basis of the existing documents, it cannot be concluded with certainty whether Carathéodory's silence was due to his unwillingness to support Einstein openly, and thus appear as an opponent of the regime, or to his impression that, in the final analysis, Einstein, through his "anti-German" attitude, had himself provoked his ill-treatment by the academies. On 30 May 1933 Einstein wrote to Born from Oxford: "I believe you know that I have never thought especially favourably about the Germans (in a moral and political sense). But all the same I have to confess that they have rather surprised me by the degree of their brutality and cowardice". And under his signature he added: "I have been promoted to being an evil beast in Germany and they have taken all my money. But I console myself with the thought that I would soon run out of it anyway".[28]

The 53-year-old Paul Ehrenfest, a professor of physics at the University of Leiden, Netherlands, who had been Carathéodory's student in Göttingen in 1906, committed suicide.

In 1931, the Zionist Adolf Abraham Fraenkel had gone to the Hebrew University of Jerusalem where he shared the direction of the mathematical institute with Michael Fekete. In the spring of that year he invited Carathéodory, then in Athens, to visit the Hebrew University, but the invitation was delayed in reaching him. Carathéodory expressed his regret at having missed the opportunity to visit Palestine but believed that he would often travel to Greece in the future and would be able to take such an opportunity to visit the Hebrew University. In his opinion, there were "many common points with Greece".[29] The Hebrew University was conceived as a centre of spiritual, cultural and scientific rebirth of the Jewish people. The idea of founding a university in Jerusalem was presented in 1901 to the First Zionist Congress and was published in 1902 in a pamphlet authored by Martin Buber, Berthold Feiwel and Chaim Weizmann.[30] Its foundation in Palestine on 24 July 1918 was a central programmatic point of the Zionist movement aiming at the cultivation of a Jewish awareness, which was considered a condition for the parallel satisfaction of material needs. Hebrew, the only language allowed at the university, was intended to be the common element binding immigrants together with the purpose of creating their identity. In this respect, the Hebrew University presented striking similarities with the University of Smyrna to be founded two years later (see Chapter 3). Just as Herzl's political Zionist endeavours were complemented by the cultural project of the foundation of a Hebrew University, Venizelos's Greek nationalist endeavours were complemented by the cultural project of the foundation of a Greek University. In both cases the universities served a national aim. The Hebrew university opened with the faculties of microbiology, chemistry and Jewish studies in 1925; the Ionian University was to open with the Institute of Hygiene and the library before the catastrophe. In the 1930s, the Hebrew university housed many Jewish scholars who had fled Germany. Adolf Abraham and Wilma Fraenkel visited Carathéodory in Munich on 7 September 1931. In April 1933, Fraenkel asked for his pension from the University of Kiel and his chair there had to be occupied anew in the summer of 1933. The university attempted to bridge the gap by bringing either Kurt Reidemeister or Werner Rogosinski, both professors in Königsberg and at that time on forced leave, the former on political, the latter on racial grounds.

The astronomer Erwin Freundlich resigned his post in Potsdam at the age of 48 and emigrated to Istanbul, where for the following four years he contributed to the reorganisation of the university and the creation of a modern observatory.[31]

Kurt Hensel, one of Carathéodory's professors in Berlin who, according to Nazi laws, was a twenty-five percent "Aryan", was deprived of his authorisation to teach at the age of 72.

The philosophy professor Edmund Husserl, born into a Jewish family, was subject to abuse and harassment from 1933. His title of professor was revoked in 1936 and he died at the age of 79, two years later.

At the age of 56, Edmund Landau, one of the first professors of the Hebrew University, who in the 1920s had added the name of the great Prague rabbi and Talmud scholar Jecheskel to his first name, became the victim of a deeply offensive provocation organised by Nazi students. Landau was not dismissed, for he had been a civil servant before the 1st of August 1914. He was granted "leave" and took premature retirement in 1934. Contrary to almost everyone surrounding him, he was at first optimistic with regard to the future of the Jews under the Nazis. For a long time he had represented the Zionist point of view, that it would be better for Germans and Jews to live as two peoples recognisable in their own particularity. In 1935–1936 he became the Rouse Ball lecturer of mathematics at the University of Cambridge, England,[32] and died of heart failure in Berlin three years later.

Apart from Landau, the German Students' Association (*Deutscher Studentenbund*) also boycotted Hilbert's personal assistant Paul Bernays, and campaigned against Neugebauer, accusing him of being a communist.

The 55-year-old Leon Lichtenstein was deprived of his chair at the University of Leipzig and died of heart failure in the Polish mountain resort Zakopane in the same year.

By reason of his Jewish origin, the 50-year-old Richard Edler von Mises was forced to withdraw from the Society for Applied Mathematics and Mechanics (*Gesellschaft für angewandte Mathematik und Mechanik*) and give up his Berlin professorship; he accepted a professorship tied to the head of the mathematical institute at the University of Istanbul.[33] In a letter to Erhard Schmidt on 21 October 1933, he proposed Erich Trefftz, then in Dresden and known for his hostile political attitude towards the Nazis, as his successor. When Anni Trefftz visited Carathéodory on 25 February 1934, Erich Trefftz was still in Dresden. He took on the editorship of the *Zeitschrift für angewandte Mathematik und Mechanik* (Journal for Applied Mathematics and Mechanics)[34] from von Mises, but only from volume 14 of 1934 until volume 16 of 1936. Trefftz died before his 50th birthday.

The 51-year-old German-Jewish mathematician Emmy Noether was not allowed to lecture at the University of Göttingen after 1933. She took a post as a guest professor at Bryn Mawr Women's College, Pennsylvania, and also gave lectures at the Institute for Advanced Study, Princeton. She spent the summer of 1934 in Göttingen and she even wished the 36-year-old Helmut Hasse much success in his efforts to rebuild the great mathematical tradition of Göttingen. Hasse had become the director of the mathematical institute replacing the 48-year-old Hermann Weyl, who had succeeded Hilbert in 1930. Emmy Noether died in 1935 from postoperative complications following apparently successful surgery for the removal of a tumour.

After abortive negotiations for positions offered to him in the USA in 1927 and 1928, and fearing discrimination against his children and his wife Hella, who was of Jewish origin, as well as the end of his own academic career, Hermann Weyl decided to accept an offer from the Institute for Advanced Study in Princeton in 1933. He worked there up to his retirement in 1951. According to Born, Weyl was one of the last great mathematicians to have also engaged in theoretical physics and astronomy and produced significant contributions in these fields.[35] With Weyl's departure, a total of eighteen mathematicians had either been dismissed from their offices in Göttingen or left the town.

Being forced to resign his Berlin professorship, the 42-year-old philosopher Hans Reichenbach emigrated to Istanbul.[36] Shortly before, on 20 March 1933, he visited Carathéodory probably to discuss his situation.

The 40-year-old Kurt Reidemeister, the former student of Landau and Hecke's assistant in Göttingen, who was at the periphery of the Vienna Circle and whom the Nazis had classified as "politically unreliable", was dismissed from his professorship in Königsberg. The violent expulsion and the disintegration of the Vienna circle had already begun at the beginning of the 1930s, was intensified after the elimination of democracy in Austria in 1933–1934, and reached its high point after the *Anschluss* in 1938. Carathéodory and sixteen other mathematicians signed a petition organised by Wilhelm Blaschke on 22 June 1933 against Reidemeister's dismissal. They were trying to persuade the authorities that Reidemeister's retirement at such a young age would be a serious loss to teaching and research in Germany. The petition was successful in that Reidemeister was then appointed as Hensel's successor at the smaller and less significant University of Marburg.

Robert Erich Remak, whose attempt to habilitate in Berlin had been supported by Carathéodory, was dismissed from his Berlin position at the age of 45. He was arrested and sent to Sachsenhausen concentration camp following the *Reichskristallnacht* violence of November 1938. He returned to Berlin two months later and, with provisional permission to stay in Holland, was able to flee to Amsterdam. Pursued and arrested there, he was deported to Auschwitz where he was murdered. The last time that he was reported alive was in 1942.

Erwin Schrödinger left Germany at the age of 46. In November 1933, supported by the Imperial Chemical Industries, he arrived at Oxford where he became a fellow of Magdalen College. That year, he shared the Nobel Prize for physics, with Paul Dirac, for his work on wave mechanics leading to the equation that bears his name.[37]

In 1933, Hertha Sponer, the 38-year-old professor of physics at Göttingen University and later second wife of James Franck, started preparations for her emigration. The following year she accepted a position in Oslo, and in 1936 in Durham, North Carolina, where she built the physics institute at Duke University. She visited Carathéodory in Munich on 8 August 1938.

Otto Toeplitz, who was descended from the oldest Jewish family of Germany, was dismissed from his professorship in Bonn at the age of 52 and then devoted himself to work for the Jewish community.[38] From 1935, as head of the university department of the official organisation, the National Representation of the Jews in Germany (*Reichsvertretung der Juden in Deutschland*), he organised the emigration of Jewish students to the USA.

Artur Rosenthal, Dean of the Faculty of Natural Sciences in Heidelberg, resigned his office.

The position of Perron was discussed between the Prussian and Bavarian ministries of education as to whether he might be a suitable candidate for retirement according to the Law for the Restoration of the Civil Service, but this discussion was soon forgotten.[39]

The emigration of scientists who were not discriminated against on racial or political grounds seems to have been rather the exception. Those who decided to stay in Germany belonged to three broad groups: convinced National Socialists, those who saw an opportunity to improve their careers under that regime and shamefully used every sort of denunciation of their colleagues, and those who might have wished to leave but believed they would have had no other opportunity elsewhere. Carathéodory decided to remain in Germany, but he belonged to none of these three categories. Although he had been dismissed from the post of governmental commissioner in Greece the year before, and regarded the USA as "the other end of the world", he could have taken steps to leave Germany. But it seems that he viewed his decision to remain as his "patriotic duty" and believed that he could get by in life under the Nazis as best he could and, moreover, he would be able to have an influence on affairs. The tragedy of his case lies in the fact that, in the end, he gained nothing from this decision but gave away too much. From that point on, his freedom of choice would become increasingly limited and the course of his life would be determined almost entirely by events beyond his control.

5.3
Member of the "Reform Committee"

The reform debate concerning higher education was dominated by Nazi positions although it was stimulated by various motives and intentions. Radical measures to reform higher education were taken in the early summer of 1933 and were completed in the spring of 1935. These included a purge of academics, beginning with the Law for the Restoration of the Civil Service, the reform of the university constitution through the abolition of self-administration and the introduction of the *Führerprinzip* and, finally, the institutionalisation of authorities of political control that would guarantee a recruitment practice and the politicisation of scientific disciplines (see 5.18).

According to the *Führerprinzip* implemented at the German universities, the Minister of Education would appoint the rector and the rector would subsequently appoint his deputy, the senate members and the deans; the senate would be significantly diminished and transformed into a body that would advise the rector, who would bear the title *Führer* of the University. All of the senate's tasks, authority and rights, except for disciplinary matters concerning students, would pass over to the rector. In Bavaria, this reform was met with some annoyance and a few protests but nothing more serious. An undated protest note against the abolition of an election for the rector was attributed to the Rector of Munich University, the dermatologist Leo von Zumbusch, a right-wing conservative, who during the weeks of increasing enthusiasm for the Nazi movement had behaved rather reservedly.

A rector's letter of 7 June to the senate members shows that the university had officially accepted the reform discussion. The rector considered it appropriate to set up a committee for reforms at Munich University and proposed twelve persons, who were consequently confirmed as committee members in the senate's session of 28 June. The reform at the university had already been the intended subject of a discussion in Hugo Bruckmann's house in Munich on 8 May 1933. Bruckmann was a publisher, a Nazi member of the *Reichstag* and politically active in cultural affairs. At that meeting, the University of Munich was represented by six professors, among them the geopolitician Karl Haushofer, and two assistant professors.[40]

On 11 July 1933, the senate of Munich University stated that "the programmatic declaration of Herr Reichskanzler Adolf Hitler to the Reich governors on 6 July this year gives us the welcome opportunity to assure that we are joyfully prepared to co-operate in the tasks of the national socialist state to the best of our ability."[41]

On 28 and 29 July 1933, the Bavarian Ministry of Education made a telephone request for consultation to take place within the senate and for a proposal for a new rector in the following academic year. On 28 August 1933, the minister announced that the rector was the *Führer* of the university. Thus, abolition of elections and of university self-administration was complete in Bavaria in the summer of 1933.

In the autumn of 1933, committees (*Arbeitsausschüsse*) were established in the two sections of the Philosophical Faculty of Munich University to work out reform proposals on specific matters of university organisation as, for example, a reduction in study time, the system of tuition fees, study and examination regulations and reform of individual disciplines. However, the committess could not even begin with their substantive work because they were kept in the dark about the course of the reform. Not even the NSLB representative was aware of its aim. This was also connected with the fact that the process of centralisation had not been completed at these early stages of the Nazi state and uniform directives for the entire *Reich* were still being prepared. By the end of the winter semmester 1933–1934 it became clear to the supporters of the new political system that, despite a mood for reform within the university and despite corresponding NS propaganda, an effective co-operation with the universities was neither desired by the authorities, nor was possible within the new political system prescribing the *Führerprinzip*.[42]

The reform committee of the Philosophical Faculty, Section II, was set up on 26 October 1933 by the botanist Fritz von Wettstein, appointed dean of the faculty according to the *Führerprinzip*, but nothing became known of its work. Carathéodory was a member, together with professors Karl Haushofer, Walther Gerlach and Walter Sandt.[43] Haushofer, a former army officer who resigned from office as a major-general after World War I, had become a professor in 1921 and was appointed full professor in 1933. He was the founder of the science of geopolitics and had been a significant supporter of Hitler since 1922, when his student and friend Rudolf Hess had introduced him to Hitler. In the succeeding years, Haushofer acted as Hitler's unofficial consultant on foreign affairs. It was Haushofer's term *Lebensraum* that Hitler took up in *Mein Kampf*. Another member of the reform committee, Walther Gerlach, famous even then for the Stern–Gerlach experimental proof of space quantisation of atoms in a magnetic field, had been a full professor of experimental physics at Munich University since 1929. He was to be entrusted later, in 1943, with the head of the physics branch in the Reich Research Council and, from 1944, he was responsible for the secret German uranium project as the "authorised representative of the Reich Marshal [Hermann Göring] for nuclear physics" in Berlin. Gerlach firmly believed that Germany had to be victorious in the war. Finally, Walter Sandt had been an associate professor of botany since 1930 and an NSDAP member from 1 May 1933.

In 1933, none of the full professors was a party member and no member of the Philosophical Faculty, Section II, was an open sympathiser of national socialism. A full professor of the faculty did join the party later in 1937. Among the assistants there were already several young and active Nazis, who easily rose to the leadership of the Lecturers' Corporation (*Dozentenschaft*) and the National Socialist German Lecturers' Association (NSDDB = *Nationalsozialistischer Deutscher Dozentenbund*). A group of them at the state observatory later tried to impose one of their own candidates as Carathéodory's successor and, failing to do this, put obstacles in the path of the appointment procedure. The group was, however, faced with a united opposition, especially from Carathéodory, Perron and Tietze, who did their best to oppose external intervention in matters concerning academics and the consequent infringement of the faculty's rights.[44]

5.4
Three "Incorrigible" Opponents

Within the Philosophical Faculty, Section II, Carathéodory, Perron and Tietze were considered to be "incorrigible" opponents of the Nazis[45] and comprised a kind of self-appointed triumvirate with pre-agreed allocation of roles in their actions. From the very beginning of the Nazi regime, they acted together in educational matters and defended their positions and areas of authority, which they saw to be threatened.

Carathéodory, Perron and Tietze rejected decisively a request in April 1934 from the Faculty Students' Corporation (*Fachschaft*) and the *Deutsche Studentenschaft* of Bavaria to organise a lecture for future teachers of secondary schools on the theory of teaching and methodology in the instruction of the natural sciences. The three professors asserted that didactics and methodology could be learned only by in-service training. They feared that, otherwise, lecturing would be made accessible to professionally incompetent persons. According to Aloys Fischer[46] who belonged to the reform committee of Section I, the problem lay in the lack of qualified lecturers, namely of lecturers proved by habilitation. Educational theorists and philosophers, such as him, stressed the necessity of a training oriented to profession, and of didactics oriented to a specific subject, and supported the organisation of lectures on the theory of teaching and methodology in mathematical instruction in their expert report of 1934.[47] The political argument that lay behind the triumvirate's position was that the introduction of lectures on the theory of teaching and methodology could create posts for those whom they perceived as incompetent persons. In fact, the characterisation "incompetent persons" was a circumlocution for Nazis and the three professors were convinced that political criteria could, and would, override scientific quality.

Another example of the triumvirate's joint action was their participation in the mathematical colloquium. A circle of mathematicians bearing the name "mathematical colloquium" or "Munich mathematical circle" already existed when Carathéodory took up his appointment in Munich. It was a monthly meeting with a speech and a closing dinner. Full professors of the Technical University and the University of Munich regularly met there with assistants, lecturers and invited students. Friedrich Hartogs, Alfred Pringsheim and the retired Heinrich Liebmann took part, along with Perron, Tietze and Carathéodory. Pringsheim was considered the heart of the colloquium. In Carathéodory's guest-book there are entries of colloquium members on 6 March 1931[48] and 31 July 1931.[49] After 1933, the Nazis started to keep a close eye on the colloquium and, to hinder its existence, they attempted to bring it into disrepute and demanded the expulsion of Pringsheim, Liebmann and Hartogs. Of course, continued participation in the colloquium by other members could have indicated a sign of solidarity with their persecuted Jewish colleagues, or even have been a demonstration of some resistance to the regime, but it seems that the colloquium was left to fade away. After the war, when the colloquium was revived under the patronage of professor I. A. Barnett from the US military government, Carathéodory gave the first lecture on 11 July 1946, entitled "Über Schwarz'sches Spiegelungsprinzip und Randwerte meromorpher Funktionen" (On the Schwarz Reflection Principle and Boundary Values of Meromorphic Functions).[50] The last time that he gave a speech at the colloquium, and his last speech ever, was on 16 December 1949 when, under the title "Länge und Oberfläche" (Length and Surface),[51] he reported on recent and earlier works from all over the world based only on head-words, which he had noted earlier. The talk was given without notes, except for a list of bibliographical references he used as an *aide-mémoire*.

Apart from the mathematical colloquium, the Lecturers' Corporation suspected the Kant Society, with which Carathéodory seems to have had contact, of harbouring opposition "nuclei". Under the presidency of the associate professor of philosophy Aloys Wenzl, the society comprised a relatively homogenous group of like-minded people who, in addition to their purely scientific personal interest, cultivated liberal fundamental principles.[52] However, at the end of 1934, the Society became enmeshed in the wheels of the Rosenberg Office (an authority authorised by Hitler to survey the schooling of the NSDAP), the Ministry of Science, the Reich Literature Chamber (an institution of the Ministry of Propaganda) and the Ministry of the Exterior. In addition, it was ridiculed by the international press and affected by competition from philosophical societies abroad. Unable to do well on the international scene, and after stopping the publication of the *Kant Studien* (Kant Studies) in the beginning of 1937, the society dissolved at the end of April 1938.[53]

Acting together, Carathéodory, Perron and Tietze rejected unreasonable political demands during faculty meetings and also made unfavourable political remarks about the regime during their lectures. Carathéodory is even said to have used his stick to strike Nazi students who were boycotting Sommerfeld's lecture on relativity.[54]

Such boycotts, as well as student demands to resist anything "Jewish", marked the beginning of an aggressive anti-Semitism which brought terror to lecture halls all over Germany.

5·5
Recommending Ernst Mohr

The increasing control of the government and Nazi sympathisers over university appointments also affected Carathéodory, as the case of Ernst Mohr illustrates. Mohr had gained his doctorate at Göttingen in 1933, with a work on group representation theory suggested by Weyl.[55] On Mohr's request of 21 July 1934 for an assessment of his "scientific disposition", Carathéodory wrote directly to Hasse at Göttingen and recommended Mohr: Mohr had been known to Carathéodory since his student days in Munich, he had much love for mathematics and also a "certain talent", he was a "thoroughly decent" man and in the beginning of that semester he had given a good lecture on the theory of group representation at the Munich colloquium. "If you would like to help him a little bit, he will get ahead", Carathéodory assured Hasse.[56] Two days later, Hasse replied that it was completely impossible to lift the prohibition of Mohr's access to the mathematical institute. According to Hasse, Mohr was "a weak character" and had played a "most unpleasant role" in "the unpleasant procedures" there in the "critical days of the end of May", of which Carathéodory would be occasionally informed by F. K. Schmidt, "towards whom Herr Mohr had behaved especially wickedly".[57] Friedrich Karl Schmidt was then a lecturer of mathematics at the University of Göttingen and Mohr had taken on a student assistant position with him in the winter semester of 1933. In April 1934, the Prussian and Reich Ministry of Education had agreed to the faculty's proposal to offer Weyl's chair to Helmut Hasse. So Hasse, treasurer of the DMV, became the head of the Göttingen mathematical institute. Mohr was interested in obtaining an assistant position with Hasse, but F. K. Schmidt did not consider him suitable for this job or for an academic career in general and Mohr knew it. In any case, it would be Hasse who had to decide who was going to become his assistant. Through Mohr's various discussions at the university about Hasse coming to Göttingen with a list of assistants put together in advance, the impression was created that, under Hasse, "the old spirit" would continue to exist in Göttingen. An assistant of the mathematical institute, Werner Weber, and the leader of the students of the mathematical faculty, Heinz Kleinsorge, went to the Berlin ministry to initiate an action against Hasse. Kleinsorge even started a petition against Hasse's appointment at the institute. Hasse informed the rector of the Göttingen University about this action, the rector listened to Weber and Mohr's accusations, Weber refused to hand Hasse the key to the institute and Hasse returned to Marburg. The Berlin ministry dismissed Weber temporarily and broke off negotiations with Mohr about an assistant position at Göttingen. An internal examination had begun on 1 June. Based on its outcome, the ministry confirmed Hasse's appointment, but brought also Erhard Tornier, a dedicated Nazi, to Göttingen to occupy Landau's position. Whereas Weber was merely rebuked, Tornier denied Mohr access to the mathematical institute. Mohr's motive for his involvement in the protest against Hasse seems to have been his fear that Hasse would retain the old faculty at the institute and, consequently, his own career would be blocked, rather than any ideological opposition to Hasse. Carathéodory might have known of the incident from Mohr himself and with his recommendation he might have tried to restore the young man's academic career which, as Mohr himself believed, was at risk because of that youthful folly. But Hasse was obviously not willing to submit to Carathéodory's request and Mohr, pursuing his career elsewhere, was lucky to receive the post of a scientific fellow for stream research at the Technical University of Breslau on 1 November 1934.[58]

5.6
The Reich Ministry of Education and the Lecturers' Corporation

As soon as the Reich Ministry of Science, Education, and People's Culture (REM = *Reichsministerium für Wissenschaft, Erziehung und Volksbildung*) was founded by decree of the Reich President and the Reich Chancellor on 1 May 1934, it took over the responsibility for various scientific institutions from the Ministry of the Interior. Bernhard Rust was appointed Reich Minister of Education and would head the REM, which had emerged from the Prussian Ministry of Science, Art and People's Culture, until the end of the war. Rust was a former secondary-school teacher, who cared little about university policies. He had been severely injured in World War I, was loyal to Hitler and had belonged to the NSDAP and the SA from 1925. His political appointments included being *Gauleiter* of North Hannover from 1925 to 1928, *Gauleiter* of Süd-Hannover–Braunschweig from 1928 to 1940, and a member of the *Reichstag* from 1930 and the Prussian Minister of Education from April 1933. In this last office, he announced that he would not interfere with academic freedom. This promise was soon forgotten, but he was always cautious in his attitude towards university professors. He was generally regarded as a weak and unstable person and not capable of maintaining his own position in conflict with others. Other state institutions, as well as institutions directly under the control of the Nazi Party, intervened in the affairs of the Reich Ministry of Education, which became a battlefield of rival Nazi factions. The ministry had sections responsible for primary, secondary and university education, Rust's personal team, the central office, as well as the office for physical education. At the outbreak of World War II, the Reich ministry was financing university institutions and academic teachers to carry out projects for the arms industry.[59] The foundation of the University Committee of the NSDAP (*Hochschulkommission der NSDAP*) by Hitler's deputy Rudolf Hess followed shortly after the foundation of the REM. It was an authority which not only collected information from persons in positions of trust and gave them orders, but also demanded the right to a share in decisions made by the local ministries of education and later also in decisions by the REM.

In Bavaria, on 24 March 1934, a new organisation for university staff the *Dozentenschaft* was set up as the "only organisation of the rising academic generation acknowledged by the state" and "as part of the university constitution". It had political aims and was given the special task of educating and training the new academic generation. It was also granted some rights of participation and involvement within the senate and the faculties. Membership of the *Dozentenschaft* was compulsory. Its members were assistants, assistant professors and associate professors, who could thus occupy influential positions in academic bodies. In this respect, they were considered to be the main political power in university policies and provided a political judgement to any decision that related to academic matters. On 1 December 1934, Hans Schemm ordered the foundation of a National Socialist Lecturers' Association (NSDB = *Nationalsozialistischer Dozentenbund*) within the Expert Corporation of University Lecturers (*Reichs(fach)schaft Hochschullehrer*) of the Nazi Teachers' Association, a "unit of shock troops of the movement" in the university. However the NSDB was dissolved by a new order of 24 July 1935.

5·7
Persecutions and Resignations in 1934

In 1934 more of Carathéodory's friends suffered a cruel fate. At the age of 50, the physicist and engineer Ludwig Hopf, cousin of the mathematician Heinz Hopf, was dismissed, because of his Jewish origin, from his professorship of mathematics and mechanics in Aachen, a year after he had been given "leave of absence".[60]

One of the leading logicians in mathematics, Paul Isaak Bernays, who had worked with Hilbert since 1917, a descendant of the famous Hamburg Talmud scholar Chacham Bernays, was dismissed at the age of 46 and was forced to go to Zurich and work at the ETH in a position that in no way corresponded to his significance as a scholar. At his own expense, Hilbert maintained Bernays in Göttingen for a while. Bernays was able to continue his research at the School of Mathematics of the Institue for Advanced Study, Princeton with a stipend for the academic year 1935–1936.

The 42-year-old pacifist Hans Rademacher, a member of the International League for Human Rights and the chairman of the Breslau chapter of the German Society for Peace (*Deutsche Friedensgesellschaft*) was dismissed from his professorship by paragraph 4 of the Law for the Restoration of the Civil Service.[61] He fled Breslau and accepted an invitation to take up a visiting Rockefeller fellowship at the University of Pennsylvania. When, later, J. R. Kline, mathematics professor at the University of Pennsylvania and secretary of the committee preparing the first post-war International Congress of Mathematicians in the US, addressed an invitation to Carathéodory, he reminded Carathéodory of the Rademacher connection: "I remember with a great deal of pleasure your visit to the University of Pennsylvania in January 1928, at the time you were Visiting Lecturer of the American Mathematical Society. We have another very strong tie with you in that Rademacher has been with us for more than fifteen years. He has had a tremendous influence on the work of our graduate school. It will be a real pleasure to renew the contacts with you."[62] Indeed, when Rademacher's Rockefeller fellowship expired, the University of Pennsylvania offered him an assistant professorship, which he accepted. He remained loyal to that institution which had enabled him to emigrate from Germany, despite the fact that he had better offers from other universities.

In Swarthmore, where he was living, Rademacher became close friends with professor Arnold Dresden,[63] chairman of the department of mathematics at Swarthmore College, a former Dutch immigrant. Dresden had visited Carathéodory on 20 May 1931 and he was the one to translate van der Waerden's *Ontwakende Wetenschap* (Science Awakening) into English, since he was "fully familiar with the English and the Dutch language and mathematical terminology".[64]

Rademacher may have also been connected with Isaak Schoenberg's later appointment (in 1941) as an assistant professor of mathematics at the College of Arts and Sciences of the University of Pennsylvania. Schoenberg shared many mathematical interests with Rademacher and had met him earlier in Germany, where he had spent several years with Landau in Göttingen. In 1931, he had visited the United States as an International Research Fellow affiliated to the University of Chicago where he held the position of a research assistant. He then became a member of Princeton's Institute for Advanced Study, where he worked with John von Neumann and Einstein. The head of the institute considered him a good teacher and a man of great personal charm, whose "attractive wife" was Landau's daughter. Schoenberg also held a teaching appointment as an acting assistant professor at Swarthmore College in 1935–1936, when Dresden was on leave of absence. Schoenberg visited Carathéodory in Munich on 27 August 1935.[65]

Gabor Szegö, professor in Königsberg since 1926, emigrated to the United States to avoid Nazi persecution and taught at Washington University in St. Louis from the autumn of 1934 until June 1938. During the summer of 1935 he had a visiting appointment at Stanford. It was not until 1938 that he became the chairman of the mathematics department, a position he held until 1953. In 1946 he was with Rademacher at the university set up in the south of France for US war veterans.[66]

The Hungarian mathematician Georg Pólya, who had taught at the ETH from 1940, joined Gabor Szegö at the Stanford department of mathematics in 1942. Gabor Szegö had been one of Pólya's fellow students at the university of Budapest.

Schmidt-Ott and the other members of the directorship and the executive committee of the *Notgemeinschaft* (see 4.24) were prepared to give up their offices in the summer of 1933, however new elections did not take place. In the dispute concerning the structure of the *Notgemeinschaft*, Schmidt-Ott's authoritative ideas had prevailed, whereas Fritz Haber's ideas about a stronger self-administration had only been partly considered. This facilitated the passing of the *Notgemeinschaft* into the Nazi state, and Haber, who was of Jewish descent, was ousted from all his offices. On 17 July 1934, the new Reich Ministry of Education intervened in the *Notgemeinschaft* and put Johannes Stark in the post of acting director. Stark, a proponent of "Aryan physics", was the Nobel Prize winner for physics in 1919 and had been president of the Physical-Technical Reich Institution (*Physikalisch-Technische Reichsanstalt*) since 1933. According to the statutes of the *Notgemeinschaft*, such an action required the approval of the members' assembly. However, the members did not convene and approval for a change of statutes was pushed through on their behalf by application of the *Führerprinzip*. Of the 57 members of the *Notgemeinschaft* including academies, universities and some societies, 47 agreed obediently (the four academies and the University of Munich rejected Stark). This majority was enough to remove Schmidt-Ott from the *Notgemeinschaft* once and for all. Ironically, in the anticipation of a revival of monarchist ideas under national socialism, Schmidt-Ott had, in 1933, shown sympathy for the Nazis.

5.8
Under Observation and Judgement

At this time, the mathematical seminar of Munich University was under close observation, in fact by the students themselves. On 11 November 1934, the organised student group for mathematics and physics in Munich (*Mathematisch-physikalische Fachabteilung der naturwissenschaftlichen Fachschaft der Studentenschaft der Universität München*) reported to the leader of the mathematics section of the German Students' Corporation (*Reichfachabteilungsleiter Mathematik der DSt*) on an enquiry concerning Oskar Perron that had arrived four days earlier. The informer was a 24-year-old doctoral student from the physics institute, who had been in the SS since the summer of 1933 and in the Hitler Youth from as early as 1924–1925. He reported on the mathematical seminar, the office and special subjects of the participants and also about their attitude towards the National Socialist regime. He mentioned that Dr. Fritz Lettenmeyer, an associate professor, had volunteered for the academic camp, that "the full professor Dr. Friedrich Hartogs (a Jew), who did not belong to the seminar [held] lectures alternately on synthetic and descriptive geometry", that Perron held lectures on algebra, Carathéodory on function theory and the calculus of variations and Tietze on number theory, and that Lettenmeyer was a head assistant and Boerner an assistant.[67]

In a report on Lettenmeyer written in the final months of 1935, Gustav Doetsch held Carathéodory responsible for the fact that Lettenmeyer had not been offered an appointment. The alleged reasons for him not being appointed were ascribed to Lettenmeyer's scientific interests, which followed Perron's research interests and so made Lettenmeyer irrelevant to Carathéodory. But even more significant for Doetsch was the fact that Lettenmeyer was a "hard core German, a Bavarian man", whereas Carathéodory was a "cosmopolitan". Doetsch reproached Carathéodory for sharpening the "race conflicts" by always praising and giving preferential treatment to the "Polish Jew" Bochner. Doetsch, nonetheless, valued Carathéodory as an "outstanding scholar".[68]

5.9
A Catholic or an Orthodox?

In 1934, all civil servants were required to furnish the state with information regarding the racial orgin of their wives. The Carathéodorys filled out a questionnaire "On the Aryan Descent of the Wife" and signed it on 24 August 1934. Euphrosyne gave an affirmative answer to the "Aryan descent" of her parents and grandparents according to the strict interpretation of the "Aryan descent" in the first decree to the Law for the Restoration of the Civil Service: "One is considered to be a non-Aryan, if one descends from non-Aryan, especially Jewish parents and grandparents. It suffices if one of the parents or the grandparents is a non-Aryan. This can be particularly assumed, if one of the parents or grandparents had belonged to the Jewish religion. Decisive, however, is not the religion, but the race to which the 4 grandparents belong." For her parents and grandparents Euphrosyne stated (with the exception of Stephanos Carathéodory for whom she gave no information in this respect) that they were Greek Catholics.[69]

There is a discrepancy between what Carathéodory and his wife stated as their own and their parents' religion and what the documents issued by the church authorities testified: for the Nazi authorities they stated they were Catholics, for the Eastern Orthodox church they were Orthodox. Maybe the Carathéodorys just presented themselves as Catholics, because the main distinction line between Christians and Jews was drawn by the Nazis between Catholics and Jews.

5.10
In Pisa

In the middle of all this turmoil, Carathéodory endeavoured to maintain his international connections. So, on 26 January 1934, he wrote to the Bavarian Ministry of Education and Cultural Affairs to say that he had been invited by the *Scuola Normale Superiore* in Pisa to give a series of lectures at the university there in the second half of March 1934. The visit was presumably at the initiative of Leonida Tonelli, then professor of analysis at Pisa. In his letter to the ministry, Carathéodory signed himself as university professor,[70] something that he had not done before, probably to stress the significance of his title, which seemed to have lost its meaning. Accompanied by his daughter, he travelled to Pisa on 20 March to give four lectures.[71]

5.11
Honorary President
of the Inter-Balkan Congress of Mathematicians

The Inter-Balkan Congress of Mathematicians (*Congrès Interbalcanique de Mathématiciens*) took place in Athens from 2 to 9 September 1934. There, N. Hatzidakis communicated Carathéodory's paper *Sur les equations de la Mécanique* (On the Equations of Mechanics),[72] and Carathéodory himself was elected honorary president of the congress. Documents relevant to the congress do not disclose whether Carathéodory had in fact attended in person. It should be noted that the Greek Mathematical Society had opposed the convocation of an inter-Balkan congress and finally voted against it under the pretext of lack of information about it, lack of time for preparation and a not clearly defined purpose and also under the pretext of favouring a Panhellenic congress instead. In Greece, universities, faculties, the Academy of Athens, and the majority of professors were against what had been declared by the very few organisers of the inter-Balkan congress as a purely scientific event.

ACTES

DU

CONGRÈS INTERBALCANIQUE

DE MATHÉMATICIENS

ATHÈNES, 2-9 SEPTEMBRE 1934 (I)

ATHÈNES

IMPRIMERIE NATIONALE

1935

Congress Files of the Inter-Balkan Congress of Mathematicians.
Courtesy of the Bavarian Academy of Sciences, Archive.

5.12
Nuremberg Laws and New Measures

The peak of the Nazi racial laws were the so called Nuremberg laws passed in 1935. According to the Blood-Protection Law (*Blutschutzgesetz*) the marriage of Jews to persons of "German or related blood" was forbidden. Marriage by a German to a "non-Aryan" against this prohibition was punished as a "betrayal of race" and extra-marital sexual intercourse as a "racial disgrace". The Reich Citizens' Law (*Reichs-bürgergesetz*) of 15 September 1935 and the implementation order of 14 November 1935 made a distinction between citizens of the Reich, who possessed full rights, and citizens of the state (nationals). Persons of mixed descent were classified as Jew-ish if at least three grandparents were Jews, or if two grandparents were of Jewish descent and practising Jews, or were married to practising Jews. Persons with three or four Jewish grandparents could only be citizens of the state and not of the Reich; on 31 December 1935, such citizens were dismissed from office as civil servants. These new laws now applied to all Jews in Germany without exception. The earlier exemptions allowed for those who had served as front-line soldiers and their rela-tives, as well those who had been civil servants before the war, were abolished. The director of the Institute for Advanced Study in Princeton, Abraham Flexner, wrote to Einstein on 7 September 1935 that the "German Government has sunk to the depths of the Medieval Ages."[73]

On 17 March 1935, the Bavarian Ministry of Education presented the names of those professors who had answered the questionnaire on racial origin and submitted evidence of the "Aryan origin of their wives" to the rector of Munich university. Carathéodory was listed as having satisfied the necessary criteria. Of "non-Aryan origin" were the wives of his colleagues Kasimir Fajans, Aloys Fischer, Wilhelm Prandtl and Alfred Pringsheim.[74]

That year Carathéodory lost many close friends and colleagues to the Nuremberg laws. Max Dehn, whose adoption of the Christian faith had not brought him the expected acceleration of his career,[75] was sent off to retirement from his post at Frankfurt University at the age of 57. He was then at the height of his career and had made remarkable contributions to the foundations of geometry, and to topology and group theory.

The 64-year-old associate professor at the University of Frankfurt, Paul Epstein, gave up his teaching commission for geodesy, didactics and history of mathematics. In November 1938, he was to be deported to a concentration camp but was left behind, since he too ill to be moved. In the summer of 1939, following a summons from the secret police, he committed suicide by a lethal dose of the barbiturate Veronal.

The shyly Jewish professor lecturer in descriptive geometry at Munich, Friedrich Hartogs, who had been Carathéodory's guest many times, was forced into retirement at the age of 61. In 1943 he, too, committed suicide.

Hans Ludwig Hamburger, who had habilitated under Carathéodory in Berlin, was dismissed from his professorship in Cologne at the age of 46. He moved to Berlin from where he was able to emigrate in 1939.[76]

Felix Hausdorff retired in Bonn at the age of 67. He committed suicide together with his wife and her sister to avoid deportation to a concentration camp in 1942.

Ernst Hellinger, who was descended from a Jewish trader family from Silesia and had attended lectures by Carathéodory in Göttingen at the beginning of the century, was dismissed at the age of 52 from his full professorship at the University of Frankfurt. He was arrested on 13 November 1938, temporarily interned in the Festival Hall in Frankfurt and then transported to Dachau. Six weeks later he was released from Dachau on condition he leave Germany.[77]

Artur Rosenthal, full professor since 1922 and director of the mathematical institute of the university in Heidelberg, was forced to ask for retirement after a student boycott of his lecture. He was interned for two months in Dachau, in November and December 1938, emigrated to the Netherlands in July 1939 and was just able to escape to the USA in March 1940. The Harvard university professor William Graustein recommended Rosenthal in 1940 by citing Carathéodory: he "was one of the first to understand the ergodic problem solved by Birkhoff (his work is about 15 years before that of Birkhoff)".[78]

Rosenthal's colleague at the University of Heidelberg, the 61-year-old Heinrich Liebmann, also victim of a student boycott, resigned his post and moved to Munich in the following year.

The 60-year-old Russian-Jewish mathematician Issai Schur fell victim to the racial laws at the end of the year and was forced into early retirement, mainly by Bieberbach. Three years later, he was thrown out of the committees of the Prussian Academy of Sciences, to which he had belonged as a member since 1922, due to a defamation campaign launched by Bieberbach. In the meantime, Bieberbach had become editor of the journal *Deutsche Mathematik*, which was published in 1936 by the mathematician Theodor Vahlen, Richard von Mises' successor to the headship of the Berlin Institute of Applied Mathematics in 1934 and head of the Department for Science (*Abteilung für Wissenschaft*) at the REM soon afterwards. This journal, which started with high circulation figures and massive financial support to publish mathematics along with ideological papers, was compelled to close down in 1943. Already in 1937, when Johannes Stark fell from his post as the acting director of the *Notgemeinschaft*, financial contributions were decreased. In September 1933, Blichfeldt remarked to Veblen that Schur was too old to be brought out of Germany and the best solution for him would be a pension in that country. Mentally and physically weakened, Schur emigrated to Palestine in 1939 and died two years later in Jerusalem. He spent the last years of his life as a honorary member of the Hebrew University in Jerusalem.

The 48-year-old physicist and chemist Kasimir Fajans, a full professor at Munich university since 1925 and director of the institute of physical chemistry since 1930, was deprived of all his positions because of his Jewish confession. He emigrated to the USA via Great Britain the following year.[79]

Against Ernst Zermelo, the ministry opened disciplinary proceedings aiming to deprive him of his teaching authorisation and remove him from the University of Freiburg where he had occupied a honorary professorship since 1926. The reason for such an action was Zermelo's refusal of the Nazi salute and his derogatory remarks about Hitler and the "Third Reich". In anticipation of the inevitable outcome, Zermelo resigned his professorship.[80] In both of his 1935 letters to Zermelo,[81] Carathéodory addresses his friend as "Zerline", the name of the innocent peasant girl in Mozart's *Don Giovanni*, an opera which deals with sexual conquest, betrayal and death. The Don, the incarnation of a force of nature without any sense of morality and responsibility, applies his charm to Zerline, who almost yields to his seduction. The implication is clear.

Another step taken by the Nazis, this time against university professors in general, was a reduction in their supplementary salaries. The supplementary salaries of the budgeted full and associate professors at Bavarian universities and the Technical University of Munich were fixed anew with effect from 1 January 1935. While the upper limit of the supplementary salary still amounted to RM 3800 per year and those supplementary salaries that were less under RM 1000 were not reduced, the supplementary salaries of full professors were reduced by twenty percent and those of associate professors by ten percent. As a consequnce of this measure, Carathéodory's supplementary salary was reduced to RM 1920.[82]

The National Socialist Teachers' Association leader, Hans Schemm, died in spring 1935 and Hess separated the National Socialist Lecturers' Association, which was in his responsibility, from its umbrella organisation. All party members among academics, from full professors up to assistants, were included in the new National Socialist German Lecturers' Association (NSDDB) which exercised a strong influence on Munich University, particularly since this Reich authority had its seat in Munich and the new leader of the NSDDB was the doctor and party member Walter Schultze, director of the medical service at the Bavarian Ministry of the Interior since 1933, and an honorary professor of public hygiene at Munich University since 1934.

5.13
In Bern and Brussels

On 12 March 1935, Carathéodory told the dean of the Philosophical Faculty II that he had been invited to give a scientific lecture in Bern on the occasion of the 25th anniversary of the foundation of the Swiss Mathematical Society (*Schweizerische Mathematische Gesellschaft*). The celebration was to take place two months later, on a Sunday, so that his Munich lectures were not going to be affected. Carathéodory stated that the contents of his speech would be scientific and that the occasion would be a celebration.[83] He was granted permission to attend the event to talk on *Examples particuliers et théorie générale dans le calcul des variations* (Particular Examples and General Theory in the Calculus of Variations).[84]

On 3 June 1935, Carathéodory asked the dean again to determine whether any reservations had been expressed regarding his acceptance of an invitation to the celebration of the centenary of the *École Militaire de Belgique*. The meeting was to take place on 7 July and Carathéodory had been invited by the commander of the school, General J. Neefs.[85] Carathéodory was, indeed, granted permission to leave Germany and on 14 July he wrote to Zermelo:

In Berlin I had a two-hour session that was presided over by the beetroot [G. Hamel?]. I have spoken with Schmidt about Scholz; it seems that Bi[eberbach] makes difficulties. Last week I was in Brussels at the centenary of the *École Militaire*. There – among others – I also became acquainted with the commander of the Paris *École Polytechnique*. On my return journey, I was in Aachen and also visited Schlot. Unfortunately, he is terribly aged, does not smoke any more and has become quite silent. However, I have not seen him since that memorable day when you, my wife and I went almost astray in Alpgrüm (1910). Do you remember that exciting occurrence? This condition of Schlot must have been caused by the death of his brother, who had poisoned himself with Veronal a few years ago. The wife is very beautiful and nice.[86]

On 1 August 1934, Gustav Doetsch, a full professor of mathematics at the University of Freiburg, had tried to draw the attention of Wilhelm Süss, one of his colleagues, to Heinrich Scholz, that "petty bourgeois" and "nobleman", who had gone to Kiel to give a lecture, where he was talked of as E.Tornier's successor by Tornier himself. Doetsch remarked that with Scholz's teaching the "poor Kieler" "would be taken for a ride".[87] From Carathéodory's letter to Zermelo it seems that Bieberbach was placing obstacles in Scholz's way and obviously shared Doetsch's opinion in this matter. Scholz, however, had asked for an examination of his case and Süss had answered a request from the *Dozentenschaft* about the matter hoping that the situation could be sorted out.[88]

On his return from Belgium, Carathéodory found Born's book, whose German version he only knew by repute but whose English translation he was reading daily with growing enthusiasm. "It is a splendid book. I admire the skill with which you have structured the material and the ease with which you handle the most complicated things. I hope to learn a lot from it."[89] Carathéodory was obviously referring to Born's *Moderne Physik* (Modern Physics),[90] which proved to be a financial disaster for the publishers, thanks to the Nazis, but a great success in Anglo-Saxon countries. The book contained Born's lectures given to a large audience at the Technical University of Berlin-Charlottenburg in March and April 1932. When Born fled to England, the publishing company Blackie and Son, Ltd. in Glasgow showed great interest in the book and published it under the title *Atomic Physics*.

5.14
Member of the International Commission of Mathematicians

On 9 April 1935, the dean's office informed the Philosophical Faculty, Section II, of Carathéodory's election to the International Commission of Mathematicians (ICM). The faculty regarded the fact that one of their members would become a member of the commission as a great honour. The dean, Privy Councillor Schmauss, declared: "In my opinion there should be no reservations against the acceptance of the election, since it concerns a purely scientific matter. Besides his appreciated scientific standing, Herr Carathéodory is, through his verbal skill, especially qualified to represent Germany with dignity."[91]

Carathéodory was authorised by decree of the Bavarian Ministry of Education and Cultural Affairs to take up the appointment. The ICM had been set up to look again at the question of permanent international collaboration, following a crisis in the International Mathematical Union (IMU) after World War I, and to present its recommendations to the 1936 International Congress of Mathematicians. The story of German participation in that international forum is revealed by Carathéodory's report to the Reich minister on 18 May 1941 in reply to the latter's request twelve days earlier:

By resolution of the International Congress of Mathematicians in Zurich (5–12 September 1932) the *Union mathématique internationale* was conclusively wound up.

From a remark of Prof. Severi in one of the transcripts at my disposal, I gather that the financial assets of the Union were also immediately liquitated.

Likewise, the attempt to create an international organisation of mathematicians based on equal [national] representation to replace the union failed.

The Zurich congress resolved in fact to form an international commission 'in order to study the relations among mathematicians of various countries anew and to propose the reorganisation of these relations at the next congress'.

The commission formed then included one representative of each member country, Italy (chairmanship), Germany, France, Spain, Belgium, Denmark, Hungary, Poland, Russia, England and America. A three-membered committee was selected out of this commission consisting of prof. F. Severi (Rome), H. Weyl (Göttingen) and G. Julia (Paris), who were to prepare the planned proposal.[92]

Germany's representation on this latter committe changed several times. After Weyl had gone to America, Severi asked me by a letter on 16 November 1933 to take Weyl's position in the commission. I rejected this request on the grounds that Weyl had been elected to the commission as the President of the German Union of Mathematicians and, therefore, one ought to attempt to substitute him by someone from the executive committee of the union. Then, with the consent of the German government, Prof. Dr. Blaschke (Hamburg) was admitted to the commission; by the way, this person was also elected Chairman of the German Union of Mathematicians in September 1934. But, already in January 1935, Blaschke resigned this chairmanship and expressed the wish also to resign from the commission. In a letter of 4 February 1936, the new chairman of the German Union of Mathematicians, Prof. G. Hamel (Technical University of Charlottenburg) asked me to join the Severi commission and particularly since Severi himself has only command of the Italian and French languages. By decree no. V, 30357 of 13 June 1935, I was authorised by the Bavarian Ministry of Education and Cultural Affairs to accept the appointment to the International Commission of Mathematicians.[93] In the meantime, the Severi committee has had two meetings, in which Prof. Blaschke participated. The first meeting took place in Rome in March 1934, the second in Paris in February 1935. A third meeting scheduled for Hamburg in August 1934 fell through at the last minute for reasons beyond the committee's control.

A questionnaire designed to outline and establish the purpose of the organisation, which was to be founded anew, was worked out in Rome. This questionnaire was sent to the eight members of the commission who were not present. But a year later, it turned out in Paris that two of these members had not answered the questionnaire at all and that a number of the others had taken a basically opposing position.[94] Therefore, it was decided to hold no meeting of the committee until the international congress in Oslo. I was member of the German delegation at the congress in Oslo. I took part in the session of the eleven-membered commission on Monday, 13 July 1936. Because the Italians were not represented in Oslo, Prof. Severi had passed on the chairmanship to Prof. Julia (Paris). But only six out of the eleven commission members were present. Right at the beginning of the session, it became obvious that Prof. Mordell (England) and Prof. Veblen (USA) regarded every international organisation of mathematicians, of whatever kind it might be, as useless and therefore they rejected it in principle.

After a long discussion, it was decided to consider the commission's task as having been done and to make a relevant statement at the final session of the congress. The wording of this statement, which I edited together with Prof. Julia, was approved in a second session, which took place on the 15th and lasted only a few minutes, and is on page 47 of volume I of the congress minutes.[95]

Unfortunately, two inaccuracies, though of an entirely secondary nature, occurred in the editing of the minutes. Firstly, the title *L'Union mathématique internationale* [attributed to the Union], which does not exist anymore, is incorrect. Secondly, a five-membered committee that has never existed has been constructed out of the three-membered Severi committee.

C. Carathéodory [96]

In the period 1932–1936 between the two international congresses of mathematicians, views differed as to whether the IMU had in fact been dissolved in 1932 or temporarily suspended. At the Zurich congress, Weyl had resolutely condemned the IMU policy up to then and stressed the need to keep politics away from mathematics. With his participation in the Severi committee, to which the Bavarian Ministry of Education consented in 1935, Carathéodory represented German mathematicians on the international scene. Why did Carathéodory accept the nomination after the devastating purge of mathematicians from German universities? Carathéodory's decision becomes even more surprising in view of the fears expressed by various member countries that Germany would attempt to increase its influence within the union. Carathéodory's participation might have been motivated by his concern that, with the Nazis in power, contacts between German mathematicians and their foreign colleagues would be ended; on the other hand, he may have been encouraged by the hope that the IMU would be resurrected after a return to normal political conditions. In either case, Carathéodory was convinced of the absolute necessity of international scientific co-operation.

Within Germany, on a smaller scale, things were also being reordered in mathematics. In the spring of 1935, for example, Professor Dr. Georg Feigl (Breslau) asked Süss (Freiburg), Hamel (Berlin), Koebe (Leipzig), Kowalewski (Dresden) and Tornier (Göttingen) to co-operate within the Mathematics Department of the Reich Research Council (*Reichsforschungsrat*) to produce a new journal, *Neue Deutsche Forschungen* (New German Researches). The invited collaborators were asked to identify doctoral and post-doctoral theses appropriate for publication in this collection.[97] Feigl found the fact that the journal *Compositio* was publishing a series of works by "non-Aryans" at that time to be embarrassing.[98] Feigl's friend, Carathéodory, was not included in the planning of the journal. However, invited by Blaschke, he did give a lecture at the mathematical seminar at Hamburg University on 15 June 1935 and was paid RM 60 for his travel expenses.[99]

5.15

Protest

On 22 June 1935, 250 academics refused to participate in the celebration of the foundation of Munich University. Various excuses were offered for non-attendance. Some boycotted the event because of their opposition to the current political, racial or sexual discrimination; others because of their dislike of the main speaker and the ideologically determined policies of the ministry, which had chosen him against the will of the faculty. Others declined to attend, simply because of the obligatory character of the celebration. The invited speaker was the philosopher Dr. Wolfgang Schultz, who had been appointed professor in Munich in 1934 against the will of the philosophical faculty and whose topic at the celebration was Nietzsche's Zarathustra and his historic relevance. Schultz was neither an acknowledged philosopher nor an entertaining lecturer. In addition, for the first time, lecturers appeared at the university in SA uniform and a changed seating plan allowed lecturers and the members of the *Studentenbund* to sit together. Carathéodory, just like Perron, Sommerfeld, Müller-Erzbach and Rehm, abstained from the celebrations as a political protest against policies of the ministry regarding appointments.[100] Being a faculty member, Carathéodory felt obliged to intervene in appointment policies in favour of the qualitatively best candidates and thus preserve high scientific standards, which were threatened by increasing Nazi political influence on the university. So, on 4 December 1935, together with Arnold Sommerfeld, Heinrich Wieland and Walther Gerlach he addressed a letter to the dean's office supporting Heisenberg's candidacy to succeed Sommerfeld. In the appointment committee's opinion, Heisenberg was by far the best candidate.[101] The eminent quantum physicist and former doctoral student of Sommerfeld had received the Nobel Prize for physics in 1933 and was at that time professor of theoretical physics at the University of Leipzig. In the event, Heisenberg was not appointed, but later would lead Germany's main reactor experiments in Leipzig and Berlin.

5.16
Carathéodory's View of Damköhler

Wilhelm Damköhler, together with N. Kritikos, A. Duschek and A. Rosenthal, helped Carathéodory with the corrections of his 1935 book on the variational calculus. Damköhler had gained his doctorate under Carathéodory and Perron on 18 July 1933 with the dissertation *Über indefinite Variationsprobleme* (On Indefinite Variational Problems). On Hasse's inquiry of 28 January 1935 about Damköhler's scientific performance and personality,[102] Carathéodory recommended him for an assistant position in Göttingen: "He is an excellent mathematician, but up to now he has engaged himself almost only with the geometric theory of sets. But he is very intelligent and keen-sighted, as well as very eager to learn everything possible. As a student he was very conceited and a bit peculiar, but he became more modest the more he learned and, personally, I have never had to complain about him. I consider it to be out of the question that he could be involved in intrigues, since his only interest is mathematics."[103] Either Carathéodory himself considered the organisation of academic mathematics to be an entirely innocent matter or he was representing it as such in order to support the appointment of his doctoral student in Göttingen.

5.17
Despina Leaves Munich for Athens

In 1935, Carathéodory sent his daughter Despina to Athens to continue her studies. The specific reason for her breaking off her studies in Munich at this time is not clear.

There may have been an entirely personal reason for her father to decide to send her away from Munich or it may have been as a result of a combination of circumstances. The political and social conditions in Germany were, in any case, growing worse at the time, as this account will explain.

On 3 March 1936, the Ministry of Finances received the following request from the Bavarian Ministry of Education for the speedy approval of a child benefit allowance for Carathéodory's daughter: "Despina Carathéodory has remained alien according to the reservation expressed in the employment resolution for her father of 30 April 1924, no. V 15507. Before his appointment to Munich her father was professor at the University of Athens; now he again has an invitation to hold guest lectures there. [...] There is no doubt about the positive attitude of professor Dr. Carathéodory to the present German Reich." [104]

The relevant part in the employment resolution reads: "Acquisition of [German] nationality for Professor Dr. Caratheodory, which is bound to employment in Munich according to paragraphs 14 and 16, section I of the nationality law of 22 July 1913 for the Reich and state subjects, does not apply to the wife of Professor Dr. Carathéodory and to his children, to whose legal representation he is entitled by virtue of his parental authority, according to the expressed reservation under paragraph 16 section II of the quoted law." [105]

Despina was registered at the law faculty of Munich University up to the winter semester of 1934–1935. [106] She had not wanted to study law and would rather have chosen languages and linguistics, but her father had insisted on law studies for her. Despina considered cultural studies to be an opportunity to gain general knowledge before getting married rather than as a preparation for a professional career. She entered the university in a year when female students made up 19% of the entire student population. At the end of March 1935, her father sent her to Athens to follow

Despina, Munich, ca. 1933. Courtesy of Mrs. Rodopoulos-Carathéodory.

law studies at the university there but she was unable to conclude them there. In Athens, the political situation was again unstable. Pressure for the restoration of the monarchy was increasing. Alarmed by the prospect of a purge in the army, and with the silent approval of Venizelos, a group of officers loyal to him launched an abortive coup on 1 March 1935. Stephanos Deltas was arrested two days later and, together with his wife Penelope, was put under house arrest up to 16 April. Georg von Streit belonged to the anti-Venizelist People's Party and it was in his house that Despina was staying in Athens. A vivid description of the Greek conditions at the time is given by Penelope Delta in her diary entry of 30 April 1935: "The entire Greece a Dante's hell: murders, betrayals, dishonesty, cowardice, prostitution, defections, plunder, filth, faint-heartedness, renunciations, bloodshed, baseness, rottenness and stench, vileness and criminality, emetic disgust, bragging and cowardice, threats, filth, filth, filth, ... ".

Conditions for study deteriorated in Germany after the rise of the Nazis and by the time Carathéodory decided there was no future for his daughter in Germany, things had become decidedly worse, even though some of the most stringent measures affecting students of both sexes did not apply to foreign nationals. For example, all students had to carry out community-service work in the period between 1 April and 30 September 1935. However, foreign students could be exempted, so a possible unwillingness to carry out community-service work could not in itself have been the reason for the interruption of Despina's studies in Munich.

In 1935, measures were taken to reduce the number of students in German universities. According to a decree issued by the Reich Ministry of Education on 30 March 1935, ordering the maximum number of students for universities of big cities, the figure for the University of Munich was fixed at 5000. Within the faculties, the number ought not to exceed 70% of that in the summer semester 1934. As a consequence, the number of students decreased from 8065 in the winter semester 1934–1935 to 5480, somewhat more than was allowed, by the summer semester 1935. Foreign students were not, in fact counted as being part of the overall total of students, so again this decree would not have forced Despina's departure from Munich.

The measures for the reduction of student numbers were preceded by the REM decision of February 1935 to abolish the *numerus clausus* and to open up the universities to students without the *Abitur*. The REM expected advanced students to participate in the *Fachschaft* which organised a kind of national socialist parallel study.

As early as November 1933, the head of the Central Office VI of the German Students' Corporation (*Hauptamt VI der DSt*) had issued detailed guidelines to local officers, by which all female students would have to complete an extensive compulsory programme during their studies. This included, among others things, regular physical education during the first semesters, education in the so-called *Frauendienstkursen* (courses for women's services) on air-raid protection, first-aid, news service, participation in collections for the Nazi relief organisation *Winterhilfswerk* and also cultural activities running under the name of *Gemeinschaftspflege* (community care), such as folk dance, folk songs, poetry of the homeland, etc.

Such an adulteration of the standards of university education, together with a suffocating atmosphere of Nazi control might well have decided Carathéodory to interrupt his daughter's studies in Munich. But there were also specific measures by the Nazi state against women who practised a profession which made studies for women seem hopeless. Those affected most by these measures were potential jurists, such as Despina, who were very few even before 1933. In 1933 there were in Germany just 252 female lawyers and notaries (1.3%) and 36 female judges and public prosecutors (0.3%). Female students were also put at a disadvantage with respect to financial aid, scholarships and other supporting benefits. Particularly after 1935, female law students could no longer receive any financial grants, because their studies led to a profession from which the Nazis had debarred women. In addition, a firm belief in nazism became a prerequisite for gaining a credit or scholarship. In law studies a period of at least six but at most ten semesters was followed by the first state examination carried out by the Ministry of Justice. In a decree of 1934 students of law, when registering for their first state examination, were obliged through "written confirmation of the appropriate authorities", to present proof that they had pursued "physical training and affinity to other folk groups". After passing the examination, the students had to undertake a three-year in-service training programme before the final state examination. Law studies, in which Jewish students were strongly represented, were restructured in the autumn of 1934 by the Nazi Karl August Eckhardt, who was given a post at the REM for this very purpose. The regulations he proposed came into force in January 1935 under the title Guidelines for the Studies of the Science of Law (*Richtlinien für das Studium der Rechtswissenschaft*) and were prefaced by "basic ideas" of the "NS revolution". These regulations had the primary aim of breaking with the traditional system of the science of law and the secondary purpose of introducing a beginner's series of lectures in the first two semesters to acquaint students with the *völkisch* foundations of the science of law. New examination courses, such as the Law of the German Farmer (*Recht des deutschen Bauern*) and the Law of Work (*Recht der Arbeit*) were introduced, as can be seen in the examination regulations for the two state examinations issued by the Reich Ministry of Justice in July 1934. In the curriculum vitae, which the student had to submit to be allowed entry to the first state examination, information had to be given about the student's personal world-view and political development, as well as about his attitude towards the science of law. These attempts to test the future jurist politically carried on through the in-service training period and into the final state examination. The reform discussion provided for a heavier emphais of German law at the expense of Roman law.

In Munich, the law faculty reduced the number of obligatory lectures on Roman law to a total of 5 hours per week for the winter term 1933–1934. Traditionally, professors at law faculties had the task of mediating and explaining the law in force to their students. However, the more this law was distorted by Nazi laws and orders, the more Nazi ideology infiltrated Germany's lecture halls. A jurist who taught family law, for instance, could not avoid having to lecture on the Nuremberg laws. Furthermore, between 1933 and 1939, all professors of the Munich law faculty were replaced, and by men who had either been NSDAP members before 1933 or had joined in 1937. Despina remembers the return of Leopold Wenger, professor of Roman and German civil law, to his homeland of Vienna in the spring of 1935, as being a dismissal. He was the president of the Bavarian Academy of Sciences and a highly respected scholar, who presented his students with a critical assessment of Caesar's personality. Wenger did not wish to offer lectures in other areas of law, despite the decreasing number of his students on account of the reduced estimation of Roman law. Wenger's place was taken by Mariano San Nicolo, former rector of the German University of Prague, who enjoyed the approval of the Reich authorities.

As far as all this affected Despina, it could well have been that the non-existent prospects of a professional career, together with the pervasive Nazi indoctrination at the university were sufficient grounds for Carathéodory to believe that a continuation of his daughter's studies in Munich was senseless.

However, Carathéodory's insistence on Despina's departure from Germany might also have been simply due to his wish to remove her from the Nazi student ambience. This view seems to be supported by what his daughter herself has said of her father's aversion to the Nazis. Under their rule, he forbade his children to speak German at home and, if they did so, he punished them severely. He could not tolerate what could be considered positive remarks about Hitler and reacted violently when confronted with remarks with even a hint of admiration towards the regime. By 1934, when Carathéodory had probably taken the decision to send his daughter to Athens, the Nazi's hold on power was absolute. The street terror of the Brown Shirts in 1933–1934 and the chaotic disturbances that brought about the end of the Weimar government, following the *Röhm-Putsch* of 30 June 1934 and the death of Hindenburg on 2 August, had now given way to a Nazi state of all-pervasive power, and no one could any longer doubt the intentions of Hitler and the Nazis.

5.18
"On the Present State of the German Universities"

Those remaining liberal spirits within German universities in the 1930s were now becoming subject to an increasingly hostile political atmosphere. There was certainly considerable inflexibility in the system and power was concentrated in the hands of a relatively small number of professors, but the best of the universities had preserved the ideal of liberal disinterested scholarship. All this was to change.

The view of Schultze, head of the *NS-Dozentenschaft*, was that the sins of the German educational system before 1933 were to be found in a misdirected education, embracing the ideas of individualism and an internationalism which contradicted the racial ideal, in colourless objectivity, in the most vulgar utilitarianism and in the fragmentation of scientific disciplines that had been exploited by the Jewish political ambition for power. To this can be added that Hitler himself had stated that objectivity was merely a slogan invented by the professors for the protection of their own interests. The power of a full professor, that in the Weimar Republic had been characterised by C. H. Becker as almost absolutist, was now being threatened, if not completely destroyed. University professors were forced to suffer humiliations and local actions were being taken against them by Nazi students. Nazi antipathy towards "intellectualism" as well as continuous polemics in the press against the "mendacious educated arrogance" of the bourgeoisie took its toll on the academic community.

A lecture, "On the Present State of the German Universities", by Georg Dahm, a jurist specialised in criminal law and rector of the University of Kiel in 1935, is absolutely revealing as to the situation prevailing in German universities. In his words, before the era of national socialism, university learning and knowledge that belonged to a guild-like fraternity reflected a mentally fragmented society and the intellectual expression of a many-party system. As a remnant of a past age, the university had no real part in the bright development of the present and, worse, was an instrument of intellectual dissolution under the influence of an increasingly stronger Jewry. The outlook of the universities lacked spiritual orientation and therefore the basis for real scientific research. The way out of this situation had been sought in positivism and specialisation or – what was even more dangerous in Dahm's opinion – in a meaningless neutral world of concepts, stripped of real content. The aim now was a new *völkisch* science that would expound the German spirit as a race-totality and reality of life and would bring it to its self-consciousness; a national science that would focus on the history of the German people, its world-view and its right. Contemporary *völkisch* science should not be judged by practical usefulness alone; stricter methods, technical tools and the construction of technological concepts of a genuine scientific thought were required; it had to overcome positivism and combat rationalistic ways of thinking; it had to turn to a holistic, concrete and intrinsic way of looking at things that would bring all sciences closer to philosophy and closer to each other; it could never be comprehensible to the layman but it would aspire to a closeness to the people, it would be *volkstümlich*; it had to give attention to the specialist, who would see his life task in performing clean painstaking and detailed work, and must clarify individual questions. Faultless technology would still be necessary and meaningful on the basis of a national science. Besides this kind of science, the other aim was a new university built by people who would be called to expound the new intellectual life. The means to the achievement of this goal would be the removal of inappropriate lecturers, the elimination of Jewish influence on German university, the production of an appropriate new academic generation, the arrangement of the habilitation procedure in such a way as to offer young lecturers the possibility to achieve their aim within a short time and secure them a financial and institutional position without themselves having to turn to a practical profession, the cultivation of collectivism and the sense of comradeship, the strengthening of the community of young lecturers and students by participation in academic camps and study groups.[107]

5.19

Carathéodory Meets Tsaldaris at Tegernsee

After the abortive Venizelist military coup of March 1935, the anti-Venizelists showed no greater respect for the constitution than had their opponents. Elections were held in June 1935 while the country was under martial law. The Venizelists' abstention from the election ended in a victory for the People's Party: 65% of the popular vote gave them 96% of seats in the parliament. Despite the fact that the communists gained 10% of the votes, the most they ever received in the inter-war period, an electoral system that favoured the majority prevented their participation in the parliament.

Panayis Tsaldaris became Prime Minister. He had been exiled by the Venizelists in 1918, and had several times been a minister of the People's Party government. A royalist and anti-Venizelist himself, he enjoyed the reputation of a law-abiding, moderate and conciliatory person. Tsaldaris had gained his doctoral title from Athens University in 1889 and studied further in Göttingen, Leipzig and Paris. In October 1935, a group of high-ranking officers were to demand that he either push for the restoration of the monarchy or resign. Tsaldaris's response was to resign and he was replaced by General Georgios Kondylis who, although a former supporter of Venizelos, had defected to the anti-Venizelist camp in 1928. Kondylis declared the abolition of the republic and shortly afterwards a farce-like plebiscite approved the restoration of the monarchy.

In the summer of 1935 Carathéodory visited Tsaldaris, who was staying in Bad Wiessee on the north-western coast of Tegernsee, the *Lago di Bonzi*, where so many Nazi officials had bought villas and where a year earlier the so-called *Röhm-Putsch* had been suppressed. He showed the Prime Minister a letter from the vice-rector of the Technical University of Athens, Konstantinos Georgikopoulos, discussed with him various matters concerning the Technical University of Athens and its professors, and promised to provide him with a report before his departure from Germany.

Carathéodory provided the Premier with his report, dated 19 August 1935. In his opinion, the Technical University of Athens had a very good reputation abroad. A way had to be found to enable those professors who had been removed in the academic year 1934–1935 to return to their posts. The dismissals had occurred at a difficult time and the reasons for the dismissals had not been investigated. Carathéodory saw two alternatives for the reinstatement of the removed professors: they should either be re-elected or, even better, the minister in charge would have to accept responsibility for their dismissal which would have to be revoked by decree. For any new appointments, the government ought to show confidence in the professors' own judgement. Carathéodory argued in favour of the appointment of Caesar Alexopoulos, one of the best students of Paul Scherrer, to a chair of experimental physics at the Technical University, and also argued in favour of other Greeks who had studied in Munich and wished to follow an academic career in Greece. Finally, Carathéodory enclosed a German work praising the Greek cartographic service.[108]

Alexopoulos' and Kujumtzelis's names were entered by hand under the title "deputy professors" in the 1936–1937 list of academics of Athens University, but not of the Technical University; however, they do not appear in the programme of studies. The following year, their names are printed in both the list and the programme. Alexopoulos was appointed as a full professor of physics at the University of Athens only in 1939 and Kujumtzelis ten years later.[109]

On 9 October 1936, the Reich and Prussian Minister of Science, Education and People's Culture wrote to Carathéodory, then at Wisconsin State University, Madison, through the rector of Munich University:

From a report of the German legation in Athens, which reached me through the Ministry of the Exterior, I assume that the Technical University plans to appoint three German academic teachers for one or two years as full professors at the university. According to this report, the rector of the Technical University should have checked with you and asked you to make the necessary enquiries in Germany.

I ask you to report to me in detail about the facts as soon as possible. [110]

Carathéodory replied almost immediately:

Last March, during my stay in Greece, I was informed by the rector of the Technical University in Athens that there was a plan to appoint three German academic teachers for one or two years at this institution. I also had a long relevant discussion with the Greek Minister of Transport, who is responsible for the Technical University. Before my departure from Greece, I reported orally on the matter to the German envoy and after my arrival in Munich I also informed the rector of Munich University, Herr Professor Kölbl.

Because there is no final decision on the subjects that are going to be taught yet, I asked the rector of the Technical University in Athens, Herr Professor Sinos, to send me a letter with the necessary details to Munich, before I take further steps in Germany. As I learned from a colleague in Athens, shortly before my departure for America, this letter had in fact been written but never reached me. I myself have not pursued this matter further, since I think that, after the death of the Greek prime minister, Professor Demertzis, [111] the interest of the Greek government in the expansion of the Technical University would not be the same as before.

Today, I immediately wrote to Athens again to become informed about the present state of the whole matter and, after receipt of the reply, I will report on this further.

Madison, Wis. 30 October 1936.

C. Carathéodory [112]

There was considerable German interest in the National Technical University of Athens. The author of *Das neue Hellas* (New Hellas) wrote that

[the National Metsovian Technical University] enjoys a reputation as one of the best technical universities in the world. Not only its scientific performance but also the unusual work ethos of all of its academics and students is remarkable. Examinations at the Technical University are dreaded, the engineers produced there are not only respected and highly paid in Greece but also highly sought-after in the Balkans, in Asia Minor and in the Near East. The Technical University includes six faculties, for engineers, electrical-engineers, flight engineers, chemists, architects and mechanical engineers, as well as various subordinate institutions, among which is the almost independent 'Academy (literally: ´σχολή´) of Fine Arts'. The prerequisites for admission to study at the Technical University are the possession of the Abitur certificate and the successful completion of an admission examination. Studies last four to five years with compulsory intermediate examinations every two months and annual final examinations. The course leads to the award of 'diploma engineer' following a final examination. Although the number of students is not limited by law, on account of the demanding studies and examinations, it is quite small. (Of an average number of 300 candidates who register annually for the admission examination, no more than 25 satisfy the requirements). The Technical University possesses excellent teaching material. [113]

Euphrosyne's nephew, John Argyris, had studied at the National Metsovian Technical University. However, unable to graduate from there and having come into conflict with two of his professors, he was sent to Munich to complete his studies at the Technical University there, under Carathéodory's supervision. [114]

Leopold Kölbl, whom Carathéodory had informed of the planned appointment of German academics at the Technical University of Athens, was an Austrian born in Vienna in 1895, and had been since 1932 a member of the NSDAP and SA, and a former director of the Geological Institute of the College for Soil Cultivation (*Geologisches Institut der Hochschule für Bodenkultur*) in Vienna. Against the expressed wish of the Philosophical Faculty, Section II of Munich University, he was given full professorship for general and applied geology in the summer semester of 1934. He was appointed as a full professor in October that year, this time with the faculty's support. In July 1935 he was nominated dean of the faculty, and in December 1935 he was appointed rector of Munich University. Accused of *Fortgesetzte Vergehen der Unzucht zwischen Männern* (repeated sexual offences among men), Kölbl was arrested by the Gestapo on 10 February 1939, but acquitted by the Munich regional court I, and an appeal was lodged against the verdict. In August 1941 he was sentenced by the Munich regional court II to two years and three months imprisonment taking account of the one year and three months already spent in custody. He was then dismissed from the NSDAP and SA. It seems that Kölbl had been popular with both opponents and sympathisers of the Nazi regime.[115]

5.20
Corresponding Member of the Austrian Academy of Sciences

On 2 April 1935, W. Wirtinger, Ph. Furtwängler, E. Schweidler, St. Meyer and H. Mache proposed Carathéodory's election to the Class of Mathematics and Natural Sciences of the Austrian Academy of Sciences. Their supporting proposal contained several inaccuracies as to Carathéodory's life and work that were indeed astonishing considering the fact that the authors were members of an academy of science: they recommended him because of his technical studies in Paris (not his military studies in Brussels) and his professorship at the University (not the Technical University) of Breslau, as the Greek minister of education after his time in Smyrna (not as a refugee from Smyrna gaining his former position at Athens University); they mentioned that he had been at Munich University since 1928 (not since 1924), that his book on the theory of real functions had been published for the first time in 1916 (not in 1918), etc. However, they stressed that he belonged among the most respected mathematicians in Germany, enjoyed international recognition and combined outstanding scientific achievements with experience of the world. On the basis of this proposal, Carathéodory was appointed corresponding member of the Academy on 28 May 1935.[116]

5.21
Expecting the War – On the Political Situation in Europe and Greece

On 27 August 1935 Carathéodory wrote to Kalitsounakis: "I see that the clouds covering Europe do not bother you. For the moment I hesitate to travel to Greece. I do not doubt that we are moving towards an Anglo-Italian war. England, which is not prepared militarily, is of course interested in delaying the outbreak [of war]. But there is a danger that the Italian will provoke the situation, since at present he possesses military superiority which he could easily lose. Next week will undoubtedly be critical. If things take a turn for the worse, the Italians will occupy various harbours in Crete and possibly Samos as well. [...] Let us hope that the English can delay. Then, a good politician in Greece could reduce the danger."[117]

At that time Italy was ill-disposed towards the national socialist German Reich. Since October 1935 it had been carefully preparing for war against Abyssinia, which broke out despite sanctions imposed by the League of Nations under England's leadership. The war led to the subjugation of Abyssinia and its unification with the colonies of Eritrea and Somalia and to an Italian East Africa under a viceroy.

After the King's return to Greece and the abolition of the Republic, Carathéodory asked Kalitsounakis on 7 December 1935:

What do you think about the unexpected change of the situation in Greece? *The Daily Telegraph* has sent a correspondent to Athens, who daily telegraphs details and interesting things. I am not entirely convinced that the King [George II] and Venizelos have already come to an understanding. Nevertheless, it does not seem to be completely impossible. In the summer, a Greek from London told me that Sir John Stavridis and Lady Crosfield had attempted to bring the King into contact with Venizelos. But I had not imagined that the results would be so crystallised. "C'est très bien, pourvou aché cel doure" [C'est très bien, pourvu que cela dure], Napoleon's mother has said after the coronation of her son. After all, maybe the King will achieve a real reconciliation of Greeks. I wish that he could and I hope for it. However, the uncompromising persons of both parties would be filled with despair.

In great friendship K.[118]

In exile Venizelos had indeed appealed to his supporters to accept the King. On 18 March 1936, Venizelos died and Penelope Delta wrote in her diary for that day: "Finis Greciae ... ". The National Schism had lost its meaning since both sides, royalists as well as republicans, saw their task as preserving the bourgeois status quo which was threatened by a strengthening Greek Left. Even the value of parliamentary democracy was being questioned by a considerable part of the population who looked with admiration at Hitler's Germany and Mussolini's Italy. The evaluation of the situation by the royalist Hondros in Athens is expressed in his letter to Sommerfeld on 1 February 1935:

We have all waited for the outcome of the plebiscite with great suspense and the result was greeted in Greece with real jubilation. It proves after all that no agreement among the wielders of power of the moment can strangle the innate right of a nation to self-determination. Many Greeks suffer under the English or Italian yoke and indeed in North Epirus, for the benefit of Italian world-power politics, a compact Greek population, the only moving force of culture, became subject to the Albanians, a primitive people who did not even possess a written language 20 years ago.

Our main weapon in the struggle for self-preservation has always been the ancient Greek culture and its instrument, the Greek language. In the same way, Germany owes its re-annexation of the Saar region to the dear mother language.

We feel your joy to be our own joy.[119]

5.22

4 August 1936: Dictatorship in Greece

King George II, doubting the capabilities of the politicians with whom he had to co-operate, chose the royalist General Ioannis Metaxas, a graduate from the Evelpides Officers' School and the Prussian Military Academy, as his prime minister on 25 April

1936. Metaxas had been a brilliant officer of the general staff during the Balkan Wars and had participated in the negotiations for peace with Turkey and in the conclusion of the peace treaty with Serbia. Chief of the general staff in 1915, he resigned his post having disagreed with the Greek involvement in the Dardanelles operation during World War I. Being one of the basic organisers of the paramilitary terror groups of the royalists, he contributed to the escalation of political conflicts and was exiled to Corsica in 1917. He returned to Greece in November 1920 after the royalists won the elections but refused to head the army in Asia Minor in the spring of 1921, being among the very few to have advised the government against a military expedition. That year he founded the Freethinkers' Party (*Κόμμα των Ελευθεροφρόνων*), a small nationalist party, which he led during the years of the republic, with no attempt at hiding his contempt for the political establishment. In the elections of 1928 he was not even elected as a member of parliament and his party gained only 5.3% of the votes. After the suppression of the Venizelist military coup in 1935, Metaxas drew on the support of the mob to demand the government's resignation and the use of the death penalty for the instigators of the coup. After the death of Premier Demertzis, King George II swore in Metaxas as president of the government without previous consultation with any of the other political leaders. Metaxas also took on the foreign ministry portfolio. The Liberals, led by Themistoklis Sophoulis, supported Metaxas with a vote of confidence and so avoided the need for a coalition government with the People's Party. Sophoulis had married Euphrosyne's cousin, Loukia, in 1930 and thus was John Argyris's stepfather. Confronted with a growing movement of the working class, Metaxas suspended key articles of the constitution, especially those referring to personal freedom, and declared martial law on 4 August 1936. He founded the dictatorial "Regime of the Fourth of August 1936" that was to last up to his death, three months before the Wehrmacht's invasion of Greece in April 1941. Metaxas referred to the dictatorship as the "Third Hellenic Civilisation", a term supposed to demonstrate the character of his regime in combining the pre-Christian elements of the first civilisation of ancient Greece with the Christian values of the second civilisation of medieval Byzantium as well as making reference to an ideological affinity with Hitler's "Third Reich". Metaxas copied the outward signs of a fascist state, such as fascist salutes, as well as political organisations, such as the National Youth Organisation (EON = *Εθνική Οργάνωση Νεολαίας*), and a brutal secret service. Metaxas's deputy minister of public order, the notorious Konstantinos Maniadakis, was also an ardent admirer of the Third Reich. However, it should be pointed out, that the Greek dictatorial regime was not actively racist and its remarkable difference from the Italian or German states was the absence of a mass movement of support for the dictator in Greece. The source of Metaxas's power was his relation to the King, who, like Metaxas himself, never questioned Greece's association with Britain. The decision of Metaxas to turn towards England should be seen in the context of a general move of the anti-Venizelists towards England and France at that time. Metaxas suppressed any opposition, especially the communists, with torture, imprisonment and exile. England was informed of his Gestapo-like activities, but attempted to strengthen his regime by eliminating those characteristics which made it unpopular and, consequently, vulnerable. In the absence of popular support, the EON with the motto "King, fatherland, religion, Greece's rebirth and the regime of the 4th of August", was created as a way of institutionalising the dictator's power.

In the disciplinary proceedings against Kakridis in 1942, Georgios Papandreou reproached the University of Athens as a whole "apart from a few honourable exceptions" both for inactivity throughout Metaxas's dictatorship and for submissiveness to Metaxas. From their official position, university professors had "repeatedly praised slavery and celebrated submissiveness and the humiliation of the intellect." [120] Metaxas's Minister-Governor of Athens, Kostas Kotzias, contributed to a German publication of the year 1938, entitled *Unsterbliches Hellas* (Immortal Greece), with an article *Die griechische Jugend durch die Jahrhunderte* (The Greek Youth through the Centuries) [121] using rhetoric reminiscent of the Great Idea: "Since the 4th of August 1936, efforts are being made for present Greece again to find the way which centuries have prescribed for her. The Greek people, one of the strongest and most vital of the European community of nations, educated through the centuries-long work of civilisation, donor of the highest intellectual culture to the whole world and above all teacher of the peoples, would commit a crime if she did not herself, at all costs, seek to find the right way and especially seek to inspire herself with the flame her ancestors have lit." The editors of *Unsterbliches Hellas* were the press officer at the Royal Greek Legation in Berlin, Major Dr. Charilaos Kriekoukis, who also wrote the foreword, and the Reich Head at the Foreign Political Office of the NSDAP (*Reichsamtsleiter im Außenpolitischen Amt der NSDAP*), Karl Bömer. Prefaces to the book were written by Alexander Rizos-Rangavis, Greek ambassador in Berlin, and Alfred Rosenberg, who held the title of Official authorised by the Führer for the surveillance of the entire intellectual and world-view schooling and education of the NSDAP (*Beauftragter des Führers für die Überwachung der gesamten geistigen und weltanschaulichen Schulung und Erziehung der NSDAP*). Franz Baron von Weyssenhoff, Government Councillor at the Reich Ministry of People's Education and Propaganda (*Regierungsrat im Reichsministerium für Volksaufklärung und Propaganda*), wrote the introduction. Metaxas himself dedicated one of his portraits "*Für das Buch 'Unsterbliches Hellas'*". Many Greek scientists, among them Carathéodory's friends Ioannis Kalitsounakis, then professor at Berlin's Foreign University (*Ausland-Hochschule*) and full professor at the University of Athens, Periklis Vizoukidis, full professor of law at the University of Thessaloniki, and Vasileios Vojatzis, the director of the Greek Social Security Foundation IKA contributed to the publication. [122] In G. D. Daskalakis's contribution to the *Geschichte und Bedeutung der Universität Athen* (History and Significance of Athens University), [123] Carathéodory's intervention for the reorganisation of Athens University, whose centenary was to be honoured with this publication, was not mentioned. Carathéodory did not contribute to the book and maybe he was not even asked to.

The dictatorial regime of the 4th of August overthrew the last remnants of the educational reform of Venizelos's government of 1928–1932 (see 4.31) and introduced radical changes in the educational system. Structure, function and programme had now to serve political and social aims.

5.23
The Oslo Congress: awarding the First Fields Medals

In 1936, when Erhard Schmidt was president of the DMV, Carathéodory took part in the International Congress of Mathematicians in Oslo as a member of the German delegation under Walter Lietzmann, Klein's former close assistant in educational matters, now honorary professor of didactics of the exact sciences at Göttingen University. The largest groups of participants came from the Anglo-Saxon world and the language of the congress proceedings was French. Total participation was smaller than in the two previous international congresses. Somewhat more than 500 mathematicians accompanied by 200 family members represented 35 nations. Soviet mathematicians were absent and Italian mathematicians were prevented by their government from going to Oslo because of international sanctions imposed on their country due to its invasion of Abyssinia. German participation included supporters of the Hitler regime as well as emigrants. Abraham Fraenkel represented both the Hebrew university and the British mandate government of Palestine, Richard Courant came from New York, Landau and Neugebauer were there, but Hilbert was absent. The Reich Minister of Education provided Lietzmann with the names of mathematicians who had permission to participate in the congress. Delegates were to conduct themselves strictly in accordance with Lietzmann's instructions, so that "the German delegation appears united and effective during the congress." Rust had also asked the foreign ministry to allow the equivalent of 400 Swiss francs for the German contribution to the International Commission for Mathematical Instruction. The exact titles of the speeches, which had to be limited to 15 minutes, were to be given to the secretary's office by 15 April. Summaries were to be published in the congress files and the manuscripts themselves had to be submitted before the end of the congress.[124] The price for a congress card amounted to 40 Norwegian crowns (half for a family member). Participants were offered travelling within Norway before and after the congress at reduced fares.[125] The DMV in Berlin thanked the president of the organisation committee, Professor Dr. Carl Störmer, and announced that its official representatives would be Erhard Schmidt and the union's treasurer, Helmut Hasse.[126] Professor Dr. Wilhelm Lorey from the University of Frankfurt believed that the DMV ought to honour Abel and Lie by laying a wreath on their graves in Norway, since both mathematicians had connections with Germany: Abel through his relation to Crelle, Lie as Klein's successor in Leipzig.[127] Hecke was asked to give a forty-five minute plenary speech on a subject from the theory of modular functions.[128]

The congress took place from 13 to 17 July at the University of Oslo and was presided over by Carl Störmer, who was elected at the opening ceremony. The Norwegian government was represented by the Minister of Education and the Foreign Minister. In addition to the formal business of the conference, there were the usual excursions and receptions, including one given by the King and Queen and another by the City of Oslo. The Oslo congress was also important for hosting the first awarding of Fields medals. In the absence of Severi, Cartan announced the decision of the Fields medal committee to award the first two medals to Lars Ahlfors and Jesse Douglas. The other members of the Fields medal committee, in addition to Severi and Cartan, who had been elected by the executive committee of the Zurich international congress in accordance with Fields' memorandum, were George D. Birkhoff, Carathéodory and Takagi. Élie Cartan delivered a plenary address at the congress and Carathéodory, chairman of the Fields medal committee, presented the work of the two Fields medal recipients.[129] Both of them were planning to go to Harvard to represent European mathematics at the university's tercentenary celebration the following month.

In the presence of the King of Norway, Cartan presented the first medal to the 29-year-old Lars Valerian Ahlfors. The ten years' older Jesse Douglas, though in Oslo, was absent from the opening ceremony and his medal was accepted on his behalf by Norbert Wiener. Carathéodory established the tradition of giving a rather specific account of the medallists' work at the congress. Carathéodory's role in this event is interpreted by Behnke as a sign of his neutrality before the outbreak of World War II. The prestige of these medals can be judged by the standing of the mathematicians of the committee which awarded them.[130]

The Fields medal had been endowed by John Charles Fields, professor of mathematics at the University of Toronto and chairman of the Canadian committee of the International Congress of Mathematicians in Toronto in 1924. His motivation is usually explained by the strong internationalist sentiment he had developed after the exclusion of mathematicians from Germany and the former enemy countries from the 1924 Toronto congress and by the lack of a Nobel Prize for mathematics. Fields' memorandum, sketching out the procedure, principles and philosophy of the award, was accepted by the Canadian congress committee – still working on behalf of the congress – at its meeting in January 1932. The International Congress of Mathematicians in Zurich in 1932 had readily adopted the proposal conveyed by J. L. Sygne, then a faculty member at the University of Toronto, and the Fields medals were awarded for the first time at the international congress in Oslo. Actively campaigning for the establishment of the medal, made possible by a surplus of funds from the Toronto congress, Fields travelled to Europe to gain the support of distinguished personalities for his idea. For this purpose, two years before his death, he visited Carathéodory in Munich on 3 September 1930.[131] In accordance with Fields' wish that the medals should honour existing work as well as the promise of future performance, it had been agreed that the medals should be awarded to mathematicians under the age of forty who would also receive a cash prize. Recipients had to be of international and objective character, as far as that could be determined. At least two but no more than four medals should be awarded every four years. Since the Oslo congress, the Fields medal is awarded regularly at the international congresses of mathematicians and, despite differing from the Nobel Prize in many respects, it is considered equivalent to the latter.

At Oslo, Ahlfors was honoured for research on covering surfaces related to Riemann surfaces of inverse functions of entire and meromorphic functions, which opened up a new chapter in analysis, that of "metric topology".[132] He had studied at Helsingfors University under Ernst Lindelöf and Rolf Nevanlinna. He spent the years 1930–1932 in Paris as an International Research Council Fellow of the Rockefeller Foundation and he also visited other European centres. He got his doctorate

19.

1937

103 Bemerkungen zu den Strahlenabbildungen der
geometrischen Optik.
Math. Ann. 114, Heft 2 (1937) S. 187-193.
[München, Sommer 1936].

104 Besprechung v. Oseen C.W.: Une méthode nouvelle
de l'optique géométrique. Svenska Vetensk. z Kad. Hdl.
s. 15, Nr 6, 1-41 (1935)
Zentralblatt 1937, Bd. 16, S. 90.
[in Madison verf. Wisc. verfasst].

105 Bericht über die Verleihung der Fields medaillen.
C. R. du Congrès des mathématiciens Oslo (13-17
Juillet 1936) Tome I pp. 308-314.
Oslo, A. W. Brøggers 1937.

106 A generalization of Schwarz's Lemma.
Bullet. Amer. Math. Soc. April 1937, pp. 231-241
[From an address delivered before the Soc. under the title
Bounded analytic functions Nov. 27. 1936 at Lawrence, Kans].

107 The most general Transformations of plane regions
which transform circles into circles.
Ibid. August 1937, pp. 573-579.
[English: on the pacific Ocean betw. Los Angeles & Panama]

108 On Dirichlets Problem.
Amer. Journ. of Mathem. Vol. 59 N≘ 4 October 1937
pp. 709-731.
[Aus e. Seminar in Madison Wis. im Winter 36-37]

*Page 19 of Carathéodory's hand-written chronological list
of his mathematical works, including his* Report on the Award of the Fields Medals.
Courtesy of Bavarian Academy of Sciences, Archive.

EXTRAIT DES

COMPTES RENDUS
DU
CONGRÈS INTERNATIONAL
DES MATHÉMATICIENS
OSLO 1936

C. Carathéodory:

BERICHT ÜBER DIE VERLEIHUNG
DER FIELDSMEDAILLEN

A. W. BRØGGERS BOKTRYKKERI A/S

OSLO 1937

Carathéodory's Report on the Award of the Fields Medals.
Separate print from the publication of the International Congress of Mathematicians in Oslo.
Courtesy of the Bavarian Academy of Sciences, Archive.

in 1932 and lectured at Helsingfors in the succeeding years. In 1935, Carathéodory, who had met Ahlfors in Germany, recommended him to Harvard University, where Ahlfors lectured in the academic year 1935–1936. Ten years later, he became the first William Caspar Graustein Professor of Mathematics there. In 1936, Ahlfors became a member of the Finnish Academy of Sciences (*Societas Scientiarum Fennica*), while Carathéodory himself became an honorary member of the same academy in February 1943.

Jesse Douglas had gained his doctorate from Columbia University in 1920. In 1928–1930 he visited the universities of Princeton, Harvard, Chicago, Paris and Göttingen as a National Research Council Fellow of the Rockefeller Foundation. In the period 1930–1934 he was an assistant professor and in 1934 an associate professor at the Massachusetts Institute of Technology. A member of the editorial board of the *American Mathematical Society Transactions* and a member of the AMS Council, he became AMS colloquium lecturer in 1936. The Fields medal was awarded to Douglas for his treatment of Plateau's problem. The problem, whose solution marked a very important step in the calculus of variations, was to find a surface of least area bounded by one or more given closed curves. He and Tibor Radó[133], independently and almost simultaneously, were the first to establish the necessary and sufficient conditions for the existence of a solution to the problem. Douglas used a daring new method to arrive at a general result, whereas Radó employed a method of conformal mapping.

In 1940, Carathéodory would engage with Plateau's problem in two reviews. He remarked that in Schiffman's *The Plateau problem for non-relative minima*, a proved theorem represents the first example in the theory of variation problems, in which the existence of an extremal with prescribed boundary is derived without the assumption of an absolute or a relative minimum.[134] Also, he noted that in a paper on *The existence of minimal surfaces of given topological structure under prescribed boundary conditions*, Courant had found the solution of Plateau's problem, not in the case of fixed boundary curves (that case was set and solved by Douglas), but for free boundary curves.[135]

The Oslo Congress decided that the committee to award the Fields medal in the following international congress would be composed of Hardy, the president, Alexandrov, Hecke, Julia, Levi-Civita, as members, and Lefschetz and Nevannlina, as deputy members. On 15 July, the bust of Sophus Lie was unveiled at Oslo University and, on that occasion, Cartan gave his lecture *Quelques aperçus sur le rôle de la théorie des groupes de Sophus Lie dans le développement de la géométrie moderne* (Some Observations on the Role of Sophus Lie's Group Theory in the Development of Modern Geometry). The general assembly decided to call the next congress in America in 1940, on the invitation of the American Mathematical Society, whose delegates in the Oslo congress were G. D. Birkhoff, H. F. Blichfeldt, L. P. Eisenhart, S. Lefschetz, H. C. M. Morse, V. Snyder, O. Veblen and N. Wiener. Greek participation in Oslo included K. Papaioannou for the Greek government, N. Sakellariou for the University of Athens and F. Vasileiou for the University of Thessaloniki. The last two and Lucas Caesar also represented the Greek Mathematical Society. None of them

Ahlfors's signature on page 13 of Carathéodory's guest-book.
Courtesy of Mrs. Rodopoulos-Carathéodory.

had received any financial support, either from the Greek state, the university, or the Greek Mathematical Society.

5.24
Against an International Congress of Mathematicians in Athens

A controversy emerged among the Greek participants on the third day of the Oslo congress, when Papaioannou urged the delegation members to try to have the next congress convened in Athens. Carathéodory, who obviously also felt some responsibilty for the performance of the Greek delegation, opposed the initiative, arguing that the Americans had already been working on the preparation for the next congress and that the decision had in fact been taken in favour of it taking place in America. Carathéodory argued that the Greek mathematicians ought not to act hastily to invite anyone without having first secured either the approval of the government and the Greek universities, or the necessary funds for hosting the congress.

The fact was that an amount of $15000 towards the expenses of a possible international congress of mathematicians in America had already been secured and the delegates of the American Mathematical Society were empowered "to issue an invitation to the Congress to convene in America in 1940, provided it appears ... that no other nation has a prior right to hold this Congress." [136]

Sakellariou expressed the opinion that it would be better if the Greeks arranged unofficial meetings with other participants and persuaded them of the case for holding the congress in Athens; in this way, a proposal for holding the 1940 congress in Athens could come from another delegation and be accepted by the Greeks. Carathéodory agreed with these tactics but he did nothing to promote such a solution. It should be noticed that P. Zervos, rector of Athens University, was explicitly opposed to the holding of an international congress in Athens. [137] Sakellariou and Caesar had visited Carathéodory in Munich on 9 July prior to the Oslo congress and may well have discussed with him there the possibility of inviting the ICM to convene in Athens. N. Kritikos, the former secretary of the Bureau for the Organisation of Smyrna University, had resigned his membership in the executive committee of the Greek Mathematical Society in 1936, obviously as a protest against political interference in the society.

Carathéodory returned from Oslo to Munich on 23 July 1936. [138] As Marston Morse wrote to Dr. Flexner on 24 September 1936, "The Mathematical Congress at Oslo voted to accept the invitation of the AMS to hold an International Mathematical Congress in America in 1940. No other nation was able at that time to present a formal invitation for the next Congress." [139]

5.25
Invitation to the University of Wisconsin

Carathéodory was invited to become the Carl Schurz Memorial Professor at the University of Wisconsin for the winter semester of 1936–1937. The history of the Carl Schurz professorship dates back to 1911, when German-American citizens of Wisconsin, wishing to commemorate their fellow citizen Carl Schurz, donated the sum of $ 30 570 to the regents of the University of Wisconsin for the purpose of establishing and maintaining a chair at the university in the state capital, Madison, to be occupied by visiting professors from German universities. About half of the original donations were of small amounts given by alumni of the university.[140]

Carl Schurz had left Germany after the revolution of 1848 and arrived in Madison in 1855. He joined the Republican Party, supported Lincoln for president, served as Minister to Spain in 1861, but resigned later that year to fight in the Civil War where he obtained the rank of major general. He was elected US senator from Missouri in 1869 and served president Hayes as Secretary of the Interior (1877–1881). He became known for his efforts at civil service reform and for the protection of national forests. In the early 1880s Schurz edited the *New York Evening Post* and *The Nation*.[141]

The holders of the Carl Schurz chair before Carathéodory had come from all parts of the German Reich and from the scientific disciplines of German language and literature, political economy, physics, art history, philosophy, chemistry, Egyptology and sociology. Carathéodory was the first to teach mathematics. Sommerfeld had held the same professorship for physics in 1922.

The Carl Schurz Professorship Committee required the prospective candidate for the chair to be of "first rate standing as a recognized authority in his field" and to have "attractive personal qualifications and [be] sufficiently at home in English to lecture in this language with relative ease". The successful nominee was engaged at a salary which would normally amount to $ 3500 for the semester.[142]

The mathematics department was very interested in Caratheodory's candidacy as Carl Schurz Professor for the first semester of 1936–1937. A member of the Carl Schurz Professorship Committee, A. R. Hohlfeld, of the Department of German, called on Carathéodory while in Munich in the middle of August 1935 and inquired of him, whether he would be interested in such an invitation. Hohlfeld got "the impression that a visit to the United States would be very attractive to him indeed, provided of course that he could secure the necessary leave of absence from his home authorities. The latter he did not doubt, but at the same time he felt that it might take some little time." Hohlfeld had found Carathéodory, with whom some members of the mathematics department were personally acquainted, to be "a most interesting man in regard to a number of subjects of general appeal". In his opinion, Carathéodory's political affiliations made him fit well in any American university and his repeated residence in English-speaking countries had provided him with a fluent command of conversational English. Hohlfeld would be pleased to support his candidacy in the committee.[143]

The chairman of the mathematics department, Mark Ingraham, was more than enthusiastic about the prospect: "There is no one in Germany at present whom we would rather have here on a Carl Schurz professorship more than Professor Caratheodory". "The recollection of his short visit to Madison would make us all anticipate with delight the possibility of a longer stay, and I feel that he would win many friends not just in our department but throughout the university", he wrote to Hohlfeld.[144]

Representing the committee, Hohlfeld suggested a semester salary of $ 4000 for Carathéodory and sang his praises to the president of the university, Glenn Frank: "I found him to be a man of great charm, of personality and broadest international interests and outlook, due largely no doubt to the fact that he was born in Greece, and on account of his father's position in the Greek consular service, grew up and received his education in various countries of Europe before he definitely settled in Germany and attained to his present position of prominence in the University of Munich."[145]

Apart from some inaccuracy in Hohlfeld's information regarding Carathéodory's place of birth and education and his father's diplomatic position, which had in fact been in the service of the Sultan, his general impression coincided almost completely with that of friends and colleagues who were attracted by Carathéodory's personality and appearance. Here elements of a myth around the Carathéodory family were introduced to support the family's Greekness and their services to Greece. This was, of course, during the period of national socialism in Germany.

A letter of invitation to Carathéodory was sent by the chairman of the mathematics department and dated 18 January 1936. "Our whole mathematical group at Wisconsin as well as in the neighboring universities will consider it the highest of privileges to have you with us. If, as I hope, you accept, please do not hesitate to call upon me for any assistance I can be in making arrangements for you", Mark Ingraham assured Carathéodory.[146]

On 2 March 1936, Carathéodory informed Hohlfeld that he was "extraordinarily pleased" with the "honorable" invitation to Madison and asked him to convey his thanks to the president of the university. He acted quickly and had a meeting with the rector of Munich University, who promised Carathéodory he would try to speed up the decision process of the Reich Ministry of Education, if that were possible. Carathéodory considered that the Bavarian Ministry of Education was "favorably inclined towards the invitation" and he ascribed this Bavarian attitude to the prestige associated with the awarding of the Carl Schurz professorship to a Munich professor. He viewed the fact that he had been asked to consider the possibility of a substitute during his absence as a sign of an expected positive outcome. From then on, he could do nothing else but wait for the decision from Berlin.[147]

Following Carathéodory's request for permission to accept the Carl Schurz professorship for the semester 1936–1937, the new dean of the Philosophical Faculty, Section II, Friedrich von Faber, wrote to Carathéodory on 16 April 1936 to ask him for his curriculum vitae, a list of publications, a questionnaire about his "Aryan" origin, also that of his wife, certified copies of his own birth certificate and the marriage certificate of his parents (copies of the same certificates also for his wife) and a statement about membership of lodges, all to be provided in duplicate.[148] A copy of Carathéodory's christening certificate and the marriage certificate of his parents were certified by the Munich police authority two days later.[149]

It was only on 10 June that Carathéodory was able to tell Glenn Frank that he had been unofficially informed that he was going to be granted leave. Ingraham immediately notified R. G. Richardson, the dean at Brown University, that if he wished "to approach Carathéodory in connection with any portion of the mathematical program for the meetings next year", it would be the right time to do so.[150]

On 6 July 1936, there was still no official confirmation of permission for Carathéodory to leave. Despite that, Mark Ingraham decided to inform Carathéodory about the obligations of a Carl Schurz professor. Carathéodory had to give one course of three lectures a week and then either a second course or a two-hour seminar once a week. "The students that we have will not be very advanced but most of them will have had complex variable and will probably be taking a course in real variables. They will have had all the ordinary undergraduate work in mathematics. Some of them will be candidates for the master's degree and quite a large group for the doctor's degree. We have wondered somewhat whether you would not like to give a semester course in the calculus of variations but we would not wish you to do this if there was some other topic you would rather lecture on. It is probable that your seminar would be for a small group and could be, of course, on whatever topic you would desire. Some of our staff would probably wish to attend." The classes were planned to start on 23 September. Ingraham suggested the University Club as the best place for Carathéodory to stay if he were going to come alone, as it was close to the campus and would offer him the opportunity of meeting all the faculty members. Ingraham had already asked the manager to reserve a room with bath for Carathéodory at a rental of $33 a month up to the middle of August but he was planning to talk in more detail to Carathéodory in Cambridge where he was going for the mathematical meetings.[151]

While in Oslo, at the ICM, Carathéodory received a telegram to tell him that his leave had been approved but he still had no official letter. He informed Ingraham that the matter had been definitely settled. The proposed schedule of a three-hour course and a two-hour seminar per week suited him, but he would not like to lecture on the calculus of variations, of which he confessed that he was tired. "But as you write that most of your students have already taken a course in complex variables, I would suggest that I give my lectures about this matter, perhaps under the title 'Methods in the Theory of Functions'", he wrote to Ingraham. "This would have several advantages: first of all I could make out during the first week what kind of a lecture would best suit your students and adapt myself to their needs. Secondly I have everything prepared to write a small pamphlet on this matter and I should be exceedingly pleased if this little tract could appear as the result of my work at Madison. As for the subject of the Seminar we could speak about it in Cambridge and certainly it would be safer to make it depend on the wishes of the students who will attend it." Carathéodory asked Ingraham for rooms outside the faculty club, but near the university or the library and with facilities for cooking breakfast. He and his wife were going to take the M. S. *St. Louis* of the North German Lloyd Steamship Company, sailing from Hamburg and due to arrive in New York on 26 August, and then travel on to Boston on the following day. Carathéodory also asked Ingraham for a small payment in advance to enable him to buy their tickets from Boston to Madison and to leave him with a sufficient sum to spend between 1 September and the opening of the academic year. Because of the German exchange regulation, he was not allowed to take enough money out of the country and had arranged with Graustein to help him for the first few days of his stay in the United States.[152]

In the meantime, Ingraham and Hohlfeld had found what they believed to be "a very satisfactory apartment" for the Carathéodorys on the 5th floor of Kennedy Manor, "perhaps the best apartment house in Madison". The apartment consisted of a bath, a kitchenette, a large dressing closet, a small dining room convertible into a bedroom and a living room with two inner door-beds folding into a closet. The apartment was reserved for Carathéodory from approximately 10 September to 10 February, for a total rent of $400 including heat, gas, electricity, full furnishing with dishes, cooking utensils, bed and table linen and also maid service. The Manor itself met Carathéodory's requirements by presenting a very pleasant view of the lake, being about a mile from the mathematics building and less than that from the library and possessing a restaurant.[153]

A State Treasurer's cheque for $500 "payable to the order of Dr. Constantin Caratheodori, drawn on the First Wisconsin National Bank of Milwaukee and representing an advance payment on the Carl Schurz Professorship" was going to be delivered to Carathéodory on his arrival in New York by Ingraham, who had been authorised to do so by the regents of the university.[154]

In Munich, Carathéodory was replaced by Hermann Boerner, who gave the four-hour lecture *Grundlagen der Geometrie* (Foundations of Geometry) in the winter semester 1936–1937.[155] Boerner had habilitated in 1934–1935 under Carathéodory with a work *Über die Extremalen und geodätischen Felder in der Variationsrechnung der mehrfachen Integrale* (On the Extremals and Geodesic Fields in the Calculus of Variations of Multiple Integrals) and in 1936 he had taken on the assistant position formerly occupied by Lettenmeyer. He had belonged to the SA since 1934 but was dismissed in 1937, because he did not renounce his Jewish friends and relatives. He was branded by the Nazis as a *Judenknecht* (Jew slave), an *überkultivierter Schwächling* (overcultivated weakling) and a politically unreliable person.[156] Despina remembers him as a "handsome blonde young man in leather trousers", who often visited them. She still has the impression that he was a Jew. After the war, he was described by Threlfall as a youthful personality holding clear and formally perfect lectures.[157]

5.26
Carl Schurz Professor at the University of Wisconsin

On 17 September 1936, Ingraham notified students and staff of a change in the advanced courses of mathematics. The course on the calculus of variations would not be offered and the credit courses to be added were the second course in the theory of functions, requiring the theory of analytic functions as a prerequisite, together with a seminar in potential theory which, though dealing with the most recent aspects of the subject, did not require more than a first course in the theory of functions. Carathéodory was going to give these courses as the Carl Schurz Memorial Professor for the first semester of the academic year 1936–1937. The chairman of the mathematics department was proud to offer the students the "rare opportunity" to attend Carathéodory's lectures and some senior members of staff would have "the privilege of auditing the work in this seminar". Ingraham introduced Carathéodory as "one of the leading mathematicians of the world [...] particularly noted for his work in the branches of mathematics" in which he was going to give courses that year.[158]

As well as his lecturing, Carathéodory took part in meetings of the mathematics department from September to December 1936, gave talks, engaged in correspondence and produced papers. Despite the fact that he wished to avoid the calculus of variations, he gave a lecture on "Some Geometrical Features of the Calculus of Variations" on 23 October. At the same event professor Gilbert Ames Bliss of the University of Chicago[159] spoke on "The Problem of Bolza in the Calculus of Variations".[160] Later, at the 24th annual meeting of the Ohio Section of the Mathematical Association of America on 8 April 1939, Bliss gave a lecture on "The Hamilton–Jacobi theory in the calculus of variations and its sources". He showed how the Hamilton–Jacobi theory had influenced the development of the calculus of variations and its applications to geometry and mechanics, especially by comparison with problems in optics and dynamics. Bliss's paper was suggested by studies of recent formulations of the Hamilton–Jacobi theory for parametric problems in the calculus of variations by Carathéodory and Teach.[161]

On 27 October 1936, Carathéodory wrote to his Athens colleague Panagiotis Zervos that the really rapid progress of Greek scientists in the eight years since his last visit to America had impressed him the most. Not only was the number of Greeks whom he met at the universities large, but some of them occupied the best positions, he wrote.[162]

Following Carathéodory's departure from the University of Wisconsin at the end of the semester, his colleagues and the Carl Schurz Professorship Committee were left with lasting impressions of his time there. In the spring of 1937, the committee must have also included Carathéodory in their view that, among the holders of the chair in the past, "the most valuable and most lasting impressions on the University community at large have been made by those men who were able to speak English with fair ease and fluency".[163]

5.27
Support for Blumenthal

While in Madison, Carathéodory tried to organise a lecture tour for Blumenthal who had been dismissed from his professorship in Aachen and expressed the wish to lecture in the United States, since he saw no opportunities for himself in Europe.[164] Carathéodory approved of Blumenthal's choice of subjects from the history of mathematics, but suggested he include the additional subject *Mathematik in Göttingen im 19. Jahrhundert* (Mathematics in Göttingen in the 19th Century) in his programme. Blumenthal followed Carathéodory's advice. At von Kármán's request, to whom Blumenthal had expressed his anguish, Carathéodory would contact the American Mathematical Society about this. Blumenthal counted on Carathéodory's support, because he considered him to be reliable.[165]

On 2 September 1936 Kármán sent the whole list of lecture titles to Carathéodory. He believed that the way to proceed would be to ask the AMS to send a circular letter to the universities to announce that Blumenthal was aiming to hold lectures on the subjects listed and to ask them whether the faculties would be interested in hearing Blumenthal lecture on one or more subjects. Kármán promised to help in the organisation and was very eager to meet Carathéodory and learn from him his views on the events of the previous five years.[166]

Carathéodory first approached Salomon Lefschetz, Fine Professor of Mathematics at Princeton, President of the AMS for the year 1935–1936 and editor of the *Annals of Mathematics*, in order to arrange the matter. Lefschetz, however, indicated the existence of a specific committee of the society in charge of visiting lectureships consisting of Birkhoff (chairman), Bliss and Hedrick and suggested that Carathéodory should approach Birkhoff. He himself could do nothing, since he was "only the President of the American Mathematical Society (until the end of this year) and not its Fuhrer [*sic*]!"[167] With this irony, Lefschetz was obviously expressing his opposition not to non-democratic procedures as his wording might suggest, but to those academics who had willingly remained in Germany during the Nazi era, including Carathéodory.

Carathéodory sent Lefschetz's reply to Kármán and remarked that he did not at all like the fact that Blumenthal's tour was to be considered by a committee of the AMS and that they had to be very careful in order not to be turned down. Carathéodory believed that if he met Birkhoff personally he would be able to influence him in favour of Blumenthal's lecture tour. However, he wanted to avoid writing to him, since Birkhoff would be sure to reply that the lectures had already been arranged for the coming year. But he was prepared to send Birkhoff a letter if Kármán thought they ought to take the risk.[168]

Kármán had now to change his plans. His new proposal was that the AMS, or a group of prominent American mathematicians, would inform mathematics departments of many universities by letter about Blumenthal's intention to hold lectures on the history of mathematics. The letter would be followed by the list of subjects as indicated by Blumenthal, a short description of the lecturer's records and abilities and a proposal for a minimum fee. Kármán wished to have the lectures wholly or partly sponsored by the AMS but, in his opinion, it would be best if the expenses could be covered by the honoraria for the lectures that were to be given in the autumn of 1937 or the spring of 1938. He believed that the most probable sponsors would be universities employing scientists who knew Blumenthal personally or by name, but not the main centres of mathematics employing scientists who were more interested in up-to-date problems than in general lectures.[169]

It is obvious that Carathéodory and von Kármán's attempts did not bear fruit and the course of events thereafter proved disastrous for Blumenthal. He remained the main editor of the *Annalen* until 1938 and emigrated to the Netherlands a year later. In 1943 he was arrested there by the Gestapo and deported to the concentration camp of Theresienstadt where he died in November 1944. His wife Mali, born Ebstein, died in a transit camp in Holland.

5.28
Pontifical Academician

On 15 November 1936, when still at the mathematics department of the University of Wisconsin, Carathéodory asked the Reich Ministry of Science, Education and

People's Culture for permission to accept the academic honour of being appointed a member of the newly founded *Pontificia Accademia delle Scienze* and Pontifical Academician by Pope Pius XI.

Pius XI became Pope in 1922, on the eve of Benito Mussolini's "march to Rome", through which Mussolini seized governmental power. The Lateran Treaty signed on 11 February 1929 between the Holy See and Mussolini solved the Roman question through a single financial settlement and recognition of the Vatican City as an independent state, the seat of the Holy See. Thus, what was considered the prelude to Pacellis' Concordat with Hitler in 1933, terminated the hostility that had lasted since 1870 between the Holy See and the Italian state. The Lateran Treaty was most heartily welcomed by Hitler. After it had been signed, the Vatican suggested that Catholics withdraw from politics and encouraged the priests to support the Fascists. One of the problems discussed during the preliminary talks prior to the treaty was certainly the question of the two academies, the *Reale Accademia d'Italia*, which was founded by the head of the fascist government by means of a decree on 7 January 1926, with the aim of providing a submissive organ for Mussolini's cultural and political purposes, and the pre-existing *Regia Accademia dei Lincei* or *Reale Accademia dei Lincei* which, according to an article of the 1926 decree, would remain unaffected. The new *Reale Accademia d'Italia* was required not only to "promote Italian intellectual development in the field of the sciences, letters and the arts" but also to "keep pure the character of the race, according to its genius and tradition and to encourage the expansion and influence [of such genius and tradition] beyond the frontiers of the state".[170] *Reale Accademia dei Lincei* was the name given to the Italian National Academy that had grown out of the *Pontificia Accademia dei Nuovi Lincei* in 1870, following the proclamation of the Kingdom of Italy and the decline of the Pontifical State. The *Pontificia Accademia dei Nuovi Lincei* continued to exist for a certain period with its headquarters in the Vatican and in 1936 it became the *Pontificia Accademia delle Scienze*. The activities of the *Accademia d'Italia* were inaugurated on 28 October 1929. The academy was housed in the Villa della Farnesina with frescoes by Raffael and his disciples while, across the street, Palazzo Corsini was the headquarters of the *Reale Accademia dei Lincei*. On 12 December 1929 Carathéodory was nominated *socio straniero* (foreign member) of the *Reale Accademia dei Lincei*.

In 1933 the Pope negotiated a concordat with the German Reich. Signed on 20 July 1933, it gave the Pope the authority to establish new ecclesiastical laws and granted Catholic schools and clergy generous privileges. In return, political Catholicism in the shape of the Centre Party was dissolved and Hitler's dictatorial regime could be established without any opposition from Catholics. In the Concordat Hitler saw a significant opportunity for the "advancing struggle against the international Jewry" and a sufficient guarantee that the Catholics would place themselves unreservedly in the service of the Nazi state. Belatedly, the Pope condemned the aggressive neopaganism of the "Third Reich" in his encyclical *Mit brennender Sorge* (With Burning Anxiety) of 14 March 1937 concerning the plight of the Church in Germany. He did not, however, explicitly condemn anti-Semitism, national socialism and Hitler.

Since 15 February 1934 Carathéodory had been a corresponding member of the *Pontificia Accademia dei Nuovi Lincei*, which preserved the continuity and tradition of the official Academy of Sciences of the Pontifical State. Its members adhered to the Statutes of 1847 and affirmed their loyalty to the Pope. A new charter enforced by decree on 11 October 1934 obliged all members of the *Pontificia Accademia dei Nuovi Lincei* – now subject to the general reform applied to all the academies recognised by the state – to swear fidelity to the fascist regime. This reform provided for the head of the government, together with the Minister of Education, to assume the right to appoint the academy's president and vice-president. The national members were also to be appointed by the government, each selected from lists of three candidates elected by members of the academy.

The difficult task of renewing the *Pontificia Accademia dei Nuovi Lincei*, so as to reflect the world-wide interests of the Holy See, was given to a physics professor, Father Giuseppe Gianfranceschi, who worked under the direct supervision of Pope Pius XI. Gianfranceschi died in 1934 but his successor continued his work so that two years later, on 28 October 1936, the Pope issued his *Motu proprio, In multis solaciis* (By one's own motion, in many consolations) thereby establishing the *Pontificia Accademia delle Scienze* not only as a team of scientists, but also as a kind of scientific senate of the Holy See. According to Pope John Paul II, the intention of his predecessor was to surround himself with a select group of scholars relying on them to inform the Holy See in complete freedom about developments in scientific research, and thereby to assist him in his reflections.[171]

The new academy consisted of seventy internationally chosen scholars. Five of them, Abderhalden in Halle, Debye in Dahlem, Guthnik in Neubabelsberg, Planck in Berlin and Carathéodory in Munich, were professors of German universities. Apart from Carathéodory, six other mathematicians were members of the Pontifical Academy, namely U. Amaldi and V. Volterra in Rome, G. Birkhoff in Cambridge, Massachusetts, Ch.-J. de la Vallée Poussin in Louvain, E. Picard in Paris and E. Whittaker in Edinburgh.[172]

Vito Volterra had refused to take the oath of allegiance to Mussolini and was forced to leave the University of Rome in 1931 and all Italian scientific academies the following year.[173] Sympathy for him had been expressed in the International Congress of Mathematicians in Oslo. Volterra and Picard had favoured the exclusion of Germany from the International Research Council (IRC) and its Unions.

Emile Picard was President of the IRC in 1919–1931 and Honorary President of the IMU in 1920–1932. At the time of the 1928 International Congress he still assumed an anti-German attitude.

Charles-Jean de la Vallée Poussin had served as president of the IMU in 1920–1924 and as honorary president in 1924–1932. In 1918–1920 he had been one of the main opponents of mathematical co-operation with Germany, however later, at the International Congress of Mathematicians in Toronto, he distanced himself from this policy. He was appointed to represent the IMU on the Executive Committee of the IRC for the three-year period 1922–1925, but still held the position when, with his consent, William Henry Young replaced him in 1930.

At the conclusion of the Oslo International Congress, George D. Birkhoff was appointed by the American Mathematical Society as President-Designate of the congress to be held 1940 in Cambridge, Massachusetts, which finally took place after the war, ten years later than originally planned.

Edmund Whittaker was elected Fellow of the Royal Society in 1905, became Astronomer Royal of Ireland in 1906, professor of mathematics in Edinburgh in 1912 and set up the Edinburgh Mathematical Laboratory. He was a member of the London Mathematical Society being its President in 1928–1929 and a Fellow of the Royal Society of Edinburgh serving it as President for most of the years of World War II. He was knighted in 1945.

Ugo Amaldi was President of the Faculty of Sciences in Rome (1944–1946), President of the Italian Society of Mathematical and Physical Sciences *Mathesis* (1941–1943), Secretary of the *Pontificia Accademia delle Scienze*, a member of the National Committee of Applied Physics and Mathematics of the National Research Council and a member of many Italian academies of arts, literature and sciences.

All mathematicians of the Pontifical Academy were powerful intellectuals and almost all had an international involvement, reflecting the Vatican's philosophy of universalism. Carathéodory's nomination as a member of that academy was the greatest international honour he had ever received and he valued his membership as a great privilege, particularly since he shared it with only a few others.

On 30 November 1936, the rector of Munich university asked the dean of the Philosophical Faculty II and the head of the lecturers for their opinion about Carathéodory's appointment.[174] Two weeks later, von Faber announced to the rector that the faculty warmly supported the acceptance of this honour by Carathéodory.[175]

On 26 April 1937, Carathéodory asked the rector of Munich University for the Reich Minister's permission to attend the inauguration of the Papal Academy of Sciences on 31 May and 1 June.[176] On 10 May 1937, the rector permitted him to accept his appointment and travel to Rome. At the same time he notified him that the Reich and Prussian Minister of Science, Education and People's Culture, the Foreign Ministry, the Foreign Organisation of the NSDAP (*Auslandsorganisation der NSDAP*) in Berlin and the office in Rome of the German Academic Exchange Service (*Deutscher Akademischer Austauschdienst*) had been informed of his visit. During his stay Carathéodory would have to contact the latter and also the relevant German Foreign Mission (*Auslandsvertretung*) that would support him in carrying out his programme and in analysing and evaluating his stay abroad. Carathéodory was advised by the rector to find the address of the German Foreign Mission in the *Handbuch für das Deutsche Reich* (Handbook for the German Reich, Berlin, Karl Heymann-Verlag) and to ask the German Foreign Misssion for the address of the office in Rome of the German Academic Exchange Service. For the necessary foreign exchange he had to turn to the German Central Office for Conferences (DKZ = *Deutsche Kongreßzentrale*) in Berlin, which had received a copy of the rector's "order" to Carathéodory. During his stay abroad, Carathéodory ought also to contact the local Foreign Organisation of the NSDAP which had also been informed of his trip. He would receive more precise information in this respect from the German Foreign Mission. A visit to the Foreign Organisation in Berlin before starting his trip was desirable, but that should not incur any costs for the public purse. In addition, Carathéodory had to submit a report of his journey in triplicate to the rector.[177]

All this was the usual procedure for a scientist travelling abroad at that time and this absolute surveillance and control did not only apply to Carathéodory.

However, on 25 May, the rector revoked his permission and did not allow Carathéodory the visit to Rome.[178] In the middle of 1937, Nazi authorities were instructed to collect information about the activity of the churches, their organisations, and their leaders and to report on the contents of the preaching and the reactions of the congregation. Carathéodory's participation in the Papal Academy was obviously no longer seen in a favourable light.

The two Academies, *Reale Accademia d'Italia* and *Pontificia Accademia delle Scienze* coexisted, with similar activities up to June 1939, when the fascist regime decided to combine them by decree, so that the Pontifical Academy and all its activities were absorbed into the *Reale Academia d'Italia*. Some of the original Lincei members were co-opted onto the new Academy. The only Academy to exist during World War II was the *Reale Accademia d'Italia*.[179]

On the occasion of the 10th anniversary of the Lateran Treaty, Pope Pius XI intended to issue an encyclical on Nazi racism and anti-Semitism which was never released. Part of its text is full of traditional Catholic hostility against the Jews and reproaches them for being themselves responsible for their fate. Another part refers

Carathéodory's completion of the staff questionnaire for the year 1936:
Carl Schurz Professor, Papal Academician and member of the Pontifical Academy
of Sciences. Courtesy of the Bavarian Academy of Sciences, Archive.

to spiritual risks arising from contact with the Jews as long as their hostility against Christianity lasts. After the Pope's death, on the eve of World War II, his successor Pius XII, alias Eugenio Pacelli, suppressed the encyclical. Pacelli was an Apostolic Nuncio in Munich and Berlin in 1917–1929, then cardinal permanent secretary and, finally, Pope from 1939 to 1958. Traditionally anti-Jewish, anti-modern and anti-communist, he had come to terms with national socialism through the Reich Concordat of 1933. He had a solid, commonly shared prejudice against the Jews. The Munich revolution of 1919 was for him a result of the Jewish atheistic world conspiracy. Communism posed for him a greater danger than national socialism. From the beginning of his career he had shown aversion to the Jews and his diplomatic activity in Germany contributed extensively to the dissolution of Catholic associations that would possibly be able to challenge Hitler's regime and thwart the "final solution". As Pope, he kept silent about the mass murder of European Jews, of which he had been informed since 1942, and also about the deportation of the Jews of Rome. Only in his Christmas address of 1942 might one recognise a hint of scepticism in his reference to the hundreds of thousands "destined to death or to progressing impoverishment without being themselves guilty, sometimes only because of their descent."[180]

5.29
Geometric Optics

5.29.1
The Book

Carathéodory's book on *Geometrische Optik* (Geometric Optics) was published as issue 5 of the 4th volume of *Ergebnisse der Mathematik und ihrer Grenzgebiete* by Neugebauer in Berlin in 1937. At that time, Neugebauer was working at the mathematical institute in Copenhagen and was supported by Harald Bohr and the Rockefeller Foundation. He addressed the following lines to Carathéodory on 3 September 1937:

many thanks for your last two cards. I have given the appendix on galley proof 55 to be printed; after all, your book has rather the character of a general account than a mere report on the literature.

The Damianos translation seemed to me excellent. It is indeed extremely peculiar to see how such semi-teleological principles appeared so early. Did Fermat know for instance this 'Ptolemaic'-Heronian optics?[181]

In his *Geometrische Optik*, Carathéodory, starting from Huygens' principle, instead of Fermat's principle, proved the mathematical equivalence of the two principles[182] with the help of Cauchy's theory of characteristics, thereby reversing the way which Hamilton had used. In Carathéodory's own words, his method had the advantage that the theorem of preservation of H. Poincaré and É. Cartan's integral invariants, which does not only substitute but, above all, completes the famous Malus theorem, arises almost naturally. Carathéodory treated the image of rays with the help of the eiconal and the coupling of individual line elements of optical spaces in different chapters in order to make clear that the two problems, which are usually treated together, are actually different.

Carathéodory explained that Fermat, in his attempt to establish the study of dioptrics in 1657, had the idea of applying a minimum principle similar to that used by Heron of Alexandria[183] for the treatment of reflection. In the case of reflection, the light ray remains in the same medium and it suffices to postulate that the usual length of the light path should be as short as possible. However, in the case of refraction, light passes through two different media and Fermat prescribes that the length, when evaluated in both media under consideration of their different specific gravities, should again produce a minimum. Heron's text *De Speculis* (On the Mirrors),[184] which was preserved only in a Latin translation of the 13th century, had for long been attributed to Ptolemy. It was established in the 19th century, however, that it is the work of Heron. The testimony of Damianos (4th century AD) is significant in this respect. In chapter 14 of his book *Main Themes of Optics* (Κεφάλαια των οπτικών υποθέσεων)[185] Damianos, also citing Heron,[186] reviews Heron's minimum principle and adds verbatim: "At the end of his [Heron's] proof he says: if nature does not lead the light of our eyes in vain, then it must reflect it in equal angles." Greek physicists believed that light comes from the eye and not from the object being viewed and therefore Damianos uses the expression "the light of our eyes".[187]

5.29.2

The Schmidt Mirror Telescope

After the end of the winter semester in Madison, Carathéodory took the opportunity to travel to the West Coast. On 6 February 1937 the Carathéodorys left Chicago for California, visiting Santa Fe and the Grand Canyon on the way and arriving in Los Angeles on 12 February. Carathéodory was anxious to meet Kármán in Pasadena and was also in contact with Hedrick.[188] At that time Kármán was the director of the Aeronautical Laboratory at the California Institute of Technology and engaged in research on the use of rockets to boost the performance of a conventional aircraft, especially for assisting its take off.

In the foreword to his booklet *Elementare Theorie des Spiegelteleskops von B. Schmidt* (Elementary Theory of B. Schmidt's Mirror Telescope),[189] Carathéodory noted that he had heard of the new telescope for the first time in the winter of 1937 during his visit to Pasadena. An instrument of this kind had just been set up in South California and W. Baade from the Mount-Wilson observatory had remarked to Carathéodory that someone ought to try to work out the theory of the new telescope on the basis of the spherical symmetry of its mirror. That was the opportunity for Carathéodory to produce his booklet in 1940.

Bernhard Schmidt's mirror telescope is based on the idea of eliminating the residual spherical aberration for light parallel to the axis by putting a corrector plate (also called a Schmidt plate) in the centre of a spherical mirror. A wide-field telescope was invented by Schmidt of Hamburg Observatory in 1930. Without any further optical element, the spherical aberration of the mirror would produce stellar images of poor quality: light falling near the edge of the mirror would be focused too close to

ELEMENTARE THEORIE DES SPIEGELTELESKOPS VON B. SCHMIDT

VON

C. CARATHÉODORY
IN MÜNCHEN

MIT 7 ABBILDUNGEN IM TEXT

1940

LEIPZIG UND BERLIN

VERLAG UND DRUCK VON B. G. TEUBNER

Carathéodory's booklet on the
Elementary Theory of B. Schmidt's Mirror Telescope.
Courtesy of the Bavarian Academy of Sciences, Archive.

it, and light falling near the centre of the mirror would be focused a little too far away. But the corrector plate is designed as a weak meniscus lens with one aspheric optical surface, so that it bends the rays precisely the amount needed to bring them from the various (radial) zones of the mirror to a single focus. The optical combination of lens and mirror creates an efficient telescope giving sharp stellar images of uniform quality over the full 5 degree field. For this reason, the Schmidt mirror telescope is used for astronomical observations.[190]

Following W. Baade's suggestion, Carathéodory worked out the theory of Schmidt's telescope and, in 1940, he published the theory of the off-axis aberrations of the Schmidt telescope, proving that they were symmetrical. To his surprise, he found not only that the equations defining the form of the correction plate could be established almost without calculations but also that the method created by Euler, Lagrange and Laplace for the establishment of celestial mechanics could be applied to the theory of the Schmidt telescope in a simple way. "In the first chapter of this small text", Carathéodory wrote in the foreword to his booklet, "the unperturbed problem is treated; in the second, the first-order perturbations are calculated. In this way, one obtains a system of formulae which enables the thorough calculation of rays from an arbitrarily large angle of incidence and thus masters the details of light transition through the Schmidt telescope with a completeness that can only rarely be achieved, even in the case of the simplest optical apparatus."

Carathéodory wrote to Herglotz on 16 January 1940: "I was able to calculate the B. Schmidt peculiar mirror telescope (the aberration of a spherical mirror is cancelled through a correction plate and the useful field of view amounts to 15°) in a very elementary way. The errors of the 3rd order are all zero and the aberration of the aperture due to the main error of the 5th order [...] is very simple to calculate."[191]

Eleven days later he wrote to Dehn: "I myself have dealt with the new telescope of Schmidt a few months ago; it is in many respects a peculiar instrument. I hope that this small thing will be published soon and I will send it to you then."[192]

Carathéodory lectured about the Schmidt telescope at Sommerfeld's colloquium on 31 May 1940.[193] Congratulating Carathéodory for the "newest booklet on B. Schmidt's mirror telescope" on 4 November 1940, the president of the German Union of Mathematicians expressed his admiration for Carathéodory's ability to "master the whole theory of this telescope with such elementary means."[194]

Carathéodory's booklet was reviewed in the annual report of the DMV in 1943. The reviewer also found that Carathéodory had obtained the theory of the instrument in an elementary way and his presentation was pleasant and could be easily read.[195]

Carathéodory continued his engagement with properties of optical intruments in reviews and articles, even throughout the war years.[196] Hondros supposes that Carathéodory's interest in the theory of optical instruments was stimulated by the construction of a large diameter parabolic mirror by the firm Zeiss. Carathéodory had told him that this mirror, whose axis was always and automatically turned towards the sun with a photoelectric arrangement, concentrated the radiation of the sun in such a small space that if someone touched its focus with the edge of a coin half the coin melted before its other edge got hot in his hand.[197]

5.29.3
Correspondence with the Imperial Chemical Industries
on the Schmidt Mirror Systems

Carathéodory's work on the Schmidt mirror telescope was taken up after the war by the Imperial Chemical Industries Ltd. (ICI) (Great Britain). On 19 May 1949, G. R. Petrie, from the Optical Development Department of the Plastics Division of ICI, wrote to Carathéodory to ask him for a copy of his "elementary theory of Schmidt's mirror telescope".[198] Carathéodory replied on 31 May that he would send him Sommerfeld's copy, since his own copies had been destroyed by fire at Teubner's "precincts" during the war, but the copy had to be returned directly to Sommerfeld after being photocopied.[199] Petrie informed Carathéodory on June 14 that he had sent the booklet back to Sommerfeld that day and hoped to correspond further with Carathéodory "on this most interesting subject".[200]

On 20 July Carathéodory gave Petrie details of his modified mathematical treatment of the Schmidt telescope.[201] Petrie replied on 3 August:

We have not yet had time to fully appreciate the advantages of this approach, but we are particularly interested because the analysis involves the radius of the image sphere. In our work we are interested in the use of this system for projecting television pictures with a magnification of about 8 times to a finite distance of about 1 metre and so far we have found analytical methods difficult to apply because all the rays leaving the corrector plate do so at angles which are varying with the distance from the axis. We are also interested in the effect of moving what we call the neutral zone nearer to the axis or further away from the axis, and moving the corrector plate away and closer to the mirror, and of aspherising the mirror itself. By a suitable combination of these variables we hope to be able to reduce greatly the small residual higher order aberrations. In order to get reasonable results in the past we have found it necessary (especially in the case of very large aperture systems of F/.8) to rely on trigonometrical ray tracing, and we have made every effort to find an analytical method, such as your own, which would simplify the work, but so far without success.

We hope this note will help you to understand the work we are doing, and if you have, yourself, studied systems in which the conjugate distances are finite we shall be very grateful to hear what results you obtained.[202]

Petrie wrote again to Carathéodory on 30 August:

In your first letter of August 11th you say that you would be afraid to aspherise the mirror in a Schmidt system because it so nearly satisfies Abbe's condition. We have done a great deal of computing with this type of system and have found that even at large apertures the trigonometrically calculated coma is invariably very small if the system is designed to satisfy Abbe's condition, but for off axis pencils the most serious defect is spherical aberration which becomes intolerably bad over 15° off the axis. We have therefore endeavoured to design systems in which a more favourable compromise has permitted a small amount of aberration to remain on the axis so that the spherical aberration off the axis has been considerably reduced. The magnitude of these aberrations is, of course, relatively small, and as I have already stated, we have not yet found an analytical method which could easily take account of these small quantities, but it may be that your own approach will enable us to obtain more accurate and useful results.[203]

In Carathéodory's papers at the Bavarian Academy of Sciences there are two of his manuscripts, one written at the beginning of August 1949, the other during 17–19 August 1949, which deal with Schmidt's mirror for objects at a finite distance. Numerous leaves of paper attached to those manuscripts contain calculations

IMPERIAL CHEMICAL INDUSTRIES LIMITED
PLASTICS DIVISION

Telephone: Welwyn 460

Telegrams: Iciplast, Telex, Welwyn Garden City

Our ref. Optical/GRP/JC

Your ref.

OPTICAL DEVELOPMENT DEPT.

THE HALL

WELWYN

HERTS

19th May, 1949

Professor C. Caratheodory,
Rauchstr. 8,
Munich,
Germany.

Dear Sir,

We are interested in the analytical theory
of Schmidt telescopes, and would like to obtain a
copy of your paper entitled, "Elementare Theorie
des Spiegelteleskops von B. Schmidt", in the
Hamburg Math. Einzelschr. 28 (1940). We have
written to Hamburg University, as we believe they
published this journal, but if you could help us
to secure a copy we should be most grateful.

Yours faithfully,
IMPERIAL CHEMICAL INDUSTRIES LIMITED
PLASTICS DIVISION

For Manager
Optical Development Department

I.C.I./1625/P/W 11.47 K71864

Letter of the Imperial Chemical Industries
asking for a copy of Carathéodory's booklet on Schmidt's mirror telescope.
Courtesy of the Bavarian Academy of Sciences, Archive.

of the coefficients of the associated power series. This is explained in a note to the paper *Der B. Schmidtsche Projektionsapparat* (The B. Schmidt Projection Apparatus),[204] which was prepared by Frank Löbell in Munich and was based on these two manuscripts. The paper obviously contains Carathéodory's painstaking work which was sent to the ICI and is indispensable to the understanding of a letter from the Optical Development Department of the Plastics Division of ICI to Carathéodory on 3 October 1949:

Dear Prof. Carathéodory,

Thank you for your letters of 30th August and 8th September.

We have studied in detail your letters dealing with Schmidt systems working at finite conjugates and agree with your method of arriving at the corrector plate equation bearing in mind, of course, the stipulations of equivalence of basically spherical and basically plane corrector plates and of adherence to the Sine Condition. Referring to your letter of 23rd August, however, we are sorry to have to point out that you made a slip in formulating equations 2.1, 2.2, 2.4, 2.5 and 2.7 as you assume in these that a and a' are measured from the centre of the mirror whereas to satisfy equations 1.1, 1.2 and 1.3 these distances must be measured from the pole of the mirror.

We followed through your working after having corrected these equations [...] and appreciate your kindness now even more than before. The radius of the mirror, R, was introduced as an extra check to keep the dimensions consistent through our working but we doubt whether the extra complication is worth while as the only satisfactory check is for two people to carry out the calculations independently. Therefore, before we actually apply the formulae to calculating a corrector plate we are forwarding our results so that you can compare them with those you get when you correct your equations. These results are attached on a separate sheet.

We shall be grateful if you will let us know whether you agree with the attached formulae.

Yours sincerely

G. R. Petrie [205]

The Optical Development Department of the Plastics Division of ICI wrote to Carathéodory again on 9 November 1949:

Dear Prof. Carathéodory,

I am sorry we have been so long in replying to your letters of 14th and 19th October, but Mr. Petrie has temporarily joined another department of ICI and at present I shall have to carry on this correspondence for him.

I can only explain our not appreciating the equivalence of the formulae

$$\frac{1}{x} + \frac{1}{x'} = \frac{1}{f} \quad \text{and} \quad \frac{1}{y} + \frac{1}{y'} = \frac{1}{f}$$

by the fact that we are so used to measuring the conjugates from the pole of the mirror that we assumed without further thought that you must have made a slip and I therefore want to send you our apologies.

We were most interested, too, to see the tests you gave for the consistency of the various formulae and also your development of $\sin \psi$ in terms of a and n.

We are very occupied on other projects just now but as soon as we can we shall calculate a Schmidt plate by your method and let you know how closely it agrees with one computed by the normal trigonometrical ray tracing and numerical integration method.

Yours very sincerely,

D. Davidson [206]

Two days later, on 11 November 1949 Carathéodory reported on the development of B. Schmidt's mirror systems at the session of the Mathematics-Natural Sciences Class of the Bavarian Academy of Sciences.[207]

5.30
Nazi Measures and Laws in 1937

The "German Law Concerning Civil Servants" (*Deutsches Beamtengesetz*) of 26 January 1937 allowed entry to the civil service only to persons of "German or related blood" and this had also to apply to wives. The discrimination therefore was extended to those persons described as "Jewish by marriage". Four of Carathéodory's colleagues were now affected by this law: Aloys Fischer was dismissed from his post at Munich and died on 23 November that year; Johann Oswald Müller was deprived of his authorisation to lecture at the University of Bonn and died three years later; the probationary professor at Munich higher schools and former doctoral student of Carathéodory, Rudolf Steuerwald, was sent into retirement; the classical scholar Rudolf Helm, a professor at the University of Rostock, was forced to retire in 1937.[208] Ten years later, he was again able to teach for a short time in Rostock and Greifswald until he moved to Berlin. Fischer's wife died in Theresienstadt concentration camp and his son was killed in the war.[209]

All "non-Aryans" were soon to be excluded from scientific institutions and academies and driven out of state organisations and associations. At a session of the Mathematics-Natural Sciences Class of the Bavarian Academy of Sciences on 3 April 1937, its secretary Heinrich Tietze spoke of the "incalculable damage that an action against either foreigners or native citizens (corresponding or full members) would have as a consequence". However, the question of who of the corresponding members were "non-Aryans" was put and Alfred Pringsheim, Richard Willstätter and Heinrich Liebmann decided to withdraw their participation once and for all.[210] It is true that academicians willingly fell into line with the supposedly legal measures of the government. In Einstein's view, they comprised a "group of intellectuals who, in front of nasty criminals, take the prone position and, to a certain degree, even sympathise with these criminals".[211] In March 1936 the Bavarian academy had not opposed the imposition of its new president by the Reich Ministry of Education. In November 1938 the "non-Aryan" members declared their enforced resignation. Through pressure of the executive committee the "Jewish by marriage" members left the academy "voluntarily". On 18 March 1940, the district leader of the Lecturers' Association (*Gaudozentenbundführer*), Dr. Otto Hörner, described Tietze as "an absolutely incorrigible reactionary for whom, even today, national socialism at the universities is unworthy of discussion"; he described Sommerfeld and Perron as belonging to a "small likewise reactionary clique who, just like Tietze, reject and sabotage every national socialist demand", also as academicians who try to shield themselves from national socialism with parliamentary-like tactics. Carathéodory's name is not, however, mentioned together with the others. In the same year, seven of the seventeen newly elected full members of the Mathematics-Natural Sciences

Class were Nazi Party members, two years later one of the seven elected full members belonged to the Nazi Party (the minister did not confirm Herglotz's election in 1942 as a corresponding member of the Bavarian Academy of Sciences), in the elections of 1943 and 1944 no Nazis were elected and at the end of the war, on 1 May 1945, six out of thirty-nine full members of the Mathematics-Natural Sciences Class remained Nazis. On 16 August 1945 the Academy members who had been elected before 1940 convened to confirm or revoke the results of elections carried during the war years. Following this, only two scientists who had belonged to the Nazi Party remained in the Mathematics-Natural Sciences Class.[212]

The NSDAP local section in Bogenhausen, the fashionable residential quarter of Munich where both Perron and Carathéodry lived, believed that "although a legal behaviour should be expected" from Perron, he was not "exactly a strongly loyal supporter of the movement." He was very religious but "nothing to his disadvantage" had furhter become known about him.[213]

At the end of 1937 the British journal *Nature*, in which Carathéodory's book on the calculus of variations had been reviewed, was banished from German libraries. *Nature* condemned national socialism as the most destructive totalitarianism against science: "Of all forms of checks on reason, that on the universities is the most dangerous and destructive, for it strikes many of the best equipped to determine social values." Further, the journal reported suppression of scientific literature, loss of the potential output of German scientists, the burning of the books, degradation of science and the subservience of German scientists to the engineers of a ruthless political machine. There was also a reference to Rust, exalting the "racial instinct" over the disinterested pursuit of scientific truth, in a speech at the University of Heidelberg on 30 June 1936.[214]

In 1937, and for the third time since 1933, questionnaires had to be completed by academics. The one completed by Carathéodory on 15 May 1937 shows that he did not belong to any political party and was only a member of the air-raid protection group.[215] This membership should not be considered as something unusual, since, in the late 1930s, as international tension increased, air-raid drills were ordered in Munich and "block wardens" were put in charge in each residential block.[216] On 7 July 1937, Dr. Joachim Pelekanidis, archimandrite and pastor of the chapel of the Royal Greek Legation in Berlin, confirmed in writing that Carathéodory's wife, Euphrosyne, was a "Greek citizen, of Greek Orthodox confession and Aryan origin", "for the purpose of submitting [the confirmation] to the relevant authorities". He also confirmed the same about her parents and grandparents.[217] It is remarkable that Greek authorities certified Greek Orthodoxy as the family's confession once again, whereas his religion in his personal file at today's federal archive of Germany is declared as Greek Catholic. This was the time when Rupert Mayer, a Jesuit father in Munich, turned with great decisiveness against Nazi violations of the Concordat and ignored the prohibitions on preaching.[218] It was also two months after Carathéodory was prevented from attending the inauguration of the Papal Academy of Sciences.

5.31
"The Wandering Jew"

Munich had about 765 000 inhabitants and was the fourth largest city of the German Reich in 1937. From 2 August 1935 it bore the title of Capital of the Movement (*Hauptstadt der Bewegung*), whereas earlier, on 15 October 1933, when the foundation-stone of the House of German Art (*Haus der Deutschen Kunst*) was laid, it was named the City of German Art (*Stadt der deutschen Kunst*). In 1937 Munich became ignominiously associated with two monumental exhibitions: on 18 July the House of German Art at the Prinzregentenstrasse was inaugurated with the first Great German Art Exhibition (*Große Deutsche Kunstausstellung*) by Hitler himself. Apart from shaping public opinion about contemporary art, "Bolshevist" artists, "Jewish" traders and "culture-corrupting" art historians, there was little to learn about form and content of the new "*völkisch* racial" art. The following day, in a run-down building in the *Hofgarten*, near the House of German Art, the disgraceful exhibition "Degenerate Art" (*Entartete Kunst*) opened with 650 works of the German avant-garde, many of which were confiscated by the Nazis and later sold abroad at high prices. On 8 November 1937 the library of the *Deutsches Museum*, also in Munich, presented the dreadful show "The Wandering Jew – Great Political Show" (*Der ewige Jude – Große politische Schau*, actually "The Eternal Jew – Great Political Show") with the intention of legitimising anti-Semitic Nazi politics.[219] Henry Morgenthau, the man to whom, as Carathéodory had written to Einstein, 1.5 million Greeks owed their lives, was counted among those influential Jews portrayed in the exhibition who, according to the Nazis, had promoted Germany's decline. Besides Morgenthau the exhibition included the names of Benjamin Disraeli and even Franklin Roosevelt. Disraeli was the British delegate at the Congress of Berlin, in which Carathéodory's father-in-law, helped by his father, represented the Ottoman Empire. Roosevelt was the American President when Carathéodory was lecturing in Madison, Wisconsin.

5.32
Graeco-German Relations Before the War

In 1937 a book bearing the title *Das neue Hellas* (New Greece),[220] and based almost exclusively on Greek literature, was published in Berlin. The introduction to the book, revealing a German interest in understanding current developments in that part of south-eastern Europe, reads:

The strong and lively bonds that connect us with the art, philosophy and history of ancient Hellas make the new Greece of our days – strangely – appear to us very often only as the heir and guardian of the ancient Greek tradition, and, even if antiquity might be a bridge to the present, it does not allow us to gain a clear-cut picture of present-day Hellas. Compared with our knowledge about the nature and character of Classical Greece, our knowledge of the newly risen Hellas is in many respects poor. However, we associate with it many diverse and significant relationships of the most varied kind, which even if they do not reach the great significance that Hellenism has gained for the cultural development of the European space, they still make the country on the edge between Orient and Occident in its present appearance no less interesting and remarkable for us. [...]

German penetration of Greek economy grew speedily in the 1930s. The economies of the two countries complemented each other. In 1937, Germany imported tobacco, raisins, fresh fruit, olive oil and other mainly agricultural products from Greece and exported hard coal, machines, medical, scientific and technical instruments and apparatus, chemical and pharmaceutical products and motor vehicles to Greece. Numerous German companies maintained permanent agencies in various branches of economy, as for example Siemens & Halske PLC and the Siemens–Schuckert works PLC in Athens, Thessaloniki and Volos. Exceptional contributions to the promotion of Graeco-German commerce were made by the German-Greek Chamber of Commerce in Athens and the German-Greek Association in Hamburg. Greece and Germany were also bound to each other by a clearing agreement of 16 August 1932 that was renewed with some changes on 4 March 1936.[221]

This penetration of the Greek economy however, did not bring an increased German political influence to bear on Greece, despite the fact that on the German side there was always a determination to show friendship towards Greece. And there was of course a very material reason for Greece's unconditional dependence on Great Britain, the strongest power in the Mediterranean. Greece herself had a fleet of world significance and, with 1.9 million tons gross, she occupied, after Italy and France, third place in the Mediterranean, and in fact was the world leader in 1938. The Greek trade fleet had always worked predominantly for British interests and Greek capital was mostly invested in England. At the end of October 1940 Greece joined in the war on the side of Great Britain against Italy.

Although Nazi ideology contained the idea of German-Greek relations rooted in a common descent, real relations between Germans and Greeks had started only in 1821 when German Philhellenes showed enthusiasm for the Greek War of Independence. A strong link between Greece and Bavaria developed following the decision of the Great Powers to install the Bavarian Otto von Wittelsbach as King of Greece in 1832, with the consequence of the settlement of thousands of Bavarians in Greece. Bavarian rule in Greece ended however with the revolution of 1843, leading to Otto's abdication in 1862. The second half of the 19th century was marked by the brilliant work of German scholars in Greece, most of all in archaeology and the natural sciences. In World War I, friendly relations between the two countries broke down, but in the inter-war years German nationals began again to take up positions within Greece and by 1940 there were some 3000 of them living there. Three groups of Germans played significant roles in the connection between Greece and Germany: businessmen, interested in the revival of trade, technicians, who massively participated in the quickly expanding Greek industry, and educators, who were invited to Greece more and more often as a consequence of the Greek preference for the German language.[222]

5.33
Archaeological Interest

In 1937 Carathéodory's involvement in Greek affairs was limited to his paper *About the Curvature of the Stylobate of the Parthenon and concerning its Intercolumnia* (Περί των καμπυλών του στυλοβάτου του Παρθενώνος και περί της αποστάσεως

ΠΕΡΙ ΤΩΝ ΚΑΜΠΥΛΩΝ ΤΟΥ ΣΤΥΛΟΒΑΤΟΥ

ΤΟΥ ΠΑΡΘΕΝΩΝΟΣ

ΚΑΙ ΠΕΡΙ ΤΗΣ ΑΠΟΣΤΑΣΕΩΣ ΤΩΝ ΚΙΟΝΩΝ ΑΥΤΟΥ

υπο

Κ. ΚΑΡΑΘΕΟΔΩΡΗ

ΑΝΑΤΥΠΩΣΙΣ ΕΚ ΤΟΥ ΠΑΝΗΓΥΡΙΚΟΥ ΤΟΜΟΥ

ΤΗΣ ΑΡΧΑΙΟΛΟΓΙΚΗΣ ΕΦΗΜΕΡΙΔΟΣ 1937

1938

Separate print of Carathéodory's paper
About the Curvature of the Stylobate of the Parthenon and concerning its Intercolumnia.
Courtesy of the Bavarian Academy of Sciences, Archive.

Parthenon curvature: (above) *platform and entablature;* (below) *temple platform.*
Plate 20 from Orlandos, Anastasios: Η Αρχιτεχτονική του Παρθενώνος
(The Architecture of the Parthenon) (Athens, Archaeological Society, 1986).

Parthenon, east side. Photograph: S. Georgiadis.

των κιόνων αυτού)[223] for the centennial celebration of the *Archaeological Paper* (*Αρχαιολογική Εφημερίς*). By a purely historical argument he explained why the curvature of the steps could be nothing else but a circle, the only known curve at the time of Iktinos, one of the architects of the Parthenon. Based on topographic measurements by Lampadarios, a professor of geodesy at the University of Athens, which were communicated to Carathéodory by the Acropolis restorer Balanos, he found that the curves of the eastern and western side of the stylobate could be represented with great precision by circles of 1850 m radius, and the curves of the northern and southern side by circles of three times that radius. By comparing his theoretical values for the distances between the columns with the measured distances and without taking into account the most ruined southern part of the temple, he came to the conclusion that the architects of the Parthenon had started from the middle of every side of the stylobate in order to determine the place of each column and, in doing so, they made small mistakes during the building of the temple.

The *Archaeological Paper* was published by the Greek Archaeological Society, one of the most significant institutions of the Greek state, owing its foundation in 1837 to private initiative, namely to the German archaeologist, Philhellene and professor at Athens University, Ludwig Ross, to the prominent Phanariot and Greek Minister of Education, Iakovos Rizos Neroulos, to the first secretary of the society and later professor of archaeology, Alexander Rizos-Rangavis and to Kyriakos Pittakis, who dedicated himself to the supervision of excavations and the editing of the *Archaeological Paper*. Carathéodory was himself a member of the Greek Archaeological Society.

German interest in Greek archaeology had also been expressly indicated by Hitler, who at the 11th Olympic Games in Berlin in 1936 announced that the German Reich would continue the excavations in Olympia.

5.34
A "Symbol" of German-Greek Contact

Apart from the archaeological paper referred to above, Carathéodory was not especially productive in this period. On 3 November 1937 he accepted an appointment to the jury of the Alfred Ackermann-Teubner Memorial Award for the Promotion of Mathematical Sciences (*Alfred-Ackermann-Teubner-Gedächtnispreis zur Förderung der Mathematischen Wissenschaften*). The purpose of the Memorial Award was to recognise work in analysis. The prize was an endowment of the University of Leipzig established by the engineer Alfred Ackermann-Teubner in 1912. The decision to invite Carathéodory to join the jury, together with Erhard Schmidt, had been taken at the 1937 conference of the DMV in Bad Kreuznach. The rector of Leipzig University proposed Koebe for chairmanship of the committee.[224] Further, in the name of the Bavarian Academy of Sciences, Carathéodory congratulated Wilhelm Wirtinger, a university professor in Vienna, on the 50th anniversary of his doctorate, but the relevant text was not published.[225] Carathéodory wrote that Wirtinger became known to mathematicians through the publication of his studies on theta functions in which Riemann function theory had been combined with ideas of Klein more than forty years earlier. The investigation of Abel functions and the general theta function could have occupied an entire life. But Wirtinger was not the kind of specialist who would engage with one problem alone. He had a feeling for what was essential in his science, studied various problems and was interested in the history of science and the philosophy of the foundations of mathematics. He combined geometric fantasy with the rare mastery of mathematical symbolism in his works.

Blaschke had gained his doctorate under Wirtinger in 1908. The executive committee of the Greek Mathematical Society asked the Greek Minister of Education for his approval to invite Blaschke, who had shown an interest in visiting Greece, to give a series of lectures to the society in the spring of 1938. Blaschke's invitation was indicated by the minister to be of great value, "expedient and also nationally useful". It seems that Blaschke was invited by Sakellariou and that Carathéodory had not been officially involved in the invitation. Blaschke's journey to Greece was preceded by a visit to Messina. Besides enjoying the warm Italian sun at the Hotel Royal, Blaschke, who openly expressed his approval of the *Anschluss*, earned his living by giving lectures in poor Italian. He intended to be in Athens from 28 March onwards and, from Messina, he provided Wilhelm Süss, the president of the DMV, with Sakellariou's address in Athens for contact.[226] Towards the end of March 1938, he told Süss that he would stay on in Athens until 16 April.[227] Behnke believes that Blaschke was confronted with mistrust both within Germany and abroad. He had acquaintances everywhere but nobody desired to be his friend. And therefore his effect as a German abroad was considerably limited.[228]

Blaschke gave two lectures at the University of Athens and two at the Technical University of Athens, all on various areas of geometry. He lectured three more times: on Leonardo da Vinci, on his own scientific impressions of the Far East, and on the contribution of Ancient and Modern Greece to the promotion of mathematics. This last lecture was held on 13 April 1938 and attended by various representatives of Greek authorities as well as the German envoy in Athens. Blaschke said that Greek and German mathematicians co-operated in the most satisfying way in the competition for the promotion of mathematics and that Carathéodory was a lively symbol of this co-operation. Blaschke was introduced by N. Hatzidakis in the first and final lecture and he was received very warmly by Athenian colleagues and authorities. A great part of his lectures was to be later published in Greek translation in Athens.[229]

As soon as he returned to Germany, Blaschke visited Carathéodory in Munich, on 21 April 1938, and invited him to the mathematical seminar of Hamburg University. Carathéodory accepted the invitation to give a lecture, for a fee of RM 100, and stayed in Hamburg from 16 to 21 May 1938. There, he met the American mathematician Morse who had been invited from Princeton to give three lectures at the seminar.[230] Morse's work on the calculus of variations in the large in the 1930s is counted among the most significant contributions to the subject since that of Kneser and Bolza.

From left to right: James Alexander, Albert Einstein, Frank Aydelotte, Oswald Veblen, Marston Morse. Late 1940s, photographer unknown. Courtesy of the Archives of the Institute for Advanced Study, Princeton.

5.35
Release from the Civil Service – Flexible in Surviving

On 7 April 1938, Carathéodory asked the rector of Munich University, professor Dr. Kölbl, to excuse his absence from that day's rally because he had to travel to Freiburg following the bereavement of a close relative.[231] It could well be that he was looking for an excuse to avoid an obligatory attendance at some event, since none of his relatives, either close or remote, had died in Freiburg at that time. But he may also have wished to contact DMV president Wilhelm Süss, a full professor at Freiburg University since 1934, to find out how he might be further engaged in mathematics after his retirement.

Wilhelm Süss, a geometer, and SA member from 1934 to 1937, became the president of the DMV in 1937 and joined the Nazi Party in 1938.[232] The union had elected him president and, at the same time, had voted for a change in the statutes which enabled his re-election until the end of the war. Süss became rector of Freiburg University in 1940 and kept the post until 1945. He was an extraordinarily skilful and successful person in matters of mathematical policies, with a pronounced civil servant mentality. He was on good terms with the Reich Minister of Education, Bernhard Rust, and acted in close co-operation with politicians concerned with matters of research during the war.[233]

Carathéodory belonged to the union, but had not had any particular involvement in its activities up to that time. From then on, he began to become more actively engaged and, on 9 June 1938, contacted Süss (in common with Nazi practice, now referred to as leader – *Führer* – of the DMV), whose role was that of mediator between the regime and German mathematicians.[234] In a letter to Süss, Carathéodory referred to his younger colleagues Grunsky, Unkelbach and Petersson: he stated that he was fond of Grunsky's works but had never met Grunsky personally, so was unable to say anything about how Grunsky lectured; Unkelbach at Munich was one of the younger people to be encouraged; Petersson worked on really bad mathematics under Hecke's influence. Carathéodory added that young people had to realise that concrete problems were more significant than general theories.[235]

Carathéodory's letter contained several inconsistencies. His judgement about Hecke is incomprehensible since, ten years earlier, he had seen in Hecke a guarantee of quality for the *Annalen*. Later, he had also proposed Hecke as his successor in Munich. Hecke, who had been to Princeton from 28 January to 18 May 1938 but was not permitted by the Reich ministry to visit Harvard, was known as a convinced adversary of national socialism.[236]

Carathéodory's alleged aversion to "general theories" and his favouring of "concrete problems" is odd, because it was he who had been reproached (as was the case during his appointment procedure in Hannover) for remoteness from practical things and finding satisfaction in pure mathematics. But in the pre-war years, utilitarianism was considered a virtue in Nazi ideology and an inclination towards "general theories", from which he now wished to distance himself, was something of which the Nazis accused the Jews. Helmut Grunsky habilitated in 1938 at Berlin University and he was employed afterwards as a scientific collaborator at the Foreign Ministry. He belonged to the group of very talented young mathematicians, who were not given regular posts at the mathematical institute of Berlin University by Bieberbach but worked for the *Jahrbuch über die Fortschritte der Mathematik*, which Bieberbach continued to edit on behalf of the Prussian Academy of Sciences in the Nazi era.

Hans Petersson, who had gained his doctorate under Hecke and habilitated at Hamburg University in 1929, held at that time an associate professorship there. He belonged to both the SA and the NSDAP.[237] Helmut Unkelbach had forced the dissolution of the Munich mathematical circle because of opposition to Pringsheim's participation.[238]

Carathéodory concluded his letter with the Nazi salutation, *Heil Hitler*, as was then prescribed when addressing official authorities. A refusal to use the salutation did not lead to automatic sanction with respect to the civil servant status, but it could cause the writer to be labelled as non-sympathetic to the regime.

It is obvious that, through Süss, Carathéodory was making preparations for his role after retirement and attempted by this letter to show himself loyal to the regime, the DMV and to Süss personally. This time, Carathéodory did not wait to be invited by a political personality to take on a duty in the service of education or educational policies, as had been the case with Venizelos's call to Carathéodory to come to the aid of the Greek government. Here, he himself took the initiative in directly approaching a person of power. No obvious reason forced him to act as he did. Even if he was determined to remain in Germany so as preserve his pension, he could have done so and remained inactive.

On 27 May 1938, according to the decree of 29 June 1935, the Bavarian Ministry of Education sent a proposal for the release from service of Bavarian academics born in the period between 2 April and 1 October 1873, together with a similar proposal for one academic who had asked for his release after completing his 60th year, to the Reich Ministry of Education. Postponement of the release from service was not requested for any case. The accompanying text concerning Carathéodory reads: "With regard to par. III of the implementation regulations of 12 July 1937 (Government Gazette I p. 771) concerning the appointment of civil servants and the ending of this permanent status, I establish that [...] the above professor for mathematics at the Faculty of Natural Sciences of the University of Munich, Privy Councillor Constantin Carathéodory, has proved his worth in the best manner in the long years of his academic activity. He did not belong to any political party. He had proved his and his wife's Aryan origin (both of them are of Greek descent). He had never been convicted of any offence and he had not belonged to any lodge."[239] This formulation was a circumlocution for stating the fact that Carathéodory had not been purged but was lawfully retired. (The Philosophical Faculty, Section II had been renamed as the Faculty of Natural Sciences.[240])

After a total of thirty-five and a half years, and with a salary corresponding to the group A1c of the Bavarian pay scales according to paragraph 2 of the law "Concerning the Special Legal Status of Teachers in Permanent Status at the Universities" (*Über die besonderen Rechtsverhältnisse der beamteten Lehrer an den wissenschaftlichen Hochschulen*), Carathéodory was released from service because he had reached the age limit.[241] From Berchtesgaden Hitler signed Carathéodory's release certificate on 12 August 1938 and in the name of the German people he acknowledged Carathéodory's successful academic effectiveness and his loyal services to the German people.[242]

A law of 21 January 1935 had set the retirement age at 65. In the case of special interests or needs, and with the permission of the Reich Ministry of Education, academics could retire later. According to a decree issued by the REM on 15 May 1935, the rector could decide whether an emeritus professor with suitable political views was allowed to continue lecturing or not, since retirement also meant the loss of the right to lecture. From 1938 such an arrangement was only possible with the consent of the Führer's deputy Rudolf Hess. It is a fact that Carathéodory continued lecturing after his retirement.

5.36
Honorary Professor of the University of Athens

On 20 October 1938, Carathéodory was informed in a letter from the Greek Minister of Education and Religious Affairs, K. Georgakopoulos (the former deputy minister at Metaxas's political office), that he had become honorary professor of the University of Athens by royal decree of 20 July 1938 published in the Government Gazette no. 128 on 13 September, namely on his 65th birthday. [243]

At that time, honorary professors became either those retired professors who had been lecturing for at least ten years at the university, or those personalities who were elected by the general assembly of professors for this honour and were appointed by decree of the Minister of Education and Religious Affairs. [244] Carathéodory had been proposed for the election by the Faculty of Natural Sciences and Mathematics of Athens University. At the 16th session on 20 May 1938, [245] the dean had asked for a debate concerning the award of the honorary professorship to Carathéodory. Ten days later, [246] the faculty unanimously voted that "the former full professor C. Carathéodory become an honorary professor." At the general assembly which nominated him, K. Maltezos, the dean of the faculty, and P. Zervos, N. Sakellariou, and D. Hondros spoke about his scientific work. That year Carathéodory was also elected honorary president of the Greek Mathematical Society. [247]

The Greek Mathematical Society chose two professors at American Universities, Aristotle Michal and Marston Morse, to be its representatives at the semi-centennial celebration of the American Mathematical Society to be held at Columbia University on 6–9 September 1938. [248] Morse's participation was possibly recommended by Carathéodory, who had met him in Hamburg in May. Birkhoff was designated as delegate of the DMV. He was very happy to receive that honour, "especially since there has always been a close bond between the two societies." [249]

5.37
The Fate of the Last Remaining Friends

The last step in the systematic deprivation of rights of all those considered to be Jews, in the racial sense of the term, was taken on 9 November 1938, on the so-called *Reichskristallnacht*. Two days later, Rust sent a telegraphic order to the rectors forbidding Jewish participation in lectures and Jewish entrance to the university. In addition, according to a decree of the Reich Ministry of Education in 1938, only Reich citizens were allowed to be members of the German Union of Mathematicians, and so its Jewish members were now obliged to leave the DMV.

Jews were not merely excluded from DMV; the Nazis attempted to eliminate them from the history of the union, as if they had never existed. Wilhelm Lorey, for example, was entrusted by the University of Jena with the writing of the history of mathematics at that institution from 1558. In 1940, he announced to E. Sperner, a member of the union's executive committee, the editor of the *Jahresbericht* and a full professor at the University of Königsberg, that the foundation of the union was to a considerable extent due to the "non-Aryan" Cantor and this fact could not be concealed.[250] With this in mind, Sperner tried to defame Lorey by informing Süss, Müller and Hasse that Lorey had been excluded from the Reich Literature Chamber (*Reichsschrifttumskammer*) because he was unreliable.[251]

On 22 June 1939, the deputy minister of the interior in Bavaria, State Counsellor Dr. Boepple, an old party member, reported to the Reich minister Rust that Erich von Drygalski ought to have left the Bavarian Academy of Sciences, because his wife was a "half-breed of first grade"; although von Drygalski was "Jewish-related", the Nazi Party had permitted him to remain at the Academy because of his world-reputation as a geographer. Von Drygalski had been an emeritus professor of geography since the spring of 1934. He was highly admired by Carathéodory.[252] Boepple informed the minister that von Drygalski had not participated in academy sessions since December 1938 and asked him whether or not he shared the NSDAP view on this matter.

The 28-year-old Curt Fulton (born Kurt Freudenthal) emigrated, with Carathéodory's help and mediation, from Munich to Bogota, Colombia where he lectured until 1946 before going to the USA.[253]

Aloys Wenzl, an associate professor of philosophy at Munich University since 1933, was deprived of his teaching authorisation at the age of 51. In 1946, he became a full professor and a member of the board of the Philosophical Institute at the same university.

The woman mathematician Toni Stern, Carathéodory's guest on 11 May 1926, was expelled from the DMV.

Patron of the arts, Alfred Pringsheim, after many years of harassment and negotiations, obtained permission to sell his precious majolica-collection in London on condition that he surrender the greatest part of the income to the Nazis. In addition, his house on Arcisstrasse had to be "sold" to the NSDAP. At the end of 1939, at the age of 89, he could at last emigrate with his wife to Zurich where he died one and a half years later.

In 1939 Richard Willstätter, Nobel-prize winner for chemistry in 1915, and a friend of the president of the Zionist world-organisation, Chaim Weizmann, also fled to Switzerland and died there three years later. Carathéodory's unshakeable belief in institutions is confirmed by a story from the Nazi period in Munich told by Willstätter in which he juxtaposed his two friends, von Drygalski and Carathéodory: A lady asked the two what that big building was which they had just left. The mathematician C. answered: "That is the university". Von D. corrected: "That was a university".[254] Both Willstätter and von Drygalski lived in Bogenhausen, in Carathéodory's neighbourhood.

The 62-year-old Werner Deetjen died in Weimar in 1939.

In the summer of 1938 Mussolini issued a "law" according to which all Jews who had immigrated to Italy since 1919 had to leave within six months. Austria's annexation in 1938, together with that of Bohemia and Moravia in 1939, provoked new waves of expulsion, especially in Vienna and Prague. In 1938, the 54-year-old

Philipp Frank was forced to leave the German University of Prague, where he had succeeded Einstein in 1912, and had worked as full professor and head of the Institute for Theoretical Physics since 1917.[255]

In 1939, the 53-year-old Paul Funk lost his full professorship at the Technical University of Prague where he had been full professor for the previous ten years.

The 46-year-old function theoretician Karl Löwner, also a full professor at the German University since 1934, who had contributed to Frank and von Mises' book together with Carathéodory, was expelled from Prague in 1939.[256]

Ferdinand von Lindemann, Carathéodory's predecessor in the chair of mathematics in Munich, died on 6 March 1939. Carathéodory referred to his colleague's mathematical performance at a session of the Department of Mathematics and Natural Sciences of the Bavarian Academy a year later.[257]

At that time Pauli, who feared that, in the event of a German occupation of Switzerland, he would be subject to intimidation and treated like a Jew, intended to flee to France and get arrested by the French so that he would be able to move on to the United States with the help of French colleagues.[258]

Debye, the director of the Kaiser Wilhelm Institute for Physics in Berlin, was ousted by the German authorities to free his institute for secret war research.[259] It was there that the German army established its main reactor research project.

5.38
Dispute about Carathéodory's Successor

Among the various state and party authorities concerned with university policies during the Nazi period, such as the REM, the NSDDB, the NSDStB, the DSt, the state ministries of education, etc., none was powerful enough to exercise a dominant influence and so able to create university policies corresponding to its own ideas alone. There was always conflict with respect to the areas of competence of these authorities.

5.38.1
The Persons Involved

The dispute over Carathéodory's successor lasted almost six years. It began with a list submitted by the faculty in 1938 and ended in 1944 with a list that was a compromise between mathematics professors and lecturers. Up to January 1941 faculty and lecturers could reach no agreement as regards the candidates. A strong involvement and interest in the occupation of the chair by a competent mathematician was shown by Carathéodory, Tietze and Perron, who opposed absurd procedures and partly senseless proposals mainly submitted by the astronomer Bruno Thüring, the astronomer and assistant professor of astrophysics at the University of Munich Wilhelm Führer, the botanists Ernst Bergdolt and Reinhard Orth, the geologist Karl Beurlen, the mathematics lecturer Hugo Dingler[260] and the holder of the Munich chair of theoretical physics Wilhelm Müller. The astronomer Wilhelm Rabe, a full professor since 1935 and director of the observatory of Munich University, although not very actively, supported the attempt of these persons to impose candidates with a Nazi ideology and inclined towards applied mathematics, whereas mathematics professors strove for the chair to be occupied by a candidate of Carathéodory's mathematical prestige.

The dean of the Faculty of Natural Sciences was the botanist Friedrich von Faber, a Nazi who gave great support to active Nazis like Thüring and Führer. With von Faber's appointment in 1936 the harmony between the dean and the majority of the faculty had been destroyed. The dean behaved in the way the leadership of the Lecturers' Corporation expected and sometimes ignored the faculty in his considerations. In April 1941 von Faber became the vice-rector and Wilhelm Müller took his place as the new dean. Müller had succeeded Sommerfeld on 1 December 1939. He was known as the representative of the so called "German physics" and his main qualification was his polemic booklet on *Judentum und Wissenschaft* (Jewry and Science), in which he sharply criticised the theory of relativity as a specific and typical Jewish phenomenon.[261] Perron had often come into conflict with him.

Walther Wüst held the rector's office from 1941 to 1945 and was the curator of the SS Ancestral Heritage (*SS-Ahnenerbe*), an organisation of the SS set up by Himmler with the aim of finding ontological evidence for the alleged superiority of the "Aryan Race" and thus "scientifically" establishing Nazi racial ideas. It therefore carried out research projects in prehistory and archaeology and also performed secret experiments, mostly on prisoners in concentration camps, during the later years of the war.

The persons who played the role of a buffer between Munich mathematics professors and lecturers, and were also greatly involved in the appointment policies, were the president of the German Union of Mathematicians, Süss, himself a candidate for Carathéodory's chair, Hasse, who was informed of developments by Wilhelm Führer, corresponded with Süss, named candidates and gave his opinions, and Blaschke, who attempted to avert Behnke's appointment and invited the candidates Strubecker on 4 March 1941 and Hopf on 17 July 1941[262] to lecture at the mathematical seminar of Hamburg University.

In the six years preceding the appointment of his successor, Carathéodory was active in proposing or blocking candidacies. For a whole year 1939–1940 he tried to persuade Herglotz to accept the appointment until his candidacy was overruled in November 1940. In addition, Carathéodory consulted with Hecke in 1942[263] and vehemently opposed Steck's candidacy and objected to Teichmüller's candidacy; with one exception in March 1941, when he was excluded from a session, Carathéodory participated together with Perron, Tietze, Rabe, Eduard Rüchardt (physics), Bergdolt and Müller in the appointment committee.

Carathéodory started lecturing again in the winter semester 1939–1940 after a one-year interval. On 25 May 1939, he asked von Faber for permission to give a four-hour lecture (Monday, Tuesday, Thursday, Friday 11–12 h) on potential theory in the winter semester following, as he had been asked to do by the Students' Corporation.[264] He gave course lectures during the following four years. In April 1941 Carathéodory was to be replaced in geometry courses by the emeritus Friedrich von Dalwigk,[265] a former professor of applied mathematics at Potsdam's geodesic institute. But von Dalwigk could not take on teaching because he was ill, and so a special teaching commission for geometry was arranged for Max Steck, an assistant professor at the Technical University of Munich, for whom Müller had received a favourable report from Vahlen. Steck lectured from 1942 to 1944.

5.38.2

The Lists Submitted

The first list, with Gustav Herglotz (Göttingen) and Bartel van der Waerden (Leipzig) *ex aequo* in the first place, followed by Carl Ludwig Siegel (Frankfurt) was proposed by the faculty and communicated to the rector by the dean, von Faber, on 15 July 1938. This list of names had to be abandoned, as we shall see, but meanwhile, Carathéodory was in contact with Herglotz.

Herglotz was reluctant to take up the appointment, anticipating difficulties in finding a house in Munich. Ten days after the outbreak of war Carathéodory tried to persuade him: "If you come to Munich soon and you do not possess a home, I would like to draw your attention to the villa of Frau Eidam, Rauchstrasse 5, opposite me, where my daughter was living with her husband two weeks ago. There are wonderful rooms and an excellent bathroom. If you write to me, I can ask whether the room is still free. In the past foreigners of every nationality have been living there. It is a sort of guest-house but much more similar to a private house. Bogenhausen is surely less endangered than the centre of the city, as well. I can think that it could be convenient for the first weeks."[266] On 22 September 1939 Carathéodory informed Herglotz that he had visited the dean, who promised him to move Berlin to accelerate the appointment procedure.[267] The Reich Ministry informed Herglotz on 29 September 1939 that his appointment was temporarily put aside.[268] On 19 May 1940 Carathéodory wrote to Herglotz that he had recently learned from Frau Hartogs that the ground floor flat of their house was rented by the military administration but she would prefer to give it to Herglotz. The period of notice in the tenancy agreement was three months, so it was probable that Herglotz would get the flat, if he decided to go to Munich. "But you should do that in any case; in Göttingen you have hardly any good friends anymore and at present it is surely better to live in a bigger city. It would be a great disappointment for us, if you could not decide" Carathéodory assured him and repeated his proposal for the guest-house.[269] On 25 May 1940 he advised Herglotz to proceed as follows: Herglotz ought to explain to the minister that he had been promised a house in September, now rented by the military administration, since his appointment had been stopped; that he could get the house, if it became free; finally, that he would accept the appointment, if he were able to get the house. Carathéodory thought that the minister would help Herglotz in this matter and that he himself could, in turn, move the dean to get the minister's support going. Carathéodory informed Herglotz further that the time schedule for the lectures could be arranged in such a way for him so as to avoid the overcrowding in the tram on his way to Munich. He also gave Herglotz information about the housing market emerging from confiscated property in Munich and reassured him that if new orders were issued, it would be a matter for the ministry to remove the official difficulties.[270] By use of the word 'official' he indicated that there were indeed no substantial difficulties standing in the way of Herglotz's appointment.

On 11 November 1940 the dean of the Faculty of Natural Sciences, von Faber, wrote to the rector that since Herglotz hesitated to accept his appointment in Munich, the faculty could no longer wait for the occupation of the chair and would therefore submit a new proposal.[271] Sixteen days later, the Reich minister told the Bavarian minister that Herglotz's appointment was no longer possible.

As for van der Waerden, life was extremely difficult. When Holland was attacked without declaration of war on 10 May 1940 and then occupied by Germany, many Dutch citizens resident in Germany were interned. Van der Waerden, a Dutch civilian, had to report to the police where it was explained to him that he would be considered an enemy because of his nationality but he would enjoy a privileged position, because of the German civil-servant oath he had sworn; he must stay in Leipzig and report daily to the police. However, on 22 May, he was told that the oath was not enough to diminish his status as a hostile foreigner and, consequently, he was interned. Behnke suspected that some "dear anonymous friend" had learned that van der Waerden had expressed some "inappropriate" remarks, the SS got knowledge of it and this provoked van der Waerden's bad treatment. Behnke asked Süss to help the Dutch mathematician: "We completely destroy mathematical life in Germany if we do not protect such people. I am too weak for it. But you should be able to achieve something in this case", he wrote to Süss on 24 May 1940.[272] Van der Waerden was released soon afterwards and he asked for permission to lecture again.[273] The permission was granted to him by telephone by the senior civil servant Dames on 11 June 1940.[274]

The third candidate, Carl Ludwig Siegel, had been Landau's doctoral student in 1920 and later Courant's assistant in Göttingen, and from 1922 a full professor for mathematics at the University of Frankfurt. He had, in fact, close contacts, not only with Dehn and Hellinger, but also with Paul Epstein. He knew about the attempts of the first two to emigrate and also the plans of the third to commit suicide. Siegel himself left Germany in 1940 for Norway and two days before the German invasion he fled for a self-imposed exile in America. With the help of Oswald Veblen and the Bohr brothers he arrived in the United States and began working at the Institute for Advanced Study in Princeton.[275]

The reason why the list of candidates was rejected was because of a report by Bruno Thüring made on 6 September 1938 on the political suitability of the candidates: Herglotz did not belong to the party, was not active in any of its sections and associated units and could neither be described as an opponent nor as a supporter of the regime. He was only a scientist who considered the service of international science to be his duty. Thüring doubted van der Waerden's honesty and believed in addition that the latter was decisively philo-Semitic and held anti-Semitism to be pointless. Van der Waerden was further a close contributor of the *Zentralblatt für Mathematik* and the *Mathematische Annalen* and, in Thüring's words, these journals were run by the Jew Neugebauer and the *Mathematische Annalen* was still publishing articles and obituaries of Jews. Thüring reported finally on van der Waerden's attitude towards national socialism: the Dutch did not show the will to understand the current political development in Germany and, therefore, he ought to be suspected.

Thüring concluded that van der Waerden belonged to the type of academic teacher not desired any more. Siegel was, according to Thüring, the type of person who could be characterised as a "mere scholar", too friendly towards the Jews and with friends among them. He had suggested that Ernst Hellinger and Max Dehn ought to be allowed to participate in the mathematical colloquium again.[276]

It now became necessary to prepare another list with the name of another candidate in addition to Herglotz, given that van der Waerden and Siegel had been ruled politically undesirable. Carathéodory, Perron and Tietze named Hecke. They based their proposal on Hecke's excellent performance in number theory and his successful lecture tour in the USA in the previous year.[277]

Meanwhile Thüring proposed the candidacy of Anton Huber in September 1938. Huber was a professor in Freiburg, Switzerland, who had already heard that he would be appointed to Vienna. He realised that the three Munich mathematics professors were using the prestige of Carathéodory's chair to offer an appointment to a significant mathematician, who, however, had no interest in national socialism and the Nazi Party.[278]

At von Faber's request, Karl Strubecker's name was put forward by Theodor Vahlen, Ludwig Bieberbach and Alfred Klose of Berlin but he was rejected by the mathematics professors in Munich early in 1939. Strubecker had been a lecturer at the University of Vienna since 1934, but changed to the Technical University there in 1938 and became an associate professor. For von Faber, Strubecker did not enjoy Herglotz's reputation but he was one of the most well-known younger German researchers in geometry. Von Faber's intention to add the applied mathematician Alfred Klose to the list failed, because Perron had characterised Klose as an especially naive and ignorant person and, consequently, would compromise the scientific reputation of the faculty.[279] Süss valued Strubecker especially because of his pleasant personal attitude but he did not find him suitable for Carathéodory's chair because of a certain narrow-mindedness, not combined with great success, which Strubecker had to overcome before he could be considered for an appointment in Munich.[280]

Finally, on 13 March 1939, a second list was submitted to the rector by von Faber, containing just the names of Herglotz and Strubecker.[281] It was not until 27 November 1940, as we have noted, that it became known that Herglotz's appointment would not be considered any more, although when Dean von Faber told the rector earlier, on 11 Novemeber 1940, that a new list would be submitted, he no longer mentioned the name of Strubecker.

On 24 January 1941, with Herglotz's appointment having failed, the three Munich mathematics professors proposed a third list with new candidates: (a) Heinrich Behnke, a professor in Münster, "one of the best specialists in the theory of analytic functions of several variables"; (b) Hellmuth Kneser, a professor in Tübingen, "a multifaceted and thorough researcher" with notable work on the theory of functions of several complex variables; (c) Franz Rellich, "a successful researcher, full of ideas", who became professor in Dresden the following year; (d) Herbert Seifert, a regular associate professor in Heidelberg whose work in topology qualified him for the Munich chair. All of them were considered by the Munich professors as being not as significant as the earlier proposed candidates, but still entirely appropriate candidates for a full professorship in Munich.[282]

On 27 January 1941, Müller recommended Max Steck to be added to the list, mainly because of Steck's contrary attitude towards Einstein. Carathéodory considered him to be a completely incompetent mathematician.[283] Steck's candidacy was strongly supported by Thüring. Süss, on the other hand, had equally strongly refused to support Steck as a candidate, even for less significant chairs, and called him "a figure of the fifth order".[284] Süss believed that Müller, Dingler and Thüring, whose group was occasionally joined by the astronomer Vogt in Heidelberg, and Führer, then with a post at the REM which allowed him considerable influence on academic life, regularly promoted persons like Steck.[285] "If one would really wanted to hang something on this Munich group, the main thing to do would be to first of all make Dingler and Müller explain themselves", Süss wrote to Kneser. Dingler suddenly appeared as a "Jack-in-the-box" [*Stehaufmännchen*] in the procedure without the Munich faculty having either any previous knowledge or having been consulted. This happened after the forestry students had rejected Dingler's teaching commission in mathematics for forestry students. Dingler was arbitrarily declaring himself the present holder of a chair in the philosophy of the natural sciences at the mathematics faculty. As for Müller, he had just had his "rubbish" on Jewish and German physics published.[286] Süss believed that this group was using every sort of intrigue to impose Steck as Carathéodory's successor.[287] Even at the end of May 1941, and without the knowledge of the faculty, Müller was still trying to obtain academic expert reports in support of Steck's candidacy.[288]

Wilhelm Führer, now having connections within the Bavarian Ministry of Education, informed Steck that he was no longer considered as a candidate for Carathéodory's chair. While Süss considered this to be pleasing, he found the fact that an adviser in the REM corresponded directly with Steck about this matter to be strange. Süss had received overwhelming material about Steck from Sperner and Bachmann.[289] An NSDAP district commissioner had produced a devastating judgement about his former comrade Steck resulting in the strong advice to encourage him to give up his career as a lecturer.[290] Later, in a letter written to the military government in 1945, Blaschke declared: "When professor Carathéodory was forced into retirement (in about 1940?) and the incapable "Dozent" (lecturer) Steck was to become his successor, I wrote a very direct letter to the Ministry of Education in Berlin (among others: "Want of knowledge in algebra is not sufficient to guarantee knowledge of geometry"). After this letter Mr. Steck was told by the Ministry that he could not count on becoming professor Carathéodory's successor. Since that time I was no longer – or at least very seldom – asked for my advice by the Ministry of Education."[291]

On 22 February 1941, von Faber asked professors Wirtinger in Vienna, Konrad Knopp in Tübingen and Otto Haupt in Erlangen to submit reports on the candidates of the third list. Wirtinger seems to have favoured Kneser. Knopp praised all persons on the list, but wrote that also other mathematicians could equally be considered for Carathéodory's chair. Haupt expressed a positive judgement about all four candidates without, however, recommending anyone in particular.[292]

Süss was ready "to realise the necessity – presented by Thüring as the reason for Steck's candidacy – of appointing, alongside Perron and other 'reactionary' members of the faculty, a politically flawless man in Munich." Since this man had to be a geometer, Süss named Hellmuth Kneser, who was "a party member and an SA man".[293]

Asked by Müller, the new dean, on 2 July 1941, Paul Koebe (Leipzig) proposed the university professors Eberhard Hopf (Leipzig), Konrad Knopp (Tübingen), Gustav Doetsch (Freiburg) and Erich Kähler (Königsberg, but at the time in the navy) as possible candidates for Carathéodory's chair.[294]

In the same month, Hasse suggested other names: Kneser, Georg Nöbeling, an associate professor at Erlangen University since 1940, Kähler, Friedrich Lösch, a full professor at Rostock University since 1939, Wolfgang Krull, a full professor at Bonn University since 1938, and Georg Aumann. He wrote to Süss: "Dr. F[ührer] will travel to Munich in the near future in order to thoroughly discuss the matter with the authorities in question, since he fears that an acceptable proposal could not be reached in any other way. When I also named Aumann in this respect, Dr. F. waved it aside, because there was something racially not in order with the wife."[295]

In fact, Georg Aumann was an associate professor in Frankfurt am Main from 1936 to 1946. The racial descent of his wife was also the reason why Aumann, despite the fact that he was a party member since 1937, had not been appointed professor in Erlangen in 1939, and in Posen in 1942–1943; it was also the reason he had not received promotion in the Wehrmacht, where he belonged to the anti-aircraft artillery from 1939 to 1941, and to the code department from 1941 to 1945.[296]

Führer proposed Oswald Teichmüller and Hellmuth Kneser in a letter to von Faber, whom he still believed to be the dean, on 1 August 1941. Führer asked von Faber to cross Steck off the list and include the two new candidates instead. Führer recommended Teichmüller as an old Nazi, free from any kind of authoritarian behaviour, and Kneser as a party comrade and an active SA member.[297] Hasse believed that Teichmüller "would be convenient to the people of the Lecturers' Association and at the same time find the approval of seriously thinking people. Through the direction of his work he [...] was particularly suitable to become Carathéodory's successor."[298] Oswald Teichmüller, a militant party and SA member but a brilliant mathematician, had led the students' boycott against Landau in Göttingen under the pretext of the racial incompatibility of students and their lecturer. He volunteered for the Eastern Front in 1943 and was killed there.

On 3 November, Müller sent an intermediate report to the rector, in which he mentioned the names of Kneser, Teichmüller and Behnke. On 19 November 1941, the appointment committee met, consisting of Müller, Perron, Tietze, Carathéodory, Rabe, Rüchardt, von Faber and Bergdolt. The next day, Müller sent a report to the rector containing the third list with the names of Hecke, van der Waerden, H. Kneser, Behnke, in which also Oswald Teichmüller was mentioned as recommended by a person of renown from outside the faculty. Müller remarked that the first two candidates had to be rejected because of their political attitude, whereas the third was favoured by his membership in the SA and his positive attitude towards national socialism. Koebe's positive opinion about Teichmüller was mentioned together with the negative opinion of the Munich mathematicians, and especially that of Carathéodory. Carathéodory complained in his report about the imperfect reasoning in a treatise of Teichmüller.[299] Also on 13 November 1941 Carathéodory wrote to Süss that Teichmüller was not sufficiently mature for one of the main posts at a great German university and that the dean had rejected the names of the other candidates, except for Teichmüller, Behnke and Kneser.[300] In the meantime, Behnke learned from the dean that he had been put on the list of candidates and was therefore tremendously pleased, since he considered his listing as the greatest acknowledgement he could receive for his work. Of course, he had no expectation of achieving the appointment and assumed that the affirmation of his qualities by the Munich faculty would be the maximum he could achieve.[301]

Müller asked the rector to consider Kneser and Teichmüller for the occupation of the chair and distanced himself from the mathematics professors, who wished to include Hecke and van der Waerden in the list. On 3 December 1941, he proposed the redesignation of Carathéodory's chair to a chair for applied mathematics and named Robert Sauer, a full professor of applied mathematics and descriptive geometry at the Technical University of Aachen since 1937,[302] Walter Tollmien, a full professor at the Technical University of Dresden since 1938,[303] and Friedrich-Adolf Willers, a senior teacher in Erfurt since 1938, as appropriate mathematicians to occupy it.[304]

In March 1942, in a letter to the rector, the Lecturers' Corporation supported Teichmüller's candidacy putting Kneser in the second place. In a senate session that same month it was revealed that the head of the lecturers had altered documents in Teichmüller's favour, so Teichmüller was crossed off the list. From the faculty, Rector Walther Wüst demanded a new list of three candidates excluding both Teichmüller and "politically unreliable" candidates.[305]

On 12 June 1942 Carathéodory, Tietze and Perron proposed Behnke and Kneser together in the first position on the list of candidates, Wilhelm Süss, the rector of Freiburg university, in the second place and Eberhard Hopf, since 1942 on leave from Leipzig University and a conscript at the German Research Institution for Gliding (DFS = *Deutsche Forschungsanstalt für Segelflug*) in Ainring near Freilassing, Upper Bavaria, in the third place.

In the middle of July 1942, the appointment committee came together with the participation of Carathéodory, Tietze, Perron, Rabe, Beurlen (soon to become the new dean) and Orth. Beurlen and Orth were representing Bergdolt and von Faber. The compromise solution was a fourth list with Kneser, Süss and Hopf's candidacies, which was submitted to the rectorate and, through the Bavarian ministry, to the Reich Ministry of Education at the end of July.[306] Then, the Berlin adviser at the REM, Dr. Führer, was sent to the front and thus another cornerstone of the Müller–Steck group disappeared, so that even Süss started to hope for an "objective" settlement of the Munich affair. However, as he wrote to Tietze, news from Perron concerning this matter had made him deeply ashamed and took him by surprise.[307] In his letter, Süss did not specify what he meant by Perron's news, but he could have been referring to Behnke's candidacy.

What happened with Behnke's candidacy can only be assumed from Perron's letter to Blaschke on 12 December 1945. Referring to the events of 1942, Perron wrote that Carathéodory, Tietze and himself had thought of Behnke as a candidate, after their first list with the names of van der Waerden, Hecke and Siegel had been rejected for political reasons; they all valued Behnke highly, not only as a scholar and member of the editorial board of the *Annalen*, but also as a man of character and stature who could be used to counterbalance the influence of Wilhelm Müller and other "vicious" Nazis. But the lecturers, although having initially accepted Behnke as a possible candidate, suddenly raised objections. Without revealing their sources, they accused Behnke of having kept his child from his first marriage with him, whereas he should have turned that "Jewish bastard" out of his house. Perron referred also to Blaschke's role in Behnke's candidacy: Blaschke had ironically mentioned in a letter that, if it were true that Munich mathematicians were thinking of Behnke as Carathéodory's successor, he would not like to miss the opportunity to congratulate the faculty for that worthy supplement to the Müller–Steck circle. Perron and his colleagues had been extremely alarmed about that remark, however they could find nothing hinting at Behnke's affinity to the Müller–Steck attitude of mind.[308] Behnke later accused Blaschke[309] of having blocked his appointment at Munich University in a letter, of which Süss is supposed to have said, "one could only pick it up with pliers."[310]

The lecturers reported to the rector on 28 July 1942 that they did not object to the list proposed by the Faculty of Natural Sciences; that Kneser, who belonged to the SA, had always supported the aims and demands of national socialism; that Süss, who could be characterised as an excellent mathematician, stood on the soil of national socialism and was politically reliable; that Hopf could be factually characterised as very good and nothing was known to his disadvantage. This estimate was followed by a report on the qualifications, curriculum vitae, and mathematical performance of each candidate.[311]

Kneser's suitability as a candidate for various vacancies in mathematical chairs was the subject of discussion between Süss and the Berlin adviser at the REM, a certain Dr. Fischer.[312] Although Süss was very pleased by the trust placed in him, he found that Kneser was by far the best candidate for Munich because of the latter's productive abilities and the depth and extension of his knowledge.[313] But Kneser rejected the offer in November 1942 and Süss himself, who considered the occupation of Carathéodory's chair as an honourable but difficult affair,[314] did the same in February 1943. Feigl congratulated him in December 1942 for the "exceptionally honourable offer of appointment" as Carathéodory's successor in Munich stressing that the decision would surely not be easy to take.[315]

5.38.3
The Successful Candidate

In July 1943, Hopf declared himself willing to accept the appointment and, at the same time, asked for the creation of an additional chair for applied mathematics. Finally, he was appointed as full professor on 1 April 1944, but later went to the USA. He became known through his significant contribution to ergodic theory, topological dynamics, partial differential equations and the calculus of variations. In Munich he held lectures on differential and integral calculus. Fleming values Hopf as "a truly worthy successor to Carathéodory in Munich. It is unfortunate for German mathematics, but fortunate for US mathematics that Hopf remained in Munich only 4 years after his appointment in 1944."[316]

In the case of Hopf, scientific qualification was set above considerations of political liability. Hopf was neither a party member nor known for his sympathy for the Nazis.[317] He had taken up an appointment in Leipzig in 1936, succeeding the dismissed Leon Lichtenstein, and left his position of assistant professor at the MIT in Boston. In the opinion of Norbert Wiener, who was Hopf's colleague at the MIT and a cousin of Lichtenstein, Hopf's opposition to Hitler and sympathy for the victims of the regime were expressed initially but, later, strong family influence dragged him to the Nazi side. In the eyes of the emigrants, Hopf belonged to that part of an element in Germany, which would at least build an acceptable foundation for the re-establishment of academic health after the war.[318]

So it may be assumed that Carathéodory's successor was expected to play exactly the same role as Carathéodory himself. Except that Carathéodory was then dreaming of his return to Greece. In the middle of November 1943, he wrote to Kalitsounakis: "I long for Athens and Greece and I hope to come there after the end of the war. Of course, literature and science will then be reborn and I hope that we will see the eve of this rebirth."[319]

5.39
Despina's Wedding

On 20 February 1938 Artur Rosenthal wrote to Carathéodory:

I was very very happy about both of your letters, particularly because I have heard nothing directly from you since our meeting in Oslo. I was very pleased to read in your letter of the 16th that you again enjoyed your stay in America so much. But above all, I would like to express my most hearty congratulations to you and your dear family for the marriage of your daughter to Herr Scutaris. I can imagine that moving to such a faraway country was not a quite easy decision; but, presumably, the young pair will time and time again come to Europe and be able to otherwise give you the chance of seeing each other again.[320]

Despina, Carathéodory's daughter, had married Theodoros Scutaris, a civil engineer with a diploma from the ETH and the owner of a sisal plantation in the British colony of Tanganika. Despina did indeed visit Munich frequently and found herself in her father's house, together with her one-year old baby son, one day before the war broke out. She left Munich for Switzerland together with her husband and baby on 31 August 1939 having been informed by a Wehrmacht officer that Germany was going to invade Poland the following day. She managed to smuggle out of Munich a painting rolled up in a thermos flask, with her mother's golden bracelet from Constantinople hidden in its bottom. This painting, *The struggle of Tobias with the fish*, had been bought by her grandfather Stephanos from the Galerie du Vicomte Du Bus de Gisignies in Brussels and was obviously an abridged copy of the right panel of the triptych *The Miraculous Draught* painted by Rubens for the

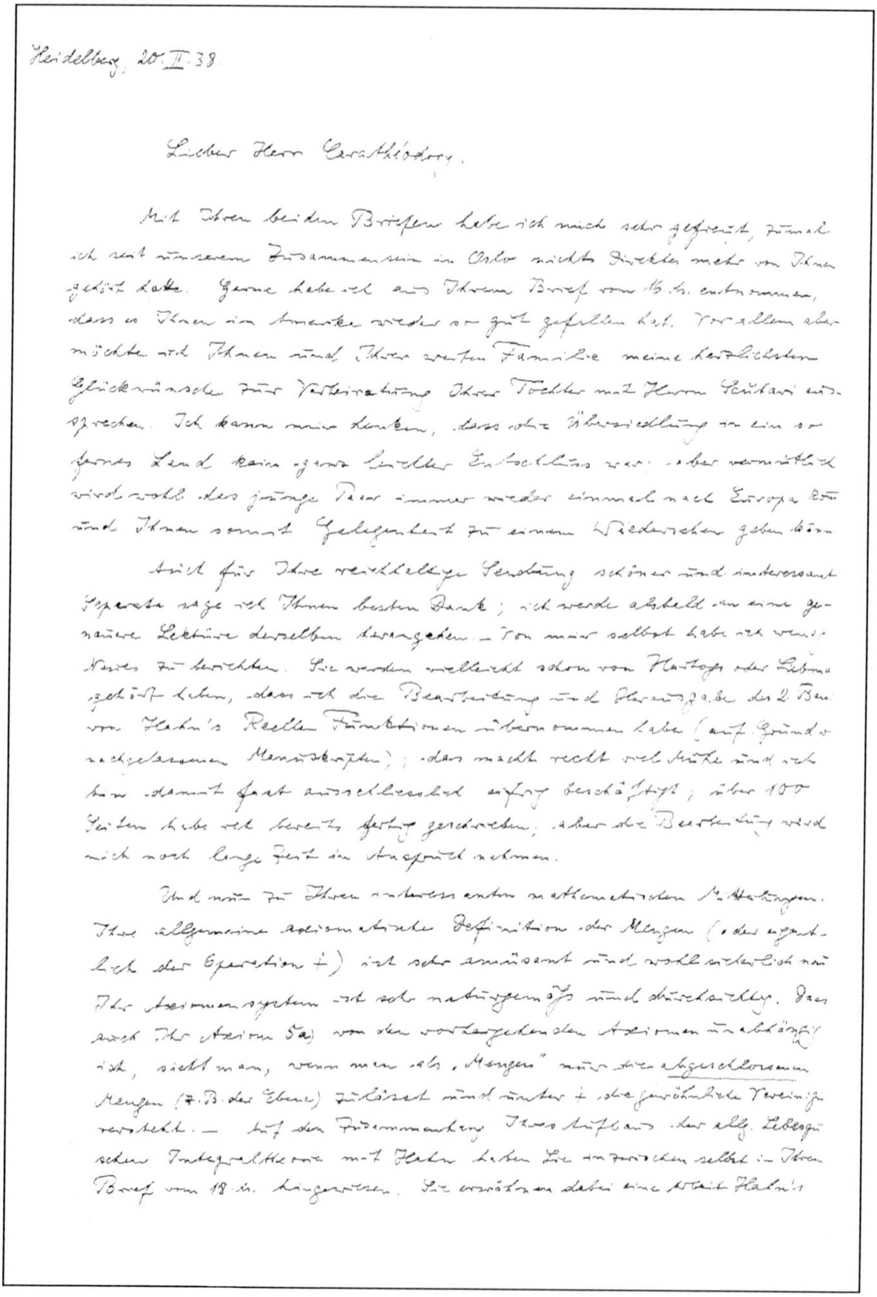

Heidelberg, 20.II.38.

Lieber Herr Carathéodory,

Mit Ihren beiden Briefen habe ich mich sehr gefreut, zumal ich seit unserem Zusammensein in Oslo nichts direktes mehr von Ihnen gehört hatte. Gerne habe ich aus Ihrem Brief vom 15.2. entnommen, dass es Ihnen in Amerika wieder so gut gefallen hat. Vor allem aber möchte ich Ihnen und Ihrer werten Familie meine herzlichsten Glückwünsche zur Verheiratung Ihrer Tochter mit Herrn Scutari aussprechen. Ich kann mir denken, dass die Übersiedlung in ein so fernes Land kein ganz leichter Entschluss war; aber vermutlich wird wohl das junge Paar immer wieder einmal nach Europa kommen und Ihnen somit Gelegenheit zu einem Wiedersehen geben können.

Auch für Ihre reichhaltige Sendung schöner und interessanter Separata sage ich Ihnen besten Dank; ich werde alsbald an eine genauere Lektüre derselben herangehen. – Von mir selbst habe ich wenig Neues zu berichten. Sie werden vielleicht schon von Hartogs oder Liebmann gehört haben, dass ich die Bearbeitung und Herausgabe des 2. Bandes von Hahn's Reellen Funktionen übernommen habe (auf Grund der nachgelassenen Manuskripte); das macht recht viel Mühe und ich bin damit fast ausschliesslich eifrig beschäftigt; über 100 Seiten habe ich bereits fertig geschrieben, aber die Bearbeitung wird mich noch lange Zeit in Anspruch nehmen.

Und nun zu Ihren interessanten mathematischen Mitteilungen. Ihre allgemeine axiomatische Definition der Mengen (oder eigentlich der Operation +) ist sehr amüsant und wohl wesentlich neu. Ihr Axiomensystem ist sehr naturgemäss und durchsichtig. Dass auch Ihr Axiom 5a) von den vorhergehenden Axiomen unabhängig ist, sieht man, wenn man als „Mengen" nur die abgeschlossenen Mengen (z.B. der Ebene) zulässt und unter + die gewöhnliche Vereinigung versteht. – Auf den Zusammenhang Ihres Aufbaus der allg. Lebesgueschen Integraltheorie mit Hahn haben Sie inzwischen selbst in Ihrem Brief vom 18.2. hingewiesen. Sie erwähnen dabei eine Arbeit Hahn's

First page of A. Rosenthal's letter to Carathéodory with congratulations on Despina's wedding and with comments on Carathéodory's paper Entwurf für eine Algebraisierung des Integralbegriffs *(see 2.17.3). Courtesy of the Bavarian Academy of Sciences, Archive.*

Despina's wedding at Kiphisia, December 1937.
Courtesy of Mrs. Rodopoulos-Carathéodory.

Church of Our Lady across the River Dyle at Mechelen in 1618–1619, which still exists there. From Switzerland Despina travelled on to Athens where she remained until the German invasion of Greece in April 1941. During the Graeco-Italian war that preceded the German invasion, she cared for injured Greek soldiers in various hospitals in the capital, driven there by the surgeon Marinos Geroulanos, Metaxas's personal doctor and the father-in-law of Ioulia's daughter, Despina Streit. As the Germans were marching to Greece, Despina was able to flee the country and make her escape to East Africa. There she managed her husband's plantation and arranged the transportation of the sisal, via the Tanganika railway built by Holtzmann, and on to Bristol for finishing. Despina's home was a villa in Dar-es-Salam overlooking the Indian Ocean[321] near the villa of the third Aga Khan, Ali Khan, the spiritual leader of the Muslim religious community of Ismaelites, and his wife, Begum Khan. Her eventual decision to give up the plantation and return to Europe may well have been connected with the so-called Mau Mau uprising in Kenya against British colonial rule in the 1950s, which was a potential threat to the lives of all Europeans in East Africa.

The miraculous draught: Rubens, 1618. Right panel: The struggle of Tobias with the fish.
Courtesy of Mrs. Lieve Lettany, Museum at Mechelen, Belgium.

5.40

Two Trips Cancelled Because of the War

Just before and during the winter term of 1939–1940, Carathéodory tried unsuccessfully to travel out of Germany. On 13 June 1939 he asked the rector for permission to accept an invitation from the Institute of Higher Studies of Belgium (*Institut des hautes études de Belgique*) to give four lectures there in the winter. He supported his request with the argument that the Institute was well-known and had been established for almost half a century and also that famous German university professors had given talks there the previous winter. He attached the letter of invitation and the programme of the Institute for the year 1938–1939 and signed it with the Nazi salutation.[322] Carathéodory's request was supported by the dean, von Faber.[323]

On 7 February 1940, Bergdolt, head of the lecturers, asked the rector of Munich University to postpone an exit permit to Carathéodory "until a victorious peace settlement", because he "was unable to guarantee the political reliability of professor Carathéodory."[324] However, a month later, the Minister of Education and the Foreign Minister permitted Carathéodory to give his lectures at the Belgian institute. His exit permit was bound to a series of formalities and conditions similar to those required earlier for his permission to travel to Rome. This time Carathéodory had to submit his passport to the Foreign Ministry and his speech in its exact wording if that was of a political nature. Speeches dealing with the philosophical-religious confrontation in Germany were not allowed. Presence at receptions of a foreign head of state or a minister of a foreign state had to be avoided in every way.[325]

In the meantime, the war had broken out with the invasion of Poland on 1 September 1939. The same day all universities closed down. Together with the universities of Berlin, Leipzig and Jena, the University of Munich was among the first to reopen ten days later. Many of the male students were immediately conscripted, although a few volunteered for the war.[326] On 10 May 1940, Belgium was attacked by the Germans without declaration of war and two weeks later its King, Leopold III, capitulated. In his obituary of Carathéodory, Perron remarks that the trip to Brussels was denied to Carathéodory,[327] which was incorrect. The truth was that Carathéodory was trapped inside Germany because of the war.

Carathéodory had also been given permission by the Reich minister to participate in the IX Volta–Congress that was to take place in Rome from 22 to 28 October 1939. Other delegation members were Blaschke (Hamburg), Wirtinger (Vienna) and G. Doetsch (Freiburg), while Hasse (Göttingen) had been chosen as delegation leader.[328] Blaschke, fearing that appointment by the ministry of Bieberbach (Berlin) would cause trouble among German participants, and also between Germans and Italians, had attempted to recommend Vahlen or Schmidt as delegation leaders, although he himself did not object to Bieberbach's participation in the congress.[329] On 20 August 1939, Carathéodory expressed to Hasse the wish to have with him the equivalent of about RM 250, i.e. about 2000 liras, since his wife would accompany him to Rome.[330] He intended to talk about the theory of measure and integral there but, due to the war, the congress did not take place.[331]

5.41
Decline in Quality

At the beginning of 1940 university studies were intensified by the introduction of trimesters instead of semesters, however only Perron, Tietze, Carathéodory and the actuarial mathematician Friedrich Böhm were still offering lectures in Munich. This measure was cancelled in the summer of 1941 in favour of a return to the traditional two semester pattern.[332]

During the winter 1940–1941, Süss, as president of the German Union of Mathematicians, began to collect evidence of the inadequate state of studies and the decline in performance in mathematics, in order to submit a memorandum, accompanied by proposals for improvement, to the Reich Ministry of Education. The ministry did not intend to produce a syllabus for mathematics and physics but Süss thought it would be appropriate for the DMV to work out a syllabus that would be recommended to his colleagues. Training for the teaching profession had to be cut down to eight semesters.[333]

Despite the war, and the resulting decline in the quality of study, dissertations were still being submitted and doctorates were awarded. Carathéodory for instance reported on the contents of the six loosely connected parts of a thesis by Artur Bischof[334] concluding that, because of the author's conscription, the work had to be hastily written and was therefore sketchily written and consequently was not easily readable.

5.42
Carathéodory and the Cartan Family – Germany Occupies France

In March 1939, in Paris, a group of friends, colleagues, students and scholars formed an international committee to organise the celebration of the 70th birthday of the prominent and world-respected French number theorist, topologist and geometer Élie Cartan. Cartan was a member of the French Academy and a professor of mathematics at the Sorbonne. The committee intended to publish a volume of some 250 to 300 pages of his scientific works followed by a selection of reminiscences, and also to organise a party for the jubilee on 8 May. Furthermore, all the issues of the *Journal de Mathématiques* for one year would be dedicated to Cartan. Carathéodory, Blaschke, Erhard Schmidt, Siegel and van der Waerden were the German participants on that committee.[335] Cartan was honoured with a celebratory symposium which praised his many mathematical achievements. Although he retired from the Sorbonne in 1940, he remained an honorary professor there until his death in 1951.

On 3 September 1942, Cartan's son Louis, professor of physics and mathematics at Poitiers, was arrested along with physicians, jurists and clergymen and thrown into the German jail at Poitiers. Some were released but the rest had to appear before the German military court in Poitiers, who decided they did not have to be held there. On 12 February 1943 they were brought to the jail of Fresnes in Paris and from there moved, presumably, to Berlin.[336] Behnke asked Süss to do something on behalf of the Cartans, who were themselves helpless, since they were connected with German mathematicians, demonstrably so through Henri Cartan's promise to contribute, together with his father, towards a gift for Carathéodory's 70th birthday (see 5.51) and also to try to enlist the support of others. Although Süss maintained that he did not know the reasons for the measures taken by the German authorities in France, he did accuse the Cartans of showing a rejectionist attitude towards Germans and considered that Élie Cartan retained strong connections with Soviet mathematicians; he also considered the fact that Henri Cartan had shown an interest in contributing towards a gift for Carathéodory, not as a willingness to co-operate with German mathematicians but, rather, as an indication of an internationalist attitude.[337] In Behnke's opinion, Süss's political report on the Cartans was objectively wrong. Behnke argued that the Cartans, *père et fils*, were Catholics in the strict sense of the term and, therefore, bourgeois and anti-communist and, secondarily, that the elderly Cartan was a French patriot.[338] Süss decided to take a different line and asked for a "relatively gentle treatment of the Cartan case" in the still "pending trial" so that it would result in a positive effect on German-French co-operation.[339] In the midst of war the Germans still desired their enemy's scientific co-operation.

On the basis of Süss's reasoning in the Cartan case, it is easy to understand why the Germans needed to hang on to Carathéodory. They needed him desperately to preserve their international scientific relations, presenting him as a famous internationalist mathematician representing Germany. After the dismissals of Jews from the universities and scientific unions in Germany there were hardly any mathematicians worthy of the role.

Henri's brother Louis Cartan was imprisoned for his activities in the French Resistance and executed in 1943 but his family did not learn of his death until 1945. In May 1944 Henri, the only surviving son of Élie Cartan (the other son Jean, a composer, had died of tuberculosis at the age of 25), now a professor of mathematics at the Sorbonne, asked the German mathematician Scholz to intervene to save the life of Jan Cavaillèr, a professor of basic mathematical research in Paris, who had been transported via Compiègne to Germany and whose life was in danger. Scholz had turned for help to Freiherr von Weizsäcker, Jr., and Behnke had also sought help from Süss. Behnke argued that one might intern politically unreliable professors, but should not treat them as hostages and prisoners if one valued German-French co-operation.[340]

Inadequately armed, France had capitulated in June 1940. After the signing of an armistice on 22 June and the occupation of northern France by German troops, Pétain set up a Nazi-collaborating regime in Vichy in the non-occupied zone. Hasse, who with the rank of a lieutenant-commander in charge of department FEP III of the Navy Ordnance Office (*Marinewaffenamt*), was in Paris in the autumn of 1940 and contacted Julia in Versailles. He judged that Julia was absolutely aware of the extent of the national disaster for France. Julia asked for Hasse's help to obtain a permit to visit his older sons in "Free France" which was completely segregated from occupied France and Hasse was able to do this through the German consulate in Paris. Julia asked him further to enable mathematical literature be sent to two French mathematicians imprisoned by the Germans in the officer's camp XVII A, where Julia's brother Roger, an electrical engineer, was also interned. Julia also requested Roger's release through the help of the Armistice Commission with the argument that his brother was indispensable to French industry.[341] In 1942 German troops occupied the whole of mainland France.

5.43
Favouring Weizsäcker's Appointment in Munich

Clashes between the Munich Faculty of Natural Sciences and the Ministry of Education reached their peak at the beginning of the war, not always to the advantage of the professors. Carathéodory, Arnold Sommerfeld and Heinrich Wieland, members of the faculty committee for the appointment of Sommerfeld's successor, intended to remove Wilhelm Müller, whom they reproached for having lectured only in mechanics in the year 1939–1940, and considered who might take up the Chair of Theoretical Physics. Since a permanent appointment made during the war could lead to difficulties later, they wished to provisionally appoint Carl Friedrich von Weizsäcker.[342] C. F. von Weizsäcker, a son of the head of the Political Department at the German Foreign Ministry, was a Heisenberg student researching astrophysics and lecturing at Berlin University at that time. He was also a member of the Kaiser Wilhelm Institute for Physics in Berlin. He had studied in Berlin, Copenhagen and Leipzig and was counted among modern theoretical physicists. The first reaction against his appointment came from the Bavarian Ministry of Education: not only was Müller not to be removed but Ludwig Glaser, a supporter of "Aryan physics" who, with the help of

Müller, had occupied the post of the former Sommerfeld's assistant Heinrich Welker, was to be given an associate professorship for theoretical physics.[343]

Carathéodory wanted to bring Weizsäcker to Munich at a time when Weizsäcker was working together with Heisenberg under the supervision of the Army Ordnance Office (*Heereswaffenamt*) for the German uranium project, which had started at the end of 1939 and was aimed at the production of an atomic weapon. In experiments lasting two years and exclusively concerning the release of atomic energy, Heisenberg had established conclusively that a critical reactor with natural uranium and heavy water could be built in Germany in a few years, in which transuranic elements could be produced for an atomic bomb.

The two atomic physicists visited the Institute for Theoretical Physics in occupied Copenhagen in September 1941 and had discussions with Chr. Møller and Niels Bohr. According to documents released in February 2002 from the Bohr archive (the originals can be seen on its web site) the visit had occurred in connection with an atomic-physics conference at the German Institute in Copenhagen, an establishment of the occupying power, in which none of the leading Danish physicists had participated. From conversation with Heisenberg and Weizsäcker, Bohr gained the impression that, in Germany, great military importance was being attributed to possibilities opened in the field of atomic research, and that Heisenberg believed that the war would be decided with atomic weapons if it lasted for long. Moreover, Bohr understood that German scientists did nothing to prevent such a development and Heisenberg und Weizsäcker wanted to force Danish scientists to collaborate with the Nazis.

In an interview in the *Süddeutsche Zeitung* on 8 February 2002, Carl Friedrich von Weizsäcker vehemently contradicted Bohr's presentation of the facts. He maintained that he went with Heisenberg to Copenhagen on a diplomatic mission, wishing to convince Bohr that the Germans would not be able to arm themselves with atomic bombs in the foreseeable future. Thus, they would have a common interest in helping to bring about an end to the war with conventional weapons.

The scope of Heisenberg and Weizsäcker's visit to Copenhagen might have been to learn about the Allied effort from Bohr. Until 1941 the German atomic physicists believed that either the atomic bomb could not be constructed or that the Germans would be the first to do so. This changed at the end of March 1941, when American physicists confirmed that plutonium was fissionable and thus usable for a bomb. Further progress was blocked in Germany because of the immense expenditure required by heavy industry and the shortage of resources; these became available in the USA some years later. Bohr fled Denmark in 1943 and worked on the Manhattan Project in America.

The American assessment of the German project is given in the following letter of 23 March 1940 by Veblen to the President of the Institute for Advanced Study, Frank Aydelotte:

The Dutch physicist, P. Debye, who has been Director of the Physics Institute of the Kaiser Wilhelm Gesellschaft in Berlin (supported by the Rockefeller Foundation) [. . .] made no secret of the fact that his work is essentially a study of the fission of uranium. This is an explosive nuclear process which is theoretically capable of generating 10,000 to 2,000,000 times more energy than the same weight of any known fuel or explosive. There are considerable deposits of uranium available near Joachimstal, Bohemia, as well as in Canada. It is clear that the Nazi authorities hope to produce either a terrible explosive or a very compact and efficient source of power. We gather from Debye's remarks that they have brought together in this Institute the best German nuclear and theoretical physicists, including Heisenberg, for this research – this in spite of the fact that nuclear and theoretical physics in general and Heisenberg in particular were under a cloud, nuclear physics being considered to be 'Jewish physics' and Heisenberg a 'White Jew'.

There is a difference of opinion among theoretical physicists about the probability of reaching practical results at an early date. This, however, is a well-known stage in the pre-history of every great invention. The tremendous importance of the utilization of atomic energy, even if only partially successful, suggests that the matter should not be left in the hands of the European gangsters, especially at the present juncture in world history.

Work of the sort which the German physicists are supposed to be doing has been going on for some time at Columbia University under Professor Fermi and Dr. Szilard, but at a slow rate because the expense of the experiments exceeds a normal departmental budget. Some effort, not entirely successful, has been made to enlist the help of the United States Government, but this process is slow and cumbersome and has met serious obstacles. It seems to us therefore, that the problem is one which might well be brought to the attention of the Rockefeller Foundation, which would be in a position to act in a simple and direct manner. We are not prepared to suggest a very definite way of attacking the problem, but suppose that if the officers of the Foundation were interested they would consult with physicists who are familiar with the practical questions involved.

We have quoted Professor Debye rather freely in writing to you, but obviously we should have to be very cautious in using his name any further. In any case his only role was unintentionally to stimulate us to bring up a question which we have had on our minds for several months, without knowing exactly what, if anything, to do about it. [344]

5.44
Sommerfeld's Successor

Sommerfeld seems to have been in despair with regard to his own successor. Two years later, in December 1942, he sought Hondros's support: "I turn to you with this printed matter because I know that, as my old friend and student, you take the warmest interest in the fate of Germany." [345]

Referring to Stark–Müller's booklet on *Jüdische und Deutsche Physik* (Jewish and German Physics) [346] whose "completely unqualified first part is due to Müller", [347] Hondros replied that Sommerfeld's influence and his great achievements, as well as the nature of his personality, had shaped the developments of modern physics in such an inextinguishable way that one could only watch certain attempts with a tiresome smile. He assured Sommerfeld that appreciation of his work and person was not confined within the narrow limits of his country. [348]

According to von Faber, dean of the Faculty of Natural Sciences, theoretical physics as conceived by Einstein and Sommerfeld, were no longer taught at the University of Munich and Müller's appointment had the purpose of bringing about this substantive change. [349] Because of his insistence on teaching the theory of relativity, Sommerfeld had been defamed by the representatives of "Aryan physics" as a "White Jew", the words denoting "Jewish according to character" as distinct from "Jewish according to descent". [350]

DIE NATURWISSENSCHAFTEN

31. Jahrgang   3. Dezember 1943   Heft 49/50

Arnold Sommerfeld zum 75. Geburtstag.

Der 2. Band seiner Vorlesungen über theoretische Physik.

Die Veröffentlichung des zweiten Bandes der Vorlesungen über die theoretische Physik[1]) von ARNOLD SOMMERFELD trifft mit seinem 75. Geburtstage, den er am 5. Dezember feiert, zusammen.

Diese Reihe von Büchern ist geeignet, das Rätsel der unerhörten und ununterbrochenen Lehrerfolge, die während so vieler Jahrzehnte die Laufbahn SOMMERFELDS begleitet — man möchte fast sagen, wenn man nicht eines besseren belehrt wäre, gekennzeichnet — haben, zu lösen. Für diejenigen, die, wie der Schreiber dieser Zeilen, den Vorzug haben, die Denkart SOMMERFELDS genau zu kennen und durch einen regelmäßigen Verkehr mit ihm darüber unterrichtet zu sein, wie er die mathematische Physik und die Behandlung ihrer Probleme auffaßt, ist der Einfluß, den er auf die besten seiner Schüler, von denen so viele unter die allerersten Physiker unserer Zeit zählen, ausgeübt hat, immer als etwas Selbstverständliches erschienen. Dieser Einfluß, der sich auf privaten und intimen Verkehr gründet, kann weder auf weite Personenkreise übertragen noch durch den Druck vervielfältigt werden. Daneben war aber für die Unterrichtsart SOMMERFELDS charakteristisch, daß er auch die weniger Begabten auf ein Niveau brachte, das man bei Durchschnittsstudenten nur selten antrifft. Dieses Resultat ist das Ergebnis der Kunst, mit welcher er den Stoff zu gliedern wußte und durch die wunderbarsten Beispiele illustrierte, einer Kunst, die erst durch die jetzige Publikation in ihrem vollen Glanze erscheint; und diese Seite seiner Fähigkeiten auch für seine besten Freunde als eine Offenbarung gelten muß. Wir müssen ihm dankbar sein, daß er sich entschlossen hat, die Vorlesungen, die in seinem Geiste (im Gegensatz zu seiner Atomphysik) nur für junge Studenten bestimmt waren, dem großen Publikum zugänglich zu machen, so daß diese weite Bücher auch in der Zukunft weite Kreise von jüngeren und älteren Physikern anregen und befruchten werden.

Vor einigen Monaten ist an dieser Stelle[2]) eine Besprechung des ersten Bandes der SOMMERFELDschen Vorlesungen durch HEISENBERG erschienen, in welcher die Hauptmerkmale dieser Vorlesungen in knappen Worten so trefflich charakterisiert sind, daß es schwer fällt, diesem endgültigen Urteil noch wesentliche Züge hinzuzufügen. Nur diejenigen Empfindungen, die bei einem gänzlich Unbeteilig-

ten durch die erste, unvermutete Begegnung mit den SOMMERFELDschen Vorlesungen hervorgerufen werden, müssen selbstverständlich bei HEISENBERG fehlen, der den ganzen Turnus dieser Vorlesungen von seiner Studentenzeit her kennt. Dazu kommt, daß der erste Band, der die Mechanik von Systemen mit endlich vielen Freiheitsgraden behandelt, gewissermaßen nur die Ouvertüre der ganzen Komposition bildet, während wir uns beim zweiten Band mit der Mechanik der Kontinua schon mediis in rebus befinden.

Man kann vielleicht sagen, daß dieser Band sogar einer der wichtigsten des ganzen Zyklus sein wird, da er das Fundament für die mathematische Behandlung der makroskopischen physikalischen Erscheinungen einschließlich der MAXWELLschen Elektrizitätslehre enthält. In einem Augenblick, wo wir uns bei der Entwicklung der Physik gewissermaßen innerhalb einer Diskontinuitätsschicht befinden, von der sich viele Wirbel unverhofft ablösen, kann man schwer sagen, welcher Teil der später zu behandelnden Gegenstände nach 50 Jahren noch seine Gültigkeit beibehalten wird. Aber hier befinden wir uns nach auf dem klassischsten Boden: die ganze Anlage des SOMMERFELDschen Buches ist von derjenigen der *Hydrodynamica* DANIEL BERNOULLIS, die vor mehr als 200 Jahren erschienen ist, gar nicht so sehr entfernt, als es ein oberflächlicher Vergleich oder die Diskrepanz des Titels vermuten lassen könnte. Denn BERNOULLI, der das Wort Hydrodynamik geprägt hat, hat darunter die ganze Mechanik der Kontinua verstanden. Und der Vergleich mit dem Hauptwerke des Begründers der mathematischen Physik, dessen ungeheuren Einfluß auf die besten Köpfe seiner Zeit heute leider nur noch ganz wenigen Historikern bewußt ist, scheint mir kein übertriebenes Kompliment für das SOMMERFELDsche Buch zu sein. Hier wie dort finden wir die weiseste Mischung der mathematischen Theorien mit der experimentellen Wirklichkeit, während z. B. allen mir bekannten Darstellungen des 19. Jahrhunderts die mathematischen Entwicklungen zu sehr in den Vordergrund gestellt wurden.

Was dem Mathematiker bei der SOMMERFELDschen Behandlung der schwierigsten Prozesse der Analysis am meisten imponiert, ist die Art, mit der er durch Streichung aller unwesentlichen Einzelheiten ein Gerüst stehen läßt, welches für das Verständnis des Zusammenhanges und für die Benutzung der Formeln am geeignetsten ist, und das jeder durch Anbringung der fehlenden Stützpfeiler tragfähig machen kann. Man lese z. B. unter diesem

[1]) ARNOLD SOMMERFELD, Vorlesungen über theoretische Physik, Bd. II: Mechanik der deformierbaren Medien. Leipzig: Akademische Verlagsgesellschaft Becker u. Erler 1944.
[2]) Naturwiss. 1943. Heft 29/30, S. 350.

Carathéodory's review of Sommerfeld's
Mechanik der deformierbaren Medien *in* Die Naturwissenschaften.
Courtesy of the Bavarian Academy of Sciences, Archive.

Gesichtspunkte im § 27 des besprochenen Buches die Behandlung der BESSELschen Funktionen, im § 15 die Ableitung der McCULLAGHschen Gleichungen, im § 17 die Einführung in die Kapillarität, eine Liste, die sich beliebig erweitern läßt.

Nur noch einige Worte über den Inhalt des Buches selbst. Dabei muß ich aber der Versuchung widerstehen, zu zeigen, wie die lose Aneinanderreihung der verschiedensten Gegenstände einer höheren Systematik gehorcht, und wie der Faden der Ariadne, der durch das ganze Werk läuft, auch bei den kühnsten Sprüngen von einem Problem zum anderen, nie aus der Hand fallen gelassen wird. Den Auftakt bildet ein erstes Kapitel über die Kinematik (oder vielmehr die Geometrie) der deformierbaren Medien. Es folgt ein Kapitel über die Statik dieser Medien, in welchem nacheinander die inkompressiblen Flüssigkeiten, die Gase, die elastischen Körper, die reibenden Medien (laminare Strömungen) behandelt werden. Im dritten Kapitel werden die EULERschen und die BERNOULLIschen Gleichungen mit einem Exkurs über ihre Ableitung durch das HAMILTONsche Prinzip, die Anwendung dieser Gleichungen auf akustische Fragen und die Dynamik der elastischen Medien besprochen. Darauf folgt das schon erwähnte Äthermodell von McCULLAGH (und W. THOMPSON) und ein mit außerordentlicher Kunst geschriebener Abriß des Turbulenzproblems. Das vierte Kapitel enthält eine ziemlich vollständige, aber nicht zu weitläufige Darstellung der HELMHOLTZschen Wirbeltheorie. Das fünfte Kapitel handelt von Schwerewellen in tiefem und seichtem Wasser und auch von Kapillarwellen, wobei die Übergangserscheinungen nicht vergessen werden und die Theorie der Gruppengeschwindigkeit ganz meisterhaft auseinandergesetzt wird. Das sechste Kapitel enthält Randwertprobleme, u. a. wird auch die v. KÁRMÁNsche Wirbelstraße behandelt. Das siebente Kapitel ist „Ergänzungen zur Hydrodynamik" betitelt. Dort finden wir z. B. das STOKESche Gesetz und die Theorie der Schmiermittelreibung, die von SOMMERFELD selbst herrührt. Im achten Kapitel werden u. a. die Elastizitätsgrenzen untersucht, bei welchen das lineare Gesetz aufhört, und auch die Elastizität von makroskopischen Einkristallen. Den Schluß bildet die berühmte Sammlung von Aufgaben, fast das wichtigste am ganzen Werke.

C. CARATHÉODORY, Universität München.

Warum hat der diploide Zustand bei den Organismen den größeren Selektionswert?

Von FRITZ v. WETTSTEIN, Berlin-Dahlem.

Es sind alte Fragen von hohem Interesse, warum im Pflanzenreich viele Gruppen als Haplonten mit einfachem Genom begannen und im Laufe der phylogenetischen Entwicklung über Formen mit antithetischem Generationswechsel zu reinen Diplonten mit doppeltem Genom gelangt sind? Warum im Tierreich mit wenigen Ausnahmen überhaupt fast nur diploide Organismen zu finden sind? Und warum mit dieser Entwicklung vom haploiden zum diploiden Zustand eine höhere Differenzierung verbunden ist, so daß die Haplonten meist nur über einfache Thallusbildungen verfügen, während die Diplonten mit komplizierter Organisation überall die Stufen der Entwicklung mit höchster Organgestaltung einnehmen? *Was ist der Vorteil des diploiden Zustandes, was ist sein Selektionswert im Laufe des Evolutionsgeschehens?*

Verschiedenes kann man zur Beantwortung dieser Frage heranziehen. Zunächst wäre daran zu denken, daß vielleicht die diploide Zelle eine größere Widerstandsfähigkeit besitzt, oder daß zum Aufbau komplizierterer Gewebe und Organe diploide Zellen besser befähigt sind. Man könnte daran denken, daß die doppelte Anlagenmenge für verschiedene Leistungen notwendig oder günstiger wäre. Einige Beobachtungen an experimentellem Material lassen sich hier auswerten.

Schon vor längerer Zeit wurden vergleichende Beobachtungen an haploiden und diploiden Moosgametophyten unter verschiedensten Kulturbedingungen durchgeführt. Die Pflanzen unterschieden sich nur durch die Genomzahl, die Genome selbst waren identisch. Vielfach fand sich eine größere Variabilität der diploiden Gametophyten, was vielleicht auf eine plastischere Reaktionsfähigkeit den Außenbedingungen gegenüber schließen läßt. Wesentliche Unterschiede in der Organbildungsfähigkeit fanden sich aber im allgemeinen nicht. Nur bei dem Laubmoos *Phascum cuspidatum* zeigte sich, daß der diploide Gametophyt direkt auf vegetativem Wege Sporogone bilden kann, während die haploide Pflanze dies in keinem Falle zustande bringt. Bei ihr erfolgt die Sporogonbildung erst über die Befruchtung, wenn durch die Kernverschmelzung der Gameten wieder der diploide Zustand erreicht ist.

Andere Beobachtungen dieser Art betreffen die sog. haploiden Blütenpflanzen. Wir kennen seit 2 Jahrzehnten in immer zahlreicheren Fällen von Blütenpflanzen haploide Individuen, während normalerweise die Blütenpflanzen Diplonten sind. Man kultiviert sie neben den diploiden Formen in den Experimentalkulturen. Sie zeigen normale Organbildung, normale Differenzierung in Stamm, Blatt und Wurzel, in Sproß und Blüte. Diese Pflanzen sind zwar in allen Teilen entsprechend nach der Kernplasmarelation verringerten Zellgröße kleiner, zeigen aber keinerlei Hemmnis, die normale Sporophytengestaltung durchzuführen. Es kann nach den Erfahrungen in unseren Kul-

Amidst all the discontent over Müller's appointment, some normal academic work continued and, for example, Carathéodory co-operated with Müller in supervising Hans Weber's dissertation *Über analytische Variationsprobleme* (On Analytic Problems in the Calculus of Variations)[351] and reviewed it.[352]

The publication of the second volume of Sommerfeld's lectures on theoretical physics in 1943 coincided with his 75th birthday. Carathéodory referred to this happy event at the start of his review of the *Mechanik der deformierbaren Medien* (Mechanics of Distortable Media)[353] and predicted that the volume would be one of the most significant of the whole cycle since it contained the foundation for the mathematical handling of macroscopic physical phenomena including Maxwell's theory of electricity. Carathéodory recognised in Sommerfeld's book the presence of classical mechanics and found its conception not very far removed from that of Daniel Bernoulli's *Hydrodynamica* written more than two centuries earlier.

5.45
Greece under German Occupation (1941–1944)

The Metaxas fascist government, brought about by the coup of 4th August 1936, should have been a natural ally of Germany and Italy. The irony is that the Italian invasion of October 1940 and the subsequent German invasion of April 1941 brought about the collapse of Metaxas's regime. Germany's imperialistic interests proved, in the end, to be much stronger than any sympathy that Hitler and Goebbels had towards Greek fascism.

During the winter of 1940–1941 Mussolini tried to invade Greece but the Italians were pushed back to North Epirus, into Albanian territory, by the Greek army. However, the Greek success proved to be temporary. Field Marshal List's Wehrmacht army launched its assault through Yugoslavia and Bulgaria at dawn on 6 April 1941. Three days later, German units entered Thessaloniki. Mayor Merkouriou and the brother of the Greek ambassador in Berlin, General Nikolaos Rangavis, who had been appointed military governor of Thessaloniki by Maniadakis, delivered the city to the Germans.

Appointed then by the Germans as Governor General of Northern Greece, N. Rangavis resigned six months later protesting against mass reprisal killings and village burnings by German forces in Greek Macedonia. His son, a civil engineer from Hannover with a German mother, was accused of spying for the British, and executed by the Germans in Crete. For the continuous, intensive and unhindered supply of Rommel's army in North Africa, the Germans had constructed a large airport at Tymbaki, Crete, by recruiting thousands of workers and engaging their army's whole technical apparatus. In mid February 1942, as soon as the landing strips were ready, RAF bombers attacked them causing a critical delay in Rommel's supply from the air. Rangavis, Jr., whom the Germans had entrusted with the construction of the airport, belonged to the armed resistance group *Midas 614*, led by the officer Ioannis Tsigantes, with direct association with the General Staff of the Allies' forces in the Middle East, and had marked targets for British planes. He was captured by the head of the airfield security and executed some day later, as Major Hartmann, head of the German counter-intelligence in Heraklion, reported to the Philhellenic German plenipotentiary in Athens, Günther Altenburg, who had been sent by Rangavis's German relatives to Heraklion in an attempt to save the young man's life.[354] The Germanophile Greek ambassador in Berlin, Alexander Rizos-Rangavis, had himself been informed by the head of the political department in the German Foreign Ministry, Ernst Freiherr von Weizsäcker, of the German plan to invade Greece. Rangavis's daughter Elmina was married to Euphrosyne's nephew, John Argyris.[355]

On 18 April 1941, shortly after the German invasion, the Greek Prime Minister Alexandros Koryzis shot himself in despair; three days later he was succeeded by Emmanouel Tsouderos. Tsouderos had been the president of the State Bank of Greece from 1931 to 1940. He was known for his contacts with the British embassy and Crown Prince Paul, and for his repeated announcements that he was going to replace Metaxas with the King's approval. The German Luftwaffe attacked harbours and cities. Wealthy families prepared to leave the country to avoid the consequences of occupation although among the propertied class there were those who showed admiration for Germany.[356] Despina and her baby son left the country for Tanganika just as the Germans were marching into Greece. They embarked in Piraeus and she paid the captain of the ship in gold.[357] People fled the cities, taking refuge in the mountains. General Tsolakoglou signed the final surrender document in Thessaloniki on 23 April 1941.

On Sunday 27 April, German troops reached Athens and went to the Acropolis to hoist the swastika. Unable to face this humiliation, Penelope Delta poisoned herself that day and died after five days of suffering. She had asked to be buried in the garden of her house and demanded a funeral with no priests or memorial service. In front of the Reichstag Hitler announced on 4 May 1941: "We are filled with sincere compassion for the defeated unfortunate Greek people. They are victims of their King and his small blinded leadership group. Still, they have fought so bravely that they cannot be denied the respect of their enemies".[358] Hitler's "compassion" found no echo among the Greeks. Two young men took down the swastika from the Acropolis on 31 May 1941 and their action marked the start of the resistance against the German occupation. At first, only sporadic acts of resistance were carried out by isolated individuals or small groups and had the character of sabotage or intelligence gathering. But as the occupation took hold, a mass movement headed mainly by the Communist Party grew. Its leader, Nikos Zachariadis, was deported to the concentration camp of Dachau where he was interned for the duration of the war.

In contrast, the so-called old political world was apathetic, reluctant, and full of fear and doubt about the value of armed resistance. After the fall of Crete to German airborne troops at the end of May, the King and his government fled the island, where they had taken refuge with British forces, and formed a government in exile in the Middle East. They were almost sure that the future of post-war Greece would be decided outside the country. Tsouderos fled from Crete to Alexandria. The naval Governor of Crete in 1940–1941 was Ioannis Kasimatis, who in 1940 had been called back to military service from which he had been dismissed[359] because of his participation in the Venizelist coup of 1935. Kasimatis had visited Carathéodory in Munich seven months before the war broke out, on 30 January 1939. From 1943 until his retirement in 1946 he represented the Greek air-force ministry in London.

In Greece itself a government of collaboration with the occupiers, initially headed by General Tsolakoglou, was established. About the end of October 1941, there were rumours that Admiral Wilhelm von Canaris, Philhellene and head of the German counter-intelligence, would return as Tsolakoglou's successor to Greece, the "land of his ancestors", and take on the state leadership.[360] By the beginning of June 1941 Greece was suffering under a tripartite occupation by Germany, Italy and Bulgaria. The Germans occupied the two largest cities, Athens and Thessaloniki, the island of Crete and a number of the Aegean islands as well as the border with Turkey. Poverty and hardship among the Greek population were the consequences of a harsh German occupation regime. The Germans plundered the country's agricultural resources and its industry; demonstrating extreme perversity, Germany demanded from Greece payment for the costs of the occupation. In the autumn of 1942, Marshal Göring boasted in Berlin that Germany would be able to feed itself from occupied countries. This was at the very moment when Greece's granaries, Eastern Macedonia and Thrace, were occupied by Bulgaria and cut off from Greece, when much of the land remained uncultivated and the system of international communication had been destroyed by the invasion. Collapse of the national economy was inevitable. The Red Cross estimated that 250 000 people died directly or indirectly as a result of famine between 1941 and 1943, among them, in January 1942, Carathéodory's friend, the Greek mathematician Hatzidakis. German propaganda was employed to attract Greeks to volunteer for work in the Reich but only a few responded, in a desperate hope of avoiding starvation. In Germany, these migrant Greeks openly expressed their anti-German sentiments, did not comply with the rules, demonstrated their opposition to national socialism outside their work place and even claimed that Germany was about to lose the war. On 8 April 1942, Goebbels admitted that the Germans could not conquer Greece morally at that moment.[361]

Within Greece the Wehrmacht reacted violently to organised resistance through anti-guerrilla operations, reprisal murders and the destruction of villages, as well as through the arrest of civilians. Against Jews they proceeded as in any other European country to implement the "final solution". Thus 96 percent of the Sephardic Jewish community of Thessaloniki, comprising one-fifth of the city's population, was deported to the death camps of Auschwitz within a few weeks in early 1943. The National Liberation Front (EAM = *Εθνικό Απελευθερωτικό Μέτωπο*), which had been created in September 1941 by the Communist Party of Greece, with the co-operation of other small socialist parties, with the aim of organising the resistance and preparing for a political change on the eventual liberation of the country, approached Georgios Papandreou in the winter of 1943–1944 and invited him to head a provisional government in Free Greece. However, the liberal and militantly anti-communist politician rejected the offer. In April 1944, at a conference in Lebanon attended by representatives of all resistance groups and political forces, the British backed Papandreou to build the Government of National Unity. However, this fragile government in exile fell apart as hostility grew between the communists and the democrats until it culminated in clashes between units of the military branch of the EAM, the National People's Liberation Army (ΕΛΑΣ = *Εθνικός Λαϊκός Απελευθερωτικός Στρατός*), and British troops for control of the Greek capital.

Carathéodory's friend Kritikos and his acquaintances Kitsikis and Papapetrou, who joined the resistance movement, were later dismissed from the Technical University of Athens because of their leftist political ideas. They belonged to those intellectuals with a German educational background and strong sympathies towards German culture, who comprised four-fifths of the Technical University professors and half of the Athens University professors.[362] The Wehrmacht took over many scientific institutes for their own use, but could not count on co-operation since the "attitude of Greek academic circles – also those who had been very friendly towards Germans before – had become much more reserved."[363] Because of these reservations it was not until 1942 that a contract concerning the foundation of a German-Greek joint institute for biology between the *Kaiser-Wilhelm-Gesellschaft* and the Greek government was signed. A Greek physician, one of the older alumni of a German University, had suggested that the Greek Medical Association decline the German proposal for its establishment. He was later arrested by the German occupation authorities.[364]

N. Kritikos provides a portrait of Carathéodory at that time: "Through the maturity of years, his historically and liberally oriented spirit had attained such power and depth which none of his dissenting discussion partners could comprehend. I remember, for instance, with what authority he had predicted the defeat of Germany at the beginning of the first great war, how he had foreseen Hitler's aggression and Hitler's total destruction before the last war, such that those educated Germans who were impressed by their Führer's achievements were shaken. For his homeland, Greece, he felt an infinite love and I can imagine with what inner turmoil he had pursued her poverty and hardship during the German occupation. But he also had an invincible optimism for the future and he knew how [...] to pass it on to the numerous Greeks who visited his hospitable home before the recent war."[365]

5.46
International Science Restructuring

When the Wehrmacht occupied Belgium, the archive and library of the *Union des Associations Internationales* fell into the hands of the German Central Office for Congresses, the DKZ. Thus, the NS state obtained a wealth of information about international congresses and organisations world-wide. By the autumn of 1940 a victorious outcome of the war seemed likely for Germany. Plans for the immediate post-war world under German leadership were constructed in meetings between the relevant ministries. The "importance of scientific co-operation within the international organisations for future foreign cultural policies of the Greater German Reich" was discussed. The restructuring of international organisations, and the influence of Germany on these institutions, was not considered to be an independent problem, but only part of the foreign cultural policy. Of central importance to this scheme were a series of secret meetings held at the REM in the winter of 1940–1941, which dealt with the question of transferring international organisations from Brussels to Germany.[366] In Rust's presence a session of the German Union of Mathematicians had taken place in Berlin in the end of February or the beginning of March 1941.[367] The Reich Minister's request to Carathéodory on 6 May 1941 – shortly after the German invasion of Greece – to report on his participation in the International Commission of Mathematicians should be set within the context of the plans for the restructuring of international science:

In the above matter, in order to investigate the significance of the Union with regard to the International Congresses, I have addressed the enclosed letter to Professor Dr. Blaschke in Hamburg on 7 March 1941. Then, Professor Dr. Blaschke had reported by a letter of 17 March 1941 that the information contained in my letter was essentially correct. From 1933 to 1936 Professor Dr. Blaschke had belonged as representative of Germany to the International Commission under the chairmanship of Francesco Severi (Rome), which aimed at excluding the Conseil[368] from the organisation of the International Congresses of Mathematicians. In 1936 Professor Blaschke transferred this membership to you and also passed the files on to you.

Therefore, I ask you, on the basis of the documents attached to my letter to Professor Blaschke of 7 March 1941, to report especially on your activity in the Commission and at the same time to express your opinion on the question, to what extent a connection between the Union and the International Congresses still exists.[369]

On May 18 Carathéodory answered that all the data contained in the minister's letter to Blaschke concerning the International Mathematical Union were correct, but that the assumption that the IMU might still exist did not correspond to the facts.[370]

Aspects of Carathéodory's report,[371] considerably distorted and without reference to his name, are included in a note of 5 May 1942 which concerned the department meeting at the REM on 23 March 1942 and which was entitled "The International Congress of Mathematicians and the International Mathematical Union".[372]

Wilhelm Süss had been present at that meeting. He had described it as "a session at the Reich Ministry of Education with the International Commission on Mathematical Instruction as an item on the agenda". Earlier Süss had spoken to Rudolf Mentzel, head of the Department for Science at the REM, who was of the opinion that a separate department in the Reich Research Council could not accommodate a discipline such as mathematics and that State Counsellor Esau was responsible for physics and mathematics together and had no intention of changing anything in this respect. Süss believed that German mathematicians ought to turn to the Reich Research Council more intensively than before and, if they did not desire to turn directly to the council or to Esau, he himself would be prepared to mediate.[373] Esau was Gerlach's predecessor in nuclear administration and president of the Physical-Technical Reich Institution which had responsibility for the tasks of a Bureau of Standards. In 1942 Süss was appointed head of the Mathematics Work Group (*Arbeitsgemeinschaft Mathematik*) and, in January 1945, of a separate department for mathematics at the Reich Research Council.[374]

On 22 June 1941 the German invasion of the Soviet Union, Operation Barbarossa, began. Just two weeks earlier the Reich Minister for Science, Education and People's Culture had informed Carathéodory that the Russian journal *Mathematical*

Collection (*Matematicheskii Sbornik*) intended to celebrate the 75th anniversary of its foundation in December that year and, on this occasion, the Academy of Sciences of the USSR was to publish a special issue of the journal to include works by foreign scientists. According to the German Embassy in Moscow, this information came from the USSR Society for Cultural Relations with Foreign Countries. The editorial board of the journal addressed a letter to Carathéodory, which the Reich Minister sent on to him, together with a reminder that if he intended to contribute to the issue he must submit the work through the minister's hand.[375] Carathéodory was also required to submit a copy of his contribution to the rector of Munich University.[376] Nothing more came of this, presumably as a result of Hitler's decision to tear up the German-Soviet non-aggression pact, and to set in motion the invasion of Russia.

In October 1941 Carathéodory seems to have taken part in a working conference of the DMV in Jena. He wrote to Süss that he was very happy to have met him again there and that he had found their common trip to Berlin pleasant and enjoyable; that he had a very good hotel room, although on the ground floor, on Dorotheenstrasse, but he was not disturbed by noise; that on his return, he managed to get a wagon-lit and he was even alone in the compartment.[377]

Only a couple of days after Carathéodory's euphoric letter to Süss, in mid November 1941, the first deportation of Jews from Munich to Theresienstadt was carried out in broad daylight but the non-Jewish population of Munich hardly took any notice of it.[378]

5.47
Mediating for Saltykow's Release

On 17 December 1941, Carathéodory sent Helmut Hasse a copy of a letter dated 8 December 1941 and addressed to him by Dr. S. Saltykow, a full professor for pathologic anatomy in Zagreb. Saltykow was asking Carathéodory on behalf of his brother, the mathematician Nikolaus Saltykow, and his brother's wife, to intercede on behalf of Nikolaus and to approach the general administrative councillor Turner in Belgrade for his release. Nikolaus, together with other Serbian university professors, had been taken into custody by the German authorities. Saltykow feared for his brother's life, since Nikolaus was old, in poor health and was being exposed to severe conditions while in detention.

Carathéodory tried to put forward reasons to Hasse why Nikolaus Saltykow was not of great significance as a hostage. He wrote that Saltykow was not a Serb but an emigrant Russian tsarist, who had fled the Bolsheviks in 1919 and went, first to Brussels, and then to Belgrade where he became a professor in 1922. Carathéodory found this fact to be in Nikolaus' favour. He concluded that the emigrant could not have been deeply rooted in Serbia and referred to Poggendorff's lexicon for Saltykow's biography. He then proceeded to Saltykow's mathematics: It is "not first class but at least he has found some interesting things in Lie theory and the integration of partial differential equations of the first order. Five years ago, he visited me and reported in detail about these things."[379]

On 7 January 1942, Hasse replied that he had collected information about Saltykow and it seemed to him not permissible to take any step in favour of the

Serb mathematician. But he told Carathéodory that the right place for such a matter would be either a Reich authority or the German Union of Mathematicians. Hasse wrote that Saltykow had never had connections with German mathematicians, that he had only published in French or Russian journals in the Russian language up to recent years and that he had grasped every opportunity to disparage the performance of Lie and his student Engel without himself contributing anything of value to the problems of the Lie and Engel school. Finally Hasse warned, or indirectly threatened, Carathéodory: "One should not devalue one's own authority by supporting aims that do not deserve it completely." [380]

Indeed, Carathéodory took Hasse's advice and turned to Süss with a successful outcome. Very soon, on 28 January 1942, he reported to Süss the positive outcome of the efforts to rescue Saltykow: "It's really kind that you have sought to help Saltykow and it was also my opinion that the matter would be treated better in a private way. [...] Last week I was at Hilbert's birthday in Göttingen and, on my return home, I found a letter from N. Saltykow himself from Belgrade in which he informs us that he is home again. This matter is consequently settled." [381]

5.48
Unable to Rescue Schauder

On the eve of World War II some fifteen thousand Jews were living in Drogobych (Drohobycz in Polish), Galicia, comprising over forty percent of the town's population. Galicia was the part of eastern Poland that had come under Soviet rule following the partition of Poland in 1939. Drogobych remained in Soviet hands from 1939 to 1941. In September 1939 hundreds of Jews who had escaped from German-occupied Poland joined the local Jews of Drogobych. When Nazi Germany attacked the Soviet Union on 22 June 1941, Galicia was one of the first areas to be taken from the Soviets. Young Jews tried to flee to the east and, in the attempt, many were killed by German air attacks or by Ukrainian nationalists. German forces entered the city on 30 June and Ukrainians, assisted by Wehrmacht soldiers, murdered over three hundred Jews in a three-day pogrom which began the following day. In July various measures against the Jews were introduced, including the confiscation of apartments by the Germans and the wearing of the Jewish Badge. Jews were also taken for slave labour. The district of Galicia was incorporated into the General Government, as were the Cracow, Warsaw, Lublin, and Radom districts. From the five districts of the General Government the Jews were transported to the extermination camps of Belzec, Treblinka and Sobibor, during Operation Reinhard which was aimed at the extermination of Polish Jews and named after the head of the Security Police (Sipo = *Sicherheitspolizei*) and the Security Service (SD = *Sicherheitsdienst*), Reinhard Heydrich. In September and October 1941, several dozen Jewish intellectuals were tortured and then murdered in a forest near the town. On 30 November, over 300 Jews were murdered in the Bronica Forest outside Drogobych. That winter many Jews died of starvation and a typhoid epidemic. At the end of March 1942, 2000 Jews were sent to their death to Belzec, and from 8 August to 17 August, 2500 more were sent there. Only Jews with essential jobs in the local oil industry were spared.

Germans and Ukrainian collaborators hunted down Jews in hiding, and over six hundred were killed in Drogobych itself. At the beginning of October 1942, a ghetto was established and 10 000 Jews were confined in it. Another *Aktion* took place on 23 and 24 October, during which 2300 Jews were sent to Belzec and three hundred patients in the Jewish hospital were killed.

On 29 October 1942 the Polish mathematician Julius Schauder wrote to van der Waerden from Drogobych.

Dear colleague,

about a year ago I left Lemberg [now Lvov] and consequently the university. I live here under the most miserable conditions together with my family (wife and child) selling my own clothes which, by the way, run low. [. . .] As you know, I am not of Aryan origin and therefore I am exposed to various dangerous situations, although I hardly have any connection with the Jewry. I am not treated as a scholar, on the contrary, I am counted among the grey mass. I fear that, without belongings and property, I will be pushed into the unknown together with my family and we believe that, under the prevailing conditions, we would hardly be able to see the end of the war. For this reason I ask you and my other acquaintances to take me into your care, since only the relevant protection and security would allow me to escape the ever newly arising dangers. In short, it's about saving life and already at the eleventh hour. I can give you the following information. I am a front-line soldier of World War I (the Italian front). In my youth I taught at the German gymnasium in Lemberg in Poland. Further, the greater part of my work is in the German language and has appeared in German journals. For five years I was an examiner of the *Zentralblatt* (Springer). Many German scholars are counted among my acquaintances. From all this, it is evident that I have am inwardly connected with German mathematics. I was never active politically and I live only for my science. I ask you to enable me to continue dedicating myself to mathematics in peace. Of course, it would be the best if I could go away to a neutral foreign country (Switzerland or Sweden). But I am aware of the impossibility of achieving this in war time. Consequently, only a milder treatment could be considered. Relevant steps should be taken within *one* week and made known here.

With many regards   yours obediently
                                                       Js Schauder

PS. Could you give me the addresses of Hölder, E. Köthe, Dötsch, Perron, Carathéodory and finally Heisenberg? Admittedly, I do not know Heisenberg personally, but he belonged to the jury of the international prize awarded to me for my work on the area of partial differential equations. I would be very obliged to you if you could confirm the receipt of this letter immediately.

Address: Prof. Dr. Julius Schauder, Drohobyez, Bergstrasse 4
General Government   District of Galicia [382]

Van der Waerden sent Carathéodory and also Heisenberg a copy of that "horrifying" letter on 4 November 1942. Being without influence himself, he wrote, he could do nothing for Schauder. But then the tone of his letter changed: "I do not know whether Schauder is a Jew or a half-Jew. As is known, to help Jews is severely punished. By the way, the letter is not only deeply distressing but also peculiar and full of contradictions. According to the sentence added later ["Relevant steps should be taken within *one* week and made known here"] help should be forthcoming within a week but in the postscript he asks for addresses, something which does take a longer time. Of what use should an "immediate confirmation" be to him is also a mystery to me. Is it possible that these two postscripts are dictated by an authority that intends to trap us?"[383]

Carathéodory immediately (5 November 1942) sent the copy of Schauder's letter to Süss hoping, as he said, that some favourable outcome might be possible, like the one in Saltykow's case of the previous year. But he took van der Waerden's view and commented that, on the grounds of the incomplete data, it was extremely improbable that someone would be able to undertake anything.[384] Of course, as Schauder's letter reveals, Carathéodory knew Schauder personally and maybe through Lichtenstein, whose student Schauder had been in Leipzig in 1931–1932. Also, as van der Waerden's letter reveals, Carathéodory and Heisenberg had the kind of connections that would allow an intervention in Schauder's favour. At least that was what van der Waerden believed.

What were Carathéodory's motives for passing the letter on to Süss? He cannot have been so naive as to have thought that any help could be provided for Schauder, who was a Jew according to Nazi classification. Why should he expose himself to Süss by trying to help a Jew if he did not firmly believe that Süss would be able to undertake something to save Schauder? That he referred to "incomplete data", obviously a reference regarding Schauder's racial descent, may be interpreted as an attempt to stress that, if Schauder were not a Jew, he could possibly be left in peace to survive. But perhaps Carathéodory was panicked by van der Waerden's intimation that the letter could be trap and by that time Carathéodory's actions might have been dictated by fear. If Carathéodory did not pass on the letter and the letter was a trap, he would be behaving "illegally" by withholding information. But what would be the consequences for him? Probably minor and insignificant. Just as Carathéodory had written to Süss, Scherrer had written to Heisenberg, and Heisenberg in his turn to Scholz, but everyone turned Schauder away. Süss received Carathéodory's letter on his return from a trip on 19 November. The letter was brought to the REM and registered the following week. Schauder's case was discussed later and the information was probably passed over to the SD.[385]

During the whole month of November a further *Aktion* was carried out at Drogobych. Ten days after it had begun, one thousand Jews were deported to Belzec and few days later several hundred more. Hundreds of Jews were killed in the ghetto as well. At the end of 1942 and early in 1943, the Jewish oil workers were sent to work camps. On 15 February 1943, 450 Jews from the ghetto were killed in the Bronica Forest. The liquidation of the ghetto began on 21 May 1943 and was completed by 10 June. Many of the ghetto houses were set on fire so that the Jews were driven out. The last Jews found in the ghetto were also killed in the Bronica Forest. The destruction of the ghetto was followed by the murder of the Jews in the work camps, where only absolutely essential workers were kept alive. On 27 August 1943 all the Jews employed in the cement factory were killed. In September 1943, Schauder was shot by the Gestapo. When the Soviet forces entered Drogobych in August 1944, four hundred Jews emerged from their hiding places. As a result of Operation Reinhard almost two million Polish Jews were gassed in the course of 1942–1943.

Were Carathéodory in Greece, he might have acted differently. As the *Philhellene* reported in 1944, "dispatches from the Near East and reports coming out of

Greece tell of the wonderful way in which the people of Greece, though starving and suffering brutal treatment by the Nazis, have come to the aid of their Jewish fellow countrymen."[386]

5.49
Papal Audience in Rome

An international conference of mathematicians was held in Rome from 9 to 12 November 1942. Although Hasse and Carathéodory had not submitted an application to participate through official channels, on 19 October 1942 the Reich Minister of Science, Education and People's Culture granted them permission to take part, on condition that the rector, the dean and the local head of lecturers of the relevant universities raised no objection.[387]

The conference took place against the height of the war – the very morning of the opening of the conference coincided with the British-American landing in Morocco and Algeria under the command of Eisenhower – and was not, of course, international in any real sense. Hasse, the head of the German group, in his report of 3 February 1943 to Wilhelm Süss, attempted to clarify the official description of the event.[388] In his opinion, the use of the words *convegno* instead of *congresso* and *di* instead of *dei*, in the expression *Convegno Internazionale di Matematici*, implied that the event was not an international congress in the usual sense but rather an Italian congress with invitations to foreigners. Wilhelm Blaschke, Carathéodory and himself from Germany, participants from Bulgaria, among them K. Popoff, Norway, Rumania, Sweden, Switzerland (Fueter and Speiser), and Hungary (v. Kérékjártó) as well as about one hundred Italian mathematicians comprised the "genuine and noble assembly".[389] Hasse was given special attention everywhere as *Capo della Delegazione Tedesca* (Leader of the German Delegation).[390]

The conference was organised by the *Reale Istituto Nazionale di Alta Matematica* in Rome and was hosted by its president, professor Dr. Francesco Severi. Carathéodory knew Severi from the International Congress of Mathematicians in Rome in 1908, where Severi had been awarded the *Guccia Medaglia*, a prize for his many investigations into the geometry of algebraic surfaces. Severi was later described by Süss as the head of the so-called Italian school of geometry and an internationally acknowledged mathematician, at the forefront of the whole organisation of sciences and, especially, of mathematics. Severi owed his "papal-position" to his own performance and personal qualities and he deserved to be publicly honoured.[391]

The conference opened with a reception given by the Governor of Rome, Prince Borghese, in the Capitol. After several official welcoming speeches, Severi gave a comparative overview on the relations between mathematics and their physical applications in the 19th and 20th centuries. Hasse provided the official reply in the name of the foreign guests. He stressed that most mathematicians in Germany were either soldiers like himself or employed in various tasks of importance to the war. He attributed the enjoyment of the scientific programme in Rome to the "brave soldiers on all fronts who protect us from the enemy onslaught". He thanked the fascist government in that they "uphold the values of the spirit even in the most difficult of times", as well as Severi personally for having taken care of all those details that provide for a successful conference. Finally, he expressed his conviction that Italy would take a privileged position in a future Europe after successfully passing a hard crucial test. Hasse's speech could hardly have been bettered if it had been given by a representative of the Reich Propaganda Ministry itself. From 1939 to 1945 Hasse was on war leave from Göttingen, working in Berlin on ballistic problems for the navy. Representatives of the National Ministry of Education, of the Senate, of the Chamber of Unions and Corporations, of the fascist party, of the Church, as well as the president of the Royal Academy of Italy and the governor of Rome came to the opening of the event. The under permanent secretary at the Italian Ministry of Education attributed the reason as to why fascism treated sciences, arts and education with such "reverence" to Mussolini's "crusade against Bolshevism" which, in his opinion, meant setting human personality above materialism and determinism lacking morality.

Carathéodory was unable to attend the opening ceremony, for he arrived in Rome in the afternoon of the opening day, travelling in a wagon-lit which he had managed to get only "through great patronage".[392] Also absent was the head of the NSDAP section for Italy in Rome, Dr. Erich. Hasse later visited Erich, who claimed that he had not been informed of the event.[393]

During their stay in Rome, the participants were received by President Federzoni of the *Reale Accademia d'Italia* and shown around the impressive building complex of Villa Farnesina. They were also received by President Di Marzio of the *Confederazione Fascista Professionalisti e Artisti*, who was a member of the National Council and by Severi, to whose opening address Speiser, Fueter and Carleman replied in the name of the foreign guests. Finally, there was a reception by Professor de Francisci, the rector of the University of Rome. Foreign participants received hospitality from mathematicians living in Rome. The daily newspapers constantly reported about the conference.

The scientific programme of the conference consisted of eleven reports providing an overview of the state, methods and scope of the most significant current branches of modern mathematics. The lectures, with one exception, were delivered at the mathematical institute of the University of Rome, which was housed in the same building as the *Reale Istituto Nazionale di Alta Matematica*, where also Blaschke gave one of his lectures aside from the official programme.[394] Hasse explained in his report which methods and results modern number theory owes to the arithmetic theory of algebraic functions. Carathéodory spoke in French on "Problems of Analytic Functions of One Variable"[395] and his lecture made a "particular impression" on Hasse. Carathéodory's lecture took place at the Papal Academy on 12 November 1942. Just before he spoke, the Pope received the participants at a private audience and were introduced to him by Severi. Pius XII, in greeting them, said that the study of mathematics was based on the "most real of all truths" and that, more than any other science, it approached the science of God; he added that all mathematicians of various nations could unite in the science of God. He ended his speech with a wish that the scientists should contribute to a future in which all peoples could find their righteous demands satisfied. The participants then passed through the Raffael loggia to the Sistine Chapel and then to the garden house of Pius IV, the seat of the Papal

Academy of Sciences. There they gathered in the conference room and were received by the academician Professor Lombardi and the Chancellor Dr. Salviucci. Severi thanked Professor Lombardi and especially Father Professor Gemelli, the president of the academy, for the friendly reception, and he read a reverential telegram of tribute to the Pope and to Father Gemelli. The conference convened under the chairmanship of the senate member Lombardi and Carathéodory delivered his speech. At noon they all had buffet lunch in the *Casa dello Studente*, organised by the rector of the University of Rome. Blaschke, rector of Hamburg University, who had been on a lecture-trip in Italy earlier that year, and had received an honorary doctorate[396] from Padua University, spoke there again.[397] Other contributors were: F. Conforto, on the state of the theory of equivalent systems and the theory of correspondences between algebraic manifolds; A. Speiser, on modern problems in the theory of abstract groups; G. Ricci, on old problems and new results in additive number theory; K. Popoff, on ballistics and the theory of integral and differential equations; O. Onicescu, on modern problems of the probability calculus; A. Signorini, on recent progress in the theory of finite thermoelastic transformations; R. Fueter, on recent viewpoints in the theory of analytic functions of two variables; G. Sansone, on recent problems in the theory of ordinary differential equations and some types of partial differential equations; B. von Kerékjártó, on the theory of continuous groups.[398]

The conference ended with a detailed introduction to the works of the *Istituto per le Applicazioni del Calcolo* headed by professor Dr. M. Picone. The institute operated under the aegis of the *Consiglio Nazionale delle Richerche*, a council similar to the Reich Research Council. Judging from the number of its personnel, the limited field of its work, mainly on numerical integration of differential equations and related studies, and the space it occupied, Hasse found that the institute did not create the impression of being a "central institute of applied mathematics".[399]

5.50

Why Should Every Philistine Know who Hilbert was?

Carathéodory used to visit Hilbert and offer him birthday congratulations. Probably on such an occasion in 1942, Herglotz reserved a room in hotel Krone for Carathéodory, who wished to be there together with Hecke,[400] apparently to discuss with him problems about the occupation of the Munich chair. Carathéodory informed Süss about an accident Hilbert had suffered, which compelled him to lie motionless, and recounted his impressions of the trip:

Two days before his birthday, Hilbert fell down the stairs and broke his left arm. Despite this, he received a few people but he was very quiet and apparently he still had pains, he was also tired because of the rector, the mayor, etc., who had been there earlier to visit him and, therefore, I saw him only very briefly. However, the fracture should be of a simple kind, but since he went again to the clinic on Saturday to get a new bandage, I spoke in detail only with his wife before my departure. Hecke and van der Waerden were also there and some younger people (Zassenhaus, Deuring & Bachmann); the last two had given quite nice lectures at the institute on Saturday. Particularly, Bachmann's lecture, in which he substituted the ordering axioms by a completion of the congruence axioms, would interest you.

Unfortunately, the trip [Munich-Göttingen-Munich], especially on my return, was not very pleasant. Overcrowding, delays etc. We only arrived with a five-hour delay at just 3 o'clock in the morning. But oddly enough, I was not at all tired and found the way in the snow through the abandoned Munich very romantic and beautiful! Yesterday, however, I completely caught up on the sleep that I was missing.[401]

On 14 February 1943 Hilbert died of complications arising out of his accident. Hecke represented the Hamburg Faculty of Mathematics and Natural Sciences at Hilbert's funeral. From Munich only Sommerfeld attended. Carathéodory was ill and the speech he had prepared in the name of German mathematicians was read at Hilbert's grave by the emotionally touched Gustav Herglotz. In his text, Carathéodory stressed Hilbert's uniqueness and integrity.[402] Therefore, he reacted with annoyance when Süss asked him for a dictum of Hilbert that might provide an appropriate engraving on a stone that the German Union of Mathematicians had decided to place on Hilbert's grave:[403]

Dear Herr Süss, I thank you very much for your letter, which I also made known to Perron and Tietze and discussed with them. We do not believe that a citation out of Hilbert's works would increase the effect of his gravestone. Apart from the fact that our generation is much too close to him to decide or to only predict which one of his achievements, let's say in 100 years, will be regarded as the greatest. And, after all, a gravestone must be for centuries. I personally have felt the grave of Gauss, which, except for the name and the date, bears no other inscription, to be particularly beautiful and impressive. But if you wish, one could still add another title, such as 'professor of mathematics in Göttingen 1896–1943'. One could also write 'the great mathematician' or maybe even better 'the mathematician'. But why should every Philistine who passes by the grave, know who Hilbert was?[404]

It is said that when Carathéodory was himself once asked why he did not give his title when signing his name, he replied that the reader of his work would not be interested in knowing who Carathéodory was but would be interested in knowing what Carathéodory wrote. This statement is consistent with his reply to Süss. Carathéodory obviously identified a person with the person's actions and not with the person's origin or institutional position.

As to the obituary of Hilbert for the Göttingen Mathematical Society, Carathéodory discussed with Sommerfeld the refusal of the Göttingen mathematicians to write it. Both of them assumed that the reason for the refusal was to be found in the behaviour of Hasse, Hilbert's successor. Sommerfeld was prepared to write the obituary himself, only if Herglotz desired it, and only as Hilbert's oldest student and friend, but not as an expert in mathematics. Besides, he did not wish to intervene in Göttingen's internal affairs.[405]

Herglotz assured Sommerfeld that Hasse himself had refused to write the obituary under the pretext of being engaged with military activity and he implied that the most appropriate person for the task would be someone who could write about Hilbert's personality. If Hasse could not be persuaded, Hecke should be asked, since he had been Hilbert's assistant for many years. However, Hecke had proposed that it would be better to entrust a commission under Carathéodory with the obituary. Finally Herglotz asked Sommerfeld to provide the obituary which was to be written as the most dignified memory of Hilbert in earlier years.[406] On 14 October 1944

Ansprache bei dem Begräbnis
Hilberts.

[NB. Hilbert ist in der Nacht vom 14 zum 15 Februar 1943 in der Chirurgischen Klinik zu Göttingen gestorben, nachdem er kurz nach seinem 81ten Geburtstag sich bei einem unglücklichen Fall in seiner Wohnung das Bein gebrochen hatte. Der Sarg wurde in der Frühe vom Donnerstag den 18ten in seine Wohnung gebracht, wo in Gegenwart seiner 78-jährigen Frau, die nicht mehr ausgeht, die Trauerfeier um 12 Uhr Mittags stattgefunden hat. Ein Geistlicher war nicht zugegen. Die Reihenfolge der Redner war: Der Rektor der Georgia-Augusta, der Oberbürgermeister von Göttingen, dessen Ehrenbürger Hilbert war, der Dekan der Naturwissenschaftlichen Fakultät Correns (Mineraloge), Sommerfeld (München) der als alter Königsberger von seiner frühesten Jugend mit beiden Hilberts befreundet war, Carathéodory, ein mathematischer Student, zuletzt Herglotz (Göttingen). Dabei hat Correns auch im Namen der Ges. d. Wiss. Göttingen & der Preussischen Akademie gesprochen, Sommerfeld im

Carathéodory's notes for his address at Hilbert's funeral.
Courtesy of the Bavarian Academy of Sciences, Archive.

Carathéodory wrote to Herglotz: "I am sending you a small obituary of Hilbert, which I have written for the academy here."[407]

That Carathéodory was considered Hilbert's inheritor in the unofficial hierarchy of the remaining German mathematicians is confirmed by another event. On 4 May 1943, *Hofrat* Wirtinger, Mayrhofer, Kruppa, Grey, Schweidler, Graff, Saliq and others proposed Carathéodory for honorary membership of the Class of Mathematics and Natural Sciences of the Austrian Academy of Sciences. They supported their proposal with the argument that after the death of their honorary member David Hilbert, Carathéodory, "one of the mathematicians of outstanding merit", was entitled to be honoured highly on his 70th birthday.[408]

5.51
Summer Vacations in the Black Forest

Carathéodory spent the summer vacations of 1942 and 1943 in the Black Forest near Freiburg. Wilhelm Süss, then rector of Freiburg University, and Georg Feigl, a full professor of mathematics at Breslau University since 1935, received the Carathéodorys in a particularly friendly and generous way. Feigl, one of Koebe's doctoral students and, since 1919, a budgeted assistant at the mathematical institute of Berlin University directed by Schmidt, had been entrusted with the editorship of the important mathematical journal *Fortschritte der Mathematik* by the Prussian Academy of Sciences in 1925. From this position he moved to Breslau University as a full professor.[409] In the spring of 1941 he was elected a member of the committee of the German Union of Mathematicians.[410] Together with Behnke, Süss and Hamel, he was in charge of the instruction commission of the DMV.[411] Feigl worked as the head of a group engaged in aeronautical research during the war.[412] After 1 July 1943 his group bore the name "Mathematical Institute".[413]

In the Black Forest, a large company of colleagues gathered daily during lunch and dinner to discuss new theories, new and old books, and the life and work of friends on the other side of the war-front, all enlivened by anecdotes told by Carathéodory. Politics was an excluded topic of conversation.[414] Through Feigl, Carathéodory had reserved a room in the hotel Kyburg from 20 August 1942. He intended to stay there for one to two weeks[415] and he was probably not willing to work on anything else in mathematics except his own projects.[416]

Carathéodory's 70th birthday was celebrated in the Black Forest on 13 September 1943 with a party organised by the DMV, paid for by them and "another party". Carathéodory was very amused and satisfied and he thanked Süss very much for the honour.[417] Süss had just returned from the conference of rectors of German universities (Salzburg 26 August 1943) where he had demanded the optimal use and the complete mobilisation of Germany's scientific potential for warfare.[418]

Carathéodory was then staying in hotel Luisenhöhe in Horben near Freiburg. Munich University certified that he was on vacation for relaxation from the 1 to the 21 September 1943.[419] Via Süss, Fueter and Finsler sent two presents for Carathéodory as a greeting from Switzerland.[420] Sperner confessed that it was "a pity not to be able to come to Cara's celebration".[421] Carathéodory had called upon Feigl and his wife, but they were unable to attend.[422] Tietze, who was also invited to Cara-théodory's birthday,[423] did not go. Only Kneser and Behnke[424] could travel to the celebration, which cost the union RM 85.50, and included a basket of flowers, lunch, wine and coffee for Carathéodory and his wife.[425]

In mid-November 1943 Carathéodory wrote to Kalitsounakis: "In September I completed the 70th year of my life. Schmidt wrote a beautiful article in *Forschungen und Fortschritte*,[426] where he also mentions my activity in Greece. It would interest you if you could find it in Greece. It would be somewhere there, since, through the Foreign Ministry, the Rector of the Technical University sent me a telegram, which has greatly moved me. For the time being we are all well and I would ask you to send us your news every now and then. With great love and friendship. K."[427]

The Nazi Party paper *Völkischer Beobachter* No 256 of 13 September dedicated an article to Carathéodory entitled *A Significant Mathematician*.[428] It referred to his origin, his academic career and his Smyrna University project, but omitted to mention his work on the reorganisation of Greek universities in the 1930s.[429] Schmidt, similarly, had also omitted a reference to Carathéodory's intervention in Greece in the 1930s, and failed to mention Venizelos as well. Carathéodory himself might have considered his Greek project of the 1930s a great failure. Though appointed by Venizelos himself, he had been dismissed and naturally avoided trumpeting the fact in Germany. It is worth mentioning that, with respect to that extremely significant effort of his, his daughter still expresses the opinion that "one does not always succeed". However, the Greek universities have recently shown signs of restoring Carathéodory's reputation.

5.52

An Unrealised Plan to Visit Finland and the Rosenberg Report on Carathéodory

On 11 January 1943, Carathéodory asked for the rector's permission to travel to Finland to deliver a series of lectures in April or May that year. He had been invited by Lindelöf, who had been a member of the German Union of Mathematicians since 1900 and was counted, according to Carathéodory, among the best mathematicians in Europe. Carathéodory praised him as "the only foreigner, of whom, as far as I know, a mathematical textbook has been translated into German and published here in recent years".[430]

Carathéodory's invitation was probably connected with his appointment as an honorary member of the Finnish Society of Sciences. On 26 January 1943, von Jan, a department head at the Bavarian Ministry of Education, reported to the Reich Minister of Education: "Taking into account the opinion of the former head of the lecturers at the University of Munich, then shared also by the rector, which was submitted together with the marginal report of the Rector of the University of Munich dated 9 February 1940 No 619, I would ask you to examine whether Professor Carathéo-dory's further travelling abroad is desired during the war. I cannot approve of such trips in view of the internationalist attitude of Professor Carathéodory. (Cf. also my report dated 27 November 1942 No V 55108 concerning honouring of retired university lecturers)."[431]

However, with a ministerial decree of 16 February 1943, the Reich minister informed Carathéodory that he reserved the final decision until the conclusion of the investigation regarding Carathéodory's person which could take longer. Therefore, Carathéodory would be able to travel in May 1943 the earliest.[432]

On 19 March 1943, the Reich minister informed the Bavarian minister: "In response to the application of 20 February 1943 – No V 7332 – I grant permission to the full professor of the University of Munich, Privy Councillor Dr. Constantin Carathéodory, to accept his election to an honorary member of the Finnish Academy of Sciences. The Foreign Ministry and the German Central Office for Conferences in Berlin have been informed."[433] It becomes evident, therefore, that Carathéodory was suspect to the Nazis in Munich but still enjoyed the trust of the Nazis in Berlin. Despite Berlin's decision, however, the authorities in Munich insisted on blocking Carathéodory's trip.

Two months later, the Nazi Party office (*Partei-Kanzlei*) in Munich asked the opinion of the NSDDB, which on 27 May 1943 sent them the following warning, a copy of which was communicated to the party member Härtle of the party's Science Department (*Hauptamt Wissenschaft*) in Berlin:

To your question of 19 May 1943 if we could agree that Professor Dr. Constantin Carathéodory, Munich, who on behalf of the Finnish Mathematical Association has been invited for a lecture by Professor Lindelöf, travels to Helsinki, we should inform you that it is urgent to warn you of a further trip of Professor Carathéodory abroad. Unfortunately, his trip to Rome in November 1942 could not be prevented in time. Dr. Constantin Carathéodory is a Privy Councillor and a professor of Mathematics at the University of Munich since 1 April 1913. He was born in Constantinople, as the son of the Turkish envoy in Brussels, and educated there at the Belgian Military Academy. Later, he went to Egypt as an engineer, was employed there for the building of railways, studied mathematics there and became a professor in Hannover and at the University of Berlin. After the war he was invited by Venizelos, whose intimate friend he was, to found the University of Smyrna. Here and in Athens he "organised" the Greek university relations, that is he moulded them to the way of thinking of Venizelos, who is a half-Jew. After expulsion by the Turks, he became a professor at the University of Munich. Carathéodory himself and his family members are Greek citizens. He maintains a large international correspondence and relationships and, according to reports here, he is particularly friendly to the Jews. Further, it has once been claimed that he provided the Jewish mathematician A. Rosenthal with an appointment to Giessen. His partiality to the Jews is especially cultivated. So, he decisively supported the appointment of the dreadful East-Galician Jew Salomon Bochner to Munich University (around 1928). To be precise, this enterprise was then rejected by the ministry but, despite that, on the grounds of mediation from Carathéodory who described Bochner as a "Jewish genius", Bochner gained permission to lecture at the University of Munich.

In what way Professor Carathéodory has further taken advantage of his extensive international relationships is not entirely clear. However, his quite cosmopolitan attitude compels us to observe his relationships with the strongest distrust. Also the fact that the group participating in the mathematicians' congress in Rome (Nov. 42), whose leader was professor Hasse (Kiel), and in which also Professor Carathéodory participated, paid a visit to the Pope and the Papal Academy then, is worthy of particular attention. This visit probably took place through Professor Carathéodory's mediation, since he had been appointed to the Papal Academy a short while before. Further, the fact that Professor Carathéodory obtained a sabbatical in the summer term of 1937 to give guest lectures in America should be a warning.

It is to be feared that professor Carathéodory will be active, and also report in a manner harmful to the German Reich in Finland, where a clique, Jewish-friendly and closely sympathising with the liberal science of the democratic countries, indeed exists to a certain percentage.

Therefore, we raise the sharpest doubts against Professor Carathéodory's trip abroad.

Heil Hitler!

Signed: Hr. Hiltner [434]

This warning, a villainous denunciation of a respected and world-wide famous scholar, reveals how incompetent the NSDDB was; none of their information about Carathéodory was precisely correct, whereas they ignored things which might have been harmful in their opinion. On the other hand, it reveals their hatred towards a person who possessed power because of his intellectual capacity and who was therefore identified with those considered to be enemies of the Reich. As to the errors in the letter, the following points can be made.

In November 1929, the list of candidates for the vacant chair of Ludwig Schlesinger, the retiring full professor of mathematics at the University of Giessen, included Gustav Doetsch, a full professor at the Technical University of Stuttgart, Artur Rosenthal, an associate professor at the University of Heidelberg, and Gerhard Thomsen, an associate professor at the University of Rostock. Doetsch rejected his appointment on 2 March 1930, Artur Rosenthal on 21 May 1930. On the basis of the documents regarding the appointment and the negotiations following the appointment of Schlesinger's successor, Harald Geppert, it can not be confirmed that Carathéodory had supported Rosenthal's appointment.[435] However, Carathéodory may have been favourably disposed towards Rosenthal, as he had been in the case of the Kiel professorship in 1928.

Bochner was employed at the University of Munich until he emigrated. At the beginning of the 1930s, he obtained a teaching commission for analytic geometry. His lectures on *Fourier integrals* published in 1932 were a precursor to distribution theory.[436]

The NSDDB seems not to have not been aware of the potentially more damaging Carathéodory–Morgenthau connection or of Carathéodory's Ionian multinational vision, which embraced local Jews. They had, therefore, to invent Venizelos's half-Jewishness in order to denounce Carathéodory. Internationalism and co-operation with the enemies of the German people after 1914 were ascribed by the Nazis to the Jews[437] and Carathéodory was apparently accused of both and thus, by Nazi logic, indirectly identified with the Jews.

The Pope, despite his indifference to the holocaust, of which he was well informed, presented another centre of power to Hitler's regime and therefore Carathéodory's visit to him drew special attention.

Dr. Wolfgang Erxleben, since November 1941 director of the Office of Observation and Judgement of Science (*Amt Wissenschaftsbeobachtung und -wertung*), commonly known as the Rosenberg Office (*Amt Rosenberg*), handled the extensive correspondence with the party office, the NSDDB, and expert advisors. But Heinrich

Härtle, who was his deputy since 1942, was one of the real heads of the organisation whose name appears only in exceptional cases. On average, this office produced sixty to eighty "expert reports" per month in the period between 1941 and 1943, according to a report by Erxleben. In the statements of the NSDDB and the Rosenberg Office, which were sent to the party office, the borders between expert advice and ideological judgement were blurred.[438]

On 2 June 1943, and on Dr. Erxleben's instructions, the Office of Observation and Judgement of Science asked, in confidence, the party member Dr. Graue of the Kaiser Wilhelm Institute for Physical Chemistry in Dahlem, Berlin to examine "this case" once more.[439]

On 19 July 1943 Dr. Georg Graue answered to Dr. Erxleben that

> Professor Carathéodory is a Greek and not a German, which at first should be taken into account for the judgement. In my opinion, Munich is starting from false assumptions in the objection. According to my information, C. is one of the most significant living mathematicians, who, as a consequence, certainly maintains numerous scientific and personal relations. He has studied mathematics in Germany and he was earlier an engineer in the Turkish service. His father was a Turkish ambassador. When after World War I a part of Asia Minor seemed to come under Greek rule, C. resigned his Berlin full professorship and put himself at the disposal of the Greek government for the building of a greater Greece. That was an act springing from his national feeling. C. shares the destiny of being appointed to the Papal Academy – as the Munich expert report says – with a great number of German, also non-Catholic colleagues. Planck is also a member of the 'Papal Academy'. In my opinion, one cannot draw any conclusions regarding his political attitude from this fact. As far as I know, C. has always stood loyally to German interests, also after 1933. I do not believe that, in Finland, he will make what we consider to be unfavourable remarks. Certainly, he would also be confronted with the personality of the rector of the University of Helsinki, who is, of course, the moving force of the invitation. This man adopts an unambiguously clear position. He is the mathematician *Nevanlinna*. Admittedly, one of the mathematicians in Helsinki, Herr Ahlfors, is married to an American Jewess. In the case of C., I have not noticed the particularly Jewish-friendly attitude mentioned in the Munich expert report.
>
> Heil Hitler!
>
> Dr. Graue [440]

Despite some inaccuracies in details, Graue's reply was objective, or even sympathetic to Carathéodory. Given that Graue was then the head of the office of war economy at the Reich Research Council and Süss was the head of the mathematics work group there,[441] it is very probable that they had exchanged views on Carathéodory's attitude.

Rolf Hermann Nevanlinna had studied under Lindelöf at the university of Helsingfors in 1913–1919. He became professor in Helsingfors in 1926, appointed to the recently established second chair of mathematics. He met Hilbert, Courant and Emmy Noether in Göttingen at Landau's invitation. When Weyl left Zurich, his chair was offered to Nevanlinna, who refused it. Nevanlinna continued to work in Helsingfors and became rector of the university there in 1941. At that time he was considered for a professorship of mathematics at Göttingen. He is known for his invention of harmonic measure in 1936 and his theory of value distribution for single-valued meromorphic functions in 1925. Nevanlinna's "unambiguously clear position", of which Graue spoke in his report, can be read in his letter to Hasse on 25 March 1941:

You know, dear Herr Hasse, your remarks about the hypocritical and stupid 'moral indignation of Western politicians, who try to hide their envy and their hate against Germany under the mantle of nice phrases', correspond completely to what we feel here and say to ourselves daily. You know those deeply rooted sympathies which connect us Finns with Germany and these bonds are today stronger than ever now that the easily understandable irritation caused by our difficult time a year ago has died down. [...]
It is absolutely clear to us that only a strong and powerful Germany, the heart of Europe, is capable of forming the fate of the European community in the way, which the interest of all European nations of culture demands. Personally, I am firmly convinced thereof and I believe to see a total justification of this conviction in European history, namely that Germany is today summoned not only to save European culture, which already happened in 1933, but to lead it to an undreamt-of blooming. The world-historic significance of the present hour is immense. Certainly, in European history, there was hardly a moment of powerful decisions depending on an organic, internal, intellectual, social and cultural development in the heart of Europe that coincided with the moment of great power-political decisions to the same degree as at present. I do not doubt for a moment about the happy outcome of this struggle. It is surely sad and horrible that this battle demands tremendously heavy sacrifices; but certainly it has to be like that. This time is particularly hard for small countries. Despite the feeling of powerlessness, which threatens to overwhelm a small country today, it is a great happiness if, like the Finnish people, it can be said that we have not failed.[442]

On 23 July 1943 the head of the Office of Observation and Judgement of Science (Braun) sent to the party member Starke of the party office in Berlin a confidential report referring to the letter dated 27 May 1943 of the NSDDB to the party office of Munich. The Rosenberg Office said that they had gathered information about Carathéodory and repeated more or less Dr. Graue's answer to Dr. Erxleben. They underlined that their judgement concerning Carathéodory deviated considerably from that of the NSDDB, to which they also sent a copy of their letter to Starke.[443]

In the meantime, Carathéodory became impatient. On 5 August 1943 he demanded the return of his passport from the dean's office of the Faculty of Natural Sciences of Munich University: "Since a lecture trip to Finland is now out of the question, I would like to request the return of my passport which is in the files in Berlin.

    C. Carathéodory
    (invited by Prof. Dr. Ernst Lindelöf)"[444]

Carathéodory stressed that he had been invited by Lindelöf and not by Nevanlinna, who was described in Dr. Graue's answer to the Rosenberg Office as the "moving force of the invitation". Ernst Lindelöf, whom Carathéodory had met for the first time, together with Lindelöf's father, at the international congress of mathematicians in Heidelberg in 1904, had been a full professor at the University of Helsingfors up to his retirement in 1938. Since 1907 he was one of the editors of the *Acta Mathematica*. He gave up his research to dedicate himself to teaching and the writing of textbooks.

The head of the NSDDB continued to press for the report on Carathéodory from the Rosenberg Office. On 25 November 1943 he wrote imperatively: "You wrote to us on 8 October that the relevant documents are in your evading place on the grounds of air-raid security. We assume that, in the meantime, you have been able to obtain the files and we expect your quick response." [445]

For the third time the request was repeated on 5 February, 1944: "We remind you of our letters of 1 October and 25 November 1943 and once more we ask you urgently to send us the requested report on Professor Carathéodory." [446]

The Rosenberg Office (Braun) replied on 14 February 1944 that they had attached their judgement about Carathéodory to their letter without sending the report on which that judgement was based. They also said that they were not informed whether Carathéodory had travelled to Finland. They asked the NSDDB to provide an indication if, in the meantime, the association had received new material supporting the one or the other opinion. [447]

The head of the NSDDB repeated his request for the fourth time on 24 February 1944: "Unfortunately, the enclosure mentioned in your letter of 14 February 1944 was not in your letter. We ask you for the immediate submission of your position regarding Professor Carathéodory." [448]

Finally, on 6 March 1944, Braun sent the requested document to a party member of the NSDDB, Dr. Glara, saying that it had been left out in error. [449]

The older Carathéodory.
Courtesy of Mrs. Rodopoulos-Carathéodory.

The NSDDB had several times attempted to draw up directives for the ideological orientation of individual disciplines, the first time in 1938. Being a party authority, Rosenberg denied the NSDDB the right to engage in philosophical (world-view) scientific research. This led to a long-year conflict between the NSDDB and Rosenberg. [450]

5.53
Munich in Wartime – Contact with Leipzig and Freiburg

By the end of 1942 the Greek Orthodox Church in Munich faced serious problems. The parish priest Archimandrite Meletios Galanopoulos, was captured in Leipzig on 23 November 1942 and accused of receiving a "hostile broadcasting station". He was sent to Dachau concentration camp. Pressure on the Greek Church increased when on 8 December 1942 the Russian Orthodox community of St. Nikolaus in Munich applied to the Bavarian Ministry of Education for the shared use of the church with the Greek Orthodox community. The minister approved the application on 19 February 1943, with permission of use to run until further notice. Members of the Greek Parochial Church Council, among them Carathéodory, its vice-president and a church donor, refused to enter into an agreement with the Russian Orthodox community on this matter.[451] Carathéodory attended the Greek Church of the Redeemer every Sunday[452] and observed all the customs and traditions of Orthodoxy, which he tried to pass on to his children.[453]

The year of 1943 started with terror and a deterioration of living conditions for the inhabitants of Munich. On 18 February 1943 Hans and Sophie Scholl, organisers of the pacifist student resistance to nazism, were arrested in Munich. On the 10th anniversary of the Nazis' taking control in Munich, during the night of 9 to 10 March, the British launched a severe air attack resulting in 205 dead, 435 wounded and 8975 homeless. Public and private air-raid shelters were overcrowded. The Bavarian State Library, the old and the new gallery, the sculpture gallery and the "brown house" were among the severely damaged buildings. The university library was extensively destroyed between 1942 and 1945, while the library building of the Technical University was turned into a pile of ruins. Bombardments continued throughout the year. According to an official German estimation, the allied attacks led to a loss of about three million volumes of scientific literature in August 1943. All journals of the Bavarian State Library and almost all journals of the mathematical seminar of the university were burnt.[454]

Carathéodory managed to retain contacts with his remaining friends and with the DMV. On 12 October 1943 he wrote to Zermelo:

Dear Zermelo, thank you for your card. I have just found one copy of the recurrence theorem, which I am sending you. The night of 2 to 3 October was really bad here. I admit that we had only pieces of broken glass but the city looks terrible. Maybe even worse than what it is in reality.

Many regards,

Yours

Cara

We are very sorry that you are so far away![455]

On 4 December 1943 the centre of Leipzig was bombed, including van der Waerden's house. Fifty-eight out of a total of ninety-two university institutes and clinics were hit that day. Van der Waerden, then professor at the University of Leipzig, and his family, went to Dresden where they stayed in the house of Franz Rellich,[456] his brother-in-law, just for a night and then travelled on to Bischofswerda near Dresden. They remained there until the end of 1944 and returned to Leipzig when the city was heavily bombed. On 11 December 1943 Carathéodory wrote to Frau van der Waerden: "For the time being I still possess my books and separata and I like to help him with them in the best possible way. He should write me what he needs." [457]

On 6 January 1944 Carathéodory wrote to Süss:

About the extent of the Leipzig catastrophe you are probably better informed than me. But the fact that a firm like Fock has lost almost all of its stock especially the Aldines[458] of the 16th century (supposedly, 1.5 million volumes of this company have been burnt) is a misfortune for the whole civilised world. And that Harassowitz, the centre of world trade with oriental books, is also destroyed is just as bad. Teubner's destruction is fatal as well, because the value of his publishing house was just in the stock of those works which had *not* been purchased and which represent the hundred year activity of German philologists and it was exactly these works that he will not want to publish. Besides, does this mean, for example, that the manuscript, the galley-proofs and the corrections of my book have been burnt? By the way, it seems that he will let it be set again some time. With respect to my clock, I think that you should not wait that long to send it. Because as soon as the first signs of the planned attack in the West become apparent, an enormous supply of services over Karlsruhe and maybe also over Freiburg will start and who knows for how long the connection with Munich will be disrupted. Therefore, I believe that you can risk to let a clock maker, known to you, wrap the clock up in an expert way and then send it here as a registered letter insured for 500 or 1000 marks. Then, *you* will also get rid of the worry of which you speak.[459]

5.54
Endeavours to Save "German Science"

5.54.1
In Favour of van der Waerden's Stay in Germany

Amidst all the disturbances and confusion of the war, some semblance of mathematical activity continued. Van der Waerden was offered a post in Utrecht. He said later that he had not wished to accept it, since he would have been appointed by the German Reich Ministry. However, from the following letter, it becomes apparent that he seriously considered the offer. Seeing the end of the war approaching, Carathéodory wrote to him on 25 March 1944:

With regard to U[trecht] I very well understand your point of view. But not everything is hopeless yet and it would be regrettable, if you could not arrange it in such a way as the possibility to move there remains open for some more time. If you could and wished to visit me in Munich one day, we could find together many books in my house, which I do not need and would be useful to you and which I would enjoy to give them to you as a present. But we would have to choose them together; by letter, this is difficult.[460]

After Germany's capitulation van der Waerden did not wish to return to Leipzig and was employed by an industrial enterprise in Holland.[461] The Queen of Holland had refused to sign his appointment at the University of Utrecht, because he had spent the war time in Germany and that was reason enough to assume collaboration with the Nazis in his case. While in Leipzig, van der Waerden attended the course of lectures given by the prominent philosopher Hans-Georg Gadamer on Plato's *Republic*. It is very probable that Plato was a subject of discussion between him and Carathéodory, who in a letter to van der Waerden remarked:

Derkyllides is rather convincing. I have kept him so long because I wanted to exactly compare what he says with the passage of Πολιτεία [*The Republic*]. Unfortunately, I haven't had the chance. Today I have given the wonderful chapter III of your book to Rehm and asked him at the same time, how it conforms with the Zodiac. He told me that, at first, only individual constellations of the Zodiac came from Babylon to Greece and that probably it was not known in Greece that it was a full circle. The whole circle became known for the first time in the 6th century; one of its constellations – I believe it is Aries – is not of Babylonian origin.[462]

Carathéodory was writing here about van der Waerden's book *Science Awakening*,[463] in which the author refers to the *Republic*: "The study of mathematics develops and sets into operation a mental organism more valuable than a thousand eyes, because through it alone can truth be apprehended."[464] Chapter III of his book bears the title: Babylonian mathematics. Rehm's information, passed on to him by Carathéodory, was used by van der Waerden in his book, as we can see from the remark that "the names of the signs of the Zodiac, with which the Greeks became acquainted around the year 550 [BC] through Cleostratus of Tenedos, are also derived from Babylon."[465]

Van der Waerden's interest in the history of mathematics goes back to the time of his studies when he was attended the course on the subject delivered by Hendrik de Vries. In Göttingen van der Waerden had attended Neugebauer's lectures on Greek mathematics. Later, he visited Neugebauer in Copenhagen, who aroused his interest in Babylonian astronomy.[466]

5.54.2
Von Laue's Acknowledgement

On 3 February 1944 Carathéodory and others proposed Max von Laue, the Nobel Prize winner for physics in 1914 for the diffraction of X-rays in crystals, for corresponding membership of the Class of Mathematics and Physics of the Bavarian Academy of Sciences.[467] Laue had received his doctorate with Max Planck in 1903. At that time he was Vice-Director of the Kaiser Wilhelm Institute for Physics in Berlin. An outspoken anti-Nazi, he had protested against disciplinary proceedings of the Berlin Academy against Einstein in 1933. For his scientific views he had been a target of "Aryan physics".

5.54.3
Steck's Exclusion from Lambert's Edition

A major issue of scientific prestige proved to be the planned edition of Johann Heinrich Lambert's collected works. Lambert was an 18th century Alsatian scholar, who is today regarded as a physicist, geometer, statistician, astronomer and philosopher and a representative of German rationalism. His performance is acknowledged in phenomenology and in the continuation of Leibniz's language-philosophical approach in semiotics. Among the achievements of Lambert as a physicist and a mathematician are the discovery and measurement of luminous intensity; the formulation of the laws governing light absorption, and thereby the establishment of photometry;

the formulation of a law for the motion of comets or planets. He was among the first to appreciate the nature of the Milky Way; he established several theorems in non-Euclidean geometry, developed De Moivre's theorems on the trigonometry of complex variables and introduced the hyperbolic sine and cosine functions. He proved the irrationality of both π and π^2, created a general theory of errors and, finally, was the first to express Newton's second law of motion in the notation of the differential calculus.

With considerable vigour Carathéodory tried to prevent an edition of Lambert's work by Max Steck. He wrote to Süss in April 1944:

Lambert, together with Euler and Lagrange, is not only the most important mathematician of the Friederician Academy [Prussian Academy of Sciences] and, due to his *Novum Organon*, the most important German philosopher[468] between Leibniz and Kant, but he was also an astronomer, a physicist and the earliest mathematician of the first rank, whose performance in applied mathematics counterbalanced his purely theoretical works. For example, in astronomy, the theorem which he proves in the work 'insigniores orbitae cometarum proprietates' has become classical. Against that, his cosmological letters (Augsburg 1761) are almost forgotten today; they caused a great sensation then and contain a prophetic representation of the spiral-nebula world as we know it today. In physics, pure 'photometria sive de mensure de gratibus luminis, colorum umbrae' (Augsburg 1760) is pioneering; likewise, the pyrometry which Johann III Bernoulli published after Lambert's death.

In applied mathematics, not only perspective but, most of all, his cartographic achievement should be mentioned.

[...] The difficulty is to find an editor who would do justice to such a universal genius. Dr. Steck is not even able to supervise the edition of the purely mathematical works of Lambert. If he were given the task, the whole matter would result with deadly certainty in a great disgrace and a disaster that could not be recovered.[469]

A day later, Sperner, the editor of the *Jahresbericht*, expressed strong doubts about Steck's suitability for the Lambert edition on the grounds of Steck's publications up until then.[470]

A month later, Hellmuth Kneser, professor at the mathematical seminar of Tübingen University, who had been informed that Switzerland had given Steck an award for his Lambert work and entrusted him with the task of preparing the whole Lambert edition, intervened in the discussion and argued that Steck ought to accomplish the edition because this would ensure a Reich German co-operation and prevent a purely Swiss Lambert edition.[471]

Tietze, on the contrary, believed that national considerations were of secondary importance with respect to the project of the Lambert edition, whereas the decisive criterion had to be the scientific accuracy of the edition. The multidisciplinary character of the work required the participation of scholars from various fields. As to Steck, whose overall scientific performance was more than questionable, Tietze thought that he was completely unsuitable for the task.[472]

Even more substantial was the critique coming from the historian Lucky, who blamed Steck for having misinterpreted Lambert's texts on perspective and of having maltreated the original work through unnecessary modernisation of the language.[473]

For the review of Steck's Lambert book in the *Jahresbericht*, Süss proposed Charles Pisot, who had given a public lecture on Lambert and was privately engaged in research in Lambert.[474]

On completely different grounds, Heinrich Scholz, justified a "holding action" against Steck. In his opinion, a Lambert edition of that order would at best be approved of by a necromancer but certainly not by a historian and he asserted that Lambert, despite his two-volumed *Organon*, was trivial as a logician and in the mathematical logic his only merit was that, of all his contemporaries, he was the only one willing to continue Leibniz's approach. But also in that area he had achieved almost nothing traceable.[475]

In the end, Lambert's *Opera mathematica* was edited by Andreas Speiser, appearing in 1946–1948. As to Lambert's significance, it is Carathéodory's view, and not that of Scholz, that has been established.

5.54.4
In the Jury for a Prize in Geometry

Together with Blaschke and three other mathematicians of Leipzig University, Carathéodory was appointed by the committee of the German Union of Mathematicians to the jury of the 1944 Ackermann-Teubner Prize for excellent work in geometry. He had been informed of his appointment by Süss six days after his 70th birthday[476] and accepted it two days later.[477] This procedure was met with Sperner's opposition, who, although abstaining from objections against Carathéodory's nomination as a jury member, recommended Süss to prefer younger members in such cases.[478] There was a shift in Sperner's attitude in this respect, for he himself had mentioned the emeritus Carathéodory – together with Perron, E. Schmidt, Hecke, Kneser, Haupt and Doetsch – as a possible candidate to write the report on analysis for the *Jahresbericht* after Knopp's refusal.[479]

5.55
Bombardments of Munich

In the spring of 1944 American bombers started carrying out air raids during the day while the British continued bombing in the night. The first American air attack on 18 March 1944 caused 172 dead, 296 wounded and 4085 homeless, as well as the damage or complete destruction of a series of cultural monuments. The air attacks were blamed by the *München-Augsburger Abendzeitung* on the "Jewish brains" who, allegedly, were planning the war in Washington. The mathematical institute was destroyed by a direct hit. Carathéodory and Tietze suffered no personal damage,[480] but Tietze complained about air raids, alarms, loss of time and side-effects on his health.[481] On the night of 24–25 April, the Bavarian Academy building on Neuhauserstrasse was bombed and completely destroyed. The Academy was thus forced to be housed and meet in the university.

In air attacks on 11, 12, 13 and 16 July 1944 a total of 1709 retail trade firms of various kinds were hit in Munich and the surroundings, 368 of them were totally destroyed and 102 severely damaged. Numerous hotels and restaurants suffered a similar fate. Many industrial facilities and plants were hit, so that production stopped for a period. There was a power failure, while heavy damage of the public transport caused temporary traffic interruption and traffic restrictions.[482]

Carathéodory provided an account of the dramatic situation to Süss:

It's almost a wonder that nothing happened to us. We have no water, no gas, no windows, no telephone; at the beginning no light, as well, no tram is en route but we are healthy and for the present unhurt! We also have had a small fire, but, since we were at home, it could be extinguished before causing any damage. Perron's house is fairly damaged through a high-explosive bomb but his furniture and books are saved. Also, as far as I know, nothing has happened to the rest of the mathematicians but it seems 19 colleagues have been bombed out, of whom I know 5 or 6. The greatest part of the university is burnt down. Lectures must be transferred to schools etc. [483]

Apart from Süss, Tietze had also learned from Carathéodory about the devastation of Perron's house. The furniture had to be transported, Perron's ill wife had to stay in the house of M. Müller. Five houses out of nine on the Trautenwolfstrasse, where Tietze himself was living, had been destroyed and others were damaged. [484]

5.56
Denunciations

5.56.1
Mohr

While the Allies were bombarding German cities, the Nazis dealt particularly severely with those trying to obtain information about the developments at the war-front from the allied side, especially when this information came from the BBC. In this respect, Ernst Mohr's, treatment is typical.

In Whitsun 1944 Feigl knew that Mohr, then an associate professor of applied mathematics at the German Charles University in Prague, [485] had been arrested by the Gestapo accused of having received and propagated news from a foreign broadcasting station. [486] Mohr was arrested in the hotel Béranek in Prague on 12 May 1944 having been denounced by one of his wife's friends. As he himself said, he did not belong to any resistance group. But he did associate with a small group of professors in Breslau, such as Georg Feigl and Hubert Cremer, who were clearly opposed to Hitler and his terror regime. [487] After his interrogation in Prague, Mohr was taken to a prison near Frankfurt an der Oder and from the 26th of July was held on remand. The Reich's chief prosecutor at the People's Court brought severe charges against both Mohr and his wife. Mohr himself was accused of having received a foreign broadcasting station, of having made a parody of Hitler, of having described the war situation as hopeless and the extermination of Jews as a great mistake, of having considered the representation of Stalin by the German press to be incorrect, of having drawn parallels between national socialism and Bolshevism on the basis of their dictatorial forms, and of having described a soldier at the front, the informer's husband, as "yet another idiot". The only accusation that Mohr admitted was that he had listened to the BBC. Even so, he was found guilty of having sided with the enemies and of having favoured them systematically through his words. He continued to be held on remand and the main hearing was fixed for 24 October 1944 at which the verdict of the People's Court was that he be sentenced to death. The argument which Mohr had presented in defence of himself, a remarkable performance in his research for the German arms, especially the air force, was not taken into account in mitigation of the sentence.[488]

Mohr's detention alarmed Feigl, who asked Süss for information about what had happened. In his answer, Süss reproached Mohr for rashness, carelessness and reck-lessness inconceivable for a member of a university. Moreover, he attributed Mohr's acts to the bad influence of his wife, who had also been arrested. Süss believed that this "childish behaviour" would surely affect Mohr's academic career in an especially negative way.[489] Probably after the verdict against Mohr, Süss intervened[490] with the intention of being able to use Mohr at the mathematical department of the institute for scientific military research of the Waffen-SS and the police. This institute belonged to the *Ahnenerbe* office of the personal staff of Heinrich Himmler and was housed in the Sachsenhausen concentration camp so as to exploit the expert knowledge of interned scientists.

It was probably due to Hans Rohrbach, a full professor at the German University of Prague, and a scientific assistant at the Reich Foreign Ministry, that Mohr was finally brought to Plötzensee on 18 December 1944. He had been provided with a research task connected with the development of the V-weapons in Peenemünde. In this way Mohr gained a decisive six month postponement of his execution and continued to work in Plötzensee until he was freed during the Russian advance. On 1 January 1946, he took up the Chair of Pure and Applied Mathematics at the Technical University of Berlin, whose former holder, Werner Schmeidler, had been dismissed.[491]

5.56.2

The Hopf Family

Other victims of denunciation were Eberhard Hopf and his wife. On 27 July 1944 Carathéodory asked Süss if he could do anything to help Frau Hopf, the wife of Eberhard Hopf, who had been chosen to succeed Carathéodory in Munich. Frau Hopf, he said, lived in Bayerisch Gmain "in great distress". The couple had lost their house and the entire household in Leipzig; Frau Hopf had to take care of her eight-year-old child and her father, the well-known music scientist Johann Wolf, who had come from Berlin to live with her after having lost everything he possessed there. Wolf's 75th birthday had been celebrated in Salzburg officially a few months before. Frau Hopf had no home help and she herself had been conscripted half a year earlier to carry out half-day civilian work and had to travel every day by train to Ainring at seven o'clock in the morning and leave her child alone. Frau Hopf fell ill and was released from conscription at the German Research Institution for Gliding on the orders of the head doctor of the hospital in Bad Reichenhall, but only temporarily, until the 1st of August. Carathéodory feared that she might not be released the next time she fell ill although her suffering was worsening. Hopf and his wife had been constantly plagued for more than a year by denunciations to all possible authorities by a "malevolent" fellow occupant of their house in Leipzig, not for any substantial reason but only because they could not tolerate each other.[492]

The denunciation of the Hopfs was rather an intrigue based on a personal con-troversy and was also treated as such by the authorities. The procedures taken against him cannot be considered a proof that Hopf was a heroic Nazi opponent. Indeed he himself had not asserted that he was; but Hopf was definitely not a Nazi friend.[493]

Carathéodory thought that, if rumours of their denunciations became known to the uncomprehending villagers of Bayerisch Gmain, the situation for the Hopf family would become doubly difficult. However, because of the child, Carathéodory did not consider Munich to be the proper place for the family to move to, even Hopf spent half of the week there. Therefore, he asked Süss if he could find a lodging for Frau Hopf, her child and, possibly, her father in Freiburg or its surroundings.[494]

A month later, Süss replied to explain that conditions did not allow him to fulfil Carathéodory's wish: he could house only people for whom he could certify significant war research in the region and others could not stay longer than three days in a hotel; in most cases, the authorities had confiscated space given by farmers and house owners asserted that their houses were overcrowded. Süss had offered Carathéodory help when Munich suffered air attacks but, at that time, he had in mind empty student-rooms in the end of the semester; since then these had been registered by the authorities and confiscated. Süss did, however, promise to let Carathéodory know, in case he learned something new in this respect.[495]

5.57
A Reich Institute for Mathematics

Süss further informed Carathéodory "confidentially"[496] that the Reich Research Council had instructed him to begin to set up a Mathematical Reich Institute with immediate effect. The establishment of such an institute had been proposed on 2 August 1944 by Walther Gerlach, who saw it as a remedy for a very perceptible and significant deficiency and pointed out that similar institutes, as for instance the one in Rome, existed and worked with great success in other countries.[497]

Süss saw his first task in setting up a functioning mathematical library in a place secure from air raids. For the whole duration of the war, he was authorised to borrow one or more libraries that had been removed from institutes. He thought that this measure of securing the stocks would be very welcome to some university institutes. He had already found an "almost ideal accommodation for the first seed of the institute" in the Black Forest and wished to act as soon as possible. He asked Carathéodory for his opinion about offering parts of the library of the Munich institute for that purpose after consulting with Tietze and Perron. With Carathéodory's mathematical treasures in mind, he informed Carathéodory that "perhaps the moving of a mathematical private library would also be possible."[498] In September the Reich Institute was set up in Oberwolfach in the Black Forest, housed in a building put at Süss's disposal by the state of Baden and received "the greatest viable mathematical library". Süss settled there a number of colleagues in ideal research positions, and he found a task there for himself, guaranteeing his family a shelter. This was some consolation to him after all his work as rector of Freiburg University during the previous years had been brought to nothing by the severe destruction of the university in air attacks, and he thought that the foundation of the Reich Institute would signal a bright spot in the dark future of German science.[499]

In the summer of 1944, the French number theorist Charles Pisot belonged to the armaments detachment of Süss's Research Institute. Together with Wilhelm Maak and Hans Schubart, Pisot worked on military commissions of high priority.[500] Maak researched in aerodynamics.[501] Pisot and his family had lost everything they possessed during the air attacks on Freiburg[502] at the end of November 1944. Carathéodory tried to establish contact with him in December of that year but he did not know whether Pisot had received his letter.[503] Two years earlier Carathéodory had recommended Pisot for an appointment in Germany, but Behnke doubted as to whether the official authorities would give their consent.[504]

5.58
Munich in the Autumn of 1944

On 14 October 1944 Carathéodory politely refused to accept Süss's invitation to visit Oberwolfach where, consequent to Göring's ministerial decree, the new institute was being built up.

Admittedly, it is completely out of the question always to leave Munich, because I would give up the whole meaning of my existence if I did so. After some very unpleasant weeks, when my house stood open from all sides, I can now peacefully work here again. I would not have been able to sort the house out properly, if I were not continuously there and had not immediately exploited every opportunity offered. These are things one should not entrust to others. [...] Now that travelling conditions have become much more difficult and, particularly after the last air attack on Munich, which completely destroyed the railway area, a change of place is not even to be contemplated.

Carathéodory wished "that the future will not become as bad as it sometimes appears."[505]

On the day he declined Süss's invitation, Carathéodory sent Herglotz a "small obituary of Hilbert" and asked him: "Do not throw it away, I have only one copy left and, at present, it is better if everything were not in one place. Munich looks awful. There are completely destroyed houses very near us, as well. The observatory is severely affected. Fortunately, Rauchstrasse was spared."[506] Carathéodory must have sent the same obituary to Süss, for the latter thanked him for receiving it on the day on which the city and university of Freiburg was severely hit by a "terror attack".[507]

On 26 November 1944 Carathéodory and his wife congratulated the 73-year-old Zermelo on his wedding. "We are both of the opinion that you could do nothing more reasonable. Please, greet your wife from both of us." Then, they wished Zermelo "not to be forced to leave Freiburg because of the war circumstances." "But if it comes to that", Carathéodory continued,

you should not think of the nearest or further surroundings of Munich. This area is already so overcrowded and the state of nutrition here so tense that it has basically to be ruled out from the very beginning. On the other hand, I think that a shelter could be found in the city itself under the assumption that the situation will not get worse in the next weeks, for example through access from outwards or through new destruction. We had very considerable damage in our near surroundings in May and June. Also the outer skin of our house, especially the roof and the windows, had then suffered considerably. With great effort we could somewhat repair these damages and since the middle of July we have calm in our street. By comparison, the devastation in other parts of the city is greater and, in recent weeks, we experienced a more or less heavy attack every few days. Especially in the surroundings of the railway stations, the city looks like a dessert and Munich's connection to the outer world becomes more and more complicated.[508]

While Carathéodory's letter to Zermelo was on its way to Freiburg, Freiburg University was heavily hit on 27 November 1944. Almost all the institutes of natural sciences and medicine and the clinics were then out of order for a long time to come. The humanities departments could at least save their books. The main building received two direct hits and two university professors were killed. The number of dead was estimated at about 5000, on the second day of the air raid 2400 bodies were recovered. Not a single house existed on the main street joining the city centre to its northern part, the cathedral roof and the area around the railway station were damaged, but not the railway facilities. Water, gas, light, telephone went out of order and they were still not functioning in the middle of December.[509] The mathematical institute of the university was burnt down and the French mathematician Professor Roger working there, whose home had also been hit by a bomb, was taken by Süss together with the remaining contents of the mathematical institute to the Research Institute in Oberwolfach.[510]

5.59
"In the Interest of the Union"

In 1944 Carathéodory was elected by the committee and executive committee of the German Union of Mathematicians for membership of the committee for a period of three years until 31 October 1947. Süss asked him to accept this election in the interest of the DMV and hoped for successful co-operation with Carathéodory to the benefit of German science in general.[511] It is obvious that the union and its president had realised by then that the Allies were advancing and thought of using Carathéodory and his authority to save the union and to lend weight to its independence as an institution and thus prevent its enforced dissolution.

Carathéodory rejected this election but the reason he gave for his refusal was hardly credible. He was seeking a way of maintaining his connections with the union without insulting Süss:

There are many reasons therefore, most of which may remain unmentioned, since they are not necessarily of a permanent nature. However, the main reason is the following: when I resigned from the editorial board of the *Mathematische Annalen* in 1928, I resolved never to expose myself again in such a way as to ever get caught in a situation similar to that from which I had just released myself. Since then, I have consistently refused to become a member of an editorial board or of a committee; and that was true for both Germany and Greece, also for America, once even for such a remote country like India. That my feelings for the German Union of Mathematicians are not touched by this decision is self-evident.[512]

Carathéodory was apparently not prepared to become exposed for the sake of the union. He therefore used a spurious argument, which would be transparent to anyone who knew his activities after 1928, as for instance his commitment as a member of the editorial board of the *Rendiconti del Circolo Matematico di Palermo* and his office as the Greek government's commissioner for the Greek universities in 1930–1932.

Almost a month later, on 5 February 1945, Süss addressed him again trying to win his favour in a repugnant way:

> The respect and high esteem of your work have hindered me in more peaceful times to ask you for a favour or even to offer you an office that would have obstructed your productive activity. But in this time, when so many other things and, maybe, also the life of the individual and, likewise, of some of the community are at stake, the president of the German Union of Mathematicians has to make a last extreme attempt in order to rescue the union entrusted to him in the hour of need. That was the reason why, for the cause of the union, I had to try to gain a personality, whose weight both as a scholar and as a character is unlimitedly acknowledged at home and abroad. This reasoning for your election would be significant enough for me to anew attempt to care that, exceptionally, you give up your principle in view of the extraordinary present conditions. Only the severe illness of your wife and the hard fate to which you are exposed in Munich do not allow me to express the request that I would otherwise express and so, I am afraid, that the German Union of Mathematicians will really have to do without your activity in the committee.

Süss continued by saying that he was convinced that Carathéodory, with his judgement and weight, would continue to be active in defending the good cause of the union and science in Germany.[513]

5.60
An Unlikely Captive

After Germany's capitulation in May 1945, Walter Blume, now with the rank of *SS-Oberscharführer* (commando leader), was taken prisoner by US forces in Austria and was tried in Dachau on 5 August 1947 as case 000-50-5-24: "United States versus Josef Bartl et al". He was sentenced on 16 April 1948 to ten years imprisonment, in his own words "on the grounds of a general charge", but in fact on the grounds of violation of the laws and usages of war.[514]

On 28 July 1948 Blume sent a letter to the *Syndikus* (legal adviser) of Munich university. He wrote that being a sergeant of the Luftwaffe, he was transferred to the Waffen-SS with the rank of commando leader in September 1944 and ordered to the guard detachment of the Mauthausen sub-camps Melk and Linz III. During an air raid, an older inmate at Melk had begun a discussion with him, from which Blume learned that the prisoner was a former professor of mathematics in Munich. The name of this gentleman was not known to Blume, but he learned it from some "fellow-sufferers" and former students of the professor, who were detained later together with Blume in prison in Landsberg am Lech. Blume explained further that the inmate had been forced to leave the university in 1933 and expressed the hope that the professor, whose name was "Karatheodori", was now well and could remember meeting him. The 48-year-old Blume was now interested in obtaining a statutory declaration from Carathéodory so as to use it in the revision of his sentence.[515]

However, the documents of the review of US army war crimes trials in Europe in 1945–1948 show that no petitions for review were filed; only petitions for clemency were filed by the accused on 15 August 1947 and 13 October 1947. His trial had revealed that, as a sergeant of the guard, Blume escorted the inmates of the Mauthausen sub-camp Linz III to and from work at the Hermann Göring Plant. During the escort, Blume engaged in the practice of beating inmates with the butt of his rifle. He also formed and supervised the guard chain around the plant during the time the inmates were at work.[516] In Melk, the companies Steyer, Ostmark and Steyr-Daimler-Puch were involved in the "Quarz project", the building of an underground factory planned for the interests of the Luftwaffe. Therefore, soldiers of the Luftwaffe were engaged as guards at the sub-camp Melk. The building and installation firms employed mostly small groups of specialised workers, engineers and technicians, who supervised the slave work of the camp inmates. These inmates came from at least twenty-six different countries, the largest national groups being those of Poles, Hungarians, French, Soviets, Germans, Italians, Greeks and Yugoslavs.[517]

On 3 August 1948, the *Syndikus* sent Carathéodory's address to Blume without, however, commenting on whether Carathéodory had actually been in Melk.[518]

There are hints that might support the story of Carathéodory's internment in a concentration camp for a very short period of time: the smear campaign against him resulting in the report of the NSDDB and a letter by the president of the Bavarian Academy of Sciences to Carathéodory on 25 December 1944, in which the former referred to a "written confirmation" that would protect Carathéodory "from further trouble" as to enable him to "carry out [his] so significant work in peace".[519] But, if Carathéodory was really sent to the concentration camp, why should that be Melk and not Dachau near Munich? The only plausible answer could be that he had been forced to work as an engineer to solve problems arising from a tunnel construction in Melk, but Steyr-Daimler-Puch are convinced that their archives contain no such information. On the other hand, even worse denunciations had not led to internment in concentration camps in other cases, and neither Carathéodory himself nor any of his colleagues had ever said anything about such an event. His daughter, who met him after the war in Zurich and Munich, has no knowledge of an internment. Dr. Leonhard Weigand contends that, if Carathéodory had been arrested or deported, he would have told him.[520] Blume's information about the professor's dismissal in 1933 does not apply to Carathéodory. The conclusion is that Carathéodory's internment is highly improbable.

Then the question that naturally arises is as to why Blume wanted to engage Carathéodory as a witness for the defence in the review of his case. Either Blume had got wrong information about the interned professor, or he was acting on behalf of a former student seeking an exoneration testimony from Carathéodory and not daring to contact the university directly.

Walter Blume was set free on 14 December 1951.

5.61
Euphrosyne's Illness and Air Raids

Carathéodory's wife suffered a light stroke in the morning of Sunday, 7 January 1945. In the evening of that "ominous" day a double air raid on Munich destroyed almost everything that had been left undamaged.[521] Hopf told Tietze that Euphrosyne could not descend the stairs to the cellar and Carathéodory stayed with her in the house.[522] Carathéodory wrote to Zermelo in January the following year: "A year ago my wife was critically ill. It was the time of daily alarms and we were living in the cellar of our house in the last months of the war."[523] Tietze had very high fever and could not carry his wife, who had recently had an operation, to the cellar. They left the windows of their house open to protect them from being shattered, the central heating broke down and he took long walks to various authorities to secure some of the rationed heating material for a small stove in their house; they had had no gas from July until the beginning of December and again no gas in January. In Tietze's room the temperature was $-3°$ C, in his wife's $+5°$ C.

The university did not reopen as scheduled on January 10. Damages to windows and doors had to be repaired and rooms with heating had to be found.[524]

In March 1945, Carathéodory announced to Süss that their district had been spared the most recent dreadful air attacks and that his wife was feeling comparatively well and had regained her mobility but could still not climb the stairs.[525] Gustav von Bergmann, the famous professor of medicine who with his *Funktionelle Pathologie* (Functional Pathology) had created the prerequisites for psychosomatic medicine, visited the Carathéodorys on 16 August 1946, probably to examine Euphrosyne's state of health.

5.62
Collected Mathematical Writings

Perron explains why the Bavarian Academy took the decision to publish Carathéodory's collected works: they "are dispersed in many journals of many countries, because all editorial boards were very keen on getting contributions from him. Thus, much is very difficult to access today. Therefore, the Bavarian Academy set about the edition of his collected treatises, to which he himself was able to carry through the first preliminary works"[526]

Furthermore, in a letter to Süss of 1 May 1944, Tietze stressed the value of Carathéodory's own co-operation in such a task:

In recent years I was thinking how appropriate it would be if a complete edition of Carathéodory's publications could come into being and how valuable it could be still if such an edition were to be organised with his co-operation. A thought which, of course, I still have not expressed to anyone except once to the executive committee of the Bavarian Academy (when I still belonged to it) when suggestions had to be made for the use of available means. But as was to be expected from the president at that time, it has not been discussed further. The new president who, at last, has been recently nominated, surely has other worries now that the academy building is burnt out.[527]

Notwithstanding the pressures of other more urgent matters, less than two months later, on 16 June, the executive committee of the Bavarian Academy of Sciences

Der Präsident
der Bayer. Akademie der
Wissenschaften

München 22, den   21.Juni 1944.
Ludwigstraße 17

Sehr verehrter Herr Geheimrat,

     Die überragende Bedeutung Ihres wissenschaftlichen
Lebenswerkes hat bei der Akademie den berechtigten Wunsch
entstehen lassen, im Interesse der Wissenschaft und zum
Ruhme unserer Körperschaft eine Gesamtausgabe Ihrer weit-
verzweigten, oft in schwer zugänglichen Zeit-und Gelegen-
heitsschriften veröffentlichten Forschungsergebnisse zu
veranstalten. Wir sind dabei der Meinung, daß eine solche
Sammlung nicht allein einen höheren Wert, sondern auch
ihre richtige persönlich Note erhalten würde, wenn ihre
Gestaltung und organische Gliederung von Ihnen selber
vorgenommen werden könnte.

     Im Anschluß an den Beschluß des Vorstandes vom 16.Juni d.J.
habe ich daher die Ehre und Freude, die Bitte an Sie zu
richten, eine Gesamtausgabe Ihrer mathematischen Abhandlungen und Aufsätze für unsere Akademie in Angriff nehmen
zu wollen. Trotz der Ungunst der Zeit glaube ich die
äußeren Voraussetzungen für die Veröffentlichung zusichern
zu können und bin jederzeit gerne bereit, das Nähere mit Jhnen
besprechen.

     In der Hoffnung, daß Sie sich meiner Bitte nicht ver-
schließen werden und den Wunsch der Akademie, der Sie seit
Jahren eine Zierde sind, erfüllen werden, verbleibe ich
mit vorzüglicher Hochverehrung und

                     H e i l   H i t l e r !
                     Ihr sehr ergebener

An
Herrn Geheimrat Univ.Professor
Dr.Constantin   C a r a t h e o d o r y

     M ü n c h e n.

*San Nicolo's letter to Carathéodory announcing the Academy's decision
to publish Carathéodory's work.
Courtesy of the Bavarian Academy of Sciences, Archive.*

took the decision to publish Carathéodory's works. On 21 June 1944, its president Mariano San Nicolo, asked Carathéodory, "the adornment of the Academy", to put in hand the lay out and organic structure of the publication of his collected writings, and he assured Carathéodory that he would secure the external prerequisites for the publication.[528] San Nicolo, a jurist and a Nazi Party member, had become a full professor of Roman and German civil law in 1935 and was the head of the Munich institute for papyrus research and history of law in the antiquity. He was elected to a corresponding member of the Philosophical-Historical Class of the Bavarian Academy of Sciences in 1935 and to a full member in 1936. From the end of 1943 he was president of the academy. He resigned his post on suggestion of the Bavarian Ministry of Education in October 1945 to prevent his removal by the military government.[529]

Carathéodory replied warmly, thanked San Nicolo and the executive committee for the extraordinary honour and consented to the project. He signed his letter to San Nicolo with the Nazi salutation.[530]

On 7 July 1944 the Department of Mathematical and Natural Sciences of the Bavarian Academy of Sciences was informed of the executive committee's intention to publish Carathéodory's collected writings.

Carathéodory started the preliminary work immediately and made sketches for the allocation of the material comprising the *Apanta* (Ἅπαντα), his Greek name for the collected works. At the beginning of 1945, having achieved an overview of its size and structure, he signed a contract with the academy committing the academy to financial support of the work. But then, paper was scarce and Carathéodory doubted the prospect of realising the edition.[531] With rare exceptions, paper had been under a ban by the Ministry of Propaganda since January 1943,[532] but not, of course, for persons like Dingler and Steck whose "new marvellous volume" had just appeared early in 1944.[533] Van der Waerden gave the reason for his refusal to publish Popoff's work in the *Annalen* in the spring of 1944 to both the shortage of paper and the shortage of space due to the coming publication of works dedicated to Carathéodory's celebration.[534]

On Christmas Day 1944, San Nicolo wrote to Carathéodory that the contract with Beck had been completed a week before. He wished that the printing would begin as soon as possible. "Our Academy is proud to be able to get this significant publication under way in the present time."[535]

The terms of the contract between the Bavarian Academy of Sciences and the publisher Beck, signed with Carathéodory's agreement, provided for the publisher to undertake the printing and sales of the collected works in an edition which would amount to 1000–1200 copies in the size and typography of the *Sitzungsberichte der Akademie* (Session Reports of the Academy). The publisher would remunerate the author with 12 percent of the retail price for every sold copy. The academy would pay the publisher RM 90 per 16 sheets of paper without expecting a return. The author would receive 15 free copies, and the academy 25. Both the author and the academy would have the right to obtain more copies at a reduced price by 25 percent. The retail price would be defined by the publisher according to academy rulings regarding publications. The publisher would have the right to print up to ten percent further

C. H. BECK'SCHE VERLAGSBUCHHANDLUNG MÜNCHEN UND BERLIN
(13b)   MÜNCHEN 23 · WILHELMSTRASSE 9 · FERNRUF 34 622
              12. Januar 1945

Herrn
Geheimrat Prof. Dr. C. Carathéodory
M ü n c h e n
Rauchstraße 8

Sehr verehrter Herr Geheimrat!

Soeben erhalte ich von meiner Druckerei eine Um-
fangberechnung dessen, was Sie an Satzvorlagen
für Band I und II Ihrer "Gesammelten Schriften"
meinem Verlag bisher übergeben haben. Ich ge-
statte mir, Ihnen diese Aufstellung zu überrei-
chen. Hoffentlich kann ich in absehbarer Zeit die
ersten Korrekturbogen folgen lassen. Wie mir mei-
nene Druckerei schreibt, hatdie Durchsicht der
Satzvorlagen keine unerwarteten Schwierigkeiten
ergeben. Sie glaubt technisch für diese Aufgabe
gut gerüstet zu sein und wünscht nur, es stünden
ihr so viele leistungsfähige Setzer zur Verfügung
wie in Friedenszeiten. Wie es gerade damit be-
stellt ist, möchte ich nicht weiter ausführen.
Ich bin mir ganz klardarüber, daß die ehrenvolle
Aufgabe, die hier die Academie meinem Verlag gestellt hat, unter den gegenwärtigen Umständen
nicht leicht zu bewältigen ist, aber ich hoffe,
alle Hemmungen werden sich nach und nach überwin-
den lassen. Sorge machen mir im Augenblick die
Figuren, denn die Kunstanstalt, die für die An-
fertigung der Zeichnungen und Klischees vorgese-
nen war, soll bei dem letzten Angriff auf München
sehr schwer getroffen worden sein. Welche Möglich-
keiten in dieser Hinsicht noch bestehen, konnte
bisher nicht geklärt werden.

Bei dieser Gelegenheit darf ich Sie auch bitten,
den Vertrag gutzuheißen, den die Academie über
die Herausgabe Ihrer "Gesammelten Schriften" mit

Beck's letter to Carathéodory asking him to approve the contract
for publishing his "Collected Writings."
Courtesy of the Bavarian Academy of Sciences, Archive.

meinem Verlag geschlossen hat. Ich sende Ihnen
3 Ausfertigungen. Wenn Sie unsere Vereinbarungen
billigen, so haben Sie bitte die Freundlichkeit,
alle 3 Ausfertigungen zu unterzeichnen. Eine Aus-
fertigung soll bei Ihnen bleiben, die anderen
zwei erbitte ich zurück. Die dritte, auf der
noch die Unterschrift des Herrn Präsidenten
fehlt, werde ich dann der Akademie zuleiten.

Ihre "Gesammelten Schriften" betrachte ich als
einen der Grundsteine, auf denen nach dem Kriege
die Verbindung der deutschen Wissenschaft mit
dem Ausland wieder aufgebaut werden soll. Möchte
es gelingen, das Werk ohne einschneidende Störungen fertigzustellen und dann seiner Bestimmung
zuzuführen.

Mit verbindlichen Empfehlungen
Ihr sehr ergebener

H. T Bena

copies for free distribution or for review purposes, these copies being free of royalties. If by the end of war printing had not been completed and the contract had not been fulfilled because of altered conditions, a new agreement for the continuation of the work would need to be signed.[536]

This contract was followed by a letter from Beck to Carathéodory on 12 January 1945. Beck wrote that the honourable task set by the academy to the publisher would not be easy to deal with under the prevailing circumstances but he hoped that all restrains would be gradually overcome. He asked Carathéodory to approve the contract. "I regard your collected texts as one of the foundation stones on which the connection of German science with foreign countries should be rebuilt after the war. I do hope that the completion of the work will be achieved without drastic disruptions and will then be devoted to its purpose."[537]

The printing of the collected texts began during Carathéodory's lifetime. With the help of professor Rudolf Steuerwald, an honorary professor of mathematics at Munich University from 1946, he was even able to read the galley-proofs of some of his treatises. After Carathéodory's death, at a session of the Class of Mathematical and Natural Sciences on 10 March 1950, a commission consisting of the elected members, professors Georg Faber, Josef Lense, Frank Löbell[538] and Robert Sauer from the Technical University of Munich, Oskar Perron, Heinrich Tietze and Arnold Sommerfeld from the University of Munich, Otto Haupt from the University of Erlangen and Erhard Schmidt from the University of Berlin, was appointed to continue the work. This commission convened on that day and entrusted the dealing of current details to an editorial board consisting of Faber, Lense, Perron, Tietze and Steuerwald. In the name of Carathéodory's heirs, his son Stephanos gave his written consent to the editorial work of the experts' commission.[539]

5.63
Denazification

Beginning in September 1944, the entire German Reich was gradually occupied by allied troops. On 8 May 1945 the military heads of Nazi Germany signed their unconditional surrender. On 5 June 1945 the USA, Great Britain, France and the USSR announced that they were assuming authority in defeated Germany. Hitler's war had ended in total defeat.

Among the foremost aims of the occupying powers were the punishment for crimes committed under Hitler's regime and the removal of active Nazis from their posts. However, the Allies managed to adopt a common line only as regards the treatment of the major war criminals. The actual denazification was approached differently in each zone. The Soviet Union considered it primarily a part of the communist reconstruction process, whereas in the Western zones, and especially in the American sector, it was treated as a question of individual guilt. Trying to carry through denazification, the Americans sent questionnaires to all inhabitants of their occupied zone in order to discover as much as possible about earlier and present political activities. Persons suspected of a Nazi past had to take responsibility for it in front of the so-called denazification courts. These courts were administered by Germans with a clean record, who had to establish the degree of individual involvement and guilt of the accused persons and pronounce judgement.

General George Patton took up his office as the military governor of Bavaria in July 1945. He attempted not only to put a brake on the automatic dismissal of former Nazis but also to bring some of them into his own civil staff. After Patton's release from office that year, the occupying force continued to treat former Nazis cautiously, with the double aim of re-establishing a well-functioning administration and, moreover, of reconstructing Germany as a bulwark of the West in view of the beginning Cold War. Denazification policies, which, initially, had been rigorously pursued, were finally abandoned in 1948. The comeback of the old elite, which had been deprived of power, was no longer oppposed. In July 1948, eighty-five percent of those removed from their former posts by the Western powers, were accepted again. The leadership in public administration and private economy showed themselves immune against denazification. The increasing domination of Germany's educational system and its bureaucracy by the Nazis during the previous years was the reason why the Germans lacked a corps of trained and politically acceptable individuals who could be relied upon to assume responsibility for the tasks of civil administration and re-education.

Carathéodory answered the questionnaire of the military government of Germany on 30 October 1945. His entries read as follows: 1 m 70 cm tall, with dark blonde hair and blue eyes, of Greek and German nationality and Greek Orthodox confession, with neither Nazi nor military past, from 1923 to 1945 author of more than 90 articles and books on mathematics and related disciplines in German, Greek, French and English which were published in German, Italian, Swiss, Greek, Swedish, Norwegian Hungarian, English, Indian and American journals. Carathéodory filled in the remark that none of those publications was in any way related to politics. He gave very precise information about his academic career and educational policies, his knowledge of languages, his memberships in various academic institutions, his sources of income (salary, literary work and assets) and, finally, he answered the question whether he had been a trustee for Jewish property in the negative.[540]

An even less stringent denazification took place in Austria. On 11 February 1949 the general secretary of the Austrian Academy of Sciences, Professor Dr. Josef Keil, addressed the following letter to Carathéodory: "The Austrian Academy of Sciences is preparing to write the almanac for the year 1948 and feels obliged to examine the state of its members and to adapt it to the statutory regulations. Accordingly, less incriminated former party members ('mere supporters' according to the usual terminology in Germany) can remain honorary members of the Academy. The academy's committee would be very grateful to you if you could inform the academy whether these regulations apply to you or not." Further, the academy asked Carathéodory to send his autobiography together with a list of his texts and his photograph for its files.[541]

From the private clinic Josephinum Carathéodory replied to Keil on 17 February 1949: "I would like to inform you that I have never been a party member and, as a consequence, I am not affected by the so-called denazification law". He promised to send Keil the requested documents when he returned home. "I am an emeritus professor at the University of Munich, a full member of the Bavarian Academy of Sciences and the Papal Accademia Scientiarum in the Vatican city, as of the Academy of Athens. In addition, I am a honorary member of the Finnish Academy in Helsingfors and a corresponding member in Berlin, Göttingen, Rome (Accademia dei Lincei) and Bologna. Finally, I am a honorary professor of the University and the Technical University of Athens."[542]

Persons with whom Carathéodory had personal or professional relations were subject to the process of denazification:

Aumann was dismissed from the University of Frankfurt am Main in March 1946, lodged an appeal and expected a political rehabilitation as the best outcome, but not regaining his professorship after the planned amalgamation of Giessen and Frankfurt universities.[543] According to Tietze, Aumann, like Boerner, was dogged by bad luck since, for example, in the matter of Siegel's successor at Frankfurt, Aumann had managed to exclude National Socialists from an appointment. In addition, he had not given in to the anti-Semite claims of the Nazi regime, either in Munich, where he participated in the Pringsheim colloquium, or in America, where he associated with Bochner and Walther Mayer.[544] Georg Aumann, who had gained his doctorate under Tietze in 1931, was a Rockefeller fellow at the Institute for Advanced Studies in Princeton in 1934–1935 and an associate professor at the University of Frankfurt from 1936, but failed to obtain various appointments of full professorships during the war, being identified as a "politically unreliable person". He had produced the paper *Ein Satz über die konforme Abbildung mehrfach zusammenhängender ebener Gebiete* (A Theorem concerning the Conformal Mapping of Multiply Connected Plane Domains)[545] together with Carathéodory in Christmas 1933. Tietze held Aumann in high esteem as a scientist and a person with a "concise, rather taciturn than talkative manner. After the war, he had, of course, found little support and goodwill, especially from his Frankfurt colleagues."[546]

On 1 October 1946, the denazification court Munich X imposed the highest penalty of RM 2000 on Boerner, who was found to be a "supporter". The court of appeal decreased this amount to RM 500, which was believed to correspond to the small degree of Boerner's responsibility.[547] In Threlfall's expert report of 2 January 1947 regarding the occupation of a full professorship of mathematics at Freiburg University, Boerner, then a unbudgeted associate professor at the University of Munich, is mentioned as a Carathéodory student, a man of the highest education and culture.[548] On the same occasion, Tietze asserted that, although Boerner had often spoken approvingly of national socialism, he was rather harmed by the head of the NSDDB in Munich because of his confrontational attitude.[549]

Hans Petersson, who belonged to the SA since October 1933 and the NSDAP since 1937, left his chair in Strasbourg and returned to Hamburg at the end of the war. He was dismissed by the British. On 19 October 1945, E. Hecke and the rector of Hamburg University, E. Wolff, favoured his re-employment. Wolff, Bredemann and Zassenhaus excused him for having joined the SA: he had done this to promote his career which was endangered, because his wife was a member of the socialist party and had a Jewish grandparent. They emphatically recommended his reinstatement. Petersson was re-employed as a lecturer at Hamburg University on 1 June 1947.[550]

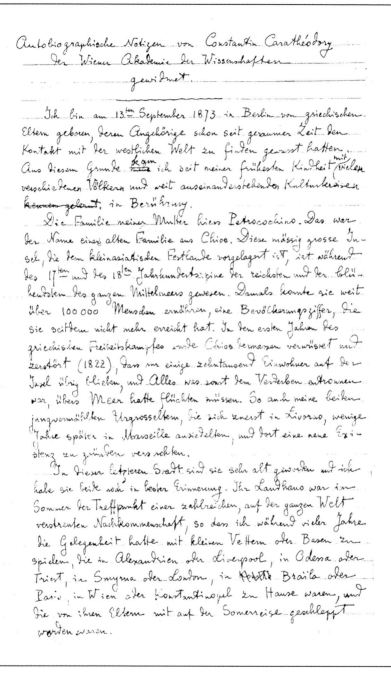

First page of Carathéodory's hand-written incomplete Autobiographical Notes
dedicated to the Austrian Academy of Sciences.
Courtesy of the Bavarian Academy of Sciences, Archive.

Gerlach, Heisenberg and Weizsäcker were interned by the Americans and British from May 1945 to January 1946. Afterwards, Gerlach held a chair at Bonn in the British occupation zone for two years before being able to return to Munich at the beginning of 1948.[551] From 1946 Heisenberg became director of the Max-Planck Institute for Physics in Göttingen, where C. F. von Weizsäcker became a scientific co-worker.

Bieberbach, who had joined the SA in 1933 and the NSDAP in 1937, was taken prisoner by the Americans and interned in a camp near Darmstadt. Prior to his arrest, the Russians had made him an offer which he rejected. According to his wife, his library, the greatest private library in Germany, was confiscated by the Americans. Bieberbach expressed the wish to work under the Americans and be thus subjected to political control. His wife told Süss that Erhard Schmidt, Andreas Speiser and Heinz Hopf were prepared to give information about his scientific qualities and his character.[552] In June 1946, Bieberbach was freed and registered as a "scientist of mathematics", obtained a food card and could work scientifically in Berlin, where he was not bound to any institution. Because of his older books, the Americans were convinced that he was a prominent mathematician.[553]

In 1945, Brouwer was temporarily suspended because of his connections with the Nazis, especially during Holland's occupation by the Germans, but was later reinstated.[554]

Blaschke, whose sympathies for fascism had earned him the nickname Mussolinetto in Hamburg and who, in his own words, had been "a Nazi in his heart",[555] was dismissed in August 1945. Although he had come under attack from the Nazis in Bad Pyrmont in 1934, he made approaches to the Nazi Party from 1936 and managed to become one of the most influential German mathematicians in matters of appointments.[556] At the beginning of the war he accused Severi for siding with the Polish viewpoint concerning Copernicus, because, in a lecture on Galileo, Severi had mentioned Copernicus as an "*astronomo polacco*".[557] Hecke, Wolff and Peschl in Bonn had spoken against Blaschke after the war and criticised the way in which he had exercised his influence in matters of appointments. A main accusation was that he had influenced the publisher Springer in favour of the Nazis and to the disadvantage of mathematicians.[558] Ernst Jacobsthal mentions Blaschke in a letter to the Springer publishing company in 1950 as the person, who "still cannot give up his Nazi methods".[559] Carathéodory had written to rector Wolff that, if bearable relations were to be restored at the mathematical institute, one had to attempt everything to cancel Blaschke's dismissal.[560] Accusations against Blaschke by his colleagues could not really be proved, Blaschke appealed against his dismissal and was re-employed by the military government at his earlier position in October 1946.

Helmut Hasse, a member of the National Socialist People's Charity (*NS-Volkswohlfahrt*), one of the largest NS mass organisations after 1936, the NSDAP from 1938, and the *NS-Altherrenbund* (a Nazi association of older men) since 1938,[561] returned to Göttingen after the war, but was dismissed by the British occupying power and lost the right to lecture. The decision for his dismissal was "obviously taken by the highest authority".[562] His colleagues might have supported his case but the attitude of the university senate and the purging committee was decisive.[563] The British

Military Government had rejected his appeal and he was conclusively dismissed in May 1946. His only alternative was a plea for clemency through official channels. Süss asked Carathéodory to support such a step through a letter to the *Kurator* in Göttingen.[564] In the end of 1946, Hasse took on a research position at Berlin Academy, in 1949 he was appointed professor at Humboldt University and on 11 October 1950 professor at Hamburg University. There, he gave his inaugural lecture at the Faculty of Mathematics and Natural Sciences on *Mathematik als Wissenschaft, Kunst und Macht* (Mathematics as Science, Art and Power)[565] on 20 January 1951.

On a request of the new rector of Freiburg university, Dr. A. Allgeier, regarding Süss's attitude, Dr. Ferdinand Springer replied that Süss had denounced him to the Reich Ministry of Education because of his connections with German Jews abroad, also that Süss had attempted to hinder Friedrich Karl Schmidt's trip to America. Schmidt was the editor of the *Grundlehren der mathematischen Wissenschaften* (Foundations of Mathematical Sciences), a mathematical collection earlier edited by Courant. He had been sent by Springer to America to secure the endangered relations with numerous authors there and especially to block, or at least moderate, the planned American competition against the German *Zentralblatt für Mathematik*. Süss had then denounced Schmidt to the ministry and Schmidt's travel permit was refused. When Springer managed to reverse the decision and obtain permission for Schmidt to travel by pointing out the significance of the visit for the relations of German mathematicians abroad, Süss attempted to arrange for F. K. Schmidt's arrest on the steamer.[566] Konrad Knopp found Springer's view of the events concerning Süss highly strange and unpleasant and suspected that Spinger wished to impede Süss's planned establishment of the new institute of mathematics in Oberwolfach, thus satisfying a possible feeling of revenge.[567] Schmidt, who had been to America, returned to Jena at the end of November 1945.[568] After a short review of his case, Süss was fully reinstated and led both the Research Institute in Oberwolfach and the mathematical institute of Freiburg University.

Thüring, W. Müller, Hugo Dingler had to give up their academic positions, some of whom were brought before denazification courts.

Von Laue, Sommerfeld and Heisenberg spoke against Johannes Stark, who had joined the NSDAP on 1 April 1930. The denazification court sentenced him to four years forced labour on 20 July 1947. In his appeal he was able to obtain a reduction and deferral of sentence.

In Strasbourg, the investigation of Pisot's case lasted a whole year, which he spent there. He became an associate professor in Bordeaux in the summer of 1946[569] and professor there in 1948. Later, in 1966, he gained a prize from the Paris Academy of Sciences.[570]

Rust committed suicide in Berne, Oldenburg on 8 May 1945, the day of the capitulation.

Karl Haushofer committed suicide in 1946. He had been sent to Dachau concentration camp in 1944 having lost his influence on Nazi politics, especially after R. Hess's flight to England. His son, the geographer, diplomat and author Albrecht Haushofer, a former consultant to the Ribbentrop Department, a member of the Information Department of the Foreign Ministry from the beginning of the war and an associate professor for geopolitics at the University of Berlin since 1940, flew to Bavaria after the attempt on Hitler's life but was arrested there and executed in Berlin on 24 April 1945.

5.64
A *"Reasonable" Compromise*

Carathéodory's antipathy towards the Nazis is never in doubt. Most of his dear colleagues and friends had emigrated or been dragged off to concentration camps, experienced denunciations, suffering, loss of property, or had been murdered or committed suicide. His own appreciation of excellence was threatened by the "levelling down" policy in education promoted by the Nazis, his civil-servant status became an object of contempt, the liberal political culture, of which he was part, was swept away by the destructive rage of national socialism, his finances deteriorated, his library was put in danger, his Christian convictions were ridiculed, his children were exposed to an unbridled violation of moral taboos, the scientific institutions, to which he belonged, declined morally.

Carathéodory experienced isolation, he was not even a subject of scientific discussion among his former colleagues who had been able to emigrate, but rather an object of their compassion. Despite all the demoralisation, Carathéodory had stayed in Munich and his motives should be examined. He had always seen Germany as the basis of his material existence. He himself was not persecuted and would not deliberately risk losing his salary or his pension after 1938. He had his residence in Munich and, what mattered even more to him, his remarkable library, which he was not prepared to abandon in any way. He had been able to get his daughter out of Germany, hoping that the rest of the family would cope in some way with the conditions. Although he despised the Nazis,[571] participation in active resistance against the regime was beyond his horizon. He considered himself as a servant of science alone, pretending to ignore its political connections. His basic concern in the Nazi era was to find a balance between a "reasonable" compromise with the regime and the autonomy of his profession, a utopian enterprise under the conditions of a dictatorial system. Persons who knew him personally have tried to describe his attitude. Perron, for example, writes in his obituary of Carathéodory:

He himself attempted, as far as that was possible, to lead a withdrawn life in the years of political pressure, of course without making concessions but also without willingly exposing himself. He observed the "Third Reich" with the eyes of a historian and constantly drew parallels between it and dictatorships of past times, or also with the eyes of a foreigner, whose attention in a foreign country is attracted by many strange rituals, which he could simply accept as facts without having to be ashamed of. But whenever possible, he had tried to relieve the hardship and suffering originating from these customs and, through his world-wide relations, he also managed to mediate a possibility of emigrant existence for some of the "non-Aryan" colleagues.[572]

Contrary to Perron's assertion, Carathéodory's life was anything but withdrawn during the Nazi dictatorship. He had continued lecturing at the university, although he had the right to leave it completely in 1938; he was in close contact with the German Union of Mathematicians and its president, a member of the Nazi Party; he kept travelling abroad on scientific occasions on his own initiative, whenever he was allowed to do so; he was a member of German delegations enjoying the consent of the Reich Minister of Education; his help for persecuted colleagues and friends always followed official channels. Presumably plagued, as Born, by the depressing "feeling that our science which is such a beautiful thing in itself and could be such a benefactor for human society has been degraded to nothing but means of destruction and death",[573] he firmly believed, despite it, that nazism would be a transitory phenomenon. Consequently, he had to survive and follow the imperative to save what could be saved in science and human life as to secure continuation after the war. But national socialism and war resulted in the loss of Jewish talent, the loss of young people as war casualties and the disruption of the educational process and academic careers of those who did not suffer the first two fates.[574] Carathéodory might have thought that political systems come and go but institutions had to be preserved and therefore supported. Continuation in the sense of keeping a mechanism going, as was the case of the DMV, turned out also to the policy of the Allies in Germany after the war. But the centre of mathematical research and initiative had been shifted to the USA, which experienced a boost to science.

Carathéodory's second close colleague at Munich University, Tietze, recalls:

We, his colleagues at the university and the academy, know how much Carathéodory also in another way rendered outstanding services to our scientific life at that time when it was essential to stem inimical scientific influences, as far as that was possible: In particular, his involvement in our academy with respect to efforts threatening its aim and existence should remain unforgetable.[575]

Behnke describes Carathéodory's moral attitude and his warm heart:

In the stormy time from 1933 onwards he was absolutely left in peace at first. But as a man standing among many nations, he was frequently asked for help. He gave it silently, tactfully and warm-heartedly. But he did not drag himself into any political statements. Yet, he knew exactly where he was standing. He has unsparingly supported his Greek countrymen, as well as the Academy of Sciences in Munich, which, because of his international high standing, he could defend better than other member against science-hostile attempts. [...] But during the war the material conditions of life necessarily became more difficult. He had no relationships with people in the countryside, who could have helped him and his family. And just the idea of how he, who always used a refined language, could have negotiated with a Bavarian farmer over foodstuffs, seems to be odd.[576]

An acknowledgement of Carathéodory's support to Greeks can be read in his guest-book. A Greek lawyer from Volos, Georgios Tamvakis, signed it on 10 July 1945 with the remark that one of the very few pleasant surprises he had was to meet "a glory of Greece, the professor" as when he came to Munich from Memmingen, where he had been kept as a prisoner of war. In November 1945, Carathéodory received a mission of the Greek Red Cross. One of the three women visiting him was Eli Adosides, whose husband Anastasios, a journalist who had studied in Paris, had been a member of Venizelos's Provisional Government in 1917.[577] Signatures in Carathéodory's guest-book are missing from the period 28 August 1941 to 10 July 1945.[578]

CHAPTER 6

The Final Years

6.1
Consequences of War

At the Potsdam Conference, held between 17 July and 2 August 1945, the powers of the anti-Hitler coalition found solutions only in compromises. Even the agreement to treat Germany as a single economic unit, and to establish a democratic system, was deliberately vague and reflected divergent aims and conflicting interests. Economic unity fell victim to disputes over reparations and political unity collapsed through the attempt of the occupying powers to establish their own political system in the areas of their control. The relationship between the four-power administration and the commanders-in-chief of the zones failed to produce a common occupation policy.

Carathéodory seems to have been more concerned about the fate of his friends than about answering the dilemma of most Germans, whether the new situation in Germany was a liberation from Hitler's dictatorial regime or an occupation. However, he sought to co-operate with the Americans in Bavaria, to improve his living conditions, to help mathematicians in need of work, to reunite himself with his daughter and to renew his international contacts as soon as possible. His opinion in matters of university policies was again requested and respected, he himself was even proposed for a post in Berlin.

On 12 September 1945 he wrote to Süss: "Maybe it is interesting to know that Erhard Schmidt can be found in Neustadt in Holstein. [...] I have not heard anything about Feigl since February; he was then in Wechselburg, not far from Leipzig. Cremer is in Miesbach and has visited me repeatedly. I have heard from Hamburg that Hecke was operated on and was very ill but that he is getting better. [...] I have repeatedly tried to write abroad, especially to my daughter, but without results until now. All traces of my wife's serious illness have almost disappeared, but she should have different living conditions in order to recover fully." [1]

From the autumn of 1944 Schmidt was in Striegau, near Breslau. He was then working with Feigl on a book on the differential and integral calculus,[2] the main subject of Feigl's lecturing activity, whose single manuscript was looted by the Poles in the county of Glatz (today Kłodzko in Poland).[3] In 1945 Carathéodory attempted to get Schmidt to come provisionally to Heidelberg, since he believed that Rosenthal would return from the United States to occupy his former position. Seifert, in Heidelberg since December 1945, wanted to have Threlfall instead and accused Carathéodory, though indirectly, of not being objective because of his friendship to Schmidt.[4] Seifert stressed that the consequent high pension costs and an increase in the proportion of elderly academics should be avoided.[5] Threlfall's candidacy was also supported by Süss, who recommended him to the faculty of natural sciences and mathematics at Heidelberg University, as a descendant of a leading family of the English gentry with acknowledged scholars, a deeply and extensively educated man with multifaceted interests, reliable and truth loving, from whose acquaintance everyone could profit.[6]

But Schmidt's presence was also desired in Berlin, where the three mathematical chairs earlier occupied by Bieberbach, H. Geppert and Klose were vacant.[7] As Carathéodory informed Süss, Schmidt himself declared his intention to return to Berlin on a trial basis and live in the mathematical institute there.[8] But his attempt to return there in April 1946 failed because of difficulties with the entry permit and this had caused him much inner suffering.[9] Süss wrote to Carathéodory that with the aid of Hopf in Zurich, Schmidt could get an invitation to Switzerland, just like Threlfall who had been invited by Wavre to the University of Geneva.[10]

Schmidt visited Carathéodory in August 1946. This visit was probably connected with the attempts by Berlin University to gain scholars from all over the country after it had resumed its activities again on 29 January 1946. The managing director of Friedrich-Wilhelm University in Berlin (located in Unter den Linden in the Soviet sector and in 1949 renamed Humboldt University) was, indeed, sent on a trip to the Western zones of Germany in December 1946 in order to "negotiate the conditions under which several gentlemen, who have been called to the university or to the Academy of Sciences, are willing to come to Berlin."

The negotiations with Carathéodory, Hamel, M. Geppert and Deuring came to nothing.[11] In 1948 and with decisive American aid, the Free University (FU) was founded in Dahlem, West Berlin, after the division of Berlin, by a group of students and young academics who broke away from East Berlin's Friedrich-Wilhelm University to seek academic freedom.

In February 1944 Feigl had been called up as an anti-aircraft gunner, but he asked Süss to apply in the name of the Reich Research Council for exemption.[12] In October 1944 he stated that he would be grateful for co-operation in work for the military between his institute, with about twenty trained students and possessing a large amount of equipment, and the Reich Institute. A very heavy air attack on 7 October 1944 damaged the town centre of Breslau and the industry works in the north. The official number of dead amounted to sixty-eight, but Feigl's mathematical institute suffered no more damage than four broken panes of glass.[13] In the spring of 1945 Feigl's team, fearing the advancing Russian army, moved the mathematical institute and its members from Breslau to Wechselburg in Sachsen, where he himself died on 24 April that year.

Carathéodory considered himself responsible for the fate of Feigl's students after his death. So he sent one of them, Ullrich Kühnel, with a message to Herglotz on 28 October 1945. He asked Herglotz to provide Kühnel with good advice or help him in any other way. In addition, he informed Herglotz that Tietze had been very tired and weakened in the summer and that he had received a letter from his mother whom he had thought to be missing. Carathéodory added that he was very happy to have saved his books. From the seminar library, only what had not been sent away still existed; everything else, all journals and collected works, had been burnt. Perron still had his books, but they were not usable.[14]

In the summer of 1945, another of Carathéodory's friends, Heinrich Behnke, was living among the ruins of his half-destroyed house in Münster. His wife had been suffering a heart condition since August of that year; his son and daughter were away from home. He was the dean of the faculty at Münster University and chairman of the political committee (information committee) in the British zone and he understood his task as mediation, help and consolation. Under the British, the practice of removing Nazis from their posts was very gentle.[15]

On 31 January 1946 Carathéodory wrote to Zermelo, whose reputation had been retrieved at the University of Freiburg in 1945: "Now she [my wife] is feeling well again and the house is only very lightly damaged. Two weeks ago we received the first news since April 1941 from my daughter Despina from Central Africa. She is well. The matter with your pension should not upset you; if you take the right steps at the right time and especially if you have patience, you will be an immensely rich man one day. I would be happy if you came to Munich soon."[16]

At Munich University, Albert Rehm worked as the acting rector in close co-operation with the American Military Government of Bavaria in the academic year 1945–1946. With the explicit permission of the administration, Carathéodory was working on a scientific report for the Pontifical Academy in Rome. For this reason, he and the philosophy student Weygand, were granted access to the mathematical seminar.[17]

At the same time, Carathéodory was actively involved in the appointment policies at German universities. At the beginning of December 1945 he was officially asked by the dean of the Faculty of Natural Sciences at Heidelberg to propose candidates to succeed Udo Wegner, who had occupied that position since 1937. He informed Süss that he was a bit surprised to learn that the decision for immediate occupation of that chair was final, since he had heard "from the American side that A. Rosenthal was thinking of returning to Europe" but "did not consider it appropriate to mention anything about it, even as a suggestion."[18] Not surprisingly, Rosenthal did not return.

Only very few of the German emigrants wanted to return to their homeland. In Süss's opinion, this unwillingness was partly due to the situation in Germany and partly to their own psychological reasons, and he was prepared to understand their attitude.[19] Carathéodory's standpoint on this matter is revealed through a note from him, Sommerfeld and Wieland on 27 March 1946: Clusius was prepared to give up his chair if Fajans wanted to return to Munich, but in view of the destruction of the chemistry department at Munich, and of the excellent work relations that Fajans had found in Ann Arbor, Michigan, his return was not expected and did not occur.[20]

6.2
Carathéodory and the Mathematical Institute in Oberwolfach: Reconstruction

After the war, the German Union of Mathematicians, having been characterised as a non-Nazi organisation, was refounded under the presidency of Erich Kamke. Its first conference was held in Tübingen from 23 to 27 September 1946.[21] Almost a year earlier, Behnke had advised Süss: "We can no longer delay, leaving the organisation to drag itself to the earth. New address lists, an overview of colleagues having no money, an attempt to obtain posts for the politically acceptable, and to obtain private support for the dismissed, as long as they have not been politically aggressive, and last but not least, negotiations with the persecutors are important. All these have got to be done this year and you, Herr Süss, have the responsibility for this as long as there is no new executive committee."[22] In January 1947 the DMV existed as before, but it still did not have the necessary permission to develop practical activities in all zones.

All of the twenty-three members of the Oberwolfach Research Institute had survived and the institute itself was not damaged. The first French troops had passed through the area and the French military governor in Freudenstadt intended to send an officer to inspect the institute. From the outcome of that inspection Süss expected to secure the future of the institute and to establish a connection with the central authorities in charge of cultural affairs. He also expected the French mathematician Roger, who was going to travel personally to Freudenstadt on 27 April 1945, to act in favour of the institute.[23]

With the help of two officers of the occupying force, G. E. H. Reuter and John Todd, both mathematicians, the Mathematical Research Institute in Oberwolfach looked to being saved after the war,[24] and in 1946 it was already receiving international guests (Heinz Hopf, Ehresmann, Henri Cartan), according to the initiative of the French military government. The institute tried to represent the general interests of mathematicians and did not demand membership contributions, but based its work on voluntary contributions.[25] The French contact officer responsible for Freiburg University was Captain Lacant, who took the mail to Süss.[26] Süss was worried about the lack of knowledge of foreign literature, about the impossibility of publications and hoped to at least receive the Mathematical Reviews, although without being able to offer anything in exchange.[27] Seifert, who would take on his professorship in Heidelberg again at the end of 1945, and Threlfall, who would be appointed a full professor at Heidelberg University in place of the dismissed Udo Wegner, co-operated with the institute, while Behnke, Bol, Görtler, Kamke, Kneser, Lösch, Maak, Magnus, Erhard Schmidt, Schneider, Günther Schule, Sperner and Walter were active either as guests or as collaborators.[28]

Ehresmann, a professor at Strasbourg University, sent four volumes by Bourbaki and the book by Élie Cartan to the Institute in August 1946. At the same time, the *Section d'Information Scientifique* in Offenburg represented by Colonel Cagniard and Captain Guillien, a friend of Pisot, entrusted the institute with the writing and editing of the *FIAT Review* for mathematics, which was intended to report on the German mathematical research results since 1939. Süss would edit two volumes devoted to pure mathematics. French colleagues like Ehresmann and Cartan were going to visit Oberwolfach and co-operate with the institute on this work.[29] Süss invited Isidore A. Barnett, Ehresmann, Mandelbrojt (from *Collège de France*)[30] and expected Schouten's visit from Switzerland, where the latter was invited for a series of lectures in the middle of October 1946. Schouten had reported the establishment of a new "mathematical centre" in Holland, which aimed at scientific co-operation abroad.[31]

Süss headed the Oberwolfach institute up until shortly before his death in 1958. He presented himself as the one who had helped colleagues in need or even saved their lives during the war. On 19 February 1948, he wrote to Eduard Wildhagen that he had been able to save a few dozen colleagues through the whirl of events, among them French colleagues, prisoners of war, who were granted leave under the pretext of being needed for research at his institute.[32]

On 30 January 1946 Süss sent Carathéodory addresses of colleagues that had become known to him over previous years and in return he asked Carathéodory for news of colleagues that he intended to gather at the Research Institute. Süss knew that Carathéodory was gathering all his information about that institute from Boerner, so he did not give him any additional news. He told him that he himself had been "reintegrated" and that lectures had restarted at the university of Freiburg; he was lecturing in Freiburg and working at the "Reich Institute". He was also trying to get some help for colleagues who were "suffering want and deprivation without being guilty" and had turned therefore through some Swiss authorities to the organisation Swiss Help for Germany (*Schweizerhilfe für Deutschland*) but so far without results. From Swiss colleagues like Hopf and Speiser he received only scanty news because of an official break-down of the mail. Süss asked Carathéodory to put at his disposal any particular way of communication he might know, for which many of his colleagues would be certainly grateful.[33]

On that same day, 30 January 1946, Carathéodory informed Süss of the fate of three mathematicians. Werner von Koppenfels, a professor at the Technical University of Brünn, had died in a military hospital in Astrakhan (Russia) at the end of August. Carathéodory gave Süss the address of the latter's widow in Thüringen. Ludwig Berwald was one of the first to have been sent from Prague to Poland. He and his wife had lived there for only a few months. Paul Funk (in Prague, where he had been suspended from his professorship in 1938) had to wear the "star" despite the fact that he was married to an "Aryan" woman. (It is astonishing how easily Carathéodory had adopted the Nazi terminology of racial separation.) Funk spent the last months of the war in Theresienstadt (where he had been deported). At the beginning of August he was appointed at the Technical University of Vienna. Carathéodory gave Süss Funk's address in Vienna.[34]

In fact, Berwald, a full professor at the German University of Prague from 1927 to 1939, had been deported to the first large Polish ghetto in Lodz (Łódź in Polish) in October 1941 and died of hunger there in March 1942. Hunger, disease and slave labour had claimed many victims in the Polish ghettoes. Added to this, constant deportations from the ghettoes to the extermination camps had started in the summer of 1942. The ghetto in Lodz, was the last to be liquidated, for it provided skilled Jewish workers for the German armaments industry. In the summer of 1944 the last remaining residents, approximately 80 000, were deported to Auschwitz. Berwald was ten years younger than Carathéodory and had been a student at Munich University from 1902 to 1908, when he gained his doctorate under A. Voss. Funk became a member of the Austrian Academy of Sciences in 1950 and remained at the 2nd chair of mathematics as a full professor at the Technical University of Vienna up to 1957, when he was succeeded by Hans Hornich.

Most probably at the beginning of February 1946, Carathéodory informed Süss of the fate and addresses of colleagues from the universities and technical universities of Graz and Vienna. He had obtained this information from Lothar Eduard Koschmieder,[35] a professor at the Technical University of Graz, who had visited him on 5 February 1946. Carathéodory described with great intensity the fate of Lothar Schrutka, who was killed in a cellar during an air raid; before dying he had been knocking on the wall. Tonio Rella, Carathéodory continued, was killed by a grenade; Anton Huber and Karl Mayrhofer were to be dismissed shortly, as Carathéodory noted "confidentially".[36] In fact, Huber and Mayrhofer, both full professors at the University of Vienna, had already been dismissed in 1945. Huber had taken on his position in 1939. Mayrhofer, a student of Perron, Sommerfeld, Blaschke and K. Knopp had succeeded Wirtinger in 1936 and was a member of the Austrian Academy of Sciences since 1941 and chairman of the Vienna Mathematical Society from 1941 to 1945. Lothar Schrutka and Tonio Rella had been full professors at the Technical University up to 1945.[37]

By providing this information, Carathéodory helped Süss to assemble an address list of mathematicians. In this list, the names of those who had been expelled from the German East as part of the post-war enforced migration (German East means former German territory in the east, and largely in Poland today, and is not the same as the German Democratic Republic (DDR) which did not then exist) were, apart from a few exceptions, not accompanied by their titles, unlike the names of their colleagues in the West. This way of listing names could lead to the misinterpretation that the refugees had been dismissed from office. Professor Dr. Hubert Cremer, then in Miesbach (Upper Bavaria), and a refugee himself from Breslau where he had worked at the Technical University since 1940, was trying to attract Süss's attention to this matter by saying that he was allowed and should use the title of a full professor at the Technical University of Breslau, and that he had also answered the official questionnaires as a professor. In Cremer's opinion, giving up the titles would be to add another step to the *fait accompli* that the Polish had attempted to create.[38] At the Potsdam Conference the areas to the east of the Oder-Neisse Line had been put under Polish administration, and the northern part of East Prussia under Soviet administration.

In the autumn of 1946 Carathéodory travelled to Switzerland. Before his trip, he asked Süss whether he could do anything for him during his stay in Switzerland and Süss replied that "obtaining foreign literature, especially from the war years belongs to our greatest problem children". Süss intended to make his Research Institute the central place to provide colleagues in Germany with the desired literature through photocopies and later through microfilms. Despite several discussions on this subject with foreign visitors to the institute, nothing had been sent up until the middle of April 1946. Süss asked Carathéodory to arrange in Switzerland that at least the *Mathematical Reviews* and some Swiss literature be sent to his institute.[39]

On 22 May 1946, Carathéodory wrote to Süss that he had arranged that professor Barnett should bring his private copy of the *Mathematical Reviews* to Munich. The complete series from January 1940 up to that date had arrived in Munich the previous day. Carathéodory asked Süss to care for the organisation and financing of many photocopies of the whole series and make them accessible to all, or at least the main, mathematical seminars in Germany. He considered the establishment of that first contact with foreign literature produced during the war years to be great progress. On his side, he would find out whether this work could be done in Munich.[40]

Süss found photocopying in Freiburg very expensive (at that time 45–75 pfennigs per page). He would like to get a complete copy for the research institute and feared that he would have to pay at least RM 200 for the copies of issues of each year. The costs would be too high for most seminars and institutes; only Göttingen, Hamburg and Münster could afford such expense. Professor Barnett was responsible for the Higher Education Section of the Military Government of Bavaria and was viewed as a "mild-mannered mathematics professor". Süss asked Carathéodory if he would be interested in editing one or more monographs of a series that enjoyed the consent of the Military Government in Freiburg.[41]

On 9 December 1946, in the name of the faculty commission for the appointment of a mathematician in place of the dismissed Gustav Doetsch in Freiburg, Süss asked Carathéodory to propose appropriate candidates for a list of three. The same request went to Sommerfeld, Tietze and Perron.[42]

Behnke announced to Süss that the first issue of the *Annalen* was planned for January 1947.[43] At the beginning of August 1946, Erhard Schmidt, having met Knopp in Tübingen, visited the Institute in Oberwolfach, probably to consult with Süss about the publication of mathematical journals. Knopp hoped that Springer would facilitate this matter considerably.[44] Leaving Freiburg, Schmidt visited Carathéodory on 14–17 August, as mentioned earlier. In the autumn of 1946 he was head of the new research institute attached to the Academy in Berlin.[45] A year later he carried out another round-trip in the West.[46] On his advice, Teubner decided to publish again the *Jahresbericht der Deutschen Mathematiker-Vereinigung* under the interim title *Jahresbericht für die Deutschen Mathematiker*.[47]

Süss's research institute was publishing at that time a series of small textbooks for students, with about 200 printed pages each. Its programme included a book on analytical geometry by Bol, a book on analytical geometry and linear algebra by Sperner, a book on function theory by Kneser, a book on ordinary differential equations by Max Müller, a book on descriptive geometry by Strubecker, a book about real functions by Lorentz,[48] the *FIAT Review* "organised on pure mathematics" by thirty colleagues[49] working for the *Section d'Information Scientifique* (French FIAT, CNRS) in connection with commander Guillien,[50] and two mathematical journals,[51] the *Archiv der Mathematik* (Archive of Mathematics) and the *Mathematisch-physikalische Semesterberichte* (Mathematical-Physical Semester Reports).[52]

Mathematicians formerly employed at universities and now working at the *FIAT Review* for mathematics were not paid a royalty, despite promises from their French colleagues.[53] Süss invited mathematicians from Germany and abroad to give guest lectures at the institute. Fueter, for example, expressed his wish to visit the institute "in order to establish again the long-interrupted connection with the German colleagues", but he wanted first to see travel in Germany, especially by car and tram, normalised again. German colleagues whom he met in Zurich were Carathéodory, Behnke and Threlfall.[54] Süss invited Besicovitch to participate in scientific discussions in March or April 1948[55] and expected German colleagues as well as Burckhardt and Stiefel from Switzerland to visit.[56] However, even in April 1948 no connection by car from Freiburg to Wolfach could be guaranteed.[57]

One of the journals published by the institute, the *Archiv der Mathematik*, was sent to Tietze. He found the fragmentation of mathematical journalism in Germany to be regrettable though unavoidable and also had some objections to the production of specialised textbooks.[58]

Boerner acted as a contact between the institutes in Oberwolfach and Munich. Süss hoped to learn from him the situation in Munich, about which the most contradictory news and rumours had been occasionally heard.[59] The institute was then still under the protection and control of the military government and working harmoniously with the official of the United Nations responsible for it.[60]

6.3
In Zurich: Family and Friends

A major edition of the works of Euler was planned by the Euler Commission of the *Schweizerische Naturforschende Gesellschaft* and Carathéodory had been commissioned with the editing of two volumes on the calculus of variations. He wrote a complete historical introduction to the topic.[61] The work took him three years, from 1942 to 1945. He submitted his manuscript while in Switzerland in the autumn of 1946.

Carathéodory had been invited by Karl Rudolf Fueter, the president of the Euler Commission, and Andreas Speiser, its general director, to go to Zurich in May or June 1946, as he informed Süss "confidentially".[62] Fueter was a colonel and artillery commander of the Swiss army and a member of high military commissions abroad. Among German mathematicians he was reputed to be anti-German; In Behnke's opinion he was precipitous and inwardly ponderous. He represented the conservative, old-established circles of Switzerland.[63] Speiser, a full professor at the University of Zurich, was descended from a very old and extraordinarily wealthy family. He had studied in Göttingen and was an assistant professor in Strasbourg before obtaining an appointment in Zurich. He was conspicuous because of his broad education which allowed him to publish works on connections between mathematics and philosophy, as well as purely mathematical works. According to Behnke, his lectures and texts did not possess the clarity that German mathematicians were used to.[64]

Carathéodory did not travel in the spring or early summer as was planned, but in October 1946. In Zurich he was surprised and happy to meet his daughter, who had just returned from Tanganika. For nearly five years neither of them had known whether the other was still alive.[65] Only in the middle of January 1946 had the Carathéodorys received the first news since 1941 from Despina from Central Africa.[66] She was living part of the year in Tanganika and the rest of the time in Athens, where her son was at school. Despina did not obtain permission to go to Munich until 1948, despite the intervention of the American ambassador in Athens. She was then able to visit her father twice. Carathéodory's son Stephanos was very active, despite the after-effects of infantile paralysis, and had helped his father the whole time.[67] Despina remembers the mathematician L. C. Young, who visited Carathéodory on 30 December 1948,[68] which means she must have been in Munich during Christmas 1948, when denazification was officially at an end.

6.4

Attempts to Leave Germany for Greece

Carathéodory discussed with his daughter how he might be able to return to Greece at their meeting in Zurich in 1946. While in Switzerland, he wrote about this to his nephew, Gabriel Krestovic:[69]

My dear Gabriel,
since February I have been trying hard to get permission to travel to Switzerland for a few weeks. Without the help of Admiral Fokas, and of Greek-American and American colleagues, I would not have achieved it. The situation in Germany is, not only for Germans but also for us foreigners, much worse than the Swiss, and – I think – you in Athens as well, imagine. I have got the impression that it could even worsen and in 1–2 years staying there would be impossible. Here I am all right, since I have secured three thousand Swiss francs through the work I do for the Eulerkommission, which will allow me to spend a few months here together with Froso [Euphrosyne] in case of emergency. However, a definite solution should be found and I see no other who could help me than you. I have lost all my savings I had abroad. But I have rescued my house in Munich, my library and various works of art of great value. I have also saved my scientific reputation and, of more significance, the friendship of American colleagues, as I can realise from their letters. I have saved all my diligence and my intellectual powers. And though at the beginning of the war I lost about a quarter of my weight, my bodily strength has not decreased as much as one might assume. I think that I can still offer many services to our homeland, especially regarding scientific connections with foreign countries. All this is a capital which I should harvest in order to bear fruit. Unfortunately, there are great, insuperable difficulties. The American authorities – in accordance with the laws – do not differentiate between Germans and foreigners who live in Germany. They do not make any exemptions even for their own civilians. Therefore, the Greek government should become convinced that Greece has an interest in taking diplomatic steps aiming at the transfer of my fortune abroad, and particularly of the most valuable part of my library, to Athens. I ask you to immediately go and see Mr. Ioannis Kalitsounakis, Palaion Faliron, 66 Poseidon Street, who will tell you whether the Academy can intervene. It would be good if you could talk to Maximos, whose connections with the palace would be useful to us. What I seek, is a combination that would allow me to live together with Froso comfortably up until the end of my life. This would be facilitated if I had the opportunity to cash in on some valuable works of art, if the Academy and the Technical University were in the position to pay me a salary in stable currency in exchange. In order to realise this plan, one could perhaps find a Greek-American sponsor. One should ask Stephanos Ladas (Langner, Parry, Card and Langner, 120 East 41st St. New York City), who is an old friend and – as I have been told – has made a great career and knows all the Greeks of America. At this moment, my ideas for realising this aim can be nothing else but vague. Therefore, I give you only the general direction. You must also use your own imagination in order to see whether there is at least a willingness to help in Athens, before I attempt to use different ways to leave Munich. Write to me immediately to the above address. I will stay in Zurich up to 20–25 October. Froso was very ill in August and has not yet recovered. However, I have left her in a better condition and I hope that the lack of my care would not cause any further damage to her health. Therefore, I am in a hurry to return. But because of various

reasons I have to spend a few weeks here. Froso does not yet know the news of Smaragda's death. Stephanos is very well. He has set up a whole school in our house and since most of his pupils pay him in edibles, we have enough to eat. I have received a letter from Despina from Cairo, where she visited Katina in June. I sent a telegram to Kinamba yesterday. I kiss you and thank you in advance for your efforts. Your loving uncle Kostias. Unfortunately, I can not imagine that Ypsilantis, who is a member of the Greek military mission in Berlin, could be useful to us.[70]

Ypsilantis would not be useful to Carathéodory because he belonged to those royalist officers who had been persecuted, exiled, imprisoned and condemned to a life sentence under Venizelist rule in World War I. Besides, he was not considered particularly intelligent[71] and did not have the necessary diplomatic skills needed to accomplish Carathéodory's wishes. Ypsilantis's signature in Carathéodory's guest-book follows that of D. Fokas, who signed on 1 August 1946 as vice-admiral, head of the Greek military mission in Berlin. Fokas graduated from the Greek Royal Navy school in 1905, fought in the Balkan Wars from 1912 to 1913 as a sub-lieutenant, and in World War I as a lieutenant-commander from 1917. He participated in the Graeco-Turkish War in Asia Minor and in the so-called Revolution of 1922 as a member of the triumvirate of the Revolutionary Committee leading the anti-dynastic tendencies of the younger officers, who demanded action to take place to purge the shame of defeat. In 1935 he retired with the rank of a rear-admiral. In World War II he resided in Greece and wrote the *Report on the Activities of the Royal Navy During World War II*. He became a member of the Athens Academy in 1960.

In Carathéodory's guest-book there are also the signatures of Stephen P. Ladas and Christine Ladas, who visited him on 9 June 1931. In 1943 S. Ladas had spoken of the "occupation of Athens by the new barbarians" and attributed the death of the Greek poet Kostis Palamas in Athens to the German occupation.[72]

Dimitrios Maximos[73] would later lead a Greek coalition government of the right-wing People's Party and the centre-right-wing heirs of the old Liberal Party from 24 January to 29 August 1947, the main task of which would be to combat the Communists.

Smaragda was one of Euphrosyne's sisters. Two years younger than Euphrosyne, she was born in Constantinople, married the industrialist and banker Alexandros Siniossoglou and had three sons and one daughter.

In his letter Carathéodory calls Euphrosyne "Froso" and her niece Aikaterini "Katina". Aikaterini, who had married a Greek industrialist in Cairo, was a daughter of Euphrosyne's brother Stephanos Carathéodory.

As we have seen, Carathéodory remained in Germany through the entire war where, far from leading a retiring life, he insisted on teaching and intervening whenever possible in educational policies, as well as doing his best to enable the continuity of scientific life under Nazi Germany. On the other hand, there is no evidence whatsoever to demonstrate Carathéodory's loyalty towards Greece, which during those years suffered bitterly under German occupation. Immediately after the war Carathéodory intensified his efforts for the restoration of mathematics, the mathematical faculties and the organisations of mathematicians in Germany, i.e., for the normalisation of academic life in his field.[74] Carathéodory's behaviour corresponded with that of those "patriotically" minded German bourgeois, who having "done their duty" during the war strove for the re-establishment of Germany within the post-war community of the civilised democratic nations. Now, in his letter to his nephew, he classified himself as a foreigner, not a German. Although he possessed the German passport, he suddenly did not want to belong to the German nation anymore and decided to seal this decision formally. In Switzerland, he visited the Greek ambassador in Bern to renew his Greek passport. His daughter reports that his reception at the embassy was not what he had expected and disappointed him extremely. The ambassador did not receive him immediately, but left him waiting outside the ambassador's office just like any other visitor. When the two men finally met, the atmosphere was cool, distanced and not particularly friendly.

Carathéodory's fears about the worsening situation in Germany were soon to come true. A lack of raw materials and transport problems, rationing and the black market, and the especially severe winter of 1946–1947, had a devastating effect on the population.

On 7 November 1946, Klaus Clusius, full professor for physical chemistry and director of the Institute of Physical Chemistry at Munich University since 1936, applied to the Economy Office for Fuel (*Wirtschaftsamt für Brennstoffe*), for an allocation of fuel to Carathéodory, assuring them that Carathéodory was a member of the university, confirmed in his office as such by the American government. As a consequence of the lack of space and inadequate heating at the damaged university, he wrote, Carathéodory was forced to prepare his exercises and lectures exclusively at home. The fuel quota applied for was destined for Carathéodory's workroom, which was about 80 m^2.[75] In August 1948, the university again expressed a wish for rapid allocation of a coal quota for Carathéodory, who had still not received anything.[76]

Meanwhile Carathéodory's plans to return to Greece were under way. On 28 January 1947, that is four days after Maximos had become Prime Minister, he wrote to Kalitsounakis:

Dear friend, Mr. Kalitsounakis, through a Greek-American friend who leaves for Paris, I have the opportunity of writing you a few words that will arrive in Athens quickly. My daughter, Despina Scutari, has written to me that, on the grounds of a memorandum from the Academy [of Athens], the Greek government negotiated with Berlin and Washington in order to allow me to transport my books and household. I have already corresponded with Ypsilantis in Berlin, but I am afraid that he might not persist if confronted with difficulties. In any case, I thank you heartily, and also the Academy, to whose decision, I assume, you have contributed. Maybe it would be good if you could review the situation with my daughter, together with whom I spent several weeks in Zurich in October. Without great help from the Greek government we will not achieve anything; therefore, I have also written to Maximos. Maybe later, when everything proceeds a little bit, I will come to Athens for a few weeks. At present everything is nebulous. In great friendship. K. Carathéodory.[77]

According to Kalitsounakis, an old colleague of Carathéodory was willing to help him leave Germany: Albert Einstein. "During my visit to Einstein in Princeton on 13 June 1947", Kalitsounakis writes, "we spoke of Carathéodory, his old acquaintance and friend. He was surprised that Carathéodory was still in Munich but he told me that, if necessary, he would be prepared to write to the American authorities in order to facilitate Carathéodory's return to Greece. 'He is a fine man', he said describing in this way his character, his manners and his excellent education."[78] The question is, whether Carathéodory's behaviour corresponded to Einstein's idea of what should an individual's moral attitude be: "The feeling for what shall be and what shall not be grows and dies like a tree, and no kind of fertiliser will be able to get very far with it. What the individual can do is to set a clear example and to have the courage to seriously support moral convictions in the society of cynics."[79] Kalitsounakis may well have met Einstein, since he was the delegate of Greece representing the Academy of Athens at Princeton's bicentennial anniversary of its foundation on 16 June 1947.[80] But whether Einstein really was willing to help Carathéodory cannot be confirmed with any certainty.

Carathéodory told Kármán later that the project of moving to Athens had been postponed and this depended much on the state of his health.[81]

6.5
Contacts with Americans

Stephen Duggan, who as director of the Institute of International Education had helped the Displaced German Scholars in 1933, expressed his hope that America would help Greece's reconstruction: "The Greeks have suffered dreadfully from the barbarism of the Germans. Starvation and exposure have reduced their numbers and their fields have been laid waste. If any of the peoples overrun by the Germans deserve well of UNRRA [United Nations Relief and Rehabilitation Administration], it is this small nation in which resides so unquenchable a love of freedom. God grant that we Americans will not fail them."[82]

After the Battle of Athens in December 1944, between units of the National People's Liberation Army and British forces, where the latter gradually won the upper hand, Prime Minister Papandreou was replaced by General Nikolaos Plastiras, the veteran of the Venizelist coups of 1922 and 1933.

In his speech to Congress, the American President Truman declared that "Poland, Czechoslovakia, Yugoslavia, Greece and Albania, on the other hand, not only have suffered greatly at the hands of the enemy in the course of the war but they are almost entirely without foreign exchange or credit resources. Consequently to date they have been the chief objects of UNRRA's activity."[83]

At the end of 1945, a government under Themistoklis Sophoulis, leader of the Liberal Party, was established in Greece with British intervention. It appeared to be more stable, but was not necessarily so. Sophoulis announced that elections were to be held on 31 March 1946, after ten years of dictatorship, war and occupation. The Communists decided to abstain, fearing that the elections would not be genuine under such politically anomalous circumstances, and so did some members of the Sophoulis cabinet. The elections brought into parliament a vengeful and repressive right-wing coalition, dominated by the People's Party, led by the revanchist Dinos Tsaldaris, a cousin of Panayis Tsaldaris. A plebiscite on the constitutional issue, held on the basis of out-of-date electoral registers on 1 September 1946, resulted in a 67 percent vote for the return of the King. Former ELAS partisans returned to the mountains, and in October 1946 the Communists announced the formation of the Democratic Army (Δημοκρατικός Στρατός).

The country was now in the grips of a civil war that was to last up until the autumn of 1949. One of the first consequences of the conflict was the extension of American influence in Greece. Great Britain, to a great degree responsible for that situation, declared itself unable to continue its engagement in Greece. Carathéodory, like all moderate liberals, must have felt very uneasy, caught between an extreme left and an extreme right. On the death of King George II on 1 April 1947, his brother King Paul succeeded to the throne. His wife was Queen Frederica, a German art historian and a very controversial figure because of her strong will and continuous political interventions. Carathéodory's daughter, Despina, asked for her father's approval to become Frederica's lady-in-waiting. But Carathéodory refused categorically, with the argument that royals come and go, but democracy survives.[84] Sophoulis did not participate in the government under Maximos. He formed the opposition, together with the Communist Party, and demanded a policy of calming things down, the return of those who had been exiled and a general amnesty under a government of his own. In July 1947 following EAM's request to take the initiative for averting the civil war, he consulted Maximos and both came to the conclusion that a political solution was still possible for Greece.

As a result of post-war agreements between the allies, Greece passed into the area of influence of the West. As Stalin himself stated, the communist cause in Greece was condemned to failure. But also there was a shift of power within the country, as the British left the Greek arena to the Americans. Massive support through US economic and military aid in the form of equipment and training finally enabled the victory of the National Army (Εθνικός Στρατός) over the Democratic Army. Under American pressure, and the threat that American aid would be stopped, Sophoulis and Tsaldaris formed a coalition government on 4 September 1947. At the end of 1947 the Communist Party, which formed the Provisional Democratic Government (Προσωρινή Δημοκρατική Κυβέρνηση), was declared an illegal organisation.

For Carathéodory, the time had come to reactivate his American contacts. On 2 February 1947 he wrote to von Kármán, then a member of the US Air Force Scientific Board at the Pentagon. Apart from family news he informed him that he was unable to get in touch with Budapest but he received a note from Fejér, who had lost all his possessions. Carathéodory's letter was forwarded to Kármán by Lieutenant John Lang Guyant of the US Air Corps, Military Attaché at the US Embassy in Paris, who had visited Carathéodory in Munich the day the letter was written and was going to meet him again about a month later.[85] Guyant signed Carathéodory's guest-book with the comment: "My respects, and a pleasure to be with you – and to meet Kármán's friend."

On 12 March 1947, the American President announced the Truman Doctrine, a programme of military and economic help for Greece and Turkey, which marked the beginning of a policy aiming, in his own words, at supporting all free people resisting subjugation attempts by armed minorities or external pressure, or in a more accurate interpretation, aiming at containing communism. As Mazower writes, "the way forward for the Greek bourgeois elite, as Tsaldaris was among the first to sense, was to profit from the tensions of the Cold War: anti-communism would not operate in the absence of foreign economic support, as had been the case in the 1930s; it would instead secure it – and on an unprecedented scale. The Truman Doctrine and the Marshall Plan proved him right and maintained the old political world in power." [86]

On 10 December 1948, Carathéodory wrote from his hospital bed to Radó, who was professor at the Ohio State University:

Dear Radó,

Your charming letter reached me yesterday. I greatly appreciate your offer to send me a food parcel and thank you very much for it. But you need not to bother about me because my daughter who is living partly in Tanganika and partly in Greece is sending me everything I want. On the contrary I think it would be very nice of you if you would send something to Prof. Perron (München 27, Friedrich Herschelstr. 11) and to Prof.Tietze (München 22, Trautenwolfstr. 7) who both are in my judgement a little starved. But I would suggest not to send CARE parcels, [87] which during the last year have deteriorated in quality but to send smaller packages with things that you have chosen yourself.

Some weeks ago I had a long visit from Szegö who had been nearly 2 months in Budapest and who told me very interesting details about Fejér and Riesz and about the general situation of the Hungarian mathematicians. He had also visited Szeged where the scientific life seems to be thriving quite a lot. [88]

Through the International Red Cross, Despina was indeed sending her father parcels with coffee and tea from Tanganika and Kenya butter in a can, but unfortunately, as she admitted, no cigars.

Carathéodory also wrote to Fueter of Szegö's visit. Szegö had come directly from Budapest and probably brought him news from Fejér. Fejér was doing rather well, also financially, since he had received a great prize at the centenary celebration of the Hungarian state in the previous summer. [89]

Radó replied to Carathéodory immediately:

Dear Professor Caratheodory:

[…] I shall contact Professor Perron at the address you gave me. In 1945, when I was in Germany working for the American Air Force, I hoped to be able to see you in Munich, but I was called back to London before I could make the trip to Munich. Since the end of the war, we were in a constant state of worry about our closest relatives in Hungary. Except for my wife and two children, all my relatives live in Hungary, and the same holds for my wife. [90]

At the beginning of 1949 Radó wrote again from Ohio:

In 1945, I spent four months in England, working for the Army Air Force, and I made a trip to Gottingen. I found it impossible to get through to Munich. In the Fall of 1946, I was made chairman of our department of mathematics. At that time, our veterans returned in mass to the colleges and universities of the country, and enrolment figures soared. With the splendid cooperation of my friends in the department, we somehow managed to take care of the heavy administrative work and to continue our research. Effective 1 January 1949, I was made a research professor and I was relieved of the chairmanship. In my present position, I am free to do as I please, with no fixed duties as regards teaching and administration.

Both of my children are grown and in fact both will get married this year. My son will finish his studies next March as a chemical engineer, while my daughter obtained last June a Master's degree in Biochemistry and is now working in the University Hospital of the University of

Michigan. Myself, Mrs. Rado, and our two children are in good health and good spirits, except for our constant anxious preoccupation about the fate of our relatives and friends in Europe.[91]

Following a conference between six major powers (the United States, Britain, France and the Benelux countries) held in London at the beginning of June 1948, at which the foundations of the future West German state were laid, the Federal Republic of Germany was officially proclaimed on 23 May 1949. After the currency reform in the Western zones and in the Soviet occupied zone (1948) Germany's economic segregation was complete; the foundation of the Federal Republic of Germany (BRD), followed by the German Democratic Republic (DDR), completed the political division of Germany.

6.6
Widowed and Fatally Diseased

In 1947 Carathéodory was stricken by the disease that finally caused his death. He had to stay in the private Josephinum clinic for six months because of uraemia caused by prostatitis.

He had not been able to see his wife more than two or three times before she died on 29 July 1947 after her last very short crisis.[92] The funeral took place three days later. Sommerfeld informed Heisenberg: "I come today from a sad thing, the burial of Frau Carathéodory; he is also very ailing (bladder, prospect of operation)."[93] On 11 September 1947 Süss wrote Carathéodory a letter of sympathy for his grief. Süss had been informed of the tragic event by Erhard Schmidt. He wished Carathéodory to get better himself and to find comfort in his work, the only thing that might help him.[94]

On 12 January 1948 Kármán and his sister sent Carathéodory their condolences from Paris, telling him that they loved Euphrosyne with all their heart.[95] Ten days later Carathéodory confessed to Kármán: "My wife often told me: if you ever need anything, your best friend in America is Kármán."[96]

Towards the end of 1947 Carathéodory was able to return home, but he had to return to the Josephinum severely ill again at Easter in 1948. After an initial improvement in his health, he worsened in July and August. He was feeling better and started taking walks for the first time in October. But still, he had to stay in the Josephinum where he was nursed and fed exceptionally well and was under the care of an outstandingly good doctor. He was beginning to take an interest in mathematics again and promised to send Süss anything that he would soon produce. At the moment, he wrote, his pigeon-holes were empty just like his head.[97]

Kalitsounakis recounts his last visit to Carathéodory in the Josephinum on 3 December 1948: "He was not lying in bed, he was allowed to take a short walk in the city. But his face gave the impression of a suffering man, though his psyche and his moods were in a good state, almost as in earlier times. The most recent Athens afternoon newspaper, which was daily sent to him by a relative or a friend from Athens by air mail was on the table there in the clinic. A massive, clearly written manuscript of a mathematical book that was going to be printed in Switzerland

[...] was also on the table. My visit lasted long, on his request the longest time possible, and almost continuously there was talk about the situation in Greece which he followed with undiminished interest." [98]

A week after Kalitsounakis's visit, Carathéodory wrote to Süss: "I am really well and I hope, if it stays like that, that I can have the operation next month. If all continues to go well, I will be able to come to Oberwolfach some day." [99] That same day, Carathéodory also wrote to Radó that he had been very ill during the past summer but in the meantime had recovered all his power. He was still in hospital hoping that the doctors would be able to operate on him in the following month so that he could "again become an active member of the society". In the postscript he mentioned: "I don't know if you are aware that I lost my dear wife 1 ½ years ago. You can imagine how much this affected me." [100]

On 21 December 1948, still in the Josephinum, he wrote to Fueter: "I am quite well again. For the moment I am in a state of health that is much better than that during all these last years. Kielleuthner, who had always repeated to me that in my case an operation was absolutely out of the question, said a few weeks ago, after having performed a bladder speculum examination, that one can now risk this operation. Although initially he stressed the risks connected with my case (because of the abscesses that I had in the summer and the scar formed out of them), I have told him straight away that I would not wish to have the handicap under which I now suffer as a permanent state. Thus, in all probability, I will be operated on at the beginning of January. If everything goes as I want, I will be released from the clinic after 5–6 further weeks." [101] Carathéodory also expressed the wish to meet Fueter again in 1949.

Carathéodory was operated on at the beginning of March 1949. On 24 March 1949 he informed Born:

In the last years I have had great misfortune. Two years ago I had suddenly a very severe prostate disease and shortly after that I lost my wife, a blow from which I cannot recover. So many complications appeared during my disease that it is a wonder that I have got over it. Up until recently it was questionable whether I could be operated on at all. For one year I am under treatment from professor Kielheuthner, who has looked after me with such care that I have gained all my powers again and he could carry out the very difficult operation at the beginning of this month. But the follow-up treatment is lengthy. Up to now everything has gone well and there is the prospect that I will be healthy again in a few weeks! [102]

Carathéodory had not joined any health insurance scheme [103] and depended on financial aid from the state for health care. On 12 July 1949, Walther Gerlach reminded the Bavarian Ministry of Education and Cultural Affairs that it had approved a contribution of DM 600 towards hospital and operation costs for Carathéodory on 5 May 1949. [104] Gerlach, the elected rector of Munich university for the period 1948–1951, struggled for the rights of the university members uncompromisingly and powerfully against all resistance he met in authorities and ministries. [105] In 1921 Born had described him as very splendid, vigorous, knowledgeable, adroit, helpful. [106]

Gerlach had acknowledged Carathéodory's value for the university and its members. He addressed Carathéodory on his 75th birthday with the words: "Your devotion to teaching and your deep interest in the rising generation, your openness to all scientific and human problems, were exemplary to all who had the luck to come closer to you." [107]

With the help provided by the ministry, but mainly using his own funds, Carathéodory managed to pay the invoice for DM 7235.50, except for a remaining debt of DM 840, which he was not able to pay because his expenses for the necessary care at home and for special dietary food were so high. But the Josephinum still remained to be paid, and the fact that he could not live up to his obligations depressed him strongly. Under consideration of the extremely severe disease and the great significance that Carathéodory had for the university and for German science, Gerlach applied for approval of another contribution to Carathéodory for half the remaining debt, namely DM 420.

In support of his appeal, Gerlach wrote that Carathéodory had left the hospital and had taken part again in faculty sessions and lectures. When he was not in the clinic, Carathéodory attended the sessions of the Bavarian Academy of Sciences and the mathematical colloquium. He thanked Gerlach on 5 November 1949:

Dear Magnificence, I must thank you very much for the great interest. I had never believed that you could achieve such a brilliant result. Thank God, it was worth making the great expenses for my health; I am fully fit again and I intend to lecture in the next semester. With many regards also to your wife. Your grateful C. Carathéodory [108]

Eight days later he announced to Süss: "I admire the medical art that put me back on my feet again. If it continues like that, I will surely be able to visit you in Oberwolfach in summer and thus fulfil a very old wish." [109] On 23 November 1949 Carathéodory wrote to the rector's office of the University of Munich, asking for a loan:

I have suffered from the consequences of a severe prostate hypertrophy since the spring of 1947. The situation as it developed in April 1947 demanded permanent treatment in a urological hospital immediately. My stay in hospital was interrupted for only five months in the winter 47/48. Since 26 March 1948 I have been in the Josephinum under the care of Professor Kielleuthner, who through the most careful and prudent treatment has improved the state of the disease so far that an operation would be possible soon. However, the continuous stay in the hospital brings great expenses. Up to autumn 1948 I could cover them to the greater part through my income as an emeritus full professor. But its cut since October 1948 does not allow me to cover the monthly hospital expenses (around DM 750) from the current income anymore. In addition, there is also the doctor's bill for the treatment in the second half-year of 1948 as well as the coming costs for the operation. I further inform the authorities in charge about the extent of the cut. The salary in September 1948, still DM 802, was reduced in November and December 1948 to DM 495. In January 1949 it was about DM 650. [110]

On 4 January 1950 he wrote to Zermelo:

In the last years I have gone through a lot. Two and a half years ago I lost my wife. I myself was then already suffering a severe prostate disease and lying in a clinic. For more than two years I have been in bed in the clinic most of the time. Finally, the operation could be carried out about ten months ago and it was more successful than we had all expected. I have been home since last summer together with my son again and entirely healthy. [111]

6.7
Theory of Functions and Carathéodory's Last Doctoral Student

"I have written lately a somewhat longish theory of functions, which will be printed in Basel during the coming months", Carathéodory announced to Radó in December 1948.[112] Three and a half months later he told Born that he had "written a rather extensive function theory which will be published by Birkhäuser in Basel. The first 100 pages are already set."[113] Karl Leonhard Weigand wrote the foreword to *Funktionentheorie* (Theory of Functions).[114] He had gained his doctorate under Carathéodory with the dissertation *Über die Randwerte meromorpher Funktionen einer Veränderlichen* (On the Boundary Values of Meromorphic Functions of One Variable) on 11 June 1947. His second doctoral father was Tietze, who mediated the contact between Weigand and Carathéodory in 1945. Since that time, Weigand had been working on his dissertation in Carathéodory's house. Their personal relationship developed into a sort of father-son friendship.[115] When the manuscript of the Theory of Functions was completed, Weigand, who had helped him continuously during its production, a time of "great external disorder and change", had the impression that Carathéodory had exercised considerable effort until he achieved his aim and then obvious exhaustion set in.[116] Weigand had taken on the task of going over the text in a critical way, but also as a person who would learn from it. To his amusement, Carathéodory often let friends and colleagues look at the manuscript during its production. Thus Weigand had felt that this work was receiving Carathéodory's special love. "I have written to Birkhäuser to send you the corrections to my book. Of course, it is for me extremely valuable that you and Nevanlinna look at the text of my book; on the other hand, I would not like to burden you both so much,"[117] Carathéodory wrote to Fueter. To Kármán he announced on 22 January 1948: "Last week I sent Fueter the manuscript of a book about the abstract theory of measure that will be published in Zurich and at this moment I am finishing a function theory that will be published in Basel."[118]

In the foreword to the publication, Weigand indeed thanked Professor Fueter, who not only read all galley proofs but also personally conducted all the negotiations with the publisher and was responsible for sending the manuscript to the press. He also thanked Nevanlinna, who had studied and presented single-valued analytic functions in his well-known book, for his lively participation and support. Erhard Schmidt had checked through the greater part of the manuscript and A. Ostrowski had honoured it with a detailed critique. "For the moment I have two books in print, which are to be published by Birkhäuser in Basel. One of them is a function theory, the other, which was already set and burnt at Teubner in Leipzig, a theory of measure and integral in Boolean rings. I hope that both books will appear by the summer,"[119] Carathéodory wrote to Zermelo on 4 January 1950. So both of his books were to be published in Basel.

6.8

Born's Natural Philosophy of Cause and Chance

The vice-chancellor of Oxford University, Professor Tizard, known through his conflict with Lindemann-Cherwell about the technical conduct of the war, had invited Born to deliver the Waynflete Lectures at Magdalen College in 1948. Born accepted the invitation, which meant a lot of work for him, and the lectures he gave appeared under the title *Natural Philosophy of Cause and Chance*.[120] This contained a report about the advances in the concept of probability in deterministic physics that culminated in quantum mechanics. Born quoted from several of Einstein's letters, causing Einstein to grumble: "You have, dear Born, put my carelessly written remarks on window-display."[121]

While Einstein revealed some misinterpretations of his remarks in Born's citation of his letters,[122] Carathéodory was obviously very satisfied when Born sent the book to him as a gift:

Dear Born,

You have given me great pleasure by sending me your book. [...] I have learned very much from the book. For example, much was new to me in your treatment of Planck's discoveries and in the role that Einstein had played on that occasion. It is really worthy of admiration that you were able to assimilate such an enormous amount of material in such a small space.[123]

Einstein's role in Planck's discoveries, mentioned in Carathéodory's letter to Born had consisted in showing that wherever oscillations occur in atomic systems, their energy follows Planck's radiation law.[124]

Planck's interpolation (with the help of entropy) between the Rayleigh-Jeans formula for the radiation density, valid for high temperatures, and Wien's law, valid for low temperatures, was in Born's modern interpretation the first successful and complete attempt to bridge the gap between the wave aspect and the particle aspect of a system of equal and independent components, whatever these may be, photons or atoms. Planck's reasoning was presented by Born in a slightly different form due to Einstein, where entropy did not appear explicitly, and a statistical argument led Einstein to the conclusion that the Wien part of the energy fluctuation was accounted for by quanta behaving like independent particles and consequently leading to the well-known interpretation of the photo-electric effect in 1905. The Indian physicist, Bose, showed that Planck's law of radiation could be obtained by regarding the radiation as a "photon gas", provided that the photons were not treated as individual recognisable particles but as being completely indistinguishable. Einstein transferred this idea to material atoms. Their so-called Bose–Einstein statistics was later realised to be a straightforward consequence of quantum mechanics.

6.9

The First Post-War International Congress of Mathematicians

After his critical operation in 1949, Carathéodory believed that he had recovered. He announced lectures again and accepted an invitation to the first post-war international congress of mathematicians that would take place in Cambridge, Massachusetts, in 1950, and would be of great significance for the reintegration of Germany into the international scientific scene. "I am now 4 months out of hospital and quite fit. I hope to be able to go to America for the congress", he wrote to Radó.[125]

J. R. Kline, general secretary of the international congress had sent him the following letter dated 8 November 1949:

Dear Professor Carathéodory: I wish to enclose a more personal letter in addition to my acknowledgement of your acceptance of the invitation to deliver an address at the Congress. I wish to assure you in as strong terms as possible that we are extremely anxious to have as large a delegation of German mathematicians as possible at the 1950 International Congress. From the very beginning of our negotiations for the resumption of the plans for the Congress, we have insisted that all mathematical groups, irrespective of national and geographic origins, must be invited to the Congress. This is a fundamental principle and we are most anxious to have it carried out in full details. Please assure all of your colleagues that we will welcome them most heartily. I remember with a great deal of pleasure your visit to the University of Pennsylvania in January 1928, at the time you were Visiting Lecturer of the American Mathematical Society. We have another very strong tie with you in that Rademacher has been with us for more than fifteen years. He has had a tremendous influence on the work of our graduate school. It will be a real pleasure to renew the contacts with you.[126]

Carathéodory thought that the letter contained a passage that should be widely known to German mathematicians – obviously that they would be welcomed in the United States – and allowed Süss to use it without mentioning Carathéodory's name. "Of course, very few would have the possibility to finance such a trip. To be precise, I have accepted the invitation [...] but I have ways that others do not have and I do not know whether my plan will succeed."[127]

Radó was eager to meet Carathéodory again: "Preparations for the Congress next year are going forward in this country. I am sure that our guests from Europe will enjoy the best in hospitality. I am very happy to know that you plan to come to the Congress."[128]

The Congress was held to replace the one that had been planned for 1940 in the United States but made impossible by the war. A new committee to make the selections for the Fields medallists was appointed in 1948 since the one appointed in 1936 had never been able to function. The chairman of the new committee was Harald Bohr. He followed the precedent of 1936 by announcing the winners and giving a description of their work at the opening session of the congress. Unlike the presentation given by Carathéodory in 1936 though, he ignored the biographical information about the recipients, giving instead only a concise exposition of their mathematical achievements.[129] After Carathéodory's death, Tietze and Perron declined on various grounds to take his place as participants in the IMC. Rector Gerlach announced their refusal to Kline on 22 April 1950 adding that "we all mourn in Carathéodory an outstanding scholar, an excellent teacher and a reliable friend."[130] Sommerfeld does not seem to have participated either, despite the fact that he was planning a trip to the America.[131] On 2 May 1950 John Todd, then at the National Bureau of Standards, US Department of Commerce, wrote to Süss: "We are very glad to know that you will have the opportunity of visiting the Congress."[132]

6.10
Death

At the beginning of 1950 Carathéodory believed he still had a long life ahead of him and he had planned to lecture again. So he wrote to Zermelo:

I can work again and I have announced a lecture for the summer. [...] Apart from us only Thorschreiber still remains from our old company.

With the best regards to you both

Your old

C. Carathéodory [133]

This is perhaps the last letter written to one of his friends from the early days of his scientific career in Germany. Carathéodory died at 12.15 on 2 February 1950.[134] He was buried on Monday, 6 February 1950 at 2 pm in the Waldfriedhof cemetery in Munich. The location of his grave is marked by an Ionic column with his name and dates, and those of his beloved wife and his son.

Carathéodory's grave. Photograph: M. Georgiadou.

The rector of the Technical University of Munich, the rector of the Federal Institute of Technology in Zurich, Prof. Dr. F. Stüssi, and the rector of the University of Bern, Prof. Dr. A. Amonn, sent Gerlach their sympathy for Carathéodory's death in the name of their universities on 6, 7, and 15 February 1950, respectively. The

Das ordentliche Mitglied der Akademie
(Mathematisch*naturwissenschaftliche Klasse)

Herr

Dr. Konstantin Carathéodory

Geh. Reg. Rat

o. Professor der Universität München

ist am 2. Februar 1950 verstorben.

Die Beisetzung findet Montag, 6. Februar 1950,

14 Uhr im Waldfriedhof statt.

München, 3. Februar 1950.

BAYERISCHE AKADEMIE
DER WISSENSCHAFTEN

Der Präsident:
H. Mitteis

Announcement of Carathéodory's death by the Bavarian Academy of Sciences.
Courtesy of the Bavarian Academy of Sciences, Archive.

Süddeutsche Zeitung published an obituary by Tietze on 6 February, the *Münchner Merkur* published one from Georg Faber a day later.[135]

Børge Jessen sent Stephanos Carathéodory, Despina Scutari and her husband his condolences:

I met him only once, at the Congress in Zurich, which was for me a special delight, since through his masterly book on real functions he had been my teacher in a way. For all of us in Copenhagen it is a grief, that we were not granted his company among us for a while and with this I would like to express my sincere sympathy.[136]

Behnke found that Carathéodory was "mourned by mathematicians in all the world, but also by all intellectuals in two nations, the Hellenes and the Germans, who both counted him among their best representatives."[137]

Süss expressed to Stephanos and Despina his condolences also in the name of his wife and of his colleagues in Freiburg, Gerrit Bol, Heinrich Görtler and Georg Lukas Tautz.[138] He wrote to Tietze that Carathéodory's death would be an especially painful loss and would trigger thoughts and feelings accordingly in every place where Carathéodory was known. Süss asked Tietze to write a two-page obituary of Carathéodory, characterising his scientific personality as a whole, and to provide him with or suggest a selection of appropriate photographs. He would like to publish it in the 4th issue of the 2nd volume of the *Archiv der Mathematik*.[139]

Für die vielen Beweise aufrichtiger Teilnahme am
Hinscheiden unseres geliebten Vaters sprechen
wir unseren herzlichsten Dank aus.

Stephanos Carathéodory u. **Despina Scutari,** geb. Carathéodory

MÜNCHEN im Februar 1950

Expression of thanks by Stephanos and Despina.

Tietze spoke with Despina about obtaining some photographs on the 21st of February, as she was leaving Munich for Athens; Stephanos was entrusted with the task. Tietze doubted that an original photographic plate could be found after all the damage caused during the war, but he thought that if necessary photographs could be copied photographically. For the obituary he needed time to acquire an overview of all of Carathéodory's works, although two pages could only give a sketch of his life's work and not a full image.[140]

On 23 April 1950 Tietze told Süss that the obituary was not what Süss had wished; it had turned out to be twenty-one double spaced pages. He was sure that what he had produced would not be what an expert would have expected, but he was influenced by his wish to record information about Carathéodory's life and personality, which might have been less known to many people. What he produced was a speech for a broader circle and not appropriate for publication in the *Archiv*.[141]

From his correspondence with Süss regarding the obituary it is clear that Tietze was trying by all means to avoid a publication in the *Archiv*.[142] Süss, however, seriously considered the publication of Tietze's obituary: "We will dedicate a few lines to our friend Threlfall, who was most closely affiliated to the foundation of the research institute and its work, and Carathéodory's death is such a case for which everyone would understand the exception. But the character of the journal does not allow us to exceed the given size. The *Jahresbericht* will apparently reappear shortly, and extensive acknowledgements should find their place there. I would be very grateful if you sent us something as soon as possible so that we could put the acknowledgement in the issue after the next (the next one is already printed)".[143]

The cashier's office at the university asked the rector on 4 March 1950 to apply to the Ministry of Education for an order by which Carathéodory's children would be paid the amount of money to which they were entitled after the death of their father. Since 17 August 1948 Carathéodory had been paid 75 percent of DM 16 536 (i.e. DM 12 402), consisting of his basic salary of DM 12 600, his additional salary of DM 1920 and a housing benefit of DM 2016.[144] In addition, Stephanos applied to the University for the reimbursement of his expenses for his father's death mask, for the printed cards of the announcement of the death and for the notes of thanks. This request was rejected in accordance with paragraph 36 of the financial-assistance law. Stephanos also applied for an orphan's benefit, but this application was rejected as well, since he was capable of earning an income for himself and his disability did not prevent him from teaching.[145]

6.11

Carathéodory's Library

Carathéodory had contributed *A Proof of the First Principal Theorem on Conformal Representation*[146] to the anniversary volume presented to Courant on his 60th birthday on 8 January 1948. On 19 June 1950, Courant wrote to Stephanos from Germany and expressed his regret that he had not written sooner. When he got the news, after some delay, he was so deeply shocked and distressed that he did not write until it was too late. "You know, how close I was to your father as a student, an admirer and a friend. What his loss means for the individual mathematician, is surely clear to you." Courant asked Stephanos if he intended to sell his father's scientific library, hoping that Stephanos would not think the question was tactless. Courant would be very interested in acquiring such a library for his institute in New York and he asked Stephanos to send him a short telegram if the answer was positive, so that he could go to Munich and discuss the arrangement directly.[147] Stephanos did sent him a telegram.[148]

Courant visited Stephanos but no agreement between them seems to have been reached. In a letter to Stephanos, Courant mentioned that he had spoken with Erhard Schmidt who was thinking exactly like him in this matter, namely that Stephanos ought to pursue the financially most advantageous solution. Courant also referred to an offer from Würzburg, which he believed would possibly be the best, especially if Stephanos sold the books and separata separately.[149] However, the university library at Würzburg and the faculty for mathematics and computer science there contain no documents to prove an offer to buy Carathéodory's library. It is almost certain that no books were bought. The faculty library then contained only about 100 books and an increase in the number of books would not have gone unnoticed.[150]

An idea of what Carathéodory's library contained, apart from mathematical works, can be gained from two testimonies. Tietze wrote that "Carathéodory's marvellous private library, which he worked with passion to extend, contains the greatest rarities and the most precious objects, especially from early Greek literature."[151] Sakellariou, who had visited Carathéodory's library in Göttingen when Carathéodory was a professor there, describes the Greek part as extensive and rich in historical, archaeological and literature books. He also says that it was surrounded by works of art that played the role of decoration.[152]

According to Despina, the library consisted of books belonging to a variety of fields and topics. The oldest and perhaps most valuable piece was the first edition of Aristotle's works published by Aldus Manutius in Venice in the years 1495–1498,

Bust of Carathéodory by the Greek sculptor Apartis.
Courtesy of Mrs. Rodopoulos-Carathéodory.

followed by an edition of Pausanias's works of 1516. Part of the library contained reports and diaries by English travellers to Greece, especially by those who had undertaken the "Grand Tour" during the 18th and 19th centuries. Despina remembers Eduard Dodwell's *Views of Greece*,[153] with 34 copperplate engravings and 13 wood-cuts, among these books. Also the monumental multivolumed work by James Stuart and Nicholas Revett, *The Antiquities of Athens*,[154] which crucially contributed to the dissemination of Greek classical style in British architecture, was owned by Carathéodory. The *Description de l'Egypte*, the fruit of multifaceted research carried out in the course of Napoleon's expedition in Egypt had been bought by one of the Carathéodorys in Florence. It comprised nine volumes of text and eleven large format pictorial volumes published in Paris in the years 1809–1822. Philosophy and literature from the Enlightenment to Romanticism were further categories of the library. Works by Voltaire, the Phanariots' favourite author, and Lamartine, as well as Goethe's Collected Works, were among them. One of the twenty books by Alphonse de Lamartine, the French Romantic writer and statesman, bore the author's personal dedication to Carathéodory's father, Stephanos. Jules Verne's books which Carathéodory had received as prizes in mathematical competitions during his Belgian years had been carefully preserved. Valuable works by Greek authors completed the library. In addition, Carathéodory's collections included three boxes of etchings and copperplate engravings.

Carathéodory cared for his books with the love of a collector. He always put loose leaflets in each of them documenting its history. On these leaflets he wrote mathematical notes, information on the contents of the book and the place and time of its purchase. He also taught his children from a very young age how to hold a book and how to turn its pages so that it would not be destroyed.

There is also evidence that he attended auctions. In November 1933, for instance, he bought books from the library of Marcus Fugger in an auction in Munich. Referring to Fugger as Fuggerus, in a letter to Kalitsounakis, he mentioned that Fuggerus had set up his library around 1550, that all books were bound by Fuggerus himself, and provided with his own signature, and that the only relatively expensive book (because it was very richly bound) was the first volume by Eustathios. Another book that he bought was Moschopoulos (published by Robert Estienne) for RM 15.[155] The Fuggers were a German mercantile and banking dynasty that had dominated European business during the 15th and 16th centuries, developing capitalistic economic concepts and influencing continental politics. Jakob Fugger the Rich, the merchant and banker of Augsburg (1459–1525), and his nephew, Anton Fugger, had amassed great properties from their mercantile activities; their descendants, however, acquired a humanistic education at European universities. Marrying within their class, they spent most of their lives on their estates, where they established valuable libraries and built magnificent residences.[156] Markus was Anton's oldest son and Carathéodory could, of course, be proud of his acquisition from the auction.

"Unfortunately, I was not able to find a copy of this book which is in Göttingen", Carathéodory wrote to Kalitsounakis on another occasion, referring to an edition of the Iliad translated by Th. Gaza into the "common dialect" (or spoken language).[157]

In March 1958 Courant asked Stephanos to introduce Mrs Jane Richtmyer to Athens. She was the "wife of a famous mathematical physicist" and "an excellent painter" who was to come from New York to Europe. Courant added that if Stephanos could do that, it would be in the interest of both.[158] One may assume that this interest had to do with works of art in Stephanos' possession.

There is also evidence that there was an archive kept in Carathéodory's library, namely that of Cyparissos Stephanos, professor for mathematics at the University and the Technical University of Athens. Carathéodory and he had belonged to the editorial board of the *Rendiconti del Circolo Matematico di Palermo* since 1909 and Carathéodory remained a member up until 1933.

Cyparissos Stephanos had obtained his doctoral degree from the University of Athens in 1878 and went on to further study under Darboux, Jordan, Hermite to Sorbonne, Paris, with a stipend from the Greek state. He returned to Athens in 1884 with a doctorate from the Sorbonne, where he had submitted his dissertation *Sur la Theorie des Formes Binaires et sur l'Elimination* (On the Theory of Binary Forms and on the Elimination) and was appointed as a professor at the University of Athens. C. Stephanos represented Greece in international mathematical congresses, belonged to a number of mathematical societies and to the editorial board of many journals, published in renowned European and American journals and was the founder and director (1901–1905) of the first school of commerce in Athens. A scientifically precise, enthusiastic and pleasant lecturer, he published around twenty-five original works in European journals on topology, higher algebra, mathematical analysis and theoretical mechanics, and left notes on original works that he had not published. After his death at the age of 60 in 1917, his family put his unpublished works at

Carathéodory's disposal. Nothing has become known of them since that time.[159] The Greek Mathematical Society, presided over by N. Sakellariou, decided on 3 March 1935 to entrust Carathéodory and professor Varopoulos of the University of Thessaloniki with the collection of Stephanos' scientific work, so as to enable a publication in the society's bulletin. There is no such publication in the following issues.

After Carathéodory's death, his son Stephanos inherited his father's mathematical books and his daughter Despina the rest. Mathematical books from Carathéodory's library have appeared on the market in Germany at various times. An auction of Carathéodory's personal library was held by Dr. Helmut Tenner in Heidelberg in May 1984. Interested bidders increased the prices for some of the books, for example, Minkowski's Collected Works of 1911, the seven volumes of Weierstrass' Collected Mathematical Works 1894–1924 and Königsberger's biography of Hermann von Helmholtz, to 130–200 percent of their estimated value.[160]

Carathéodory's personal archive, including his mathematical correspondence, no longer exists. An indication that it has been sold to archives and auctions is found in the Einstein Archive. Also several dozen non-mathematical books from Carathéodory's library, bearing his father's ex libris, the head of the goddess Athena, with Stephanos' name and very nicely bound, have appeared on the market, in Munich around 1976, for example. They were mostly belletristic literature in French. Among them were F. D. Déhèque's modern Greek-French dictionary[161] (which is impressive because it includes demotic Greek throughout), Pierre Loti's *Aziyadé*[162] and Charles DuFresne DuCange's glossary of 1688 for authors of common and base Greek.[163]

The ex libris *of Carathéodory's father depicting goddess Athena's head and the Ancient Greek maxim* Μηδέν Άγαν *(Nothing Excessive). Courtesy of Johann Strauss.*

Part of the library was kept by Carathéodory's daughter in Athens. A large number of books has been donated to the Academy of Athens and are displayed in its central hall.[164]

Epilogue

Most historians of modern Greece engage in investigations of the so-called generation of the 1930s, to which many of Carathéodory's younger friends belonged and which comprised what may be called the golden age of Greek art and scholarship. In contrast, the generation of Greek intellectuals of Carathéodory's age still remains historiographically untouched. Additionally, Carathéodory's mature years were marked by two major traumatic national experiences: the Asia-Minor disaster and the German occupation of Greece. Both influenced his own fate in an ambiguous way which has put further obstacles in his portrayal as a personality, as well as in the account of his life and work. But also historians of mathematics have not engaged in broad studies concerning Carathéodory. Where histories of mathematicians are considered from an essentially national standpoint, this failure can be easily explained by the fact that Carathéodory and his career developed beyond the demarcation of national boundaries. However, beyond this explanation for the absence of a historiography of Carathéodory, Carathéodory's decision to remain in Nazi Germany has always raised the simplistic question of whether he had belonged to "the good or the bad guys". The answer to this question seems to be crucial in respect to any attempt at a biographical approach.

Maybe one of the most remarkable qualities of Carathéodory's intellectual character was his highly developed awareness of history. He perceived history not only as an autonomous scientific discipline, but also as a field of knowledge with an instrumental value, indispensable for meeting current demands. Furthermore, he believed that a man of his class and stature had both the self-evident justification and the responsibility to intervene in the historical process and influence or even determine its course. His *Autobiographical Notes* reveal a self-perception of his own role and that of his family as eminently historic, whereas the frequent historical references in the introductions to his mathematical papers clearly disclose his proclivity to define his own contributions as a sequence of a broader continuum within the history of mathematics.

Carathéodory was aware of the ephemeral character of mathematical knowledge and of knowledge in general. As he once remarked to Süss, nobody could know how mathematicians would view Hilbert in the future. But these words did not bear the slightest trace of underestimation or misjudgement either with respect to the famous mathematician or to the role of the individual within history. Carathéodory held, on the contrary, that the privilege of articulating scientific discourse had to be left exclusively to an aristocracy of the mind. He extended the same elitist views to the field of politics, considering that exercising leadership was an ability given to only a few. He admired the "men of the French Revolution" and he held Venizelos and his skills as a political leader in high esteem. He conceived the slogan *liberté* of the French Revolution rather as expressing the will to national independence than as a guarantee of civil rights.

According to Carathéodory, permanent progress was the general trend of history, which was nevertheless characterised by the relentless occurrence of breaks and revolutions, this applying equally to science, politics and society. However, Carathéodory always tended to locate his own deeds and actions in an intermediate field, avoiding extreme positions, conflict and fanaticism, as far as that was possible. Thus, the graduate of the *École Militaire* of Brussels never participated in a war, although he did not reject the necessity of wars as a means to pursue national goals. He was a Phanariot and a Greek nationalist, a Greek and a German, an Orthodox and a Catholic, he was liberal and conservative, political and apolitical, nationalist and internationalist at the same time. As such, he always strove for mediation between cultures, mentalities, states and institutions, thus earning respect and honours from everywhere.

Carathéodory's one-man project in Smyrna was aimed at bridging the gap between the Orient and the Occident and simultaneously at confirming Greece's role as the Western outpost in the Orient; his efforts to modernise Greek Universities in 1930 were aimed at the academic integration of Greece into Europe; the occasional support of American ambitions in Greece in the field of education was meant by him as an antidote to Greek isolation. Notwithstanding the completely different conditions, Carathéodory's actions in Germany were directed by a quite similar spirit. After the two world wars he displayed a strong commitment to prevent Germany from falling into international isolation. During World War I he continued contributing to the *Mathematische Annalen*; in 1936 he participated with the German delegation in the Oslo International Congress of Mathematicians; during World War II he took part in the procedures of the Bavarian Academy of Sciences. He did not get involved in the movement for national socialism, but he did have connections with Nazi Party members. He never openly mentioned the holocaust or the Nazi crimes against Greece. However, he took great pains to re-establish mathematics as an academic discipline in Germany after the war and thus to contribute to the reintegration of this country into the community of civilised nations.

Carathéodory exhibited the same moderate behaviour with respect to his academic affairs. He avoided, for example, joining Hilbert's group and participating in the International Congress of Mathematicians in Bologna in 1928. Yet, at least twice in his life – during the quarrel concerning the *Mathematische Annalen* and in the controversy about his own succession to the Munich chair – he was unable to stay away from heated academic in-fighting.

Carathéodory had been brought up with the idea of peaceful co-existence of the peoples in the Balkans and the Near East. His studies outside the Greek Kingdom prevented him from submitting to a petty nationalism of the kind which was strongly promoted through education in all of the newly created Balkan national states. Competence, multilinguality, high social status and material wealth made him a cosmopolitan, able to bear himself confidently either in the Orient or the Occident, and this mobility determined his attitude towards the Foreign and the Other. His ideas about the role of Greece in Asia Minor, as these are expressed in his memorandum to Venizelos, are indicative of his wish for the creation of a state of equal opportunities for all people and the care for intellectual development, a precondition for prosperity and progress. Carathéodory believed that ideas and their formation were culturally determined, but also possessed the capacity to transcend the specific context of their emergence and serve the solution of problems in different conditions of space and time. As a *homo politicus*, he was a decisive supporter of Greek nationalist ambitions, an ardent advocate of the Great Idea and he believed passionately in the superiority of Greek culture. His perception of Greece was that of a "Westerner abroad", almost comparable to the perception of a German Philhellene, as were the Bavarian administrators of King Otto's court about a century before Carathéodory's last Greek educational project came to an end. His contributions to Greek affairs were never accompanied by a clear and definite decision to remain in that country, and he always considered others to be more suitable than himself to offer their services to national matters within Greece itself. But he wished to be a "Greek" and demonstrated his "Greekness" whenever he considered it necessary and not inappropriate. In Germany he stressed his "Greekness" by speaking German with a hint of a foreign accent. Probably having met him in Smyrna, Toynbee very accurately pointed at the discrepancy between Carathéodory's national voluntarism and his real Western identity by labelling him a "fish out of water". By identifying himself as "Constantin Carathéodory from Constantinople" in his Göttingen dissertation of 1904, he publicly declared his belonging to the Greek powerful elite of the Ottoman capital, the bearer of "Greekness" within the Empire. The way in which some fifteen years later he implemented his Smyrna university project is indicative of his faith in Greek supremacy; he innately believed that the Greeks were predestined to carry out the "civilising mission" in the Orient and he also knew that a radiant Greek university in a Greek-occupied Smyrna would be inconceivable without an unconditional Greek victory in Anatolia. His Hellenomania reached a peak as he saw the Greeks of the modern era as "inheritors of a sacred tradition that mingles with their history". In this feverish daydream the idealised Greeks possessed an almost metaphysical dimension. Thus, it may not seem strange that, in our days, his name is used by some dubious political circles to support ultra nationalist or even anti-Semitic theories. These attempts culminate in absurdities, such as the denouncement of Einstein as the alleged plagiarist of the theory of relativity, and its attribution to its supposed real inventor, Carathéodory! Of course, Carathéodory himself had never questioned Einstein's authorship of relativity. Besides, he belonged to the Phanariot heirs, who attributed exceptional talents to the Jews and were their rivals within the Ottoman Empire, particularly for financial influence. However, this rivalry had never resulted in anti-Semitism. The victory of totalitarianism in Germany and the succeeding World War II put a dreadful end to any sort of idealisation.

When the Nazis came to power in 1933, Carathéodory could hardly conceive how this could happen in a country with the cultural traditions of Germany. He initially tended to view the Hitler regime with a somewhat overconfident contempt, whereas later, when Hitler gained absolute power, he was incapable of resistance.

His behaviour in the Nazi era was, in fact, identical with that of the so-called German *Bildungsbürgertum*, i.e. of those educated bourgeois who, despite their humanistic background, in their overwhelming majority abstained from any opposition against Hitler's dictatorship, and especially Hitler's war, and thus dramatically failed to exercise their historic responsibility towards both Germany and humanity as a whole. Yet, according to a German perception of the *Bildungsbürgertum*, by not resisting Hitler's world-domination plans, these people had simply "done their duty for the German fatherland". In any case, Carathéodory was honestly concerned about the continuation of German culture and especially of German scientific tradition, but he kept silent in the face of crimes that violated any idea of human decency, accepted the authority of an illegal state, made his compromises and submitted to the expulsion of Jews from scientific institutions. After his retirement in 1938, Carathéodory took the decision to continue to lecture at the university, which in the meantime had lost its autonomy and reflected the new ideological, political and social order under the Nazi totalitarian regime, with engineered discipline having replaced its former intellectual openness. The deterioration of science, the suppression of scientific exchange and literature, the abandonment of the belief in reason, accompanied by racial ideology and national self-sufficiency had resulted in the general decline of Munich University. Carathéodory attempted to survive academically within this context presenting himself as an apolitical specialist, having the promotion of science as his sole objective.[1] Yet, under Nazi rule everything was political and the assumption that science and rationality alone could avert complicity in crimes against humanity was an illusory belief. Einstein would remark on this kind of attitude: "It *is* a tradition of German scholarship not to bother about public affairs. In this respect German scholars *do* differ from their colleagues in the United States and in England. [...] Has the great mass of the German scholars acquiesced in the vagaries of the Nazis? [...] They were quite willing to say 'Heil Hitler', and to ignore much that was going on around them, so long as submission in this sense allowed them to continue their study of the electron or, say, of the early history of the Spanish language. Most of my colleagues in Berlin showed this attitude rather than positive acceptance of Hitlerism or an undue preoccupation with their income. I need not emphasize that I cannot approve. Indeed, here lie the greatest dangers of particularistic expertism among the scholars. [...] Without any regard for the scientist's comfort, there are a few things which one can say in *favor* of the Ivory Tower. The difficulty is to indicate where the right balance lies."[2]

The persons with whom Carathéodory was in contact during the Nazi period were mainly Hasse, Blaschke and Süss. All three were significant mathematicians, all three belonged to the Nazi Party, and all three exercised decisive power in university politics throughout Germany, power under whose shield Carathéodory could function. He could do so because the Nazis themselves have never doubted his loyalty to their regime (the villainous dozents of Munich University being in this respect a rather insignificant exception). Carathéodory was never subject to any persecution in Germany.

If Carathéodory were to be judged by Greek conventions, one could say that at the most critical time of war between Greece and Germany, he did not distance himself sufficiently from the enemy. He did not even behave as a member of the Greek bourgeoisie who, during the war, were planning the future of their country through an understanding with the Allies. His intermediate position, which had always been his own choice and which had earned him the respect of both supporters and opponents of the Nazi regime in Germany, could not be sustained *de facto* in this extreme situation, despite Carathéodory's continuing internationalist attitude as expressed through his participation in international forums, such as the International Mathematical Commission, the International Congresses of Mathematicians and the Papal Academy. But even in these forums he represented Germany with the approval of the Nazi regime. That he had not made a fortunate choice can be clearly seen in the incident with Schauder's letter, or in his correspondence after the war, in which a sadness and a loss of self-confidence are evident. Carathéodory's value for the Nazi state lay precisely in the fact that he nowhere and never appeared to favour nazism. On the contrary, he presented an image of Germany which, under the Nazis, had *de facto* ceased to exist. All testimonies about his alleged resistance against the regime are oral, come mostly from close relatives and, naturally, appear after his death. That he had stayed in Germany and claimed a role for himself during the Nazi era is surely problematic; but because he insisted on playing this role on his own terms, the consequences of his decision were politically negligible and he was saved from defamation. Carathéodory's plight during these years also demonstrates the decomposition power of national socialism which proved so able to quickly destroy any elite.

Fair, brilliant and reliable in his academic activities, Carathéodory was loved and respected by the majority of his German students and colleagues, and he also enjoyed international esteem; and was the most honoured mathematician after Hilbert. He felt comfortable among a very large circle of friends and relatives, close and distant. In many instances, his German colleagues saw in him the *deus ex machina*, but at the same time a curiosity. They rarely expressed an opinion about his political preferences. Only his closest colleagues hinted at his aversion to dictatorships and his elitism. The embarassement German historians feel towards this exceptional intellectual may be explained by the fact that for them Carathéodory has always been the Foreign and the Other with whom they are not prepared to deal.[3]

Carathéodory viewed mathematics as a cultural good which could give meaning to his life, a retreat in moments of external pressure, or compensation for senseless socialising. He drew satisfaction from his engagement with mathematics and therefore had a positive attitude in his life, something that Kritikos had wrongly interpreted as optimism. The idea of acquiring status and social prestige by means of academic activity was not at all alien to him, but it was surely not his first priority. His mathematical engagement, although a keen personal interest first of all, resulted in the social will to improve things. He kept his own critical opinion in the hope that the best argument would win, as in the case of the Hilbert–Pringsheim debate in his Breslau period or in the *Annalenstreit*. He favoured academic exchange and the promotion of the best scholars without fearing competition. He conceived his own cultural vocation as being education, the frame within which mathematics ought to develop. He intervened in educational policies, articulating a powerful discourse and mediating between the university as a state institution and the state itself. He believed in the unlimited potentialities of education to improve the ethical character of the individual and of society as a whole. Although fascinated by secret and occult sciences as a phenomenon of cultural history, he was, at the same time, convinced of the unshakeable power of reason.

The use of Carathéodory's name as an attribute to theorems, problems and concepts attests to his rich and original mathematical contributions. He was extremely productive up to World War I, but developed, completed and improved his theories throughout his lifetime, even up to his death. To judge from his age, he again found himself in an intermediary position between two generations, the generation of his teacher, friend and colleague Hilbert, and the generation of his fellow students. And although he bore the characteristics of a Hilbert student, namely both the devotion to Hilbert's mathematical tradition and to Hilbert's person, he was mentally nearer to Hilbert than anyone else of the younger generation. Blumenthal and Schmidt, both three years younger than Carathéodory, might be comparable to him. But Hellinger, Weyl, Haar, Courant, Hecke and Funk belonged to the notable younger generation of mathematicians. Carathéodory's *modus operandi* was rather classical; he traced the historical development of an idea in earlier works of famous mathematicians, he recognised general behaviour from thoroughly worked out single cases, he aimed at completeness, precision and smoothness in his theories. He sought to simplify proofs, to perfect his theories, and to employ axiomatisation in an impressive way. His ways, however, were not always that simple and he seemed to sometimes ignore the possibilities of modern approaches in his science. Furthermore, the problems he treated were complex and he did not hesitate to increase their complexity as, for example, in the calculus of variations, where his investigations extended to spaces of many dimensions. The simplicity he praised did not apply to himself. Parallel to the calculus of variations, he engaged in studies in complex and real analysis. He enriched the theory of functions with new results in problems concerning Picard's theorem, in problems arising from Schwarz's lemma, in coefficient problems appearing in power-series expansions. He advanced the theory of functions of several variables, simplified the proof of the main theorem of conformal representation of simply connected domains onto the unit circle and worked out the theory of boundary correspondence. From 1918, starting with his *Vorlesungen über reelle Funktionen*, up to the end of his life, he endeavoured to find a proper definition of length, surface and space measures.

Carathéodory was mostly occupied with the foundations of mathematical notions, but he was deeply interested in mathematical applications, although he kept some distance from them in his work. He produced what Mehrtens described as "applicable mathematics", i.e. mathematics concerned with areas whose usefulness is considered but which, in fact, aim at the development of general knowledge without regard for specific applications.[4] Carathéodory's mathematical career had been facilitated by his engineering background: that was one of the reasons why Klein took an early interest in him, and this also played a role in his various appointments until Göttingen. His effective communication with Born, von Kármán and Sommerfeld was maybe due to their common interest in the applicability of mathematics. Carathéodory engaged in optics, mechanics, the motion of planets and, above all, thermodynamics. However, his book on geometric optics was not received with the same enthusiasm as his books on first-order partial differential equations, on conformal representation and on real variables. Perron mentions that geometric optics, tightly bound in Carathéodory's case to the calculus of variations, and later to applications, cast its spell over Carathéodory in such a way that he did not shrink from extensive numerical calculations. His thermodynamics, on which with great mastery he applied the axiomatic method, is still cited today but was not established within the culture of physics, although it earned him acknowledgement from renowned physicists. A reviewer remarks that one might say that this kind of engagement with applications was typical of the Göttingen school of applied mathematics with Carathéodory as one of its pioneers; this approach arrived in the United States with Courant, Friedrichs, John, Levy, and others, including Moser, after the war, and evolved into the applied mathematics of the Courant Institute. A "Göttingen school" was also mentioned by Lindemann, this term, however, denoting a circle of scientists trying to have their favourite candidates appointed to mathematical chairs all over Germany.

On the other hand, Carathéodory endeavoured to develop mathematical aspects of physics in an autonomous way. Thus, he effectuated a kind of feedback effect that could result in the coupling of pure and applicable mathematics. However, his canonical form, which does not always reduce to the standard form, made it difficult to apply to physical theories.

Apart from mathematics itself, Carathéodory was also interested in its history, but he was not solely concerned with establishing historical facts. Moreover, he used history to find approaches for an exact treatment of the problem under consideration. Kritikos summarises the characteristics of Carathéodory's mathematical work as follows. "It is always closely connected with older mathematics. It preferably treats the great problems, which mathematical analysis, urged by applications, has presented. The treatment is handled with generality and precision, which were not even required or were hardly possible at that time. In particular, one establishes an excellent combination of geometric concepts with logically strict numerical or axiomatic methods of contemporary mathematics. On the other hand, what is impressive is the continuous productivity and the gradual ripening of mathematical thought, as well as a strong individuality, which assimilates foreign elements within a whole, admirable for its unity. Finally, the work presents an exceptional beauty of form. An intensive, I would say artistic, mood is disclosed everywhere not only through the selection of subjects and methods but also through refinement: through the 'resplendent face', the clear plan, the weighted development of the parts and the careful elaboration of the details." [5]

Despite his friendship with Zermelo, Fraenkel and Brouwer, Carathéodory did not seem to have been interested in philosophical discourse on mathematics and he used non-technical terms in his discussions on the character of mathematics or in the obituaries of mathematicians he wrote. His daughter, for instance, mentioned that he used to compare the structure of a mathematical proof with the structure of a fugue for several voices and that he viewed mathematics as a Bach composition full of dynamics and beauty. That was exactly Hasse's statement in his speech on *Mathe-*

matics as Science, Art and Power.[6] There is a series of examples in Carathéodory's use of language in which beauty, elegance, prettiness and charm are attributed to mathematical proofs, qualities which, in accordance with the usual stereotypes, he also credited to women. On the other hand, he also recognised dynamics in mathematics, a male feature in his opinion, and in this regard he deviated from the common view of mathematics as order above all. However, he was prepared to accept order as a principle of organising mathematical business. Thus, he described Klein as a commander riding in front of a brilliant staff, who devised and applied a tight order to mathematical instruction. Carathéodory also viewed mathematical creation as an aesthetic pleasure born out of pain. We may assert that he was an eclecticist in his science as in his general performance. No wonder then that his life may be subject to several interpretations.

Appendix I

Some Explanations concerning the Text

1. Calendar

The new (Gregorian) calendar was introduced in Greece by the state on 18 January 1923. The new calendar was 12 (now 13) days ahead of the old calendar, which is still faithfully followed by the monks of Mount Athos, by the Orthodox Church in the Holy Land and by some other Orthodox churches in Eastern Europe. In the text, a double date is employed to show this difference wherever documents were dated according to the old (Julian) calendar: old date [new date]. Otherwise, the new calendar is used throughout the text.

2. Language

The original language of the quoted texts is mentioned in the notes. All texts have been translated into English by the author. In all Greek quotations, also in historical text, modern intonation rules are employed.

3. Although the Ottoman Empire was dissolved in 1923, the Europeans used to call it Turkey after the end of the Crimean War in 1856.

4. The names and places are given in the form used at the time of the events described – Smyrna (Izmir). Local spellings are used – Hannover, Marseille (not Hanover, Marseilles), except where the name might not be recognised – Cologne (Köln).

5. The following German monetary units are mentioned in the text:

The *Goldmark* (abbr.: M) (100 Pf = 0.3584 gr. fine gold) denotes the unit of the Reich's golden currency (9 July 1873–1923).

The *Rentenmark* (fixed securities mark) was introduced by law on 13 October 1923 to stabilise the German currency in the inflation following World War I.

The *Reichsmark* (abbr.: RM) was introduced by law on 30 August 1924 to replace the completely devalued *Mark*, after the currency was stabilised through the *Rentenmark*.

The *Deutsche Mark* (abbr.: DM) denotes the monetary unit after 1948, which was introduced with the currency reform for the three Western zones and West Berlin.

6. Abbreviations used in the notes:

6.1 Journals and Proceedings

Abhandl. Math. Sem. Hamb. Univ.: Abhandlungen aus dem mathematischen Seminar der Hamburgischen Universität
Acta Math.: Acta Mathematica
Ann. d. Phys.: Annalen der Physik
Ann. d. Mat. pura e. appl.: Annali di Matematica pura ed applicata
Ann. École Norm. Sup.: Annales de l'École Normale Superieure
Ann. Scuola Norm. Sup. Pisa, Sci. Fis. e Mat.: Annali della R. Scuola Normale Superiore di Pisa, Scienze Fisiche e Matematiche
Ann. of Math.: Annals of Mathematics
Archiv d. Math. u. Phys.: Archiv der Mathematik und Physik
Boll. Un. Mat. Ital.: Bolletino dell' Unione matematica Italiana
Bull. Amer. Math. Soc.: Bulletin of the American Mathematical Society
Bull. Calcutta Math. Soc.
Can. Math. Bull.: Canadian Mathematical Bulletin
Centaurus
Comment. Math. Helv.: Commentarii Mathematici Helvetici
C. r. Acad. Sci. Paris: Comptes rendus des Séances de l'Académie des Sciences Paris
DMV-Mitt.: DMV-Mitteilungen
Forsch. u. Fortschr.: Forschungen und Fortschritte
Fund. Math.: Fundamenta Mathematicae
Ges. Math. Schr.: Gesammelte Mathematische Schriften
Handb. d. Phys.: Handbuch der Physik
Jber. Deutsch. Math.-Verein.: Jahresbericht der Deutschen Mathematiker-Vereinigung
Jbuch d. Bayer. Akad. d. Wiss.: Jahrbuch der Bayerischen Akademie der Wissenschaften
J. f. reine u. angew. Mathematik: Journal für reine und angewandte Mathematik
Math. Ann.: Mathematische Annalen
Math. Nachr.: Mathematische Nachrichten
Math. Z.: Mathematische Zeitschrift
Mitt. Ges. Deutsch. Chem.: Mitteilungen der Gesellschaft Deutscher Chemiker
Monatshefte f. Math. u. Phys.: Monatshefte für Mathematik und Physik
Nachr. Kgl. Ges. Wiss. Gött.: Nachrichten der Königlichen Gesellschaft der Wissenschaften zu Göttingen
Norske Mat. Tidsskrift: Norske Matematisk Tidsskrift
NTM-Schriftenr. Geschichte Naturwiss., Technik, Med.: NTM-Schriftenreihe zur Geschichte der Naturwissenschaften, Technik, Medizin
Phys. Rev.: Physical Review
Phys. Z.: Physikalische Zeitschrift
Proc. Cambridge Philos. Soc.: Proceedings of the Cambridge Philosophical Society
Proc. Roy. Soc. London: Proceedings of the Royal Society of London

Proc. Nat. Acad. Sc.: Proceedings of the National Academy of Science

Quart. J. Math.: Quarterly Journal of Mathematics

Rend. Circ. Mat. Palermo: Rendiconti del Circolo Matematico di Palermo

Sitzungsber. Bayer. Akad. Wiss. München: Sitzungsberichte der Bayerischen Akademie der Wissenschaften München

Sitzungsber. Kgl. Preuß. Akad. Wiss. Berlin: Sitzungsberichte der Königlichen Preußischen Akademie der Wissenschaften Berlin

Trans. Amer. Math. Soc.: Transactions of the American Mathematical Society

Verh. Deutsch. Phys. Ges.: Verhandlungen der Deutschen Physikalischen Gesellschaft

Verh. Schw. Naturf. Ges.: Verhandlungen der Schweizerischen Naturforschenden Gesellschaft

Z. f. angew. Math. u. Mech.: Zeitschrift für angewandte Mathematik und Mechanik

Z. f. Deutsch. Kulturphil.: Zeitschrift für Deutsche Kulturphilosophie

Z. f. techn. Phys.: Zeitschrift für technische Physik

Zentralbl. f. Math.: Zentralblatt für Mathematik und ihre Grenzgebiete

6.2 Abbreviations used for archival sources

Barch: Bundesarchiv

BDC: Berlin Documentation Center

BayHstA: Bayerisches Hauptstaatsarchiv

LMU: Ludwig-Maximilian University

Appendix II

A Short Biographical Sketch of the Carathéodory Family[1]

The Carathéodorys' descent can be traced back to mid-18th century Adrianople (now Edirne, Turkey). Their ancestor Theodore (1740–1789), also known as Cara-Theodore because of his dark complexion, was a trader in fruit and vegetables. His wife Helene (1759–1827) gave birth to two daughters, Sultana and Sophia, and two sons, Antonios Carathéodory (Adrianople 1765–Bucharest 1815) and Stephanos Carathéodory (1789–1867), who generated the two branches of the Carathéodory family. Our character, Constantin Carathéodory, is descended from Antonios and his wife, Anna Sotirchou, the niece of the Oecumenical Patriarch Kyrillos VI. Constantin's wife, Euphrosyne, is a descendant of Stephanos and his wife Loukia Mavrocordato. We also give details of the life of Carathéodory's relative Konstantin St. Carathéodory.

Stephanos Carathéodory
Adrianople 1789–Constantinople 1867

Elementary education at the Greek School of Adrianople.

1805–1809, general education (elements of philosophy, elements of physics and mathematics, Italian and French) at the Greek School of Kydonies (Ayvalik).

1809, in Livorno recommended by Benjamin Lesvios to Greek traders there.

1809–1819, studying philosophy, mathematics and medicine, attending lectures in physics by Volta at the University of Pisa; 1819, diploma in medicine.

1820, headmaster of the Greek communal school of Adrianople; there until 1821; simultaneously, private physician and author of speeches for the church.

1822–1824, private physician in Adrianople and the region.

1824, in Constantinople.

1825, in Pazardzhik visiting Philippopolis (Plovdiv).

1826, private physician in Constantinople.

1827, personal physician to Sultan Mahmud II; later also to Sultan Abdülmecid; at the court's medical service until 1861.

1828–1867, professor at the Imperial School of Medicine.

1830, marries Loukia Mavrocordato; 7 children: Smaragda (1830–1899); *Alexander* (1833–1906), Helene (1836–1914); Zoe (1838–1889); *Konstantin* (1840–1922); Sophia (1843–1863); Maria (1845–1927).

Alexander St. Carathéodory Pasha
Constantinople 1833–Constantinople 1906

Private education and education in communal schools.

At the Translation Chamber.

1855–1860, studying law, especially French civil law, in Paris; also mathematics; 1860 thesis entitled *De l'erreur en matière civile d'après le droit romain et le Code Napoléon* (On the Error in Civil Cases According to Roman Law and the Napoleonic Code).

1860, civil servant at the Ottoman Ministry of Foreign Affairs.

1862, President of the Admiralty Court in Constantinople.

1862–1864, with his mother in Paris.

1865–1866, in Paris and Berlin.

Lawyer in Constantinople.

Kapikâhya of Chios for 20 years.

October 1867, with Grand Vizier Âli Pasha in Crete, negotiating with Christian revolutionary leaders.

1871, deputy Minister of Foreign Affairs.

1874, at the Brussels Conference for the laws and usages of war.

1874–1876, ambassador of the Ottoman Empire in Rome.

1876, member of the constitutional committee.

1876–1877, deputy Minister of Foreign Affairs; secretary at the Constantinople Conference.

13 June–13 July 1878, first Ottoman plenipotentiary at the Congress of Berlin; head of the Council of Public Works.

1878–1879, Minister of Foreign Affairs.

1884, member of the Council of Public Works.

1885–1894, Prince of Samos.

March 1895–March 1896, Governor General of Crete.

1896, First Translator of His Imperial Majesty the Sultan; later, secretary at the Department for Administrative Cases of the Council of State.

22 June 1869, marries Cassandra Mousouros; 7 children: Stephanos (1870–1941); Anna (1872–1946); Pavlos (1874–1934); Loukia (1876–1933); Smaragda (1882–1946); Roxane (1883–1948); *Euphrosyne* (1884–1947).

Konstantin St. Carathéodory
Constantinople 1840–Athens 1922

Private education in Constantinople.

Studies in engineering with specialisation in bridge construction at the Paris *École Polytechnique*.

Member of the Council of State, head of the Council of Public Works.

Representative of the Ottoman Empire in the European Commission of Danube.

March 1896, representing the Ottoman government in the enthronement of Czar Nicolas II in St. Petersburg.

1906–1907, Prince of Samos.

Marries Aikaterini Photiadi; 6 children: Alexander Carathéodory (1884–1908); Maria; Sophia; Ioanna; Loukia; Rallou.

Constantin Ant. Carathéodory
Adrianople 1802–Constantinople 1879

1807–1818, primary and secondary education in Adrianople, church music.

1819, graduation from the Bucharest School.

1820–1821, teacher at the School of Adrianople.

1821–1822, private teacher to Greeks and Bulgarians in Adrianople.

1822–1823, studying the German language in Vienna.

1824–1827, studying medicine, surgery and gynaecology at the University of Pisa; doctorate from the University of Pisa in 1827.

1827–1829, studying anatomy and surgery, gynaecology, paediatrics, pathology, ophthalmology and chemistry in Paris; working at hospitals.

1829, qualifying as an eye-specialist in London; working at hospitals; treatise on eye-diseases.

1830, private physician in Constantinople.

1830–his death, professor at the Imperial School of Medicine.

1831–1839, personal physician to Sultan Mahmud II.

From 1832, chief physician at the Kız Kulesi Military Hospital for Plague [and later Cholera] Patients.

1836, member of the newly founded Council of Hygiene.

1839–his death, chief physician at the surgical clinic of the Imperial School of Medicine.

1839–1861, personal physician to Sultan Abdülmecid.

1842–1844, professor for obstetrics at the midwife school at Galatasaray.

From 1843, at the polyclinic of the Imperial School of Medicine for out-patient surgery.

1835, marries Eriphyli Aristarchi; one son: *Stephanos* (1836–1907).

1841, marries Euphrosyne Charitonos from Adrianople; four children: Anna (died in 1939); Telemachos (1845–1927); Aglaia (1846–1919); Alexander (1848–1932).

Stephanos Carathéodory
Constantinople 1836–Brussels 1907

Primary and secondary private and millet education in Constantinople.

1852–1860, studying law in Berlin; in parallel, secretary of the Ottoman delegation there; 1860, doctorate entitled *Du droit international concernant les grands cours d'eau* (International Law Concerning the Great Waterways).

Secretary of the Ottoman embassies in Berlin, Stockholm and Vienna.

From 1866, replacing the ambassador of the Ottoman Empire in St. Petersburg.

1871, recalled to Berlin; first secretary of the Ottoman legation in Berlin.

1874–1875, on leave in Constantinople.

1875–1900, ambassador in Brussels; there up to his death.

29 August 1872, marries Despina Petrocochino (1851–1879); two children:
   a) **Constantin** (Berlin 1873–Munich 1950); b) Ioulia (1875–1942).

Appendix III

Chronology

1 [13] September 1873, birth in Berlin as the first child of Stephanos Carathéodory and Despina Petrocochino.

13 [25] January 1874, christening at the Russian Orthodox church in Berlin.

1875, birth of his sister Ioulia in Brussels.

1879, death of his mother at Cannes; death of his grandfather Constantin Carathéodory in Constantinople.

1881–1883, at the private school of Vanderstock (Belgium).

Winters of 1883–1884 and 1884–1885, vacations with his father on the riviera, in Bordighera and San Remo (Italy).

1884, birth of his aunt and future wife Euphrosyne Carathéodory.

1886–1891, at the Belgian grammar school *Athénée Royal d'Ixelles*; graduation in 1891.

October 1891–May 1896, at the Military School of Belgium; 1891–1895, 57th Year of Artillery and Engineering; 1893–1896, School of Applications.

1895–1897, in the Ottoman Empire, partly on Mytilene assisting his cousin Ioannis Aristachis to design the road network of Samos.

August 1895, in Chania visiting Alexander Carathéodory Pasha, Governor General of Crete (1895–1896).

1897, in Athens, Trieste, Semmering, Vienna, Dresden, Brussels, London, Paris.

Winter 1897–1898, in London.

1898–Easter 1900 assistant engineer at the Assiut dam in Egypt with a salary of 240 British pounds.

1900, dismissal of his father Stephanos from the post of the Ottoman ambassador in Brussels.

1901, publication of his book *Egypt* by the newly founded (1899) Society for the Promotion of Useful Books.

Summer semester 1900–winter semester 1902, mathematical studies at the University of Berlin.

Summer semester 1902–summer semester 1904, mathematical studies at the University of Göttingen; publications on function theory.

1903, member of the German Union of Mathematicians.

Easter 1903, in Isthmia, near Corinth.

22 January 1904, at the celebrations of the 200th colloquium of H. A. Schwarz.

8–13 August 1904, at the third International Congress of Mathematicians in Heidelberg; afterwards with H. Hahn and W.Wirtinger via Munich to Achensee.

Autumn 1904, to Edinburgh with his father and his brother-in-law, Georg von Streit; visit to the castle of Lord Rosebery.

1 October 1904, doctor title (Dr. phil.); dissertation *Über die diskontinuierlichen Lösungen in der Variationsrechnung* (On the Discontinuous Solutions in the Calculus of Variations).

Christmas 1904, in Brussels working on his habilitation.

5 March 1905, *venia legendi* for his work *Über die starken Maxima und Minima bei einfachen Integralen* (On the Strong Maxima and Minima in the Case of Simple Integrals).

Easter 1905, vacations in Isthmia with his father, sister and uncle Telemachos.

Easter 1905–Easter 1908, unpaid lecturer at Göttingen University; publications on function theory.

December 1906, in Grand Hotel, Paris.

End of 1907, death of his father Stephanos in Brussels.

February 1908, new habilitation at Bonn.

1 April 1908–1909 April, unpaid lecturer of applied mathematics and geodesy at Bonn University; living at Venusbergweg 32; work on the isoperimetric problem with E. Study.

6–11 April 1908, at the IV International Congress of Mathematicians in Rome.

Summer semester 1908, leave of absence.

Easter 1908, vacations in Athens and Corinth.

July–September 1908, in Constantinople, Kuru Çesme on the Bosphorus working on thermodynamics.

3 October 1908, "professor".

5 [18] February 1909, marriage to Euphrosyne in Constantinople, Kuru Çesme.

1 April 1909–30 September 1910, budgeted professor of Higher Mathematics at the Royal Technical University of Hannover, annual salary M 4500 (including a housing benefit of 660 marks) plus lecturing fees plus examination fees; living at Heinrichstr. 34 and Alleenstr. 9.

1909–1933, on the editorial board of the journal *Rendiconti del Circolo Matematico di Palermo*.

7 November 1909, birth of his son Stephanos in Hannover.

1 October 1910–30 March 1913, budgeted Professor of Higher Mathematics at the Technical University of Breslau also Vice-Rector, annual salary M 9000; living at Scharnhorstr. 30, Breslau XVIII; works on classical function theory.

1911, committee member for appointments to the Faculty of Natural Sciences and Mathematics of Athens University; enables the appointments of the physicist Dimitrios Hondros and the mathematician Georgios Remoundos.

13 October 1912, birth of his daughter Despina at Breslau.

1 April 1913–30 September 1918, full professor of the Philosophical Faculty of Georg-August University in Göttingen, joint head of the Physics-Mathematics Seminar, annual salary M 7200 plus M 720 housing benefit plus lecture fees plus other fees related to his university post; living at Friedländer Weg 31; work on functions of real functions, especially on the theory of measure.

1914–1928, on the editorial board of the *Mathematische Annalen*.

1916, correspondence with Einstein regarding the theory of canonical transformations and the problem of closed time-lines.

September 1917, with his sister in St. Moritz, Switzerland.

1918, publication of his *Vorlesungen über reelle Funktionen* (Lectures on Real Functions, later translated as: Functions of Real Variables) by Teubner.

Winter semester 1918–31 December 1919, full professor at the University of Berlin, joint head of the Mathematical Seminar; living at Pension Ludwig, Markgrafenstr. 33, Berlin W8.

October 1918, two visits to his son Stephanos in Halle.

12 December 1918, full member of the Prussian Academy of Sciences, confirmed 10 February 1919.

31 March 1919, corresponding member of the Mathematical-Physical Class of the Academy of Sciences in Göttingen, proposed 22 February 1919.

3 July 1919, introduced to the Prussian Academy of Sciences by Planck during a public session.

September 1919, meeting with Venizelos in Paris.

20 October 1919, submission of his memorandum concerning the foundation of a new Greek university to Venizelos.

1 January 1920–30 April 1924, in Smyrna and Athens.

20 March 1920, consultation with Venizelos, Leonidas Paraskevopoulos and Apostolos Psaltoff on a battle ship in the harbour of Smyrna, Asia Minor.

2 [15] June 1920, appointment as Professor of Analytical and Higher Geometry (with civil-servant status) to the Faculty of Mathematics and Natural Sciences of Athens University.

28 July 1920, appointment as organiser of the Ionian University and full professor of mathematics, monthly salary 4000 drachmas.

29 July 1920, honorary member of the Prussian Academy of Sciences.

August 1920, consultation with Venizelos and the High Commissioner of Ionia, Aristidis Stergiadis, on a battle ship in the harbour of Smyrna.

4 October 1920, travel from Athens to Smyrna.

December 1920, dismissal as a Venizelist from his Athens professorship by decree of the Minister of Education (royalist government).

June 1921–end of October 1921, in Europe for the organisation of the Ionian University.

8 September 1922, flight from Smyrna on a Greek battleship.

2 [15] September 1922, reappointment as a full professor of Higher Mathematical Analysis at the University of Athens.

1922, death of his great uncle Konstantin Carathéodory in Athens.

1923, full professor at the Technical University of Athens.

1 May 1924–August 1938, full professor of mathematics at the Philosophical Faculty, Section II of Munich University, initial annual salary of 6171 golden marks; living at Rauchstr. 8, München 27; works on variational problems on open surfaces, theory of optical instruments, theory of functions of several variables.

26 July 1924, contact with R. A. Millikan for a foundation of an institute of pure and applied physics in Athens.

21 February 1925, full member of the Bavarian Academy of Sciences, proposed 7 February 1925.

26 November 1925, associate member of the Prussian Academy of Sciences.

16 May 1926, corresponding member of the *Accademia delle Scienze dell Istituto Bologna*.

Summer 1926, in Thrace and Macedonia; report to Henry Morgenthau, Sr., on the settlement of refugees.

13 December 1926, full member of the Academy of Athens, proposed 26 November 1926.

1927, second edition of his *Vorlesungen über Reelle Funktionen*, with minor changes, by Teubner.

21 December 1927, Privy Councillor.

1927, death of his father's half-brother Telemachos Carathéodory.

1928, first Visiting Lecturer of the AMS; January, travel with his wife to the States on the *Aquitania*; 10 January, lecture at the University of Pennsylvania; visiting professor at Harvard in the 2nd semester; 21 March, William Lowell Putnam speech at Harvard; 7 May, approval of a possible permanent appointment for him by Harvard's tenth mathematics division; in Washington, DC, New Orleans, Austin, San Antonio, Los Angeles; 2–20 July, lectures on the calculus of variations and selected subjects from analysis at the University of Berkeley, payment $ 1000; with Euphrosyne at the observatories of Mt. Wilson and Mt. Hamilton, the Yosemite National Park; mid-August, in the Rocky Mountains of Canada; 20 August, in Portland, Oregon, excursion along the banks of the Ohio; 21 August, in Seattle; 22 August, in Vancouver; return to New York on the *Canadian Pacific* railway; 7 September, embarkation at New York for Europe.

14 February 1929, appointment as a full professor of mathematics at Stanford University, annual salary of $ 8000; 12 May, rejection of the offer; 17 May, rise in his supplementary salary in Munich from RM 1960 to RM 2400.

1928–1929, work with Radó in Munich.

September 1928, in Berlin, probably viewing Schwarz's papers.

October 1928, honorary member of the Calcutta Mathematical Society.

November 1928 – January 1929, active involvement in the *Annalenstreit* (*Annalen* quarrel).

January 1929, negotiating works of art in Munich for the Benakis collection in Athens.

12 December 1929, nomination to *socio straniero* (foreign member) of the *Reale Accademia dei Lincei*.

15 March – 15 May 1930, on paid leave in Greece, with half of his salary in May, to reorganise the Greek universities.

May 1930, in Thessaloniki.

7 April 1930, honorary member of the National Technical University of Athens.

Summer 1930, honorary president of the Greek Mathematical Society; report on *The Reorganisation of the University of Athens* (*Η Αναδιοργάνωσις του Πανεπιστημίου Αθηνών*).

24 September 1930, in Budapest with Haar, König, Riesz, Tangl and Fejér.

End of September – 3 October 1930, in Thessaloniki; 4–30 October 1930, in Athens.

25 March – 30 April 1931 in Athens; his Bavarian salary to Bochner, Lettenmeyer and an assistant.

3 October 1931, appointment to Commissioner of the Greek Government.

Spring 1932, in Athens and Thessaloniki.

From 1 May 1932, five-month leave of absence from Greece.

26 July 1932, dismissal from the post of Commissioner of the Greek Government.

1932, publication of his book *Conformal Representation* by the Cambridge University Press.

1932, death of his sister's husband, Georg von Streit.

18 July 1933, listed as professor of "Aryan" descent.

26 October 1933, appointment as a member of the "reform committee" of the Philosophical Faculty, Section II of Munich University.

15 February 1934, corresponding member of the *Pontificia Accademia dei Nuovi Lincei*.

20 March 1934, travel with his daughter to Pisa to give for four lectures at the *Scuola Normale Superiore*.

24 August 1934, submission of a completed questionnaire on the "Aryan descent of the wife".

28 August 1934, oath of allegiance to Hitler.

September 1934, honorary president of the Inter-Balkan Congress of Mathematicians in Athens (2–9 September).

1934, death of his brother-in-law Pavlos.

Spring 1935, Despina's departure for Athens.

May 1935, in Bern for the 25th anniversary of the Schweizerische Mathematische Gesellschaft.

28 May 1935, acceptance of a corresponding membership of the Austrian Academy of Sciences.

13 June 1935, appointment to the International Commission of Mathematicians per decree of the Bavarian Ministry of Education.

7 July 1935, in Brussels at the celebration of the centenary of the *École Militaire de Belgique.*

Summer 1935, meeting with the Greek Prime Minister Tsaldaris at Tegernsee.

19 August 1935, submission of a memorandum concerning the National Technical University of Athens to Tsaldaris.

1935, publication of his book *Variationsrechnung und partielle Differentialgleichungen erster Ordnung* (Calculus of Variations and Partial Differential Equations of the First Order) by Teubner.

March–May 1936, in Greece.

14 May 1936, speech on "A Completion of the Schwarz Lemma" at the Academy of Athens.

13–17 July 1936, at the International Congress of Mathematicians in Oslo; awards the first Fields medals to Lars Ahlfors and Jesse Douglas.

23 July 1936, return to Munich.

August 1936, on the *St. Louis* steamship to the United States: 26 August, in New York; 27 August, in Boston; 31 August, speech on "The Beginning of Research in the Calculus of Variations" at the AMS meeting on the tercentenary of Harvard University.

September 1936–February 1937, Carl Schurz Memorial Professor at the University of Wisconsin at Madison with a payment of $4000.

23 October 1936, lecture on "Some Geometrical Features in the Calculus of Variations" at the University of Wisconsin.

28 October 1936, Papal Academician and full member of the *Pontificia Accademia Scientiarum.*

27 November 1936, speech on "Bounded Analytic Functions" to the AMS at Lawrence, Kansas.

6 February 1937, departure from Chicago to California; through Santa Fe and Grand Canyon at L. A. on 12 February.

1937, paper *About the Curvature of the Stylobate of the Parthenon and concerning its Intercolumnia* for the centennial celebration of the Greek Archaeological Paper.

3 November 1937, acceptance of his appointment to the jury of the Alfred Ackermann–Teubner Memorial Award for the Promotion of Mathematical Sciences for work in analysis.

1937, publication of his book *Geometrische Optik* (Geometric Optics) by Springer.

December 1937, Despina's marriage to Theodoros Scutaris.

May 1938, lecture at Hamburg mathematical seminar; meeting with Morse.

20 July 1938, honorary professor of the University of Athens; honorary president of the Greek Mathematical Society.

12 August 1938, retirement; 1938–1944, dispute over his successor.

March 1939, member of an international committee for the celebration of Élie Cartan's 70th birthday.

Winter semester 1939–1940, lecturing after a one-year pause.

1939, publication of his *Reelle Funktionen, Bd. 1: Zahlen, Punktmengen, Funktionen* (Real Functions, vol. 1: Numbers, Point Sets, Functions) by Teubner.

18 May 1941, report on his participation in the International Commission of Mathematicians to the Reich Minister of Education.

7 June 1941, request of the Minister of Education to Carathéodory to contribute to the special issue of the journal *Matematicheskii Sbornik*, published by the Academy of Sciences of the UdSSR, for the celebration of the 75th anniversary of the journal's foundation in December 1941.

October 1941, at the work conference of the German Union of Mathematicians in Jena.

December 1941, mediation for the release of the Serb mathematician Nikolaus Saltykow.

23 January 1942, at Hilbert's birthday in Göttingen.

End August 1942, vacation in the Black Forest, Hotel Kyburg.

5 November 1942, mediation for the salvation of the Polish mathematician Schauder; Schauder's execution by the Gestapo in September 1943.

9–12 November 1942, at the International Conference of Mathematicians in Rome; 12 November, at the Papal audience and speech on "Problems of Analytic Functions of One Variable" at the Papal Academy.

1942, death of his sister Ioulia.

11 January 1943, request to travel to Helsinki, Finland for lectures in April or May 1943.

February 1943, election to honorary member of the Finnish Academy of Sciences; 24 March, acceptance of honorary membership.

4 May 1943, proposal for honorary membership of the Austrian Academy of Sciences; 29 July 1943, acceptance of honorary membership.

27 May 1943, the NSDDB report on Carathéodory to the Nazi Party office in Munich.

19 July 1943, Dr. Graue's report on Carathéodory to the Rosenberg Office.

1–21 September 1943, vacation in the Black Forest, hotel Luisenhöhe in Horben near Freiburg.

13 September 1943, his 70th birthday party by the German Union of Mathematicians.

21 September 1943, acceptance of appointment to the jury for the award of the Ackermann–Teubner Prize of 1944 for excellent work in geometry.

1943, destruction of his book on *Reelle Funktionen, Bd. II* at Teubner's during the bombardments of Leipzig.

16 June 1944, decision of the executive committee of the Bavarian Academy of Sciences for the publication of his collected works.

7 January 1945, Euphrosyne's minor stroke.

12 January 1945, rejection of election to member of the committee of the German Union of Mathematicians for three years.

30 October 1945, completed the questionnaire of the Military Government of Germany.

November 1945, scientific report for the Pontifical Academy with the explicit permission of the American Military Government of Bavaria.

February 1946, attempts for an exit permit to Switzerland.

May 1946, Professor Barnett brings his private copy of *Mathematical Reviews* to Munich, through Carathéodory's mediation.

11 July 1946, first post-war lecture "On the Schwarz Reflection Principle of Boundary Values of Meromorphic Functions" at the Munich mathematical colloquium.

October 1946, in Switzerland; submission of the manuscript on the historical introduction to Euler's works; meeting with his daughter and planning to return to Greece.

1946, publication of his *Reelle Funktionen, Bd. 1: Zahlen, Punktmengen, Funktionen* by the Chelsea Publishing Company.

1946, death of his sisters-in-law Smaragda and Anna.

Spring 1947, prostate hypertrophy.

29 July 1947, death of Euphrosyne; burial at Waldfriedhof cemetery, Munich, three days later.

End of 1947, return home from the private clinic Josephinum.

26 March 1948, back to the clinic; treatment by professor Kielleuthner.

From 17 August 1948, pension of DM 12 402.

1948, publication of his *Vorlesungen über reelle Funktionen* by the Chelsea Publishing Company.

1948, death of his sister-in-law Roxane.

December 1948, Despina's visit to Munich.

Beginning of March 1949, operation at the Josephinum.

16 December 1949, last speech ever on "Length and Surface" at the Munich mathematical colloquium.

2 February 1950, death at 12:15.

6 February 1950, burial at Waldfriedhof cemetery at 14:00.

Appendix IV

Carathéodory's Fields of Study and Contributions bearing his Name

I *Calculus of Variations*

I.1 Discontinuous Solutions

I.2 General Theory

Carathéodory theorem on convexity

Carathéodory, Constantin: Variationsrechnung und partielle Differentialgleichungen erster Ordnung. Leipzig, B. G. Teubner, 1935, p 197.

I.3 Geometric Methods

I.4 Multiple Integrals

II *Thermodynamics*

Carathéodory principle (book section 2.2)

Carathéodory C (1909) Untersuchungen über die Grundlagen der Thermodynamik. Math. Ann. 67:355–386 and Ges. Math. Schr. vol II, pp 131–166, axiom II on p 140.

III *Geometric Optics*

IV *Mechanics*

V *Theory of Functions*

V.1 Picard's Theorem

V.2 Coefficient Problems

Carathéodory class of functions (book section 2.6.2)

Carathéodory C (1907) Über den Variabilitätsbereich der Koeffizienten von Potenzreihen, die gegebene Werte nicht annehmen. Math. Ann. 64:95–115 and Ges. Math. Schr. vol III, pp 54–77.

Carathéodory class of functions; Carathéodory lemma; Carathéodory–Toeplitz theorem (book section 2.6.2)

Carathéodory C (1911) Über den Variabilitätsbereich der Fourierschen Konstanten von positiven harmonischen Funktionen. Rend. Circ. Mat. Palermo 32:193–217 and Ges. Math. Schr. vol III, pp 78–110.

Carathéodory–Fejér theorem (book section 2.6.2)

Carathéodory C, Fejér L (1911) Über den Zusammenhang der Extremen von harmonischen Funktionen mit ihren Koeffizienten und über den Picard–Landauschen Satz. Rend. Circ. Mat. Palermo 32: 218–239 and Ges. Math. Schr. vol III, pp 111–138.

V.3 Schwarz's Lemma

V.4 Conformal Mappings

V.4.1 Existence Theorem

V.4.2 Variable Domains

Carathéodory convergence theorem (book section 2.6.4.1); *Carathéodory domain* (book section 2.6.4.2)

Carathéodory C (1912) Untersuchungen über die konformen Abbildungen von festen und veränderlichen Gebieten. Math. Ann. 72:107–144 and Ges. Math. Schr. vol III, pp 362–405.

V.4.3 Mapping of the Boundary

Carathéodory–Osgood theorem (book section 2.6.5)

Carathéodory C (1913) Über die gegenseitige Beziehung der Ränder bei der konformen Abbildung des Inneren einer Jordanschen Kurve auf einen Kreis. Math. Ann. 73:305–320 and Ges. Math. Schr. vol IV, pp 3–22.

Carathéodory extension theorem (book section 2.6.4.3)

Carathéodory C (1913) Über die Begrenzung einfach zusammenhängender Gebiete. Math. Ann. 73:323–370 and Ges. Math. Schr. vol IV, pp 23–80.

V.5 Normal Families

V.6 Functions of Several Variables

Carathéodory (Riemannian) manifold or Carathéodory hyperbolic manifold (book section 2.6.6)

Carathéodory C (1932) Über die analytischen Abbildungen von mehrdimensionalen Räumen. In: Verhandlungen des Internationalen Mathematiker-Kongresses. Zürich 1932, vol I, pp 93–101 and Ges. Math. Schr. vol IV, pp 234–246.

VI *Functions of Real Variables*

VI.1 Theory of Measure

Carathéodory measure (book section 2.17.1)

Carathéodory C (1914) Über das lineare Maß von Punktmengen – eine Verallgemeinerung des Längebegriffs. Nachr. Kgl. Ges. Wiss. Gött., math.-phys. Kl. 1914, pp 404–426 and Ges. Math. Schr. vol IV, pp 249–275.

Vitali–Carathéodory theorem

Carathéodory, Constantin: Vorlesungen über reelle Funktionen. Leipzig, Berlin, B. G. Teubner, 1918, p 304.

VI.2 One-to-One Mappings

VI.3 Algebraic Theory of the Integral

VII *Geometry*

VIII *Partial Differential Equations*

IX *Book Reviews*

X *Historical-Biographical Papers*

Appendix V

A List of Carathéodory's Students

with name, thesis title and year of title award, as well as some biographical notes, wherever possible.

Carathéodory "had not published many of his ideas; they result in others' works, especially in those of the numerous students who were introduced by him to the spirit and ways of scientific research and who partly themselves occupy university chairs today."

Perron O (1952) *Constantin Carathéodory. Jber. Deutsch. Math.-Verein.* 55:39–51, here p 51.

Göttingen: Doctorates

1. *Hans Rademacher:* Eindeutige Abbildungen und Meßbarkeit (Unique Mappings and Measurability), 1916.

(Wandsbek near Hamburg 3 April 1892 – Haverford, Pa 7 February 1969)

1911–1915 stud. Göttingen, 1916 Dr. Göttingen, 1919 habil. lecturer univ. Berlin, 1922 assoc. prof. univ. Hamburg, 1925 full prof. univ. Breslau, 1933 emigr., 1934 visit. Rockefeller fellow, 1936 assoc. prof. and 1939 prof. univ. of Pennsylvania, Pa.

2. *Paul Finsler:* Über Kurven und Flächen in allgemeinen Räumen (On Curves and Surfaces in General Spaces), 1918.

(Heilbronn a. N. 11 April 1894 – Zurich 29 April 1970)

1912–1918 stud. Stuttgart and Göttingen, 1918 Dr. Göttingen, 1921 ass. univ. Cologne, 1922 habil. and 1925 teaching commission univ. Cologne, 1926–1944 assoc. prof. of applied mathematics and 1945 full prof. univ. Zurich, 1959 emer.

Berlin: Doctorates

1. *Erich Bessel-Hagen:* Über eine Art singulärer Punkte der einfachen Variations-probleme in der Ebene (About a Kind of Singular Points in Simple Variational Problems on the Plane), 1920, Carathéodory/Schmidt.

(Charlottenburg 12 September 1898 – Bonn 29 March 1946)

1917–1920 stud. Berlin, 1920 Dr. Berlin, 1921–1924 Klein's private ass., 1925 habil. and lecturer Göttingen, 1927 new habil. and lecturer Halle, 1928 new habil. and lecturer Bonn, 1931 assoc. prof. univ. Bonn, 1939 non-budgeted prof.

Berlin: Habilitations

1. *Hans Hamburger:* Über eine Verallgemeinerung des Stieltjesschen Momenten-problems (About a Generalisation of the Stieltjes Problem of Moments), 1919, Schmidt/Carathéodory.

(Berlin 5 August 1889–Cologne 14 August 1956)

1907–1914 stud. univ. Berlin, Lausanne, Göttingen and Munich, 1914 Dr. (A. Prings-heim) univ. Munich, 1919 habil. and lecturer Berlin, 1922 assoc. prof. univ. Berlin, 1924–1935 full prof. univ. Cologne, 1935 dismissed, Berlin, 1939 emigr., 1941–1947 lecturer univ. college in Southampton, 1947–1953 full prof. univ. Ankara/Turkey, 1953–1956 full prof. univ. Cologne, 1954–1955 guest prof. at Cornell univ. Ithaca (USA).

2. *Hans Rademacher:* Über partielle und totale Differenzierbarkeit von Funktionen mehrerer Variabeln (About Partial and Total Differentiability of Functions of Several Variables), 1919, Carathéodory/Schmidt.

Munich: Doctorates

1. *Ludwig Häusler:* Über das asymptotische Verhalten der Taylor-Koeffizienten einer gewissen Funktionenklasse (About the Asymptotic Behaviour of Taylor Coefficients of a Certain Class of Functions), 1929, Perron/Carathéodory.

2. *Hans Wolkenstörfer:* Probleme der Erweiterung von topologischen Abbildungen ebener Punktmengen (Problems of Extension of Topological Mappings of Plain Point Sets), 1929, Tietze/Carathéodory.

(12 February 1898), Dr., ass. TU Munich.

3. *Wladimir Seidel:* Über die Ränderzuordnung bei konformen Abbildungen (About Frontier Classification in the Case of Conformal Mappings), 1930, Carathéodory/Perron.

(1906 Odessa, Ukraine), 1933 Ph.D. at Harvard, paper on the cluster values of an analytic function at a boundary point of its disk domain of definition, Wayne State University, Detroit, MI, 1952–1953 member of the School of Mathematics at IAS Princeton.

4. *Hans Rügemer:* Die absoluten Störungen für die Planeten der Jupitergruppe (The Absolute Disturbances for the Planets of the Jupiter Group), 1930, Alexander Wilkens/Carathéodory.

(Nuremberg 25 April 1905–22 June 1977)

stud. univ. Munich and Leipzig; 1928 student ass. Munich observatory; 1929 study interrupted for financial reasons; doctorate with cum laude; 1 August 1930 ass. Bamberg observatory; May 1933 joins the NSDAP; although denouncing his head Zinner, remains ass. up to 31 July 1935; 1 August 1935–31 July 1936 recipient of a DFG scholarship, works at Munich observatory; 1 October 1936 ass. Munich observatory; from August 1939 at the Luftwaffe weather station; since August 1942 *Observator der Bayerischen Erdmessungskommission*; after 1945 in Neustadt, Waldnaab at the Johannes Kepler private observatory.

5. *Georg Aumann:* Beiträge zur Theorie der Zerlegungsräume (Contributions to the Theory of Decomposition Spaces), 1931, Tietze/Carathéodory.

(Munich 11 November 1906–4 August 1980)

1925–1929 stud. univ. Munich, 1929 state exams, 1930–1932 probationary teacher, 1931 Dr. univ. Munich, 1932–1936 ass. TU Munich, 1933 habil. Munich, lecturer univ. and TU Munich, Christmas 1933, *Ein Satz über die konforme Abbildung mehrfach zusammenhängender ebener Gebiete* (A Theorem concerning the Conformal Mapping of Multiply Connected Plane Domains) together with Carathéodory, 1934–1935 Rockefeller fellow at Princeton, NJ, 1936–1946 assoc. prof. univ. Frankfurt a. M., NSDAP member from 1937, 1939–1941 Wehrmacht anti-aircraft gun, 1941–1945 Wehrmacht code department, March 1946 dismissed from the univ. Frankfurt a. M., 1948 secondary-school teacher Munich, teaching commission univ. Munich and phil.-theol. college Regensburg, 1949 prof. univ. Lahore/Pakistan, 1949 full prof. univ. Würzburg, 1950–1960 univ. Munich, 1960–1961 visit. prof. univ. of Idaho USA, 1961 full prof. TU Munich, 1966–1967 visit. prof. univ. UCLA.

6. *Ernst Peschl:* Über die Krümmung von Niveaukurven bei der konformen Abbildung einfach zusammenhängender Gebiete auf das Innere eines Kreises. Eine Verallgemeinerung eines Satzes von E. Study (About the Curvature of Curves on the Plane in the Case of Conformal Mapping of Simply Connected Domains onto the Interior of a Circle. A Generalisation of a Theorem by E. Study), 1931, Carathéodory/Perron.

(Passau 1 September 1906–9 July 1986)

1925–1931 stud. univ. and TU Munich, 1929 state exams Munich, 1931 Dr. univ. Munich, 1931–1938 ass. univ. Jena and univ. Münster, 1935 habil. univ. Jena, 1936 lecturer univ. Jena, 1937 deputy prof. univ. Bonn, 1938 assoc. prof. univ. Bonn, 1941–1943 military service, 1943–1945 on leave as scientific researcher at Research Institution for Aeronautics (*Luftfahrt-Forschungsanstalt*) Braunschweig-Völkenrode, 1948 full prof. univ. Bonn, 1955–1968 head of the Rhine-Westf. Inst. for Instrument. Math. univ. Bonn, 1968 head of Inst. for Math., Society for Mathematics and Data Processing (*Gesellschaft für Mathematik und Datenverarbeitung*), 1974 emer.

7. *Wilhelm Damköhler:* Über indefinite Variationsprobleme (About Indefinite Variational Problems), 1933, Carathéodory/Perron.

(Klingenmünster/Rheinpfalz 25 February 1906)

1925–1933 stud. Munich, 1933 Dr. univ. Munich, 1938–1945 lecturer univ. Jena, 1944 Helmholtz-Institute Landsberg, 1946 teaching commission univ. Munich, 1948–1952 prof. univ. Tucumán/Argentina, 1952 prof. univ. Potosi/Bolivia.

8. *Josefa v. Schwarz:* Das Delaunaysche Problem der Variationsrechnung in kanonischen Koordinaten (Delaunay Problem of the Calculus of Variations in Canonical Co-ordinates), 1933, Carathéodory/Tietze.

Expert Allianz insurance PLC.

9. *Ta Li* (Chinese student from Macao): Über die Stabilitätsfrage bei Differentialgleichungen (On the Stability Question in the Case of Differential Equations), 1933, Perron/Carathéodory.

1934 Tsing-Hua univ. Peiping, China, 1937 Tung-Chi univ. prov. Kiangsi, China. Ta Li published his work on the *Neue Beweise zu den Carmichaelschen Sätzen über die Reihe* $\Omega(x) = \sum_{n=0}^{\infty} c_n g(x+n)$ (New Proofs of the Carmichael Theorems about the Series $\Omega(x) = \sum_{n=0}^{\infty} c_n g(x+n)$ in the *Journal für reine und angewandte Mathematik* 169 (1933). In the first footnote to that work, Ta Li explained that he had spoken about this subject at the mathematical seminar of Munich University on 18 February 1932 and was grateful to the lecturers for their consent. A common statement by Carathéodory, Hartogs, Perron and Tietze reads: "In issue 169 of this journal, pp 87–91 appeared a work by Herr Ta Li, of which the first footnote could easily lead the reader to wrong conclusions. The lecturers participating in the Munich mathematical seminar feel therefore obliged to make the statement that the relevant work was not shown to them before publication and that, consequently, they are not responsible for the contents in any way." Below in hand-writing: "In additon Ta Li should write in correction: 'footnote is based on a misunderstanding' ".
(Niedersächsische Staats- und Universitätsbibliothek Göttingen – Abteilung für Handschriften und seltene Drucke. Cod. Ms. H. Hasse 33:2).

10. *Josef Mall:* Grundlagen für eine Theorie der mehrdimensionalen Padéschen Tafel (Foundations for a Theory of the Padé Table of Several Dimensions), 1934, Perron/ Carathéodory.

11. *Erna Zurl:* Theorie der reduziert-regelmäßigen Kettenbrüche (Theory of Reduced-Regular Continued Fractions), 1934, Perron/Carathéodory.

12. *Rudolf Steuerwald:* Über Enneper'sche Flächen und Bäcklund'sche Transformation (About Enneper's Surfaces and Bäcklund's Transformation), 1935, Carathéodory/Tietze.
(Munich 7 November 1887 – Alzing/Upper Bavaria 25 July 1960)
1906–1910 stud. univ. Munich, Heidelberg, Freiburg, 1910 state exams, 1919–1937 higher schools in Munich, probationary professor, 1935 Dr., 1937 retirement, 1946 honorary prof. univ. Munich, teaching commission univ. Munich.

13. *Ahmet Nazim* (since 1938: *Nazim Terzioğlu*): Über Finslersche Räume (About Finsler Spaces), 1936, Carathéodory/Perron.
(Kayseri 1912 – 1976 Silivri)
stud. univ. Göttingen, Munich, 1937 assist. math. dpt. univ. Istanbul, 1942 lecturer univ. Istanbul, 1943–1944 math. prof. univ. Ankara, 1944 math. prof. univ. Istanbul, 1950–1952 dean math.-nat. sci. faculty, founder of geophys. inst. and inst. for hydrobiology in Istanbul, co-founder of a space research inst. in Uludağ, 1953 director math. instit. Istanbul, 1965–1967 foundation rector Karadeniz (Black Sea) Techn. Univ., 1969–1971 and 1971–1974 rector univ. Istanbul, 1971 established the Mathematical Research Institute, a library with 2000 math. vols., organised summer courses on modern mathematics for teachers of secondary schools, 1973 fellow of

the Hahnemann Medical Society of America, 1974 *Grosses Verdienstkreuz* from the Federal Republic of Germany.

Well-known publ: Über den Satz von Gauss–Bonnet im Finslerschen Raum. Univ. Istanbul. Fac. Sci. (1948), 26–32.

Research on Apollonius, editor of:

Das Vorwort des Astronomen Banî Mûsa b. Şâkir zu den Conica des Apollonios von Perge (The Foreword of the Astronomer Banî Mûsa b. Şâkir to the Conics of Apollonius of Perga). Istanbul 1974.

Das achte Buch zu den Conica des Apollonios von Perge rekonstruiert von Ibn al-Haysam (The Eighth Book to the Conics of Apollonius of Perga Reconstructed by Ibn al-Haysam). Istanbul 1974.

Kitâb al-Mahrûtât, Das Buch der Kegelschnitte des Apollonios von Perge (Kitâb al-Mahrûtât, The Book of Conic Sections of Apollonius of Perga). Istanbul 1981.

14. *Süe-yung Kiang* (*née* Zee): Über die Fouriersche Entwicklung der singulären Funktion bei einer Lebesgueschen Zerlegung (About the Fourier Development of Singular Functions in the Case of a Lebesgue Decomposition), 1940, Carathéodory/Schmauss.

15. *Hans Weber:* Über analytische Variationsprobleme (About Analytic Variational Problems), 1942, Carathéodory/Müller.

Müller's assistant, dismissed by Müller end of March 1943.

16. *Paul Armsen:* Über die Strahlenberechnung an einer einfachen Sammellinse (About the Calculation of Rays on a Simple Collecting Lens), 1943, Carathéodory/Rabe.

(16 December 1906 Reval)

1925–1943 stud. Dorpat, Hamburg and Munich, 1943 at the Institute for National Psychology of the Heydrich Foundation, Prague, 1943 Dr. univ. Munich, 1943–1945 ass. with teaching commission German univ. Prague, 1947–1953 teaching commission phil.-theol. college Bamberg, Erlangen, 1953 statistician Johannesburg, South Africa.

17. *Karl Leonhard Weigand:* Über die Randwerte meromorpher Funktionen einer Veränderlichen (About the Boundary Values of Meromorphic Functions of one Variable), 1947, Carathéodory/Tietze.

(30 October 1916 Regensburg)

1936–1940 stud. Munich and Göttingen, 1940 state exams, 1947 Dr., 1948–1951 ass. univ. Munich, 1951 higher-school service.

The names, subjects of dissertations and exact dates of title awards are listed in Biermann (1988), pp 359 and 367, for Berlin and in Toepell (1996), pp 450–452, for Munich.

Information about *Rügemer* in: Texte und Abhandlungen zur Geschichte der Mathematik und der Naturwissenschaften, vol XXX: Litten F (1992) *Astronomie in Bayern 1914–1945* (Astronomy in Bavaria 1914–1945). Franz Steiner, Stuttgart, p 250.

Information about *Seidel* in: J. Laurie Snell "A Conversation with Joe Doob", Web document, and in: IAS Community of Scholars, 1930–1980 (Communication from Lisa Coats, IAS Archives).

Information about *Nazim Terzioğlu* in: *Bilim Tarihi* (History of Science) 16/1993, pp 11–19 (Communication from Professor Dr. Nuran Yıldırım, Istanbul University).

Notes

I *Origin and Formative Years*

1. The appearance of the modern Greek settlements, both those created in Europe in the 18th century and those created in the Orient since the 19th century, is, according to Psyroukis N (1983) *Το Νεοελληνικό Παροικιακό Φαινόμενο* (The Modern Greek Phenomenon of Settlements). "Επικαιρότητα", Athens, directly connected to the birth and evolution of the international capitalist and colonialist market and division of labour. These settlements resulted in the transformation of Greece into a source of and market for agricultural products and raw materials for the capitalist metropolises, but also in integrating Greece into the Western capitalist world. Hasiotis I K (1993) *Επισκόπηση της Ιστορίας της Νεοελληνικής Διασποράς* (Overview of the History of Modern Greek Diaspora). Βάνιας, Thessaloniki, defines the modern Greek settlements as those groups of the Diaspora Greeks whose particular character distinguishing them from their surroundings is due to the common geographic and ethnic origin of their members and does not require an active participation of their members in collective functions. Until Greece's territorial integration the modern Greek settlements were outside the Greek Orthodox Orient. The organisation of their members in national-religious groups created the so-called *κοινότητες* (communities).

2. On the Modern Greek Enlightenment see:
Dimaras K Th (1983) *Νεοελληνικός Διαφωτισμός* (Modern Greek Enlightenment). "Ερμής", Athens.
Kondylis P (1988) *Ο Νεοελληνικός Διαφωτισμός – Οι Φιλοσοφικές Ιδέες* (Modern Greek Enlightenment – The Philosophical Ideas). "Θεμέλιο", Athens.

3. On Chios see: Calvocoressi P: The Anglo-Chiot Diaspora. Greece and Great Britain during World War I. In: Institute for Balkan Studies, King's College (eds) First Symposium organised in Thessaloniki on 15–17 December 1983. Institute for Balkan Studies–202, Thessaloniki, 1985, pp 247–257.

4. Iorga N (1989) *Το Βυζάντιο μετά το Βυζάντιο* (Byzantium after Byzantium). Gutenberg, Athens, p 72. Original title: (1935, 1971) *Byzance après Byzance: Continuation de l'Histoire de la Vie Byzantine*. Bucharest; also: (1992) *Byzance après Byzance: Continuation de l'Histoire de la Vie Byzantine*. Balland, Paris; (2000) Byzantium after Byzantium. Center for Romanian Studies in cooperation with the Romanian Institute of International Studies, Iasi, Oxford.

5. The list of hostages is given in: Vlasto A M (1913) *XIAKA*, or "The History of the Island of Chios from the Earliest Times down to its Destruction by the Turks in 1822". Privately printed, London. See also: Haniotis I (1972–1973) *Η Χίος του 1822* (Chios in 1822). Privately printed, Athens.

6. Carathéodory to Kalitsounakis on 26 June 1935. Letter in Greek. Vovolinis (ed) (1962), p 520.

7. Continental System is a term denoting Napoleon's measures for the embargo against Great Britain that was introduced by the Berlin decree of 21 November 1806. Great

Britain replied in 1807 by prohibiting neutral ships from entering French harbours and with blockading the harbours of France and its allies. In this economic war Great Britain gained a monopoly in the grocery trade and conquered the South-American market. But on the whole, the British economy was severely damaged. On the European continent some industrial branches (the textile and beet sugar industries) could develop well because of the elimination of competition, but the negative consequences predominated, for instance in eastern Germany, which was dependent on the British market for grain. Russia's resistance (since 1810) against the Continental System contributed to the outbreak of war in 1812.

8. Psyroukis N, as note 1, here p 70.
   The countries around the eastern Mediterranean and especially the coasts of Asia Minor, Syria, and Egypt were denoted by the Italian word Levante, meaning East. The Greeks used the same word to describe the Near East.

9. Shaw & Shaw (1977), p 17.

10. Carathéodory C (1957) *Autobiographische Notizen* (Autobiographical Notes). *Ges. Math. Schr.* vol V, pp 389–408, here p 389.

11. The Phanariot rule in Moldavia (since 1709) and Wallachia (since 1715) which lasted up to the beginning of the Greek War of Independence (1821) entered the Rumanian historiography as the Phanariot Epoch. In the Greek historiography, the 18th century is known as The Century of the Phanariots. From the beginning of that century up to the beginning of its last quarter, the intellectual and political life of the Greeks was a matter practically determined by the Phanariots alone.

12. The term Great Idea was first used in 1844 by Greece's Premier Ioannis Colettis in his speech to the National Assembly of the Greeks, in which he introduced three aspects of the Great Idea, namely the desire for a state unity of the Greeks, the realisation of this state unity and the civilising mission of the Greeks. The state unity was realised stepwise: without blood shed with the integration of the Ionian islands in 1863, the annexation of Thessaly and part of Epirus in 1881 and the union with the Dodecanese in 1947. Macedonia and part of Thrace were united with the national corpus as a consequence of the two victorious Balkan Wars in 1912–1913, as well as the Aegean islands that were, however, ceded to Greece only through the Treaty of Sévres in 1920. Crete came to Greece in 1913.
    The official Greece has never defined territorial demands connected with the Great Idea in a clear way. These followed according to the concrete conditions each time and according to the criterion of the chances for success. Irredentism and expansionism merged together within the Great Idea and, due to the dispersion of the Greeks, the two national aims could hardly be kept apart.

13. See: Alexandris A (1980) Οι Έλληνες στην Υπηρεσία της Οθωμανικής Αυτοκρατορίας 1850–1922 (Greeks in the Service of the Ottoman Empire 1850–1922). *Bulletin de la Société Historique et Enthologique de la Grèce* 23:365–404.

14. See: Georgiadou M (2000/2001) *Vom ersten zum zweiten Phanar und die Carathéodorys* (From the First to the Second Phanar and the Carathéodorys). *Südost-Forschungen* 59/60:164–217.
    Georgiadou M: Expert knowledge between tradition and reform – The Carathéodorys: a Neo-Phanariot Family in 19th Century Constantinople. In: Institut Français d'Études Anatoliennes (ed) *Médecins et ingénieurs ottomans à l'âge des nationalismes* (Ottoman Medical Doctors and Engineers in the Age of Nationalism). Maisonneuve & Larose, Paris, 2003, pp 243–294.

15. Carathéodory, as note 10, here p 392. In the period during which Carathéodory's family was living in Constantinople, new national states emerged and national conflicts developed in the Balkans and a reform programme known as *Tanzimat* was put forward to enable modernisation in the Ottoman Empire. Despite the fact that as regards politics,

the military and economy the Ottoman state declined to the extent that it became known as the Big Sick, the development of three branches of the economy, namely trade, manufacture, and the monetary and credit system, resulted in the emergence of an Ottoman upper middle class, mainly consisting of Armenians, Greeks, and Jews, i.e. non-Muslim subjects. The Ottoman Greeks embraced the ideas of modernisation because it served their economic insterests and resulted in their treatment as equal to that of the Sultan's Muslim subjects.

16. Carathéodory Et (1861) *Du droit international concernant les grands cours d'eau* (International Law Concerning the Great Waterways). Brockhaus, Leipzig.

17. He was christened four and a half months later, on 25 January 1874. His christening certificate was issued by the Russian Orthodox church in Berlin for submission to the Munich authorities in July 1933 as was then demanded for the proof of "Aryan" descent. The Russian church, following another calendar, gives 1 September 1873 as the birth date and 13 January 1874 as the christening date. A copy of this document is kept in: BayHStA MK 35403, Laufzeit: 1924–1950.

18. Carathéodory, as note 10, here p 393.

19. In November 1900 Stephanos C. learned from the newspapers that he was to be replaced. The author of the libel, a former translator at the French embassy in Constantinople, *Comte* Emile de Kératry (1832–1904), had written it to extort money. Though Stephanos bargained him down in his attempt to dissuade him from publishing, the Sublime Porte declined the deal and sacrificed their envoy instead. His dismissal was attributed by him and Alvensleben to the intrigues of his successor to the Brussels post, the Turkish ambassador to Paris and palace protégé, Salih Münir Pasha. Cf. Findley C V (1989) Ottoman Civil Officialdom – A Social History. Princeton University Press, Princeton, p 229.

20. Carathéodory, as note 10, here p 393.

21. Gustav Karl Ludwig Richter (Berlin 1823–Berlin 1884), portrait, genre and Orient painter and lithographer with studies in Berlin, Paris and Rome. Since 1860 member of the Berlin Academy of Arts. Painter of the high society.

22. Giacomo Meyerbeer (Berlin 1791–Paris 1864), actually Jakob Liebmann Meyer Beer. Composer of operas; 1816 he went to G. Rossini in Venice and ten years later he settled down in Paris, where he cultivated the French Great Opera with E. Scribe; in 1842 he became general conductor of the Berlin opera. Meyerbeer's operas combine German, Italian and French stylistic elements and are a mixture of the classical lyric tragedy (5 act tragedies with historical themes) and the *opéra comique*. His works: *Robert der Teufel* (1831); *die Hugenotten* (1836); *der Prophet* (1849); *die Afrikanerin* (1864).
    Die Afrikanerin (The African Lady) is the nickname of Carathéodory's daughter.

23. Carathéodory C: Μαθηματικά (Mathematics). In: Drandakis (ed) Μεγάλη Ελληνική Εγκυκλοπαίδεια (Great Greek Encyclopaedia), vol 16. "Πυρσός", Athens, 1931, pp 465–466. Translated from the Greek by Stephanos Carathéodory as *Mathematik* (Mathematics) for *Ges. Math. Schr.* vol V, pp 233–239, here p 238.
    Gaspard Monge (1746–1818): French mathematician and physicist, born in Beaune, known as the founder of descriptive geometry. He became Professor of Mathematics at Mézières in 1768, and in 1780 Professor of Hydraulics at the Lycée in Paris.
    "The earliest paper of any special importance which he communicated to the French Academy was one in 1781, in which he discussed the lines of curvature drawn on a surface. These had been first considered by Euler in 1760, and defined as those normal sections whose curvature was a maximum or a minimum. Monge treated them as the locus of those points on the surface at which successive normals intersect, and thus obtained the general differential equation. He applied his results to the central quadrics in 1795. In 1786 he published his well-known work on statics." In 1792 he became minister of the marine, but soon took charge of the national manufacture of arms and gunpowder

assisting "the committee of public safety in utilizing science for the defence of the republic." When the terror regime was established, "he was denounced, and escaped the guillotine only by a hasty flight. On his return in 1794 he was made a professor at the short-lived [...] [*École Normale*], where he gave lectures on descriptive geometry;" his *Leçons de géométrie descriptive*, in which he stated his principles regarding the general application of geometry to the arts of construction, was published the following year (1795). In 1796 he "went to Italy on the roving commission which was sent with orders to compel the various Italian towns to offer pictures, sculpture, or other works of art that they might possess, as a present or in lieu of contributions to the French republic for removal to Paris. In 1798 Monge was sent by the *Directory* to Italy, from where he followed Napoleon to Egypt. Thence after the naval and military victories of England he escaped to France.

Monge then settled down at Paris, and was made professor at the Polytechnic school, where he gave lectures on descriptive geometry; these were published in 1800 in the form of a textbook entitled *Géométrie descriptive*. This work contains propositions on the form and relative position of geometrical figures deduced by the use of transversals. The theory of perspective is considered; this includes the art of representing in two dimensions geometrical objects which are of three dimensions, a problem which Monge usually solved by the aid of two diagrams, one being the plan and the other the elevation. Monge also discussed the question as to whether, if in solving a problem certain subsidiary quantities introduced to facilitate the solution become imaginary, the validity of the solution is thereby impaired, and he showed that the result would not be affected. On the restoration he was deprived of his offices and honours, a degradation which preyed on his mind and which he did not long survive. Most of his miscellaneous papers are embodied in his works, *Application de l'algèbre à la géométrie*, published in 1805, and *Application de l'analyse à la géométrie*, the fourth edition of which, published in 1819, was revised by him just before his death. It contains among other results his solution of a partial differential equation of the second order."

In 1805 Monge, who was a counsellor and friend of Napoleon, was made a senator and Count of Pelusium, but lost both dignities on the restoration of the Bourbons.

Within a few years from its introduction, descriptive geometry was being taught in French scientific and technical schools and had spread to several other countries. Monge viewed descriptive geometry as a powerful tool for discovery and demonstration in various branches of pure and infinitesimal geometry.

Joseph Fourier had also participated in Napoleon's Egyptian expedition, became secretary of the *Institut d'Egypte* and compiled the *Description de L'Egypte*.

Citations from: A Short Account of the History of Mathematics (4th edition, 1908) by W.W. Rouse Ball. Web document.

24. The European Commission of Danube consisted of the Ottoman Empire, Prussia, Austria-Hungary (i.e. countries through which Danube flows), as well as of Britain, France, Sardinia and Russia. The navigation of the upper Danube was administered by the International Commission of Danube (CID = *Commission Internationale du Danube*) which included delegates from Austria-Hungary, Bavaria, the Ottoman Empire, Württemberg, two commissioners of the two Danube Principalities and a commissioner of the Principality of Serbia, the last three being employed by the Ottoman Empire. The two commissions had to control the implementation of the internationalisation of navigation, the equal treatment of all flags, the construction of technical works for the facilitation of navigation and, finally, the supervision and regulation of navigation.

25. C. Carathéodory to Stephanos Al. Carathéodory on 11 [23] April 1897. Letter in Greek. Vovolinis (ed) (1962), p 470f.

Maybe the reason why Carathéodory did not participate in the war was his sense of Greek national identity and his Turkish nationality.

26. C. Carathéodory to Stephanos Al. Carathéodory on 28 April [10 May] 1897. Letter in Greek. Vovolinis (ed) (1962), p 471.

27. Historical Archive of Samos, Greek Ministry of National Education and Religious Affairs.

28. Jordan C (1882–1887) *Cours d'analyse de l'école polytechnique*. 3 vols, Gauthier-Villars, Paris.

29. Behnke H: *Carathéodorys Leben und Wirken* (Carathéodory's Life and Deeds). In: Panayotopoulos (ed) (1974), pp 17–33, here p 20.
The same text is published as: Behnke H (1974) *Constantin Carathéodory 1873–1950 – Eröffnungsrede zur Centenarfeier (am 3. September 1973) in Athen* (Constantin Carathéodory 1873–1975 – Opening Speech at the Centenary Celebration (on 3 September 1973) in Athens). *Jber. Deutsch. Math.-Verein.* 75:151–165.

30. In 1859 Wilhelm Fiedler proposed to George Salmon to work out the conic sections. Thus began his 45-year long friendship with the English theologian and mathematician (Rev. Dr. hon. of Cambridge and Oxford, 1819–1904, since 1866 Professor regius of divinity in Dublin), but also his tireless work for the propagation of algebraic-geometric methods that had been developed in England and Germany.
A first fruit of these studies was his dissertation of 1859, *Die Zentralprojektion als geometrische Wissenschaft* (The Central Projection as a Geometric Science), which brought him the doctoral title of the University of Leipzig on A. F. Möbius' recommendation, although he was autodidact. Since 1857 in Chemnitz lecturer of descriptive geometry, to which he initially had not felt particularly attracted, he released already in 1860 the *Analytische Geometrie der Kegelschnitte mit besonderer Berücksichtigung der neueren Methoden, frei bearbeitet nach G. Salmon* (Analytical Geometry of Conic Sections with Special Consideration of Recent Methods, Freely Treated According to G. Salmon) (7th edition in 1907), then, in 1862, the independent work *Die Elemente der neueren Geometrie und die Algebra der binären Formen* (The Elements of Recent Geometry and the Algebra of Binary Forms) and in 1863 the *Vorlesungen über die Algebra der linearen Transformationen frei bearbeitet nach Salmon* (Lectures on the Algebra of Linear Transformations, Freely Treated According to G. Salmon) (3rd edition 1879). Although he began to elaborate on Salmon's Treatise on *Analytic Geometry of Three Dimensions* (1862) (3rd edition 1879, from the 4th only the first part appeared in 1898) about the same time, only later, in 1873, he related to it the *Analytische Geometrie der höheren ebenen Kurven, frei bearbeitet nach G. Salmon* (Analytical Geometry of Higher Plain Curves, Freely Treated According to G. Salmon) (2nd edition 1882).
Voss A (1913) *Wilhelm Fiedler. Jber. Deutsch. Math.-Verein.* 22:97–113, here 98f.

31. Carathéodory, as note 10, here p 397f.

32. Carathéodory C (1901) *Nouvelles mesures du mur sud de la grande galerie de la grande pyramide de Cheops. Académie Royale de Belgique, Bulletin de la classe des sciences*. Hayez, Bruxelles, pp 31–41 and *Ges. Math. Schr.* vol V, pp 273–281.
On page 4f of the same *Bulletin* one can read under "Rapports": "Sur la proposition de M. Ch. Lagrange, le *Bulletin* renfermera une note intitulée: *Nouvelles mesures du mur sud de la grande galerie de la grande pyramide de Cheops*; par C. Caratheodory, ancien élève de l'École militaire de Belgique."

33. *Φιλική Εταιρεία* was the name of the secret society founded in Odessa in 1814 with the single aim of liberating the Greeks from the Ottoman yoke through an armed revolt. Strongly influenced by Freemasonry its membership grew rapidly, particularly in the Diaspora settlements. Merchants were forming the largest single category within the Friendly Society.

34. Emmanuel Benakis (1843–1929). Member of Parliament, elected deputy for Attico-Boeotia in 1910. Minister of National Economy 1910, Minister of Agriculture, Trade

and Industry 1911, Mayor of Athens 1914. In 1920 he went to France but returned to Greece in 1924.

35. Eleutherios Venizelos (Mournies/West Crete 1864–Paris 1936). Greek liberal politician, jurist. As leader of the revolutionary movement for the union of Crete with Greece and because of his opposition to Prince George he enjoyed great national respect. From 1910, when he was called to Athens by the Στρατιωτικό Σύνδεσμο (Military League) of the "Goudi revolutionaries", until his exile in 1936, he dominated the Greek political life: he was prime minister in 1910–1915, 1917–1920, 1924, 1928–1932, 1932 and 1933. In 1910 he founded the Liberal Party and was its president since then. From his party emerged the later parties of the Centre in Greece. Venizelos was closely related to Lloyd George. His foreign policy aiming at the union of all Greeks in one state became a synonym of the realisation of the Great Idea.

36. Carathéodory C (1901) *Η Αίγυπτος* (Egypt). Σύλλογος προς διάδοσιν ωφελίμων βιβλίων, Athens.

37. John Horváth to Vicki Lynn Hill on 22 October 2001. Communication to the author from Vicki Lynn Hill on 08 November 2001.
   For the proof of Schwarz's result Horváth refers to "Rademacher–Toeplitz, ‚Von Zahlen und Figuren' [About Numbers and Figures], or Fejér's Collected Works, Vol. 2".

38. Friedelmeyer J-P: August Leopold Crelle, 1780–1855. In: Begehr H G W, Koch H, Kramer J, Schappacher N, Thiele E-J (eds) Mathematics in Berlin. Birkhäuser, Basle, Boston, Berlin, 1998, pp 27–32.

39. Cf. Birt Th (1900) *Deutsche Wissenschaft im 19. Jahrhundert – Eine Rede gehalten am 9. Januar 1900* (German Science in 19th Century – A Speech held on 9 January 1900). Marburger akademische Reden 1900, no 1. Elwert, Marburg.

40. Jakob Steiner's work on *Einige geometrische Betrachtungen* (Some Geometric Considerations) which dealt with circles was published in 1826 in the new *Journal für die reine und angewandte Mathematik* and marked the beginning of his important research in geometry. He contributed sixty-two articles in total to this journal. This work together with his most famous work, the *Systematische Entwicklung der Abhängigkeit geometrischer Gestalten voneinander* (Systematic Development of the Dependence of Geometric Figures on Each Other) of 1832 and his posthumously published *Vorlesungen über synthetische Geometrie and Allgemeine Theorie über das Berühren und Schneiden der Kreise und der Kugeln* (Lectures on Synthetic Geometry and General Theory on the Contact and Cutting of Circles and Spheres) dealt with elements of what has come to be called projective geometry.
   For an extensive presentation of Steiner's work see: Begehr H, Lenz H (1998) Jacob Steiner and Synthetic Geometry. In: Begehr H et al (eds), as note 38, here pp 49–54.
   In his most famous work, *Geometrie der Lage* (Geometry of Position) of 1847, von Staudt was the first to do projective geometry without reference to measurements or lengths. His *Beiträge zur Geometrie der Lage* (Contributions to the Geometry of Position), that was published three times between 1856 and 1860, supplement his 1847 book. Here, he applied his discoveries in projective geometry to synthetic geometry and he also introduced the formulae for one-, two-, or three-dimensional complex projective spaces. Notable Mathematicians.

41. Carathéodory C: *Hermann Amandus Schwarz*. In: Verband der Deutschen Akademien (ed) *Deutsches Biographisches Jahrbuch*, vol 9. Deutsche Verlags-Anstalt, Stuttgart, 1927, pp 236–238 and *Ges. Math. Schr.* vol V, pp 57–59.

42. Introduction to: Carathéodory C (1932) Conformal Representation. Cambridge University Press, London.

43. Behnke, as note 29, p 20f.

44. Born (1978), p 91f.

45. Biermann (1988), p 154f.

46. Ibid, p 155.

47. Gray JJ (1984) Fuchs and the theory of Differential Equations. Bull. Amer. Math. Soc. 10:1–26, here p 1 and p 19.

48. Biermann (1988), p 141.

49. Hensel. K. (1899) *Neue Begründung der Theorie der algebraischen Zahlen.* In this work K. Hensel had the idea of creating an analogue of the tried and tested power-series method in the case of function sets applied to the case of number sets.

50. Wilczynski E J (1902) Lazarus Fuchs. Bull. Amer. Math. Soc. 9:46–49, here p 46.

51. The *Seminar für Orientalische Sprachen* was working since 1887 for the personnel of the Foreign Ministry and the German economic interests abroad.
    Wende (1959), p 64.

52. Solomonidis (1966), p 46f.

53. Cf. Ellwein Th (1997) *Die deutsche Universität – Vom Mittelalter bis zum Gegenwart* (The German University – From the Middle Ages to the Present Day). Fourier, Wiesbaden, p 227f.

54. Ralf Haubrich speaks of departure from the neo-humanistic ideal (meaning: idealism) due partly to the transformation of the institutional and educational system of German mathematics around 1890 under Klein's influence.
    Haubrich R (1998) Frobenius, Schur, and the Berlin Algebraic Tradition. In: Begehr et al (eds), as note 38, pp 82–96.

55. Carathéodory C (1900–1901) *La Géométrie synthétique. Revue de l'Université de Bruxelles* 6:11 pages and *Ges. Math. Schr.* vol V, pp 263–272.

56. Fraenkel (1967), note 55 on p 149.

57. Reid (1970), p 97.

58. Born (1978), p 91. Born also refers to Hans Müller, a mathematician who was awarded the doctor title in 1903.

59. Reid (1970), p 97.

60. Carathéodory to Frau Young on 14 August 1927. Letter in German. Special Collections and Archives, University of Liverpool Library, ref. D.140/9/6.

61. Peckhaus V (1990) *"Ich habe mich wohl gehütet alle Patronen auf einmal zu verschießen" – Ernst Zermelo in Göttingen* ("I took good care not to fire all cartridges at once" – Ernst Zermelo in Göttingen). History and Philosophy of Logic 11:19–58, here note 44 on p 27.

62. Carathéodory C, Sommerfeld A (1943) *Zum Andenken an David Hilbert; Ansprachen im Trauerhause* (In Memory of David Hilbert; Addresses in the House of Mourning). *Die Naturwissenschaften* 31:213–214 and *Ges. Math. Schr.* vol V, pp 96–100, here p 99.

63. Behnke, as note 29, here p 21f.

64. Reid (1976), p 42f.

65. Born (1978), p 90.

66. Born (1978), p 82.

67. H. Minkowski, *Über die Körper konstanter Breite.* In: Hilbert D, Speiser A, Weyl H (eds) *Gesammelte Abhandlungen von Hermann Minkowski* (Collected Treatises of Hermann Minkowski), vol II. Teubner, Leipzig and Berlin, 1911, pp 277–279. This work appeared in Russian in the journal *Matematicheskii Sbornik* (Mathematical Collection). Moscow, vol 25, pp 505–508.

68. Carathéodory C (1935) *Schlußwort zur Encyklopädie der mathematischen Wissenschaften* (Closing Remarks about the Encyclopaedia of Mathematical Sciences). In: Akademie der Wissenschaften zu Göttingen (ed) *Encyklopädie der Mathematischen Wissenschaften mit Einschluß ihrer Anwendungen* (Encyclopaedia of Mathematical Sciences

and their Applications), 6 vols. Teubner, Leipzig, 1898–1935, here pp V-VIII of index volume for vol IV: *Mechanik* (Mechanics). Also: (in Hungarian translation) *Matematikai és Fizikai Lapok* 42:102–106 and *Ges. Math. Schr.* vol V, pp 79–83.

69. Zassenhaus H J (1975) On the Minkowski–Hilbert dialogue on mathematization. Canad. Math. Bull. 18(3):443–461, here p 454.

70. Carathéodory, as note 10, here p 401f.

71. Meschkowski (1980), p 160.

72. Klein F (^{2}1911–1914) *Elementarmathematik vom höheren Standpunkte aus* (Elementary Mathematics from an Elevated Point of View), 2 parts (I: *Arithmetik, Algebra, Analysis,* II: *Geometrie*). Teubner, Leipzig, here part II, cited in: Meschkowski (1980), p 160.

73. Hilbert D (1899) *Grundlagen der Geometrie* (Foundations of Geometry). Teubner, Leipzig. Translated as: (1902) Foundations of Geometry. Open Court, Chicago.

74. *Teilnachlaß Carathéodory* (Part of Carathéodory's Papers), *Bayer. Akad. d. Wiss.*

75. Carathéodory C (1935) *Glückwunschschreiben der Bayerischen Akademie der Wissenschaften zum 50-jährigen Doktorjubiläum von David Hilbert. Verfaßt im Auftrag der Akademie* (Congratulation Letter of the Bavarian Academy of Sciences on David Hilbert's Fifty-Year Doctoral Anniversary – Authored on the Academy's Commission). *Jbuch d. Bayer. Akad. d. Wiss.* 1934–1935, pp 128–129 and *Ges. Math. Schr.* vol V, pp 77–78.

76. Steck M (1942) *Das Hauptproblem der Mathematik* (The Main Problem of Mathematics). Lüttke, Berlin.

77. Süss to Thär on 8 February 1943. Universitätsarchiv Freiburg, Nachlaß Wilhelm Süss, C89/79.

78. Carathéodory to Klein on 24 December 1906. Letter in German. Niedersächsische Staats- und Universitätsbibliothek Göttingen – Abteilung für Handschriften und seltene Drucke. Cod. Ms. F. Klein 8, 461.

79. Mylonas V (undated), pp 36–38.

80. Klein to Carathéodory on 26 December 1906. Niedersächsische Staats- und Universitätsbibliothek Göttingen – Abteilung für Handschriften und seltene Drucke. Cod. Ms. F. Klein 8, 461/Anl.

81. Behnke, as note 29, here p 21.

82. Zassenhaus, as note 69, here p 454.

83. Weyl H (1944) David Hilbert and his mathematical work. Bull. Amer. Math. Soc. 50:612–654, here p 615.

84. Reid (1970), p 49.

85. Web document of the Mathematical Faculty of the Georg-August University of Göttingen about Hilbert.
     Half-ironically Einstein described Göttingen as *Eldorado der Gelehrsamkeit* (Eldorado of erudition). Cf. his letter to Born and Franck on 18 January 1922. *Albert Einstein – Hedwig und Max Born. Briefwechsel 1916–1955* (1972), p 75.

86. Carathéodory to Weyl on 17 April 1928. Letter in German. ETH-Bibliothek Zürich, Wissenschaftshistorische Sammlungen, HS 91:498.

87. Carathéodory, Sommerfeld, as note 62, here p 99f.

88. Weyl, as note 83, here p 653.

89. Cf. Fraser C G: Calculus of Variations. In: Grattan-Guinness (ed) Companion Encyclopaedia of the History and Philosophy of the Mathematical Sciences, vol 1. Routledge, London, 1994, pp 342–350, here p 343.

90. Zermelo E (1894) *Untersuchungen zur Variationsrechnung*. Doctoral thesis, Berlin.

91. Kneser A (1900) *Lehrbuch der Variationsrechnung*. Vieweg, Braunschweig.

92. Weyl, as note 83, here p 617.

93. Carathéodory C (1919) *Antrittsrede in der Preußischen Akademie der Wissenschaften* (Inaugural Speech at the Prussian Academy of Sciences). *Sitzungsber. Preuß. Akad. d. Wiss. Berlin* 33:564–570 and *Ges. Math. Schr.* vol V, pp 179–186, here p 182f.

94. The Corinth Canal was laid out in 1881–1893 through the isthmus to connect the Gulf of Corinth with the Saronic Gulf.

95. Carathéodory C (1904) *Zur geometrischen Deutung der Charakteristiken einer partiellen Differentialgleichung erster Ordnung mit zwei Veränderlichen. Math. Ann.* 59: 377–382 and *Ges. Math. Schr.* vol V, pp 35–41. A separate print of the *Mathematische Annalen* exists in the Göttingen university archive.

96. Hans Hahn, Gustav Herglotz, Heinrich Tietze, and Paul Ehrenfest, all students of mathematics at the Technical University of Vienna, were known as "the inseparable four". Hahn would later belong to the nucleus of the Vienna Circle, a group of scientists and philosophers, created in Vienna around 1923, who developed the neo-positivism or logical positivism based on 19th-century positivism. It was due to Hahn's initiative that Moritz Schlick was appointed to the University of Vienna in 1922. (See 4.4)

97. The Lagrange problem is a problem of the calculus of variations in an $(n+1)$-dimensional space with p ordinary differential equations as side conditions.
Carathéodory's definition in: (1926) *Die Methode der geodätischen Äquidistanten und das Problem von Lagrange* (The Method of Geodesic Equidistant [Surfaces] and the Problem of Lagrange). *Acta Math.* 47:199–236 and *Ges. Math. Schr.* vol I, pp 212–248, here p 212.

98. Carathéodory, as note 10, here p 405.

99. The Weierstrass E function in the calculus of variations is a function which isolates the main part of the increment of a functional as the extremal is varied, using a local (needle-shaped) variation for a given value of its derivative, at a given point of the extremal.
V. M.Tikhomirov, Encyclopaedia of Mathematics.
For Carathéodory's definition of the Weierstrass E function see: *Ges. Math. Schr.* vol I, p 328.

100. Carathéodory C (1904) *Über die diskontinuierlichen Lösungen in der Variationsrechnung.* Doctoral thesis, Göttingen University and *Ges. Math. Schr.* vol I, pp 3–79.

101. Carathéodory, as note 10, here p 405.
Karl Schwarzschild (Frankfurt am Main 1873–Potsdam 1916). Astronomer, professor and director of the observatory belonging to the University of Göttingen, since 1909 director of the astrophysical observatory in Potsdam, although he had rejected the requirement for this post, namely to be christened. His scientific works, around one hundred papers in the period 1890–1916 extend to pure and applied mathematics, theoretical physics, meteorology, fine mechanics and construction of apparatus, but most of all theoretical and observed astronomy and astrophysics. During his time in Göttingen (1901–1909) he published on electrodynamics and geometric optics. In 1906 he studied the transport of energy through a star by radiation. He was introduced to his theoretical research mainly by Seeliger (Munich) under whose supervision he gained his doctoral title and his habilitation in 1899. In his later time he mainly engaged with problems of the newly created general theory of relativity, but his relevant works (partly written in the battlefield) remained fragmentary. He died on 11 May 1916, the same day on which his last significant work *Zur Quantenhypothese* (On the Quantum Hypothesis) appeared in the *Sitzungsberichte der Preußischen Akademie der Wissenschaften*. He was spirited, enthusiastic about art and poetry. Fraenkel (1967), p 86.
Schwarzschild used to remain absolutely strict in the most popular presentations, but at the same time he had an elegant and intellectually stimulating style that hardly existed among scientists. Freundlich E (1917) *Schwarzschild. Über das System der Fixsterne* (On the System of Fixed Stars) Leipzig 1916. *Die Naturwissenschaften* 40:627.

102. *Gedächtnisrede des Hrn Einstein auf Karl Schwarzschild* (Commemorative Speech of Herr Einstein on Karl Schwarzschild). Presented 29 June 1916, published 6 July 1916. *Sitzungsber. Kgl. Preuß. Akad. d. Wiss. Berlin* 1916, pp 768–770.

103. Carathéodory to Sommerfeld on 12 June 1949. Letter in German. LMU München, Institut für Theoretische Physik, Nachlaß Arnold Sommerfeld.

104. Minkowski's major works in the four-dimensional treatment of electrodynamics are: (a) an address given on 21 September 1908: Minkowski H (1909) *Raum und Zeit* (Space and Time). *Phys. Ztschr* 3:104–111 and *Jber. Deutsch. Math.-Verein.* 18 (1909) 75–88) and (b) *Die Grundgleichungen für die elektomagnetischen Vorgänge in bewegten Körpern* (The Basic Equations for the Electromagnetic Processes in Moving Bodies). *Nachr. Kgl. Ges. Wiss. Gött., math.-phys. Kl.* 1908, pp 53–111.

105. Tietze H (1950), *Dem Andenken an C. Carathéodory* (In Memory of C. Carathéodory). *Sitzungsber. Bayer. Akad. d. Wiss. München, math.-nat. Kl.* 1950, pp 85–101, here note 2 on p 99. Note 25 of Zermelo and Hahn's article in the *Encyklopädie der Mathematischen Wissenschaften mit Einschluss ihrer Anwendungen*, Band II A, pp 626–641, here p 633, reads: "According to oral communications from *C. Carathéodory*, discontinuous solutions of the variation problem which are still available to appropriately modified Weierstrass' methods, can occur in such exceptional cases."

106. Funk P (1958) *Nachruf auf Prof. Johann Radon* (Obituary of Prof. Johann Radon). *Monatshefte f. Math.* 62:189–199, here 193.

107. Perron O (1952) Constantin Carathéodory. *Jber. Deutsch. Math.-Verein.* 55:39–51, here p 42.

108. Carathéodory, as note 100, here *Ges. Math. Schr.* vol I, p 57.

109. Perron, as note 107, p 42.

110. Carathéodory C (1945) *Basel und der Beginn der Variationsrechnung. (Meinem alten Freund Andreas Speiser zum sechzigsten Geburtstag in Erinnerung an viele in Göttingen, Zürich und Melide verbrachte Stunden)* (Basel and the Beginning of the Calculus of Variations. (To my old friend Andreas Speiser in memory of the many hours spent in Göttingen, Zurich and Melide)). In: *Festschrift zum 60. Geburtstag von Prof. Dr. Andreas Speiser* (Festschrift on the Occasion of Prof. Dr. Andreas Speiser's 60th Birthday). Orell Füssli, Zurich, pp 1–18 and *Ges. Math. Schr.* vol II, pp 108–128. Perron wrongly dates the print with 1700.

111. Carathéodory C (1910) *Oskar Bolza. Vorlesungen über Variationsrechnung (Leipzig 1909) – Jacques Hadamard. Leçons sur le calcul des variations recueillies par M. Fréchet (Paris 1910)* (Oskar Bolza, Lectures on the Calculus of Variations (Leipzig 1909) – Jacques Hadamard, Lectures on the Calculus of Variations Selected by Mr. Fréchet (Paris 1910)). *Archiv d. Math. u. Phys.*, 3rd series, 16:221–224 and *Ges. Math. Schr.* vol V, pp 304–308, here p 305.

112. Königsberger is known for both this biographical Festschrift for Jacobi and for his biography of Helmholtz in 1902.

113. *III. Internationaler Mathematiker-Kongress in Heidelberg 1904* and *Programm des III. Internationalen Mathematiker-Kongresses in Heidelberg 1904* (3rd ICM in Heidelberg 1904 and Programme of the 3rd ICM in Heidelberg 1904). Universitätsarchiv Heidelberg, Akte der nat.-math. Fakultät 1903–1904, Bd. 1.

114. Carathéodory, as note 10, here p 391f.
    Carathéodory was refering to the *Traité du Quadrilatère attribué à Nassiruddin-el-Toussy, d'après un manuscrit tiré de la bibliothèque de S. A. Edhem Pacha, ancien Grand-Visir, traduit par Alexandre Pacha Carathéodory, ancien Ministre des Affaires Etrangères. Par Autorisation du Ministère Impérial de l'Instruction Publique.* (Treatise on the Quadrilateral Attributed to Nassiruddin-el-Toussy, According to a Manuscript Drawn from the Library of S. A. Edhem Pasha, Former Grand Vizier, Translated by Alexander Carathéodory Pasha, Former Minister of Foreign Affairs. Authorised by the

Imperial Ministry of Public Instruction). Constantinople 1891. The book was written in Arabic and French.

Nassiruddin-el-Toussy was a Muslim scholar in fact of the 13th century and one of the greatest mathematicians of his time. In his *Treatise on the Quadrilateral*, spherical geometry was dealt for the first time as a separate subject and not as an auxiliary science of astronomy. Edhem Pasha (1818–1893) had donated part of his collections for lectures on natural sciences to the first Ottoman university and was an honorary member of the Greek Philological Society of Constantinople.

Information about Nassiruddin-el-Toussy by Johann Strauss, communicated to the author by Klaus Kreiser on 23 December 2000.

115. Carathéodory C: *Wilhelm Wirtinger*. In: Bayer. Akad. d. Wiss. (ed) *Jbuch d. Bayer. Akad. d. Wiss.* 1944–1948 (Annual of the Bavarian Academy of Sciences). Beck, München, 1948, pp 256–258 and *Ges. Math. Schr.* vol V, pp 175–176.

116. Carathéodory, as note 10, here p 406.

117. This story is told by his daughter as well as by persons who had known Carathéodory personally and heard it from him.

118. Carathéodory C (1906) *Über die starken Maxima und Minima bei einfachen Integralen.* *Math. Ann.* 62:449–503 and *Ges. Math. Schr.* vol I, pp 80–142.

119. Dean W. Fleischmann to *"Euerer Hochwohlgeboren"* (Your Honour) on 8 March 1905. Universitätsarchiv Göttingen, Königliches Universitäts-Kuratorium Göttingen, Philosophische Fakultät, Privatdozenten, Dr. Carathéodory 4JVc 242.

120. Ibid.

121. Carathéodory C (1949) *Länge und Oberfläche.* Disposition of a speech at the Munich Mathematical Colloquium on 16 December 1949. *Ges. Math. Schr.* vol V, pp 292–293.

122. Carathéodory's habilitation is mentioned in Klein's circular letter to the editorial board of the *Mathematische Annalen* on 23 April 1906 as a work evoking concern because of its extension as compared to its content: *Andere Arbeiten erscheinen im Vergleich zu ihrem gedanklichen Inhalt zu umfangreich [. . .] auch die lange Ausdehnung von Caratheodory macht mir Sorge.* (Other works compared to their intellectual content appear to be too extensive [. . .] also the great extent of Carathéodory's work worries me).
Tobies R (1987) *Zu Veränderungen im deutschen Zeitschriftenwesen um die Wende vom 19. zum 20. Jahrhundert (Teil II).* (On the Changes in the German Journal Business around the Turn of the 19th to the 20th Century (Part II)). In: *NTM-Schriftenr. Geschichte Naturwiss., Technik, Med.* 24, Leipzig, pp 31–49, here p 46.

123. *Lebenslauf Professor Carathéodory* (Professor Carathéodory's Curriculum Vitae), 6. März 1909. Niedersächsisches Hauptstaatsarchiv, Akten betreffend Professor Dr. Carathéodory (Files Concerning Professor Dr. Carathéodory) (1909/aus. 30.09.1910, Abth. II No 13) der Königlichen Technischen Hochschule zu Hannover, Signatur: Hann. 146 A Acc. 109/79 Nr. 36.

124. Tietze, as note 105, p 93.

125. Communication to the author on 17 January 2000.

126. Ahrens W (1917) *Lorey, Wilhelm. Das Studium der Mathematik an den deutschen Universitäten seit Anfang des 19. Jahrhunderts, Abhandlungen über den mathematischen Unterricht in Deutschland veranlaßt durch die internationale mathematische Unterrichtskommission, herausgegeben von F. Klein. Bd. III, Heft 9, Leipzig und Berlin. B.G. Teubner. 1916* (Lorey, Wilhelm. The Study of Mathematics at German Universities Since the Beginning of the 19th Century – Treatises on the Mathematical Instruction in Germany on Orders of the International Commission of Mathematical Instruction, edited by F. Klein. Vol. III, issue 9, Leipzig and Berlin, Teubner, 1916). *Die Naturwissenschaften* 16:259–262.

127. Address to Klein on his 60th birthday by Hilbert. In: Rowe D E (transl) (1986) David Hilbert on Poincaré, Klein, and the World of Mathematics. The Mathematical Intelligencer vol 8 no 1:75–77.

128. Carathéodory, as note 10, here p 407f.

129. See: Wind Tunnels of NASA, chapter 2: The European Tunnels. NASA History Office on the Web.

130. Süss to Prandtl on 3 February 1945. Universitätsarchiv der Albert-Ludwigs-Universität Freiburg, Nachlaß Wilhelm Süss, C89/4.

131. Carathéodory sent Toeplitz regards through Born, with whom Toeplitz was close friends. Cf. Carathéodory to Born on 9 October 1907. Staatsbibliothek zu Berlin – Preussischer Kulturbesitz, Handschriftenabteilung, Nachl. Born 105.

132. Zermelo E (1904) *Beweis, daß jede Menge wohlgeordnet werden kann*, E. Zermelo, *Beweis, daß jede Menge wohlgeordnet werden kann* (Proof that Every Set can Become Well-Ordered). *Math. Ann.* 59:514–516.

133. Carathéodory, as note 111, here *Ges. Math. Schr.* vol V, p 305.

134. According to the axiom of choice, for any family T of disjoint non-empty sets there is a subset S of the union of T such that S has exactly one element in common with each member of T.
    Zermelo proposed his axiom of choice to solve an antinomy in set theory and argued that it was obvious and needed no proof. However, many of the most prominent mathematicians of his time, including É. Borel, J. H. Poincaré and H. Lebesgue, disputed it. Notable Mathematicians.
    Fraenkel developed an axiom system in which he proved the independence of Zermelo's axiom of choice. His system was modified by the Norwegian logician Thoralf Skolem to produce what is known as Zermelo–Fraenkel–Skolem system.
    The axiom of choice was proved in 1940 by Kurt Gödel to be consistent with other axioms of set theory, but Paul Cohen showed in 1963 that it is independent of the other axioms in a certain system of set theory, thus demonstrating that it cannot be proved within this system.
    Cohen P J (1963) The Independence of the Continuum Hypothesis. Proc. Nat. Acad. Sc. 50:1143–1148; (1964) 51:105–110.

135. Zermelo E (1908) *Neuer Beweis für die Möglichkeit einer Wohlordnung. Math. Ann.* 65:107–128.

136. Zermelo E (1908) *Untersuchungen über die Grundlagen der Mengenlehre I. Math. Ann.* 65:261–281.

137. Carathéodory to Zermelo on 30 August 1907. Postcard in German. Universitätsbibliothek Freiburg, Nachlaß Zermelo.

138. Peckhaus, as note 61, p 38f.

139. For the description of the incident see: Born (1978), p 99ff.

140. Carathéodory to Königsberger on 22 July 1907. Letter in German. Staatsbibliothek zu Berlin, Preussischer Kulturbesitz, Handschriften-Abteilung, Slg Darmst. H 1907(15): Carathéodory.
    Influenced by Fuch's function theory, Königsberger worked on differential equations.

141. Carathéodory to Königsberger on 22 July 1907, ibid.

142. Carathéodory to Königsberger on 11 December 1907. Letter in German. Staatsbibliothek zu Berlin, Preussischer Kulturbesitz, Handschriften-Abteilung, Slg Darmst. H 1907(15): Carathéodory.

143. Ibid.

2 *Academic Career in Germany*

1. The Faculty to the Minister of Education, Dr. Studt, on 20 June 1907. Universitätsarchiv Bonn, Akte "Professor Philipp Furtwängler" der Philosophischen Fakultät der Rheinischen Friedrich-Wilhelms-Universität zu Bonn.

2. Carathéodory to the dean of the Philosophical Faculty of Bonn University on 29 January 1908. Universitätsarchiv Bonn, Akte "Professor Constantin Carathéodory" der Philosophischen Fakultät der Rheinischen Friedrich-Wilhelms-Universität zu Bonn.

3. E. Study to the dean of the Philosophical Faculty of Bonn University on 31 January 1908. Ibid.

4. Ibid.

5. The dean of the Philosophical Faculty to the rector of Bonn University on 21 February 1908. Ibid.

6. Krull W (1970) *Das Bonner Mathematische Seminar 1904–1927* (The Bonn Mathematical Seminar 1904–1927). In: *150 Jahre Rheinische Friedrich-Wilhelms-Universität zu Bonn 1818–1968. Bonner Gelehrte – Beiträge zur Geschichte der Wissenschaften in Bonn – Mathematik und Naturwissenschaften* (150 Years Rhine Friedrich-Wilhelm University Bonn 1818–1968. Scholars of Bonn – Contributions to the History of Sciences in Bonn – Mathematics and Natural Sciences). H. Bouvier u. Co Verlag, Ludwig Röhrscheid Verlag, Bonn, pp 40–48, here p 43.

7. He submitted his request from Constantinople on 23 June 1908. Universitätsarchiv Bonn, Akte "Professor Constantin Carathéodory", as note 2. The dean sent Carathéodory's application to the faculty to decide on 26 June; it was approved by 28 members. Ibid.

8. Born (1978), p 127ff.

9. Carathéodory C (1909) *Untersuchungen über die Grundlagen der Thermodynamik.* Math. Ann. 67:355–386 and *Ges. Math. Schr.* vol II, pp 131–166.

10. Communication of the board of trustees of the Philosophical Faculty of Bonn University to Carathéodory on 6 October 1908. On 15 October 1908 the dean sent that communication to the faculty members. Universitätsarchiv Bonn, Akte "Professor Constantin Carathéodory", op. cit.

11. Tietze, as note 105, chapter 1, p 91.

12. Landé A (1926) *Axiomatische Begründung der Thermodynamik durch Carathéodory.* Handb. d. Phys. 9:281–300.

13. In: Zemansky M (1957) Heat and Thermodynamics. McGraw-Hill, New York, Toronto, London, p 172f, the author gives a brief outline of Carathéodory's ideas, based on H. A. Buchdahl's presentation in the American Journal of Physics, January 1949. Zemansky presents the mathematical theorem which Carathéodory proved and which led him to the formulation of the second law as follows:
"Imagine a space of three dimensions with rectangular coordinates x, y, z. In the neighbourhood of any arbitrary point P_0 there are points which are not accessible from P_0 along solution curves of the equation
$$A(x, y, z)\mathrm{d}x + B(x, y, z)\mathrm{d}y + C(x, y, z) = 0,$$
if, and only if, the equation is integrable. The equation is said to be integrable if there exist functions $\lambda(x, y, z)$ and $F(x, y, z)$ such that
$$A\mathrm{d}x + B\mathrm{d}y + C\mathrm{d}z = \lambda \mathrm{d}F.$$
[…] Consider a system whose states are determined, for the sake of argument, by three thermodynamic coordinates x, y, and z. Then the first law in differential form may be written
$$\mathrm{d}Q = A\mathrm{d}x + B\mathrm{d}y + C\mathrm{d}z,$$
where A, B, and C are functions of x, y, and z. The adiabatic, reversible transitions of this system are subject to the condition
$$\mathrm{d}Q = A\mathrm{d}x + B\mathrm{d}y + C\mathrm{d}z = 0.$$

Let us now take as our mathematical statement of the second law the following:
In the neighborhood of any arbitrary initial state P_0 of a physical system there exist neighboring states which are not accessible from P_0 along quasi-static abiabatic paths. It follows from Carathéodory's theorem that it is possible if and only if there exist functions T and S such that
$$dQ = Adx + Bdy + Cdz = TdS.$$"
Carathéodory's formulation of the second law of thermodynamics implies that the linear differential form w whose vanishing represents changes of state, in which no heat is added, must satisfy $\mu w = df$ and this gives rise to the study of the function f, which is called entropy. See: Gardner R B: The Influence of Carathéodory on Differential Systems. In: The Greek Mathematical Society (ed) C. Carathéodory, as note 29, chapter 1, pp 146–153, here p 147.

14. Carathéodory, as note 93, chapter 1, here *Ges. Math. Schr.* vol V, p 183f.

15. Philosophical Faculty of Friedrich-Wilhelm University to the Minister of Education in Berlin on 22 October 1917. Humboldt-Universität zu Berlin – Archiv –, Bestand „Philosophische Fakultät 1810 bis 1945", Nr 1467, Bl 279R.

16. Born M: *Erinnerungen und Gedanken eines Physikers* (Memories and Thoughts of a Physicist). In: Hermann A (ed) *Hedwig Born, Max Born, Der Luxus des Gewissens – Erlebnisse und Einsichten im Atomzeitalter* (The Luxury of Consciousness – Experiences and Views in the Era of Atom). Nymphenburger Verlagshandlung, München, 1969, pp 27–73, here p 34f.

17. Born M (1921) *Kritische Betrachtungen zur traditionellen Darstellung der Thermodynamik. Phys. Z.* XXII:218–224, here 219.

18. (1921) *Phys. Z.* XXII:282–286, here 282.

19. Carathéodory to Born on 9 October 1907, as note 131, chapter 1.

20. Born (1975), p 330.

21. Born (1978), p 118f. Carathéodory's father was not the Turkish ambassador in Brussels. He was an Ottoman Greek, ambassador of the Ottoman Empire in Brussels. Here, Born makes the often repeated mistake of identifying the Ottomans with the Turks.

22. *Albert Einstein – Hedwig und Max Born. Briefwechsel 1916–1955* (1972), p 61.

23. Carathéodory to Born on 24 March 1949. Staatsbibliothek zu Berlin, Preussischer Kulturbesitz, Handschriftenabteilung, Nachlaß Born 105.

24. Born M (1949) Natural Philosophy of Cause and Chance. Clarendon Press, Oxford, p 38f.

25. Truesdell C (1984) Rational Thermodynamics. Springer, New York, Berlin, Heidelberg, Tokio.
    Truesdell's first citation is on p 26, second citation on p 53.

26. See: *Vorwort zur achten Auflage* (Foreword to the 8th Edition). In: Planck M (1954) *Vorlesungen über Thermodynamik* (Lectures on Thermodynamics). Walter de Gruyter & Co, Berlin, p X.

27. Carathéodory C (1925) *Über die Bestimmung der Energie und der absoluten Temperatur mit Hilfe von reversiblen Prozessen. Sitzungsber. Preuß. Akad. d. Wiss. Berlin, phys.-math. Kl.* 1925, pp 39–47 and *Ges. Math. Schr.* vol II, pp 167–177.

28. Planck to Sommerfeld on 1 July 1923: *Leider gehöre ich nicht zu den beneidenswerten Naturen, die jede einzelne ausgesparte Minute nach ihrem freien Belieben verwenden können, indem sie ihren Gedanken sofort eine entsprechende Richtung geben. Mir wird es im Gegenteil immer schwer, einen Gegenstand, in den ich mich eingesponnen habe, schnell zu verlassen und bei günstiger Gelegenheit schnell zu ergreifen.* (Unfortunately, I do not belong to the enviable natures, who can spend every minute left free just as they like, by immediately giving their thoughts an appropriate direction. On the contrary, for me it becomes increasingly difficult to quickly leave a matter in which I have been

entangled and to quickly seize it in a more favourable moment.) Cited in: Hermann, A (1995) *Planck*, Rowohlt, Reinbek bei Hamburg, p 68.

29. Carathéodory, as note 9, here *Ges. Math. Schr.* vol II, p 158.

30. Planck, as note 26. See also E. Hölder's comments on this subject in: Hölder E (1950) *Constantin Carathéodory† – Sein Beitrag zur Axiomatik der mathematischen Physik* (Constantin Carathéodory† – His Contribution to the Axiomatics of Mathematical Physics). *Forsch. u. Fortschr.* 21/22:290–293.

31. Planck M (1926) *Über die Begründung des zweiten Hauptsatzes der Thermodynamik* (On the Foundation of the Second Main Theorem of Thermodynamics). *Sitzungsber. Preuß. Akad. d. Wiss., phys.-math. Kl.* 1926, pp 453–463.

32. Carathéodory, as note 9, here *Ges. Math. Schr.* vol II, axiom II on p 140.

33. Universitätsarchiv Bonn, Akte "Professor Constantin Carathéodory", as note 2.

34. A marriage certificate was issued by the registry office for civilians of the Oecumenical Patriarchate on 27 October 1909 (old calendar). This document translated in German by Prof. Dr. Kalitsounakis on 9 July 1937 is in: BayHstA, MK 35403, Laufzeit: 1924–1950.

35. Alexander Carathéodory's cultural contribution included numerous works on topics related to antiquity, a critique of Dante's *Hades* and *Divina Comedia*, translations of Turkish and Arabian poets in ancient Greek, treatises on the monuments of Nemrud Dag, on the history of literature, on archaeology and linguistics.

36. In the Constantinople Conference (December 1876–January 1877), England proposed respect for the sovereignty and integrity of the Ottoman Empire, Russia suggested that the implementation of the reform programme ought not to be left to the Ottoman Empire alone and demanded a foreign, especially a Russian, control. During the conference the Ottoman Government proclaimed the constitution, an adaptation to the Belgian constitution of 1831 and considered the proposed reforms to be superfluous. The Great Powers insisted on the reform programme and, when the Ottoman Government rejected both the proposal that the Powers had to be asked for the appointment of the *Valis* and the compromising proposal for the employment of mixed European commissions, the conference was broken off.

37. Andrássy's "equestrian statue was later situated near the Hungarian Parliament building. On the side of the statue was a panel with a relief illustrating the debates of the Congress of Berlin, so Fejér took the Carathéodorys to see the statue. Mrs. Carathéodory looked at the simulacre of her father and after some time hummed without too much conviction: 'it looks like him.'"
John Horváth to Vicki Lynn Hill on 22 October 2001, as note 37, chapter 1.

38. Pears E (1916) Forty Years in Constantinople – The Recollections of Sir Edwin Pears 1873–1915. H. Jenkins, London, p 82.

39. The Treaty of Berlin for the Settlement of the Affairs of the East was signed on 13 July 1878.
Bulgaria was constituted as an autonomous and tributary principality under the suzerainty of the Sultan and would acquire a Christian Government and a national militia. A province would be formed south of the Balkans with the name of Eastern Rumelia which would remain under the direct political and military authority of the Sultan under the conditions of administrative autonomy. It would have a Christian Governor General. In the island of Crete the Sublime Porte undertook to apply the Organic Law of 1868 with equitable modifications. The Ottoman Provinces of Bosnia and Herzegovina would be occupied and administered by Austria-Hungary. Britain and Germany were secretly pledged to support the occupation and administration of Bosnia and Herzegovina and Alexander Carathéodory Pasha was repeatedly put under pressure that the Porte telegraph their consent. The independence of Montenegro was recognised by the Sublime Porte. The Principality of Serbia was granted independence. Roumania was also granted

independence. The Sublime Porte undertook to carry out, without further delay, the improvements and reforms demanded by local requirements in the provinces inhabited by the Armenians, and to guarantee their security against the Circassians and Kurds.

Greece did not gain any direct privileges from the Congress of Berlin. However, thanks to French support, the way to the annexation of Thessaly and a part of Epirus was opened. On the instruction of Alexander Carathéodory Pasha, his nephew Stephanos, the secretary of the Ottoman delegation, collected pictures of the delegates, took notes of their discussions, the official and unofficial or secret meetings and kept diary even of the menus served.

40. According to an address given in Greek by the academician and university professor, Ioannis Kalitsounakis, at a session of the Athens Academy on 23 February 1950, at which Carathéodory's death was officially announced. On this occasion, the assembly of all academicians was addressed by the President of the Academy, Anastasios Orlandos, and Carathéodory's friends, Panagiotis Zervos, Georgios Ioakimoglou, Ioannis Kalitsounakis and Konstantinos Maltezos.

41. Mansel (1995), p 319.

42. Zannas (ed) (1988), p 10.

43. Mansel (1995), p 319.

44. Fazy, E (1898) *Les Turcs d'Aujourd'hui ou Le Grand Karagheuz.* P. Ollendorff, Paris, pp 137–139.

45. *ex aporriton* (εξ απορρήτων): Greek word corresponding to the Turkish *muharrem esrar kumkumasi* = a personality shrouded in forbidden mystery. This expression denoted the privy counsellor to the Sultan and derives from the office of the Great Dragoman, Alexander Mavrocordato εξ απορρήτων (1641–1710) who made his career as one of the most significant diplomatic personalities in Europe. Representing the Ottoman Empire he signed for instance the peace treaties of Carlowitz in 1699. According to Iorga, his own interests overlapped with those of the Ottomans and through his talent and skill he opened the era of an Ottoman Empire led by the Greeks, of an always victorious and everywhere ruling Phanariot spirit (Iorga (1997) vol IV, p 283ff). Alexander Mavrocordato was the founder of the Phanariot rule in the Danube Principalities and head of a family which rendered six princes and several outstanding personalities. Since 1709 and 1715, the Sultan posted Mavrocordato's sons and grandsons in the thrones of Moldavia and Wallachia respectively. They ruled up to 1758 alternately with members of two almost Hellenised local prince families.

46. The doctrine was supported by the argumentation that aim of this co-operation was the maximal privileges for the Greeks. The tactics consisted in pacifying the ruler, in exploiting his weaknesses and in taking advantage of the favourable moment, whenever that might appear. So the Greeks had to occupy positions within the Ottoman Empire and never resign them because, according to this doctrine, a great damage to Greek interests would occur, if they were substituted by Armenians, Jews, or even Turks.

47. See: Mavrogenis Sp (1885) and Tantalidis I (1868).
    Constantin seems to have been a very talented and successful surgeon. In 1843, he introduced a new method for curing patients with cataract, consisting in removing the entire lens and replacing it with an artificial lens. In 1849, he applied a treatment on a military student wounded during manœuvres by a bullet in the abdomen, without operating on him. The patient was cured. The same year he successfully performed a stone fragmentation to extract a 1.5 kg gall-bladder stone from a patient. This "masterpiece" attracted the attention of local and foreign medical circles. The *Gazette Médicale de Constantinople* of the Imperial Medical School and the *Journal de Constantinople*, a French paper in Istanbul, dedicated many of their columns to this case. This rare operation was the second of its kind. Previously, a similar operation had been performed by the famous French anatomist and surgeon Baron Guillaume Dupuytren, whose patient,

however, died. Carathéodory's patient survived with good chances of recovery. Cara-théodory became "the physician of the century", whereas *L'Union Médicale*, a French scientific journal pulished in Paris, praised him as a medical celebrity, in no way infe-rior to famous doctors in London and Berlin. In 1850, Constantin surgically removed a testicle carcinoma with success. The 1849–1851 statistics of the surgery department of the Imperial School of Medicine, often mention that his tonsillectomies, amputations of extremities, surgical treatment of cataract, cystectomy, castrations, removal of tumours, ganglia extirpation, correction of congenital malformations, as well as his treatment of conjunctivitis, keratitis, traumata, syphilitic exophthalmus (protrusion), traumatic iritis, blepharitis, trachoma, otitis, adenoids, syphilitic angina, glandular swelling, ruptures, perforations, inflammations, abscesses, furuncle, diarrhoea, gonorrhoea, epididymitis, were always successful. His techniques in opthalmology have been also applied by other doctors. Constantin's activity as a medical doctor in Constantinople is presented in the doctoral thesis of Ülman Y I (1999).

Constantin also treated military patients, soldiers of the Imperial Guard affected by the cholera epidemic of 1836–1837. These were interned in the Kız Kulesi Hospital provid-ing isolation from the city. The estimated number of deaths was 20 000–30 000. On the epidemics in Istanbul see: Yıldırım, Nuran: *Salginlar* (Epidemics). İstanbul Ansiklope-disi, vol. 6. Istanbul 1994, p 423.

As to Constantin's Philhellenism, let it be noted that he participated together with scien-tists, philosophers, artists, poets, archaeologists and Hellenists in the foundation of the *Ελληνική Εταιρεία* (Hellenic Society) in Paris in 1829, which aimed at the "propagation of the Enlightenment in Greece" through edition of texts and a Greek paper. His book *Υγιεινά Παραγγέλματα προς χρήσιν του Ελληνικού Λαού* (Sanitation Commands for Use by the Greek People) was released by the publishing company of the Philhellene Ambroise Firmin-Didot (1790–1876), a student of Korais in Paris. In his book, Con-stantin gave instructions for the use of preventive measures for health protection and tried to liberate the Greeks from superstition. In the introduction, he declared that he would be very happy if he could thus contribute to the happiness and cultural progress of the Greeks, who were soon to be counted among the cultured nations of the rest of Europe.

48. Tietze, as note 105, chapter 1, here p 92.

49. Perron, as note 107, chapter 1, here p 41.

50. Carathéodory C, Study E (1910) *Zwei Beweise des Satzes, dass der Kreis unter allen Figuren gleichen Umfanges den größten Inhalt hat. Math. Ann.* 68:133–140 and *Ges. Math. Schr.* vol II, pp 3–11.

51. Universitätsarchiv Bonn, Akte "Professor Constantin Carathéodory", as note 2.

52. Ibid.

53. With a letter to the ministry dated 8 May 1909, ibid.

54. The Minister of Education "on the reports no. 263 and no. 269 of 4 and 5 March this year" on 9 March 1908. Niedersächsisches Hauptstaatsarchiv, "Akten betreffend Professor Dr. Carathéodory", as note 123, chapter 1.

55. The head of the teaching staff of Department V for General Sciences "on the decree UIT. no. 20791 of 9 March 1908" on 26 March 1908. Ibid.

56. The Minister of Education to the rector of the Technical University of Hannover on 22 April 1908 (UIT. no. 21091). Ibid.

57. The head of the teaching staff of Department V of General Sciences "on the decree UIT. no. 21091 of 22 April 1908" on 4 May 1908. Ibid.

58. The Minister of Education on 2 February 1909 (UIT. no. 23853, "Immediately. In con-nection with the decree UIT. no. 23417 of 13 October 1908"). Ibid.

59. The majority representative of the teaching staff of Department V "on the decree UIT. no. 23853 of 2 February 1909" on 27 February 1909. Ibid.

60. Professor Dr. L. Kiepert's *Sondergutachten über die Vorschläge zur Besetzung der erledigten Professur für höhere Mathematik. Anlage A, B, C, D, E* (Special report concerning the proposals for the occupation of the professorsgip for higher mathematics. Enclosures A, B, C, D, E) on 3 March 1909. Ibid.

61. As to confirm Runge's statement on the interest in applied mathematics, Carathéodory would later ask for his opinion about an integraph that he was planning to buy for the collection of Department V, now that he had a respectable sum at his disposal. Carathéodory to Runge on 14 May 1909. Letter in German. Staatsbibliothek zu Berlin – Preussischer Kulturbesitz, Handschriftenabteilung, Dep. 5 (Nachl. Runge–DuBois-Reymond) 580.

62. Carathéodory C (1911) *Über den Variabilitätsbereich der Fourierschen Konstanten von positiven harmonischen Funktionen. Rend. Circ. Mat. Palermo* 32:193–217 and *Ges. Math. Schr.* vol III, pp 78–110.

63. The Minister of Education (UIT. no. 21017 "on the report of 16 March of the year, no. 153") to the rector and senate of the Royal Technical University in Hannover on 6 April 1909. Niedersächsisches Hauptstaatsarchiv, "Akten betreffend Professor Dr. Carathéodory", as note 123, chapter 1.

64. The Minister of Education (UIT. no. 21225) on 16 April 1909. Ibid.

65. Rector's office, Hannover, 7 June 1909. Ibid.

66. Carathéodory to Klein on 14 March 1910. Letter in German. Niedersächsische Staats- und Universitätsbibliothek Göttingen – Abteilung für Handschriften und seltene Drucke. Cod. Ms. F. Klein 8, 462.

67. Carathéodory to Klein (undated). Niedersächsische Staats- und Universitätsbibliothek Göttingen – Abteilung für Handschriften und seltene Drucke. Cod. Ms. F. Klein 8, 469.

68. Koppenfels W v (1941) *Georg Prange. Jber. Deutsch. Math.-Verein.* 51:1–14. Prange had begun his mathematical studies in Easter 1903 but because of a disease of the lungs interrupted it three years later.

69. Süss to Kneser on 29 May 1941. Universitätsarchiv der Albert-Ludwigs-Universität Freiburg, Nachlaß Wilhelm Süss, C89/63.

70. So, for example, Carathéodory wrote to Hilbert on 19 December 1908 (Letter in German. Niedersächsische Staats- und Universitätsbibliothek Göttingen – Abteilung für Handschriften und seltene Drucke. Cod. Ms. D. Hilbert 55) in order to achieve a favourable reference for Broggi: "The day before yesterday I received a letter from Broggi, who has emigrated to South America and, as it seems, has been offered a mathematical post in La Plata. He also told me that you would get an inquiry about him and asked me to write to you, since I have associated with him much and, therefore, I can also assess his knowledge a little bit. I would like to do it, although it seems to me to be quite unnecessary, since you should certainly also take the view that probably no one can be found to be better in the whole of South America. Of course, he knows much and he is more or less exact and sometimes original. His book on actuarial mathematics is really good, probably the best that exists."

71. Carathéodory's undated letter to Klein, as note 67.

72. Fraenkel (1967), p 100.

73. Dehn solved the third of the 23 problems Hilbert had posed at the International Congress of Mathematicians in Paris in 1900. Its solution shows that the Archimedian axiom is needed to prove that two tetrahedra have the same volume if they have the same altitudes and have bases of the same area.
   Magnus W (1978) Max Dehn. The Mathematical Intelligencer vol I no 3:132–143, here 134.

Dehn announced a recent finding of a text of Archimedes on the construction of $\pi/7$ to Carathéodory. Carathéodory's undated postcard to Toeplitz in Kiel. Toeplitz B, Nachlaß Otto Toeplitz. Universitäts- und Landesbibliothek Bonn, Handschriften- und Rara-Abteilung.

74. Dehn M (1910) *Über die Topologie des dreidimensionalen Raumes* (On the Topology of Three-Dimensional Space). *Math. Ann.* 69:137–168.

75. Carathéodory to Hilbert (undated), probably spring 1911. Letter in German. Niedersächsische Staats- und Universitätsbibliothek Göttingen – Abteilung für Handschriften und seltene Drucke. Cod. Ms. D. Hilbert 55.

76. Magnus W, Moufang R (1954) *Max Dehn zum Gedächtnis* (In Memory of Max Dehn). *Math. Ann.* 127:215–227.

77. Fraenkel (1967), p 127.

78. R. Gnehm to Frobenius on 19 June 1913 concerning Geiser's successor in Zurich. Frobenius' reply is mentioned in: Günther Frei G, Stammbach U (1992) *Hermann Weyl und die Mathematik an der ETH Zürich* (Hermann Weyl and Mathematics at the Swiss Federal Institute of Technology in Zurich). Birkhäuser, Basel–Boston–Berlin.

79. The Minister of Education (UIT. no. 20831) to Carathéodory on 26 March 1910. Niedersächsisches Hauptstaatsarchiv, „Akten betreffend Professor Dr. Carathéodory", as note 123, chapter 1.

80. Mylonas, as note 79, chapter 1, here p 30ff.

81. Carathéodory to Hilbert on 2 February 1911. Letter in German. Niedersächsische Staats- und Universitätsbibliothek Göttingen – Abteilung für Handschriften und seltene Drucke. Cod. Ms. D. Hilbert 55.

82. Carathéodory was obviously referring to Minkowski's papers: (a) (1888) *Über die Bewegung eines festen Körpers in einer Flüssigkeit* (On the Motion of a Solid Body in a Fluid). *Sitzungsber. Kgl. Preuß. Akad. d. Wiss.* XL:1095–1110; (b) *Theorie der konvexen Körper, insbesondere Begründung ihres Oberflächenbegriffs* (Theory of Convex Bodies, Especially Foundation of the Notion of Their Surfaces) in the 2nd volume of Minkowski's collected works. According to Zassenhaus, op. cit., p 452, Minkowski's theories on pencils of convex bodies were instrumental in the creation of integral geometry and the new trend in the geometry of numbers.
As J. Dieudonné wrote (Herman Minkowski. In: Gillispie Ch C (ed) Dictionary of Scientific Biography, vol IX. Scribner, New York 1974, pp 411–414), "Long before the modern conception of metric space was invented, Minkowski realized that a symmetric convex body in an n-dimensional space defines a new notion of 'distance' on that space and a corresponding 'geometry'."

83. Hilbert D: *Hermann Minkowski* (memorial speech at the public session of the *Kgl. Ges. Wiss. Gött.* on 1 May 1909). In: Hilbert D, Speiser A, Weyl H (eds) *Gesammelte Mathematische Abhandlungen von Hermann Minkowski* (Collected Mathematical Treatises by Hermann Minkowski), vol I. Chelsea Publishing Company, New York, 1967, pp V–XXXI, here p IX.

84. Minkowski died of a ruptured appendix.

85. Reid (1970), p 119f.

86. Carathéodory to Hilbert (undated), as note 75.

87. In a very emotional letter, Carathéodory offered his condolences to Otto Toeplitz on Emil Toeplitz's death. Emil had died in Breslau on 22 August 1917.
Carathéodory from St. Moritz to Toeplitz on 3 September 1917. Toeplitz B, Nachlaß Otto Toeplitz. Universitäts- und Landesbibliothek Bonn, Handschriften- und Rara-Abteilung.

88. Carathéodory to Hilbert on 5 May 1912. Niedersächsische Staats- und Universitäts-
    bibliothek Göttingen – Abteilung für Handschriften und seltene Drucke. Cod. Ms.
    D. Hilbert 55.

89. Uniwersytet Wrocławski, Archiwum, TH 1 – Akta ogólne rektoratu, 29 VIII 1910 – 30
    III 1912.

90. Uniwersytet Wrocławski, Archiwum, TH 98 – Rektor, Eingang, 20 IV 1912 – 26 VI
    1915.
    An official document mentions that on 5 September 1912, at 9 am, the senior civil
    servant Tidick, the treasurer, civil servant von Kunowski, the vice-rector of the Technical
    University Professor Dr. Carathéodory and the government secretary Johnsdorf had met
    on the instructions of the president at the cash registry of the Technical University in
    order to carry out a special audit of the University cashier's office bookkeeping.

91. The general Picard theorem states that an analytic function $f(z)$ of a complex variable z,
    in the neighbourhood of an essential singular point, assumes every finite value infinitely
    often with the possible exception of one value.
    The Picard theorem is substantially supplemented by the Iversen theorem and the Julia
    theorem.
    E. D. Solomentsev, Encyclopaedia of Mathematics.

92. Carathéodory, as note 10, chapter 1, here p 407. The Schwarz lemma is the name given
    by Carathéodory to the known theorem of function theory. See: Carathéodory C (1912)
    Untersuchungen über die konformen Abbildungen von festen und veränderlichen Ge-
    bieten (Studies in Conformal Mappings of Constant and Variable Domains). *Math. Ann.*
    72:107–144 and *Ges. Math. Schr.* vol III, pp 362–405, here heading of chapter 1.

93. Carathéodory C (1905) *Sur quelques généralisations du théorème de M. Picard* (On
    Some Generalisations of the Picard Theorem). *C. r. Acad. Sci. Paris* 141:1213–1215 and
    Ges. Math. Schr. vol III, pp 3–5.

94. At the first Scandinavian Congress of Mathematicians (Stockholm 1909), Lindelöf pre-
    sented an "elementary proof" of Picard's original theorem, in the sharpened form in-
    dicated by Landau and Carathéodory (Elfving (1981), p 148). Carathéodory mentions
    Lindelöf in this respect in his paper: (1912) *Sur le théorème général de M. Picard* (On
    the General Picard Theorem). *C. r. Acad. Sci. Paris* 154:1690–1692 and *Ges. Math. Schr.*
    vol III, pp 10–12.

95. Carathéodory C (1907) *Sur quelques applications du théorème de Landau–Picard* (On
    Some Applications of the Landau–Picard Theorem). *C. r. Acad. Sci. Paris* 144:1203–
    1206 and *Ges. Math. Schr.* vol III, pp 6–9.

96. Perron, as note 107, chapter 1, here p 46f.

97. Carathéodory to Schottky on 15 January 1908. Letter in German. Staatsbibliothek
    zu Berlin, Preussischer Kulturbesitz, Handschriften-Abteilung, Nachl. 282 (Schottky):
    Carathéodory.
    Schottky contributed fundamentally to the theory of functions with his work on Abel
    functions and conformal mapping.

98. Carathéodory, as note 93.

99. Schottky's 1875 Berlin doctoral thesis concerning conformal mappings of multiply con-
    nected plane domains was the origin of the mapping of a domain bounded by three
    disjoint circles, which provides an example of an automorphic function with a Cantor
    set boundary. In 1904 Schottky proved the so-called Schottky theorem, a generalisation
    of the Picard and Landau theorems, which is now a well-known result in the theory of
    functions of a complex variable of the type of distortion theorems.
    Schottky's theorem (1904) *Über den Picard'schen Satz und die Borel'schen Unglei-*
    chungen (About the Picard Theorem and the Borel Inequalities). *Sitzungsber. Kgl.*
    Preuß. Akad. d. Wiss. Berlin 2:1244–1262) can be formulated as follows: if a func-

tion $w = f(z) = c_0 + c_1 + \ldots$ is regular and analytic in the disc $D = \{z : |z| < R\}$ and does not take certain finite values a_1, a_2 in D, then in any disc $|z| \leq R$, $0 < R_1 < R$, the modulus $|f(z)|$ is bounded by a number $M(a_1, a_2, c_0, R_1)$ that depends only on a_1, a_2, c_0, R_1.
E. D. Solomentsev, Encyclopaedia of Mathematics.

100. Carathéodory C, Landau L (1911) *Beiträge zur Konvergenz von Funktionenfolgen. Kgl. Preuß. Akad. d. Wiss. Berlin, phys.-math. Kl.* 1911, pp 587–613 and *Ges. Math. Schr.* vol III, pp 13–44.

101. Theorem VI, *Ges. Math. Schr.* vol III, p 26f.

102. See: Perron, as note 107, chapter 1, here note 1 on p 47.

103. Carathéodory C (1912) *Untersuchungen über die konformen Abbildungen von festen und veränderlichen Gebieten*, as note 92.

104. Carathéodory to Paul Bernays on 23 September 1912. Letter in German. ETH-Bibliothek Zürich, Wissenschaftshistorische Sammlungen, HS 975:597.
In 1912 Bernays obtained his doctorate under Landau's supervision and started working on his habilitation near Zermelo.

105. Carathéodory to Paul Bernays (undated, later than 23 September 1912). Letter in German. ETH-Bibliothek Zürich, Wissenschaftshistorische Sammlungen, HS 975:598.

106. Carathéodory C (1920) *Über eine Verallgemeinerung der Picardschen Sätze. Sitzungsber. Preuß. Akad. d. Wiss. Berlin, math-phys. Kl.* 1920, pp 202–209 and *Ges. Math. Schr.* vol III, pp 45–53.

107. Ibid, *Ges. Math. Schr.* vol III, p 45.

108. Carathéodory C (1912) *Sur le théorème général de M. Picard*, as note 94.

109. Carathéodory, as note 106, here *Ges. Math. Schr.* vol III, p 46.

110. Riesz F (1915) *Über ein Problem des Herrn Carathéodory. Journal für die reine und angewandte Mathematik* 146:83–87.

111. Carathéodory C (1907) *Über den Variabilitätsbereich der Koeffizienten von Potenzreihen, die gegebene Werte nicht annehmen. Math. Ann.* 64:95–115 and *Ges. Math. Schr.* vol III, pp 54–77.

112. Toeplitz O (1911) *Über die Fouriersche Entwickelung positiver Funktionen. Rend. Circ. Mat. Palermo* 32:191–192.

113. Carathéodory, as note 62.
In the 4th chapter of: Grenander U, Szegö G (1958) Toeplitz forms and their applications. University of California Press, Berkeley and Los Angeles, the trigonometric moment problem is solved with the help of a Carathéodory's theorem concerning the "finite moment problem". The theorem is presented on p 58 as follows:
"Let $c_1, c_2, \ldots, c_n$ be given complex constants not all zero, $n > 1$. There exists an integer m, $1 \leq m \leq n$, and certain constants $\rho_k, \varepsilon_k; k = 1, 2, \ldots, m$, such that $\rho_k > 0$, $|\varepsilon_k| = 1$, $\varepsilon_k \neq \varepsilon_1$ if $k \neq 1$, and
$$c_\nu = \sum_{k=1}^{m} \rho_k \varepsilon_k^\nu, \nu = 1, 2, \ldots, n. \quad (1)$$
The integer m and the constants ρ_k, ε_k are uniquely determined.
This basic theorem can be formulated also as follows:
$$c_\nu = \int_{-\pi}^{\pi} e^{-i\nu x} da(x), \nu = 1, 2, \ldots, n, \quad (2)$$
where $a(x)$ is a distribution function of the finite type defined in the following way. It has jumps at those points x for which e^{-ix} coincides with one of the numbers ε_k and the corresponding jump is ρ_k."

114. Fischer E (1911) *Über das Carathéodorysche Problem Potenzreihen mit positivem reellen Teil betreffend* (On Carathéodory's Problem Concerning Power Series with a Positive Real Part). *Rend. Circ. Mat. Palermo* 32:240–256.

115. Schur I (1912) *Über einen Satz von C. Carathéodory* (About a Theorem of C. Carathéodory). *Sitzungsber. Kgl. Preuß. Akad. d. Wiss. Berlin* 1912, pp 4–15.
     Schur gave an algebraic form of Carathéodory–Toeplitz conditions in his study of analytic functions bounded in the unit disc in: Schur I (1916, 1917) *Über Potenzreihen, die im Inneren des Einheitskreises beschränkt sind, I, II* (About Power Series which are Bounded in the Interior of the Unit Circle, I, II). *Journal für die reine und angewandte Mathematik* 147:205–232 and 148:122–145.

116. Frobenius G (1912) *Ableitung eines Satzes von Carathéodory aus einer Formel von Kronecker* (Derivation of a Theorem by Carathéodory from a Formula by Kronecker). *Sitzungsber. Kgl. Preuß. Akad. d. Wiss. Berlin* 1912, pp 16–31.

117. Carathéodory's lemma: if $f(z) = 1 + \sum_{n=1}^{\infty} c_n z^n$, then $|c_n| \le 2, n = 1, 2, \ldots$
     See: Szegö G: Some Recent Investigations Concerning the Sections of Power Series and Related Developments. In: Askey R (ed) Gabor Szegö: Collected Papers, vol 2: 1927–1943. Birkhäuser, Boston, Basel, Stuttgart, 1982, pp 641–648, here p 641f.
     The Riesz–Herglotz theorem states that in order that $f(z)$ be of class C it is necessary and sufficient that it has a Stieltjes integral representation
     $$f(z) = \int_{-\pi}^{\pi} (e^{it} + z)(e^{it} - z)^{-1} d\mu(t)$$
     where $\mu(t)$ is a non-decreasing function on $[-\pi, \pi]$ such that
     $\mu(\pi) - \mu(-\pi) = 1$.
     Riesz F (1911) *Sur certains systèmes singuliers d'équations intégrales* (On Certain Singular Systems of Integral Equations). *Ann. Sci. École Norm. Sup.* 28:33–62.
     Herglotz G (1911) *Über Potenzreihen mit positiven, reellen Teil im Einheitskreis* (About Power Series with Positive, Real Parts in the Unit Circle). *Ber. Verh. Kgl. Ges. d. Wiss. Leipzig, math.-phys. Kl.* 63, pp 501–511.

118. Toeplitz O (1910) *Zur Theorie der quadratischen Formen von unendlich vielen Veränderlichen. Nachr. Kgl. Ges. Wiss. Gött., math.-phys. Kl.* 1910, pp 489–506. In this paper Toeplitz engaged with another class of quadratic forms, the Jacobi forms, or J-forms, which are directly connected with the Stieltjes theory of continued fractions.

119. Fejér L (1916) *Über trigonometrische Polynome. Journal für die reine und angewandte Mathematik* 146:53–82.

120. Szegö G (1920, 1921) *Beiträge zur Theorie der Toeplitzschen Formen I, II* (Contributions to the Theory of Toeplitz Forms I, II). *Math. Z.* 6:167–202 and 9:167–190.
     Szegö G (1921) *Über orthogonale Polynome, die zu einer gegebenen Kurve der komplexen Ebene gehören* (On Orthogonal Polynomials, which Belong to a Given Curve of the Complex Plane). *Math. Z.* 9: 218–270.

121. Meyer-König W (1960) U. Grenander and G. Szegö, Toeplitz forms and their applications. *Jber. Deutsch. Math.-Verein.* 62:42–44, here 43.

122. Carathéodory C, Fejér L (1911) *Über den Zusammenhang der Extremen von harmonischen Funktionen mit ihren Koeffizienten und über den Picard–Landauschen Satz. Rend. Circ. Mat. Palermo* 32: 218–239 and *Ges. Math. Schr.* vol III, pp 111–138.
     The work had begun in Hannover in July 1910 and was produced in Breslau and Budapest in October 1910. Fejér as well as Riesz and Haar were then working in Kolozsvár, later Fejér went to Budapest.

123. G.V. Kuz'mina, Encyclopaedia of Mathematics.

124. Carathéodory C, Fejér L (1907) *Remarques sur le théorème de M. Jensen* (Remarks on the Theorem of Jensen). *C. r. Acad. Sci. Paris* 145, Paris, pp 163–165 and *Ges. Math. Schr.* vol III, pp 179–181.

125. John Horváth to Vicki Lynn Hill on 22 October 2001, as note 37, chapter 1.
     The Blaschke product was introduced by Blaschke in the paper: (1915) *Eine Erweiterung des Satzes von Vitali über Folgen analytischer Funktionen* (An Extension of Vitali's

Theorem on Sequences of Analytic Functions). *Ber. Verh. Kgl. Ges. d. Wiss. Leipzig, math.-phys. Kl.* 67, pp 194–200.

126. Carathéodory C (1929) *Über die Winkelderivierten von beschränkten analytischen Funktionen. Sitzungsber. Preuß. Akad. d. Wiss. Berlin, phys.-math. Kl.* 1929, pp 39–54 and *Ges. Math. Schr.* vol III, pp 184–204.

127. Julia's Theorem.
If a is an isolated essential singular point of an analytic function $f(z)$ of the complex variable z, then there exists at least one ray $S = \{z : \arg(z-a) = \theta_0\}$ issuing from a such that in every angle $V = \{z : |\arg(z-a) - \theta_0| < \varepsilon\}$ $\varepsilon > 0$, that is symmetric with respect to the ray, $f(z)$ assumes every finite value, except possibly one, at an infinite sequence of points $\{z_k\}CV$ converging to a. This result of G. Julia (see: Julia G (1924) *Leçons sur les fonctions uniformes à une point singulier essentiel isolé* (Lessons on Uniform Functions of an Isolated Essential Singular Point). Gauthier-Villars, Paris.) supplements the big Picard theorem on the behaviour of an analytic function in a neighbourhood of an essential singularity.
E. D. Solomentsev, Encyclopaedia of Mathematics.

128. Carathéodory C (1936) Μία συμπλήρωσις του λήμματος του *Schwarz. Praktika de l'Académie d'Athènes* 11:276–286. Translated from the Greek by Stephanos Carathéodory as *Eine Verschärfung des Schwarzschen Lemmas* (A Sharpening of Schwarz's Lemma) for *Ges. Math. Schr.* vol III, pp 225–235.

129. Carathéodory C (1937) A Generalization of Schwarz's Lemma. *Bull. Amer. Math. Soc.* 43:231–241 and *Ges. Math. Schr.* vol III, pp 236–246.

130. G. Szegö was at Washington University in St. Louis from the autumn of 1934 to June 1938.

131. Carathéodory to Szegö on 30 November 1936. Letter in German. Stanford University Libraries, Department of Special Collections, SC 323 Gabor Szegö Papers, Box 5, Folder 15.
Carathéodory was probably referring to his work: (1936) *Über beschränkte Funktionen, die in einem Paar von vorgeschriebenen Punkten gleiche Werte annehmen* (On Bounded Functions which Assume Equal Values in a Pair of Given Points). *Monatshefte f. Math. u. Phys.* 43:225–241 and *Ges. Math. Schr.* vol III, pp 205–224, beginning with the formulation: "Since the Schwarz lemma has been put in the centre of many function theoretic investigations, so many problems of the maximum concerning bounded functions have been treated that it is hardly worthy to extend this already very broad class of questioning even further. But some time ago, I was confronted with certain problems of this kind which lead to completely unexpected results, although their solution presents no difficulties at all."
The method of combining Schwarz's lemma with appropriate linear mappings, to which Carathéodory referred in his letter, was used and generalised in various directions also by Schottky and Schur. Also, Nevanlinna, in his thesis of 1919 entitled *Über beschränkte Funktionen die in gegebenen Punkten vorgeschriebene Werte annehmen* (On Bounded Functions which in Given Points Assume Prescribed Values), treats the problem formerly treated by Carathéodory from a new and unifying point of view. (Elfving (1981), p 190).

132. Carathéodory, as note 129, here *Ges. Math. Schr.* vol III, p 242f. C. Carathéodory, *Über beschränkte Funktionen, die in einem Paar von vorgeschriebenen Punkten gleiche Werte annehmen*, as note 131, here p 219ff.

133. Carathéodory to Szegö on 3 December 1936. Letter in German. Stanford University Libraries, Department of Special Collections, SC 323 Gabor Szegö Papers, Box 5, Folder 15.
The relevant theory is extensively presented in Carathéodory's lectures on the theory of functions at the University of Wisconsin: Second Course in the Theory of Functions, Pro-

fessor Constantin Carathéodory, University of Wisconsin, First Semester, 1936–1937, pp 29a–29d. University of Wisconsin, General Files, Mathematics Department, 7/22/2, Box 12.

134. Carathéodory might have thought of:
(a) Dieudonné J (1931) *Recherches sur quelques problèmes relatifs aux polynômes et aux fonctions bornées d'une variable complexe* (Investigations into Some Problems Related to Polynomials and Bounded Functions of a Complex Variable). *Ann. Sci. École Norm. Sup.* 48:247–358.
(b) Rogosinski W (1934) *Zum Schwarzschen Lemma* (On the Schwarz Lemma). *Jber. Deutsch. Math.-Verein.* 44:258–261.

135. Carathéodory to Szegö on 8 December 1936. Letter in German. Stanford University Libraries, Department of Special Collections, SC 323 Gabor Szegö Papers, Box 5, Folder 15.
Carathéodory was obviously referring to Szegö's paper: (1936) Inequalities for the zeros of Legendre polynomials and related functions. *Trans. Amer. Math. Soc.* 39:1–17. There, in a new and somewhat improved way, by applying Sturm's method to the Legendre differential equation, Szegö derived the inequalities due to Bruns and the improved inequalities due both to Stieltjes and Markoff, which indicate the "regular distribution" of the system of zeros in the interval $(0, \pi)$ if $n \to \infty$. Szegö applied further his method to general Bessel functions and to discuss the zeros of ultraspherical polynomials and of the cosine polynomials.
Fejér's investigations, which Carathéodory did not know, were probably concerning the: (1936) *Bestimmung von Grenzen für die Nullstellen des Legendreschen Polynoms aus der Stieltjesschen Integraldarstellung desselben* (Definition of Limits for the Zeroes of the Legendre Polynomial from the Stieltjes Integral Representation of the Same). *Monatshefte f. Math. u. Phys.* 43:193–209. There, Fejér derived the same result as Szegö directly from the Stieltjes integral form.
Szegö's "note in the Bulletin" was obviously his paper: (1936) Some recent investigations concerning the sections of power series and related developments. *Bull. Amer. Math. Soc.* 42:505–522. It was a report on the development of Weierstrass' theory of power series regarding various classes of functions. As Szegö noted in his conclusion, his "intention was to set forth only the most typical results of this field, illustrating the way in which properties of an analytic function are reflected in terms or in the sections of its power series development."

136. Carathéodory to Szegö on 10 December 1936. Letter in German. Stanford University Libraries, Department of Special Collections, SC 323 Gabor Szegö Papers, Box 5, Folder 15.

137. Carathéodory, as note 129, here *Ges. Math. Schr.* vol III, p 239f.

138. Carathéodory showed Szegö how to derive the inequality. Carathéodory to Szegö on 16 December 1936. Letter in German. Stanford University Libraries, Department of Special Collections, SC 323 Gabor Szegö Papers, Box 5, Folder 15.

139. Carathéodory C (1941) *Über das Maximum des absoluten Betrages des Differenzenquotienten für unimodular beschränkte Funktionen. Math. Z.* 47:468–488 and *Ges. Math. Schr.* vol III, pp 247–272.

140. In his paper: (1907) *Über die Uniformisierung beliebiger analytischer Kurven* (On the Uniformisation of Arbitrary Analytic Curves). *Nachr. Kgl. Ges. Wiss. Gött., math.-phys. Kl. 1907*, pp 191–210, Koebe achieved the main theorem of the uniformisation theory, or, in other words, the generalisation of the Riemann mapping theorem for any simply connected Riemann surface.
He "proved the existence of an absolute constant $\kappa = 0$ for which the disc $|w| < \kappa$ is contained in the range of every function $f \in S$. The value $\kappa = 1/4$ was determined by

Bieberbach a few years later." Duren P L (1983) Univalent Functions. Springer, New York, Berlin, Heidelberg, Tokyo, p 69.

141. Bieberbach L (1968) *Das Werk Paul Koebes* (Koebe's Work). *Jber. Deutsch. Math.-Verein.* 70:148–158.

142. Koebe's distortion theorems:
    a) There exist positive numbers $m_1(r), M_1(r)$, depending only on r, such that for any $f \in S, |z| = r, m_1(r) \leq |f(z)| \leq M_1(r)$.
    b) There exists a number $M(r)$ depending only on r, such that for $f \in S, |z_1|, |z_2| \leq r, M(r)^{-1} \leq |f'(z_1)f'(z_2)^{-1}| \leq M(r)$.
    E. G. Goluzina, Encyclopaedia of Mathematics.

143. Koebe, as note 140.
    "As early as 1910, Koebe described an iterative method for constructing the mapping, without proving that it really works. (Others verified years later that it does work.) In 1920, Koebe gave a proof of existence of the mapping ["Abbildung mehrfach zusammenhängender Bereiche aus Kreisbereiche", Math. Z. 7 (1920), 235–301]. Many other proofs have been given since." Duren refers to a proof (modelled after Schiffer and Hawley (1962), a combination of potential theory with the calculus of variations, given in Peter Henrici's book *Applied and Computational Complex Analysis*, volume 3, and Henrici's quotation of a survey article by Dieter Gaier, *Konforme Abbildung mehrfach zusammenhängender Gebiete* (Conformal Mapping of Multiply Connected Domains) (*Jber. Deutsch. Math.-Verein.* (1978), 25–44). Communication to the author from P. Duren on 08 November 2001.

144. Bieberbach explains that the Riemann–Schottky mapping problem, i.e. the problem of the schlicht conformal mapping of any n-connected domain onto a domain bounded by circles, which is a generalisation of the Riemann mapping theorem ($n = 1$), appears in Riemann's papers and in Schottky's dissertation and that Schottky formulated it as a problem from the theory of linear differential equations with algebraic coefficients. (Bieberbach, as note 141, here 148f). "It is the famous *Kreisnormierungsproblem*, now the *Kreisnormierungssatz*, which has been solved, at least for finitely connected domains." Communication to the author from P. Duren on 08 November 2001. (See also note 99.)
    Poincaré and Koebe's uniformisation theorem is stated in the bibliographical notes of Carathéodory's *Conformal Representation* as follows: every simply connected Riemann surface can be represented conformally either on the complete sphere, or on the Euclidean plane, or on the interior of the unit circle. Carathéodory's results of §§ 125–130 and 157–159 could be made to yield the above theorem.

145. Second Course in the Theory of Functions, Professor Constantin Carathéodory, University of Wisconsin, First Semester, 1936–37, p 96f. University of Wisconsin, General Files, Mathematics Department, 7/22/2, Box 12.
    With regard to the Koebe distortion theorem, Duren points out that "Koebe got inequalities with *some* (unspecified) constants depending only on the point." The same applies to the so-called "Koebe one-quarter theorem". Koebe proved existence of an absolute lower bound for the radius of the disk covered by the image, but he did not find the sharp bound 1/4. The sharp forms of Koebe's theorems came out of the area theorem, first proved by Gronwall in 1914, or more directly from Bieberbach's use of the area theorem in 1916 to prove the sharp bound $|a_2| \leq 2$ for the second coefficient. That allowed proofs much simpler than Koebe's for sharp forms of the theorems.
    So in retrospect, the ideas in Koebe's original proofs may not have been worth a battle over priority, because of the subsequent discovery of much more elegant proofs of stronger theorems. But of course it was important to point out the qualitative phenomena, as Koebe is credited with doing." Communication to the author on 1 November 2001. For historical remarks see also: Duren, as note 140, here notes on p 69.

146. Koebe P (1912) *Über eine neue Methode der konformen Abbildung und Uniformisierung.* *Nachr. Kgl. Ges. Wiss. Gött., math.-phys. Kl.* 1912, pp 844–848.

147. Koebe P: *Über diejenigen analytischen Funktionen eines Arguments, welche ein algebraisches Additionstheorem besitzen, und die endlich-vieldeutig umkehrbaren Abelschen Integrale.* (On Those Analytic Functions of One Variable, which Admit an Algebraic Addition Theorem, and Finite-Valued Invertible Abelian Integrals). In: Carathéodory C, Hessenberg G, Landau E, Lichtenstein L (eds) *Mathematische Abhandlungen Hermann Amandus Schwarz zu seinem fünfzigjährigen Doktorjubiläum am 6. August 1914 gewidmet von seinen Freunden und Schülern.* Springer, Berlin, 1914, pp 192–214.

148. Carathéodory C (1914) *Elementarer Beweis für den Fundamentalsatz der konformen Abbildungen* (Elementary Proof for the Fundamental Theorem of Conformal Mappings). Ibid, pp 19–41 and *Ges. Math. Schr.* vol III, pp 273–299.

149. To the so-called Carathéodory–Koebe algorithm and to the competition between Carathéodory and Koebe see: Remmert R (1998) Classical Topics in Complex Function Theory. Springer-Verlag, New York. Heinhold explains why the osculation procedure cannot be applied as a numerical procedure in: Heinhold J (1974) *Schmiegungsoperationen in der Praxis der Konformen Abbildung* (Osculation Operations in the Practice of Conformal Mapping). In: Panayotopoulos (ed) (1974), pp 203–222, here p 210.

150. The proof of the Riemann mapping theorem through repeated square-root mappings in: Koebe, as note 146.
Koebe P (1915) *Abhandlungen zur Theorie der konformen Abbildung. I. Die Kreisabbildung des allgemeinsten einfach und zweifach zusammenhängenden schlichten Bereichs und die Ränderzuordnung bei konformer Abbildung* (Treatises on the Theory of Conformal Mappings. I. The Mapping of the Most General Simply and Doubly Connected Schlicht Domain onto the Circle and the Classification of the Boundaries in the Case of Conformal Mappings). *J. f. reine u. angew. Math.* 145:177–223.

151. Carathéodory C (1937) On Dirichlet's Problem. American Journal of Mathematics 59:709–731 and *Ges. Math. Schr.* vol III, pp 311–336, here p 334f.
A.Yanushauskas (Encyclopaedia of Mathematics) defines the Dirichlet problem as the problem of finding a harmonic function u which is regular in a domain D and which coincides with a given continuous function φ on the boundary Γ of D.
The Dirichlet problem is one of the fundamental problems in potential theory and appears also in gravitation, heat conduction and elasticity theory.
See: Gårding L (1979) The Dirichlet Problem. The Mathematical Intelligencer 2/1: 43–53.
Evans Gr (1938) Dirichlet problems. In: Semicentennial Publications of the American Mathematical Society, vol 2, pp 185–226 with reference to Carathéodory's paper On Dirichlet's Problem.

152. Carathéodory to Szegö on 16 December 1936, as note 138.

153. Carathéodory, as note 151, here *Ges. Math. Schr.* vol III, p 312.
Carathéodory was referring to Perron's paper: (1923) *Eine neue Behandlung der ersten Randwertaufgabe für* $\Delta u = 0$ (A New Treatment of the First Boundary Value Problem for $\Delta u = 0$). *Math. Z.* 18:42–54 and Radó and Riesz's paper: (1925) *Über die erste Randwertaufgabe für* $\Delta u = 0$ (On the First Boundary Value Problem for $\Delta u = 0$). *Math. Z.* 22:41–44.
Perron's method employed in the study of the Dirichlet problem, or the method of upper and lower functions, is applicable to domains D of a fairly general kind. It involves the construction of sequences of upper (superharmonic) and lower (subharmonic) functions, the common limit of which is the sought solution of the Dirichlet problem.
A.Yanushauskas, Encyclopaedia of Mathematics.

154. Editorial comments on A.Yanushauskas: Dirichlet Problem. Encyclopaedia of Mathematics.

155. In the notes of his 1932 book *Conformal Representation*.
    Dirichlet's principle states that Dirichlet's problem is solved by a uniquely determined admissible function $\varphi = u$ for which Dirichlet's integral $D[\varphi] = \int \int_G (\varphi_r^2 + \varphi_s^2) \mathrm{d}r \mathrm{d}s$ attains ist minimum value. G is a domain in the x, y-plane whose boundary consists of Jordan curves. The function u possesses continuous first and second derivatives in G and is harmonic. See: Courant C (1950) Dirichlet's Principle, Conformal Mapping, and Minimal Surfaces. Interscience Publishers, New York, p 6.

156. Carathéodory C (1928) *Bemerkungen zu den Existenztheoremen der konformen Abbildung*. Bull. Calcutta Math. Soc. 20:125–134 and *Ges. Math. Schr.* vol III, pp 300–310, here p 300.

157. Radó T (1922–1923) *Über die Fundamentalabbildung schlichter Gebiete* (On the Fundamental Mapping of Schlicht Domains). *Acta Sci. Math. Szeged* 1:240–251 here 241f.
    Carathéodory mentions a proof in Bieberbach's (1931) *Lehrbuch der Funktionentheorie* (Textbook of the Theory of Functions), vol II. Teubner, Berlin, p 5.

158. Carathéodory to Hilbert on 2 February 1911, as note 81. Haar had submitted his habilitation in Göttingen in 1910.

159. Communication to the author from Peter Duren on 30 October 2001.

160. Carathéodory to Hilbert (undated), as note 75.

161. Definition in: Carathéodory C (1912) *Untersuchungen über die konformen Abbildungen von festen und veränderlichen Gebieten*, as note 92, here *Ges. Math. Schr.* vol III, p 381f.
    Instead of the complicated mappings of polygons used in Schwarz's proof of the Riemann mapping theorem, Carathéodory uses very elementary mappings, namely only linear and quadratic.

162. Shields A (1988) Carathéodory and Conformal Mapping. The Mathematical Intelligencer 10/1:18–22.

163. Bieberbach L (1913) *Über einen Satz des Herrn Carathéodory* (On a Theorem of Herr Carathéodory). *Nachr. Kgl. Ges. Wiss. Gött. 1913*, pp 552–560.
    In 1914 Bieberbach studied polynomials now named after him which approximate a function that conformally maps a given simply connected domain onto a disc. Bieberbach conjectured that $a_n \leq n$ for the coefficients a_n of each function f belonging to the class S (the Bieberbach conjecture) in: (1916) *Über die Koeffizienten derjenigen Potenzreihen, welche eine schlichte Abbildung des Einheitskreises vermitteln* (About the Coefficients of those Power Series, which Induce a Schlicht Mapping of the Unit Circle). *Sitzungsber. Kgl. Preuß. Akad. d. Wiss. Berlin, math.-phys. Kl.* 1916, pp 940–955.
    This problem and conjecture was followed by an intensive study of the class S. For a detailed history of the progress in investigating the validity of the Bieberbach conjecture see: Duren, as note 140.

164. Walsh J L (1924) On the expansion of analytic functions in series of polynomials. Trans. Amer. Math. Soc. 26:155–170.

165. Walsh J L (1926) *Über die Entwicklung einer analytischen Funktion nach Polynomen*. Math. Ann. 96:430–436.

166. Shields, as note 162.

167. Courant R (1914) *Über eine Eigenschaft der Abbildungsfunktionen bei konformer Abbildung* (About a Property of Mapping Functions in the Case of Conformal Mappings). *Nachr. Kgl. Ges. Wiss. Gött.* 1914, pp 101–109 and (1920) *Bemerkung zu meiner Note „Über eine Eigenschaft ... "* (Remark on my Note "About a Property ... ") *Nachr. Kgl. Ges. Wiss. Gött.*, 1920, pp 69–70.

168. Carathéodory to Felix Hausdorff on 30 March 1913. Letter in German. Universitäts- und Landesbibliothek Bonn, Handschriftenabteilung, NL Hausdorff, Kapsel 61.

169. For an arbitrary set M, Hausdorff defines the φ-kernel of M as the sum $\underline{M}$ of all sets $A \leq M$ for which $A \leq A\varphi$ (to which $A = 0$ certainly belongs) or the greatest set $A \leq M$

for which $A \leq A\varphi$. He defines the φ-hull of M as the intersection $\overline{M}$ of all sets $A \geq M$ for which $A \geq A\varphi$ (to which $A = E$ [the whole space which may be taken to be a pure set] certainly belongs) or the smallest set $A \geq M$ for which $A \geq A\varphi$.
Hausdorff F (1957) Set Theory. Chelsea Publishing Company, New York, p 189f.

170. Weyl, as note 83, chapter 1, here p 638.

171. Hausdorff F (1914) *Grundzüge der Mengenlehre*. Von Veit, Leipzig.
Translated from German as *Set Theory*, as note 169.

172. Carathéodory to Hilbert, apparently in the spring of 1912. Letter in German. Niedersächsische Staats- und Universitätsbibliothek Göttingen – Abteilung für Handschriften und seltene Drucke. Cod. Ms. D. Hilbert 55.

173. "Sweeping out" process is the translation used by Carathéodory himself in his paper On Dirichlet's Problem to characterise what he had denoted as Poincaré's *Ausfegeverfahren* in this letter.

174. Carathéodory C (1913) *Über die gegenseitige Beziehung der Ränder bei der konformen Abbildung des Inneren einer Jordanschen Kurve auf einen Kreis. Math. Ann.* 73:305–320 and *Ges. Math. Schr.* vol IV, pp 3–22.

175. Carathéodory to Hilbert on 5 May 1912, as note 88.

176. Riemann B: *Grundlagen für eine allgemeine Theorie der Functionen einer veränderlichen complexen Grösse*. Dissertation, Göttingen 1851. In: Weber H, Dedekind R (eds) *Bernhard Riemann's Gesammelte Mathematische Werke und Wissenschaftlicher Nachlass* (Bernhard Riemann's Collected Mathematical Works and Scientific Papers). Teubner, Leipzig, 1876, pp 3–47.

177. In 1858 Dedekind defined a real number as a partition, or "cut" of the rational numbers into two sets such that each member of one set was less than all members of the other. His cuts gave a model for the continuous number line, since they filled all the gaps in the rationals.

178. Carathéodory to Hilbert on 5 May 1912, as note 88.

179. Carathéodory C (1913) *Über die Begrenzung einfach zusammenhängender Gebiete. Math. Ann.* 73:323–370 and *Ges. Math. Schr.* vol IV, pp 23–80. Carathéodory included this work together with his preceding paper (as note 174) in a special issue beginning with his dedication *Herrn Hermann Amandus Schwarz zum siebzigsten Geburtstag 25. Januar 1913 in tiefer Verehrung gewidmet vom Verfasser* (In deep reverence dedicated by the author to Herr Hermann Amandus Schwarz on the occasion of his 70th birthday on 25 January 1913). P. Duren refers to Carathéodory's extension theorem in the special case most commonly quoted, as saying that a conformal mapping from one Jordan domain to another has a homeomorphic extension to the closures (Communication to the author on 10 November 2001). Carathéodory's theorem XV in his paper *Über die Begrenzung einfach zusammenhängender Gebiete* reads: "In the case of every conformal mapping of the interior of two simply connected domains G and H onto each other, the prime ends of these domains correspond to each other one-to-one."

180. Mentioned in: Remmert, as note 149, p 172.

181. Carathéodory had arrived to these results already in summer 1911, presented them to the Naturalists' Assembly in September 1911, made them known to Hilbert through the two manuscripts and dedicated them to H. A. Schwarz.

182. Study E (1912) *Konforme Abbildung einfach zusammenhängender Bereiche*. Teubner.

183. *Ges. Math. Schr.* vol IV, p 24.

184. Elfving (1981), p 148.
For a more recent treatment of the theory see: chapter 9 of Pommerenke Chr (1975) Univalent Functions. Vandenhoeck & Ruprecht, Göttingen.

185. Carathéodory to Hausdorff on 30 March 1913, as note 168.
     In the same letter, Carathéodory congratulated Hausdorff on his appointment to Greifs-
     wald.

186. Foreword by Lars V. Ahlfors, Harvard University to: Rassias Th M (ed) Constantin
     Carathéodory – An International Tribute vol I. World Scientific, Singapore, New Jersey,
     London, Hong Kong, 1991.

187. Young L C (1981) Mathematicians and Their Times – History of Mathematics and
     Mathematics of History. North-Holland Publishing Company, Amsterdam, New York,
     Oxford, p 243f.

188. Perron, as note 107, chapter 1, here p 49.

189. Carathéodory C (1928) Remark on a Theorem of Osgood concerning convergent Series
     of analytic Functions. Bull. Amer. Math. Soc. 34: 721–725 and *Ges. Math. Schr.* vol
     IV, pp 91–95. According to Carathéodory's own listing, this work was written in Har-
     vard. However, in his *Gesammelte Mathematische Schriften* (Collected Mathematical
     Writings), California appears as the place of its emergence.

190. Osgood W F (1901–1902) Note on the functions defined by infinite series whose terms
     are analytic functions of a complex variable; with corresponding theorems for definite
     integrals. Ann. of Math. 2nd ser, 3:25–34.
     Carathéodory presents the theorem of Osgood as follows: a sequence of analytic func-
     tions which converges in a region R must converge uniformly in at least one subregion
     R_1 of R (*Ges. Math. Schr.* vol IV, p 91).

191. See: Szegö G (1939) Orthogonal Polynomials. Amer. Math. Soc. Colloquium Publica-
     tions vol 23, New York, p 364.

192. Carathéodory C (1932) Conformal Representation. Cambridge Tracts in Mathematics
     and Mathematical Physics no 28, CUP, London, p 86.

193. Radó T (1948) Length and Area. Amer. Math. Soc. Colloquium Publications volume
     30, New York, p 44.

194. *Ges. Math. Schr.* vol IV, note 13, p 117.

195. Carathéodory C (1929), *Stetige Konvergenz und normale Familien von Funktionen.*
     Math. Ann. 101:515–533 and *Ges. Math. Schr.* vol IV, pp 96–118.

196. Carathéodory uses the term *stetige Konvergenz* (continuous convergence) with which Du
     Bois-Reymond had denoted the uniform convergence of a sequence of functions (*stetige
     Convergenz*). In (1964) The Theory of Integration (Cambridge Tracts in Mathematics
     and Mathematical Physics no 21, New York and London, p 44), L C Young gives the
     following definition of uniform convergence.
     Let $f_1, f_2, \ldots, f_n, \ldots$ be any sequence of functions of x, not necessarily continuous,
     and let f be their limiting function. It is assumed that all these functions are bounded
     and defined in the same interval. If a new function $F(x, p) = f_n(x) - f(x)$ is defined
     only at those points where p represents an integral value of n and vanishes elsewhere,
     then the convergence of the given sequence is said to be uniform at a point x_0, if in the
     new space, in which $F(x, p)$ has been defined, the upper and the lower limits of the
     function $F(x, p)$ at the point which corresponds to $x = x_0$ and $n = \infty$ coincide and are
     equal to zero, or, in other words, if F is continuous at that point.
     Carathéodory's exposition of the theory of uniform convergence can be found in his
     1918 book *Vorlesungen über reelle Funktionen*, pp 173–180 or in his 1939 book *Reelle
     Funktionen, Band I*, pp 165–172. The theory of continuous convergence is presented in
     the latter, pp 173–182.
     In his *Remark on a Theorem of Osgood concerning convergent Series of analytic Func-
     tions*, Carathéodory uses the term "regular convergence" to distinguish from the term
     stetige Konvergenz (continuous convergence) that had been used by: Hans Hahn (1921)
     Theorie der reellen Funktionen. 1. Bd (Theory of Real Functions, vol 1). Springer, Berlin,

p 238). Carathéodory defines the term as follows: "We shall say that a sequence of functions $f_n(z)$, meromorphic on a closed region A, is *regularly convergent* to a point z_0 of A if for every sequence of points $z_1, z_2, \ldots$, belonging to A and converging towards z_0, the (finite or infinite) limit $\lim_{n \to \infty} f_n(z_n)$ exists.

197. Perron, as note 107, chapter 1, here note 1 on p 47.

198. Montel P (1927) *Leçons sur les familles normales de fonctions analytiques et leurs applications*. Gauthier-Villars, Paris.

199. Montel P (1912) *Sur les familles de fonctions analytiques qui admettent des valeurs exceptionnelles dans un domaine* (On the Families of Analytic Functions which Admit Exceptional Values in a Domain). *Ann. Sci. École Norm. Sup.* 24:487–535, here 493.

200. Carathéodory mentions and defines the "so-called Weierstrass' double series theorem" in his 1950 book *Funktionentheorie*, vol I, p 205.

201. Carathéodory to Zermelo on 14 July 1935. Letter in German. Universitätsbibliothek Freiburg, Nachlaß Ernst Zermelo.
    In this letter, Carathéodory uses the term *Grenzschwankung* which I have translated as limiting oscillation in accordance with the terminology in his 1932 book *Conformal Representation*, p 61.

202. Perron, as note 107, chapter 1, p 49.

203. Behnke, as note 29, chapter 1, p 29.

204. Carathéodory C (1926) *Über das Schwarzsche Lemma bei analytischen Funktionen von zwei komplexen Veränderlichen. Math. Ann.* 97:76–98 and *Ges. Math. Schr.* vol IV, pp 132–159.

205. *Ges. Math. Schr.* vol IV, p 133.

206. Carathéodory C (1928) *Über die Geometrie der analytischen Abbildungen, die durch analytische Funktionen von zwei Veränderlichen vermittelt werden. Abhandl. Math. Seminar Univ. Hamburg* 6:96–145 and *Ges. Math. Schr.* vol IV, pp 167–227.

207. Carathéodory C (1932) *Über die Abbildungen, die durch Systeme von analytischen Funktionen von mehreren Veränderlichen erzeugt werden* (About Mappings Produced Through Systems of Analytic Functions of Several Variables). *Math. Z.* 34:758–792 and *Ges. Math. Schr.* vol III, pp 406–448, here p 415f.
    For an exposition of the theory of Carathéodory's metric see: Behnke H, Thullen P (1934) *Theorie der Funktionen mehrerer komplexer Varänderlichen* (Theory of Functions of Several Complex Variables). Verlag von Julius Springer, Berlin, pp 100–103.

208. Blaschke W (1927) *Zur Geometrie der Funktionen zweier komplexer Veränderlicher I: Die Gruppen der Kreiskörper. Abhandl. Math. Seminar Univ. Hamburg* 5:189–198.

209. Kritikos N (1950) Το μαθηματικό έργο του K. Καραθεοδωρή (Carathéodory's Mathematical Work). Αιών του Ατόμου (Century of the Atom) Δ΄/2–3:328–332, here 330.
    Poincaré H (1907) *Les fonctions analytiques de deux variables et la représentation conforme* (Analytic Functions of Two Variables and Conformal Representation). *Rend. Circ. Mat. Palermo* 23:185–220.
    Reinhardt K (1921) *Über Abbildungen durch analytische Funtionen zweier Veränderlicher* (On Mappings by Analytic Functions of Two Variables). *Math. Ann.* 83:211–255.

210. Carathéodory C (1927) *Über eine spezielle Metrik, die in der Theorie der analytischen Funktionen auftritt. Atti della Pontificia Accademia delle Scienze Nuovi Lincei, Anno 80*, pp 135–141 and *Ges. Math. Schr.* vol IV, pp 160–166.

211. Carathéodory C (1931) *Ein dem Vitalischen analoger Satz für analytische Funktionen von mehreren Veränderlichen. Journal f. reine u. angew. Math.* 165:180–183 and *Ges. Math. Schr.* vol IV pp 228–233.

212. Kritikos, as note 209, p 330f.

213. Carathéodory C (1932) *Über die analytischen Abbildungen von mehrdimensionalen Räumen.* In: *Verhandlungen des Internationalen Mathematiker-Kongresses.* Zürich 1932, vol I, pp 93–101 and *Ges. Math. Schr.* vol IV, pp 234–246.

Liouville's classical theorem, stating that every bounded holomorphic function on the entire complex plane is a constant function, may be proved by the distance-decreasing property of the Carathéodory distance. See: Goldberg S I: Liouville's Theorem for a Class of Quasiconformal Mappings. In: Panayotopoulos (ed) (1974), pp 154–164, here p 154.

"A Riemannian manifold (resp. complex space) *X* is called Liouville, if it carries no non-constant bounded harmonic resp. holomorphic functions. It is called Caratheodory, or Caratheodory hyperbolic, if bounded harmonic (resp. holomorphic) functions separate the points of *X*."

From Prof.Vladimir Lin's speech on "Liouville and Caratheodory coverings in Riemannian and complex geometry" at The Technion on Tuesday, 22 October 1996.

214. C. Carathéodory (1932) *Über die Abbildungen, die durch Systeme von analytischen Funktionen von mehreren Veränderlichen erzeugt werden. Math. Z.* 34:758–792 and *Ges. Math. Schr.* vol III, pp 406–448.

The main theorem is in *Ges. Math. Schr.* vol III, p 426f.

215. Carathéodory to Hilbert on 12 December 1912. Letter in German. Niedersächsische Staats- und Universitätsbibliothek Göttingen – Abteilung für Handschriften und seltene Drucke. Cod. Ms. D. Hilbert 55.

216. Hilbert D (1913) *Begründung der elementaren Strahlungstheorie* (Foundation of Elementary Theory of Radiation). *Jber. Deutsch. Math.-Verein.* 22:1–20, here p 2.

217. Hilbert had engaged with physics from 1910 to 1922 and with integral equations from 1902 to 1912.

218. Born (1978), p 122f.

219. Lummer O, Pringsheim E (1900) *Ueber die Strahlung des schwarzen Körpers für lange Wellen* (On the Radiation of the Black Body in the Case of Long Waves). *Verh. Deutsch. Phys. Ges.* Jhrg. 2:163–180.

220. Hilbert's integral eguation is:

$$\eta - \frac{\alpha}{4\pi q^2} \iiint \frac{e^{-A}}{S} \eta(x_1, y_1, z_1)\, dx_1 dy_1 dz_1 = 0.$$

221. In text in parentheses: *Ztschr. für wissenschaftliche Photographie Bd. I (1903) S 360–364; Herleitung des Kirchhoffschen Gesetzes* (Journal for Scientific Photography, vol I (1903) pp 360–364; Derivation of Kirchhoff's Law).

222. Hilbert gives the reference for Pringsheim's treatise: *Herleitung des Kirchhoffschen Gesetzes, Zeitschrift für wissenschaftliche Photographie (1903)*, as note 221, in: Hilbert, as note 216, here remark on p 16. Also his "axiomatic treatment of the elementary theory of radiation" comprises pp 16–20 of the above mentioned paper.

223. Carathéodory to Hilbert on 4 April 1913. Letter in German. Niedersächsische Staats- und Universitätsbibliothek Göttingen – Abteilung für Handschriften und seltene Drucke. Cod. Ms. D. Hilbert 55.

224. Hellinger E D, Toeplitz O (1910) *Grundlagen für eine Theorie der unendlichen Matritzen* (Foundations for a Theory of the Infinite Matrices). *Math. Ann.* 69:289–330. *M* is given in this paper.

Hellinger, a close friend of Born, Toeplitz and Courant, "remembered for his wit, strength of personality, and uncommon ability to make friends" (Rovnyak J (1990) Ernst David Hellinger 1883-1950: Göttingen, Frankfurt Idyll, and the new World. Operator Theory: Advances and Applications 48:1–41, here 1), had been Hilbert's assistant from 1905 to 1907, when he gained his doctorate under Hilbert's supervision with a work on *Die Orthogonalinvarianten quadratischer Formen von unendlich vielen Variablen* (The Orthogonal Invariants of Quadratic Forms of Infinitely Many Variables). He had attended lectures also by Carathéodory, then he habilitated in Marburg in 1909 and stayed there until 1914. Only in 1920 he became full professor at the Royal University in Frankfurt am Main.

He produced important work on Stieltjes's moment problem and after long work together with Toeplitz the encyclopaedia article on integral equations.

225. Previously mentioned important papers of 1907 and 1911.

226. Toeplitz, as note 118.

227. Hilbert, as note 216, here remark 1 on p 18.

228. Planck M (1913) *Vorlesungen über die Theorie der Wärmestrahlung.* Barth, Leipzig.

229. (1915) G P – *Η Δίκη του Ναυπλίου* (G[eorgios] P[Papandreou] – The Trial of Nauplion). In: *Δελτίο Εκπαιδευτικού Συλλόγου* 5: 194–205. Both citations in: Bouzakis (1997), pp 160–168, here p 160 resp. p 165.

230. Carathéodory to Tsaldaris on 29 August 1935. Vovolinis (ed) (1962), p 520f.

A decree published in the Government Gazette, issue 86 of 31 December 1836, provided for the establishment of a university, the University of Athens, which would be officially opened on the third day of Easter 1837 and would carry the name "Ottonian University" in honour of its founder, King Otto of Greece. The opening ceremony took place on 3 May 1837. On 20 October 1862 the Ottonian University changed its name to "National University of Greece" and on 17 July 1911, to satisfy a condition of the bequest of the "great benefactor" of the university, Ioannis Dombolis (d. Russia, 1850), to the Greek state, the "Capodistrian University" was established (instead of establishing a separate university, the Capodistrian was integrated into the National University, which obtained the new name of "National and Capodistrian University"). The National and Capodistrian University is globally known as the University of Athens.

231. Hondros to Sommerfeld on 24 December 1909. Letter in German. LMU München, Institut für theoretische Physik, Nachlaß Sommerfeld, N 89, 009.

232. Eckert M, Märker K (eds) *Arnold Sommerfeld – Wissenschaftlicher Briefwechsel. Band 1: 1892–1918* (Arnold Sommerfeld – Scientific Correspondence, vol 1: 1892–1918). Deutsches Museum, München, 2000, pp 272f.

233. The Minister of Education to the rector and the senate of the Royal Technical University in Breslau on 6 February 1913. Uniwersytet Wrocławski, TH 156 – Professoren, 18 XI 1910–24 IV 1933.

234. The Minister of Education to the rector and the senate of the Royal Technical University in Breslau on 8 March 1913. Ibid.

235. Königliches Universitäts-Kuratorium, Philosophische Fakultät, Ordentliche Professoren 1913 "Dr. Carathéodory". Universitätsarchiv Göttingen 4JVb 277.

236. Carathéodory to Hilbert on 4 April 1913, op. cit.

237. Rector Schenk on 31 March 1913. Uniwersytet Wrocławski, TH 98, op. cit.

238. Reid (1970), p 138.

239. Faber G (1959) *Mathematik. Geist und Gestalt* (Mathematics. Spirit and Form). In: *Biographische Beiträge zur Geschichte der Bayerischen Akademie der Wissenschaften vornehmlich im zweiten Jahrhundert ihres Bestehens, Bd 2: Naturwissenschaften* (Biographic Contributions to the History of the Bavarian Academy of Sciences, Primarily in the Second decade of its Existence, vol 2: Natural Sciences). Munich, pp 1–45, here p 37. Mentioned in: Toepell (1996), p 295f.

240. There is a drawing of the site plan (1:1000) and the section of the elevator (1:50) in the Architectural Museum of Wrocław. (Information supplied by Anna Korpalska and Sp. Georgiadis.)

241. Runge C (1913) The mathematical training of the physicist in the university. In: Hobson E W, Love A E H (eds) Proceedings of the 5th International Congress of Mathematicians. CUP, Cambridge. Reprint 1967, Kraus, Nendeln/Liechtenstein, pp 598–607.

242. *Und Karlchen Runge, Der hat die Passion, Daß bei seinen Formeln, Stimmt die Dimension* (And little Karl Runge, he has the passion, that in his formulae, is correct the dimension).
Bei noch so großer Hitze, Der Prandtl Kühlung find't, Er stellt an den Motor, Und fächelt sich Wind (Even in such a heat, Prandtl cools off, by turning on the motor, he takes wind thereof).
Zum Physiker drillt, Debye seinen Sohn; Am ersten Tag stopft er, Mit Quanta ihn schon (His son, Debye drills, to become a physicist, with quanta he fills, him since the hours of mist).
From a poem by Margarethe Goeb entitled *Zu Professor Carathéodorys Abschied von Göttingen am 1.8.1918* (On the Occasion of Professor Carathéodory's Departure from Göttingen on 1 August 1918), Courtesy of Mrs. Rodopoulos-Caratheodory. For information about Goeb see: Thiele R (2001) *Lobgedicht auf Caratheodory* (A Poem in Praise of Caratheodory). In: *Briefe an die Herausgeber* (Letters to the Editor), *DMV-Mitt. 2.* The poem has been published, analysed and commented by Alexander Sideras and Paraskevi Sidera-Lytra in a paper entitled *Constantin Caratheodory's Abschied von Göttingen – Ein unveröffentlichtes Gedicht der Studenten anlässlich seiner Berufung nach Berlin* (Constantin Caratheodory's Departure from Göttingen – An Unpublished Poem from the Students on the Occasion of his Appointment in Berlin). Communication to the author from Angela Kakavoutis, Deutsche Schule Thessaloniki, on 01 July 2002.

243. Weyl (1955), Vorwort p V.

244. Cf. Hilbert D (1915) *Die Grundlagen der Physik (erste Mitteilung)* (The Foundations of Physics (First Communication)). *Nachr. Kgl. Ges. Wiss. Gött., math.-phys. Kl.* 1915, pp 395–407.

245. Caratheodory C (1917) *Über die geometrische Behandlung der Extrema von Doppelintegralen. Verh. Schw. Naturf. Ges. 99. Jahresversammlung*, Zurich 1917; 2nd part, pp 127–129 and *Ges. Math. Schr.* vol I, pp 371–373.

246. Thiele, as note 242.

247. Caratheodory to Klein on 22 December 1913. Letter in German. Niedersächsische Staats- und Universitätsbibliothek Göttingen – Abteilung für Handschriften und seltene Drucke. Cod. Ms. F. Klein 8, 464.

248. *Zur frd. Kenntnisnahme und in der Bitte C's Aufnahme in die Redaktion nunmehr herbeizuführen. Herzlichste Weihnachtsgrüsse. Hilbert* (To your kind attention and with the request to now bring about Caratheodory's admission to the editorial board. Most heartily Christmas greetings. Hilbert). Niedersächsische Staats- und Universitätsbibliothek Göttingen – Abteilung für Handschriften und seltene Drucke. Cod. Ms. D. Hilbert 55.

249. Caratheodory to Hilbert on 22 December 1913. Letter in German. Niedersächsische Staats- und Universitätsbibliothek Göttingen – Abteilung für Handschriften und seltene Drucke. Cod. Ms. D. Hilbert 55.

250. Caratheodory, as note 41, chapter 1, here *Ges. Math. Schr.* vol V, p 58.

251. Mentioned in Tobies, as note 122, chapter 1, here p 33.

252. Caratheodory to Hilbert on 1 July 1925. Letter in German. Niedersächsische Staats- und Universitätsbibliothek Göttingen – Abteilung für Handschriften und seltene Drucke. Cod. Ms. D. Hilbert 55, 9.

253. Die Redaktion der Mathematischen Annalen (1925) *Felix Klein. Math. Ann.* 95:1 and *Ges. Math. Schr.* vol V, p 52.

254. Address to Klein on his 60th birthday by Hilbert, as note 127, chapter 1.

255. Caratheodory to Klein on 1 September 1919. Letter in German. Niedersächsische Staats- und Universitätsbibliothek Göttingen – Abteilung für Handschriften und seltene Drucke. Cod. Ms. F. Klein 8, 468.

256. Carathéodory to Hilbert on 14 April 1926. Letter in German. Niedersächsische Staats- und Universitätsbibliothek Göttingen – Abteilung für Handschriften und seltene Drucke. Cod. Ms. D. Hilbert 55.

257. Sommerfeld was the initiator of Blumenthal's appointment to Aachen and person- ally close to him. Blumenthal was a co-editor of the *Jahresbericht der Deutschen Mathematiker-Vereinigung* in the years 1924–1933 and on the board of chief editors of the *Mathematische Annalen* in 1924–1938.

   Pinl M (1969) *Kollegen in einer dunklen Zeit* (Colleagues in a Dark Time). *Jber. Deutsch. Math.-Verein.* 71:167–228, here p 169.

258. Carathéodory to Hilbert on 17 April 1926. Postcard in German. Niedersächsische Staats- und Universitätsbibliothek Göttingen – Abteilung für Handschriften und seltene Drucke. Cod. Ms. D. Hilbert 55.

259. Carathéodory to Einstein on 4 October 1926. Letter in German. The Jewish National and University Library, Jerusalem, Einstein Archive, call number: 8 344.

260. Certificate of Posting with stamp: 16 October 1926. Ibid.

261. Von Kármán to Carathéodory on 12 February 1927. Letter in German. Caltech Archives, Theodore von Kármán Papers, Folder 5.4.

262. Carathéodory to Pólya on 7 February 1917. Postcard in German. ETH-Bibliothek Zürich, Wissenschaftshistorische Sammlungen, HS 89:83.

263. Carathéodory to Pólya on 7 April 1917. Postcard in German. Stanford University Libraries, Department of Special Collections, SC 337 George Polya Papers, Box 1, Folder 14.

264. Jentzsch R (1918) *Über Potenzreihen mit endlich vielen verschiedenen Koeffizienten* (About Power Series with Finitely Many Different Coefficients). *Math. Ann.* 78:276– 285.

   Pólya G (1918) *Über Potenzreihen mit endlich vielen verschiedenen Koeffizienten. Math. Ann.* 78:286–293.

   In his paper, Pólya grounded function theoretical results of Jentzsch's work in a new way.

   Jentzsch's preceding work: (1917) *Untersuchungen zur Theorie der Folgen analytischer Funktionen* (Studies in the Theory of Sequences of Analytic Functions) was published in: *Acta Math.* 41:219–251.

265. Van der Waerden to Carathéodory on 19 März 1944. Letter in German. ETH-Bibliothek Zürich, Wissenschaftshistorische Sammlungen, HS 652:10610.

266. Carathéodory to van der Waerden on 25 March 1944. Letter in German. ETH-Bibliothek Zürich, Wissenschaftshistorische Sammlungen, HS 652:10661. Blaschke's evaluation of Popoff may be mentioned here: "Humanly, politically and scientifically, I hold K. Popoff in less esteem." Blaschke's report on his lecture trip to Bucharest and Sofia from 5 to 25 May 1943. Staatsarchiv Hamburg, Hochschulwesen, Dozenten- u. Personalakten IV 86, Blaschke, Wilhelm, 13.9.1885.

267. Vol 21, 1949/50. At that time the editorial board consisted of Heinrich Behnke (Münster, Westfalen), Heinz Hopf (Zurich), Franz Rellich (Göttingen), Richard Courant (New York), Kurt Reidemeister (Marburg [Lahn]), Bartel L. van der Waerden (Amsterdam).

268. Carathéodory to Born on 9 October 1907, as note 131, chapter 1.

269. Description of the incident in Born (1978), p 108f.

270. R.Tobies (1981), p 85.

   In a common letter of 4 August 1914 to H. A. Schwarz on the occasion of his golden doctoral jubilee, Landau and Carathéodory mention "the malicious behaviour of the enemies of our German fatherland, which made the first days of August the most severe that humanity has ever experienced".

   NL Schwarz, Zentrales Archiv der Akademie der Wissenschaften der DDR, Historische

Abteilung. Abschnitt II: Akten der Preußischen Akademie der Wissenschaften 1812–1945.

271. Born (1978), p 160.

272. This incident had been repeatedly discussed in Carathéodory's house and always felt as a great insult. Carathéodory has been attempting to keep it secret from his children, but his daughter Despina was aware of the accusations.

273. Cf. Wehler H-U, *Der erste totale Krieg. Woran das deutsche Kaiserreich zugrunde ging – und was daraus folgte* (The First Total War. On what the German Kaiserreich was Destroyed – and what Followed from that). *Die Zeit* no. 35, 20 August 1998.

274. Cf. Meinhardt G (1977) *Die Universität Göttingen – Ihre Entwicklung und Geschichte von 1734–1974* (The University of Göttingen – Its Development and History from 1734 to 1974). Musterschmidt Göttingen, Frankfurt, Zurich, p 86 and Hund F (1987) *Die Geschichte der Göttinger Physik* (The History of Göttingen Physics). Vandenhoeck & Ruprecht, Göttingen, p 53.

275. Carathéodory to Zermelo on 31 March 1915. Letter in German. Universitätsbibliothek Freiburg, Nachlaß Ernst Zermelo.

276. Carathéodory to Fejér on 2 July 1916. Letter in German sent to the author by Dr. A.Varga, József Attila University, Bolyai Institute, Szeged, on 19 June 2000.

277. Carathéodory from St. Moritz to Toeplitz on 03 September 1917. Toeplitz B, Nachlaß Otto Toeplitz. Universitäts- und Landesbibliothek Bonn, Handschriften- und Rara-Abteilung.

278. Carathéodory to Hilbert (undated). Letter in German. Niedersächsische Staats- und Universitätsbibliothek Göttingen – Abteilung für Handschriften und seltene Drucke. Cod. Ms. D. Hilbert 55.

279. From a poem by Margarethe Goeb, as note 242.
    The verse for Carathéodory is: *In Aegypten da nimmt man, Chinin gegen's Fieber, Cara tut's nicht, Kotelett ist ihm lieber* (One takes in Egypt, quinine against fever, Cara rejects it, a chop is him dearer).

280. Carathéodory to Hilbert on 12 August 1916. Postcard in German. Niedersächsische Staats- und Universitätsbibliothek Göttingen – Abteilung für Handschriften und seltene Drucke. Cod. Ms. D. Hilbert 55.

281. Carathéodory to Klein on 15 September 1916. Postcard in German. Niedersächsische Staats- und Universitätsbibliothek Göttingen – Abteilung für Handschriften und seltene Drucke. Klein XXII, A.

282. In 1920, Lichtenstein became a full professor at Münster, in 1922 at the University of Leipzig, where he also took on the headship of the Mathematical Institute. His scientific contribution includes works on problems of potential theory, of conformal representation, on the theory of integral equations and on hydrodynamics.
    Cf. Beckert H (1980) *Leon Lichtenstein 1878–1933. Wissenschaftliche Zeitschrift der Karl-Marx-Universität Leipzig* 1:3–13.

283. Reid (1970), p 152.

284. Carathéodory C, *Deutsches Wissen und seine Geltung. Deutsche Allgemeine Zeitung*, 12 April 1929 and *Ges. Math. Schr.* vol V, pp 198–199.

285. Hilbert, as note 244, here p 395.

286. Einstein to Carathéodory on 6 September 1916. Letter in German. The Jewish National and University Library, Jerusalem. Scient.Corr.File, Folder "C-Misc.", 8 334. (Einstein's letter is filed with the date November 6th 1916).
    Permission granted by the Albert Einstein Archives, The Jewish National and University Library, the Hebrew University of Jerusalem, Israel.

287. Einstein to Carathéodory (undated letter in German, probably written on 10 December 1916). The Jewish National and University Library, Jerusalem. Scient.Corr.File, Folder "C-Misc.", 8 343.
   Permission granted by the Albert Einstein Archives, The Jewish National and University Library, the Hebrew University of Jerusalem, Israel.

288. Carathéodory to Einstein on 16 December 1916. Letter in German. The Jewish National and University Library, Jerusalem. Scient.Corr.File, Folder "C-Misc.", 8 335.

289. By "Analytical Dynamics", Carathéodory meant: Whittaker E T (1904) A treatise on the analytical dynamics of particles and rigid bodies. CUP.

290. Carathéodory C, *Variationsrechnung. Ges. Math. Schr.* vol I, pp 312–370. Printed from: Frank–Mises, *Die Differential- und Integralgleichungen der Mechanik und Physik*, chapter 5, pp 227–279. 2nd edition, Braunschweig 1930. The first volume with contributions from Bieberbach L, Carathéodory C, Courant R, Löwner K, Rademacher H, Rothe E, Szegö G was released in 1925, the second two years later. There has been a second edition in 1930 and a third in 1945.

291. Cf. chapter entitled *Kosmologie* (Cosmology), pp 311–321 of: Born M (1964) *Die Relativitätstheorie Einsteins* (Einstein's Theory of Relativity). Springer, Berlin, Göttingen, Heildelberg.

   In general relativity theory, the concept of a gravitational force as a four-dimensional vector is absent and gravitational properties are defined by the Riemann space–time structure. Accordingly, the motion of a particle in a gravitational field (provided it is not acted upon by any non-gravitational forces) is considered as free in the general theory of relativity. The precise formulation of a geodesic hypothesis is as follows: the world line of a free test particle with non-zero rest mass is a non-isotropic time-like geodesic line of space–time; the world line of a free test particle with zero rest mass (a photon, a neutrino) is an isotropic geodesic line of space-time.
   D. D. Sokolov, Encyclopaedia of Mathematics.

292. Einstein A (1914) *Die formale Grundlage der allgemeinen Relativitätstheorie* (The Formal Foundations of the General Theory of Relativity). In: Kox A J, Klein M J, Schulmann R (eds) The Collected Papers of Albert Einstein, vol 6: The Berlin Years: Writings, 1914–1917. Princeton University Press 1996, pp 73–130, here p 122.

293. Communication to the author from D. Christodoulou on 20 November 2000.

294. Lecture on 2 November 1926. *Teilnachlaß Carathéodory*, as note 74, chapter 1.

295. Carathéodory C: Χωρόχρονος. Μεγάλη Ελληνική Εγκυκλοπαίδεια, vol 24, 1934, pp 800–802. Translated from the Greek by Stephanos Carathéodory as: *Der moderne Raum–Zeit-Begriff* (The Modern Notion of Space–Time) for *Ges. Math. Schr.* vol II, pp 402–413.

296. Carathéodory C (1914) *Über das lineare Maß von Punktmengen – eine Verallgemeinerung des Längebegriffs. Nachr. Kgl. Ges. Wiss. Gött., math.-phys. Kl.* 1914, pp 404–426 and *Ges. Math. Schr.* vol IV, pp 249–275.

297. Today, what is called Carathéodory's measure is the outer measure $\mu*$ defined on the class of all subsets of a metric space M (with a metric ρ) such that $\mu*(A \cup B) = \mu*(A) + \mu*(B)$ provided that $\rho(A, B) > 0$. If an outer measure $\mu*$ is given, it is possible to specify the class of measurable sets on which $\mu*$ becomes a measure.
   Outer measures result, in particular, from the construction of the extension of a measure from a ring R onto the σ-ring generated by it. In the classical theory of the Lebesgue measure the outer measure of a set is defined as the greatest lower bound of the measures of the open sets containing the given set; the inner measure of a set is defined as the least upper bound of the measures of the closed sets contained in the given set.
   V. A. Skrortsov, Encyclopaedia of Mathematics.
   Radó gives following definition for the Carathéodory outer measure: Let S be a metric

space, and let $\Gamma(E)$ be a finite-valued, non-negative function defined for every subset E of S (not only for Borel sets). Then $\Gamma(E)$ will be called a Carathéodory outer measure if and only if the following conditions hold.

(C_1) $\Gamma(0) = 0$, and $\Gamma(E_1) \leq \Gamma(E_2)$ whenever $E_1 \subset E_2$.

(C_2) $\Gamma\left(\sum E_n\right) \leq \sum \Gamma E_n$ for every sequence $E_1, \ldots, E_n, \ldots$ of subsets of S.

(C_3) $\Gamma(E_1 + E_2) = \Gamma(E_1) + \Gamma(E_2)$ whenever the distance $\rho(E_1, E_2)$ of the sets E_1, E_2 is positive.

Radó, as note 193, p 44.

298. Perron, as note 107, chapter 1, here p 50f.

299. Rosenthal A (1916) *Beiträge zu Caratheodorys Meßbarkeitstheorie. Nachr. Kgl. Ges. Wiss. Gött., math-phys. Kl. 1916*, pp 305–321.
     On the inner measure and comparison of inner and outer measure cf. relevant chapters of: Carathéodory C (1956) Algebraic Theory of Measure and Integration. Chelsea Publishing Company, New York.

300. A satisfactory theory of the inner measure was eventually produced by J. Ridder, but not until 1940. In: (1941) *Mass- und Integrationstheorie in Strukturen* (Measure and Integration Theory in Structures). *Acta Math.* 73:131–173, Ridder reduced the asymmetry between the theories of inner and outer measure to a minimum measure as to modify the definition of internal measure without restricting the fundamental properties of this concept which are needed in applications. Carathéodory acknowledged that Ridder had developed an axiomatic theory based on Carathéodory himself and Glivenko. Cf. *Zur Geschichte der Definition der Meßbarkeit* (On the History of the Definition of Measurability), Carathéodory's hand-written note to his work *Über das lineare Maß von Punktmengen – Verallgemeinerung des Längenbegriffs*, as note 296, and also Carathéodory's review of Ridder's paper in: (1941) *Zentralbl. f. Math.* 24:103–104 and *Ges. Math. Schr.* vol V, pp 368–369.

301. This is explained clearly on page 307 of the survey paper: Federer H (1952) Measure and area. Bull. Amer. Math. Soc. 58:306–378.

302. Wendell Fleming's report on the Carathéodory manuscript on 2 October 2001.
     Hausdorff's definition is given in: Hausdorff F (1919) *Dimension und äußeres Maß* (Dimension and Outer Measure). *Math. Ann.* 79:157–179. Hausdorff explains that he generalises Carathéodory's length measure and uses analogous notions of measure, which indirectly include non-integer values of the dimension of a point set, so that sets of fractional dimension become possible.

303. Randolph, then a member of the Institute for Advanced Study, had visited Carathéodory in Munich on 27 June 1938, when the latter was engaged in similar subjects.

304. Morse A P, Randolph J F (1945) Gillespie measure. *Fund. Math.* 33:12–26.

305. Carathéodory C (1919) *Über den Wiederkehrsatz von Poincaré. Sitzungsber. Preuß. Akad. d. Wiss. Berlin, math.-phys. Kl.* 1919, pp 580–584 and *Ges. Math. Schr.* vol IV, pp 296–301, citation on p 296.
     Poincaré H (1890) *Sur le problème des trois corps et les équations de la dynamique* (On the Problem of Three Bodies and the Equations of Dynamics). *Acta Math.* 13:1–270.

306. Carathéodory C, Some Applications of the Lebesgue Integral in Geometry. British Association for the Advancement of Science, Report of the 94th Meeting (96th Year) Oxford (1926, August 4–11) p 338 and *Ges. Math. Schr.* vol V, p 240.

307. Carathéodory's proof of Poincaré's recurrence theorem is presented by: V.V. Rumyantsev, Encyclopaedia of Mathematics.

308. Carathéodory C (1918) *Vorlesungen über reelle Funktionen.* Teubner, Leipzig and Berlin. According to his own notes, Carathéodory was writing the book from 20 August 1914 to November 1915.

309. *Ann. Sci. École Norm. Sup.* 1910.

310. Boerner H: Carathéodory, Constantin. In: Gillispie Ch C (ed) Dictionary of Scientific Biography, vol III. Scribner, New York 1971, p 62f, here p 63.

311. Lebesgue formulated the theory of measure in 1901 and, in the following year, he gave the definition of the Lebesgue integral that generalises the notion of the Riemann integral by extending the concept of the area below a curve to include many discontinuous functions. There is a letter from the Hungarian mathematician Zoárd Geöcze to Fejér dated 16 May 1916 dealing with Carathéodory's opinion about Carathéodory's definition of the area as compared to that from Lebesgue. "I do not agree with Professor Carathéodory's opinion that his definition concerning the area is simpler than that by Lebesgue; because his definition is related to a manifold of points (so that there are no double points), whereas Lebesgue's definition is related to the image of a part of the plane u, v (so it may have double points, etc). These two definitions are incomparable as are the manifold of points and some part of the plane u, v".
Zoárd Geöcze: a high-school teacher with a doctoral degree from the Sorbonne and a habilitation from Budapest. He died during World War I. His letter to Fejér reveals that he corresponded with Carathéodory in 1915. Geöcze to Fejér on 16 May 1916. Letter in Hungarian, translated by Dr. Antal Varga. The importance of Geöcze's results in modern real function theory and their connections with those of Carathéodory, Lebesgue, Riesz, Radó can be found in: Saks S (1937) Theory of the Integral. Hafner Publishing Company, New York. Communication to the author from Dr. Antal Varga.

312. Carathéodory to Fejér on 2 July 1916, as note 276.
Carathéodory's results were formulated by Hans Hahn for arbitrary metric spaces. See: chapter VI of Hahn H (1921), as note 196.

313. Akademie der Wissenschaften zu Göttingen, Pers. 20, 985.

314. Fleming's report, as note 302.
Carathéodory's 1914 article and 1918 book did not only influence the abstract formulation and development of measure theory but also the efforts to lay the foundations of probability theory. D. Kappos explains that, since probability is a normed measure on a Boolean algebra of events, Carathéodory's algebraic measure theory is appropriate to introduce the concept of probability as a strictly positive and normed measure. See: Kappos D: Generalized Probability with Applications to the Quantum Mechanics. In: Panayotopoulos (ed) (1974), pp 253–270, here p 253.

315. Schmidt E (1943) *Constantin Carathéodory zum 70. Geburtstag* (To Constantin Carathéodory on the Occasion of his 70th Birthday). *Forsch. u. Fortschr.* 23/24:249–250.

316. Fréchet R-M (1928) *Les espaces abstraits et leur théorie considérée comme introduction à l'Analyse générale* (Abstract Spaces and their Theory Considered as an Introduction to General Analysis). Jacques Gabay, Paris.

317. Carathéodory C (1938) *Entwurf für eine Algebraisierung des Integralbegriffs. Sitzungsber. Bayer. Akad. d. Wiss. München, math.-nat. Abt.* 1938, pp 27–68 and *Ges. Math. Schr.* vol IV, pp 303–342.

318. Kritikos, as note 209, p 332.

319. Carathéodory C (1939) *Reelle Funktionen, Bd. I, Zahlen / Punktmengen Funktionen.* Teubner, Leipzig and Berlin.

320. Rosenthal to Carathéodory on 20 February 1938. *Teilnachlaß Carathéodory*, as note 74, chapter 1.
In a "Request for Research Funds" submitted to the Committee on Research of the University of New Mexico on 10 September 1944, Rosenthal explained in a "brief statement of the problem to be investigated": "I am writing a book on 'Set Functions' (about 350 pages) based on a part of my German manuscript for the second volume of H. Hahn's 'Reelle Funktionen'. I shall get the coming 2nd term off (my first vacation for 2 ¼ years) and I wish to spend this time entirely on finishing the book; (about 200 more

pages have to be written). I hope that I shall be able to do so and that the book then can be printed in the 3rd term." Rosenthal File in the University Archive of the University of New Mexico.

321. Blaschke's report of 17 February 1938 on the work of W. Maak. Staatsarchiv Hamburg, Hochschulwesen, Dozenten- u. Personalakten IV 648, Maak, Wilhelm 13.8.1912.

322. Carathéodory C (1938) *Bemerkungen zur Axiomatik der Somentheorie*. *Sitzungsber. Bayer. Akad. d. Wiss. München, math.-nat. Abt.* 1938, pp 175–183 and *Ges. Math. Schr.* vol IV, pp 343–351.

323. Carathéodory to Jessen on 26 July 1938. Letter in German. University of Copenhagen, Institute for Mathematical Sciences, Archive, Børge Jessen Papers, Box 12, Folder "Carathéodory, Constantin".

324. Carathéodory C (1939) *Die Homomorphien von Somen und die Multiplikation von Inhaltsfunktionen*. *Ann. Scuola Norm. Sup. Pisa, Sci. Fis. e Mat.* 8:105–130 and *Ges. Math. Schr.* vol IV, pp 352–384.

325. Theorem 7 (main theorem) in *Ges. Math. Schr.* vol IV, p 382.

326. A torus space Q_w is, according to Jessen, the metric space whose elements are the infinite sequences of real numbers $\xi = (x_1, x_2, \ldots, x_n, \ldots)$ where $0 \leq x_n \leq 1$ for $n = 1, 2 \ldots$, the distance $\rho(\xi, \eta)$ of two points $\xi = (x_1, x_2, \ldots, x_n, \ldots)$ and $\eta = (y_1, y_2, \ldots, y_n, \ldots)$ in Q_w being defined by the formula $\rho(\xi, \eta) = \sum_n |y_n - x_n|/2n$. See: Saks, as note 311, here p 157.
For the Jessen infinite-dimensional torus see: Carathéodory, as note 299, p 320ff.

327. Jessen B, draft of an undated letter to Carathéodory in German. University of Copenhagen, Institute for Mathematical Sciences, Archive, Børge Jessen Papers, Box 12, Folder "Carathéodory, Constantin".
Jessen was spending his summer vacations together with H. Bohr, to whom Carathéodory sent his regards saying that he had no news from him for a long time.

328. Carathéodory C (1940) *Über die Differentiation von Massfunktionen* (On the Differentiation of Measure Functions). *Math. Z.* 46:181–189 and *Ges. Math. Schr.* vol IV, pp 385–396.
From properties of place functions and of the integral, Carathéodory obtained equations connected with the differentiation of measure functions. To carry out a differentiation means to determine a place function f given the two measure functions $\varphi(X)$ and

$$\psi(X) = \int_X f \, \mathrm{d}\varphi.$$

This paper was written on Perron's 60th birthday. Carathéodory wrote to Herglotz: "Knopp has found out that Perron has his 60th birthday on 7 May. He has proposed to edit an issue of the *M. Z.* full of works dedicated to Perron. You are also on the list, which is not very long, by the way. I find that it is too late to dash off something decent until the 1st of March and I have suggested to him, instead of this plan, to write an address that would come from the editorial board of the *M. Z.* and in which P.'s close friends and students could participate. What is your opinion? In the name of the Göttingen Mathematical Society we have presented a very nice address on the 70th birthday of Cantor. It would be very unpleasant for me to dash off something in a few weeks, especially, since I have recently written things that do not interest Perron at all." And Carathéodory mentioned the mirror telescope of Schmidt.
Carathéodory to Herglotz on 16 January 1940. Letter in German. Niedersächsische Staats- und Universitätsbibliothek Göttingen – Abteilung für Handschriften und seltene Drucke. Cod. Ms. Herglotz F 18.

329. Carathéodory C (1941) *Bemerkungen zum Riesz–Fischerschen Satz und zur Ergodentheorie* (Remarks on the Riesz–Fischer Theorem and on Ergodic Theory). *Abhandl. Math. Sem. Hamb. Univ.* 14:351–389 and *Ges. Math. Schr.* IV, pp 397–442.

Carathéodory formulated the Riesz–Fischer theorem, gave a short history of ergodic theory and acknowledged that the Principal Ergodic Theorem was proved by Birkhoff in 1931, whereas new proofs and generalisations of the Birkhoff theorem were owed to Khintchine and E. Hopf. A chapter entitled "Ergodic Theory" is part of his book *Algebraic Theory of Measure and Integration*. Carathéodory dedicated this paper to Herglotz on his 60th birthday.

330. Carathéodory C (1942) *Gepaarte Mengen, Verbände, Somenringe* (Pair of Sets, Lattices, Rings of Somas). *Math. Z.* 48:4–26 and *Ges. Math. Schr.* vol IV, pp 443–473.
     This paper was dedicated to Knopp on his 60th birthday on 22 July 1942. On this occasion, Carathéodory had contacted Herglotz: "You have surely heard that a special issue of the *M. Z.* is planned for Knopp's 60th birthday. The related works should be sent to Knopp himself, as usually, already before Christmas and at the same time Kamke, who would provide these works before their printing with an identical dedication, should be informed."
     Carathéodory to Herglotz on 1 November 1941. Letter in German. Niedersächsische Staats- und Universitätsbibliothek Göttingen – Abteilung für Handschriften und seltene Drucke. Cod. Ms. Herglotz F 18.

331. Carathéodory C (1944) *Bemerkungen zum Ergodensatz von G. Birkhoff. Sitzungsber. Bayer. Akad. d. Wiss. München, Math.-Nat. Abt.* 1944, pp 189–208 and *Ges. Math. Schr.* vol IV, pp 474–494.

332. Pitt H R (1943) Some Generalisations of the Ergodic Theorem. *Proc. Cambridge Philos. Soc.* 38:325–342.

333. Carathéodory to Born on 24 March 1949, as note 23.

334. Carathéodory C (1956) *Mass und Integral und ihre Algebraisierung. Lehrbücher und Monographien aus dem Gebiete der exakten Wissenschaften*, vol 10. Birkhäuser, Basel, Stuttgart.

335. In the preface found among the papers left by Carathéodory.

336. Carathéodory to Radó on 10 December 1948. Letter in English. The Ohio State University Archives, Tibor Rado papers (Record Group R6 40/76/1), "Caratheodory, C. (Professor): 1948–1949".
     Around 1930 Besicovitch from Trinity College, Cambridge, England, extended his density properties of sets to those of finite Hausdorff measure. After the war, Besicovitch was invited by Süss to visit the Mathematical Research Institute in Oberwolfach, participate in scientific discussions and report on his scientific work. (Süss to Besicovitch on 16 February 1948. Universitätsarchiv Freiburg, Nachlaß Wilhelm Süss, C89/5).

337. Referring to Besicovitch, Radó wrote in his book *Length and Area* (as note 193), p 559f: "let Σ be a simple closed surface, A its area, and V the volume enclosed by Σ. The isoperimetric inequality asserts that $V^2 \le A^3/36\pi$. If one attempts to prove the inequality in a general form, then the definition of V is an important issue. Indeed, in analogy with the Osgood curve, Besicovitch [1] constructed a simple closed surface Σ such that $|\Sigma|$ (the three-dimensional measure of Σ) is positive. If $|D|$ is the three-dimensional measure of the (open) domain D enclosed by Σ, then we should distinguish between the interior volume $V_i = |D|$ and the exterior volume $V_e = |D| + |\Sigma|$, since now $V_e > V_i$. The isoperimetric inequality may be then considered in either one of the forms

(1) $V_e^2 \le A^3/36\pi$

(2) $V_i \le A^3/36\pi$.

Now Besicovitch shows that for given $\varepsilon > 0$ and $G > 0$, his surface may be made to satisfy $V_e > G$, $A < \varepsilon$, where A is the Lebesgue area of Σ. Thus V_e is not compatible with the Lebesgue area A as far as the inequality (1) is concerned. On the other hand, it is readily seen that that in the Besicovitch example the inequality (2) *does* hold, a situation which suggests that *the concept of enclosed volume V must be properly adjusted to the concept of surface area.* As a matter of fact, in a paper to appear in the Transactions of the American Mathematical Society, the writer established the isoperimetric inequality in a very general form, using the Lebesgue area. Subsequent results of J.W.T.Youngs, as yet unpublished, contain even stronger results. Let us recall in this connection the application of the Lebesgue area in the Plateau problem [...]. The role played by the Lebesgue area in these two classical variation problems may be considered as added evidence of the relevancy of this concept of surface area."

In the bibliography to *Length and Area* (p 562), Radó mentions Besicovitch's paper: (1945) On the definition and the value of the area of a surface, Quart. J. Math. 16:86–102

338. Radó to Carathéodory on 16 December 1948. The Ohio State University Archives, Tibor Rado papers (Record Group R6 40/76/1), "Caratheodory, C. (Professor): 1948–1949". In his communication *Sur l'Aire des Surfaces Continues* (On the Area of Continuous Surfaces) to the International Congress of Mathematicians in Bologna, Radó distinguished his own parametric surfaces from Carathéodory's amorphous surfaces. Carathéodory's surfaces he said, are nothing more than assemblages of points and are not revealed as surfaces in any other way except in the way in which they behave towards some procedure of measuring the area.
See: International Congress of Mathematicians. Bologna 1928. Vol 1/2, Kraus, Nendeln/Liechtenstein, reprint 1967, pp 355–360.

339. Radó T (1947) The Isoperimetric Inequality and the Lebesgue definition of surface area. Trans. Amer. Math. Soc. 61:530–555.

340. Carathéodory C (1949) *Erhard Schmidt: Die Brunn-Minkowskische Ungleichung und ihr Spiegelbild sowie die isoperimetrische Eigenschaft der Kugel in der euklidischen und nichteuklidischen Geometrie. I* (Erhard Schmidt: The Brunn–Minkowski Inequality and its Mirror Image as well as the Isoperimetric Property of the Sphere in Euclidean and non-Euclidean Geometry. I). *Math. Nachr.* 1, Berlin 1948, pp 81–157. *Zentralbl. f. Math.* 30:76–77 and *Ges. Math. Schr.* vol V, pp 383–385.

341. Carathéodory to Radó on 2 January 1949. Letter in English. The Ohio State University Archives, Tibor Rado papers (Record Group R6 40/76/1), "Caratheodory, C. (Professor): 1948–1949".

342. In the preface to his book *Length and Area* (as note 193), Radó writes that the theory of length and area which he presents is based upon fundamental ideas of Lebesgue and Geöcze.
Radó's impulse to start working on a book on length and area came from an invitation by the American Mathematical Society to give the Colloquium Lectures in 1942. When he went to the Institute for Advanced Study in September 1944 he started revising his manuscript. Courtesy of the Archives of the Institute for Advanced Study.

343. Radó refers to chapter III.2 of his book (pp 191–220) entitled "Bounded Variation and Absolute Continuity".

344. Radó to Carathéodory on 10 January 1949. Letter in English. The Ohio State University Archives, Tibor Rado papers (Record Group R6 40/76/1), "Caratheodory, C. (Professor): 1948–1949".

345. Carathéodory to Radó on 1 February 1949. Letter in English. Ibid.

346. Carathéodory to Radó on 5 February 1949. Letter in English. Ibid.
"At the time of this correspondence, research on general continuous surfaces in 3-dimensional space was very active. Radó used the Lebesgue definition of surface area while Besicovitch used a different definition. For 'highly irregular' surfaces the definitions may disagree. The correspondence indicates Carathéodory's views on some of issues in the

Besicovitch-Radó controversy.", Communication to the author from Wendell Fleming on 4 June 2002.

347. Rademacher H (1916) *Eindeutige Abbildungen und Meßbarkeit. Monatshefte f. Math. u. Phys.* 27:183–290.

348. Carathéodory C, Rademacher H (1917) *Über die Eindeutigkeit im Kleinen und im Grossen stetiger Abbildungen von Gebieten. Archiv d. Math. u. Phys.*, 3rd series, 26:1–9 and *Ges. Math. Schr.* vol IV, pp 278–287.
   Rademacher's broad cultural interests and his disappointment due to Klein's lectures had driven him directly to the study of philosophy with Leonhard Nelson. He himself attributed his return to mathematics to Courant's influence.
   See: Grosswald E (ed) Collected Papers of Hans Rademacher, 2 vols. MIT Press, Cambridge, MA, 1974, here vol I, p XIII.

349. Cartan E (1934) *Les espaces de Finsler*. Hermann et Cie, Paris.
   A Finsler space is a differentiable manifold supplied with a metric that can be given by a real positive-definite convex function $F(x, y)$ of co-ordinates of x and components of contravariant vectors y acting at the point x.
   M. I. Voĭtsekhovskiĭ, Encyclopaedia of Mathematics.

350. Young, as note 187, p 242.

351. Humboldt-Universität zu Berlin – Archiv –, Bestand „Universitätskurator-Personalia C9", Bl 1, 2.

352. Humboldt-Universität zu Berlin – Archiv –, Bestand „Philosophische Fakultät 1810 bis 1945", Nr 1467, Bl 287R.
   Cf. also: Biermann (1988), pp 182–186.

353. The gradual development of functional analytic technique subsequently made possible to extend Schur's techniques to important groups containing infinitely many elements, including the groups of three-dimensional and of higher-dimensional rotations, and of the Lorentz group, which is important in all relativistic physical theories.

354. Fraenkel (1967), p 120.

355. Humboldt-Universität zu Berlin – Archiv –, Bestand „Philosophische Fakultät 1810 bis 1945", Nr 1467, Bl 275R.

356. Ibid, Bl 277.

357. Ibid, Bl 277 & 277R.

358. Ibid, Bl 275R & 276.

359. Ibid, Bl 279R & 280. Cf. also: Biermann (1988), pp 328–333. When Schottky retired, his full professorship was given to Schur (on 21 May 1921), who had been a full professor *ad personam* in Berlin since 29 December 1919.
   At the third general session of the Prussian Academy of Sciences on 12 January 1933, Issai Schur submitted Carathéodory's paper: (1933) on *Die Kurven mit beschränkten Biegungen* (Curves with Bounded Curvatures) (*Sitzungsber. Preuß. Akad. d. Wiss. Berlin, math.-nat. Kl.* 1933, pp 102–125 and *Ges. Math. Schr.* vol II, pp 65–92). On the same day Schur wrote to Carathéodory: "I have read it with great interest and I believe I understood the aim and character of this investigation. I find that your great presentation skill has proved to be of particular worth. I have formulated the short resumé that had to be attached as follows: the work is connected with a study begun by H. A. Schwarz and continued by Herr Erhard Schmidt in 1925. In particular, it is shown that a theorem proved by Schmidt for the curves described here as curves with bounded curvatures can be used in the clear description of this class of curves. I hope that you will receive a correction of this advance notice so that you will be able to possibly make changes."
   Schur to Carathéodory on 12 January 1933. *Teilnachlaß Carathéodory*, as note 74, chapter 1.
   In the paper: (1925) *Über das Extremum der Bogenlänge einer Raumkurve bei vorge-*

schriebenen Einschränkungen ihrer Krümmung (On the Extremal of the Arc Length of a Space Curve in the Case of Given Limitations of its Curvature). *Sitzungsber. d. Preuß. Akad. d. Wiss. Berlin* 1925, pp 485–490, Schmidt had given a simple and complete proof of Schur's theorem stating that in every *Verwindung* (a twisting that preserves curvature) of an arc on the plane the distance of its end points becomes longer.

See: Leichtweiß K: *Arbeiten Blaschkes zur Differentialgeometrie im Großen und zur Riemannschen Geometrie* (Blaschke's Works in Differential Geometry in the Large and in Riemannian Geometry). In: Burau W et al (eds) *Wilhelm Blaschke, Gesammelte Werke* (Collected Works), vol 6: *Selecta zusammengestellt von Kurt Leichtweiß* (Selected Papers Compiled by Kurt Leichtweiß). Thales-Verlag, Essen, 1986, pp 79–82, here p 79.

Schmidt's main result reads: for an arc of a curve with bounded curvature, whose length S is smaller than 2π, the relation $r \geq 2 \sin S/2$, in which r denotes the distance between the end points of the arc, is always true. Its "very clear geometric interpretation" according to Carathéodory reads: the chord r of our curve is never smaller than the chord of an equally long arc of unit radius. The equality sign is valid only when the curve is congruent with an arc of unit radius. Carathéodory showed in his own work that Schmidt's theorem can be reversed.

360. The passage is taken from the section dealing with the education of the Guardians who would receive their main mathematical training between the ages of 20 and 30, after spending two or three years for the study of music and gymnastics and as a preliminary to a five-year study of dialectics. The speakers in the dialogue are Socrates and Glaucon.

See: Goold G P (ed) Selections illustrating the history of Greek Mathematics, with an English translation by Ivor Thomas, vol I: From Thales to Euclid. The Loeb Classical Library, Harvard UP, Cambridge, MA, London, 1998 ([1]1939), pp 9, 11.

361. Carl Heinrich Becker (Amsterdam 12 April 1876–Berlin 10 February 1933), orientalist and Prussian politician. He followed Middle Eastern and oriental studies and, after long trips to Italy, Spain, Greece and the Near East, submitted his habilitation in Heidelberg in 1902. In 1908, he was appointed professor at the Colonial Institute in Hamburg, in 1913 at the University of Bonn. In 1916 he joined the Ministry of Education in Berlin as an adviser on staff matters for the Prussian Universities. Appointed permanent secretary of this ministry in April 1919, Becker was active as a permanent secretary or minister (1921 and 1925–1930 Prussian Minister of Education and Cultural Affairs) in Prussian and German culture politics without any interruption until 1930. In the autumn of 1926, he founded the German Academy of Poets which comprised the third section of the Academy of Arts, in addition to the sections of Fine Arts and Music. After his resignation in 1930 he resumed his academic activity as a full professor at Berlin University. Lecture tours brought him especially to Anglo-Saxon countries. As member of an international committee sent by the League of Nations to China for the study of the Chinese educational system, he spent there three months in the autumn of 1931. Wende (1959), p 247 and p 331.

362. Tobies R (1987) *Die Berufungspolitik Felix Kleins* (Felix Klein's Appointment Politics). In: *NTM-Schriftenr. Geschichte Naturwiss., Technik, Med.* 24/2, Leipzig, pp 43–52, here p 46.

363. Carathéodory C (1919) *Die Bedeutung des Erlanger Programms. Die Naturwissenschaften* 7 (special issue on Klein's 70th birthday): 297–300 and *Ges. Math. Schr.* vol V, pp 45–51.

364. Ibid, *Ges. Math. Schr.* vol V, pp 49–51.

365. Carathéodory C (1926) *Über Flächen mit lauter geschlossenen geodätischen Linien und konjugierten Gegenpunkten. Abhandl. Math. Sem. Hamb. Univ.* 4: 297–312 and *Ges. Math. Schr.* vol V, pp 3–21.

366. Carathéodory to Hilbert on 14 April 1926, as note 256.

367. A Liouville surface is a surface for which the equations of the geodesics admit a quadratic integral $a_{ij}du^i du^j$, where the tensor a_{ij} is different from the metric tensor g_{ij} of the surface.
     I. Kh. Sabitov, Encyclopaedia of Mathematics.

368. Paul Funk had gained his doctorate under Hilbert on 22 May 1911 with a dissertation *Über Flächen mit lauter geschlossenen geodätischen Linien* (On Surfaces with Nothing but Closed Geodesic Lines).

369. *Ges. Math. Schr.* vol V, p 5.
     The calculus of variations in the large, which makes use of topological concepts and methods to prove the existence and estimate the number of extremals, to study certain qualitative properties of the extremals and the relations between extremals of various types, crystallised in the third decade of the 20th century out of the attempt to estimate the number of closed geodesics on a closed Riemannian and more generally a Finsler manifold.
     See: D.V. Anosov, Encyclopaedia of Mathematics.

370. Toepell (1996), p 296.

371. Burau W (1969) *Mathematik* (Mathematics). In: *Universität Hamburg 1919–1969* (University of Hamburg 1919–1969). Selbstverlag der Universität Hamburg, Hamburg, p 255.

372. Carathéodory C (1935) *Einfache Bemerkungen über Nabelpunktskurven. Festschrift 25 Jahre Technische Hochschule Breslau*, pp 105–107 and *Ges. Math. Schr.* vol V, pp 26–30.

373. Monge G (1850) *Application de l'analyse à la géométrie*. Bachelier, Paris.

374. Hamburger H (1940) *Beweis einer Vermutung von Carathéodory* (Proof of an Assumption by Carathéodory), part I. *Ann. of Math.* 41:63–86 and Hamburger H (1941) *Beweis einer Vermutung von Carathéodory*, parts II, III. *Acta Math.* 73:174–332.

375. Bol G (1944) *Über Nabelpunkte auf einer Eifläche* (About Umbilic Points on an Ovaloid). *Math. Z.* 49:389–410.

376. Blaschke W (1942) *Sugli ombelichi d'un ovaloide* (About Umbilic Points of an Ovaloid). *Atti Conv. Mat. Roma* 1942, 201–208.

377. Leichtweiß, as note 359, here p 80.

378. Blaschke W (1945) *Vorlesungen über Differentialgeometrie, I: Elementare Differentialgeometrie* (Lectures on Differential Geometry, I: Elementary Differential Geometry). Springer, Berlin, pp 49, 64, 214, 224, 231, 232, 234.

379. Erich Bessel-Hagen, *Über eine Art singulärer Punkte der einfachen Variationsprobleme in der Ebene*, submitted on 27 August 1920.

380. Hans Rademacher (1919) *Über partielle und totale Differenzierbarkeit von Funktionen mehrerer Variabeln und über die Transformation der Doppelintegrale* (submitted on 15 December 1919). *Math. Ann.* 79:340–359.

381. Hans Hamburger, *Über eine Verallgemeinerung des Stieltjesschen Momentenproblems*, submitted on 16 April 1919.
     Hamburger had gained his doctoral degree under Pringsheim in 1914 with the dissertation *Über die Integration linearer homogener Differentialgleichungen* (On the Integration of Linear Homogeneous Differential Equations).
     The trigonometric moment problem can be formulated as follows: let c_n be a sequence of complex constants, $c_{-n} = \bar{c}_n$. What are the necessary and sufficient conditions in order that a distribution function $a(x)$ in $[-\pi, \pi]$ exists for which the equations of its Fourier-Stieltjes coefficients
     $$c_n = (2\pi)^{-1} \int_{-\pi}^{\pi} e^{-inx} da(x), \quad n = 0, \pm 1, \pm 2, \dots$$
     hold.

For the trigonometric moment problem see: Grenander, Szegö, as note 113, here pp 19–20.

382. Siegmund-Schultze R (1998) *Mathematiker auf der Flucht vor Hitler – Quellen und Studien zur Emigration einer Wissenschaft* (Mathematicians Fleeing from Hitler – Sources and Studies in the Emigration of a Science). *Dokumente zur Geschichte der Mathematik,* vol 10. Vieweg, Braunschweig, Wiesbaden, p 123.
Die Mathemat'sche Gesellschaft, Dem Remak mißfällt, Warum die Bonzen nie erklär'n, Wozu kriegen's ihr Geld? (Remak dislikes, the Mathematical Society, the bigwigs never say, what for they get their money). Poem by Margarethe Goeb, as note 242.

383. Cf. Biermann (1988), pp 182–186.

384. Carathéodory to Hilbert on 10 October 1918. Letter in German. Niedersächsische Staats- und Universitätsbibliothek Göttingen – Abteilung für Handschriften und seltene Drucke. Cod. Ms. D. Hilbert 55.

385. Cf. Wehler, as note 273.

386. Carathéodory to Klein on 17 November 1918. Letter in German. Niedersächsische Staats- und Universitätsbibliothek Göttingen – Abteilung für Handschriften und seltene Drucke. Cod. Ms. F. Klein 8, 467.

387. He means the *Zeitschrift für Mathematik und Physik* founded around 1856 by Oskar Xavier Schlömilch (1823–1901). Schlömilch taught at Jena and Dresden. Causchy's techniques in analysis became well known in Germany through his textbook. In 1847 he gave a general remainder formula for the remainder in Taylor series. He discovered an important series expansion of an arbitrary function in terms of Bessel functions in 1857. Klein's appointment as a professor of geometry in Leipzig in the winter semester 1880–1881 is attributed to Schlömilch, who, in the mid 1870s, left the Dresden Polytechnic for a post at the Saxony Ministry of Education and Cultural Affairs. Cf. Tobies (1981), p 47.

388. *Albert Einstein – Hedwig und Max Born. Briefwechsel 1916–1955* (1972), p 155.

389. In this way this traumatic experience is impressed in Despina's memory.

390. Hartkopf W (1992) *Die Berliner Akademie der Wissenschaften – Ihre Mitglieder und Preisträger 1700–1990* (The Berlin Academy of Sciences – Its Members and Prize Winners 1700–1990). Akademie Verlag, Berlin. See also: Thiele, as note 242 and Zentrales Archiv der Akademie der Wissenschaften der DDR, Historische Abteilung, Abschnitt II: Akten der Preußischen Akademie der Wissenschaften 1812–1945. Titel: Personalia, Mitglieder. Signatur: II-III, 37.

391. Carathéodory, as note 93, chapter 1, here p 181f.

392. Ibid, p 186.

393. Mancosu (1998), p 4.

394. Brouwer (1921) *Intuitionistische verzamelingleer* (Intuitionist Set Theory). *KNAW Verslagen* 29:797–802. Paper presented to the Royal Dutch Academy of Sciences on 18 December 1920. See: Mancosu (1998), pp 23–27.

395. Peckhaus, as note 61, chapter 1, here p 45.

396. Brouwer, *Begründung der Mengenlehre unabhängig vom logischen Satz vom ausgeschlossenen Dritten. Erster Teil: Allgemeine Mengenlehre* (Foundation of Set Theory Independently of the Logical Theorem of the Excluded Middle. First Part: General Set Theory). *KNAW Verhandelingen, 1e Sectie, deel XII,* no. 5, pp 1–43. Paper presented to the Royal Dutch Academy of Sciences in November 1917.
Brouwer, *Begründung der Mengenlehre unabhängig vom logischen Satz vom ausgeschlossenen Dritten. Zweiter Teil: Theorie der Punktmengen* (Foundation of Set Theory Independently of the Logical Theorem of the Excluded Middle. Second Part: Theory of Point Sets). *KNAW Verhandelingen, 1e Sectie, deel XII,* no. 7, pp 1–33. Paper presented to the Royal Dutch Academy of Sciences in October 1918.

397. Brouwer (1919) *Intuitionistische Mengenlehre* (Intuitionist Set Theory). *Jber. Deutsch. Math.-Verein.* 28:203–208. The German version of Brouwer, as note 394.

398. Carathéodory to Klein on 1 September 1919, as note 255.

399. Brouwer und Weyl, "have the common intention to solidly found part of analysis, to sacrifice the rest, further, they commonly take the view that only the real constructability of mathematical objects should establish their logical, free of contradictions existence. Weyl's book *Das Kontinuum* was released in 1918. Already early, Brouwer had rejected the applicability of the theorem of the Excluded Middle in matters concerning an infinity of things and with great power started, likewise around 1918, to build up those parts of analysis which can be founded without the application of this theorem. But Hilbert rejects every sacrifice of mathematical goods."
O. Blumenthal (1935) *Lebensgeschichte* (Life-Story). In: *David Hilbert – Gesammelte Abhandlungen*, vol 3, pp 388–429, here p 423.
In 1922, Hilbert indeed turned against Brouwer's and Weyl's critique to classical mathematics: "What Weyl and Brouwer do, amounts in principle to following in the footsteps of Kronecker! They seek to ground mathematics by throwing overboard all that which is troublesome and set up a dictatorship à la Kronecker. But this means that they would chop up and mutilate our science. If we were to follow such reformers, we would run the risk of losing a great part of our most valuable treasures."
Hilbert D (1922) *Neubegründung der Mathematik* (New Founding of Mathematics). *Abhandl. Math. Sem. Hamb. Univ.* 1, pp 157–177.

400. Weyl H (1921) *Über die neue Grundlagenkrise der Mathematik* (On the New Crisis Concerning the Foundations of Mathematics). *Math Z* 10:37–79.
Weyl H (1918) *Das Kontinuum. Kritische Untersuchungen über die Grundlagen der Analysis* (The Continuum – A Critical Examination of the Foundations of Analysis). Veit, Leipzig.

401. Weyl, as note 83, chapter 1, p 639.

402. Biermann (1988), pp 190–192.

403. Fraenkel (1967), p 165.

404. Born (1978), p 93 and p 95.

405. Born (1978), p 93.

406. Cf. Wenzel U J, *Denken in Beziehungen – Die ersten Bände der Simmel-Gesamtausgabe* (Thinking in Relations – The First Volumes of the Complete Simmel Edition). *Neue Zürcher Zeitung* no. 167, 21/22 July 1990.
Oesterle K, *Die Vereinheitlichung der Welt – Zurück zu Georg Simmel: zwei Bände der Gesamtausgabe, ein Porträt des jungen Philosophen* (The Standardisation of the World – Back to Georg Simmel: two Volumes of the Complete Edition, a Portrait of the Young Philosopher). *Süddeutsche Zeitung*, 19 March 1997.

407. Nelson L (undated) *Kritik der praktischen Vernunft*. Verlag „Öffentliches Leben", Göttingen. Also: 1917, Verlag von Veit & Comp. This book was the first volume of a several-volumed work concerning the foundations of ethics that was published under the title *Vorlesungen über die Grundlagen der Ethik* (Lectures on the Foundations of Ethics). It also appeared as volume 4 (1972) of: Paul Bernays et al. (eds) *Gesammelte Schriften in neun Bänden* (Collected Texts in Nine Volumes) Felix Meiner Verlag, Hamburg, 1970–1977.

408. Georg Elias Müller, the co-founder of experimental psychology, was forced to retire because of his advanced age in 1922. Müller had made his own mark by studies in psychophysics, the psychology of memory and thought and the theory of colour.

409. Unless otherwise stated, the account of the "Nelson affair" is based on V. Peckhaus, as note 61, chapter 1, and Peckhaus V (1990) *Hilbertprogramm und Kritische Philosophie – Das Göttinger Modell interdisziplinärer Zusammenarbeit zwischen Mathematik und*

Philosophie (Hilbert's Programme and Critical Philosophy – The Göttingen Model of Interdisciplinary Co-operation between Mathematics and Philosophy). Vandenhoeck & Ruprecht, Göttingen, esp. chapter 6: *Hilbert und die Philosophie: Sein Engagement für Leonhard Nelson* (Hilbert and Philosophy: his Engagement for Leonhard Nelson), pp 196–224. The event with the investigation against Nelson in May 1915 is mentioned in note 621, p 212.

410. Carathéodory to Hilbert on 10 October 1918, as note 384.

411. Wittwer W W: *Carl Heinrich Becker.* In: Treuer W, Gründer K (eds) *Berlinische Lebensbilder 3. Wissenschafts-Politik in Berlin – Minister, Beamte, Ratgeber* (Berlin Biographies 3. Science Politics in Berlin – Ministers, Civil Servants, Counsellors). Einzelveröffentlichungen der Historischen Kommission zu Berlin vol 60, Colloquium Verlag, Berlin, 1987, pp 251–267, here p 254.

412. Wende (1959), p 108.

413. Cf. Beyerchen (1980), pp 150–151.

414. Wende (1959), p 99f.

3 *The Asia-Minor Project*

1. Humboldt-Universität zu Berlin – Archiv –, Bestand „Universitätskurator-Personalia C9", Bl 4.

2. The long distances of the city lost him much time and power: Kalitsounakis's impression of how Carathéodory had felt in Berlin. Vovolinis (ed) (1962), p 484.

3. Michalakopoulos A (1911) *Το Πανεπιστήμιον* (The University). *Παναθήναια* vol IA. Mentioned in: Kyriazopoulos B D (1976) *Τα Πενήντα Χρόνια του Πανεπιστημίου Θεσσαλονίκης 1926–1976* (Fifty Years University of Thessaloniki 1926–1976). Privately printed, Thessaloniki, note 2 on p 213.

4. Northern Greece through the Balkan Wars in 1912–1913.

5. At that time, the University of Athens consisted of the National University and the Capodistrian University.

6. Kyriazopoulos, as note 3, p 42.

7. *Εφημερίς των συζητήσεων της Βουλής, Περίοδος Β΄, Συνεδρίασις 17–20 Δεκεμβρίου 1929* (Paper of Parliament Debates, Second Period, Session of 17–20 December 1929), p 276.

8. Stephanou St (ed) *Ελευθερίου Βενιζέλου – Πολιτικαί υποθήκαι ανθολογηθείσαι από τα κείμενά του* (Eleutherios Venizelos's Political Advices Selected from his Texts), vol I and vol. II. Privately printed, Athens, 1965 and 1969, here vol I, p 296.

9. For information on the Treaty of Sèvres see: Major Peace Treaties of Modern History 1648–1967 (1967), pp 2080ff.

10. Housepian (1972), p 106.

11. G. Deschamps (1907), pp 135–138.

12. Carathéodory to Delta on 17 June 1928. Letter in Greek. Leukoparidis (ed) (1956), p 475.

13. Horton (1985), p 90.
George Horton had worked as Consul and Consul General of the United States in the Near East for thirty years. A poet, literary critic, journalist of the *Chicago Herald* and author of two best-selling novels, he had been offered a consular post in Berlin which he rejected for one in Greece and was appointed Consul to Athens in 1893. In 1909 he was transferred to Thessaloniki and in 1911 he was appointed US Consul in Smyrna. From the outbreak of World War I until 1917, when Turkey broke diplomatic relations with the USA, Horton represented all Allied interests in Smyrna. He was Consul General of the

USA in Smyrna in the period 1919–1922. Horton spoke fluently the same six languages as Carathéodory, namely English, French, German, Italian, Greek and Turkish, and, just like Carathéodory, he was thoroughly familiar with the history of the Near East.

14. Solomonidis (1962), p 22 and pp 202–222.

15. Vellay Ch (1919) *Smyrne ville grecque* (Smyrna, a Greek City). Librairie Chapelot, Paris, p 24.

16. Carathéodory to Klein on 1 September 1919, as note 255, chapter 2.

17. The Greek Minister of Education and Religious Affairs to the rector's office of the Athens University on 2 June 1920 (old calendar) and protocol no. 10550. The document is signed by D. Glinos. BAyHStA MK 35403.

18. Faculty session of 23 October 1920 (old calendar). Πρακτικά Συνεδριών Σχολής Φυσικών και Μαθηματικών Επιστημών (Minutes of Mettings of the School of Natural and Mathematical Sciences), p 246f.

19. Memorandum in Greek in: Vovolinis (ed) (1962), pp 484–489.

20. *Ex oriente lux. Gespräche eines Meisters mit seinem Schüler über wesentliche Punkte des urkundlichen Christentums. Berichtet vom Schüler selbst Georg Jacob Aaron, cand. sacr. theol. Erstes Gespräch* (Ex Oriente Lux – Conversations of a Master with his Disciple about Fundamental Points of Documentary Christian Faith, Reported by the Disciple Himself, Georg Jacob Aaron, cand. sacr. theol. – First Conversation) was the title of a book edited by Georg Cantor in Halle in 1905. But here *Ex Oriente Lux* suggests the "civilising mission" of the Greeks, "to one day pass on the torch with the light of European civilisation to Asia and even farther away" (State Councillor Georg Ludwig von Maurer, 1835); "the inspiration of the Orient" (Prime Minister Colettis, 1844); "the inspiration of Europe or Asia" (poet and journalist Panagiotis Soutsos, 1846). *Ex Oriente Lux* was the peaceful interpretation of the Great Idea, alias expressed in the Greek national rhetoric as "the Orient through the Orient".

21. See for instance: Dimaras K: Η ιδεολογική υποδομή του νέου ελληνικού κράτους – Η κληρονομιά των περασμένων, οι νέες πραγματικότητες, οι νέες ανάγκες (The Ideological Foundation of the Modern Greek State – The Heritage of the Past, the New Realities, the New Needs). In: Christodoulos G, Bastias I, Simopoulos K, Daskalopoulou Chr (eds) Ιστορία του Ελληνικού Έθνους (History of the Greek Nation), vol 13. Εκδοτική Αθηνών, Athens, 1977, pp 455–484.

22. So, in 1897, Lewis Sergeant expected a "civilising role in the East" from the Greeks and acknowledged "a strong claim for the extension of their influence and authority".
Sergeant L (1897) Greece in the Nineteenth Century – A Record of Hellenic Emancipation and Progress: 1821–1897. Fisher Unwin, London, p 351.
François Lenorman realised that "wherever trade, industry and civilisation have attained a certain degree of development in Eastern countries, the honour of the fact belongs to the Greeks".
Lenorman F, *La Grèce et les Iles Ioniennes*, p 5. Cited in Sergeant, ibid, p 359.
George Horton remarked that "Smyrna will grow great again when a live and progressive western civilization once more develops in Ionia. History has demonstrated that the Greeks, from their geographical position, their industrial and economic enterprise, and their relative maritime supremacy in the Mediterranean are the people ultimately destined to carry European progress into Asia Minor."
Horton (1985), p 94.
In Versailles, the British Prime Minister Lloyd George supported Venizelos's territorial demands for Eastern Thrace and, on that occasion, he explicitly referred to the expected Greek contribution to the expansion of civilisation in that region.
Werner Zürrer (1976) *Die ‚Griechische Frage' auf den Friedenskonferenzen von 1919/20* (The 'Greek Question' at the Peace Conference of 1919/20). *Südost-Forschungen* 35: 183–246, here p 220.

Jean Gout, member of the French delegation at the Versailles Peace Conference, emphatically argued in favour of the Hellenisation of Asia Minor "in order that a natural development of civilisation is guaranteed".
Werner Zürrer, ibid, p 189f.

23. Carathéodory C (1926) The Urban Refugees in Macedonia and Thrace, Chapters I, II, and IV, here Chapter I, p 9 (English translation of Carathéodory's text). Carathéodory, writings. In: The Papers of Henry Morgenthau, Sr., container 29, reel 24 of the microfilm edition, subject file, 1868–1939. The Library of Congress, Manuscript Division.

24. 76th session of the 4th constitutional assembly. Mentioned in: Kyriazopoulos, as note 3, here p 44.

25. In 1920 Glinos worked out the project for the direct foundation of an *Ελεύθερο Πανεπιστήμιο* (Free University) in Athens that would counterbalance the dark activity of the state university. Financial reasons hindered its operation but in its place the *Ανωτέρα Γυναικεία Σχολή* (Superior Female School), "a light point within this Middle Ages darkness", as Penelope Delta said, was inaugurated in 1921. In 1924, Glinos rejected his appointment to rector of the University of Thessaloniki for ten years, which had been ordered by special decree. He himself had written the university regulations but considered that he could not accommodate his lecturing within the given frame. See: Iliou Ph, *Δημήτρης Γληνός: Από την εκπαιδευτική μεταρρύθμιση στην κοινωνική επανάσταση* (Dimitris Glinos: From the Educational Reform to Social Revolution). *Τα Νέα*, 8 February 2000.

26. Citation from a speech by Colettis mentioned in: Markezinis S (1966) *Πολιτική Ιστορία της Νεωτέρας Ελλάδος* (Political History of Modern Greece). "Πάπυρος", Athens, vol 1, p 208.
About Venizelos's intentions as regards Constantinople see:
Stephanou (ed), as note 8, here vol II, p 161.
Mazarakis-Ainian I: *Ο Ελευθέριος Βενιζέλος και οι εθνικές μας διεκδικήσεις στη Συνδιάσκεψη Ειρήνης, Α′ εξάμηνο 1919* (Eleutherios Venizelos and our National Claims at the Peace Conference, A′ Semester 1919). In: Εταιρεία Ελληνικού Λογοτεχνικού και Ιστορικού Αρχείου και Μουσείο Μπενάκη (eds) *Συμπόσιο για τον Ελευθέριο Βενιζέλο – Πρακτικά* (Symposium on Eleutherios Venizelos – Files). Privately printed, Athens, 1988, pp 245–264.
Gibbon, H A (1920) Venizelos. Mifflin, Boston, New York, p 345.
Richter H (1990) *Griechenland im 20. Jahrhundert*. Bd. 1: *Megali Idea – Republik – Diktatur: 1900–1940* (Greece in the 20th Century. Vol 1: Great Idea – Republic – Dictatorship: 1900–1940). Romiosini, Köln, p 49f.
Venizelos believed that if Greece were established in Smyrna, she would be able to hope for an annexation of Constantinople. However, if Constantinople were to be ceded to Greece, then Greece would have to resign from claims on Ionia.
See: Petsalis-Diomidis N: *1919: Τη Σμύρνη ή την Πόλη; Μια εναλλακτική λύση που ο Βενιζέλος απέρριψε μάλλον βεβιασμένα* (1919: Smyrna or Polis? An Alternative Solution which Venizelos Rejected rather Hastily). In: Dimitrakopoulos O, Veremis Th (eds) *Μελετήματα γύρω από τον Βενιζέλο και την εποχή του* (Studies on Venizelos and his Time). Stratis G Philippotis, Athens, 1980, pp 101–118, here p 105. Also: "Adrianople and Smyrna are stepping stones to Constantinople". Gibbon, op cit, p 376.

27. On the work of the High Commission in the vilayet of Aydin, see two very well informed dissertations:
a) Solomonidis V (1984) Greece in Asia Minor: The Greek Administration of Aidin, 1919–1922. Thesis submitted for the degree of PhD, King's College, University of London.
b) Theodorou G (1991) *Η άλλη όψη της μικρασιατικής εμπειρίας* (The Other Side of the Asia-Minor Experience). Thesis submitted for the degree of PhD, Philosophical Faculty, University of Thessaloniki.

28. Smith (1973), p 94.
29. Markezinis, as note 26, here vol I, Athens 1973, p 305.
30. Information from Mrs. Despina Rodopoulos-Carathéodory.
31. Housepian (1972), p 104 and Horton (1985), p 92.
32. Carathéodory was employed on 28 October (old calendar) 1920 with the High Commissioner's resolution no 34123/119/12711/12720.
   The text of the contract signed one day earlier reads:

   ### CONTRACT

   1) Between his Excellency, the High Commissioner of Smyrna and Herr C. Carathéodory, former full professor at the University of Berlin, following contract is agreed.

   2) Herr Carathéodory was called from Germany as organiser of the now being founded Smyrna University for the period of five years and took on his managing activity on 15 July 1920 [old calendar]. At the same time he was appointed full professor for mathematics at the university for an unlimited period.

   3) Since 15 July 1920 [old calendar] Herr Carathéodory is entitled to a monthly compensation of 4000 drachmas, independently of the moment of the begin of his lecturing activity. Both the professor's salary and the benefit allowance for the office of the rector, as soon as he will occupy it, are included in this compensation.

   4) Herr Carathéodory will obtain this income also for the whole period of his respective absence in Europe or elsewhere, as long as this concerns a mission connected with the organisation of the University. In implementation of the respective instructions of the High Commissioner, an additional daily compensation for travel expenses will be fixed.

   5) Herr Carathéodory is entitled to reimbursement of the half of the expenses for the transport of his furniture and books from Germany to Smyrna, as well as of the expenses for the travel of his family.

   This has been written in duplicate and signed by both sides.

   Smyrna, 27 October 1920 [old calendar]

	The High Commissioner of Smyrna
C. Carathéodory	A. Stergiadis

33. At the session of the Faculty of Mathematics and Natural Sciences on 16 December 1920 (old calendar), the dean announced the rector's document no. 2035 of 11 December 1920 (old calendar) according to which the Ministry of National Education and Religious Affairs had revoked by decree the dismissals of professors Theodoros Skoufos and Georgios Athanasiadis and the appointments of professors D. Aeginitis (professor for astronomy and meteorology since September 1896 and director of the astronomy laboratory and the National Observatory), P. Zervos, G. Papanikolaou, I. Politis (professor for botany since 1918 and director of the botanical laboratory and museum), K. Maltezos (professor for physics since 1918), C. Carathéodory and of the associate professor N. Sakellariou.
   Πρακτικά Συνεδριών Σχολής Φυσικών και Μαθηματικών Επιστημών, as note 18, p 250.
   In 1918–1920, Glinos had worked out the report for the purge against the anti-Venizelists at the university and, from his post of General Secretary of the Ministry of Education and President of the Educational Council, had attempted to promote the reform in primary education and in language teaching. A law allowing the use of demotic Greek in primary schools had been passed in 1917.
   Iliou, as note 25 and Dimaras A (ed) *Η μεταρρύθμιση που δεν έγινε* (The Reform that was not Carried Out). "Ερμής", Athens, 1974.
   After the electoral defeat of the Liberals on 14 November 1920, Venizelos and Benakis left Athens for Nice.
34. Toynbee (1970), p 203.
35. Horton (1985), p 78f.

36. Carathéodory to Klein on 11 March 1921. Letter in German. Niedersächsische Staats- und Universitätsbibliothek Göttingen – Abteilung für Handschriften und seltene Drucke. Cod. Ms. F. Klein 8, 468A.
37. Carathéodory to Klein on 17 November 1918, as note 386, chapter 2.
38. Toynbee (1970), preface to the second edition, p x.
39. Toynbee (1970), p 166.
40. The relevant files of the year 1914 concerning these expulsions exist in the Archive of the Greek Foreign Ministry and bear the register no A/21.
41. "Tuesday, 10/23 March 1915
    [...] We will really bring Germany into discord, but its victory disastrous for the Greeks, anyway. Plan for liquidation of the Greeks. Statements (by the way denied, after all unimportant in themselves) by Liman von Sanders to some kaimakam about profitable liquidation of the Greek element [...]"
    Streit G v (1965) Ημερολόγιον-Αρχείον (Diary-Archive) vol II-A. Privately printed, Athens.
42. Smith (1973), p 100.
43. Toynbee (1970), p 168.
44. Initiative of the Union of Professors at the Physical Schools of the Superior Educational Institutions (ed) Δημήτριος Χόνδρος, Καθηγητής Πανεπιστημίου Αθηνών 1882–1962 – Αφιέρωμα για τα 100 χρόνια από τη γέννησή του (Dimitrios Hondros, Professor of the University of Athens, 1882–1962 – Dedication on the 100th Anniversary of his Birthday). Privately printed, Athens, 1984, p 12.
45. In 1902 Leon Lichtenstein had joined *Siemens & Halske* as an electrical engineer.
46. Carathéodory to Stergiadis on 29 June [12 July] 1921. Letter in Greek. Vovolinis (ed) (1962), p 497f.
47. Ausserer was an expert in history and ancillary sciences at the Austrian National Library. Cf. Stummvoll J (ed) (1968) *Geschichte der Österreichischen Nationalbibliothek* (History of the Austrian National Library). Georg Prachner Verlag, Wien, p 631.
48. Carathéodory to Stergiadis on 29 June [12 July] 1921, as note 46.
49. Carathéodory C (1922) *Über eine der Legendreschen analoge Transformation. Δελτίον της Ελληνικής Μαθηματικής Εταιρείας* (Bulletin of the Greek Mathematical Society) 3:16–24. Translated from the Greek by D. Kappos for the *Ges. Math. Schr.* vol I, pp 374–382.
50. Carathéodory C (1922) *Über die kanonischen Veränderlichen in der Variationsrechnung der mehrfachen Integrale. Math. Ann.* 85:78–88 and *Ges. Math. Schr.* vol I, pp 383–395.
51. Carathéodory C (1922) *Über die Reziprozitätsgesetz der verallgemeinerten Legendreschen Transformation. Math. Ann.* 86:272–275 and *Ges. Math. Schr.* vol I, pp 396–400.
52. Solomonidis (1962), p 399.
53. Carathéodory to Stergiadis on 13 September 1921 (probably old calendar). Letter in Greek. Vovolinis (ed) (1962), p 499.
54. G. Ioakimoglou in: Πρακτικά Ακαδημίας Αθηνών (Minutes of the Academy of Athens), vol 25, p 680.
55. Cf. Engin Berber (1993) *Mütareke ve Yunan İşgali Döneminde İzmir Sancağı* (The Sandjak of Izmir During the Armistice and the Greek Occupation). Doctoral thesis, Izmir. Berber uses only Greek sources for the part of his work referring to the Ionian University. Translation of Ioakimoglou's text by Professor Dr. K. Kreiser and the *Lehrstuhl für türkische Sprache, Geschichte und Kultur* (Chair for Turkish Language, History and Culture) of Bamberg University.
56. Toynbee (1970), p 177.

57. Ibid, p 175.

58. How Stergiadis dealt with Turkish education is described in: Toynbee (1970) pp 172–177.

59. Clogg R: *Κιγκς Κόλλετζ, Λονδίνο και Ελλάδα, 1915–1922* (King's College, London and Greece, 1915–1922). In: Veremis Th, Goulimi G (eds) *Ελευθέριος Βενιζέλος – Κοινωνία–Οικονομία– Πολιτική στην Εποχή του* (Eleutherios Venizelos – Society – Economy – Politics in his Epoch). "Γνώση", Athens, 1989, pp 143–167, here p 165f.

60. Carathéodory to Klein on 4 February 1922. Letter in German. Niedersächsische Staats- und Universitätsbibliothek Göttingen – Abteilung für Handschriften und seltene Drucke. Cod. Ms. F. Klein 8, 468B.

61. The Greeks carried out excavations in order to produce evidence for the Greek character of Asia Minor. G. Soteriou, the curator of Byzantine antiquities, headed the excavation works in Ephesus. In the previous years, the British had found the location of Artemis' temple and the Austrians many public buildings of the younger city. The excavations of the Greeks aimed at localising the Byzantine church of Hagios Ioannis Theologos, which was finally found in a very bad condition. Architectonic members of the Artemis temple had been used for its construction.

62. Copy to the author from a letter in possession of Mrs. Despina Rodopoulos-Carathéodory. President of the Young Men's Christian Association of Smyrna was Professor J. Kingsley Birge, Vice-President the Honorable George Horton, Treasurer F. S. McVittie, esq., and General Secretary E. O. Jacob.

63. Horton stated that "it was all ready for business when the Turks burned Smyrna, possessing an installation similar to that of the great universities of Europe, including a good library and complete equipment of appliances. It would never have lacked money or support and would have been at the service of all classes, irrespective of creed or race." Horton (1985), p 76.

64. The myth of Stergiadis's treason is refuted by: Solomonidis V (1989) *Βενιζέλος–Στεργιάδης – Μύθος και Πραγματικότητα* (Venizelos–Stergiadis, Myth and Reality). In: Veremis Th, Goulimi G (eds), as note 59, here pp 475–536. A very good documentation from the Venizelos Archive in Benakis Museum, the Archive of the Greek Foreign Ministry, the Government Gazette, the High Commission of Constantinople, the High Commission of Smyrna, and the Foreign Office, London.

65. In: *Παιδεία και Ζωή*, April 1950. Quoted in: Vovolinis (ed) (1962), p 503.

66. Horton (1985), p 104.

67. Schramm-von Thadden, E: *Griechenland vom Beginn der Dynastie Glücksburg bis zum Frieden mit der Türkei (1863–1923)* (Greece from the Beginning of the Glücksburg Dynasty until the Peace with Turkey (1863–1923)). In: Schieder Th (ed) *Handbuch der europäischen Geschichte* (Handbook of European History), vol 6. Union Verlag, Stuttgart, 1973, pp 610–617, here p 616. About the Allied indifference to the massacre of Smyrna read: Miller H (1958) The Colossus of Maroussi. New Directions, New York.

68. Session of 6 October 1922 (old calendar). *Πρακτικά Συνεδριών Σχολής Φυσικών και Μαθηματικών Επιστημών* 6 October 1922 – March 1925, p 1.

69. Carathéodory to von Kármán on 11 December 1922. Postcard in German. Caltech Archives, Theodore von Kármán Papers, Folder 5.4.
    In his letter Carathéodory added: "What you tell about Abraham is indeed very sad – when did it happen?" This last statement refers to Max Abraham's illness. Abraham was deputising for the professor of physics at the Technical University of Stuttgart up until 1921, when he accepted a chair in Aachen, his first professorship in Germany. But he fell ill of a brain tumour and died in Munich on the way to his new post, at the age of 47 on 16 November 1922.

70. Hondros to Sommerfeld on 14 September 1922. Letter in German. Institut für theoretische Physik der Universität München, Nachlaß Sommerfeld, 1977-28A, 147/1.

In this letter, Hondros also refers to Max Abraham:

"Full of suffering I read about Herr Abraham's disease in your letter, and I feel even more obliged to stand by him as much as I can, since I had associated with him in Göttingen and was even invited by him to attend his lectures about partial differential equations, free of charge. I have immediately submitted a request to the bank consortium to receive foreign currency, a thing connected with insurmountable difficulties. [...] Now, my cousin, Herr Phrixos [Theodoridis], doctor of engineering, who was in Smyrna with Herr Carathéodory as professor of physics, luckily had 8000 marks in change and, together with Herr [Theodoridis], I send them to you for Herr Abraham hoping that I will soon be in the position to repeat my contribution."

71. Themistoklis Sophoulis (1860–1949), an archaeologist from Samos. As revolutionary leader against the semi-autonomous hegemonial regime of Samos, Sophoulis proclaimed the Ἕνωσις (Union) of Samos with Greece in 1912, during the first Balkan War. In May 1914 he was appointed Governor General of Macedonia and in May 1915 he was elected deputy for Samos with the Liberals. During the National Schism he sided with Venizelos. In 1916 he became Minister of the Interior, in the years 1917–1920 he was President of the so-called Revived Chamber. He formed a government on 10 July 1924 but resigned on 8 October that year. Deputy leader of the Liberals since 1928, he became Leader of the Liberal Party after Venizelos's death. Repeatedly Minister and President of the Parliament. During the German occupation of Greece he participated in a resistance group and had contacts with the Allied Headquarters of the Middle East. He was arrested in May 1944 and sent to prison for six months. Twice post-war Prime Minister of Greece, in 1945–1946 and in 1947–1949. Sophoulis is listed by the Grand Lodge of Greece among the free masons belonging to the Hera Lodge.

72. Carathéodory to Hilbert on 25 December 1922. Letter in German. Niedersächsische Staats- und Universitätsbibliothek Göttingen – Abteilung für Handschriften und seltene Drucke. Cod. Ms. D. Hilbert 55.

73. Carathéodory C (1926) *Die Methode der geodätischen Äquidistanten und das Problem von Lagrange. Acta Math.* 47:199–236 and *Ges. Math. Schr.* vol 1, pp 212–248.

74. Carathéodory C, Schmidt E (1923) *Über die Hencky–Prandtlschen Kurven. Z. f. angew. Math. u. Mech.* 3:468–475 and *Ges. Math. Schr.* vol II, pp 339–352.

75. Carathéodory to Frau van der Waerden on 11 December 1943. Letter in German. ETH-Bibliothek Zürich, Wissenschaftshistorische Sammlungen, HS 652:10609.

76. Carathéodory C (1923) *Sui Campi di Estremali Uscenti da un Punto e Riempienti tutto lo Spazio* (On the Fields of Extremals that Start from a Point and Take Up the Whole Space). *Boll. Un. Mat. Ital.* 1:2–6, 2:48–52, 3:81–87 and *Ges. Math. Schr.* vol I, pp 188–202.

77. There he wrote: (1923) *Über die* Enveloppen *der Extremalen eines Feldes in mehrdimensionalen Räumen* (On the Envelopes of Extremals of a Field in Spaces of Several Dimensions). Δελτίο Ελληνικής Μαθηματικής Εταιρείας 4:22–31 and *Ges. Math. Schr.* vol I, pp 203–211). In Mai 1924 he ended his work on The Method of Geodesic Equidistant [Surfaces] and the Problem of Lagrange. Also in 1924, in Kiphisia, he had started to write his work: (1924–25) *Über geschlossene Extremalen und periodische Variationsprobleme in der Ebene und im Raume* (On Closed Extremals and Periodic Variational Problems in the Plane and in Space) *Ann. d. Mat. pura e. appl.* (4) 2:297–320 and *Ges. Math. Schr.* vol II, pp 40–64, which he ended in Munich on 5 February 1925. The examples in the end of this work show that closed extremals which produce an absolute minimum in the usual sense of the word, do not give a minimum when they are considered to be "fibres" of a fibred space.

Finally, his work on real functions: (1924) *Sur les Transformations Ponctuelles* (On Point-like Transformations). Δελτίον της Ελληνικής Μαθηματικής Εταιρείας 5:12–19 and *Ges. Math. Schr.* vol IV, pp 288–295 was written in Kiphisia on 14 January 1924.

78. Anastasiadis (ed) (1973), p 4.

79. Carathéodory to von Kármán on 11 December 1922, as note 69.

80. Carathéodory's programme in: *Επετηρίς του Πανεπιστημιακού Έτους 1923–1924* (Yearbook of the Academic Year 1923–1924). Session of 6 October 1922. *Πρακτικά Φυσικομαθηματικής Σχολής.*

81. Presidents of the executive committee of the Greek Mathematical Society were N. Hatzidakis in 1918–1925, G. Remoundos in 1926-1927, K. Maltezos in 1928, N. Sakellariou from 1929 through Metaxas's dictatorship. Since 1919, the Society was publishing the *Bulletin*, which aimed at the communication among mathematicians, the promotion of mathematics, and the promotion of educational matters. It was able to publish works of Blaschke, Alfred Rosenblatt, T. Bell, E. Lagrange, A. Hoborski, G. Garcia and organise lectures (R. Archibald had held more than one), publications and a competition for the presentation of original studies on didactics or on pure mathematics.

82. Cf. Fili Chr, *Θεμελιωτές των μαθηματικών στην Ελλάδα του 20ου αιώνα* (Founders of mathematics in 20th-Century Greece). *Τα Νέα*, 14 January 2000.

83. Carathéodory C (1924) *Πρόλογος Αναλυτικής Γεωμετρίας Ν. Σακελλαρίου.* Blazoudaki, Athens. Translated from the Greek as *Vorwort zu dem Buche von Nilos Sakellariou „Elemente der analytischen Geometrie"* (Foreword to N. Sakellariou's book "Elements of Analytical Geometry") by Stephanos Carathéodory for *Ges. Math. Schr.* vol V, pp 187–190.

84. Carathéodory C (1924) *Περί των μαθηματικών εν τη μέση εκπαιδεύσει. Δελτίον της Ελληνικής Μαθηματικής Εταιρείας* 5:1–6. Translated from the Greek by Stephanos Carathéodory as *Über den Mathematikunterricht an den höheren Schulen* (About Mathematical Instruction at Superior Schools) for *Ges. Math. Schr.* vol V, pp 191–197.

85. This is a story told by Euphrosyne's nephew, John Argyris, who has experienced the events as a nine-year old boy.

86. In an article entitled *Herr Caretheodory Lectures Today on Conformal Mapping*, the student paper *The Summer Texan* (Austin, Texas, June 12, 1928) presented following information: "Herr Caretheodory [...] is known as the founder of the famous Smyrna University which was destroyed by fire by the Turks. He has studied Near East political situations and is an authority on the subject."
A Museum of Physical Sciences and Technology has been recently founded as part of the School of Positive Sciences and the Department of Pharmacology of Athens University. It is housed in the chemistry building and includes, among its exhibits, Carathéodory's personal archive and cupboard from the High Commissioner's Office in Smyrna, as well as scientific apparatus from the Ionian University.

87. Bierstadt E H (1997) *Η Μεγάλη Προδοσία – Ο Ρόλος των Μεγάλων Δυνάμεων στη Μικρασιατική Καταστροφή και στη Συνθήκη της Λωζάνης* (The Great Betrayal – The Role of the Great Powers in the Disaster of Asia Minor and in the Treaty of Lausanne). "Νέα Σύνορα", Athens. Original title: (1924) The Great Betrayal. A Survey of the Near East Problem. Hutchinson & Co, London.

88. Zannas (ed) (1988), p 138.

89. Nearing (1923).

90. Quoted from: Delta P (1995) *Περί της ανατροφής των παιδιών μας* (About the Raising of our Children). "Περίπλους", Athens, p 45.

91. The Lausanne Peace Treaty was signed on 24 July 1923 between the Allied Powers and Turkey. The frontier of Turkey with Greece was laid down as following the course of Έβρος (Maritza river). The Greek Government undertook to establish no naval base and no fortification on the islands of Mytilene, Chios, Samos and Ikaria and to forbid the flight over the territory of the Anatolian coast. The islands of Imbros and Tenedos remained under Turkish sovereignty and enjoyed a special administration which guaran-

teed rights for the native non-Muslim population. Turkey renounced all rights and title over Astypalaia, Rhodes, Chalki, Carpathos, Casos, Tilos, Nisyros, Calymnos, Leros, Patmos, Lipsos, Symi and Cos and also over the island of Castellorizo. Turkey recognised the annexation of Cyprus proclaimed by the British Government on 5 November 1914. The exchange of prisoners of war and interned civilians detained by Greece and Turkey respectively formed the subject of a separate agreement between those Powers signed at Lausanne on 30th January 1923. The provisions of an annex regulated the transit and navigation of commercial vessels and aircraft and of war vessels and aircraft in the Straits in time of peace and in time of war.

92. Morgenthau H (1929) I was sent to Athens. Doubleday, Doran & Co, New York, p 89. The whole quotation reads: "Venizelos also spoke with affectionate enthusiasm of Etienne [Stephanos] Delta, the President of the Greek Red Cross, a man of conspicuous strength of mind and benignity of character. 'You can treat him with unlimited confidence', said Venizelos, a remark that my later experience bore out fully, in my many dealings with that sagacious, benevolent, and most loveable philanthropist and patriot."

93. Greiner B (1995) *Die Morgenthau-Legende – Zur Geschichte eines umstrittenes Plans* (The Morgenthau Legend – On the History of a Controversial Plan). Hamburger Edition, Hamburg, p 87.

94. Carathéodory to Einstein on 12 May 1930. Letter in German. The Jewish National and University Library, Jerusalem, Einstein Archive, 8 345-2, 8 345-3.
Permission granted by the Albert Einstein Archives, The Jewish National and University Library, the Hebrew University of Jerusalem, Israel.
Einstein had replied to Carathéodory as can be concluded from a letter of Carathéodory to Delta on 21 May 1930 (Letter in Greek. Leukoparidis (ed) (1956), p 483). As soon as he got Einstein's letter, Carathéodory sent it to Penelope Delta for the autograph collection of her daughter Virginia, and he explained that "Herr M." mentioned in Einstein's letter was Morgenthau.

95. Carathéodory C, The Urban Refugees in Macedonia and Thrace, as note 23, here Chapter IV, p 22 (with minor editorial changes). "Concordia domi pax foris" [Harmony at home, peace abroad].

96. Seven cities in Asia Minor have been immortalised by a letter of prophesy addressed by John to the Christians living in them. These churches of the Book of Revelation were located in Ephesus (Efes in Turkish), Smyrna (İzmir), Pergamum (Bergama), Thyatira (Akhisar), Sardis (Sart), Philadelphia (Alaşehir) and Laodicea (Laodikea). It is obvious that Carathéodory identifies the churches with the cities and refers to Smyrna and Philadephia twice.

97. Carathéodory C, Influence of the refugees *on Greek life* (from an article about to be published in a Greek work). English translation of Carathéodory's text. Carathéodory, writings. In: The Papers of Henry Morgenthau, Sr., as note 23.

98. Carathéodory to Delta on 21 August 1928. Letter in Greek. Leukoparidis (ed) (1956), p 477f.

4 *A Scholar of World Reputation*

1. BayHStA, MK Reg.Sp. V Abgabe 1991 Vorl.Nr. 1376, Laufzeit: 1922–1944.
2. Ibid. Carathéodory's name has been Germanised into "Karatheodory".
3. Ibid. The same spelling of Carathéodory's name is followed here. The expression "after the collapse of these plans" could evoke the impression that Carathéodory himself had been responsible for the collapse.
4. Large (1998), p 268.
5. Ibid, p 199f.

6. BayHStA, MK Reg.Sp. V Abgabe 1991 Vorl.Nr. 1376, Laufzeit: 1922–1944.

7. As note 261, chapter 2.

8. Willstätter was Carathéodory's guest on 10 March 1931.

9. Boehm L and Spörl J (eds) *Ludwig-Maximilians-Universität 1472–1972*. Berlin, 1972, p 352. Quoted in: Volker Loseman (1977), p 29.

10. Broszat M, Fröhlich E, and Wiesemann F (eds) *Bayern in der NS-Zeit – Soziale Lage und politisches Verhalten der Bevölkerung im Spiegel vertraulicher Berichte* (Bavaria in the Nazi Era – Social Condition and Political Behaviour of the Population as Reflected in Confidential Reports). Oldenburg, München, Wien, 1977, p 429f.

11. BayHStA, MK Reg. Sp. V Abgabe 1991 Vorl. Nr. 1376, Laufzeit: 1922–1944.

12. See below: Carathéodory's negotiations to stay in Munich after receiving an appointment to Stanford.

13. BayHStA MK 35403, Laufzeit: 1924–1950.

14. Kalitsounakis's address at the session of Athens Academy on 23 February 1950, as note 40, chapter 2. Carathéodory's letter to him is also mentioned there.

15. Young L C, as note 187, chapter 2, here p 242.

16. Hilbert extensively modified the mathematics of invariants – the entities that are not altered during geometric changes, such as rotation, dilation and reflection. Hilbert proved the theorem of invariants, namely that all invariants can be expressed in terms of a finite number. In his *Zahlbericht* (Commentary on Numbers), a report on algebraic number theory published in 1897, he consolidated what was known in this subject and pointed the way to future developments.

17. *Πρακτικά Συνεδριών Σχολής Φυσικών και Μαθηματικών Επιστημών 1924–1925*, 1st session, p 199.

18. Ibid, 3rd session, p 213.

19. W. Blaschke to Captain Jackson on 17 September 1945 (in English). Staatsarchiv Hamburg, Hochschulwesen, Dozenten- u. Personalakten IV 86, Blaschke, Wilhelm 13.9.1885.

20. Hondros D (1950) *Από την ζωή του Καραθεοδωρή* (From Carathéodory's Life). *Αιών του Ατόμου* (Century of the Atom) Δ'/2–3:326.

21. See: Questionnaire of the Military Government of Germany. BayHStA MK 35403, Laufzeit: 1924–1950.

22. Toepell (1996), p 15. Toepell asserts that with these scientists, mathematics was able to retain a remarkable independence under national socialism.

23. Fraenkel (1967), p 79.

24. Toepell (1996), p 292.

25. Behnke, as note 29, chapter 1, p 29.

26. Vovolinis (ed) (1962), p 504.

27. Communication to the author on 19 December 1999.

28. Tietze, as note 105, chapter 1, p 94.

29. Young, as note 187, chapter 2, here p 242f.

30. An absolutely credible communication to the author from Mrs. Despina Rodopoulos-Carathéodory.

31. Heinhold J: *Erinnerungen an eine Epoche Mathematik in München (1930–1960)* (Memories of an Epoch of Mathematics in Munich (1930–1960)). In: Chatterjii S D, Fenyö I, Kulisch U, Laugwitz D, Liedl R (eds) *Jahrbuch Überblicke Mathematik – Mathematical Surveys*, vol 17. Bibliographisches Institut-Wissenschaftsverlag, Mannheim/Wien/Zürich, 1984, pp 177–209, here p 183.

32. See for example: Carathéodory to Delta on 4 February 1929. Letter in Greek. Leukoparidis (ed) (1956), p 480f.

33. Cf. Mayer H (1994) *Wagner*. Rowohlt, Hamburg, pp 21–28.

34. Wördehoff B, *Hitlers Wagner – Joachim Köhlers provozierende Studie „Der Prophet und sein Vollstrecker"* (Hitler's Wagner – Joachim Köhler's Provocative Study *The Prophet and his Executioners*). *Die Zeit* no. 16, 11 April 1997.

35. Hitler did not merely adore the music, but recognised its composer as the only political precursor of national socialism. It was Wagner's music that played as prisoners were marched off the trains and into the death camps. And Wagner's daughter-in-law, Winifred, was the friend who provided Hitler in the 1920s with the paper on which he wrote *Mein Kampf* while in prison. According to Rose, the extraordinary power of Wagner's music lies in its emotional appeal to violence and paganism, it is shockingly magnificent in its contempt for common humanity.
    Tim Jackson on a searching inquiry into Wagner's anti-Semitism (on the book by Paul Lawrence Rose. *Wagner: Race and Revolution*). *The ring of true contempt. The Independent*, Saturday 25 July 1992.

36. Carathéodory C, *90 Semester München – Zum 80. Geburtstag Alfred Pringsheims* (90 Semester Munich – On Alfred Pringsheim's 80th Birthday). *Münchener Neueste Nachrichten*, 31 August 1930 and *Ges. Math. Schr.* vol V, pp 60–61.
    Hedwig Pringsheim-Dohm, Alfred Pringsheim *als angehender Neunziger* (as prospective 90-year-old), Emilia Maria Pringsheim, Peter Pringsheim have signed in Carathéodory's guest-book next to the date 12 September 1930.

37. Frevert U (1995) *Ehrenmänner – Das Duell in der bürgerlichen Gesellschaft* (Men of Honour – Duel in the Bourgeois Society). Beck Verlag, München, p 237.

38. Wördehoff B, *Führer zu deutscher Art – National nach Noten* (Leader of Germanness – National by Music). *Die Zeit* no. 28, 5 July 1996.

39. Kotsowillis K: *Die Griechische Kirche zum Erlöser in München – Zum 500jährigen Jubiläum (1494–1994)* (The Greek Church of the Redeemer in Munich – On the Quincentenary (1494–1994)).
    Old Catholics are members of the independent Old Catholic Church of the Utrecht Union, separated from Rome since 1723. Relationships between the *Salvatorkirche* (Church of the Redeemer) and Catholics were locally good until about 1975, when the church was occupied by supporters of the Julian (old) calendar, which is followed by the Patriarchates of Jerusalem, Moscow, Serbia and Georgia, as well as all monasteries of the Holy Mountain of Athos. The Old Calendarists is a highly conservative group of Orthodox Christians hostile to all oecumenical ties. Today, the *Salvatorkirche*, albeit a possession of the Bavarian state, is under the jurisdiction of the Oecumenical Patriarchate.
    Communication to the author from Dr. Albert Rauch, Ostkirchliches Institut Regensburg on 22 February 2001.
    Old Catholics reject papal infallibility, celibacy and auricular confession. In Old Catholic communities, laymen have the right to elect their priests and bishops.

40. Millikan R A (1916) A Direct Photoelectric Determination of Planck's "h.". Phys. Rev. Ser. 2, 7 no 3:355–388.

41. Communication to the author from Shelley Erwin on 7 January 1999.

42. Frank W. Ober, Near East Relief representative, to professor Robert Andrews Millikan, Pasadena on 14 August 1924. Letter in English. Caltech Archives, Millikan Papers, Folder 31.4.

43. Carathéodory to Millikan, on 26 July 1924. Letter in English. Ibid.

44. H. H. Johnson to R. A. Millikan on 18 August 1924. Letter in English. Ibid.

45. Millikan to Carathéodory on 27 September 1924. Letter in English. Ibid.
    Nikolaos Politis (1872–1942): author of many works and books on international law,

lectured international law at the University of Paris. Director of the Greek Ministry of the Exterior 1914–1916, Minister of the Exterior of the National-Defence government 1916–1917 and of Venizelos's government 1917–1920, first representative of Greece in the League of Nations and, later, ambassador of Greece in Paris 1924–1925, 1927 and 1930–1931.

46. Carathéodory to Millikan, on 26 July 1924, as note 43.

47. Millikan to Andreadis on 6 October 1924. Letter in English. Ibid.

48. Andreadis to Millikan on 11 November 1924. Letter in English. Ibid.

49. Hondros to Millikan on 25 December 1924. Letter in French. Ibid.

50. Session of 24 February 1924. *Πρακτικά Συνεδριών Σχολής Φυσικών και Μαθηματικών Επιστημών* 6 October 1922–March 1925, pp 153–155.

51. Kenoyer to Millikan on 5 August 1924. Letter in English. Caltech Archives, Millikan Papers, Folder 31.4.

52. Chambers to Millikan on 20 August 1924. Letter in English. Ibid.

53. Chambers to Millikan on 2 September 1924. Letter in English. Ibid.

54. H. von Euler to Millikan on 29 June 1924. Ibid.

55. Millikan to H. von Euler on 30 December 1924. Letter in English. Ibid.

56. Millikan to L.A. Kenoyer on 30 December 1924. Letter in English. Ibid.

57. See: History of Athens College on the Web: www.haef.gr/american/ac/history

58. Carathéodory to Reichenbach on 25 June 1924. HR 016–28–01 Hans Reichenbach Sammlung, Philosophisches Archiv der Universität Konstanz im Auftrag der University of Pittsburgh Libraries, Special Collections Department – 363 Hillman Library. Cited with permission of the University of Pittsburgh. All rights reserved.

59. Reichenbach H (1924) *Axiomatik der relativistischen Raum–Zeit-Lehre*. Vieweg, Braunschweig.

60. C. Caratheodory, Goettingen (1). Scient.Corr.File, Folder "C-Misc.". Albert Einstein Archives, The Jewish National and University Library, Jerusalem.

61. Reichenbach to Carathéodory on 13 July 1925. HR 016–03–17 Hans Reichenbach Sammlung, Philosophisches Archiv der Universität Konstanz im Auftrag der University of Pittsburgh Libraries, Special Collections Department – 363 Hillman Library. Cited with permission of the University of Pittsburgh. All rights reserved.

62. *Sitzungsber. Bayer. Akad. d. Wiss. München, math.-nat. Abt.* 1925, pp 133–175, here p 133.

63. See: Hecht H, Hoffmann D (1982) *Die Berufung Hans Reichenbachs an die Berliner Universität – Zur Einheit von Naturwissenschaft, Philosophie und Politik* (Hans Reichenbach's Appointment to Berlin University – On the Unity of Natural Sciences, Philosophy and Politics). In: *Deutsche Zeitschrift für Philosophie*. VEB Deutscher Verlag der Wissenschaften, Berlin, pp 651–662, here p 655.

64. Cf. Reichenbach H: *Rationalismus und Empirismus – Eine Untersuchung der Wurzeln philosophischen Irrtums* (Rationalism and Empiricism. An Inquiry into the Roots of Philosophical Error). In: Kamlah A, Reichenbach M (eds) *Hans Reichenbach – Gesammelte Werke in 9 Bänden* (Hans Reichenbach – Collected Works in 9 Volumes), vol 1. Vieweg, Braunschweig, 1977.

65. Von Mises R (1951) Positivism. Harvard University Press, Cambridge.

66. Reid (1970), p 181.

67. Hilbert D, Neumann J v, Nordheim L (1928) *Über die Grundlagen der Quantenmechanik* (About the Foundations of Quantum Mechanics). *Math. Ann.* 98:1–30.

68. Dirac P (1926) The Elimination of the Nodes in Quantum Mechanics. Proc. Roy. Soc. London 111A:281–305.

69. First of a series of monumental papers on quantum mechanics: Schrödinger E (1926) *Quantisierung als Eigenwertproblem [erste Mitteilung]* (Quantisation as an Eigenvalue Problem [First Communication]). *Ann. d. Phys.* 81:109–140.

70. Born M, Heisenberg W, Jordan P (1925) *Zur Quantenmechanik II* (On Quantum Mechanics II). *Z. f. Phys.* 35:557–615.
    "A couple of lines about Heisenberg's new work, of which it it said that it looks mystical but it is right. Undoubtedly, it concerns the treatise (*Z. f. Phys.* 35, 1925, p 879), in which he formulates the basic thoughts of quantum mechanics and explains them by simple examples."
    Born in: *Albert Einstein – Hedwig und Max Born. Briefwechsel 1916–1955* (1972), p 94.

71. In his *Quantisierung als Eigenwertproblem*, as note 69.

72. *Albert Einstein – Hedwig und Max Born. Briefwechsel 1916–1955* (1972), p 70.

73. Einstein to Born on 30 December 1921. Ibid, p 72.

74. Ibid, p 73.

75. Pauli W (1927) *Zur Quantenmechanik des magnetischen Elektrons* (On Quantum Mechanics of the Magnetic Electron). *Z. f. Phys.* 43:601–623.

76. Dirac P (1928) The quantum theory of the electron, part II. Proc. Roy. Soc. London 118A:351–361.

77. Heisenberg W (1928) *Zur Theorie des Ferromagnetismus* (On the Theory of Ferromagnetism). *Z. f. Phys.* 49:619–636.

78. Dirac P (1929) Quantum mechanics of many-electron systems. Proc. Roy. Soc. A 123: 714–733.

79. Heisenberg W, Pauli W (1929) *Zur Quantendynamik der Wellenfelder* (On the Quantum Dynamics of Wave Fields). *Z. f. Phys.* 56:1–61 and Heisenberg W, Pauli W (1929) *Quantentheorie der Wellenfelder II* (Quantum Theory of Wave Fields II). *Z. f. Phys.* 56:168–190.

80. Carathéodory to Hilbert on 14 April 1926, as note 256, chapter 2.

81. Van der Waerden, B L (1967) Sources of Quantum Mechanics. North-Holland Publishing Company, Amsterdam.

82. Neumann J v (1932) *Mathematische Grundlagen der Quantenmechanik* (Mathematic Foundations of Quantum mechanics). Grundlehren den mathematischen Wissenschaften vol 38. Springer, Berlin.

83. Carathéodory C (1908) *Sur une Méthode directe du Calcul des Variations. Rend. Circ. Mat. Palermo* 25:36–49 and *Ges. Math. Schr.* vol I, pp 170–187.

84. Perron, as note 107, chapter 1, p 53.

85. Fleming's report, as note 302, chapter 2, with the note: "An account of Tonelli's techniques up to the early 1920s is included in: Tonelli L (1921–1923) *Fondamenti di Calcolo delle Variazioni*. 2 vols, Zanichelli, Bologna." See also: Tonelli L (1913) *Sul caso regolare nel calcolo delle variazioni* (On a Regular Case in the Calculus of Variations). *Rend. Circ. Mat. Palermo* 35:49–73.

86. Carathéodory to Königsberger on 11 December 1907, as note 142, chapter 1.

87. The mentioned example is in *Ges. Math. Schr.* vol 1, p 184.

88. Carathéodory C (1935) *Variationsrechnung und partielle Differentialgleichungen erster Ordnung*. Teubner, Leipzig.
    Maxwell's example is on p 312.

89. As note 290, chapter 2. For Carathéodory's method in the calculus of variations see also: Thiele R (1997) On Some Contributions to Field Theory in The Calculus of Variations from Beltrami to Carathéodory. *Historia Mathematica* 24:281–300.

90. Cf. Carathéodory C (1926) *Über den Zusammenhang der Theorie der absoluten opti-*
    schen Instrumente mit einem Satze der Variationsrechnung (On the Connection between
    the Theory of Absolute Optical Instruments and a Theorem of the Calculus of Variations).
    Bayer. Akad. d. Wiss. München, math.-nat. Abt. 1926, pp 1–18 and *Ges. Math. Schr.* vol
    II, pp 181–197.
    Carathéodory C (1930) *Les transformations canoniques de glissement et leur application*
    à l'optique géométrique (Canonical Transformations of Sliding and their Application
    to Geometric Optics). *Rendiconti della R. Accademia Nazionale dei Lincei, Classe di*
    Scienze fisiche, mathematiche e naturali, vol XII, serie 6a, pp 353–360 and *Ges. Math.*
    Schr. vol II, pp 198–206.

91. See: § 21 of Carathéodory C (1937) *Geometrische Optik* (Geometric Optics). Verlag von
    Julius Springer, Berlin.

92. Carathéodory C (1926) *Die Methode der geodätischen Äquidistanten und das Problem*
    von Lagrange. Acta Math. 47:199–236 and *Ges. Math. Schr.* vol I, pp 212–248.

93. *Ges. Math. Schr.* vol I, p 212.

94. Carathéodory C (1932) *Über die Existenz der absoluten Minima bei regulären Varia-*
    tionsproblemen auf der Kugel. Ann. Scuola Norm. Sup. Pisa, Sci. Fis. e Mat. 1:79–87
    and *Ges. Math. Schr.* vol I, pp 253–263.

95. Carathéodory's theorem concerning the envelope curves of geodesic lines on closed reg-
    ular surfaces, whose Gauss curvature is everywhere positive, is formulated by Blaschke
    as: *Der Ort der zu einem Punkte der Eifläche konjugierten Punkte hat mindestens vier*
    Spitzen. Blaschke mentions that he owes the knowledge of this theorem to a communi-
    cation from Carathéodory in 1912. See: Blaschke W (1945) *Vorlesungen über Differen-*
    tialgeometrie, I: Elementare Differentialgeometrie. Springer, Berlin, p 231.

96. The projective plane is a 2-dimensional manifold, which has a number of equivalent
    definitions. First of all, the projective plane is a set of all straight lines in 3-dimensional
    space, passing through the origin. Each line is uniquely defined by its intersection with
    the sphere around the origin, but two opposite points on the sphere define the same line.
    Therefore, another model of the projective plane is a sphere in 3-dimensional space, on
    which the pairs of opposite points are identified, i.e. considered as one point. We can
    always imagine that this sphere is mapped somewhere, e.g. to 3-dimensional space again,
    using a mapping, which actually glues the opposite points in one. This will provide us
    with a model of the projective plane as a surface in 3-dimensional space. Also, we can use
    only a hemisphere to represent the projective plane, and take into account that opposite
    points on its equator should be identified. Topologically, the hemisphere is equivalent
    to a disc (or rectangle), so the projective plane can be defined as a result of patching the
    opposite points on the disc.
    Information from the Web.
    For the projective plane see also V.V. Afanas'ev's definition in the Encyclopaedia of
    Mathematics.

97. Carathéodory's references concerning Darboux and Hilbert are found in: (a) *Flächen*
    mit lauter geschlossenen geodätischen Linien, as note 365, chapter 2, here *Ges. Math.*
    Schr. vol V, p 5: Darboux, *Leçons sur la théorie générale des surfaces*, T III, p 4 and
    (b) *Über die starken Maxima und Minima bei einfachen Integralen, Ges. Math. Schr.*
    vol I, p 81: Hilbert (1899) *Über das Dirichletsche Prinzip. Jber. Deutsch. Math.-Verein.*
    8:184.

98. In: *Sui Campi di Estremali Uscenti da un Punto e Riempienti tutto lo Spazio* (On the
    Fields of Extremals that Start from a Point and Take Up the Whole Space). *Ges. Math.*
    Schr. vol I, pp 188–202, Carathéodory showed that two points whose ordinates differ
    more than π, can never be connected through an extremal and that in this case the
    variational problem has no solution.

99. Carathéodory to Hopf on 14 January 1931. Letter in German. ETH-Bibliothek Zürich, Wissenschaftshistorische Sammlungen, HS 621:349.
   In this letter Carathéodory congratulated Hopf on his appointment to Zurich. Hopf had become a full professor at the ETH.
   Eleven years later, Behnke would give his impression of Hopf, whom he met during a lecture trip to Zurich: in Switzerland, Hopf had created his own scientific school and from all other scientists there, he had the greatest influence on the rising scientific generation. Since he was born as a German citizen he could have never gained any more influence in Switzerland. He was a reserve officer of World War I and an owner of the E.K. I. (iron cross I).
   Behnke on 13 June 1942: *Bericht über meine Vortragsreise nach Zürich vom 13. April (Einreise) 1942 bis 25. April (Ausreise) 1942* (Report About my Lecture Trip to Zurich from 13 April (Arrival) 1942 to 25 April (Departure) 1942). Universitätsarchiv Freiburg, Nachlaß Süss, C89/42.
   Hopf obtained Swiss citizenship in 1943.

100. Zermelo, as note 132, chapter 1, here p 514.

101. *Ges. Math. Schr.* vol I, p 130f.

102. Carathéodory C (1929) *Über die Variationsrechnung bei mehrfachen Integralen. Acta Sci. Math. Szeged* 4:193–216 and *Ges. Math. Schr.* vol I, pp 401–426.

103. Notes 49, 50, 51, chapter 3.

104. Haar A (1928) *Über adjungierte Variationsprobleme und adjungierte Extremalflächen. Math. Ann.* 100:481–502.
   The notion of the adjunct variation problem and the adjunct extremal surface enables the transfer of theorems about minimal surfaces to the calculus of variations for the solutions of problems of the form $\delta \iint f(\partial z/\partial x, \partial z/\partial y)\,dx\,dy = 0$.

105. Fleming's report, as note 302, chapter 2.

106. Rund H: The Hamilton–Jacobi Theory of the Geodesic Fields of Carathéodory in the Calculus of Variations of Multiple Integrals. In: Panayotopoulos (ed) (1974), p 497.

107. Carathéodory to Born on 24 January 1935. Letter in German. Staatsbibliothek zu Berlin, Preussischer Kulturbesitz, Handschriften-Abteilung, Nachl. Born 105.

108. Born M (1933) On the Quantum Theory of the Electromagnetic Field. Proc. Roy. Soc. London 143:410–437.

109. Born M (1933) Modified Field Equations with a finite Radius of the Electron. Nature, vol CXXXII:282.

110. Born M, Schroedinger E (1935) The Absolute Field Constant in the New Field Theory. Nature CXXXV:342.

111. Born M (1935) Quantised Field Theory and the Mass of the Proton. Nature CXXXVI: 952.

112. Carathéodory to Born on 26 February 1935. Letter in German. Staatsbibliothek zu Berlin, Preussischer Kulturbesitz, Handschriften-Abteilung, Nachl. Born 105.

113. Weyl H (1934) Observations on Hilbert's Independence Theorem and Born's Quantization of Field Equations. Phys. Rev. 46:505–508.

114. Carathéodory to Weyl on 5 March 1935. Letter in German, handwriting. ETH-Bibliothek Zürich, Wissenschaftshistorische Sammlungen, HS 91:44.

115. Weyl to Carathéodory on 23 March 1935. Typewritten copy. ETH-Bibliothek Zürich, Wissenschaftshistorische Sammlungen, HS 91:45.

116. Weyl H (1935) Geodesic Fields in the Calculus of Variation for Multiple Integrals. Ann. of Math. 36/3:607–629.

117. See: Christodoulou D (2000) The Action Principle and Partial Differential Equations. Annals of Math. Studies No 146, Princeton University Press, pp 51–56.

118. E. L. I. (1935) A Neglected Aspect of the Calculus of Variations – Variationsrechnung und partielle Differentialgleichungen erster Ordnung, von Prof. Constantin Carathéodory. Nature CXXXVI:814.

119. Fleming's report, as note 302, chapter 2.

120. Perron O, as note 107, chapter 1, here pp 43–46.

121. See: Boerner H (1953) *Carathéodory's Eingang zur Variationsrechnung* (Carathéodory's entry to the Calculus of Variations). *Jber. Deutsch. Math.-Verein.* 56:31–58.

122. For Carathéodory's method see: Hermann Boerner: *Carathéodory und die Variationsrechnung* (Carathéodory and the Calculus of Variations). In: Panayotopoulos (ed), as note 29, chapter 1, pp 80–90, here p 83ff.
    In two letters to Zermelo (Carathéodory to Zermelo on 25 April and on 3 May 1933. Letters in German. Universitätsbibliothek Freiburg, Nachlaß Ernst Zermelo) and in a letter to Blaschke (Carathéodory to Blaschke on 24 April 1933. Letter in German. Courtesy of Professor Dr. Dr. h.c. Walter Benz), Carathéodory presented solutions to a navigation problem, which had been the subject of a lecture held by Zermelo in Prague on 18 September 1929 and published the following year under the title *Über die Navigation in der Luft als Problem der Variationsrechnung* (On Navigation in the Air as a Problem of the Calculus of Variations). *Jber. Deutsch. Math.-Verein.* 39:44–48. Zermelo had formulated what was to become the famous "Zermelo navigation problem" as follows: "An airship or aeroplane moves with constant velocity k with respect to the surrounding air mass in an unlimited plane in which the wind distribution is given by a vector field u, v as a function of place and time. How should the aircraft be piloted in order to arrive from one point P_0 to another P_1 within the shortest time?" Fleming notes that although Carathéodory's solution is ingenious, this problem is by now of merely historical interest.
    Carathéodory included the solutions to this problem in his 1935 book, sections 276 and 458.
    In his second letter to Zermelo, Carathéodory noted that Zermelo's use of the word "megistochrone" had shocked him. "$M\acute{\varepsilon}\gamma\iota\sigma\tau\sigma\varsigma$ means the biggest, not the longest", he wrote, "and if one has a sensitive ear for Greek, then one has to seek a connection with $\mu\acute{\alpha}\varkappa\rho\sigma\varsigma$, consequently something like makistochrone from $\mu\acute{\alpha}\varkappa\iota\sigma\tau\sigma\varsigma = \mu\acute{\eta}\varkappa\iota\sigma\tau\sigma\varsigma$. Euclid says also $\Gamma\rho\alpha\mu\mu\acute{\eta} = \mu\acute{\eta}\varkappa\sigma\varsigma\ \alpha\pi\lambda\alpha\tau\acute{\varepsilon}\varsigma$, doesn't he?" So, Carathéodory gave Euclid's definition for the line in Ancient Greek as $\mu\acute{\eta}\varkappa\sigma\varsigma\ \alpha\pi\lambda\alpha\tau\acute{\varepsilon}\varsigma$. In his Elements, Euclid writes that $\Gamma\rho\alpha\mu\mu\acute{\eta}\ \delta\varepsilon\ \mu\acute{\eta}\varkappa\sigma\varsigma\ \alpha\pi\lambda\alpha\tau\acute{\varepsilon}\varsigma$, i.e. a line is length without breadth.
    Carathéodory underlined the word $\mu\acute{\eta}\varkappa\sigma\varsigma$ (length). Chrone is an ending coming from the Greek word $\chi\rho\acute{\sigma}\nu\sigma\varsigma$ (time). $M\acute{\alpha}\varkappa\iota\sigma\tau\sigma\varsigma$ is a Doric word used only in Ancient Greek tragedy (Sophocles, $O\iota\delta\acute{\iota}\pi\sigma\upsilon\varsigma\ T\acute{\upsilon}\rho\alpha\nu\nu\sigma\varsigma$ (Oedipus the King), line 1300) to denote $\mu\acute{\eta}\varkappa\iota\sigma\tau\sigma\varsigma$ (the longest).
    Makistochrone was thus a word constructed by Carathéodory to denote the longest time, in analogy with the word brachistochrone which denotes the shortest (least) time.
    In his contribution to Frank–Mises' book on the differential and integral equations of mechanics and physics, Carathéodory had remarked that the least-time problem, the oldest in the calculus of variations, which cannot be solved with methods of elementary geometry, had been posed by Johann Bernoulli in 1696. It was the famous problem of curves of minimum time of descent, and gave the opportunity to many respected mathematicians, above all to Euler, to publish on it. With this began the modern interest in the calculus of variations.
    The brachistochrone problem had been formulated by Bernoulli as follows: two points of a vertical plane should be connected by a curve in such a way that a material point which under the influence of gravitation starts to fall from A along the curve reaches B in the shortest possible time. The problem consists in finding the minimum of the integral $\int \sqrt{T(h-u)^{-1}}\, dt$ for a conservative mechanical system that moves along a

curved path $x_i(s)$ $(i = 1, 2, \ldots, n)$ in the generalised co-ordinate space in case when no other movement is involved except that for which the total energy is h. The extremals of this problem are then characterised by the equations $R_i = -2N_i$ (and this corresponds to Euler's theorem), where $R_i = (\mathrm{d}L_{\dot{x}_i}/\mathrm{d}t) - L_{x_i}$ are the components of the reaction, $L = T - U$ the Lagrangian of the system and $N_i = -U_{x_i} + \dot{U}T_{\dot{x}_i}(2T)^{-1}$ the components of the normal force.

The problem was solved independently by Johann Bernoulli, his brother Jakob and Isaac Newton. The basic idea to set up an integral for the total time of fall in terms of the unknown curve and then vary the curve so that a minimum time is obtained led to a differential equation, whose solution is a curve called the cycloid.

In his paper: (1933) *Généralisation d'un théorème d' Euler sur le movement brachistochrone* (Generalisation of a Theorem of Euler on the Brachistochrone Movement). *Rendiconti della R. Accademia dei Lincei, Classe di Scienze fisiche, mathematiche e naturali,* vol 17, seria 6ª:10–12 and *Ges. Math. Schr.* vol II, pp 374–376, Carathéodory considered the general problem $\int (h - U)\sqrt[\alpha]{T}\,\mathrm{d}t = \min$ and found that the differential equations for the extremals take the form $R_i = (2\alpha - 1)N_i$ from which, depending on the values of α, one could derive interesting particular cases. For example, for $\alpha = 0.5$ one gets the Maupertuis principle, or for $\alpha = 0$ a generalisation of the Galilée principle. Carathéodory closed his text with the remark "it is thus amusing enough that the Euler theorem is really invariant for conservative systems and in fact, as he himself believed, 'a general principle of nature'".

123. Carathéodory C (1937) The beginning of research in the Calculus of Variations. Osiris III:224–240 and *Ges. Math. Schr.* vol II, pp 93–107.

124. Draft of publication. General Files, Mathematics Department, 7/22/2, Box 12, the University of Wisconsin Archives. In Carathéodory's printed text (see previous note) the "charming prince" has become the "prince charming".

125. Carathéodory, as note 308, chapter 2, here p 672.

126. Carathéodory, as note 88, p 12.

127. Young L C, as note 187, chapter 2, here p 244f.

128. Pesch H J, Bulirsch R (1994) Historical Paper – The Maximum Principle, Bellman's Equation and Carathéodory's Work. Journal of Optimization Theory and Applications 80 (2):199–225.

129. Communication to the author from H. J. Pesch on 6 June 1996.

130. Fleming's report, as note 302, chapter 2, with the note: "One can see clearly from Perron's obituary article how Bellman's optimal feedback controls are obtained by Carathéodory's method (See top of page 45)."

131. As note 129.

132. Pesch H J, Bulirsch R, as note 128, here eq 76 on p 218.

133. Fleming's report, as note 302, chapter 2.

134. As note 129.

135. Fleming's report, as note 302, chapter 2.

136. Ibid, with the note: "For recent developments and applications of Hamilton-Jacobi PDEs see: M. Bardi and I. Capuzzo-Dolcetta, Optimal Control and Viscosity Solutions of Hamilton-Jacobi-Bellman Equations, Birkhauser, 1997; J. A. Sethian, Fast marching methods, SIAM Review 2 (41) 199–235."
    See also: Barrett J F: On the Relation between Carathéodory's Work in the Calculus of Variations and the Theory of Optimal Control. In: Panayotopoulos (ed), as note 29, chapter 1, pp 54–74.

137. Phokion Negris (1846–1928): scientist and politician; Minister of Finances 1898–1899 and 1901–1902; Minister of Transport 1916; Minister of the Interior 1917. Venizelist.

138. Theodoros Pangalos (1878–1932): a Venizelist with military studies at the Evelpides Officers' School and the War School of Paris. Active in the Military League of Goudi in 1909, he joined the Salonica National-Defence Movement in 1917 and fought at the Macedonian front. After King Konstantin's abdication in 1917, he became chief of the personnel department at the Ministry of War. At the beginning of the Asia-Minor expedition, he became chief of the general staff at General Paraskevopoulos' headquarters, but was sent into retirement from army service in the following year, after Venizelos's defeat in the elections of November 1920. Pangalos returned to politics with the so-called Plastiras revolution of 1922, staged the Trial of the Six and came to power through a coup d'état in June 1925. His dictatorial regime lasted fourteen months. The "ambitious and ruthless general" (M. L. Smith) abolished the freedom of press, dissolved the national assembly, exiled opposing politicians and military, persecuted communists, imposed a hypocritical prudery and the death penalty for embezzlers of public money, and abolished permanent positions in the civil service. In international relations, he appeared as a supporter of the "patriotic irreconcilability" and was seized by the persistent idea of an armed revenge against Turkey, while dreaming of a confederation with Albania.
    See: Kallivretakis L, Θεόδωρος Πάγκαλος: "Υπήρξα και εγώ μία ανωμαλία εν μέσω τόσων άλλων" (Theodoros Pangalos: "Also I Have Been an Anomaly among so Many Others"). Τα Νέα, 27 December 1999.

139. Kalitsunakis I: *Die Gründung der Akademie von Athen* (The Foundation of the Academy of Athens) In: Kriekoukis Ch, Bömer K (eds) *Unsterbliches Hellas* (Immortal Greece). Zeitgeschichte Verlag Wilhelm Andermann, Berlin, 1938, pp 129–130.

140. Dimitrios Aeginitis (1862–1934): professor of astronomy, member of the Athens Academy; Minister of Education and Religious Affairs in 1917 and 1926.

141. As note 139.

142. Carathéodory to Kalitsounakis on 21 November 1926. Vovolinis (ed) (1962), p 506.

143. Nikolaos Balanos (1860–1937): civil engineer and archaeologist; head of the archaeology department of the Greek Ministry of Education. In 1902–1930 he supervised the restoration works on the Acropolis; in 1902–1909 he carried out a fundamental restoration of the entire Erechtheion. Although his restoration resulted in a harmonious building, his unconsidered use of iron had disastrous consequences. The metal brackets and supports corroded so much in the course of time that, in many parts of the temple, the marble was destroyed.
    Cf. Scholl A (1998) *Die Korenhalle des Erechtheion auf der Akropolis. Frauen für den Staat* (The Porch of the Maidens of Erechtheion on the Acropolis – Women in the Service of the State). Fischer Taschenbuchverlag, Frankfurt am Main, p 14.
    Carathéodory was Balanos's relative: Sofia Baltatzi, a cousin of Carathéodory's wife, was married to Balanos.

144. See: Müller, W (1844) *Griechenlieder*. Brockhaus, Leipzig.

145. Hugo Ritter von Seeliger (Bielitz 23 Sept.1849 – Munich 2 Dec. 1924): doctoral degree in 1871; engaged at Leipzig observatory up to 1873; at Bonn observatory since 1873; leader of the German expedition to Auckland islands for the observation of the Venus transit in 1874; lecturer in Bonn in 1877; lecturer in Leipzig in 1878; director of Gotha observatory in 1881; director of Munich's observatory since 1882 and full professor for astronomy at the University of Munich. Von Seeliger was a member of the Bavarian Academy of Sciences and of numerous academies at home and abroad. In his works concerning the distribution of stars and astrophysics he developed a theory of stars and nebula for the creation of new stars.

146. Carathéodory to Runge on 8 December 1924. Staatsbibliothek zu Berlin – Preussischer Kulturbesitz, Handschriftenabteilung, Dep. 5 (Nachl. Runge–DuBois-Reymond) 580.

147. Communication to the author from Dr. R. Heydenreuter on 30 May 2001.

148. Carathéodory, von Dyck, von Faber, Finsterwalder, Perron, Pringsheim and Voss.

149. Toepell (1996), p 282f.

150. Alexander Wilkens, Arnold Sommerfeld, Georg von Faber, Jonathan Zenneck, Robert Emden, Walther von Dyck, Wilhelm Wien.

151. *nicht besonders glücklich gewählt, da sie ethische Analogien nahelegt, die mit der Einsteinschen Theorie nichts zu tun haben.*
Einstein's personal file in the Bavarian Academy.
Communication to the author from Dr. Michael Eckert on 22 October 1999.

152. Carathéodory to Sommerfeld on 28 March 1927. Letter in German. Deutsches Museum. Archiv NL 89, 019, Mappe 5,9. *Sommerfeld-Projekt* on the Web.

153. For example, Carathéodory intended to spend the new year's eve of 1927 together with Herglotz and Sommerfeld in Seefeld. Sommerfeld to Wilhelm Lenz on 24 December 1926. *Sommerfeld-Projekt* on the Web. Or: "I hear through Despina that you are well; she was absolutely delighted by your invitation." Carathéodory to Sommerfeld on 28 April 1928. Letter in German. LMU München, Institut für theoretische Physik, Nachlaß Sommerfeld.

154. The others were Alexander Wilkens, Anton Hauptmann, August Schmauss, Erich Kaiser, Erich von Drygalski, Heinrich Tietze, Heinrich Wieland, Hermann Anschütz-Kaempfe, Karl Frisch, Karl Ritter von Goebel, Kasimir Fajans, Rita Anschütz-Kaempfe, Richard Hertwig, Richard Willstätter, Theodor Mollison, Kollmann.
Munich, Deutsches Museum. Archiv NL 89, 019, Mappe 5,10. *Sommerfeld-Projekt* on the Web.

155. Beyerchen (1980), p 149f.

156. BAyHstA MK 35403, Laufzeit: 1924–1950.

157. Birkhoff G D (1913) Proof of Poincaré's Geometrical Theorem. Trans. Amer. Math. Soc. 14:14–22.

158. Poincaré H (1912) *Sur un théorème de Géométrie* (On a Theorem of Geometry). *Rend. Circ. Mat. Palermo* 13:355–407.

159. Mac Lane S (1994) Jobs in the 1930s and the Views of George D. Birkhoff. The Mathematical Intelligencer 16/3:9–10, here 9.
Poincaré's (Conjectured) Geometric Theorem may be stated as follows: let a ring-shaped plane surface be mapped onto itself by a topological transformation in such a way that points on the inner boundary are regressed, those on the outer boundary advanced by the transformation; then, either there exists at least one invariant point, or else there exists a transformation field whose boundary is formed by two simple closed curves corresponding to each other by the transformation, each of which separates the two boundary circles of the ring-shaped surface.
See: Kérékjártó B. de (1928) Note on the general translation theorem of Brouwer. In: International Congress of Mathematicians. Bologna 1928. Comunicazioni. Reprint 1967, Kraus, Nendeln/Liechtenstein.
In his *Proof of Poincaré's Geometrical Theorem*, p 14, Birkhoff states the theorem as follows:
let us suppose that a continuous one-to-one transformation T takes the ring R, formed by concentric circles C_a and C_b of radii a and b respectively ($a < b > 0$), into itself in such a way as to advance the points of C_a in a positive sense, and the points of C_b in the negative sense, and at the same time to preserve areas. Then there are at least two invariant points.
In his paper: (1926) An extension of Poincaré's last geometrical theorem. *Acta Math.* 47:297–311, here 298, Birkhoff states the theorem as follows:
let r, θ stand for polar coordinates in the plane, so that $r = a > 0$ in the equation of a circle C of radius a. A doubly connected ring R, bounded by the circle C and a closed curve Γ encircling C, as well as a second like ring R_1 bounded by the same circle C

and a like encircling curve $\Gamma_1(\ldots)$ are taken to be related in that a one-to-one, direct, continuous point-transformation T carries R into R_1.

160. Birkhoff G D (1917) Dynamical Systems with two degrees of Freedom. Trans. Amer. Math. Soc. 18:199–300.

161. Birkhoff G (1927) Dynamical Systems. Amer. Math. Soc. Colloquium Publications vol 9, New York, p 189.

162. "The Division agreed unanimously to invite Professor C. Carathéodory of the University of Munich to take Professor Birkhoff's place during the second half of the academic year 1927–28.
Voted: that the Chairman communicate the invitation to Professor Carathéodory"
III. Division of Mathematics, UAV 561.3, Harvard University, Minutes of Meetings, Vol. III, September 1924–June 1928.

163. Carathéodory to Delta on 1 January 1928. Letter in Greek. Leukoparidis (ed) (1956), p 474f.

164. Semicentennial Publications of the American Mathematical Society, vol I, p 21.

165. BAyHstA MK 35403, Laufzeit: 1924–1950.

166. Heinhold J, as note 31, here p 183.
Litten F (1994) *Die Carathéodory-Nachfolge in München* 1938–1944 (Succession to Carathéodory in Munich). *Centaurus* 37:154–172, here 154.
Toepell M (1996), p 227.

167. BAyHstA MK 35403, Laufzeit: 1924–1950.

168. *Provinzialismus als einer geistigen Gefahr und unfreien Gesinnung.* Large (1998), p 267.

169. Special Collections and Archives. University of Liverpool Library, ref. D. 140/9/6

170. *The Pennsylvanian* did not publish any reviews or follow-up articles on Carathéodory's visit. Two other university publications, The *Alumni Register* and *The Pennsylvania Gazette*, did not mention the visit. In the *Public Lectures of the University* there is no material on Carathéodory or his lecture. The *Alumni Records* does not contain any folder on Carathéodory.
Communication to the author from Martin Hacket and Christopher Rooney on 19 October 1999.
The *Aquitania* was built in 1914 and was at that time the biggest British steamer. It took part in the Dardanelles operation as troopship and directly afterwards it was employed for the transport of wounded soldiers. On board the *Aquitania* the newly married Venizelos and his wife travelled to the USA in October 1921.

171. Harvard University Archives, HU 20.41. Harvard University Catalogue 1927–1928. Cambridge 1927, Courses of Instruction, p 555.

172. Carathéodory to Sommerfeld on 28 April 1928, as note 153.

173. III. Division of Mathematics, Harvard University, UAV 561.3, Minutes of Meetings, Vol. III, September 1924–June 1928.

174. Helge Kragh (1990), p 21.

175. Slater J (1927) Radiation and absorption in Schrödinger's theory. Proceedings of the National Academy of Sciences 13:7–12.

176. Slater J (1929) The theory of complex spectra. Phys. Rev. 34:1293–1322.

177. John Clarke Slater (1900–1976). Information from the American Philosophical Society and Lois Beattie, Institute Archives MIT, on 29 January 1999.

178. Carathéodory to Weyl on 17 April 1928. Letter in German. ETH-Bibliothek Zürich, Wissenschaftshistorische Sammlungen, HS 91:498.

179. III. Division of Mathematics, Harvard University, UAV 561.3, Minutes of Meetings, Vol III, September 1924–June 1928.

180. Siegmund-Schultze, as note 382, chapter 2, here p 49.

181. Birkhoff G D (1938) Fifty years of American Mathematics. In: Semicentennial Publications of the American Mathematical Society, vol 2, pp 270–315, here p 246f.

182. Siegmund-Schultze, as note 382, chapter 2, here p 50.

183. Carathéodory to Coolidge on 25 July 1928. Letter in English. Harvard University Archives, UAV 561.8, 1920–1930, C. Carathéodory, Department of Mathematics, Correspondence.
     Julian Lowell Coolidge (1873–1954) was educated at Harvard and Oxford. He studied further under Corrado Segre in Turin and under Study in Bonn. He lectured mostly at Harvard.

184. As note 123.

185. Perron to Carathéodory on 6 July 1928. Letter in German. Harvard University Archives, UAV 561.8, 1920–1930, O. Perron, Department of Mathematics, Correspondence.

186. Toepell (1996), p 276.

187. III. Division of Mathematics, Harvard University Archives, UAV 561.3, Minutes of Meetings, Vol. III, September 1924–June 1928.

188. Carathéodory to Coolidge on 25 July 1928, as note 183.

189. See: Herbert U, *Aus der Mitte der Gesellschaft* (From the Middle of Society). *Die Zeit* no. 25, 1 June 1996.

190. Coolidge to Carathéodory on 29 July 1928. Letter in English. Harvard University Archives, UAV 561.8, 1920–1930, J. Coolidge, Department of Mathematics, Correspondence.

191. Siegmund-Schultze, as note 382, chapter 2, here p 307.

192. Reid (1976), p 96.

193. Habilitation title: *Konvergenzsätze für Fourierreihen grenzperiodischer Funktionen* (Convergence Theorems for Fourier Series of Almost-Periodic Functions).
     See also: Bochner S (1927) *Beiträge zur Theorie der fastperiodischen Funktionen. I. Teil, Funktionen einer Variablen* (Contributions to the Theory of Almost-Periodic Functions. Part I, Functions of One Variable). *Math. Ann.* 96:119–147. There, Bochner defined functions equivalent to Bohr's almost-periodic functions. A function $f(x)$ continuous in the interval $(-\infty, \infty)$ is said to be a Bochner almost-periodic function if the family of functions $\{f(x+h) : -\infty < h < \infty\}$ is compact in the sense of uniform convergence on $(-\infty, \infty)$.
     E. A. Bredikhina, Encyclopaedia of Mathematics.

194. Mac Lane, as note 159, here p 10.

195. Carathéodory to Frau Hilbert on 20 June 1928. Letter in German. Niedersächsische Staats- und Universitätsbibliothek Göttingen – Abteilung für Handschriften und seltene Drucke. Hilbert 772, 6.

196. Carathéodory to Delta on 17 June 1928. Letter in Greek. In: Leukoparidis (ed) (1956), p 475f.

197. According to Fritz Stern (1999), p 251, Masaryk from Czechoslovakia and Paderewski from Poland (leaders of small democracies pursuing national claims) were scientists, musicians and poets, political beings by necessity and by some special calling, who pitted the human spirit against power and authority. The word "calling" again implies a mission.
     Carathéodory might have got to know Morgenthau and also Paderewski in Paris when invited there by Venizelos.

198. *The Summer Texan.* Tuesday June 12, 1928 and Thursday, June 14, 1928. The Center for American History, The University of Texas at Austin.

In Texas, Carathéodory wrote: (1928) *Bemerkungen zu den Existenztheoremen der konformen Abbildung*, as note 156, chapter 2.

199. Carathéodory to Delta on 17 June 1928, as note 196.

200. Carathéodory to Frau Hilbert on 20 June 1928, as note 195.

201. Carathéodory to Delta on 17 June 1928, as note 196.

202. William M. Roberts, University Archivist, the Bancroft library. Berkeley, California 94720–6000.
     Bayerische Akademie der Wissenschaften.

203. Carathéodory to Frau Hilbert on 20 June 1928, as note 195.

204. Carathéodory to Delta on 17 June 1928, as note 196.

205. Carathéodory to Delta on 21 August 1928, as note 97, chapter 3.

206. Carathéodory to Delta on 17 June 1928, as note 196.

207. Carathéodory to Delta on 21 August 1928, as note 97, chapter 3.

208. Carathéodory C (1929) *Stetige Konvergenz und normale Familien von Funktionen. Math. Ann.* 101:515–533 and *Ges. Math. Schr.* vol IV, pp 96–118.
     Carathéodory uses the term *stetige Konvergenz* (continuous convergence) with which Du Bois-Reymond had denoted the uniform convergence of a sequence of functions. In *The Theory of Integration* (Cambridge Tracts in Mathematics and Mathematical Physics no. 21, New York and London 1964, p 44), L. C. Young gives the following definition of uniform convergence:
     Let $f_1, f_2, \ldots, f_n, \ldots$ be any sequence of functions of x, not necessarily continuous, and let f be their limiting function. It is assumed that all these functions are bounded and defined in the same interval. If a new function $F(x, p) = f_n(x) - f(x)$ is defined only at those points where p represents an integral value of n and vanishes elsewhere, then the convergence of the given sequence is said to be uniform at a point x_0, if in the new space, in which $F(x, p)$ has been defined, the upper and the lower limits of the function $F(x, p)$ at the point which corresponds to $x = x_0$ and $n = \infty$ coincide and are equal to zero, or, in other words, if F is continuous at that point.
     Carathéodory's presentation of the theory of uniform convergence can be found in his book *Vorlesungen über reelle Funktionen* (1918), pp 173–180 or in his book *Reelle Funktionen, Band I* (1939), pp 165–172. The theory of continuous convergence is presented in his *Reelle Funktionen, Band I* (1939), pp 173–182.
     In his *Remark on a Theorem of Osgood concerning convergent Series of analytic Functions*, Carathéodory uses the term "regular convergence" to distinguish from the term *stetige Konvergenz* (continuous convergence) that had been used by H. Hahn in his: (1921) *Theorie der reellen Funktionen*. Springer, Berlin, p 238. Carathéodory defines the term as follows: "We shall say that a sequence of functions $f_n(z)$, meromorphic on a closed region A, is *regularly convergent* to a point z_0 of A if for every sequence of points $z_1, z_2, \ldots$, belonging to A and converging towards z_0, the (finite or infinite) limit $\lim_{n \to \infty} f_n(z_n)$ exists.

209. Carathéodory to Delta on 17 June 1928, as note 196.

210. Gilbert Ames Bliss (Chicago, Illinois 1876 – Harvey, Illinois 1951): educated in Chicago and professor there from 1908 until his retirement. Known works by him: *Mathematics for Exterior Ballistics* (1944) and *Lectures on the Calculus of Variation* (1946).
     Oswald Veblen (Decorah, Iowa 1880 – Brooklyn, Maine 1960): famous for his work in projective and differential geometry as well as for his contibution to the development of modern topology. He held the Henry B. Fine Professorship of Mathematics at Princeton from 1926 to 1932, when he was appointed to the School of Mathematics of the newly founded Institute for Advanced Study at Princeton. He remained there until his retirement in 1950.

211. H. F. Blichfeldt to Dr. R. L.Wilbur on 1 November 1928. Letter in English. Department of Special Collections, Stanford University Libraries, SC64a, Wilbur Papers, Box 70, Folder: Mathematics.

212. Blichfeldt to Carathéodory on 14 November 1928. Letter in English. Ibid.

213. Carathéodory to the dean 10 December 1928. Archiv der LMU München, E–II N Car.

214. Carathéodory's letter to Blichfeldt on 26 December 1928 is mentioned in Wilbur's letter to Carathéodory on 23 January 1929 (Department of Special Collections, Stanford University Libraries, SC64a, Wilbur Papers, Box 70, Folder: Mathematics) as follows: "Professor Blichfeldt has discussed with me your letter of December 26th. I realize that it is very difficult for one placed as well as you are to think of changing to another country and to new conditions."

215. Wilbur to Carathéodory on 23 January 1929. Letter in English. Ibid.

216. Wilbur to Carathéodory on 15 February 1929. Letter in English. Ibid.

217. BayHstA MK 35403, Laufzeit: 1924–1950.

218. Note of the State Ministry of Finances no. 16640 of 11 April 1924.

219. BayHStA, MK Reg.Sp. V Abgabe 1991 Vorl.Nr. 1376, Laufzeit: 1922–1944.

220. The Princeton Mathematics Community in the 1930s. Transcript Number 6 [PMC6]. An Oral-History Project. The Trustees of Princeton University, 1985.

221. BayHstA, MK Reg.Sp. V Abgabe 1991 Vorl.Nr. 1376, Laufzeit: 1922–1944.

222. Carathéodory to Blichfeld on 12 May 1929. Letter in English. Department of Special Collections, Stanford University Libraries, SC64a, Wilbur Papers, Box 70, Folder: Mathematics.

223. Blichfeldt to Wilbur (Secretary of the Interior, Washington, DC) on 17 May 1929. Ibid.

224. Courtesy of Dr. Antal Varga.

225. Kérékjártó to Kneser on 4 May 1928. Universitätsarchiv Freiburg, Nachlaß Wilhelm Süss, C89/63.

226. Marston Morse to Doctor Aydelotte on 18 September 1944. Courtesy of the Archives of the Institute for Advanced Study.

227. Communication to the author by Dr. A.Varga.

228. In his letter of 12 May 1929 to Blichfeldt, Carathéodory mentioned that he was working together with Dr. Radó, who, thanks to a Rockefeller fellowship, was staying in Munich before visiting Harvard and the Rice Institute as an invited lecturer the following year.

229. Carathéodory C (1952) Conformal Representation. Cambridge at the University Press.

230. Van der Waerden B L (1941) *Topologie und Uniformisierung der Riemannschen Fläche* (Topology and Uniformisation of the Riemann Surface). Quoted in: *Ges. Math. Schr.* vol IV, p 125.

231. Radó T (1925) *Über den Begriff der Riemannschen Fläche. Acta Sci. Math. Szeged* 2:101–121.

232. Carathéodory C (1950) *Bemerkung über die Definition der Riemannschen Flächen, Herrn Oskar Perron zum 70. Geburtstag am 7. Mai 1950 gewidmet. Math. Z.* 52:703–708 and *Ges. Math. Schr.* vol IV, pp 125–131.

233. Carathéodory to Radó on 17 November 1949. Letter in English. The Ohio State University Archives, Tibor Rado papers (Record Group R6 40/76/1), "Caratheodory, C. (Professor): 1948–1949".

234. Carathéodory to Radó on 18 November 1949. Letter in English. Ibid.

235. Radó to Carathéodory on 28 November 1949. Letter in English. Ibid.

236. Heins M (1949) The Conformal Mapping of Simply-Connected Riemann Surfaces. Annals of Mathematics 50(3):686–690.

237. Brouwer's circular letter *An Verleger und Redakteure der Mathematischen Annalen* from Laren on 30 April 1929. Caltech Archives, Theodore von Kármán Papers, Folder 4.14.

238. Hilbert's motives are exposed in a letter that Hilbert sent to the co-editors of the *Mathematische Annalen* on 15 October 1928. This letter and citations from it are mentioned in a letter by Carathéodory to Courant on 24 December 1928. Brouwer Archief, CB MA 69.

239. Lehto (1998), p 46.

240. See: Brouwer's circular letter *An Verleger und Redakteure der Mathematischen Annalen* of 5 November 1928. Caltech Archives, Theodore von Kármán Papers, Folder 4.14. Hilbert's letter to Brouwer is given in: Blumenthal *An Verleger und Redakteure der Mathematischen Annalen* on 16 November 1928. The Jewish National and University Library, Jerusalem, call no 13–154. Hilbert refers only to the difference in fundamental matters and the authorisation by Blumenthal and Carathéodory to remove Brouwer.

241. Hilbert to Einstein on 15 October 1928. The Jewish National and University Library, Jerusalem, call no 13–139.

242. Von Dyck to editors on 18 January 1925. The Jewish National and University Library, Jerusalem, call no 6 110–2.

243. Hilbert to Einstein in the postscript of his letter of 15 October 1928. The Jewish National and University Library, Jerusalem, call no 13–139.

244. Born to Einstein on 20 November 1928. *Albert Einstein – Hedwig und Max Born. Briefwechsel 1916–1955* (1972), p 104f.

245. Einstein to Hilbert on 19 October 1928. The Jewish National and University Library, Jerusalem, call no 13–141.

246. Carathéodory to Einstein on 20 October 1928. Utrecht 18 148–1.

247. Born to Einstein on 20 November 1928. *Albert Einstein – Hedwig und Max Born. Briefwechsel 1916–1955* (1972), p 104.

248. Einstein to Carathéodory on 23 October 1928. Utrecht 18–150.

249. Hilbert to Einstein on 25 October 1928. The Jewish National and University Library, Jerusalem, call no 13–143.

250. Brouwer to Blumenthal from Laren on 2 November 1928. The Jewish National and University Library Jerusalem, call no 13 171–1.
     Also Caltech Archives, Theodore von Kármán Papers, Folder 4.14.

251. Brouwer *An Verleger und Redakteure der Mathematischen Annalen* on 30 April 1929. The Jewish National and University Library, Jerusalem, call no 13–175. Also Caltech Archives, Theodore von Kármán Papers, Folder 4.14.

252. Brouwer to Carathéodory on 2 November 1928. The Jewish National and University Library Jerusalem, call no 13–171.1. Also Caltech Archives, Theodore von Kármán Papers, Folder 4.14.

253. Blumenthal *An Verleger und Redakteure der Mathematischen Annalen* on 16 November 1928, as note 240.

254. Carathéodory to Courant on 3 November 1928. CB MA 3 Brouwer Archief.

255. Brouwer's circular letter *An Verleger und Redakteure der Mathematischen Annalen* on 5 November 1928, as note 240.

256. Springer to Courant 13 November 1928. Mentioned in: van Dalen D (1990) The War of the Frogs and the Mice, or the Crisis of the Mathematische Annalen. The Mathematical Intelligencer 12/4:17–31, here 23.

257. Blumenthal *An Verleger und Redakteure der Mathematischen Annalen* on 16 November 1928, as note 240.

258. Blumenthal to Carathéodory on 20 November 1928. Blumenthal sent a copy of that letter immediately to Courant and Springer. CB MA 26 Brouwer Archief.

259. Born to Einstein on 20 November 1928. *Albert Einstein – Hedwig und Max Born. Briefwechsel 1916–1955* (1972), p 103ff.

260. Bieberbach to Blumenthal on 24 November 1928. The Jewish National and University Library, Jerusalem, call no 13–161.

261. van Dalen D, as note 256, here 24.

262. Einstein to Blumenthal on 25 November 1928. In the postscript Einstein writes that he sends this letter also to Brouwer. The Jewish National and University Library Jerusalem, call no 13–157.

263. Blumenthal to von Kármán on 22 November 1928. Caltech Archives, Theodore von Kármán Papers, Folder 3.10.

264. Carathéodory to Courant (postcard dated 26 November 1928). CB MA 29 Brouwer Archief.

265. Carathéodory to Blumenthal on 27 November 1928. CB MA 32 Brouwer Archief.

266. Courant to Carathéodory on 30 November 1928. CB MA 37 Brouwer Archief.

267. Carathéodory to Courant on 1 December 1928. CB MA 39 Brouwer Archief.

268. Brouwer *An Verleger und Redakteure der Mathematischen Annalen* on 23 December 1928. The Jewish National and University Library Jerusalem, call no 13–170.

269. Carathéodory to Einstein on 3 December 1928. Einstein Archives, The Jewish National and University Library Jerusalem, call no 18 163.

270. Carathéodory to Courant on 12 December 1928. CB MA 47 Brouwer Archief.

271. The old contract signed by Klein, Hilbert, Einstein, Blumenthal as editors of the *Mathematische Annalen* on one side and Julius Springer as publisher on the other, in Berlin on 25 February 1920. The Jewish National and University Library, Jerusalem, call no 411058.

272. Einstein to Courant on 18 December 1928. The Jewish National and University Library Jerusalem, call no 13–169.
Einstein probably alluded to the extant *Batrachomyomachia* (Battle of Frogs and Mice), a Greek parody, written perhaps during the Persian Wars (490 and 480–479) and attributed (the latest in the 2nd century BC) to Homer. See for example: Homeri Ilias, Ulyssea, batrachomyomachia, hymni XXXII. Aldus, Venice, 1504. This parody narrates the fate of small creatures with epic grandiloquence and pathos. It describes the tragic death of mouse "breadcrumb thief" on the back of the frog King. The war that broke out as a consequence between mice and frogs had to be settled by Zeus's mediation through lightning and finally through an army of crabs. The parody had been adapted and translated. Adaptation in German: *Froschmeuseler* by Rollenhagen, 1595.

273. Courant to Carathéodory on 17 December 1928. Brouwer Archief CB MA 53.

274. Carathéodory to Courant on 19 December 1928. In the postscript Carathéodory added that he would now officially announce to Hilbert that he would not be able to join the new editorial board. CB MA 58 Brouwer Archief.

275. Courant to Carathéodory on 23 December 1928. CB MA 65 Brouwer Archief.

276. Bohr to Carathéodory on 23 December 1928. CB MA 66 Brouwer Archief.

277. Carathéodory to Courant on 24 December 1928. CB MA 69 Brouwer Archief.

278. Brouwer to Bieberbach, Bohr, Carathéodory, Courant, v. Dyck, Einstein, Hölder, v. Kármán, Sommerfeld on 23 January 1929. The Jewish National and University Library, Jerusalem, call no 173–1.

279. Carathéodory to Fraenkel on 5 November 1928. Arc 4° 1621/Caratheodory. Archive Adolf Fraenkel. The Jewish National and University Library, Jerusalem.

280. Landau to Fraenkel on 5 November 1928. Arc 4° 1621/Edmund Landau. Archive Adolf Fraenkel. The Jewish National and University Library, Jerusalem.

281. Landau to Fraenkel on 18 Cheschwan 5689. Arc 4° 1621/Edmund Landau. Archive Adolf Fraenkel. The Jewish National and University Library, Jerusalem.

282. Carathéodory C (1930) *Untersuchungen über das Delaunaysche Problem der Variationsrechnung. Abhandl. Math. Sem. Hamb. Univ.* 8:32–55 and *Ges. Math. Schr.* vol II, pp 12–39.

283. Law 4599/2 May 1930 Περί αποδοχής δωρεάς των κληρονόμων Εμμανουήλ Μπενάκη προς ίδρυσιν Μουσείου Μπενάκη (Concerning the Acceptance of the Endowment by Emmanuel Benakis's Heirs for the Foundation of the Benakis Museum) was published in issue 138 of the Government Gazette.

284. Oeconomos ex Oeconomon: a fanatic supporter of the Great Idea, since 1814 director of the philological high school of Smyrna, in 1819 nominated by the Oecumenical Patriarch of Constantinople, Grigorios V, to "General Preacher of the Great Church and all Orthodox Churches of the Greek Genus".

285. Vincent Lanza (1822–1902): an Italian political emigrant in Greece, where he fled in 1848; professor of perspective at the Σχολή Τεχνών (School of Arts), an institution preceding the Technical University in Athens.

286. Peter von Hess (1792–1871): descendant of a family of artists. Starting his career with genre and landscape painting, he owes his fame to later paintings of battle scenes from the Napoleonic Wars. In 1833 he travelled to Greece where he painted the two famous paintings "Landing of Otto in Nauplion" and "Reception of Otto at the Theseion". In 1839 he was invited by Czar Nikolaus to St. Petersburg where he became painter of the court. He was a member of the academies of arts of Munich, Berlin, Vienna and St. Petersburg.

287. Karl Rottmann (1797–1850): he started his studies at the Munich Academy of Fine Arts in 1821. After a trip to Italy in 1826–1827 he became known as a landscape painter. Later, in the service of Ludwig I, he painted frescos with Italian landscapes in *Hofgarten*, Munich. After Otto's enthronement as King of Greece he had to add new frescoes, this time with Greek landscapes. Therefore he was sent to Greece where he created his famous water-colour sketches. From these sketches emerged twenty-three paintings on canvas that are on display in the Rottmann Hall of the *Neue Pinakothek*. Rottmann is the most significant painter of Greek landscape.

288. Karl Wilhelm Freiherr von Heideck (actually Heydeck) (1787–1861): a general and a talented amateur in drawing and painting. He lived in Greece in 1826–1829, participated in various battles as member of the expedition corps sent by Ludwig I of Bavaria to insurgent Greece, and became commandant of Nauplion. He returned to Greece with King Otto, still a minor, as a member of the Regency and remained there another three years, from 1833 to 1836. Leo von Klenze mentions that Heydeck had been his teacher in oil-painting. Heydeck's works are sketches and oil paintings of landscapes and battle scenes, of historic-documentary value. In 1824 Heydeck became an honorary member of the Munich Academy of Fine Arts.

289. Information contained in Carathéodory's letters (in Greek) to Penelope Delta on 1 November 1928, 25 January 1929, 4 February 1929, 6 February 1929, 21 March 1929. Leukoparidis (ed) (1956), pp 478–482.

290. Adolf von Harnack (1851–1930): theologian; active Christian; Lutheran. Professor of theology at Berlin University, he took on the post of General Director of Berlin Royal Library at the age of 57, four years later he became president of the then founded *Kaiser-Wilhelm-Gesellschaft*, which he presided over until his death. He was a teacher and friend of Hermann Scholz. During World War I he repeatedly attempted to intervene in politics supporting moderate positions. After the war he worked together with Friedrich Naumann on paragraphs concerning church, school and science. He was a friend of Friedrich Schmidt-Ott, second vice-president of the *Kaiser-Wilhelm-Gesellschaft*.

291. Friedrich Schmidt-Ott (1860–1956): he studied law in Leipzig and Berlin. In 1888 he joined the Prussian Ministry of Education and Cultural Affairs where he worked in close co-operation with Althoff. A year later he was entrusted with the administration of the Prussian Academy of Sciences and the Royal Academy of Sciences in Göttingen. In 1895 he was granted the title of Privy Counsillor and Speaker Counsillor. He organised the exhibition of the chemical specimen of the world exhibition of 1900 that took place in Paris. Entrusted with the Department of Art in 1903; from 1905 responsible for the big museums and the building of the Museum Island. In August 1917 he was appointed State Minister of Education and Cultural Affairs but resigned in November 1918 and joined the *Notgemeinschaft* that was founded in 1920, financing research. Schmidt-Ott became its president and retained this post until 1934 when he resigned. After World War II he supported the foundation of the *Deutsche Forschungsgemeinschaft* (German Research Society), an organisation succeeding the *Notgemeinschaft* that nominated him to its honorary president. He participated in the foundation of the *Max-Planck-Gesellschaft* (Max Planck Society), a successor to the *Kaiser-Wilhelm-Gesellschaft*.

292. Archiv der Staatsbibliothek zu Berlin – Preußischer Kulturbesitz – Festschrift für Schmidt-Ott, 1930, Blatt 20.

293. Schmidt-Ott F (1952) *Erlebtes und Erstrebtes 1860–1950* (Experienced and Attempted 1860–1950). Steiner, Wiesbaden, p 182. Cited from: Treue W (1987) *Friedrich Schmidt-Ott*. In: Treuer W, Gründer K (eds), as note 411, chapter 2, pp 235–250, here p 245.

294. Wende (1959), p 94.
This was something revolutionary because, traditionally, science politics in Germany were a matter of the federal states and not of the Reich.

295. Archiv der Staatsbibliothek zu Berlin – Preußischer Kulturbesitz – Festschrift für Schmidt-Ott, 1930, Blatt 201.

296. Carathéodory C, Dyck W v (1930) *Mathematik – Aus fünfzig Jahren deutscher Wissenschaft*. In: *Festschrift, Friedrich Schmidt-Ott zur Feier seines 70. Geburtstages überreicht*. Walter de Gruyter, Berlin, pp 275–285 and *Ges. Math. Schr.* vol V, pp 62–76.

297. Carathéodory to Delta on 21 August 1928, as note 97, chapter 3.

298. Georg von Streit (1966) Ημερολόγιον-Αρχείον (Diary–Archive), vol II-B. Privately printed, Athens, p 48.

299. Zannas (ed) (1988), p 60 & p 272.

300. Price C (1917) Venizelos and The War: A Sketch of Personalities and Politics. Simpkin, Marshall, Hamilton, Kent & Co, London, pp 30–39.

301. H R H Prince Nicholas of Greece (1929) Political Memoirs 1914–1917, Pages from my Diary. Hutchinson and Co, London, p 27.

302. Mazower M (1991), p 108.

303. Carathéodory to Delta on 21 October 1928. Letter in Greek. Leukoparidis (ed) (1956), p 478.

304. Carathéodory to Delta on 24 March 1929, one day before the national celebration of the Greek War of Independence. Leukoparidis (ed) (1956), p 482.

305. Carathéodory probably alludes to Benakis's maltreatment by the anti-Venizelists during the so-called Νοεμβριανά (November Occurences) of 1916 and to Benakis's trial on the suspicion of having instigated the assassination of Ion Dragoumis. Dragoumis, Penelope's great love, was murdered on his way from Kiphisia to Athens by men of the "security" battalion on 13 August 1920 in retaliation for the attempt on Venizelos's life at the Gare de Lyon, Paris, by two retired Greek officers the previous day. Venizelos was then returning from Sèvres after having signed the Peace Treaty of Sèvres (see 3.2).

306. Venizelos's statement to the press published in Ελεύθερο Βήμα of 1 October 1929. Stephanou (ed), as note 8, chapter 3, here vol II, p 262.

307. Ibid.

308. Georgios Papandreou (Kalentzi/Achaia 1888–Athens 1968): Prefect of Lesbos 1916; Governor General of Chios 1917–1920; Minister of the Interior in Gonatas's Cabinet 1923; Minister of National Economy 1925; exiled by Pangalos' dictatorship 1926; Minister of Education 1930–1932; Minister of Transport 1933; founder of the Democratic Party (later Democratic Socialist Party) 1935; persecuted and exiled by Metaxas's dictatorship 1936; imprisoned by the Italians 1942; Greek Prime Minister in the Middle East 1944; Prime Minister in Greece, leading the Centre Union 1963–1965.

309. Venizelos's election speech reads: "However, we will support our social system even more effectively through the necessary direct and far-reaching reform of our educational system. Our future can only seem gloomy because the state obviously works to raise the future army of social revolt, since hundreds of schools of the so-called classical education still release thousands of young people from their ranks, who normally are extremely inadequately educated and essentially incapable of doing any productive work. Of course, I am a supporter of classical education but only for a small part of the school youth, for those few chosen ones who would form the leadership of tomorrow. All those who, even if they come from the most inferior strata, will stand out in primary school are counted by me amongst the chosen ones. If they are penniless, the state will provide them with grants so as to enable them to continue their education. For most of those who study at secondary schools, the so-called classical education is absolutely fruitless and unrewarding. The primary school, which should change into a work school and which we should ensure is first a six-class school and hopefully later a seven-class one, is, in my opinion, not only a school of general education. I believe that during the last years at these schools, some practical agricultural knowledge could be mediated to pupils from agricultural regions that would make them more able in farming, even if they would not have the means or the opportunity to complete their agricultural training by studies at a practical agricultural school. It is certainly natural that I touch upon this extremely complex and difficult problem only in its general features."
See: Michalakeas T, Nasiotis G (1961) Βίβλος Ελευθερίου Βενιζέλου (Book about Eleutherios Venizelos), 5 vols. Ιστορικαί Εκδόσεις, Athens, here vol V: Ιστορική Σύν-θεσις (1922–1936) (Historical Synthesis (1922–1936)), pp 628–644, chapter entitled Παιδεία (Education).

310. During Capodistria's government, 7 school buildings were built, 437 by 1910, 1045 in the period 1910–1920, and 3167 during Venizelos's government of the years 1928–1932.

311. Carathéodory to Kalitsounakis on 3 February 1930. Letter in Greek. Vovolinis (ed) (1962), p 507.

312. In Government Gazette no. 169 of 16 May 1930. D. Kalitsounakis belonged to the Themis Lodge.

313. Bouzakis (1997), p 28ff.

314. BayHstA MK 35403, Laufzeit: 1924–1950.

315. Archiv der LMU München, E–II–1054.

316. BayHstA MK 35403, Laufzeit: 1924–1950.

317. The French architect and urban planner Ernest Hébrard (1875–1933), a graduate of the *École des Beaux-Art* of Paris, came to Greece with the *Armée d'Orient* in 1917 and was appointed to its archaeological service in Thessaloniki. From this post he headed a group of Greek and foreign researchers to plan the reconstruction of the city after the great fire of the late summer of 1917, which had destroyed a great part of Thessaloniki. From June 1918 until the end of 1921 he supervised the application studies and modification of the original plans. In collaboration with A. Papanastasiou and, presumably, accompanied by his friend Aristotelis Zachos, he photographically documented the enormous efforts to reconstruct destroyed cities and villages of Northern Greece after World War I. In 1918 he was appointed professor to the newly founded architectural faculty of Athens,

participated in the organisation of the study programme and lectured on architectural composition and urbanism. He was also appointed to the Council of Public Works with the rank of an inspector. As such, he became a member of the Superior Technical Council and supervised the realisation of technical works. With a group of Greek architects he worked on a new plan of Athens in 1919. After Venizelos's electoral defeat in 1920, he was dismissed from the National Technical University together with forty-five other professors, who had been appointed by the Liberals. He was also removed from all other projects which were revoked by war in Asia Minor. In 1922, he accepted the post of the chief architect in the Government General of Indochina and left Greece for Vietnam. Towards the end of 1927, he returned to Greece and worked at the University of Thessaloniki and the Athens architectural faculty, to which he was reappointed in February 1929. As consultant to the Greek government, he participated in a series of programmes for the modernisation of the cities, for archaeological research, for protection of monuments and organisation of a contemporary architectural education. He also participated in the programme for construction of school buildings as head of the architectural department of the Ministry of Education. His directives to the architects favoured free architectural creation and offered the possibility of a flexible and functional building programme. In 1928, Hébrard was entrusted with the university buildings of Thessaloniki and submitted his drawings at the end of 1929. However, they were not executed but only kept as historical texts in the university archive. Later, in 1938 the architect N. Mitsakis planned the university complex on a confined site covering a small part of the present university site.

The *Institut Français de Thessalonique* organised the exhibition *Ernest Hébrard – Architecte & Photographe* in Thessaloniki from 13 December 2001 to 13 February 2002.

318. Dimitrios Pikionis (1887–1968): civil engineer from the National Technical University of Athens; studied painting in Munich and Paris; associate professor of decoration at the National Technical University of Athens 1925; full professor 1943; member of the Academy of Athens in 1966; representative of Greek rationalism in architecture at the beginning of the 1930s, later traditionalist.

319. Carathéodory C (1930) *Η Αναδιοργάνωσις του Πανεπιστημίου Αθηνών*. National Printing Office, Athens. See: Georgiadou (2003) The Reorganisation of the Universities of Athens and Istanbul in the 1930s – Two One-Man Projects and a Historical Coincidence. *Chronos, Revue d'Histoire de l'Université de Balamand*, vol 8. (In print).

320. BayHstA MK 35403, Laufzeit: 1924–1950.

321. There exists a note by Haar and Carathéodory to Zermelo, also signed by others, showing that Carathéodory was in Budapest on 24 September 1930:
"Dear Zermelo,
we have eaten the overleaf menu to honour Cara (who makes the best jokes). Why don't you come here?
Kind regards,
Haar
Why don't you come to Munich anymore? I will travel with my wife to Athens and return only at the beginning of November; but then you can come whenever you want.
Cara
Kind regards
Dénes König   F. Riesz   Karl Tangl
L. Fejér"
The "overleaf menu" was offered in restaurant Gundel, hotel Gellért in Budapest on 24 September 1930 and included:
"Crème Dubarry in cups, goose liver risotto, roasted partridge, lentils with lard, apple purée, fruit, a glass of Fructidor, coffee."
Universitätsbibliothek Freiburg, Nachlaß Ernst Zermelo.

322. Centre for the History of Thessaloniki (ed) *Η Θεσσαλονίκη στα χαρακτικά από τον 15ο έως τα τέλη του 19ου αιώνα.* (Thessaloniki in the Engravings from the 15th to the 19th Century). Thessaloniki, 1998.

323. Schumacher (1937), p 174.

324. Cf. Touloumakos I, *Η ίδρυση του Πανεπιστημίου Θεσσαλονίκης και ο μαθηματικός Κωνσταντίνος Καραθεοδωρή* (The Foundation of the University of Thessaloniki and the Mathematician Constantin Carathéodory). *Μακεδονία*, 16 January 2000.

325. Carathéodory to Courant on 15 July 1931. Letter in German. New York University Archives.
    De Possel proved the possibility of mapping of a multiply connected domain onto a parallel-slit domain. See: de Possel R (1931) *Zum Parallelschlitztheorem unendlich vielfach zusammenhängender Gebiete* (On the Parallel-Slit Theorem of Multiply Connected Domains). *Nachr. Ges. Wiss. Gött., math.-phys. Kl.*, pp 199–202.
    Courant defines the parallel-slit domains or, briefly, the slit domains as domains consisting of the whole plane of the complex variable $\omega = u + iv$ except for straight segments $u = $ constant, the "boundary slits".
    Courant R (1950) as note 155, chapter 2, p 45f.

326. Courant to Carathéodory on 19 July 1931, Letter in German. New York University Archives.

327. BayHstA MK 35403, Laufzeit: 1924–1950 and Archiv der LMU München, OC N 14. The papers were *Ελεύθερον Βήμα* (Free Tribune), *Πρωία* (Morning) and *Ελεύθερος Άνθρωπος* (Free Man).

328. Archiv der LMU München, OC N 14.

329. Tietze to the rector's office of Munich University on 11 September 1931. Archiv der LMU München, E-II-1054.

330. Carathéodory to Courant (postcard with stamp 18 May 1931). New York University Archives.

331. Archiv der LMU München, OC N 14.

332. BayHstA MK 35403, Laufzeit: 1924–1950. Carathéodory's name is again written as Karathéodory (in capital letters).

333. Ibid.

334. Vovolinis (ed) (1962), p 515ff.

335. BayHstA MK 35403, Laufzeit: 1924–1950.

336. Cf. Dimaras A (ed) (1974) *Η μεταρρύθμιση που δεν έγινε*, as note 33, chapter 3.

337. Bouzakis (1997), p 64ff.

338. Glinos D (1971) *Τα ιδανικά της Παιδείας και ο κ. Παπανδρέου* (The Ideals of Education and Mr. Papandreou). In: Glinos D (1971–1975) (Exclusive Pages), 4 vols. Στοχαστής, Athens, here vol 1, pp 99–113.

339. *Η Εκπαιδευτική μας πολιτική* (Our Educational Policies). In: *Ερμής*, year ΙΑ', issue 456, Athens, 4 April 1931.

340. Papandreou's address to the Greek parliament on 19 March 1932 during the debate concerning law 5343/1932.

341. Ministry of Education and Religious Affairs, 20 April 1932, register no. 23414.
    In a session on 12 January 2000, the senate of the University of Thessaloniki unanimously decided to adopt the proposal of the dean's office of the *Σχολή Θετικών Επιστημών* (School of Positive Sciences) and honour Carathéodory for his contribution to the university. They have decided to name the administration building after him and to put his bust in an appropriate place on the university site.

342. Communication from Mrs. Despina Rodopoulos-Carathéodory, who has not presented any document. The information about the Papanastasiou government of summer 1932 in Greek papers of that epoch.

343. Aristotelis Oeconomou: professor of higher mathematics at the National Technical University of Athens since 1929, a member of the Munich mathematical colloquium in the summer of 1931, author of the paper *Sur un théorème de Carathéodory et Fejér* (On a Theorem of Carathéodory and Féjer) and of the entries on *Space* and *Time* for the *Μεγάλη Ελληνική Εγκυκλοπαίδεια* (Great Greek Encyclopaedia) vol 24 (1934), for which Carathéodory had written his *Space–Time* article.

344. Dimitrios Kappos: holder of the chair I of mathematics at Athens University, known for his works and books on the theory of probability; lecturer at Erlangen University from summer semester 1947 to summer semester 1956, on leave from summer semester 1954 to summer semester 1956, giving no lectures during this time.
    Communication to the author from Petra Schwarz on 27 November 1998.
    Kappos had a teaching commission for mathematics at the Philosophical-Theological University of Bamberg for three semesters after World War II: introduction to analysis in winter semester 1946–1947, introduction to higher algebra in summer semester 1947, introduction to higher algebra, part II in winter semester 1947–1948.
    Communication to the author from Dr. John Moore on 14 December 1998.
    Kappos's book on infinitesimal calculus (Athens 1960) was dedicated to "the holy shadow of [his] ever memorable teacher Constantin Carathéodory".

345. In 1910, Hatzidakis supported the creation of mathematical faculties with the argument that whoever desired higher spiritual pleasures for himself should first of all study mathematics. He was dismissed from the university as a supporter of the demotic language but readmitted in 1912. In 1913, being the dean of the Faculty of Natural Sciences and Mathematics, he proposed the foundation of a mathematical journal, which would also accept works by non-members of the academic community, the introduction of Sunday lecturing by university professors and the foundation of secondary schools of technical education. Interested in foreign languages, literature and linguistics, he wrote poems, published in renowned journals under a pseudonym and was said to speak thirteen languages.
    Hatzidakis writtenly proposed marriage to Lise Meitner on 22 August 1915. See: Sexl L, Hardy A (2002) *Lise Meitner*. Rowohlt, Reinbek bei Hamburg, p 56.
    On 9 July 1942, Carathéodory provided the president of the German Union of Mathematicians with the following information: "Nikolaos J. Hatzidakis, born 25 April 1872 in *Berlin*, 1900–04 Prof. at the Military School in Athens, since 1904 at the University there. 1939 retired, died in *Athens* on 25 January 1942. He occupied himself especially with surface theory and kinetics and wrote, apart from those in Greek, German, French, English, Italian, also several works in Danish. He was a member of the German Union of Mathematicians since 1899. His father's name is spelled *Hazzidaki* in the mathematical literature (for example, Encyclop. III 3 p 344 or 348). This is entirely incomprehensible to me, since his father's brother, Georgios Hatzidakis, who died only ten months ago and was a known linguist, is also written Ha*t*zidaki*s* in the list of members of the Prussian Academy."
    Carathéodory to Süss on 9 July 1942. Letter in German. Universitätsarchiv Freiburg, Nachlaß Wilhelm Süss, C89/48.
    Kalitsounakis had brought Georgios Hatzidakis in contact with Carathéodory in Athens. Carathéodory, who immensely respected Hatzidakis and his endeavours for the establishment of the *καθαρεύουσα* (purified language) as the official national language, was very pleased with this acquaintance. The meeting of the two men is mentioned in: Vovolinis (ed) (1962), p 484.
    In 1919 Georgios Hatzidakis had accused the educational reformers Glinos, Triantafyl-

lidis and Delmouzos of being nationally dangerous persons who pursued the demise of religion, language, family and fatherland.

See: Iliou, as note 25, chapter 3.

G. Hatzidakis was the first rector of the University of Thessaloniki in 1926–1927; his *Einleitung in die neugriechische Grammatik* (Introduction to Modern Greek Grammar) had been published in Leipzig in 1892.

346. Triantafyllidis's most important work, his Νεοελληνική Γραμματική (Modern Greek Grammar), was published in June 1941. It is the first systematic study on phonetics, morphology, production and composition of the modern Greek language and, even now, a symbol of resistance against the lethargy and stagnation in matters of language and education ruling the Greek society.

Cf. Stavridi-Patrikiou R, Μανόλης Τριανταφυλλίδης. Διανοούμενος και αγωνιστής του κινήματος της Δημοτικής (Manolis Triantafyllidis. An Intellectual and Fighter of the Demotic-Language Movement). Τα Νέα, 30 December 1999.

347. Kakridis spent the years 1931–1939 as a lecturer and associate professor at the Philosophical Faculty of the University of Thessaloniki before leaving for Athens in 1940. At the Philosophical Faculty of Athens University, he was confronted with a conspiracy which ended in the notorious Δίκη των τόνων (Trial of Accents) in the midst (November 1941 to July 1942) of the German occupation. Accused of "criminal" and "nationally harmful" acts, Kakridis was able to survive within the hostile environment for three more years, but, in the end, he was forced to return to Thessaloniki. His only "crime" was having supported the idea that the revival of the dead body of Greek antiquity through an approach based on the use of modern Greek language was a prerequisite for the understanding and appropriation of its achievements. In his opinion, the gap between the ancient and the modern Greek worlds could not be bridged through blind imitation or borrowing of non-assimilated and anachronistic values of antiquity, but only through dynamic confrontation with antiquity that would result in a fruitful mixing, necessary to redetermine contemporary reality. The whole trial and the indictment aimed mostly at the rejection of the monotonal system that had been proposed to the Academy of Athens by Kakridis and, before him, by the renowned linguist Georgios Hatzidakis in 1929. Kakridis was accused of trying to diminish the value of ancient Greek civilisation or to evoke doubts about its usefulness. Although he was defended by numerous respected personalities of the Greek intelligentsia, such as Hondros, Triantafyllidis and Kritikos, who were also acquainted with Carathéodory, the disciplinary proceedings against him led to his temporary dismissal. Kakridis's philological production, about forty books and more than two-hundred articles in Greek and other languages, covers a vast spectrum of subjects and methods: interpretative studies of antique lyric and dramatic poets, comparative linguistic studies, comments and interpretation of modern poets and prose writers. Although he was the editor of the Ιστορικό Λεξικό της Νέας Ελληνικής Γλώσσης (Historical Lexicon of the Modern Greek Language) for the Academy of Athens, he never became a member of this institution.

Apart from his lecturing activity at the university, Kakridis served education by accepting the presidency of the Παιδαγωγικό Ινστιτούτο (Pedagogic Institute) through which the Georgios Papandreou government intended to implement the daring educational reform of 1964. Interrupting his academic career, he called for help from personalities dedicated to education, such as Kritikos, and received it. Unfortunately, the work of the institute was stopped due to the "July defection" in 1965 and officially ended because of the April coup d'etat in 1967.

Cf. Hourmouziadis N, Ι. Θ. Κακριδής: Άνδρας πολύτροπος, ωραίος ως Έλληνας (I. Th. Kakridis: A Cunning Man, Nice like a Greek). Τα Νέα, 10 January 2000.

Ioannis Kakridis and his family visited Carathéodory in Munich, together with his wife Olga on 11 April 1927 and later with his daughter Eleni on 29 July 1928; his sister, Penelope Kakridis, a mathematican with whom Carathéodory's son was in love, N. Kritikos

and Josefa von Schwarz were Carathéodory's guests on 20 July 1929.
Ioannis Kakridis's mother was a niece of the linguist Georgios Hatzidakis.

348. Othon Pylarinos continued the N. Hatzidakis's line of thought in geometry.

349. Gratsiatos, *Über das Verhalten der radiotelegraphischen Wellen in der Umgebung des Gegenpunktes der Antenne und über die Analogie zu der Poissonschen Beugungserscheinung* (About the Behaviour of Radio-Telegraphic Waves in the Neighbourhood of an Opposite Point of an Aerial and about the Analogy to Poisson's Diffraction Phenomenon), 1928.

350. The medical faculty, which by law 835 of 2 September 1937 had been replaced by the veterinary faculty, and the theological faculty began their work in 1942 to support resistance against the German occupation. During the occupation, the university lost all of its installations and a great part of its equipment, its academics and students; its administrative personnel suffered famine, persecutions, imprisonment and executions. Seminars were held in private rooms, in cinemas, in clubs, in synagogues.

351. Carathéodory to Hilbert on 19 January 1932. Letter in German. Niedersächsische Staats- und Universitätsbibliothek Göttingen – Abteilung für Handschriften und seltene Drucke. Hilbert 452c,7.

352. Goethe, *Faust II*, dritter Akt: "*Vor dem Palaste des Menelas zu Sparta*".
"*Ja, auf einmal wird es düster, ohne Glanz entschwebt der Nebel,*
Dunkelgräulich, mauerbräunlich. Mauern stellen sich dem Blicke,
Freiem Blicke, starr entgegen. Ist's ein Hof? ist's tiefe Grube?
Schauerlich in jedem Falle! Schwestern, ach! wir sind gefangen,
So gefangen wie nur je."

353. Large (1998), p 289. Large refers to the Berlin journalist Carl von Ossietzky as writing that, in 1932, the Germans took Goethe to heart "not as a poet and a prophet but, above all, as opium".
Werner Deetjen, director of the library from 1916 to 1939, endeavoured to bring the traditional tasks of a library in accord with the education of a broader circle of readers, to stress the public character of the library and to reform its administration, and also to enrich the library with works of art. In December 1918 the *Großherzogliche Bibliothek* was given the name *Weimarer Bibliothek* and in 1919 it was renamed as *Thüringische Landesbibliothek*. In the following years both of the names *Großherzogliche Bibliothek* and *Thüringische Landesbibliothek* were used to denote the library. Deetjen assessed "a certain physiognomy which bears such pronounced historic traits of a great intellectual past" as a special feature of the *Thüringische Landesbibliothek*. In the years 1928–1930, the financial means for book purchases were enormously reduced to the amount of RM 2400 and the library had the smallest budget of all German regional libraries. Deetjen's visit to Carathéodory could have also been aimed at persuading Carathéodory to donate books to the library, which in 1925–1935 had set literature, language and history as its main focus. For the history of the library, which is now called *Herzogin Anna Amalia Bibliothek* see: Knoche M (ed) *Herzogin Anna Amalia Bibliothek – Kulturgeschichte einer Sammlung* (The Library of Duchess Anna Amalia – Cultural History of a Collection). Carl Hanser Verlag, München, Wien, 1999.

354. Cited from: Rehm W (1952) *Griechentum und Goethezeit – Geschichte eines Glaubens* (Greeks and the Time of Goethe – History of a Belief). L. Lehnen, München, p 164.

355. Butler E M (1935) The Tyranny of Greece over Germany – A study of the influence exercised by Greek art and poetry over the great German writers of the eighteenth, nineteenth and twentieth centuries. CUP.

356. Carathéodory C (1931) Τα Μαθηματικά, as note 23, chapter 1.

357. Carathéodory to Courant on 24 March 1931. Letter in German. New York University Archives.

358. Carathéodory to Courant. Postcard in German with stamp 18 May 1931. Ibid.
359. Courant to Carathéodory on 20 May 1931. Letter in German. Ibid.
360. Courant to Carathéodory on 26 May 1931. Letter in German. Ibid.
361. Carathéodory to Courant on 1 June 1931. Letter in German. Ibid.
362. Courant to Carathéodory on 3 June 1931. Letter in German. Ibid.
363. Courant to Carathéodory on 19 July 1931. Letter in German. Ibid.
364. Courtesy of the Archives of the Institute for Advanced Study. Cf. (1999) American National Biography. Oxford University Press, New York, vol 16, p 302.
365. The programme of the Congress mentions:
    "Dienstag, den 6 September
    9–10 Uhr: Vortrag, C. Carathéodory, Über die analytischen Abbildungen durch Funktionen mehrerer Veränderlicher. Auditorium III." Courtesy of the ETH Archives.
366. Behnke, as note 29, chapter 1, p 31.
367. Siegmund-Schultze, as note 382, chapter 2, here p 271f.
368. *Internationaler Mathematikerkongress Zürich 1932 – Programm* (International Congress of Mathematicians, Zurich 1932 – Programme). Courtesy of the ETH Archives.
369. The Princeton Mathematics Community in the 1930s, as note 220.
370. Carathéodory C (1933) *Über die strengen Lösungen des Dreikörperproblems* (presented at the session of 8 July 1933). *Sitzungsber. Bayer. Akad. d. Wiss. München, math.-nat. Abt.* 1933, pp 257–267 and *Ges. Math. Schr.* vol II, pp 387–396.
371. Carathéodory C (1935) *Sur les Équations de la Mécanique. Actes du Congrès Interbalcanique de Mathématiciens.* Imprimerie Nationale, Athènes, pp 211–214 and *Ges. Math. Schr.* vol II, pp 397–401.
372. Carathéodory to Herglotz on 2 September 1940. Letter in German. Niedersächsische Staats- und Universitätsbibliothek Göttingen – Abteilung für Handschriften und seltene Drucke. Cod. Ms. Herglotz F 18.
373. Carathéodory C (1940) *Über die Integration der Differentialgleichungen der Keplerschen Planetenbewegung. Revue mathématique de l'Union Interbalcanique* III, 3–4:8 pages and *Ges. Math. Schr.* vol II, pp 414–425.
374. Carathéodory C (1945/46) *Über die Integration der Differentialgleichungen der Keplerschen Planetenbewegung. Sitzungsber. Bayer. Akad. d. Wiss. München, math.-nat. Klasse* 1945/46, pp 57–76 and *Ges. Math. Schr.* vol II, pp 426–445.

5 *National Socialism and War*

1. *Bekenntnis der Professoren an den deutschen Universitäten zu Adolf Hitler und dem nationalsozialistischen Staat, überreicht vom Nationalsozialistischen Lehrerbund Deutschland/Sachsen.* Dresden 1933.
    The presentation of the German university conditions in 1933 is mainly based on an *AstA-Referat für Grund- und Freiheitsrechte* (Students' Union Report on Fundamental and Civil Rights) by Michael Laux, vol X uni 7.
2. Craig G A (1982) The Germans. Putnam, New York, p 184. Translated into German by Hermann Stiehl as: (1984) *Über die Deutschen.* Gütersloh, Stuttgart.
3. Cf. Bollenbeck (1999).
4. Bieberbach's Berlin speech in 1934 (cited in Fraenkel (1967), p 166) is revealing in this respect. Bieberbach saw "German mathematics [...] rooted in blood and soil" and believed that, in practical cultural policies, mathematics should be freed from "the curse of sterile intellectualism", which burdened thinkers "alien to folk and race." In his opinion, these thinkers should not be regarded as German researchers and would not exist in the future.

5. Doetsch to Süss on 1 August 1934. Universitätsarchiv Freiburg, Nachlaß Wilhelm Süss, C89/5.
   Cf. Remmert V (1999) Mathematicians at War. Power Struggles in Nazi Germany's Mathematical Community: Gustav Doetsch and Wilhelm Süss. *Revue d'histoire des mathématiques*, 5:7–59.

6. Süss to Feigl on 3 April 1941. Universitätsarchiv Freiburg, Nachlaß Wilhelm Süss, C89/51.

7. Schappacher N: The Nazi era: the Berlin way of politicizing mathematics. In: Begehr et al (eds), as note 38, chapter 1, pp 127–136, here p 134.
   For a detailed description of the events in the DMV see: Mehrtens H (1985) *Die „Gleichschaltung" der mathematischen Gesellschaften im nationalistischen Deutschland* (The "Gleichschaltung" of Mathematical Societies in National Socialist Germany). In: *Jahrbuch Überblicke Mathematik*, as note 31, chapter 3, here vol 18, pp 83–103.

8. BArch, REM-Akte Stroux, Johannes (ehem. BDC).

9. Archiv der LMU München, E II N Car.
   At that time the term Greek Catholic properly applied to members of all dioceses of Byzantine (Eastern) rite in full union with Rome.
   Communication to the author from Dr. Albert Rauch, Ostkirchliches Institut Regensburg, on 22 February 2001.
   The main church bodies with members resident in Istanbul are:
   the Eastern Orthodox (Greek Orthodox) Churches which recognise the decisions taken in the Seven Oecumenical Councils; the Oriental Orthodox (Armenian and Syrian) Churches which separated themselves from Eastern Orthodoxy after the Third Council which took place in Ephesus in 431; the Eastern Rite Catholic Churches which include groups that united with Rome during the past 400 years (the Armenian Catholic and the Greek Catholic, for instance) and one group (Maronite) which considers that it never broke loose from Rome; the Latin (Roman) Catholic Church has some foreigner members, who are more or less permanent residents in Istanbul.
   In various documents Carathéodory presents himself either as belonging to the Greek Orthodox Church, or to the Oriental Church, or to the Greek Catholic Church.

10. BayHStA MK 35403, Laufzeit: 1924–1950.

11. Ibid.

12. Archiv der LMU München, E II N Car.

13. Cf. Mac Lane S, as note 159, chapter 4.

14. Wolff S L (1993) *Vertreibung und Emigration in der Physik* (Persecutions and Emigration in Physics). *Physik in unserer Zeit* 6:267–273.

15. Blumenthal to von Kármán on 29 September 1935. Letter in German. Caltech Archives, Theodore von Kármán Papers, Folder 3.10.

16. Born to Einstein on 31 May 1939. *Albert Einstein – Hedwig und Max Born. Briefwechsel 1916–1955* (1972), p 143.

17. Department of Public Information, Princeton University, Release 26 March 1969. Willy Feller Faculty File. Courtesy of Mudd Library, Princeton University.

18. In December 1924, before the election for the Reichstag, Born had declared his support for the German Democratic Party (DDP = *Deutsche Demokratische Partei*). Heiber (1992), p 12.

19. Born thought of his family as being "Jews only according to the present laws but they have never thought of that earlier. They lack (and so do I) every, but every emotional relation to the real Jewry, their forms, their laws. I am in nature – or I believe to be – a liberal Westeuropean with a strong cultural element". He had a rejecting attitude towards a possible appointment at the Hebrew University. "I am a nationalist just as little for Juda as for Germania, or actually less; for I know neither Hebrew nor Jewish literature,

while I have grown very fond of the German language, poetry, art." Cf. Wolff S, as note 14. "But of course now I feel very strongly, not only because I and my own are counted thereto, but because oppression and injustice provoke me to anger and resistance." Born to Einstein on 2 June 1933. *Albert Einstein – Hedwig und Max Born. Briefwechsel 1916–1955* (1972), p 122.

20. Cf. Stern F (1999), p 68f.

21. Born to Einstein on 2 June 1933. *Albert Einstein – Hedwig und Max Born. Briefwechsel 1916–1955* (1972), p 125.

22. Courant to Carathéodory on 19 July 1933. Letter in German. Courant Institute at the New York University.

23. Widmann, Horst (1973) *Exil und Bildungshilfe – Die deutschsprachige akademische Emigration in die Türkei nach 1933* (Exile and Educational Help: The German-Speaking Academic Emigration to Turkey after 1933). Herbert Lang, Bern; Peter Lang, Frankfurt am Main, p 53.
    For the reorganisation of the University of Istanbul see: M. Georgiadou, as note 319, chapter 4.

24. Cf. Beyerchen (1980), p 51.

25. Born to Einstein on 2 June 33. *Albert Einstein – Hedwig und Max Born. Briefwechsel 1916–1955* (1972), p 124.

26. Veblen to Flexner on 12 August 1933. Letter in English. Courtesy of the Archives of the Institute for Advanced Study.

27. Correspondence re. Prussian and Bavarian and Italian Academies of Sciences, File Folders 35(4) to 36(2), Box 49 in Albert Einstein Duplicate Archive, The Princeton University Library.
    For the history of this conflict see: Stoermer M (1995) *Die Bayerische Akademie der Wissenschaften im Dritten Reich* (The Bavarian Academy of Sciences in the Third Reich). *Acta historica Leopoldina* 22:89–111, here p 89f.

28. *Albert Einstein – Hedwig und Max Born. Briefwechsel 1916–1955* (1972), p 121.

29. Carathéodory to Fraenkel on 30 April 1931. Letter in German. Arc 4° 1621/Caratheodory, Archive Adolf Fraenkel. The Jewish National and University Library.

30. Buber M, Feiwel B, Weizmann Ch (1902) *Eine jüdische Hochschule* (A Jewish University). Jüdischer Verlag, Berlin.

31. In 1937, Freundlich returned to Prague to take on an appointment as a professor of astronomy at Charles University but was forced to leave the country two years later. This time he reached St. Andrews University where he founded a department of astronomy.

32. "The Rouse Ball lecturer is appointed annually by the Faculty Board of Mathematics. In 1932–1933 it was Einstein, in 1933–1934 van der Waerden, in 1934–1935 Heisenberg. The minutes of the Faculty Board are totally unrevealing: 'Dr. Landau was appointed Rouse Ball lecturer for 1935 and failing him Professor Cartan.' Cartan was Rouse Ball lecturer in 1938–1939." Communication to the author from Dr. Elisabeth Leedham-Green, Cambridge University Archives, on 21 October 1998.

33. In 1939, von Mises emigrated to the States and worked there as the Gordan McKay Professor for Aerodynamics and Applied Mathematics at Harvard University.

34. In this journal Carathéodory published his paper for von Mises' 50th birthday. He examined the example of a heavy sledge gliding on a horizontal plane without being affected by external forces and showed that it is impossible to approximate the motion of the sledge by assuming a reaction force similar to the force of friction in the case of motion in viscous fluids. On the contrary, for the non-holonomous motion examined, one could assume static friction, as is the usual case in the statics of solid bodies.
    Carathéodory C (1933) *Der Schlitten* (The Sledge). *Z. f. angew. Math. u. Mech.* 13:71–76 and *Ges. Math. Schr.* vol II, pp 377–386.

35. *Albert Einstein – Hedwig und Max Born. Briefwechsel 1916–1955* (1972), p 117.

36. Reichenbach stayed in Istanbul until 1938. He spent the following years in LA until his death in 1953.

37. In the winter semester of 1936–1937, Schrödinger accepted a post in Graz from where he was dismissed without notice on 1 September 1938. In October 1939, he went to Dublin where he became professor at the Dublin Institute for Advanced Studies in 1940. He returned to Vienna as university professor in 1956.

38. At the beginning of 1939, Toeplitz emigrated to Palestine where Salman Schocken offered him an academic-administrative position at Hebrew University. He died in Jerusalem a year later. Toeplitz had been discriminated in the past by being put back in an appointment procedure, a thing which he himself attributed to anti-Semitic motives. This assertion is mentioned in: *Albert Einstein – Hedwig und Max Born. Briefwechsel 1916–1955* (1972), p 28.

39. Cf. Litten F (1994) *Oskar Perron, ein Beispiel für Zivilcourage im Dritten Reich* (Oskar Perron, an Example of Courage of Convictions in the Third Reich). *DMV-Mitt.* 3:11–12.

40. What was actually discussed at that meeting were the attempts of the executive committee of the Association of German Universities to adjust to the new conditions.

41. Cited in Böhm (1995), p 72.

42. Ibid, p 90ff and 222f.

43. Ibid, p 223f.

44. Ibid, p 401f.

45. Ibid, p 585.

46. Aloys Fischer (1880–1937): educational theorist, psychologist, philosopher; 1907–1918 teacher of the Bavarian Princes Luitpold and Albrecht; 1910 founded the Educationalist-Psychological Institute of the Munich Teachers' Association; 1919 head of the Seminar of Education and 1929 also of the Psychological Institute of Munich University. An extremely versatile scientist, he comiitted himself to the humanisation of vocational schools.

47. Böhm (1995), p 224f.

48. Margaret D. Kennedy, Josa v. Schwarz, Gianfranco Cimmino, Orin J. Farrell, Hans Wolkenstörfer, Georg Aumann, de Possel, Ernst Peschl, S. Bochner.

49. de Possel, A. Oeconomou, Sophia A. Oeconomou, Josa v. Schwarz, S. Bochner, H. Boerner, Margaret Kennedy, G. Aumann, E. Peschl. Miss Josefa v. Schwarz gained her doctorate under Carathéodory and Tietze on 26 July 1933 with the dissertation *Das Delaunaysche Problem der Variationsrechnung in kanonischen Koordinaten* (The Delaunay Problem of the Calculus of Variations in Canonical Co-ordinates).

50. Published as: Carathéodory C (1946–1947) *Zum Schwarzschen Spiegelungsprinzip (Die Randwerte von meromorphen Funktionen). Comm. Math. Helv.* 19: 263–278 and *Ges. Math. Schr.* vol III, pp 337–353.

51. Carathéodory, as note 121, chapter 1.

52. Böhm (1995), p 584f.

53. Cf. Leaman G, Simon G (1994) *Die Kant-Studien im Dritten Reich* (Kant Studies in the Third Reich). *Kant-Studien* 85:443–469.

54. The incident is told by his wife's nephew, John Argyris.

55. Feigl to Süss on 5 July 1944. Universitätsarchiv Freiburg, Nachlaß Wilhelm Süss, C89/51.

56. Niedersächsische Staats- und Universitätsbibliothek Göttingen – Abteilung für Handschriften und seltene Drucke. Cod. Ms. H. Hasse. 1:265.

57. Ibid, Beil.

58. See the very well researched article: Litten F (1996) *Ernst Mohr – Das Schicksal eines Mathematikers* (Ernst Mohr – The Fate of a Mathematician). *Jber. Deutsch. Math.- Verein.* 98:192–212.

59. Cf. Siegmund-Schultze R (1986) *Faschistische Pläne zur "Neuordnung" der europäischen Wissenschaft. Das Beispiel Mathematik* (Fascist Plans for a "New Order" of European Science. The Paradigma of Mathematics). *NTM-Schriftenr. Geschichte Naturwiss., Technik, Med.* 23/2, Leipzig, pp 1–17.
    Beyerchen (1980), pp 82 and 87.
    Grüttner (1995), pp 87 and 157.

60. Ludwig Hopf got a research grant for Cambridge, England in 1939 and was appointed as a lecturer of higher mathematics at Trinity College, Dublin where he died on 21 December that year.

61. Personalakte Dr. Hans Rademacher. Staatsarchiv Hamburg, Hochschulwesen, Dozenten- und Personalakten, I 64, 1922–1934.

62. Kline to Carathéodory on 8 November 1949. Letter in English. Universitätsarchiv Freiburg, Nachlaß Wilhelm Süss, C89/407.

63. Arnold Dresden was a member of the mathematical faculty of the University of Wisconsin from 1909 to 1927.

64. "I am greatly indebted to Professor Dresden, Swarthmore, for the fine piece of work he performed!" Van der Waerden in the preface to the English edition.

65. Courtesy of the Archives of the Institute for Advanced Study.
    Signatures of I. J. Schoenberg, Dalli Schoenberg and Elisabeth Schoenberg in Carathéodory's guest-book. Schönberg went to Colby College in the autumn of 1936. An article in the college student paper in the autumn of 1937 notices that Dr. Schoenberg had recently addressed the mathematical colloquia of Harvard and Brown universities and that he was publishing work in the Annals of Mathematics and in the Transactions of the American Mathematical Society.

66. Süss to Behnke on 4 March 1946. Universitätsarchiv Freiburg, Nachlaß Wilhelm Süss, C89/5.

67. Barch, WI-Akte Perron, Oskar (ehem. BDC).

68. BayHStA, MK 69776. Communication to the author from Freddy Litten on 22 November 2001.

69. Archiv der LMU München, E II N Car.

70. BayHStA MK 35403, Laufzeit: 1924-1950.

71. Carathéodory to Kalitsounakis on 22 February 1934. Letter in Greek. Vovolinis (ed) (1962), p 519.

72. Carathéodory, as note 371, chapter 4.

73. Courtesy of the Archives of the Institute for Advanced Study.

74. BArch, REM-Akte Stroux, Johannes (ehem. BDC).

75. Fraenkel (1967), p 127. In January 1939, Dehn could escape to Copenhagen and later to Trondheim, Norway, in the beginnings of 1941 over the German patrolled Norwegian-Swedish border through Finland, Russia, Siberia, Japan to arrive in San Francisco. He made many stopovers in the States founding positions either not up to his merits or loading him with a lot of stress. From the University of Idaho, to the Illinois Institute of Technology in Chicago, then to St. John's College in Annapolis, Maryland, he arrived finally in 1945 in the small Black Mountain College in North Carolina, an institution having arts as the core of its curriculum.
    In a postcard dated 27 January 1940 (Postcard in German. Max Dehn Papers, Archives of American Mathematics. Collections in the Center for American History. The University of Texas at Austin. Archive notice: "Postcard from Caratheodory, distinguished

mathematician and, like Dehn, a student of Hilbert."), Carathéodory congratulated Dehn for his paper on ornamentation (Dehn M (1940) *Über Ornamentik* (On Ornamentation). *Norske Mat. Tidsskrift* 21:121–153) which he had read "with great interest. It is an original and intelligent work."
This paper provides many interesting and even unexpected examples and analyses ornamentation both as a link between mathematics and the arts and also as a branch of applied mathematics. Magnus W (1978) Max Dehn. The Mathematical Intelligencer 1/3:132–143, here p 140.
"Common roots of mathematics and ornamentics" was the title of one of the two lectures Dehn gave at the Black Mountain College in March 1944. Sher R B (1994) Max Dehn and Black Mountain College. The Mathematical Intelligencer 16/1:54–55, here p 54.

76. Hamburger lectured in the period 1941–1947 at the University of Southampton and became a full professor at the University of Ankara in 1947. He stayed there up to 1953 when he returned as professor to the University of Cologne.

77. In late February 1939 Hellinger arrived in the States where he started lecturing at the Northwestern University in Evanston, Illinois. He became a full professor there in 1946.

78. Siegmund-Schultze, as note 382, chapter 2, here p 260.
Rosenthal became a research fellow and lecturer at the University of Michigan in 1940; member of the mathematical faculty of the University of New Mexico 1942–1947; professor of mathematics at Purdue University in Lafayette, Indiana 1947–1957.

79. In the period 1936–1957 Fajans was professor of physical chemistry at the University of Michigan in Ann Arbor.

80. Only in 1946 Zermelo was reappointed as a honorary professor of the University of Freiburg.

81. Carathéodory to Zermelo on 25 April and on 3 May 1933. Letters in German. Universitätsbibliothek Freiburg, Nachlaß Ernst Zermelo.

82. BayHStA MK 35403, Laufzeit: 1924–1950.

83. Ibid.

84. Carathéodory C (1935) *Examples particuliers et théorie générale dans le calcul des variations. Enseignement mathématique* 34:255–261 and *Ges. Math. Schr.* vol I, pp 306–311.

85. BayHStA MK 35403, Laufzeit: 1924–1950.

86. Carathéodory to Zermelo on 14 July 1935. Letter in German. Universitätsbibliothek Freiburg, Nachlaß Ernst Zermelo.

87. Doetsch to Süss on 1 August 1934. Universitätsarchiv Freiburg, Nachlaß Wilhelm Süss, C89/5.

88. Süss to Feigl on 14 January 1936. Universitätsarchiv Freiburg, Nachlaß Wilhelm Süss, C89/51.

89. Carathéodory to Born on 21 July 1935. Letter in German. Staatsbibliothek zu Berlin, Preussischer Kulturbesitz, Handschriften Abteilung, Carathéodory, Constantin, Nachlaß Born 105.

90. Born M (1933) *Moderne Physik*. Verlag von Julius Springer, Berlin.

91. Archiv der LMU München, OC N 14.

92. Lehto (1998), p 58: The General Assembly authorised Fueter, President of the Congress in 1932, to assign members to the Commission to study the question of permanent international collaboration in mathematics. After consultations with E. Cartan, Severi, Veblen and Weyl, Fueter appointed F. Severi (Rome) Chairman of the Commission and the following members: P. S. Aleksandrov (Moscow), H. Bohr (Copenhagen), L. Fejér (Budapest), G. Julia (Paris), J. L. Mordell (Manchester), E. Terradas (Madrid), Ch.-J. de

la Vallée Poussin (Louvain), O.Veblen (Princeton), H.Weyl (Göttingen), and S. Zaremba (Cracow).

93. "The preponderance of German professors is conspicuous: Blaschke, Carathéodory, and Weyl." and "This choice of members presumably reflected the fact that Germany had been and would remain the key problem." Lehto (1998), p 66.
Having that in mind, the new elected president of the IMU, William Henry Young, concluded his letter to Carathéodory on 31 January 1929 as follows: "I hope my German colleagues will see in my election to the office a good omen for the future." Lehto (1998), p 53 and p 338.

94. "The outcome of these meetings was disappointing in that difficulties preventing the foundation of a new union were found to be insurmountable." Lehto (1998), p 67.

95. Lehto (ibid) writes that the fact that two meetings were held in Oslo seems to indicate that the Commission have not given up the possibility of a positive recommendation.

96. Carathéodory's report to the Reich Minister on 18 May 1941. Archiv der LMU München, OC N 14.

97. Feigl to Süss on 30 May 1935. Universitätsarchiv Freiburg, Nachlaß Wilhelm Süss, C89/51.

98. Ibid.

99. Staatsarchiv Hamburg, Z=Personalakte Blaschke, Hochschulwesen-Personalakten I128 Bd. 4. Aktenzeichen: PB 130 Z U.A.

100. For a description of the celebration see: Böhm (1995) p 567 and pp 573–579.

101. Munich, Deutsches Museum. Archiv NL 89, 004. *Sommerfeld-Projekt* on the Web.
For further proposals from Sommerfeld, Carathéodory, Wieland and Gerlach regarding Sommerfeld's succession see: Litten (2000), p 66f.

102. Hasse to Carathéodory on 28 January 1935. Niedersächsische Staats- und Universitätsbibliothek Göttingen – Abteilung für Handschriften und seltene Drucke. Cod. Ms. H. Hasse 1:265, Beil.

103. Carathédory to Hasse on 30 January 1935. Letter in German. Niedersächsische Staats- und Universitätsbibliothek Göttingen – Abteilung für Handschriften und seltene Drucke. Cod. Ms. H. Hasse 1:265.

104. BayHStA MK 35403, Laufzeit: 1924–1950.

105. Archiv der LMU München, E II N Car.
However, the previously mentioned letter of 10 February 1930 to the Speaker Councillor Dr.Tergende at the Foreign Ministry in Berlin documents that it was Carathéodory who, at the time of his appointment, had requested that the German nationality bound to his employment in Bavaria should not to apply to his wife and children.
BayHStA MK 35403, Laufzeit: 1924–1950.

106. Archiv der LMU München, E II N Car.

107. Dahm G (1935) *Zur gegenwärtigen Lage der deutschen Universität. Z. f. Deutsch. Kulturphil.* 1:211–224.

108. Vovolinis (ed) (1962), p 520.

109. See: *Επετηρίς του Πανεπιστημιακού Έτους 1936–1937*. Athens 1936.

110. Archiv der LMU München, E II N Car. The letter was signed by Scurla *im Auftrag* (on behalf). Dr. Herbert Scurla was later sent to Turkey by the *Reichsministerium für Wissenschaft, Erziehung und Volksbildung* to inspect the activities of German university professors, emigrants in their large majority, in 1939. The original of the *Scurla-Bericht* (Scurla Report) is located in the political archive of the German Ministry of Foreign Affairs. See: Klaus-Detlev Grothusen (ed) *Der Scurla-Bericht*. Dagyeli-Verlag, Frankfurt am Main, 1986.

At the Philosophical Faculty, University of Athens there were four foreign associate professors: Vincenzo Biagi, Italian Philology; Gottfried Merkel, German Philology; Alfred Romain, German Philology; Nogué J, French Philology. See: *Εθνικόν και Καποδιστριακόν Πανεπιστήμιον Αθηνών, Επετηρίς Πανεπιστημιακού Έτους 1936–1937* (National and Capodistrian University of Athens, Yearbook of the Academic Year 1936–1937). Athens 1936.

111. Konstantinos Demertzis was a full professor of civil law at Athens University. He had been minister of Venizelos's government of 1917 and was the prime minister of Greece from 30 November 1935 to 13 April 1936. His government granted an amnesty to the Venizelist officers who had organised the military coup of 1 March 1935.

112. Archiv der LMU München, E II N Car.

113. Schumacher H (1937), p 174f.

114. His own version of the story. Argyris says that the reason he left Athens for Munich was his conflict with Protopapadakis, a professor of railway engineering. He was the brother of the former Prime Minister Petros Protopapadakis, who, after the Asia-Minor disaster, had been charged with high treason at the "Trial of the Six". Argyris believes that the conflict was due to a feeling of revenge, since his family were Venizelists.

115. Cf. Heiber (1992), p 537ff and Böhm (1995), p 465f. For a balanced view of Kölbl see: Litten F (2003) *Die „Verdienste" eines Rektors im Dritten Reich – Ansichten über den Geologen Leopold Kölbl in München* (The "Merits" of a Rector in the Third Reich – Views about Geologist Kölbl in Munich). In: *NTM* 11, Basel, pp 34–46. Communicated to the author on 14 June 2002.

116. Archiv der Österreichischen Akademie der Wissenschaften, Personalakte Carathéodorys.

117. Vovolinis (ed) (1962), p 521.

118. Ibid. Lady Crosfield was Domini Eliadi (1892–1963), a Greek millionairess and wife of Sir Arthur Crosfield. Sir Arthur was a wealthy industrialist, Philhellene and Liberal MP for Warrington in 1906. It was in the Crosfields' lavish house at 41 West Hill, Highgate that Venizelos married Elena Skylitsi, a close friend of Lady Crosfield, on 2 [15] September 1921. Sir John Stavridis had been the Greek Consul in London 1903–1920 and a personal friend of Lloyd George. He was reputed to belong to the British Intelligence.

119. Hondros to Sommerfeld on 1 February 1935. Letter in German. Deutsches Museum, Archiv, HS, 1977-28/A, 147/2.
    In 1919, through the Saar statute (art. 45–50 of the Versailles Peace Treaty), the southern parts of the Prussian Rhine province (1465 km^2) and the western parts of the Bavarian Palatinate (416 km^2) had been assigned on trust of the League of Nations for 15 years. After this period, the outcome of a plebiscite had to decide to which state the Saar area would belong. France had gained ownership of the coal mines. The League of Nations had appointed a governing commission which was advised by a regional council. In 1925 the Saar area was included into the French customs area. On 13 January 1935, 90.8 % of the population declared themselves in favour of an annex to Germany. The Saar mines were bought back. The national socialist government put a Reich Commissioner in charge of the Saar area called "Saarland" thereupon.

120. Bouzakis (1999), p 261.

121. Kotzias K: *Die griechische Jugend durch die Jahrhunderte.* In: Kriekoukis Ch, Bömer K (eds), as note 139, chapter 4, pp 40–49, here p 49.

122. Kalitsounakis's contribution was *Die Gründung der Akademie von Athen*, pp 129–130, and Vizoukidis's *Das Griechische Recht im Laufe der Jahrtausende*, pp 73–82.

123. Daskalakis G D: *Geschichte und Bedeutung der Universität Athen.* In: Kriekoukis Ch, Bömer K (eds), as note 139, chapter 4, pp 117–128.

124. Universitätsarchiv Freiburg, Handakte des Schriftführers Conrad Müller E4/53.

125. Ibid.

126. Ibid.

127. Lorey's letter to Schmidt on 30 June 1936. Ibid.

128. Staatsarchiv Hamburg, Hochschulwesen, Dozenten- u. Personalakten IV 337, Hecke, Erich 20.9.1887.

129. Carathéodory C: *Bericht über die Verleihung der Fieldsmedaillen an L.V. Ahlfors und J. Douglas* (Report on the Award of the Fields Medals to L.V. Ahlfors und J. Douglas). In: *Comptes Rendus du Congrès Internationale des Mathématiciens, Oslo 1936.* A.W. Brøggers, Oslo, 1937, pp 308–314 and *Ges. Math. Schr.* vol V, pp 84–90.

130. (1978) A Short History of the Fields Medals. The Mathematical Intelligencer I/3:127–129.

131. Signature in Carathéodory's guest-book.

132. Lehto (1998), p 70.

133. Radó was engaged with this problem already earlier. Cf. for example his polemic: Radó T (1927) *Bemerkungen zur Arbeit von Herrn Ch. H. Müntz über das Plateausche Problem* (Remarks on the Work of Herr Ch. H. Müntz about Plateau's Problem). *Math. Ann.* 96: 587–596.

    Radó T (1930) On Plateau's problem. Ann. of Math. 31:457–469.

    Radó T (1930) The problem of least area and the problem of Plateau. *Math. Z.* 32:763–796.

    Douglas J (1931) Solution of the Problem of Plateau. Trans. Amer. Math. Soc. 33:263–321.

    Plateau's problem simply stated is to find a surface $z = f(x, y)$, which satisfies the partial differential equation obtained by setting the mean curvature equal to zero and which, on the boundary of a given domain in the x, y-plane, takes on a given value of z. The vanishing of the mean curvature results from the Euler equation associated with minimising the area of the surface $z = (x, y)$, which caps the given boundary curve. More generally stated, it is to find the expression of a surface of zero mean curvature (a minimal surface) which caps a given space curve, in terms of two curvilinear co-ordinates, that is to map it on a plane, but not by direct projection. The most important question is the relation of the minimal surface to the surface of least area.

    Semicentennial Publications of the American Mathematical Society, vol 2, p 221.

    An account of the work on the Plateau problem up to 1933 is given in: Radó T (1933) On the Problem of Plateau. Springer, Berlin, 1933. Jesse Douglas reviewed Radó's above mentioned account in: Douglas J (1934) Radó on Plateau's problem. Bull. Amer. Math. Soc. 40:194–196.

134. Carathéodory C (1940) Max Schiffman, The Plateau problem for non-relative minima (Ann. of Math. II (1939) 834–854). *Zentralbl. f. Math.* 23:398–399 and *Ges. Math. Schr.* vol V, pp 364–365.

135. Carathéodory C (1940) R. Courant, The existence of minimal surfaces of given topological structure under prescribed boundary conditions. Acta Math. 72 (1940) 51–98. *Zentralbl. f. Math.* 23:399–400 and *Ges. Math. Schr.* vol V, pp 366–367.

    To minimise an area it is sufficient to minimise Dirichlet's integral under prescribed boundary conditions. The vector which minimises Dirichlet's integral automatically solves the problem of least area. Douglas restricted the vectors admissible in this variational problem to harmonic vectors, so that Dirichlet's integral could be expressed only in terms of the boundary values of the vector representing the minimal surface. Consequently, the variational problem was reduced to a problem with functions of one variable. Independently of each other, Courant and Tonelli observed that a wider class of vectors allowed a considerable simplification of the problem. Their approach enabled solutions

of the general Douglas's problem for higher topological structure, of free boundary problems and of problems concerning unstable minimal surfaces.
See: Courant R (1950) as note 155, chapter 2, p 100.

136. Semicentennial Publications of the American Mathematical Society, vol I, p 20.

137. Sakellariou N (1937) *Το διεθνές μαθηματικόν συνέδριον εν Όσλο* (The International Mathematical Congress in Oslo). *Δελτίον της Ελληνικής Μαθηματικής Εταιρείας* IZ:176–181.

138. Carathéodory to Ingraham on 24 July 1936. Letter in English. General Files, Mark Ingraham, 7/22/11. The University of Wisconsin, Madison.

139. Courtesy of the Archives of the Institute for Advanced Study.

140. Living memorials (1947) p 22f. Communication to the author from James Frank Cook on 6 January 1999.

141. Communication to the author from James Frank Cook on 6 January 1999. See also: Encyclopaedia Britannica.

142. Carl Schurz Professorship Committee to the Department of German on May 8, 1937. Departmental Minutes from Series 7/22/1. The University of Wisconsin, Madison.

143. A. R. Hohlfeld to M. Ingraham on 14 October 1935. General Files, Mark Ingraham, 7/22/11. The University of Wisconsin, Madison.

144. Mark H. Ingraham to A. R. Hohlfeld on 17 October 1935. Ibid.

145. A. R. Hohlfeld to President Glenn Frank on 15 November 1935. Ibid.

146. Mark Ingraham to Carathéodory on 18 January 1936. Letter in English. Ibid.

147. Carathéodory to Hohlfeld on 2 March 1936. Letter in English translation. Ibid.

148. Archiv der LMU München, OC N 14.

149. BayHStA MK 35403, Laufzeit: 1924–1950.

150. Mark H. Ingraham to R. G. D. Richardson on 10 June 1936. General Files, Mark Ingraham, 7/22/11. The University of Wisconsin, Madison.

151. Mark H. Ingraham to Carathéodory on 6 July 1936. Letter in English. Ibid.

152. Carathéodory to Ingraham on 24 July 1936. Letter in English. Ibid.
This sea route is not in agreement with Carathéodory's own chronological record of his works. He has noted namely that he had written "The most general transformations of plane regions which transform circles into circles" in August 1936 on the Pacific Ocean between Los Angeles and Panama.

153. Mark H. Ingraham to C. Carathéodory on 6 August 1936. Letter in English. Ibid. This letter was sent to Germany and a copy to Carathéodory, c/o M.S. St. Louis of the North German Lloyd Steamship, New York City.

154. M. E. McCafrey Secretary. The Regents of the University of Wisconsin, Madison on August 7, 1936. Ibid.

155. Archiv der LMU München, E-II-1054.

156. Universitätsarchiv Freiburg, Nachlaß Wilhelm Süss, C89/5.

157. Threlfall's report *Über die Besetzung eines Ordinariats für Mathematik an der Universität Freiburg i. B.* (On the Occupation of a Full Professorship for Mathematics at the University of Freiburg i. B.) on 2 January 1947. Universitätsarchiv Freiburg, Nachlaß Wilhelm Süss, C89/37.

158. Notice of changes in advanced courses in mathematics signed by Mark H. Ingraham, Chairman, Department of Mathematics, on 17 September 1936. General Files, Mark Ingraham, 7/22/11. The University of Wisconsin, Madison.
Also in: Departmental Minutes from series 7/22/1. The University of Wisconsin, Madison.

159. G. A. Bliss had been one of the vice-presidents of the IMU in the period 1924–1932.

160. Departmental Minutes from series 7/22/1. The University of Wisconsin, Madison.

161. See: (1939). American Mathematical Monthly 46:320.

162. Vovolinis (ed) (1962), p 522.

163. Carl Schurz Professorship Committee to the Department of German on May 8, 1937, as note 142.

164. Blumenthal to von Kármán on 29 September 1935. Letter in German. Caltech Archives, Theodore von Kármán Papers, Folder 3.10.

165. Blumenthal to von Kármán on 6 August 1936. Ibid.

166. Von Kármán to Carathéodory on 2 September 1936. Letter in German. Ibid.

167. S. Lefschetz to Carathéodory on 7 December 1936. Letter in English. Caltech Archives, Theodore von Kármán Papers, Folder 5.4.

168. Carathéodory to von Kármán on 10 December 1936. Letter in German. Caltech Archives, Theodore von Kármán Papers, Folder 5.4.

169. Von Kármán to Carathéodory on 1 March 1937. Letter in German. Ibid.

170. Morghen R (1990) The Accademia Nazionale dei Lincei in the Life and Culture of United Italy on the 368th Anniversary of its Foundation – (1871–1971). Accademia Nazionale dei Lincei, Rome, p 62f.

171. Pope John Paul II. His message to the Pontifical Academy of Sciences on 22 October 1996.

172. Archiv der LMU München, OC N 14.

173. The question, if the foreign members of the academy were also obliged to swear the oath of allegiance to Mussolini, remained unanswered by the Papal Academy. The accessible period of the *Archivio Segreto Vaticano* extends up to the whole Benedict XV's pontificate, i.e. up to 22 January 1922. So there is no way to tell whether Carathéodory sworn the oath or not. However, to remain at the University of Munich, he had to take the oath of allegiance to Hitler.

174. Archiv der LMU München, OC N 14.

175. Ibid.

176. Archiv der LMU München, E-II-1054.

177. Encyclical no. 2053 of 10 May 1937. Archiv der LMU München, E II N Car.
     Alongside the Foreign Ministry, the Nazis created various instruments for their foreign policy: the *Außenpolitisches Amt der NSDAP* (A. Rosenberg), the *Dienststelle Ribbentrop*, the *Auslandsorganisation (AO) der NSDAP*, the *Deutsche Kongreßzentrale* (DKZ), etc. The DKZ was an administration authority aimed at controlling and employing international networks, although Nazi ideology in fact ruled out any positive assessment of international networks. The DKZ came to the fore as a private organisation, whereas, in fact, it was a subdivision of the *Reichsministerium für Volksaufklärung und Propaganda* (Reich Ministry for People's Education and Propaganda). Initially, through allocation of foreign exchange, it only controlled the attendance of German delegates to international congresses but gradually, by employing modern information technologies, it began to register and manipulate the expansion and composition of international networks. The DKZ was supported by the SS and, against resistance from the Foreign Ministry, acquired significant power of disposal in foreign affairs. The parallel existence of so many authorities in Nazi foreign politics confirms the polyarchic order of the "Third Reich". However, compared to the Wehrmacht and the Foreign Ministry, i.e. the traditional instruments of German power politics, these authorities started to fade in importance from 1938. Herren M (2000) Netzwerk Aussenpolitik: Internationale Kongresse und Organisationen als Instrumente schweizerischer Aussenpolitik 1914–1950 (The Network of

Foreign Politics: International Congresses and Organisations as Instruments of Swiss Foreign Politics 1914–1950). Web document.

178. Archiv der LMU München, E-II-1054.

179. The *Pontificia Accademia delle Scienze* was refounded under the name *Accademia Nazionale dei Lincei* in 1944. The term *Reale* was not used again, also because Italy became a republic in 1946.

180. Cornwell (1999), p 342f.

Cf. Cornwell J (1999) and Passelecq G, Suchecky B (1997) *Die unterschlagene Enzyklika: Der Vatikan und die Judenverfolgung* (The Suppressed Encyclical: The Vatican and the Persecution of the Jews). Carl Hanser Verlag, München, Wien. Original: (1995) *L'encyclique cachée de Pie XI. Une occasion manquée de l'Église face à l'antisémitisme* (The Concealed Encyclical of Pius XI. A Missed Opportunity of the Church in View of Anti-Semitism). Éditions La Découverte, Paris.

181. Neugebauer to Carathéodory on 3 September 1937. Letter in German. *Teilnachlaß Carathéodory*, as note 74, chapter 1.

182. According to Carathéodory, Fermat's principle represents a geometric theorem appropriate to characterise the form of light rays infiltrating an optical instrument in all cases. See: Carathéodory C (1937) *Geometrische Optik*, as note 91, chapter 4, p 12.

183. Chris Weeks draws attention to Neugebauer's argument for Heron's date:
"The problem of Heron's date used to be one of the most disputed questions in the history of Greek mathematics. All that could certainly be said was that he came after Apollonius, whom he quotes, and before Pappus, who cites him, say between 150 BC and AD 250. But in 1938 O. Neugebauer showed that the eclipse of the moon described in Heron, *Dioptra* 35 [...] as taking place on the tenth day before the spring equinox and beginning at Alexandria at the fifth (seasonal) hour of the night and at Rome at the third (seasonal) hour must have been the eclipse of 13th March AD. [...] Though strictly this establishes only an upper limit, Neugebauer's argument that Heron was referring to something in the recent memory of his readers has been generally accepted."
See: Goold G P (ed) Selections illustrating the history of Greek Mathematics, with an English translation by Ivor Thomas, vol II: Aristarchus to Pappus. The Loeb Classical Library, Harvard UP, Cambridge, MA, London, 1993 ([1] 1941), p 466.

184. In: (1899–1914) Heronis Alexandrini opera quae supersunt omnia. Bibliotheca scriptorum Graecorum et Romanorum Teubneriana, 5 vols. Teubner, Leipzig, here vol II, 1, pp 301–365.

185. Carathéodory mentions that Damianos's book Κεφάλαια των οπτικών υποθέσεων, *Haupttatsachen der Optik* (Main Themes of Optics) was published in Greek and German by R Schöne in Berlin in 1897. Carathéodory C (1937) *Geometrische Optik*, as note 91, chapter 4, here note 26 on p 7.

186. Heronis Alexandrini opera II, 1, as note 187, p 303.

187. Carathéodory's remark 26 on p 7 of his book.

188. Carathéodory to von Kármán on 6 February 1937. Letter in German. Caltech Archives, Theodore von Kármán Papers, Folder 5.4.

189. Carathéodory C (1940) *Elementare Theorie des Spiegelteleskops von B. Schmidt*. Teubner, Leipzig, Berlin, and *Ges. Math. Schr.* vol II, pp 234–279.

190. The Palomar and UK 48-inch (1.22 m) Schmidt cameras were used to make the all-sky surveys.
For the history of the development of Schmidt systems see: Christensen M (2001) Bernhard Schmidt: His Camera and Its Derivatives. Web document.

191. Carathéodory to Herglotz on 16 January 1940. Letter in German. Niedersächsische Staats- und Universitätsbibliothek Göttingen – Abteilung für Handschriften und seltene Drucke. Cod. Ms. Herglotz F 18.

192. Carathéodory to Dehn on 27 January 1940, op. cit.

193. Sommerfeld to Grimm on 5 May 1940. *Sommerfeld-Projekt* on the Web.

194. Süss to Carathéodory on 4 November 1940. Letter in German. Universitätsarchiv Freiburg, Nachlaß Wilhelm Süss, C89/48.

195. Bückner H (1943) C. Carathéodory, Elementare Theorie des Spiegelteleskops von B. Schmidt. *Jber. Deutsch. Math.-Verein.* 53:23f.

196. Carathéodory C (1937) *Bemerkungen zu den Strahlenabbildungen der geometrischen Optik* (Remarks on the Ray Images of Geometric Optics). *Math. Ann.* 114:187–193 and *Ges. Math. Schr.* vol II, pp 207–215.

C. Carathéodory (1937) *v. Oseen C.W., Une méthode nouvelle de l'optique géometrique* (v. Oseen, C.W., A New Method of Geometric Optics). Svenska Vetensk. Akad. Hdl., III. s. 15, Nr. 6, 1–41 (1936). *Zentralbl. f. Math.* 16:90–91 and *Ges. Math. Schr.* vol V, pp 356–357.

In (1940) *Das parabolische Spiegelteleskop* (The Parabolic Mirror Telescope). *Vierteljahrsschrift der Naturforschenden Gesellschaft in Zürich* 85:105–120 and *Ges. Math. Schr.* vol II, pp 216–233, Carathéodory's approach aimed at analysing the image in a small number of simpler elements from which the properties of the instrument could be assessed. He then compared his formulae with already tried results.

In Carathéodory C (1943) A. Colacevich. *Teoria del telescopio Schmidt* (A. Colacevich, Theory of Schmidt's Telescope) (Ottica, Firenze, 7, p 40–78). *Ges. Math. Schr.* vol V, pp 375–376 (according to a galley proof for the *Jahrbuch über die Fortschritte der Mathematik*), Carathéodory remarked that the author had used the inconvenient method of calculating with waves instead of light rays and his approach was based on a delusion. Moreover, despite the fact that the size of the terms to be omitted had to be estimated from the beginning so as not to influence the desired approximation, Colacevich had not made up his mind about the order of errors to be retained.

Armsen was "a repatriated Balt from Estonia" (Feigl to Süss on 19 July 1943. Universitätsarchiv Freiburg, Nachlaß Wilhelm Süss, C89/51) working at the *Institut für Völkerpsychologie der Heydrich Stiftung* (Institute for National Psychology of the Heydrich Foundation) (Feigl to Süss on 17 October 1944. Universitätsarchiv Freiburg, Nachlaß Wilhelm Süss, C89/51). He had recently gained his doctorate under Carathéodory and Rabe with a work on geometric optics. In his review of Armsen's thesis (Carathéodory C (1944) *Paul Armsen. Über die Strahlenbrechung an einer einfachen Sammellinse.* München: Dissertation 1942. 49 S. (Maschinenschrift). *Zentralbl. f. Math.* 28:269–270 and *Ges. Math. Schr.* vol V, p 378), Carathéodory wrote that the author had treated the case of a single non-spherical lens and developed a method to calculate the terms of the eiconal up to the desired approximation from the coefficients which determine the form of the lens. Armsen found that the errors of the 5th order cannot be neglected. The interesting by-product of this work was that out of twelve 5th-order errors, three depended on the Seidel coefficients alone and were almost as large as the other nine. Carathéodory concluded that Armsen's contribution was to show that these three errors had to be taken into account in relevant treatments.

With this "by-product", Carathéodory confronted the Bavarian Academy of Sciences on 5 February 1943. In: *Die Fehler höherer Ordnung der optischen Instrumente* (The Higher-Order Errors of Optical Instruments). *Bayer. Akad. d. Wiss. München, math.-nat. Abt.* 1943, pp 199–216 and *Ges. Math. Schr.* vol II, pp 290–307, he argued that the three errors of the 5th order cannot be changed without at the same time touching on errors of 3rd order. This work is part of the contents of two manuscripts in Carathéodory's papers in the Bavarian Academy of Sciences. The manuscripts were temporary sketches for a work meant to simplify the formulae for the calculation of the aberration effects in case of errors up to the 5th order. Max Herzberger presented the contents of the two manuscripts accompanied by some remarks in a text entitled *Berechnung der Diffraktionskurven aus dem Eikonal* (Calculation of the Diffraction [today: Aberration] Curves from the Eiconal). *Ges. Math. Schr.* vol II, pp 308–335. In 1932, in *Zentralbl. f. Math.* 3:88–89 (see also: *Ges. Math. Schr.* vol V, pp 350–352), Carathéodory had reviewed: Herzberger M (1931) *Strahlenoptik* (Optics of Rays). Julius Springer, Berlin.

197. Hondros D (1950) *Ο Κ. Καραθεοδωρή ως φυσικός* (Carathéodory as a Physicist). *Αιών του Ατόμου* (Century of the Atom) *Δ*/2–3:333–334, here 334.

198. *Teilnachlaß Carathéodory*, as note 74, chapter 1.

199. Ibid.

200. Ibid.

201. Ibid.

202. Ibid.

203. Ibid.

204. Carathéodory C (Aus dem Nachlaß) *Der B. Schmidtsche Projektionsapparat. Ges. Math. Schr.* vol II, pp 280–289.

205. *Teilnachlaß Carathéodory*, as note 74, chapter 1.

206. Ibid.

207. *Ges. Math. Schr.* vol II, p 281.

208. R. Helm had visited Carathéodory on 29 September 1935.

209. Böhm (1995), p 376.

210. Clause II 5 of the protocol to the session of the Mathematics-Natural Sciences Class of the Bavarian Academy of Sciences of 3 April 1937 signed by Tietze. In: Stoermer M, as note 27, here p 96f.

211. Einstein welcoming Fritz Haber's decision to emigrate: "Let us hope that you will not return to Germany any more. It is no business to work for a group of intelectuals [...]"

212. Stoermer M, as note 27, here pp 102, 105, 106 and 107.

213. "General Judgement" of the NSDAP local section in Bogenhausen on 3 February 1941. Barch, WI–Akte Perron, Oskar (ehem. BDC).
   In the "General Judgement" is mentioned that Perron was "black", this term generally used to denote a Catholic, but in Nazi terminology a "reactionary". Perron was a Protestant.

214. (1937) Freedom of Mind. Nature 139:1.

215. BAyHStA MK 35403, Laufzeit: 1924–1950.

216. Large (1998), p 376f.

217. His signature was certified by the head of the consular office of the Greek legation, the legation secretary A. Coundouriotis.
   On 9 July 1937 Kalitsounakis, who was an officially acknowledged interpreter, translated Carathéodory's wedding certificate in German.
   BayHstA MK 35403, Laufzeit: 1924–1950.

218. Large (1998), p 371.

219. Large (1998), p 339ff. Cf. Schuster, Peter-Klaus (ed) *Die ,Kunststadt' München 1937 – Nationalsozialismus und ,Entartete Kunst'* (Munich, the City of Art in 1937 – National Socialism and 'Degenerate Art'). Prestel-Verlag, München, 1987.

220. Schumacher H (1937) Das neue Hellas. Georg Stilke, Berlin.

221. Ibid, pp 128–132: *Die deutsch-griechischen Handelsbeziehungen* (The German-Greek Commercial Relationships).

222. Cf. Fels E (1941) *Griechenlands wirtschaftliche und politische Lage* (Greece's Economic and Political State). *Geographische Zeitschrift* 47:57–71.

223. Carathéodory C (1937) *Περί των καμπυλών του στυλοβάτου του Παρθενώνος και περί της αποστάσεως των κιόνων αυτού*. Archaeological Paper 1937, centenary volume, pp

120–124. Translated by Stephanos Carathéodory as *Über die Kurven am Sockel des Parthenon und die Abstände seiner Säulen* (About the Curvature of the Stylobate of the Parthenon and concerning its Intercolumnia) for *Ges. Math. Schr.* vol V, pp 257–262.

224. Conrad Müller to Carathéodory on 29 October 1937. Letter in German. Carathéodory to Müller on 3 November 1937. Universitätsarchiv Freiburg, Handakte des Schriftführers Conrad Müller, E4/53.

225. Carathéodory C (1937) *Glückwunschschreiben der Bayerischen Akademie der Wissenschaften zum 50-jährigen Doktorjubiläum von Wilhelm Wirtinger (23.12.1937).* Written on the academy's commission but published only in *Ges. Math. Schr.* vol V, pp 91–92.

226. Blaschke to Süss on 29 March 1938. Universitätsarchiv Freiburg, Nachlaß Wilhelm Süss, C89/37; Blaschke to Süss on 17 March 1938. Universitätsarchiv Freiburg, Nachlaß Wilhelm Süss, C89/45.

227. Blaschke to Süss on 29 March 1938. Universitätsarchiv Freiburg, Nachlaß Wilhelm Süss, C89/45.

228. Behnke to Süss on 18 September 1943. Universitätsarchiv Freiburg, Nachlaß Wilhelm Süss, C89/5.

229. Δελτίον της Ελληνικής Μαθηματικής Εταιρείας, vol IΘ′ (1938), p 93f. Also: Blaschke's report of 20 May 1938 on his lecture tour in Italy and Greece. Staatsarchiv Hamburg, Hochschulwesen, Dozenten- u. Personalakten IV 86, Blaschke, Wilhelm 13.9.1885.

230. Staatsarchiv Hamburg, Z=Personalakte Blaschke, Hochschulwesen-Personalakten I 128 Bd. 4 Aktenzeichen: PB 130 Z U.A.

231. Archiv der LMU München, E-II-1054.

232. Universitätsarchiv Freiburg, Nachlaß Wilhelm Süss, C 89/5.

233. Cf. Siegmund-Schultze, as note 59, here p 2 and Mehrtens as note 7, here p 93.

234. Expressed by Mehrtens, as note 7, here p 92, as: part of his compromise with the regime was that the autonomy of the German Union of Mathematicians would remain untouched.

235. Carathéodory to Süss on 9 June 1938. Letter in German. Universitätsarchiv Freiburg, Nachlaß Wilhelm Süss, C89/48.
   In May 1935 Feigl had announced to Süss that Grunsky would take over the *Fortschritte* after a transitional period and in special cases would be consulted by Feigl. Feigl to Süss on 30 May 1935. Universitätsarchiv Freiburg, Nachlaß Wilhelm Süss, C89/51. Indeed, in 1935–1939, Grunsky was editor of the *Jahrbuch über die Fortschritte der Mathematik*.

236. Staatsarchiv Hamburg, Hochschulwesen, Dozenten- u. Personalakten IV 377, Hecke, Erich 20.9.1887.

237. Staatsarchiv Hamburg, Hochschulwesen, Dozenten- u. Personalakten IV 780, Petersson, Hans 24.9.1902.

238. F. Litten's communication to the author on 28 March 2000 with the note that he was rather used by Thüring and his friends than acted on his own initiative.

239. BArch, REM-Akte Stroux, Johannes (ehem. BDC). Lists of names of free masons published by the Grand Lodge of Greece confirm that persons very close to Carathéodory were free masons: Venizelos belonged to the Athena Lodge, von Streit and Benakis to the Pythagoras Lodge, Papoulias Dimitrios to the Orpheus Lodge, Christos Gryparis, retd. Major and with Carathéodory member of the parochial church council *zum Erlöser* in 1943, to the Delphi Lodge. The Grand Lodge of Greece, on the other hand, maintain that there were also other significant personalities belonging to the free masons for whom, however, there exists no confirmation because the archives of the Grand Lodge of Greece were destroyed to a significant part directly after the German invasion of Greece in 1941. On the basis of this information it may be assumed that Carathéodory could have belonged to the free masons at least up until 1932. But taking into account

his synkretism, his independent intellect and Despina's clear negation of her father's involvement with free masonry, it would be rather advisable to believe that he really did not belong to any lodge.

240. Böhm (1995), p 401.

241. Barch, Reichsstatthalter-Carathéodory, C. (ehem. BDC)
One day before Carathéodory's official release from the civil service, Georg Prange and his wife Hildegard visited him in Munich. They signed in Carathéodory's guest-book on 11 August 1938. It was the year in which Prange's paper *Zur dreihundertjährigen Wiederkehr des Erscheinens von Galileis Diskorsi* (On the 300th Anniversary of the Release of Galilei's Diskorsi) appeared in: *Forsch. u. Fortschr.* 14:138–140.

242. Archiv der LMU München, OC N14.

243. Königreich Griechenland. Ministerium für Unterricht und Kultus. No. 83163. Athens, 20 October 1938. BayHstA MK 35403, Laufzeit: 1925–1950.

244. Daskalakis, as note 123, here pp 125 and 128.

245. 16th session of 20 May 1938. *Πρακτικά Συνεδριών Σχολής Φυσικών και Μαθηματικών Επιστημών* 1937–1938, p 132.

246. 17th session of 30 May 1938. Ibid, p 134.

247. It should be noticed that, during Metaxas's dictatorship, permission from the relevant minister was required for civil servants to be members of societies. K. Georgakopoulos was responsible for granting permission to professors of mathematics with civil servant status to belong to the Greek Mathematical Society.

248. *Δελτίον της Ελληνικής Μαθηματικής Εταιρείας*, vol ΙΘ′ (1938), p 94.
In the war, Marston Morse established a Ballistic Research Section of the US Army Ordnance Department which investigated mathematical problems arising from tactical needs. He conducted important studies in the ricochet of bombs and shells, the clearance of enemy minefields, the effectiveness of bombs and their fragments and many other subjects of vital interest to the Armed Forces. Courtesy of the Archives of the Institute for Advanced Study.

249. Birkhoff to Conrad Müller on 3 October 1938. Universitätsarchiv Freiburg, Handakte des Schriftführers Conrad Müller E4/53.

250. Lorey to Sperner on 5 August 1940; Sperner to Süss, Müller, Hasse on 10 September 1940. Universitätsarchiv Freiburg, Nachlaß Wilhelm Süss, C89/77.

251. Sperner to Süss, Müller, Hasse on 10 September 1940. Universitätsarchiv Freiburg, Nachlaß Wilhelm Süss, C89/77.

252. In a letter to Kalitsounakis on 7 December 1935 (Vovolinis (ed) (1962), p 521), Carathéodory mentioned von Drygalski as a colleague whose opinion he respected.

253. Siegmund-Schultze, as note 382, chapter 2, here p 117.

254. Richard Willstätter, *Aus meinem Leben. Von Arbeit, Muße und Freunden* (From my Life. About Work, Leisure and Friends). Verlag Chemie, München 1949, p 354.

255. After a lecture tour in the US, Frank obtained a half-time appointment as a lecturer on physics and mathematics at Harvard University.

256. Karl Löwner was brought to the University of Louisville through the efforts of the Emergency Committee in Aid of Displaced German Scholars in New York. He taught at the engineering school of the University of Louisville from 1939–1944. He left for a "temporary" appointment as research mathematician at Brown University, specifically at the National Defence Research Committee, and he did not return to Louisville. Communication to the author by Tom Owen on 14 January 1999. Löwner became professor of mathematics at Stanford University in 1951.

257. Carathéodory C (1940) *Ferdinand von Lindemann. Sitzungsber. Bayer. Akad. d. Wiss. München, math.-nat. Abt. 1940*, pp 61–63 and *Ges. Math. Schr.* vol V, pp 93–95.

258. Courtesy of the Archives of the Institute for Advanced Study.

259. Ibid.

260. Thüring, a former assistant at the observatory in Heidelberg, had gone to Munich at the end of 1936. He was the head of the Lecturers' Corporation and the Nazi Lecturers' Association and a supporter of "Aryan" physics. In 1940–1941 he became a full professor of astronomy at the University of Vienna and was replaced by Bergdolt, a party member since 1922.
Litten (1994), as note 166, chapter 4, p 161.
Führer was a party member since 1930, in the SS since 1933, the head of the Lecturers' Corporation in 1935. In July 1935 he was appointed *Gaudozentenbundführer* and thus became a party representative. Dingler was lecturing in Munich since 1938 and received an appointment to the university there in 1940. In 1924, Carathéodory and Perron supported the renewal of Hugo Dingler's teaching commission for elementary mathematics that the latter had been holding since 1920. Teaching of elementary mathematics at the university, they argued, was necessary for secondary school teachers to outgrow the subject and realise the connection with scientific research. It was also necessary to bridge the gap between higher and lower mathematics. A ministerial decree of 30 September 1924 did not approve of the commission which was however renewed on 6 April 1926.
Toepell (1996), p 274f.
This teaching commission arranged by Lindemann in 1920 had a permanent character and was attached to the supernumerary (unbudgeted) professorship for methodology, instruction and history of mathematical sciences that was created at that time. Klein's lectures *Elementarmathematik vom höheren Standpunkte aus* of 1907–1908 (as note 72, chapter 1) were for long considered to have a prototypical role in this respect.
Aufgaben- und Forschungsbereiche in der ersten Hälfte des 20. Jahrhunderts – die Ära Perron, Carathéodory, Tietze und Hartogs (Tasks and Fields of Research in the First Half of the 20th Century – Perron, Carathéodory, Tietze and Hartog's Era). Web document, LMU München.
Dingler took part in the intrigues concerning Carathéodory's successor after his teaching commission in mathematics for forestry scientists had been rejected.
Süss to Kneser on 29 May 1941. Universitätsarchiv Freiburg, Nachlaß Wilhelm Süss, C89/63.
For Dingler see also: Beyerchen (1980), p 243f.

261. (1936) *Judentum und Wissenschaft*. Theodor Fritsch, Leipzig, p 48–55. Cf. Beyerchen (1980), pp 226 and 340.

262. Staatsarchiv Hamburg, Z=Personalakte Blaschke, Hochschulwesen-Personalakten I 128 Bd. 4 Aktenzeichen: PB 130 Z U.A.

263. Evidence of their meeting in: Carathéodory to Herglotz on 14 January 1942 (letter in German) and (postcard in German) on 20 January 1942. Niedersächsische Staats- und Universitätsbibliothek Göttingen – Abteilung für Handschriften und seltene Drucke. Cod. Ms. G. Herglotz F 18.

264. Archiv der LMU München, E-II-1054.

265. Universitätsarchiv Freiburg, Nachlaß Wilhelm Süss C89/61.

266. Carathéodory to Herglotz. Postcard in German with stamp 10 September 1939. Niedersächsische Staats- und Universitätsbibliothek Göttingen – Abteilung für Handschriften und seltene Drucke. Cod. Ms. Herglotz F 18.

267. Carathéodory to Herglotz on 22 September 1939. Ibid.

268. Litten (1994), as note 166, chapter 4, p 160.

269. Carathéodory to Herglotz on 19 May 1940. Letter in German. Niedersächsische Staats- und Universitätsbibliothek Göttingen – Abteilung für Handschriften und seltene Drucke. Cod. Ms. Herglotz F 18.

270. Carathéodory to Herglotz on 25 May 1940. Letter in German. Ibid.

271. BayHStA, MK Reg.Sp. V Abgabe 1991 Vorl. Nr. 1376, Laufzeit: 1922–1944.

272. Behnke to Süss on 24 May 1940. Universitätsarchiv Freiburg, Nachlaß Wilhelm Süss, C89/42.

273. Behnke to Süss on 3 June 1940. Ibid.

274. Behnke to Süss on 12 June 1940. Ibid.

275. Cf. Siegel C L (1979) On the History of the Frankfurt Mathematics Seminar (Address given on June 13, 1964 in the Mathematics Seminar of the University of Frankfurt on the Occasion of the 50th Anniversary of the Johann-Wolfgang-Goethe-University Frankfurt). The Mathematical Intelligencer I/4:223–230.

276. BayHStA, MK Reg.Sp. V Abgabe 1991 Vorl. Nr. 1376, Laufzeit: 1922–1944.

277. Litten (1994), as note 166, chapter 4, p 157.

278. Ibid, p 157.

279. Ibid, p 158f.

280. Süss to Hasse on 17 July 1941. Universitätsarchiv Freiburg, Nachlaß Wilhelm Süss, C89/61.

281. Litten (1994), as note 166, chapter 4, p 159.

282. Ibid, p 161.

In 1943, Seifert was described by Süss as "one of the scientifically leading people, maybe more appropriate in lecturing to middle and higher semesters than to beginners." Süss to Chuboda on 26 January 1943. Universitätsarchiv Freiburg, Nachlaß Wilhelm Süss, C89/48.

He was then entrusted with war activity as a department director at the *Luftforschungsanstalt Hermann Göring* (Hermann Göring Air-Research Institution) in Braunschweig.

283. Carathéodory to Süss on 24 April 1944. Letter in German. Universitätsarchiv Freiburg, Nachlaß Wilhelm Süss, C89/48.

284. Universitätsarchiv Freiburg, Nachlaß Wilhelm Süss C89/61.

285. Süss to Sperner on 29 May 1941. Universitätsarchiv Freiburg, Nachlaß Wilhelm Süss, C89/77.

286. Süss to Kneser on 29 May 1941. Nachlaß Wilhelm Süss, C89/63.

287. Süss to Kerkhoff on 14 May 1942. Ibid.

288. Süss to Sperner on 29 May 1941. Universitätsarchiv Freiburg, Nachlaß Wilhelm Süss, C89/77.

289. Süss to Sperner on 11 June 1941. Universitätsarchiv Freiburg, Nachlaß Wilhelm Süss, C89/77.

290. Süss to Kneser on 6 June 1941. Universitätsarchiv Freiburg, Nachlaß Wilhelm Süss, C89/63.

291. Blaschke to the military government on 17 September 1945. Text entitled "Help for people persecuted out of political reasons". Staatsarchiv Hamburg, Hochschulwesen, Dozenten- u. Personalakten IV 86, Blaschke, Wilhelm 13.9.1885.

292. Litten (1994), as note 166, chapter 4, p 161f.

293. Süss to Hasse on 23 April 1941. Universitätsarchiv Freiburg, Nachlaß Wilhelm Süss, C89/61.

294. Litten (1994), as note 166, chapter 4, p 162.

295. Hasse to Süss on 29 July 1941. Universitätsarchiv Freiburg, Nachlaß Wilhelm Süss, C89/61.

296. Universitätsarchiv Freiburg, Nachlaß Wilhelm Süss C89/5.

297. Litten (1994), as note 166, chapter 4, p 163.

298. Hasse to Süss on 29 July 1941. Universitätsarchiv Freiburg, Nachlaß Wilhelm Süss C89/61.

299. Litten (1994), as note 166, chapter 4, p 163f.

300. Universitätsarchiv Freiburg, Nachlaß Wilhelm Süss C89/48.

301. Behnke to Süss on 11 November 1941. Universitätsarchiv Freiburg, Nachlaß Wilhelm Süss C89/42.

302. Sauer had succeeded Müller as dean of the faculty of general sciences at the Technical University of Aachen in 1939. He had gained his doctorate under S. Finsterwalder at the Technical University of Munich.

303. Tollmien was employed at the *KWI Strömungsforschung* (Kaiser Wilhelm Institute for Air-Stream Research) in Göttingen in 1935–38.

304. Litten (1994), as note 166, chapter 4, p 164f.

305. Ibid, p 165f.

306. Ibid, p 166f.

307. Süss to Tietze on 27 July 1942. Universitätsarchiv Freiburg, Nachlaß Wilhelm Süss, C89/79.

308. Perron to Blaschke on 12 December 1945. Universitätsarchiv Freiburg, Nachlaß Wilhelm Süss, C 89/5.

309. In his letter of 20 September 1945 to the rector of Hamburg University.

310. Blaschke to colleagues on 15 June 1946. Universitätsarchiv Freiburg, Nachlaß Wilhelm Süss, C89/5.

311. BayHStA, MK Reg.Sp. V Abgabe 1991 Vorl.Nr. 1376, Laufzeit: 1922–1944.

312. Süss to Kneser on 8 September 1942. Universitätsarchiv Freiburg, Nachlaß Wilhelm Süss, C89/63.

313. Süss to Tietze on 27 July 1942. Universitätsarchiv Freiburg, Nachlaß Wilhelm Süss, C89/79.

314. Süss to Sperner on 17 December 1942. Universitätsarchiv Freiburg, Nachlaß Wilhelm Süss, C 89/77.

315. Feigl to Süss on 23 December 1942. Universitätsarchiv Freiburg, Nachlaß Wilhelm Süss, C89/51.

316. Fleming's report, as note 302, chapter 2.

317. Litten (1994), as note 166, chapter 4, 168.

318. Siegmund-Schultze, as note 382, chapter 2, here p 137f.

319. Carathéodory to Kalitsounakis on 15 November 1943. Vovolinis (ed) (1962), p 523.

320. *Teilnachlaß Carathéodory*, as note 74, chapter 1.

321. Mrs. Rodopoulos has presented a video made from old photographs with scenes from her family's life in the plantation. In one of them her little son sits triumphantly with a weapon in his hand on the head of an elephant killed within the plantation.
Born to Einstein about Chaim Weizmann on 22 May 1948: "He could have saved many more Jews, if he had accepted the offer of the British to give him a piece of Kenya (East Africa)". *Albert Einstein – Hedwig und Max Born. Briefwechsel 1916–1955* (1972), p 180.

322. Archiv der LMU München, OC N 14.

323. Ibid.

324. Archiv der LMU München, E II N Car.

325. Archiv der LMU München, OC N 14.

326. Grüttner (1995), p 361.

327. Perron, as note 107, chapter 1, p 41.

328. Archiv der LMU München, E II N Car.

329. Blaschke to Bol on 6 July 1939 in order that the latter asks Süss. Universitätsarchiv Freiburg, Nachlaß Wilhelm Süss, C89/45.

330. Niedersächsische Staats- und Universitätsbibliothek Göttingen – Abteilung für Handschriften und seltene Drucke. Cod. Ms. H. Hasse. 21:3.

331. Carathéodory C (1943) *Maßtheorie und Integral* (Theory of Measure and Integral). Reale Accademia d'Italia, Fondazione Alessandro Volta. Istituta dalla Società Edison di Milano. Atti dei Convegni 9. Convegno di scienze fisiche matematiche e naturali, 1939, Matematica contemporanea e sue applicazioni, Roma, pp 195–208 and *Ges. Math. Schr.* vol V, pp 242–256.

332. Toepell (1996), p 313 and Grüttner (1995) p 370 & 374.

333. Süss to Feigl on 3 April 1941. Universitätsarchiv Freiburg, Nachlaß Wilhelm Süss, C89/51.

334. Carathéodory C (1942) *Artur Bischof, Beiträge zur Carathéodoryschen Algebraisierung des Integralbegriffs* (Schr. math. Inst. u. Inst. angew. Math. Univ. Berlin 5 (1941) 237–262 and Berlin: Dissertation 1940). *Zentralbl. f. Math.* 25:33–34 and *Ges. Math. Schr.* vol V, p 363.

335. Universitätsarchiv Freiburg, Nachlaß Wilhelm Süss, C89/48.

336. Behnke to Süss on 3 March 1943. Universitätsarchiv Freiburg, Nachlaß Wilhelm Süss, C89/42.

337. Süss to Behnke on 9 March 1943. Ibid.

338. Behnke to Süss on 12 March 1943. Ibid.

339. 13 October 1943. Re: person and family of Louis Cartan. The Rector of Freiburg University. Universitätsarchiv Freiburg, Nachlaß Wilhelm Süss, C89/48.

340. Behnke to Süss on 11 May 1944. Universitätsarchiv Freiburg, Nachlaß Wilhelm Süss, C89/42.

341. Hasse to Süss on 18 October 1940. Universitätsarchiv Freiburg, Nachlaß Wilhelm Süss, C89/61.

342. Their letter to Munich University is dated 1 September 1940. Munich, Deutsches Museum. Archiv NL 89, 019, Mappe 5,11 (*Sommerfeld-Projekt* on the Web). Cf. Beyerchen (1980), p 340.

343. Beyerchen (1980), p 243f. Glaser habilitated under Stark in Würzburg in 1921 and had been an NSDAP member since 1932.

344. Courtesy of the Archives of the Institute for Advanced Study.

345. Sommerfeld to Hondros in December 1942. LMU München, Institut für Theoretische Physik, Nachlaß A. Sommerfeld, N 89, 002.

346. Stark J, Müller W (1941) *Jüdische und Deutsche Physik*, Vorträge zur Eröffnung des Kolloquiums für theoretische Physik an der Universität München (Jewish and German Physics: Lectures on the Beginning of the Colloquium for Theoretical Physics at the University of Munich). Leipzig. Beyerchen (1980), p 317. See also: 1942 review of the book by Weizel W in: *Z. f. techn. Phys.* 1:25.

347. Süss to Sperner on 29 May 1941. Universitätsarchiv Freiburg, Nachlaß Wilhelm Süss, C89/77.

348. Hondros to Sommerfeld on 7 December 1942. Deutsches Museum, Archiv, HS, 1977–28/A,147/4.

349. Beyerchen (1980), p 227.

350. Ibid, p 216.

351. Hans Weber, Über analytische Variationsprobleme (10.9.1942) Carathéodory/Müller.

352. Carathéodory C (1943) *Hans Weber. Über analytische Variationsprobleme.* München: Dissertation 1941. 64 S. (Maschinenschrift). *Zentralbl. f. Math.* 27:404–405 and *Ges. Math. Schr.* vol V, p 377.
Carathéodory wrote that the author had treated variation problems whose basic function $F(x_i, \dot{x}_i)$ was analytic along an extremal or extremaloid $\xi_i(t)$ which was not necessarily an analytic curve. Weber showed that, in this case, the integral $\int F dt$ existed over curves of comparison of a certain *closer* neighbourhood and could be represented by a series development.

353. Carathéodory C (1943) *Arnold Sommerfeld zum 75. Geburtstag. Der 2. Band seiner Vorlesungen über theoretische Physik* (To Arnold Sommerfeld on the Occasion of his 75th Birthday. The Second Volume of his Lectures on Theoretical Physics). *Die Naturwissenschaften* 31:573–574 and *Ges. Math. Schr.* vol V, pp 379–382.

354. Harokopos G (1971) *Το Φρούριο Κρήτη 1941–1944 – Η κατασκοπεία και αντικατασκοπεία στην Κρήτη* (Fort Crete 1941–1944 – Intelligence and Counter-Intelligence on Crete). Georgios Evangeliou, Athens, p 79f.

355. Argyris's connection to Rangavis and his daughter is omitted from an autobiographical sketch released on the Web by the World Innovation Foundation. On the contrary, his connection to the head of the German counter-intelligence, Admiral Wilhelm von Canaris, is stressed. See: *A small view of the life of John Argyris – Engineer Extrodinaire* [*sic*]. In: The Newsletter of the World Innovation Foundation, April–July 2000, vol 4, edition 1.

356. Fleischer (1986), p 77.

357. Mrs. Rodopoulos' credible version of the story.

358. *Hitler-Reden* (Hitler's Speeches) II, 1707, cited in Fleischer (1986), p 76.

359. As chief of the general staff of the Air Force.

360. Fleischer (1986), p 188.

361. Cited in Fleischer (1986), p 178.

362. Cf. Fels, as note 223, here p 67.

363. Report of Max Hartmann, a department head at the Kaiser Wilhelm Institute for Biology who was promoting a joint research institute for biology in Piraeus, in December 1941. Cited in: Macrakis (1993), p 147.

364. The Philhellene, Bulletin of the American Friends of Greece, vol II no 5, May 1943. The scientist addressed his audience with the question: "But what purpose would the proposed institution serve at this time when Greece, bleeding and starved, is in the throes of such unspeakable misery?"

365. N. Kritikos in: *Παιδεία και Ζωή* (Education and Life), April 1950. Cited in: Vovolinis (ed) (1962), p 523.
Longing for peace, also the couple Conrad and Martha Müller noted in Carathéodory's guestbook on 28 August 41: "How nice being with you. We have spoken of past times of peace. Thank you for the nice moments."

366. Macrakis (1993), p 137f. See also: Herren, as note 177.

367. Universitätsarchiv Freiburg, Nachlaß Wilhelm Süss, C89/77.

368. By "Conseil" is meant the *Conseil international de recherches* (IRC = International Research Council), whose Constitutive Assembly was held in Brussels during 18–28 July 1919 and ratified the implementation of post-war international science policies. The Germans adopted the view that the IRC was founded by learned societies in France and Britain supported by their governments for the purpose of undermining the position of German science. Still in 1928 German academic institutions assumed a negative stand to the invitation to join the IRC despite repeated governmental requests. In 1931 the IRC was replaced by the International Council of Scientific Unions (ICSU). In the

protocols of the *Verband wissenschaftlicher Körperschaften* (Association of Scientific Corporations) of the years 1933–1937 the ICSU was not mentioned at all. See: Lehto (1998).

369. Archiv der LMU München, OC N 14.

370. Ibid.

371. Cf. Carathéodory's previously mentioned report to the Reich Minister of Education about the International Mathematical Union.

372. For example Carathéodory's expression "the assumption that the IMU might still exist did not correspond to the facts" was turned into "the question, whether the International Mathematical Union still exists, is difficult to decide"; the expression "After Weyl had gone to America" was read as "up to his appointment in America". The abolishment of the *Union mathématique internationale* by resolution of the International Congress of Mathematicians in Zurich, as well as Hamel's request to Carathéodory to join the Severi Commission and Carathéodory's rejection because of respect to democratic procedures are not mentioned in the remark. The Fields medal is mentioned as Fils medal. Cf. Siegmund-Schultze, as note 59.

373. Süss to Sperner on 18 March 1942. Universitätsarchiv Freiburg, Nachlaß Wilhelm Süss, C89/77.

374. Mehrtens H: Mathematics and War: Germany, 1900–1945. In: Forman P, Sánchez-Ron J M (eds) National Military Establishment and the Advancement of Science and Technology. Boston Studies in the Philosophy of Science, vol 180. Dordrecht, Boston, London, 1996, pp 87–134, here p 115.

375. Archiv der LMU München, E II N Car.

376. Ibid.

377. Carathéodory to Süss on 13 November 1941. Letter in German. Universitätsarchiv Freiburg, Nachlaß Wilhelm Süss, C 89/48.

378. Large (1998), p 420.

379. Carathéodory to Hasse on 17 December 1941. Letter in German. Niedersächsische Staats- und Universitätsbibliothek Göttingen – Abteilung für Handschriften und seltene Drucke. Cod. Ms. Hasse 1:265.

380. Hasse to Carathéodory on 7 January 1942. Letter in German. Ibid, Beil.

381. Carathéodory to Süss on 28 January 1942. Letter in German. Universitätsarchiv Freiburg, Nachlaß Wilhelm Süss, C89/48.

382. Copy of Schauder's letter (in German) to van der Waerden of 29 October 1942. Universitätsarchiv Freiburg, Nachlaß Wilhelm Süss, C89/48.
   Schauder published fixed point theorems for Banach spaces in 1930, his 1934 paper on topology and partial differential equations is of major importance and his last work was to generalise results of Courant, Friedrichs and Lewy.
   "Hölder is an excellently equipped analyst, with taste and fantasy, at the same time respectable and competent." Kneser to Süss on 2 July 1940. Universitätsarchiv Freiburg, Nachlaß Wilhelm Süss, C89/63.
   SD: officially, the information service and counter-intelligence of the German Reich since 1936. It provided the Gestapo with information about opponents of national socialism at home and abroad. In 1939 it was united with the Sipo within the *Reichssicherheitshauptamt* (RSHA = Reich Security Department). In World War II, operation troops of the SD and the Sipo carried through mass executions of Jews. The Gestapo had unlimited power to keep people in detention without any guaranty of their rights, to send them to concentration camps, etc. Due to its methods, including torture, it was feared and disreputable. In World War II, the Gestapo became the epitome of total rule in areas occupied by German troops. The international military tribunal in Nuremberg (1946) declared both the Gestapo and the SD to be criminal organisations.

383. Van der Waerden to Carathéodory on 4 November 1942. Letter in German. Universitätsarchiv Freiburg, Nachlaß Wilhelm Süss, C89/48.

384. Carathéodory to Süss on 5 November 1942. Letter in German. Ibid.

385. "Im REM vorgebracht 26/28.11.42 Protokoll. (illegible name) will SD zum Eingreifen angeben 14.1.43" (Presented at the REM 26, registered 28 November 1942. (illegible name) wants to report it to the SD to take action). Note in different hand-writing at the end of Carathéodory's letter to Süss on 5 November 1942, as note 384.

     Heisenberg later spoke of Schauder's murder: Schauder "had written to me and I had put out feelers in order to see what could be done. I wrote to Scholz who had something to do with Poland. Then Scherrer wrote me the following ridiculous letter saying he also had something to do with the case. He wrote: 'Dear Heisenberg, I have just heard that the mathematician Schauder is in great danger. He is now living in the little Polish town of so-and-so under the false name of so-an-so.' That came in a letter which was of course opened at the frontier. It is unbelievable how anyone can write that from Switzerland. I heard nothing more about Schauder and I have now been told that he was murdered." Bernstein J (2001), p 98. The name is wrongly spelled as "Schouder".

     Information about the elimination of Jews of Drogobych in: Encyclopedia of the Holocaust. Macmillan Publishing Company, New York, 1990.

386. The Philhellene, vol III, no 7–8, July–August 1944.

387. Archiv der LMU München, E II N Car.

388. Hasse's report to Süss on 3 February 1943. Universitätsarchiv Freiburg, Nachlaß Wilhelm Süss, C89/61.

389. Giornale d'Italia on 9 November 1942, included in Hasse's report.

390. Hasse's report to Süss on 3 February 1943, as note 387.

391. Süss to Sperner on 4 February 1943. Universitätsarchiv Freiburg, Nachlaß Wilhelm Süss, C 89/77.

392. Carathéodory to Hasse on 4 November 1942. Letter in German. Niedersächsische Staats- und Universitätsbibliothek Göttingen – Abteilung für Handschriften und seltene Drucke. Cod. Ms. Hasse 21:4.

393. Hasse's report to Süss on 3 February 1943, as note 388.

394. Ibid and Hasse H (1943) *Internationale Mathematikertagung in Rom im November 1942.* *Jber. Deutsch. Math.-Verein.* 53:21–22.

395. Carathéodory C: *Probleme der analytischen Funktionen einer Veränderlichen.* In: R. Istituto Nazionale di Alta Matematica: Atti del Convegno Matematico, tenuto in Roma dall'8 al 12 Novembre 1942, pp 209–213. Tipografia del Senato del Dott. G. Bardi, Roma, 1945 and *Ges. Math. Schr.* vol IV, pp 119–124.

396. Süss to Blaschke on 22 June 1942. Universitätsarchiv Freiburg, Nachlaß Wilhelm Süss, C89/45.

     On the occasion of the memorial celebration of Titus Livius and Galileo Galilei, Blaschke became a honorary Dr. of Padua University on 31 May 1942. Staatsarchiv Hamburg, Hochschulwesen, Dozenten- u. Personalakten IV 86, Blaschke, Wilhelm 13.9.1885.

397. Hasse's report to Süss on 3 February 1943, as note 388.

398. Hasse H, *Internationale Mathematikertagung in Rom im November 1942*, as note 394.

399. Hasse's report to Süss on 3 February 1943, as note 388.

400. Carathéodory to Herglotz on 14 January 1942 (letter in German) and (postcard in German) on 20 January 1942. Niedersächsische Staats- und Universitätsbibliothek Göttingen – Abteilung für Handschriften und seltene Drucke. Cod. Ms. G. Herglotz F 18.

401. Carathéodory to Süss on 28 January 1942, as note 381.

     "As far as I can judge, Deuring has performed excellently in his discipline." Kneser to Süss on 2 July 1940. Universitätsarchiv Freiburg, Nachlaß Wilhelm Süss, C89/63.

     Max Deuring, an assistant and lecturer in Jena since 1937, became an associate professor at the University of Posen (now Poznan in Poland) in 1943. Hans Zassenhaus was a lecturer at the Faculty of Mathematics and Natural Sciences of Hamburg University. Carathéodory himself found it odd to feel an aesthetic pleasure amidst all this misery. Yet, according to Friedrich Schiller, it is aesthetics that curbs the arbitrary use of power, the irresponsibility and the brutishness: Schiller F (1965) *Über die ästhetische Erziehung des Menschen* (About the Man's Education in Aesthetics). Reclam, Stuttgart, p 36.

402. Carathéodory C, Sommerfeld A, as note 62, chapter 1.
Reid (1970) p 213f.

403. Süss to Carathéodory on 17 May 1944. Letter in German. Universitätsarchiv Freiburg, Nachlaß Wilhelm Süss, C89/48.

404. Carathéodory to Süss on 30 May 1944. Ibid.

405. Sommerfeld to Herglotz on 26 March 1944. Niedersächsische Staats- und Universitätsbibliothek Göttingen – Abteilung für Handschriften und seltene Drucke. Cod. Ms. Herglotz F 135. Communication to the author from Michael Eckert on 22 October 1999.

406. Herglotz to Sommerfeld on 1 April 1944. Ibid.

407. Carathéodory to Herglotz on 14 October 1944. Letter in German. Niedersächsische Staats- und Universitätsbibliothek Göttingen – Abteilung für Handschriften und seltene Drucke. Cod. Ms. Herglotz F 18.
He referred to: *David Hilbert. Bay. Akad. d. Wiss. München, math.-nat. Abt. 1943*, pp 350–354 and *Ges. Math. Schr.* vol V, pp 101–105.

408. Archiv der Österreichischen Akademie der Wissenschaften, Personalakte Carathéodorys.

409. Süss's confirmation to Frau Feigl after the war. Universitätsarchiv Freiburg, Nachlaß Wilhelm Süss, C89/5. Süss writes that Feigl had been heading the *Fortschritte* since 1925. In the *Mitgliedergesamtverzeichnis der Deutschen Mathematiker-Vereinigung 1890–1990* (List of Members of the Union of German Mathematicians 1890–1990) (Toepell M (ed), IGN der LMU München, 1991) the year is mentioned as 1928.

410. Süss to Feigl on 3 April 1941. Universitätsarchiv Freiburg, Nachlaß Wilhelm Süss, C89/51.

411. Behnke to Pohl on 21 January 1943. Universitätsarchiv Freiburg, Nachlaß Wilhelm Süss, C89/42.

412. In Breslau, his aim was the building and heading of the Mathematical Institute through lectures, supervision of doctorates and habilitations, co-operation in the *Reichsverband deutscher mathematischer Gesellschaften und Vereine* (Reich Association of German Mathematical Societies and Clubs) and participation in managerial and scientific tasks of the *Deutsche Versuchsanstalt für Luftfahrt* (German Experimental Institution for Aeronautics).
Pinl M (1967) *Georg Feigl zum Gedächtnis* (In Memory of Georg Feigl). *Jber. Deutsch. Math.-Verein.* 79:53–60.

413. Feigl to Süss, undated letter. Universitätsarchiv Freiburg, Nachlaß Wilhelm Süss, C89/51.

414. Behnke, as note 29, chapter 1, p 32.

415. Carathéodory to Süss on 9 July 1942. Letter in German. Universitätsarchiv Freiburg, Nachlaß Wilhelm Süss, C89/48.

416. After consultation with Bieberbach, the senior civil servant at the Reich Research Council Dr. Kerkhoff intended to ask Harald Geppert, a full professor at Berlin University and a member of the Prussian Academy of Sciences, to write the overview report on mathematics for the *Forschungen und Fortschritte*. He assumed that he could not count on Carathéodory for the writing of this report.
Kerkhoff to Süss on 2 November 1942. Universitätsarchiv Freiburg, Nachlaß Wilhelm Süss, C89/63.

417. Süss to Sperner on 24 September 1943. Universitätsarchiv Freiburg, Nachlaß Wilhelm Süss, C89/77.

418. Mehrtens, as note 374, here p 114.

419. Archiv der LMU München, E-II-1054.

420. Süss to Behnke on 29 July 1943. Universitätsarchiv Freiburg, Nachlaß Wilhelm Süss, C89/42.

421. Sperner to Süss on 21 September 1943. Universitätsarchiv Freiburg, Nachlaß Wilhelm Süss, C89/77.

422. Feigl to Süss on 12 September 1943. Universitätsarchiv Freiburg, Nachlaß Wilhelm Süss, C89/51.
    Feigl was probably occupied with a war conference of the German Union of Mathematicians which was to take place in Würzburg from 6 to 10 September (Süss to Behnke on 29 July 1943. Universitätsarchiv Freiburg, Nachlaß Wilhelm Süss, C89/42), in which also Pisot and Schmidt were going to participate. Schmidt preferred a small event without participation of foreigners. If, however, foreigners were to come, he would take on the presidency of the conference. Schmidt had been severely ill probably because of an over-exertion.
    Feigl to Süss on 9 May 1943. Universitätsarchiv Freiburg, Nachlaß Wilhelm Süss, C89/51.

423. Süss to Behnke on 29 July 1943. Universitätsarchiv Freiburg, Nachlaß Wilhelm Süss, C89/42.

424. In September 1943 Behnke, Feigl, Flatt, Hamel and Lietzmann were comprising the Instruction Commission of the *DMV*. Universitätsarchiv Freiburg, Nachlaß Wilhelm Süss, C89/77.

425. Feigl to Süss on 7 February 1944. Universitätsarchiv Freiburg, Nachlaß Wilhelm Süss, C 89/51.

426. Schmidt E, as note 315, chapter 4.

427. Vovolinis (ed) (1962), p 523.

428. Dr. E.W.: *Ein bedeutender Mathematiker*.

429. Archiv der LMU München, E-II-1054.

430. Ibid.
    Carathéodory obviously meant: Lindelöf E (1934) *Einführung in die Höhere Analysis* (Introduction to Higher Analysis). Teubner, Leipzig, Berlin 1934. In the same letter, Carathéodory assured the rector that he was not obliged to do military service.

431. BayHStA MK 35403, Laufzeit: 1924–1950.

432. The Bavarian Ministry of Education and Cultural Affairs to the rector of Munich University.
    Archiv der LMU München, E-II-1054.

433. Archiv der LMU München, OC N 14.

434. Institut für Zeitgeschichte München, MA-116.

435. Communication to the author from Dr. Eva-Marie Felschow, Universitätsbibliothek, Universität Giessen, on 16 November 1998.

436. Cf. *Aufgaben- und Forschungsbereiche in der ersten Hälfte des 20. Jahrhunderts – die Ära Perron, Carathéodory, Tietze und Hartogs*, as note 260.

437. See for example: Stark J (1932) *Nationale Erziehung* (National Education). Eher, München.

438. Loseman (1977), pp 59 and 74.

439. Institut für Zeitgeschichte München, MA-116.

440. Ibid.

441. Universitätsarchiv Freiburg, Nachlaß Wilhelm Süss, C89/4.

442. Universitätsarchiv Freiburg, Nachlaß Wilhelm Süss, C89/61.

443. Institut für Zeitgeschichte München, MA-116.

444. Archiv der LMU München, E-II N Car.

445. Institut für Zeitgeschichte München, MA-116.

446. Ibid.

447. Ibid.

448. Ibid.

449. Ibid.

450. Grüttner (1995), p 203.

451. Kotsowillis K, as note 39, chapter 4.

452. Communication to the author from Dr. Weigand on 19 September 1999.

453. His daughter says that Carathéodory did not allow his children to clean their teeth before receiving the Holy Communion.

454. Bayerische Akademie der Wissenschaften to Mathematisches Forschungsinstitut on 11 November 1946. Universitätsarchiv Freiburg, Nachlaß Wilhelm Süss, C89/5.

455. Carathéodory to Zermelo on 12 October 1943. Universitätsbibliothek Freiburg, Nachlaß Ernst Zermelo.

456. Franz Rellich gained the greatest part of his scientific training in Göttingen where he received his doctorate under Courant with a dissertation on a subject from the theory of partial differential equations. After the war he took up the heading of the Göttingen Mathematical Institute. He informed the emigrants and Americans about the situation of the personnel and the institutes in Germany and endeavoured for scientific contacts with the outside world.
Cf. Siegmund-Schultze, as note 382, chapter 2, pp 277 and 282.

457. Carathéodory to Frau van der Waerden on 11 December 1943, as note 75, chapter 3.

458. The *Aldinen* are prints, especially from the 16th century, produced at the printing works of Aldus Manutius (~1450 to 1515) and his successors in Venice. For the Greek *Aldinen* see: Sicherl M (1997) *Griechische Erstausgaben des Aldus Manutius. Druckvorlagen, Stellenwert, kultureller Hintergrund* (Aldus Manutius' Greek First Editions. Printer's Copies, Status, Cultural Background). Ferdinand Schöningh, Paderborn, München.
By "Fock" Carathéodory means Gustav Fock's bookshop and publishing house in Leipzig.

459. Carathéodory to Süss on 6 January 1944. Letter in German. Universitätsarchiv Freiburg, Nachlaß Wilhelm Süss, C89/48.

460. Carathéodory to van der Waerden on 25 March 1944. Letter in German. ETH-Bibliothek Zürich, Wissenschaftshistorische Sammlungen, HS 652:10661.

461. Cf. for example: Hubert Cremer to Süss on 22 November 1946. Universitätsarchiv Freiburg, Nachlaß Wilhelm Süss, C89/5.

462. Carathéodory to van der Waerden on 28 June 1944. Letter in German. ETH-Bibliothek Zürich, Wissenschaftshistorische Sammlungen, HS 652:10612.

463. English translation by Arnold Dresden with additions of the author. P. Noordhoff, Groningen, Holland 1954.

464. p 148 of van der Waerden's book.

465. p 84 of van der Waerden's book.

466. Neugebauer's mathematical interest in Babylonia was continuous. When he was a member of the Institute for Advanced Study in 1945–1946, he took on the critical edition of all available lunar and planetary ephemerides of the Seleucid period under the title

Astronomical Cuneiform Texts. Courtesy of the Archives of the Institute for Advanced Study.

467. The other persons who signed the proposal were: Arnold Sommerfeld, August Schmauss, Friedrich Boas, Georg von Faber, Guido Fischer, Günther Scheibe, Heinrich Tietze, J. Ossamy, Klaus Alfred Clusius, Ludwig Föppl, Martin Näbauer, Rudolph Tomaschek, Walther Gerlach, Walther Meißner, Richard Baldus, Thomeis. Personal file von Laue in the Bavarian Academy, Munich. *Sommerfeld-Projekt* on the Web.

468. Carathéodory calls Lambert a German philosopher, although Lambert was born in Mülhausen, Alsace in 1728. Mülhausen (Mulhouse) belonged from 1515 up to the French Revolution to the *Eidgenossenschaft* (Swiss Federation).

469. Carathéodory to Süss on 24 April 1944. Letter in German. Universitätsarchiv Freiburg, Nachlaß Wilhelm Süss, C89/48. Lambert's papers are kept in the public library of the Basel University.

470. Sperner to Süss on 25 April 1944. Universitätsarchiv Freiburg, Nachlaß Wilhelm Süss, C89/77.
Sperner was obviously referring to: Steck M (ed) *Lambert, Johann Heinrich, 1728–1777: Schriften zur Perspektive* (Writings on Perspective). Lüttke, Berlin, 1943.

471. Kneser to Süss on 25 Mai 1944. Universitätsarchiv Freiburg, Nachlaß Wilhelm Süss, C89/63.

472. Tietze to Süss on 1 May 1944. Universitätsarchiv Freiburg, Nachlaß Wilhelm Süss, C89/79.

473. Kneser to Süss on 24 August 1944. Universitätsarchiv Freiburg, Nachlaß Wilhelm Süss, C89/63.

474. Süss to Sperner on 11 May 1944. Universitätsarchiv Freiburg, Nachlaß Wilhelm Süss, C89/77.

475. Scholz to Süss on 23 November 1944. Universitätsarchiv Freiburg, Nachlaß Wilhelm Süss, C89/4.

476. Süss to Carathéodory on 19 September 1943. Letter in German. Universitätsarchiv Freiburg, Nachlaß Wilhelm Süss, C89/48.

477. Carathéodory to Süss on 21 September 1943. Letter in German. Ibid.

478. Sperner to Süss on 26 June 1943. Universitätsarchiv Freiburg, Nachlaß Wilhelm Süss, C89/77.

479. Sperner to Süss on 9 September 1940. Universitätsarchiv Freiburg, Nachlaß Wilhelm Süss, C89/77.

480. Arnold Sommerfeld to Gustav Herglotz on 26 June 1944. Niedersächsische Staats- und Universitätsbibliothek Göttingen   Abteilung für Handschriften und seltene Drucke. *Sommerfeld-Projekt* on the Web.

481. Tietze to Süss on 1 May 1944. Universitätsarchiv Freiburg, Nachlaß Wilhelm Süss, C89/79.

482. Cf. *Monatsbericht des Regierungspräsidenten von Oberbayern*, 7.8.1944 (Monthly Report of the Administration Head of Upper Bavaria, 7 August 1944). Broszat et al (eds), as note 10, chapter 4, p 671.

483. Carathéodory to Süss on 19 July 1944. Postcard in German. Universitätsarchiv Freiburg, Nachlaß Wilhelm Süss, C89/48.

484. Tietze to Süss on 21 July 1944. Universitätsarchiv Freiburg, Nachlaß Wilhelm Süss, C89/79.

485. Rohrbach to Süss on 13 November 1944. Universitätsarchiv Freiburg, Nachlaß Wilhelm Süss, C89/4.

486. Feigl to Süss on 5 July 1944. Universitätsarchiv Freiburg, Nachlaß Wilhelm Süss, C89/51.

487. See: Litten, as note 58, chapter 4, here p 198f.
     Hubert Cremer: Koebe's assistant in Jena 1927–1931; assistant professor at the University of Cologne 1931; professor at the Technical University of Breslau 1940. Cremer H (1968) *Erinnerungen an Paul Koebe* (Memories of Paul Koebe). *Jber. Deutsch. Math.-Verein.* 70:158–161.

488. Litten, as note 58, chapter 4, here p 200f.

489. Feigl to Süss on 5 July 1944 and Süss to Feigl on 18 July 1944. Universitätsarchiv Freiburg, Nachlaß Wilhelm Süss, C89/51.

490. Osenberg to Süss on 13 December 1944. Universitätsarchiv Freiburg, Nachlaß Wilhelm Süss, C89/4.

491. Litten, as note 58, chapter 4, here pp 201ff and 206.

492. Carathéodory to Süss on 27 July 1944. Universitätsarchiv Freiburg, Nachlaß Wilhelm Süss, C89/48.

493. Communication to the author from Freddy Litten on 4 June 2001.

494. Universitätsarchiv Freiburg, Nachlaß Wilhelm Süss, C89/48.

495. Süss to Carathéodory on 25 August 1944. Letter in German. Universitätsarchiv Freiburg, Nachlaß Wilhelm Süss, C89/48.

496. As note 495.

497. Gerlach to the committee of the Reich Research Council on 2 August 1944. Universitätsarchiv Freiburg, Nachlaß Wilhelm Süss, C89/4.
     The story of the foundation of a national German institute for applied mathematics goes back to 1941. A prototypical role for German played the Institute for Advanced Study, Princeton and the Italian *Istituto per le Applicazioni del Calcolo*. Cf. Mehrtens, as note 373, here pp 115–118.

498. Süss to Carathéodory on 25 August 1944, as note 495.

499. Süss to F. Schmidt-Ott on 8 January 1945. Universitätsarchiv Freiburg, Nachlaß Wilhelm Süss, C89/4.

500. 4 July 1944, List of male members of the Freiburg Mathematical Institute according to their significance for the current works. Universitätsarchiv Freiburg, Nachlaß Wilhelm Süss, C89/4.

501. Staatsarchiv Hamburg, Hochschulwesen, Dozenten- u. Personalakten IV 648, Maak, Wilhelm, 13.8.1912.

502. Süss to Blaschke on 10 January 1945. Universitätsarchiv Freiburg, Nachlaß Wilhelm Süss, C89/4.

503. Carathéodory to Süss on 3 March 1945. Universitätsarchiv Freiburg, Nachlaß Wilhelm Süss, C89/4.

504. Behnke to Süss on 6 January 1943. Universitätsarchiv Freiburg, Nachlaß Wilhelm Süss, C89/42.

505. Carathéodory to Süss on 14 October 1944. Universitätsarchiv Freiburg, Nachlaß Wilhelm Süss, C89/48.

506. Carathéodory to Herglotz on 14 October 1944, op. cit.

507. Süss to Carathéodory on 12 December 1944. Universitätsarchiv Freiburg, Nachlaß Wilhelm Süss, C89/4.

508. Carathéodory to Zermelo on 26 November 1944. Universitätsbibliothek Freiburg, Nachlaß Ernst Zermelo.

509. Süss to Erich Röver, on 19 December 1944. Universitätsarchiv Freiburg, Nachlaß Wilhelm Süss, C89/4.

510. Süss to Alexander von Humboldt Stiftung on 15 December 1944. Ibid.

511. Süss to Carathéodory on 19 December 1944. Ibid.

512. Carathéodory to Süss on 12 January 1945. Ibid.

513. Süss to Carathéodory on 5 February 1945. Ibid.

514. National Archives Microfilm Publication M 1217: Review of US Army War Crimes Trials in Europe 1945–1948. Roll 5: Reviews and Recommendations for Cases 000-50-5-21 to 000-50-136.

515. 13B Landsberg/Lech, Hindenburgring 12. Archiv der LMU München, E-II-1054.

516. National Archives, as note 514.

517. Perz B (1991) *Projekt Quarz. Steyer-Daimler-Puch und das Konzentrationslager Melk.* Industrie, Zwangsarbeit und Konzentrationslager in Österreich, vol 3. Verlag für Gesellschaftskritik, Wien.

518. Archiv der LMU, E-II-1054.

519. *Teilnachlaß Carathéodory*, as note 74, chapter 1.

520. Communication to the author from Dr. Weigand on 19 September 1999.
The International Service of the Red Cross for missing persons does not provide information to non-relatives.

521. Carathéodory to Süss on 12 January 1945. Universitätsarchiv Freiburg, Nachlaß Wilhelm Süss, C89/4.

522. Tietze to Süss on 12 January 1945. Ibid.

523. Carathéodory to Zermelo on 31 January 1946. Letter in German Universitätsbibliothek Freiburg, Nachlaß Ernst Zermelo.

524. Tietze to Süss on 12 January 1945, as note 522.

525. Carathéodory to Süss on 3 March 1945. Universitätsarchiv Freiburg, Nachlaß Wilhelm Süss, C89/4.

526. Perron, as note 107, chapter 1, p 51.

527. Tietze to Süss on 1 May 1944. Universitätsarchiv Freiburg, Nachlaß Wilhelm Süss, C89/79.

528. *Teilnachlaß Carathéodory*, as note 74, chapter 1.

529. Monika Stoermer, as note 27, here pp 94, 100 and 108.

530. Munich, 24 June 1944. *Teilnachlaß Carathéodory*, as note 74, chapter 1.

531. Tietze H (1953) Foreword to *Ges. Math. Schr.* vol I.

532. Feigl to Süss on 9 May 1943. Universitätsarchiv Freiburg, Nachlaß Wilhelm Süss, C89/51.

533. Tietze to Süss on 6 Ferbuary 1944. Universitätsarchiv Freiburg, Nachlaß Wilhelm Süss, C89/79.

534. Van der Waerden to Carathéodory on 19 March 1944. Letter in German. ETH-Bibliothek Zürich, Wissenschaftshistorische Sammlungen, HS 652:10610.
Indeed, many mathematicians dedicated contributions to Carathéodory on the occasion of his 70th birthday in the *Mathematische Annalen* issue 2, vol 119 of 1944. These were E. Schmidt, Georg Aumann, Josef Lense, H.Tietze, O. Perron, E. Hecke. This issue includes still another article by M. Ghermanescu (Timisoara).

535. San Nicolo to Carathéodory on 25 December 1944. Letter in German. *Teilnachlaß Carathéodory*, as note 74, chapter 1.

536. München, den 12. Jan. 1945 C. H. Beck'sche Verlagsbuchhandlung Oscar Beck.
München, den 20. Jan. 1945 C. Carathéodory.
Teilnachlaß Carathéodory, as note 74, chapter 1.

537. *Teilnachlaß Carathéodory*, as note 74, chapter 1.

538. Baier O, Lenz H (1967) *Frank Löbell zum Gedächtnis* (In Memory of Frank Löbell). *Jber. Deutsch. Math.-Verein.* 70:1–15.

539. Tietze, as note 530.

540. BayHStA MK 35403, Laufzeit: 1924–1950.

541. Archiv der Österreichischen Akademie der Wissenschaften, Personalakte Carathéodorys. Keil, Josef (Reichenberg in Böhmen 13 October 1878 – Vienna 13 December 1963): 1904 head of Secretary's Office of the Austrian Archaeological Institute; 1920 lecturer; 1925 professor of ancient history and archaeology at the University of Vienna; 1927 professor of ancient history and archaeology at the University of Greifswald; 1936 professor of ancient history and archaeology at the University of Vienna and also joint head of the Austrian Archaeological Institute up to 1956.
Hartkopf W, as note 390, chapter 2.

542. Archiv der Österreichischen Akademie der Wissenschaften, Personalakte Carathéodorys.

543. Aumann to Süss on 13 April 1946. Universitätsarchiv Freiburg, Nachlaß Wilhelm Süss, C89/5.

544. Tietze to Süss on 16 December 1946. Universitätsarchiv Freiburg, Nachlaß Wilhelm Süss, C89/37.

545. Aumann G, Carathéodory C (1934) *Ein Satz über die konforme Abbildung mehrfach zusammenhängender ebener Gebiete. Math. Ann.* 109:756–763 and *Ges. Math. Schr.* vol III, pp 449–457.

546. Tietze to Süss on 4 June 1951. Universitätsarchiv Freiburg, Nachlaß Wilhelm Süss, C89/385.

547. Die Berufungskammer München, Ber. Reg. Nr. 562/47, Akt. Z. I. Inst. X/393/46, May 30, 1947. Universitätsarchiv Freiburg, Nachlaß Wilhelm Süss, C89/5.

548. Threlfall's report on 2 January 1947 "Über die Besetzung eines Ordinariats für Mathematik an der Universität Freiburg i.B.". Universitätsarchiv Freiburg, Nachlaß Wilhelm Süss, C89/37.

549. Tietze to Süss on 16 December 1946. Universitätsarchiv Freiburg, Nachlaß Wilhelm Süss, C89/37.

550. Staatsarchiv Hamburg, Hochschulwesen, Dozenten- u. Personalakten IV 780, Petersson, Hans 24.09.1902.

551. Helmut Rechenberg, *Walther Gerlach (1889–1979)*. Web document.

552. Frau Bieberbach to Süss on 15 May 1946 and Bieberbach to Süss on 14 June 1946. Universitätsarchiv Freiburg, Nachlaß Wilhelm Süss, C89/5.

553. Süss to Frau Bieberbach on 6 June 1946 and Bieberbach to Süss on 3 August 1946. Universitätsarchiv Freiburg, Nachlaß Wilhelm Süss, C89/5.

554. Fraenkel (1967), p 162.

555. Staatsarchiv Hamburg, Hochschulwesen, Dozenten- u. Personalakten IV 86, Blaschke, Wilhelm 13.9.1885.

556. Heiber (1992), p 103.

557. Universitätsarchiv Freiburg, Nachlaß Wilhelm Süss, C89/45.

558. Blaschke's letter to "Meine Herren Kollegen" on 15 June 1946. Universitätsarchiv Freiburg, Nachlaß Wilhelm Süss, C89/5.

559. Siegmund-Schultze, as note 382, chapter 2, p 287.

560. Carathéodory to Süss on 6 February 1946. Letter in German. Universitätsarchiv Freiburg, Nachlaß Wilhelm Süss, C89/5.

561. Staatsarchiv Hamburg, Hochschulwesen, Dozenten- u. Personalakten IV 1851, Prof. Dr. Helmut Hasse 1950–1966 (–1981).

562. Paul ten Bruggencate to Süss on 11 October 1945. Universitätsarchiv Freiburg, Nachlaß Wilhelm Süss, C89/5.

563. Süss to Paul ten Bruggencate on 31 January 1946. Universitätsarchiv Freiburg, Nachlaß Wilhelm Süss, C89/5.

564. Süss to Carathéodory on 31 May 1946. Letter in German. Universitätsarchiv Freiburg, Nachlaß Wilhelm Süss, C89/5. There is no evidence of how Carathéodory reacted.

565. Hasse H (1952) *Mathematik als Wissenschaft, Kunst und Macht* (Mathematics as Science, Art and Power). Verlag für Angewandte Wissenschaft, Wiesbaden.

566. Springer to Allgeier on 21 October 1946. Universitätsarchiv Freiburg, Nachlaß Wilhelm Süss, C89/4.

   Süss to Behnke on 7 December 1946. Universitätsarchiv Freiburg, Nachlaß Wilhelm Süss, C89/37.

567. Knopp to Süss on 5 January 1947. Universitätsarchiv Freiburg, Nachlaß Wilhelm Süss, C89/37.

568. Carathéodory to Süss on 30 January 1946. Letter in German. Universitätsarchiv Freiburg, Nachlaß Wilhelm Süss, C89/5.

569. Süss to Frau Feigl on 8 July 1946. Universitätsarchiv Freiburg, Nachlaß Wilhelm Süss, C89/5.

570. Süss to Ehresmann on 19 January 1948. Universitätsarchiv Freiburg, Nachlaß Wilhelm Süss, C89/5.

571. His last doctoral student, Dr. Leonhard Weigand, asserts that Carathéodory was a decisive opponent of the regime and a strict anti-Nazi.
   Communication to the author from Dr. Leonhard Weigand on 19 September 1999.

572. Perron, as note 107, chapter 1, p 41.

573. Born to Einstein on 15 July 1944. *Albert Einstein – Hedwig und Max Born. Briefwechsel 1916–1955* (1972), p 149.

574. According to Moser, a triple negative. A reviewer's remark.

575. Tietze, as note 105, chapter 1, p 95

576. Behnke, as note 29, chapter 1, p 32.

577. Anastasios Adosides was Governor of Macedonia at the end of World War I; later, when the islands received the first waves of refugees from Smyrna, Prefect of Samos and the Cyclades (his grandfather had been Prince of Samos under the Ottomans) and member of the Refugee Settlement Commission.

578. A possible explanation would be that he would not like to expose his guests and himself, in case that his guest-book came to Nazi hands. On the other hand, it is not sure that the copy in the author's hands is complete.

6 *The Final Years*

1. Carathéodory to Süss on 12 September 1945. Letter in German. Universitätsarchiv Freiburg, Nachlaß Wilhelm Süss, C89/5.

2. Feigl to Süss on 17 October 1944. Universitätsarchiv Freiburg, Nachlaß Wilhelm Süss, C89/51.

3. Frau Feigl to Süss on 20 June 1946. Universitätsarchiv Freiburg, Nachlaß Wilhelm Süss, C89/5.

4. Carathéodory to Süss on 6 February 1946. Letter in German. Universitätsarchiv Freiburg, Nachlaß Wilhelm Süss, C89/5.

5. Seifert to Carathéodory on 28 December 1945. Letter in German. Carathéodory attached a copy of this letter in his own handwriting to his letter to Süss on 6 February 1946. Universitätsarchiv Freiburg, Nachlaß Wilhelm Süss, C89/5.

6. Süss to the Faculty of Natural Sciences and Mathematics of Heidelberg University on 3 September 1946. Universitätsarchiv Freiburg, Nachlaß Wilhelm Süss, C89/385.

7. Süss to Carathéodory on 15 February 1946. Letter in German. Universitätsarchiv Freiburg, Nachlaß Wilhelm Süss, C89/5.

8. Carathéodory to Süss on 21 February 1946. Letter in German. Ibid.

9. Süss to Frau Feigl on 8 July 1946. Universitätsarchiv Freiburg, Nachlaß Wilhelm Süss, C89/7.

10. Süss to Carathéodory on 17 April 1946. Letter in German. Universitätsarchiv Freiburg, Nachlaß Wilhelm Süss, C89/5.

11. Siegmund-Schultze R (1998) The University of Berlin from reopening until 1953. In: Mathematics in Berlin, as note 38, chapter 1, pp 137–141, here p 137.

12. Feigl's telegram to Süss in February 1944. Universitätsarchiv Freiburg, Nachlaß Wilhelm Süss, C89/51.

13. Feigl to Süss on 17 October 1944. Universitätsarchiv Freiburg, Nachlaß Wilhelm Süss, C89/51.

14. Carathéodory to Herglotz on 28 October 1945. Letter in German. Niedersächsische Staats- und Universitätsbibliothek Göttingen – Abteilung für Handschriften und seltene Drucke. Cod. Ms. Herglotz F 18.

15. Behnke to Süss on 7 November 1945 Universitätsarchiv Freiburg, Nachlaß Wilhelm Süss, C89/5.

16. Carathéodory to Zermelo on 31 January 1946. Letter in German. Universitätsbibliothek Freiburg, Nachlaß Ernst Zermelo.

17. Rector's draft on 2 November 1945. Universitätsarchiv München, E-II-1054.

18. Carathéodory to Süss on 6 February 1946, as note 4.
Together with Udo Wegner, Carathéodory's nephew John Argyris supervised Joseph F. Gloudemann's doctoral dissertation *Zur numerischen Berechnung der linearen und nichtlinearen Differentialgleichungen mit Hermiteschen Interpolationspolynomen* (On the Numerical Calculation of Linear and Non-Linear Differential Equations with Hermite Interpolation Polynomials). Gloudemann gained the title of a *Dr. Ing.* from the University of Stuttgart in 1970.
The Mathematics Genealogy Project on the Web.

19. Süss to Behnke on 4 March 1946. Ibid.

20. Cf. Litten F (1998) „*Er half . . . , weil er sich als Mensch und Gegner des Nationalsozialismus dazu bewogen fühlte*" – Rudolf Hüttel (9.7.1919–12.10.1993) ("He Helped . . . , because as a Man and an Opponent of National Socialism he Felt Emotionally Touched" – Rudolf Hüttel (9.7.1919–12.10.1993)) *Mitt. Ges. Deutsch. Chem.* 14:78–109, here note 14, p 102.

21. Universitätsarchiv Freiburg, Nachlaß Wilhelm Süss, C89/4. Süss to Cartan on 6 September 1946. Universitätsarchiv Freiburg, Nachlaß Wilhelm Süss, C89/5.

22. Behnke to Süss on 7 November 1945. Universitätsarchiv Freiburg, Nachlaß Wilhelm Süss, C89/5.

23. Süss to Zentgraf on 26 April 1945. Universitätsarchiv Freiburg, Nachlaß Wilhelm Süss, C89/4.

24. Todd was professor at King's College, London. Süss to Behnke, Bol, Krull, Maak, Nöbeling, Rellich, Sperner, Wielandt on 15 August 1949: "The Institute owes to Herr Todd maybe its continued existence in the year 1945 and we would like to make him perceive this appreciation also by interesting company." Universitätsarchiv Freiburg, Nachlaß Wilhelm Süss, C89/5.

25. Süss to H. Epheser on 7 January 1947. Ibid.

26. Süss to Ehresmann on 8 April 1946. Ibid.

27. Süss to Todd on 26 April 1946. Universitätsarchiv Freiburg, Nachlaß Wilhelm Süss, C89/385.

28. Süss to Ehresmann on 6 March 1946. Universitätsarchiv Freiburg, Nachlaß Wilhelm Süss, C89/5.

29. Süss to Ehresmann on 23 August 1946. Ibid.

30. Süss to Barnett on 4 April 1946. Ibid.

31. Süss to Ehresmann on 9 October 1946. Ibid.

32. Süss to Eduard Wildhagen on 19 February 1948. Universitätsarchiv Freiburg, Nachlaß Wilhelm Süss, C89/398.

33. Süss to Carathéodory on 30 January 1946. Letter in German. Universitätsarchiv Freiburg, Nachlaß Wilhelm Süss, C89/5.

34. Carathéodory to Süss on 30 January 1946. Letter in German. Ibid.

35. Lothar Eduard Koschmieder (1890–1974) was a full professor at the Technical University of Graz from 1940 until 1946. His next post was professor (on contract) of the engineering faculty at the Syrian State University of Aleppo in 1948.

36. Loose leaflet in Carathéodory's handwriting. Universitätsarchiv Freiburg, Nachlaß Wilhelm Süss, C89/5.

37. See: Einhorn R (1985) *Vertreter der Mathematik und Geometrie an den Wiener Hochschulen 1900–1940.* VWGÖ Wien.
    Mayrhofer K (1971) *Nachruf von Hans Hornich.* Offprint from: *Almanach der Österreichischen Akademie der Wissenschaften,* 120. Jahrgang (1970), Wien.
    Communication to the author from Dipl. Ing. Erich Jiresch, Director of the University Archive of the Technical University of Vienna, on 29 November 1999.

38. Cremer to Süss on 22 November 1946. Universitätsarchiv Freiburg, Nachlaß Wilhelm Süss, C89/5.

39. Süss to Carathéodory on 17 April 1946. Ibid.

40. Carathéodory to Süss on 22 May 1946. Ibid.

41. Süss to Carathéodory on 31 May 1946. Universitätsarchiv Freiburg, Nachlaß Wilhelm Süss, C89/5.
    Barnett, a "mild-mannered mathematics professor": F. Litten's communication to the author on 28 March 2000.

42. Süss to Carathéodory on 9 December 1946. Universitätsarchiv Freiburg, Nachlaß Wilhelm Süss, C89/37.

43. Süss to Behnke on 7 December 1946. Ibid.

44. Knopp to Süss on 1 August 1946. Universitätsarchiv Freiburg, Nachlaß Wilhelm Süss, C89/5.

45. Bieberbach to Süss on 28 November 1946. Ibid.

46. Heisig to Sperner on 24 September 1947. Universitätsarchiv Freiburg, Nachlaß Wilhelm Süss, C89/4.

47. Reidemeister to Süss on 6 January 1948. Ibid.

48. Süss to Dr. E. Witt on 13 November 1947. Universitätsarchiv Freiburg, Nachlaß Wilhelm Süss, C89/385.

49. Süss to E. Wildhagen on 19 February 1948. Universitätsarchiv Freiburg, Nachlaß Wilhelm Süss, C89/398.

50. Süss to H. Cartan on 6 September 1946. Universitätsarchiv Freiburg, Nachlaß Wilhelm Süss, C89/5.

51. Süss to E. Wildhagen on 19 February 1948, as note 49.

52. Süss to B. Bavnik on 19 April 1947. Universitätsarchiv Freiburg, Nachlaß Wilhelm Süss, C89/5.
    The 1st volume of the *Archiv der Mathematik* was edited by the Mathematical Research Institute and published by the company G Braun in Karlsruhe in 1948–1949. The journal's board consisted of: G Bol, Freiburg; E Bompiani, Rom; P ten Bruggencate, Göttingen; J Dieudonné, Nancy; Ch. Ehresmann, Strasbourg; H Görtler, Freiburg/H Hadwiger, Bern; H Hopf, Zurich/H Kneser, Tübingen; W Magnus, Pasadena/T Nagell, Upsalla; Chr Pauc, Cape Town/J Radon, Vienna; K Reidemeister, Marburg; J A Schouten, Amsterdam; H Seifert, Heidelberg/E Sperner, Bonn; E Stiefel, Zurich.
    The 1st volume of the *Mathematisch-Physikalische Semesterberichte* was edited by Heinrich Behnke und Walter Lietzmann in connection with the Mathematical Research Institute and published by Vandenhoeck & Ruprecht in Göttingen in 1950.

53. Süss to Ehresmann on 9 October 1946. Universitätsarchiv Freiburg, Nachlaß Wilhelm Süss, C89/5.

54. Fueter to Süss on 23 December 1947. Ibid.

55. Süss to Besicovitch on 16 February 1948. Ibid.

56. Süss to Ehresmann on 19 January 1948. Ibid.

57. Süss to Ehresmann on 6 April 1948. Ibid.

58. Tietze to Süss on 13 October 1948. Universitätsarchiv Freiburg, Nachlaß Wilhelm Süss, C89/385.

59. Süss to Tietze on 26 January 1948. Ibid.

60. Süss to Alexander Dinghas on 23 January 1948. Universitätsarchiv Freiburg, Nachlaß Wilhelm Süss, C89/5. Süss to Besicovitch on 16 February 1948, Universitätsarchiv Freiburg, Nachlaß Wilhelm Süss, C89/5.
    Alexander Dinghas (1908 Smyrna–Berlin 1974) was a doctoral student of Schmidt. Dr. 1936; habil. 1939; associate at the U. Berlin 1946; full professor at the FU Berlin 1949; work on function theory in modern setting, esp. on value distribution theory, and also on isoperimetric problems, convex sets, set functions, special functions; 1952–1953 visiting professor at Columbia University, New York.

61. Carathéodory C (1952) *Einführung in Eulers Arbeiten über Variationsrechnung. Leonardi Euleri Opera omnia I 24.* Bernae, pp VIII–LXIII and *Ges. Math. Schr.* vol V, pp 107–174.

62. Carathéodory to Süss on 21 February 1946. Letter in German. Universitätsarchiv Freiburg, Nachlaß Wilhelm Süss, C89/5.

63. Behnke on 13 June 1942. *Bericht über meine Vortragsreise nach Zürich*, as note 99, chapter 4.

64. Ibid.

65. Carathéodory to von Kármán on 2 February 1947. Letter in English. Caltech Archives, Theodore von Kármán papers, Folder 5.4. and Carathéodory to von Kármán (Paris) on 22 January 1948. Letter in French. Caltech Archives, Theodore von Kármán Papers, Folder 5.4.

66. Carathéodory to Süss on 30 January 1946. Universitätsarchiv Freiburg, Nachlaß Wilhelm Süss, C89/5.

67. Carathéodory to von Kármán on 22 January 1948, as note 65, and Carathéodory to Born on 24 March 1949, as note 23, chapter 2.

68. L. C. Young, Carathéodory's former student, had announced a visit to Freiburg University during the first days of January 1949. He would come from Cape Town but, even at the end of November 1948, he had not obtained permission to travel to Germany. Süss spoke about Young's entry to Germany to the chief adjutant of the General in charge of that zone; he assured Süss that he would give permission. Every foreigner intending to travel from an occupied zone to another needed a special permit from each zone.

Süss to Carathéodory on 29 November 1948. Letter in German. Universitätsarchiv Freiburg, Nachlaß Wilhelm Süss, C89/407.

Young planned to arrive in Paris on 28 December 1948, travel to Freiburg the following day, visit Carathéodory in Munich on 30 December, arrive in Freiburg via Erlangen again on 2 January and travel further to Paris. Carathéodory thought that Young "seems to have no idea of travel difficulties within Germany."

Carathéodory to Süss on 10 December 1948 and Süss to Carathéodory on 28 December 1948. Letters in German. Universitätsarchiv Freiburg, Nachlaß Wilhelm Süss, C89/407.

69. Gabriel Krestovic had visited Carathéodory in Munich on 8 March 1930. He was the son of Euphrosyne's sister Loukia. His father, Stephanos Krestovic, had been assistant to the Governor General of Monastir in the years 1896–1909.

70. Carathéodory's letter from "Brunnen/Switzerland, c/o Prof. Dr. Fueter, Zürich 7, Klosbachstr. 75", was written on 1 October 1946. Vovolinis (ed) (1962), p 523f.

71. Cf. references to Ypsilantis in: Dragoumis Ph St (1995) *Ημερολόγιο – Διχασμός 1916–1919*. (Diary – Schism 1916–1919). "Δωδώνη", Athens and Ioannina. The author of this book, Philippos Dragoumis, was a nephew of Carathéodory's old friend Markos Dragoumis.

72. Kostes Palamas by Stephen P. Ladas. The Philhellene, vol II no 6–7, June–July 1943.

73. D. Maximos (1873–1955): economist and politician; 1921–1922 Director of the National Bank; 1922–1928 in Florence; from 1928 in Athens; in 1933–1935 MP and Minister of Foreign Affairs; 1947 Prime Minister not belonging to the parliament. He was the first Greek politician to open concentration camps for the Communists.

74. A testimony about Carathéodory's efforts in this field comes from Hölder: "When I met him for the last time (1947) on the grounds of Bogenhausen he spoke [...] very seriously about the state of our science in Germany and thought that the few still available forces should concentrate." Hölder E, as note 30, chapter 2, here p 290f.

75. Archiv der LMU München, OC N 14.

76. Archiv der LMU München, OC N 14.

77. Carathéodory to Kalitsounakis on 28 January 1947. Vovovlinis (ed) (1962), here p 524.

78. Kalitsounakis's address at the session of the Athens Academy on 23 February 1950, as note 40, chapter 2. Ioannis Kalitsounakis (1878–1966) was born in Chania, Crete. He graduated from the Philological School of Athens University in 1901 and continued his studies in philosophy and paedagogy in Jena and Berlin with the help of a scholarship from the Cretan government. In 1904 he taught at the Theological Seminar and the secondary school of Chania and the following year he took on a lecturership (and later a full professorship) of Modern Greek at the *Seminar für Orientalische Sprachen* of Berlin University, where from 1928 he was teaching Greek philology of the Middle Ages. In parallel, he was appointed full professor to the chair of Ancient Greek philology at the University of Athens. He retained both of the chairs, in Germany and Greece, until 1948. After World War II he took on lecturing at the *Freie Universität Berlin* and in 1953 he was nominated as an honorary acting professor. His studies concern Greek philology and language in antiquity, Middle Ages and modern times. A full member of the Academy of Athens from its foundation, he became its president in 1947 and was its Secretary of Publications in the years 1950–1966. In 1961 he also took on the presidency of the Philological Society *Parnassos*. He died in Bucharest while representing the Academy of Athens at a congress there in 1966.

79. Einstein to Born on 7 September 1944. *Albert Einstein – Hedwig und Max Born. Briefwechsel 1916–1955* (1972), p 153.

80. The Princeton Herald, Friday, June 20, 1947.

81. Carathéodory to von Kármán on 22 January 1948, as note 65.

82. Duggan's text is published in the Philhellene Special Issue on the Liberation of Athens, October 14, 1944.
    The "Agreement for United Nations Relief and Rehabilitation Administration", also signed by Greece on 9 November 1943, can be read in: Pamphlet no 4, PILLARS OF PEACE. Documents Pertaining to American Interest in Establishing a Lasting World Peace: January 1941–February 1946.
    Published by the Book Department, Army Information School,
    Carlisle Barracks, Pa, May 1946.

83. See: President Truman's UNNRA Message to Congress. New York Times. November 13, 1945.

84. Mrs. D. Rodopoulos-Carathéodory's version of the story.

85. Carathéodory to von Kármán on 2 February 1947, as note 65.

86. Mazower M (1991), p 304.

87. The US new course inaugurated by the Truman Doctrine was most clearly reflected in the Marshall Plan, alias European Recovery Program for the economic reconstruction of Western Europe, whose announcement denoted the intensification of the Cold War and the failure of the last allied efforts for a common policy towards Germany. The resources flowing into Western Europe, aiming primarily at purchasing raw materials, food and capital goods from the USA, amounted to about 13 billion US dollars until help came to a stop in 1952. The Co-operative for American Relief to Everywhere (CARE) was an organisation established in the USA in 1946 to organise relief consignments (CARE parcels) initially to European countries, later also to other countries, thus contributing to overcoming the desperate straits of the post-war era.

88. Carathéodory to Radó on 10 December 1948, as note 336, chapter 2. Minor editorial changes.

89. Carathéodory to Fueter on 21 December 1948. Letter in German. ETH-Bibliothek Zürich, Wissenschaftshistorische Sammlungen, HS 1227:1594.

90. Radó to Carathéodory on 16 December 1948. Letter in English. The Ohio State University Archives, Tibor Rado papers (Record Group R6 40/76/1), "Caratheodory, C. (Professor): 1948–1949".

91. Radó to Carathéodory on 10 January 1949, as note 344, chapter 2.

92. Carathéodory to von Kármán on 22 January 1948, as note 65.
    BayHStA MK 35403, Laufzeit: 1924-1950.

93. Sommerfeld to Heisenberg on 1 August 1947. Communication to the author from Dr. Michael Eckert on 22 October 1999.

94. Süss to Carathéodory on 11 September 1947. Letter in German. Universitätsarchiv Freiburg, Nachlaß Wilhelm Süss, C89/407.

95. Von Kármán to Carathéodory on 12 January 1948. Letter in French. Caltech Archives, Theodore von Kármán papers, Folder 5.4.

96. Carathéodory to von Kármán on 22 January 1948, as note 65.

97. Carathéodory to Süss on 15 October 1948. Letter in German. Universitätsarchiv Freiburg, Nachlaß Wilhelm Süss, C89/407.

98. Vovolinis (ed) (1962), p 525.

99. Carathéodory to Süss on 10 December 1948. Letter in German. Universitätsarchiv Freiburg, Nachlaß Wilhelm Süss, C89/407.

100. Carathéodory to Radó on 10 December 1948, as note 336, chapter 2.

101. Carathéodory to Fueter on 21 December 1948, as note 89.

102. Carathéodory to Born on 24 March 1949, as note 23, chapter 2.

103. BayHStA MK 35403, Laufzeit: 1924–1950.

104. BayHstA MK 35403, Laufzeit: 1924–1950.

105. Rechenberg, as note 550, chapter 5.

106. Born to Einstein on 12 February 1921. *Albert Einstein – Hedwig und Max Born. Briefwechsel 1916–1955* (1972), p 62.

107. Archiv der LMU München, E-II-1054.

108. Deutsches Museum, Archiv.

109. Carathéodory to Süss on 13 November 1949. Letter in German. Universitätsarchiv Freiburg, Nachlaß Wilhelm Süss, C89/407.

110. BayHstA MK 35403, Laufzeit: 1924–1950.

111. Carathéodory to Zermelo on 4 January 1950. Letter in German. Universitätsbibliothek Freiburg, Nachlaß Ernst Zermelo.

112. Carathéodory to Radó on 10 December 1948, as note 336, chapter 2.

113. Carathéodory to Born 24 March 1949, as note 23, chapter 2.

114. Carathéodory C (1950) Funktionentheorie. Birkhäuser, Basel.

115. Dr. Weigand's communication to the author on 19 September 1999.

116. Tietze, as note 105, chapter 1, p 95.

117. Carathéodory to Fueter on 21 December 1948, as note 89.

118. Carathéodory to von Kármán (Paris) on 22 January 1948, as note 65.

119. Carathéodory to Zermelo on 4 January 1950. Letter in German. Universitätsbibliothek Freiburg, Nachlaß Ernst Zermelo.

120. Born, as note 24, chapter 2.

121. Einstein to Born 12 April 1949. *Albert Einstein – Hedwig und Max Born. Briefwechsel 1916–1955* (1972), p 184.

122. Ibid, p 166.

123. Carathéodory to Born on 24 March 1949, as note 23, chapter 2.

124. Planck's radiation law is given by equation 8.16 in Born, as note 24, chapter 2.

125. Carathéodory to Radó on 17 November 1949, as note 233, chapter 4.

126. Kline to Carathéodory on 8 November 1949, as note 62, chapter 5. Kline's letter was sent by Carathéodory to Süss on 13 November 1949.

127. Carathéodory to Süss on 13 November 1949. Letter in German. Universitätsarchiv Freiburg, Nachlaß Wilhelm Süss, C89/407.

128. Radó to Carathéodory on 28 November 1949, as note 235, chapter 4.

129. See: (1978) A short history of the Fields Medal, as note 130, chapter 5.

130. Universitätsarchiv München, E-II-1054.

131. Communication to the author by Dr. M. Eckert.

132. Todd to Süss on 2 May 1950 Universitätsarchiv Freiburg, Nachlaß Wilhelm Süss, C89/385.

133. Carathéodory to Zermelo on 4 January 1950, as note 111.

134. BayHstA MK 35403, Laufzeit: 1924–1950.

135. Universitätsarchiv München, E-II-1054.

136. Draft of a condolence letter in German. University of Copenhagen, Institute for Mathematical Sciences, Archive, Børge Jessen Papers, Box 12, Folder "Carathéodory, Constantin".

137. Behnke, as note 29, chapter 1, here p 33.

138. Süss to Stephanos Carathéodory on 16 February 1950. Letter in German. Universitätsarchiv Freiburg, Nachlaß Wilhelm Süss, C89/407.

139. Süss to Tietze on 17 February 1950. Universitätsarchiv Freiburg, Nachlaß Wilhelm Süss, C89/385.

140. Tietze to Süss on 23 February 1950. Ibid.

141. Tietze to Süss on 23 April 1950. Ibid.

142. Ibid.

143. Süss to Tietze on 26 April 1950. Ibid.

144. BayHstA MK 35403, Laufzeit: 1924–1950.

145. Archiv der LMU München, E-II-1054.

146. Carathéodory C: A Proof of the First Principal Theorem on Conformal Representation. In: Studies and essays presented to R. Courant on his 60th birthday, Jan. 8, 1948. Interscience Publishers, New York, pp 75–83 and *Ges. Math. Schr.* vol III, pp 354–361.

147. Courant to Stephanos Carathéodory on 19 June 1950. Letter in German. New York University Archives.

148. Courant to Stephanos Carathéodory on 23 June 1950. Letter in German. Ibid.

149. Courant to Stephanos Carathéodory on 10 July 1950. Letter in German. Ibid.

150. Communication to the author from Gisela Sprenger, Universitätsbibliothek Würzburg on 16 June 1998.

151. Tietze H: Konstantin Karathéodory. *Süddeutsche Zeitung* no. 30, 6 February 1950.

152. Δελτίον της Ελληνικής Μαθηματικής Εταιρείας, vol 26 (1952).

153. Dodwell E (1821) Views of Greece, 2 vols. London.

154. Stuart J (1762–1816) Antiquities of Athens, 4 vols. London. A hugely influential publication in the history of architecture.
Stuart J (^{2}1841) The antiquities of Athens: and other monuments of Greece/ as measured and delineated by James Stuart and Nicholas Revett. Tilt and Bogue, London.

155. Carathéodory to Kalitsounakis on 22 February 1934. Letter in Greek. Vovolinis (ed) (1962), p 519.
Carathéodory might have meant Eustathios, Archbishop of Thessalonica ca. 1194, the author of: (1827–1830) *Commentarii ad Homeri Iliadem*, 4 vols. Weigel, Lipsiae.
Manuel Moschopoulos was a Byzantine grammarian who had worked on Sophokles' tragedies. The most distinguished of the Greek presses in France and Germany in the 16th century was that of the Estienne family in Paris.

156. Cf. *Fugger* in Encyclopaedia Britannica.
Lehmann P (1956–1960) *Eine Geschichte der alten Fugger Bibliotheken* (A History of the Old Fugger Libraries), 2 vols. *Schwäbische Forschungsgemeinschaft bei der Kommission für bayerische Landesgeschichte*. Mohr, Tübingen.

157. Carathéodory to Kalitsounakis on 26 June 1935, as note 6, chapter 1, here p 520.
Homer's Iliad was translated by Th. Gazis and edited in Modern and Ancient Greek by the Cypriot Nik. Theseus (printer: Karli, Florence, 1811–1812). The book might have been of interest to Carathéodory because his wife's grandfather, the Sultan's physician Stephanos, had co-operated with Theseus in the edition. Theseus was living in Marseille and working as a trader before the Greek War of Independence. During the war he participated in battles in Greece and offered all his fortune for Greece's independence. He aimed at English protection for Greece, combated Capodistria's candidacy and, in 1829, was forced to leave Greece.

158. Courant to Stephanos Carathéodory on 4 March 1958. Letter in German. New York University Archives.

159. Cf. Κυπάρισσος Στέφανος (Cyparissos Stephanos). In: Roussos M N (1986) Επιφανείς Συριανοί (Prominent Men from Syros). Published by the Movement of Catholic Scientists and Intellectuals of Greece, pp 43–51.

160. Frei G (1986) Auction of Carathéodory's personal library. The Mathematical Intelligencer 8/1:78.

161. Déhèque F D (1825) *Dictionnaire grec moderne français contenant les diverses acceptions des mots, leur étymologie ancienne ou moderne, et tous les temps irréguliers des verbes suivi d'un double vocabulaire de noms propres d'hommes et de femmes, de pays et de villes* (Modern Greek-French dictionary containing the different significations of words, their old and modern etymology and all irregular tenses of verbs, followed by a double vocabulary of male and female first names and names of countries and cities). Jules Duplessis et Cie, Paris; Chez Treuttel et Wurtz, Treuttel Jr., et Richter, London.

162. Pierre Loti (1850–1923): pseud. of Julien Viaud, French novelist, an officer in the French navy. He achieved popularity with his romances of adventure in exotic lands, such as *Aziyadé*, set in Constantinople in 1879 (Calman Lévy, Paris, [17]1891; J.Tallandier, Paris, 1977). Since 1861 he belonged to the *Académie Française* and in 1914 he did not hesitate to volunteer for the war.
    Aziyadé was translated into English by Marjorie Laurie and published by Kegan Paul International, New York, in 1989.

163. Charles duFresne Sieur DuCange (1688) *Glossarium ad scriptores mediae et infimae graecitatis* (Glossary for Authors of Mediocre and Base Greekness), 2 vols. Lyon. Unaltered edition: 1958, Akad. Druck- u. Verl. Anst., Graz.
    Information about time and content of sale by Dr. Johann Strauss, Orientalisches Seminar Freiburg, on 11 September 1996 and Département d'Études turques, Université Marc-Bloch, Strasbourg on 29 April 2000.

164. The contents are catalogued but still on hand-written pieces of paper.
    Communication to the author from Eva Kalpourtzi on 20 December 2000.

Epilogue

1. For the allegedly apolitical character of mathematics during the Nazi era see: Mehrtens, as note 7, chapter 5, here p 100.

2. Einstein to Dr. Frank Aydelotte on 29 May 1944. Courtesy of the Archives of the Institute for Advanced Study.

3. Günther Meinhardt who wrote about the development and the history of the University of Göttingen from 1734 to 1974 fails to mention Carathéodory even with one word. Friedrich Hund who wrote the history of physics in Göttingen and refers not only to Carathéodory's friends from physics but also to the Göttingen mathematicians leaves Carathéodory out of this history. Even Inge and Walter Jens, who recently wrote about the life of Katharina Pringsheim, publish part of Carathéodory's text on A. Pringsheim's 80th birthday but mention Carathéodory's name only in an appendix regarding information on the quotations and this as "Constanthéodory".

4. Mehrtens, as note 374, chapter 5, here p 91.

5. Kritikos, as note 209, chapter 2, here vol $\Delta'/4$, p 378.

6. Hasse, as note 565, chapter 5.

Appendix II

1. The biographical sketch is taken from work done by the author for the conference *Élites urbaines et savoir scientifique dans la société ottomane. XIXe–XXe siècles* (Urban Elites and Scientific Knowledge in the Ottoman Society, 19th and 20th Centuries) of the *Institut français d'Etudes anatoliennes* in Istanbul from 21 to 23 March 2002. For paper title see note 14, chapter 1.

Bibliography

A. *Carathéodory's Works*

Carathéodory, Constantin: Η Αίγυπτος (Egypt) (Athens, Σύλλογος προς διάδοσιν ωφελίμων βιβλίων, 1901). Series: Publications of the Society for the Promotion of Useful Books, no. 14.

Carathéodory, Constantin: Gesammelte Mathematische Schriften (Collected Mathematical Writings) (München, C. H. Beck, 1954–1957), vol. I: Variationsrechnung (Calculus of Variations) 1954; vol. II: Variationsrechnung, Thermodynamik, Geometrische Optik, Mechanik (Calculus of Variations, Thermodynamics, Geometric Optics, Mechanics) 1955; vol. III: Funktionentheorie (Theory of Functions) 1955; vol. IV: Funktionentheorie, Reelle Funktionen (Theory of Functions, Real Functions) 1956; vol. V: Geometrie, Partielle Differentialgleichungen, Historisch-Biographisches, Verschiedenes (Geometry, Partial Differential Equations, Historical-Biographical Texts, Various Texts) 1957.

Carathéodory, Constantin: Conformal Representation (London, Cambridge University Press, [1]1932; with an additional chapter on the theory of Poincaré and Koebe on uniformisation: Cambridge, Cambridge University Press, [2]1952). Series: Cambridge Tracts in Mathematics and Mathematical Physics, no. 28.

Carathéodory, Constantin: Elementare Theorie des Spiegelteleskops von B. Schmidt (Elementary Theory of B. Schmidt's Mirror Telescope) (Leipzig and Berlin, B. G. Teubner, 1940). Series: Hamburger Mathematische Einzelschriften, 28. Heft.
Also in GMS vol. II, pp. 234–279.

Carathéodory, Constantin: Funktionentheorie (Basel, Birkhäuser [1]1950; revidierte Aufl.: Basel, Birkhäuser, [2]1960–1961). Series: Lehrbücher und Monographien aus dem Gebiete der exakten Wissenschaften. Mathematische Reihe, Bd. 8–9.
English translation by F. Steinhardt: Theory of Functions of a Complex Variable (New York, Chelsea Publishing Company [1]1954, [2]1958–1960; [3]1964–).

Carathéodory, Constantin: Geometrische Optik (Geometric Optics) (Berlin, J. Springer, 1937). Series: Ergebnisse der Mathematik und ihrer Grenzgebiete, vierter Band.

Carathéodory, Constantin: Mass und Integral und Ihre Algebraisierung (Measure and Integral and their Algebraisation). Edited by P. Finsler, A. Rosenthal and R. Steuerwald (Basel, Birkhäuser, 1956). Series: Lehrbücher und Monographien aus dem Gebiete der exakten Wissenschaften. Mathematische Reihe, Bd. 10.
English translation by F. E. J. Linton: Algebraic Theory of Measure and Integration (New York, Chelsea Publishing Company, [1]1963, [2]1986).

Carathéodory, Constantin: Reelle Funktionen, Band I: Zahlen, Punktmengen, Funktionen (Real Functions, Vol. I: Numbers, Point Sets, Functions) (Leipzig, B. G. Teubner, [1]1939; New York, Chelsea Publishing Company, [2]1946).

Carathéodory, Constantin: Variationsrechnung und partielle Differentialgleichungen erster Ordnung (Calculus of Variations and Partial Differential Equations of the First Order) (Leipzig, B. G. Teubner, [1]1935; Leipzig, B. G. Teubner, [2]1956).
English translation by Robert B. Dean; Julius J. Brandstatter, translating editor: Calculus of Variations and Partial Differential Equations of the First Order (San Francisco, Holden-Day, [1]1965–1967; New York, Chelsea Publishing Company, [2]1965, [3]1982, [4]1999).

Carathéodory, Constantin: Vorlesungen über reelle Funktionen (Lectures on Real Functions) (Leipzig, B. G. Teubner, [1]1918; Leipzig, B. G. Teubner [2]1927; reprint: New York, Chelsea Publishing Company, [3]1948; corrected edition: New York, Chelsea Publishing Company, [4]1968, [5]1997).

Carathéodory, writings: The Urban Refugees (English translation, three pages); (July 1926) The Urban Refugees in Macedonia and Thrace, Chapters I, II, and IV (English translation, 22 pages); Influence of the refugees *on Greek life* (English translation, four pages). In: The Papers of Henry Morgenthau, Sr., container 29 (reel 24 of the microfilm edition), subject file, 1868–1939. The Library of Congress, Manuscript Division.

Projet d'une nouvelle Université en Grèce, présenté au Gouvernement Hellènique par C. Carathéodory (Project for a New University in Greece presented by C. Carathéodory to the Hellenic Government). Greek translation in: Vovolinis (Ed.), see below, pp. 484–489.

B. *Selected Bibliography*

Lexica and Encyclopaedias

Elsevier's Dictionary of Mathematics (Amsterdam, Lausanne, New York, Oxford, Shannon, Tokyo, Elsevier, 2000).

Encyclopaedia Britannica, electronic edition, 2001.

Encyclopaedia of Mathematics (An updated and annotated translation of the Soviet "Mathematical Encyclopaedia") (Dordrecht, Holland, Reidel, an imprint of Kluwer Academic Publishers, 1988–1994).

Encyclopedia of the Holocaust (New York, Macmillan Publishing Company, 1990).

Major Peace Treaties of Modern History 1648–1967, edited by Fred L. Israel (New York, Chelsea House Publishers, 1967).

Meschkowski, Herbert: Mathematiker-Lexikon (Mannheim, Wien, Zürich, Bibliographisches Institut, 1980).

Notable Mathematicians – From Ancient Times to the Present, edited by Robyn V. Young and Zoran Minderovic (Detroit, New York, Toronto, London, Gale Research, 1998).

Sturdza, Mihail-Dimitri: Dictionnaire historique et généalogique des grandes familles de Grèce, d'Albanie et de Constantinople (Historical and Genealogical Dictionary of the Great Families of Greece, Albania and Constantinople) (Paris, M.-D. Sturdza, 1999, 1983).

Books

Albers, Donald J., Alexanderson, G. L., and Reid, C.: International Mathematical Congresses. An illustrated history 1893–1986 (New York, Heidelberg, Springer, 1987).

Albert Einstein – Hedwig und Max Born. Briefwechsel 1916–1955 (Albert Einstein – Hedwig and Max Born. Correspondence 1916–1955) (München, Nymphenburger Verlagshandlung, [1]1969; Reinbek bei Hamburg, Rowohlt Taschenbuch Verlag, [2]1972).

Archibald, Raymond Clare: A Semicentennial History of the American Mathematical Society 1888–1938, Vol. 1 of the American Mathematical Society Semicentennial Publications (New York, American Mathematical Society, 1938).

Bernstein, Jeremy: Hitler's Uranium Club – The Secret Recordings at Farm Hall (New York, Springer, 2001).

Beyerchen, Alan D.: Wissenschaftler unter Hitler – Physiker im Dritten Reich (Scientists under Hitler – Physicists in the Third Reich) (Köln, Kiepenheuer & Witsch, 1980). Original title: Scientists under Hitler (New Haven, London, Yale University Press, 1977).

Biermann, Kurt-R: Die Mathematik und ihre Dozenten an der Berliner Universität 1810–1933 – Stationen auf dem Wege eines mathematischen Zentrums von Weltgeltung (Mathematics and its Lecturers at Berlin University 1810–1933 – Landmarks on the Way of a Mathematical Centre of World Importance) (Berlin, Akademie-Verlag, 1988).

Böhm, Helmut: Von der Selbstverwaltung zum Führerprinzip. Die Universität München in den ersten Jahren des Dritten Reiches (1933–1936) (From Self-Administration to the Führer Principle. The University of Munich in the First Years of the Third Reich (1933–1936)) (Berlin, Duncker & Humblot, 1995).

Bollenbeck, Georg: Tradition, Avantgarde, Reaktion – Deutsche Kontroversen um die kulturelle Moderne 1880–1945 (Tradition, Avant-garde, Reaction – German Controversies About Cultural Modernity 1880–1945) (Frankfurt am Main, S. Fischer, 1999).

Born, Max: Natural Philosophy of Cause and Chance (Oxford, Clarendon Press, 1949).

Born, Max: My Life – Recollections of a Nobel Laureate (London, Taylor & Francis Ltd, 1978). Original title: Mein Leben – Die Erinnerungen des Nobelpreisträgers (München, Nymphenburger Verlagshandlung, 1975).

Bouzakis, Sifis: Γεώργιος Α. Παπανδρέου 1888–1968 – Ο Πολιτικός της Παιδείας (Georgios A. Papandreou 1888–1968 – The Politician of Education), vol. Α′: 1888–1932 and vol. Β′: 1933–1968 (Athens, Gutenberg, 1997 and 1999).

Clogg, Richard: A Concise History of Greece (Cambridge, Cambridge University Press, 1992; reprints: 1993, 1994).

Cornwell, John: Pius XII. – Der Papst, der geschwiegen hat (Pius XII – The Pope who Kept Silent) (München, C. H. Beck, 1999). Original title: Hitler's Pope. The Secret History of Pius XII (London, Penguin, 1999).

Dakin, Douglas: Η Ενοποίηση της Ελλάδας 1770–1923 (Athens, Μορφωτικό Ίδρυμα Εθνικής Τραπέζης, [1]1982, [2]1989). Original title: The Unification of Greece 1770–1923 (London, Ernest Benn Ltd., 1972).

Deschamps, Gaston: Sur Les Routes d'Asie (On Asia's Routes) (Paris, A. Colin, [3]1907).

Elfving, Gustav: The History of Mathematics in Finland 1828–1918. In: The History of Learning and Science in Finland 1828–1918 (Helsinki, Societas Scientiarum Fennica, 1981).

Fleischer, Hagen: Im Kreuzschatten der Mächte. Griechenland 1941–1944 (Okkupation – Resistance – Kollaboration) (In the Cross Shadow of the Powers – Greece 1941–1944 (Occupation – Resistance – Collaboration) (Frankfurt am Main, P. Lang, 1986). Series: Studien zur Geschichte Südosteuropas (Studies into the History of South-Eastern Europe), vol. 2.

Fraenkel, Abraham A.: Lebenskreise – Aus den Erinnerungen eines jüdischen Mathematikers (Life Circles – From the Memories of a Jewish Mathematician) (Stuttgart, Deutsche Verlags-Anstalt, 1967).

Gondicas, Dimitri, and Issawi, Charles (Eds.): Ottoman Greeks in the Age of Nationalism: Politics, Economy, and Society in the Nineteenth Century (Princeton, The Darwin Press, Inc., 1999).

Grüttner, Michael: Studenten im Dritten Reich (Students in the Third Reich) (Paderborn, Schöningh, 1995).

Heiber, Helmut: Universität unterm Hakenkreuz. Teil II: Die Kapitulation der Hohen Schulen. Das Jahr 1933 und seine Themen (University under the Swastika. Part II: The Capitulation of Superior Schools. The Year 1933 and its Topics) (München, London, New York, Paris, Saur, 1992).

Hilbert, David – Gesammelte Abhandlungen (Collected Treatises), vol. 1: Zahlentheorie (Number Theory); vol. 2: Algebra, Invariantentheorie, Geometrie (Algebra, Theory of Invariants, Geometry); vol. 3: Analysis, Grundlagen der Mathematik, Physik, Verschiedenes, nebst einer Lebensgeschichte (Analysis, Foundations of Mathematics, Physics, Various Texts Together With a Life-Story) (Berlin, Springer, 1932–1935).

Hobsbawm, Eric J.: Nationen und Nationalismus – Mythos und Realität seit 1780 (Nations and Nationalism – Myth and Reality since 1780) (Frankfurt am Main, New York, Campus Verlag, 1991).

Hobsbawm, Eric J.: Das imperiale Zeitalter 1875–1914 (The Imperial Age 1875–1914). (Frankfurt am Main, Fischer Taschenbuch Verlag, 1995). Original title: The Age of Empire 1875–1914 (London, Weidenfeld and Nicolson, 1987).

Horton, George: Report on Turkey, USA Consular Documents (Athens, the Journalists' Union of the Athens Daily Newspapers, 1985).

Housepian, Marjorie: Smyrna 1922 – The Destruction of a City (London, Faber and Faber, 1972).

Iorga, Nicolae: Geschichte des Osmanischen Reiches (History of the Ottoman Empire), 5 vols. (Gotha, Perthes, 1908–1913; reprint: Darmstadt, Primus Verlag, 1997).

Kleanthis, Fanis N.: Η Ελληνική Σμύρνη – Σελίδες από την Ιστορία της, εικόνες από τη ζωή της και πλήρης εξιστόρηση της τραγωδίας του 1922 (The Greek Smyrna – Pages from its History, Images of its Life and the Complete Story of the Tragedy of 1922) (Athens, Βιβλιοπωλείο της "Εστίας", 1996).

Kragh, Helge: Dirac, A Scientific Biography (Cambridge, Cambridge University Press, 1990).

Large, David Clay: Hitlers München – Aufstieg und Fall der Hauptstadt der Bewegung (Hitler's Munich – The Rise and Fall of the Capital of the Movement) (München, C. H. Beck, 1998). Original title: Where Ghosts Walked – Munich's Road to the Third Reich (New York, Norton, 1997).

Lehto, Olli: Mathematics Without Borders – A History of the International Mathematical Union (New York, Springer, 1998).

Leukoparidis, X. (Ed.): Αλληλογραφία της Π. Σ. Δέλτα 1906–1940 (P. S. Delta's Correspondence 1906–1940). (Athens, Βιβλιοπωλείο της "Εστίας", 1956).

Lewis, Bernard: The Emergence of Modern Turkey (London, Oxford University Press, 1968).

Litten, Freddy: Mechanik und Antisemitismus – Wilhelm Müller (1880–1968) (Mechanics and Anti-Semitism – Wilhelm Müller (1880–1968))(München, Institut für Geschichte der Naturwissenschaften, 2000). Series: Studien zur Geschichte der Mathematik und der Naturwissenschaften (Studies on the History of Mathematics and Natural Sciences), edited by Menso Folkerts, vol. 34.

Losemann, Volker: Nationalsozialismus und Antike. Studien zur Entwicklung des Faches Alte Geschichte 1933–1945 (National Socialism and Antiquity. Studies in the Development of the Field of Ancient History 1933–1945) (Hamburg, Hoffmann & Campe, 1977).

Macrakis, Kristie: Surviving the Swastika – Scientific Research in Nazi Germany (London, Oxford University Press, 1993).

Mancosu, Paolo: From Brouwer to Hilbert – The Debate on the Foundations of Mathematics in the 1920s (New York, Oxford, Oxford University Press, 1998).

Mansel, Philip: Constantinople – City of the World's Desire, 1453–1924 (London, Penguin Books Ltd., 1995).

Mazower, Mark: Greece and the Inter-War Economic Crisis (Oxford, New York, Clarendon Press, 1991).

Mazower, Mark: Inside Hitler's Greece – The Experience of Occupation, 1941–44 (New Haven and London, Yale University Press, 1993).

Meynaud, J.: Πολιτικές Δυνάμεις στην Ελλάδα (Political Forces in Greece) (Athens, "Byron", [1]1966, [2]1974).

Miller, William: The Ottoman Empire and Its Successors, 1801–1927 (Cambridge, Cambridge University Press, 1936).

Morgenthau, Henry: I was sent to Athens, illustrations from photographs (Garden City, N.Y., Doubleday, Doran and Company, Inc., 1929). Greek translation by Kasesian, S: Morgenthau, Henry: Η Αποστολή μου στην Αθήνα – 1922 Το Έπος της Εγκατάστασης (My mission in Athens (or: I was sent to Athens) – The Epos of Settlement in 1922) (Athens, Τροχαλία, 1994).

Nearing, Scott: Oil and the Germs of War (Ridgewood, NJ, Nellie Seeds Nearing, 1923).

Panayotopoulos, A. (Ed.): C. Carathéodory International Symposium, Athens, September 1973 – Proceedings (Athens, Greek Mathematical Society, 1974).

Peckhaus, Volker: Hilbertprogramm und Kritische Philosophie – Das Göttinger Modell interdisziplinärer Zusammenarbeit zwischen Mathematik und Philosophie (Hilbert's Programme and Critical Philosophy – The Göttingen Model of Interdisciplinary Co-operation Between Mathematics and Philosophy) (Göttingen, Vandenhoeck und Ruprecht, 1990).

Psomiades, Harry John: Greek-Turkish Relations, 1923–1930 – A Study in the Politics of Rapprochement (Columbia University, PhD thesis, 1962).

Psyroukis, Nikos: Η Μικρασιατική Καταστροφή 1918–1923 – Η Εγγύς Ανατολή μετά τον πρώτο Παγκόσμιο Πόλεμο (The Asia-Minor Disaster – The Near East after World War I) (Athens, Επικαιρότητα, 1973).

Questions on German History – Ideas, forces, decisions from 1800 to the present. Historical Exhibition in the Berlin Reichstag Catalogue, 4th (updated) Edition (Bonn, German Bundestag Publications' Section, 1993).

Reid, Constance: Courant in Göttingen and New York: the story of an improbable mathematician (New York, Springer-Verlag, 1976).

Reid, Constance: Hilbert. With an appreciation of Hilbert's mathematical work by Hermann Weyl (Berlin, New York, Springer-Verlag, 1970).

Ringer, Fritz K.: The Decline of the German Mandarins: The German Academic Community, 1890–1933 (Harvard University Press, 1969).

Saks, Stanisław: Theory of the Integral (New York, Hafner Publishing Company, 1937).

Schumacher, Hans: Das neue Hellas (New Greece) (Berlin, Georg Stilke, 1937). Series: Schriftenreihe der Preußischen Jahrbücher, vol. 46.

Semicentennial Addresses of the American Mathematical Society, Vol. 2 of the American Mathematical Society Semicenntenial Publications (New York, American Mathematical Society, 1938).

Shaw, Stanford & Shaw, Ezel Kural: History of the Ottoman Empire and Modern Turkey, vol. 2: Reform, Revolution, and Republic: The Rise of Modern Turkey, 1808–1975 (Cambridge, Cambridge University Press, 1977).

Smith, Michael Llewellyn: Ionian Vision – Greece in Asia Minor 1919–1922 (London, Allen Lane, a division of Penguin Books Ltd., [1]1973; a facsimile edition, with a new introduction by the author: London, C. Hurst & Co., [2]1998).

Solomonidis, Christos S.: Η Παιδεία στη Σμύρνη (Education in Smyrna) (Athens, Mavridis's printing house, 1962).

Solomonidis, Christos S.: Σμυρναίοι Ακαδημαϊκοί – Ιωακείμογλου–Καλομοίρης–Σεφεριάδης (Academics from Smyrna – Ioakimoglou–Kalomiris–Seferiadis) (Athens, Mavridis's printing house, 1966).

Steinberg, Michael S.: Sabers and Brown Shirts: the German Students' Path to National Socialism 1918–1935 (Chicago, IL, Chicago University Press, 1977).

Stern, Fritz: The Politics of Cultural Despair; A Study in the Rise of the German Ideology (Berkeley, CA, Berkeley University Press, 1961).

Stern, Fritz: Einstein's German World (Princeton, Princeton University Press, 1999).

Svolopoulos, Konstantinos: Κωνσταντινούπολη 1856–1908 – Η Ακμή του Ελληνισμού (Constantinople 1856–1908 – The Prosperity of Hellenism) (Athens, Εκδοτική Αθηνών, 1994).

Tobies, Renate: Felix Klein (Leipzig, B. G.Teubner, 1981).

Toepell, Michael: Mathematiker und Mathematik an der Universität München: 500 Jahre Lehre und Forschung (Mathematicians and Mathematics at the University of Munich: 500 Years of Teaching and Research) (München, Institut für Geschichte der Naturwissenschaften, 1996). Series: Münchener Universitätsschriften, vol. 19.

Toynbee, Arnold: The Western Question in Greece and Turkey – A Study in the Contact of Civilisations (New York, Howard Fertig, 1970).

Trakakis, Georgios: Η Βιομηχανία εν Σμύρνη και εν τη Ελληνική Μικρασία – Οικονομική μελέτη, Σμύρνη 1920 (Industry in Smyrna and in Greek Asia Minor – A Financial Study, Smyrna 1920) (Athens, Τροχαλία, 1994).

Wende, Erich: C. H. Becker – Mensch und Politiker. Ein biographischer Beitrag zur Kulturgeschichte (C. H. Becker – A Man and a Politician. A Biographical Contribution to Cultural History) (Stuttgart, Deutsche Verlags-Anstalt, 1959).

Weyl, Hermann: Die Idee der Riemannschen Fläche (The Idea of the Riemann Surface) (Leipzig, B. G.Teubner, 1913).

Zannas, P. A. (Ed.): Αρχείο της Π. Σ. Δέλτα, Α´: Π. Σ. Δέλτα, Ελευθέριος Κ. Βενιζέλος – Ημε-ρολόγιο – Αναμνήσεις – Μαρτυρίες – Αλληλογραφία (P. S. Delta's Archive, Α´: P. S. Delta: Eleutherios K.Venizelos – Diary – Memories – Testimonies – Correspondence) (Athens, Ερμής, 1988).

Zürcher, Erik Jan: The Unionist Factor. The Role of the Committee of Union and Progress in the Turkish National Movement 1905–1926 (Leiden, Brill, 1984).

Papers on Carathéodory and his Work

Anastasiadis, M (Ed.): Κωνσταντίνος Καραθεοδωρή, Οργανωτής του Πανεπιστημίου Σμύρ-νης (Constantin Carathéodory, Organiser of the University of Smyrna) (Athens, privately printed, 1973).

Aumann G: Distortion of a Segment under Conformal Mapping and Related Problems. In: Panayotopoulos, A. (Ed.): C. Carathéodory International Symposium, Athens, September 1973 – Proceedings (Athens, Greek Mathematical Society, 1974), pp. 46—53.

Barrett J F: On the Relation between Carathéodory's Work in the Calculus of Variations and the Theory of Optimal Control. In: Panayotopoulos, A. (Ed.), ibid., pp. 54–74.

Behnke H: Carathéodorys Leben und Wirken (Carathéodory's Life and Deeds). In: Panayo-topoulos, A. (Ed.), ibid., pp. 17–33. The same text was published as: (1974) Constantin Carathéodory 1873–1950 – Eröffnungsrede zur Centenarfeier (am 3. September 1973) in Athen (Constantin Carathéodory 1873–1975 – Opening Speech at the Centenary Celebra-tion (on 3 September 1973) in Athens). Jber. Deutsch. Math.-Verein. 75:151–165.

Bieberbach L: Über einen Satz des Herrn Carathéodory (On a Theorem of Herr Carathéo-dory). In: Nachr. Kgl. Ges. Wiss. Gött. 1913, pp. 552–560.

Boerner H (1953) Carathéodory's Eingang zur Variationsrechnung (Carathéodory's Entry to the Calculus of Variations). Jber. Deutsch. Math.-Verein. 56:31–58.

Boerner H: Carathéodory, Constantin. In: Gillispie Ch C (Ed.): Dictionary of Scientific Biog-raphy, vol. III (New York, Scribner, 1971), p. 62f.

Boerner H: Carathéodory und die Variationsrechnung (Carathéodory and the Calculus of Variations). In: Panayotopoulos, A. (Ed.), op. cit., pp. 80–90.

Born M (1921) Kritische Betrachtungen zur traditionellen Darstellung der Thermodynamik (Critical Considerations of the Traditional Representation of Thermodynamics). Phys. Z. XXII:218—224; 249—254; 282—286.

Bückner H (1943) C. Carathéodory, Elementare Theorie des Spiegelteleskops von B. Schmidt (C. Carathéodory, Elementary Theory of B. Schmidt's Mirror Telescope). Jber. Deutsch. Math.-Verein. 53:23–24.

E. L. I. (1935) A Neglected Aspect of the Calculus of Variations – Variationsrechnung und par-tielle Differentialgleichungen erster Ordnung. Von Prof. Constantin Caratheodory. Nature CXXXVI:814.

Georgiadou M (1997) Die Gründung der Ionischen Universität in Smyrna – Die griechische ‚zivilisatorische Mission‘ im Orient (The Foundation of the Ionian University in Smyrna – The Greek 'Civilising Mission' in the Orient). Südost-Forschungen 56:291–317.

Georgiadou M (2003) The Reorganisation of the Universities of Athens and Istanbul in the 1930s – Two One-Man Projects and a Historical Coincidence. Chronos, Revue d'Histoire de l'Université de Balamand, vol. 8. In print.

Hölder E (1950) Constantin Carathéodory† – Sein Beitrag zur Axiomatik der mathematischen Physik (Constantin Carathéodory† – His Contribution to the Axiomatics of Mathematical Physics). Forsch. u. Fortschr. 21/22:290–293.

Hondros D (1950) Από την ζωή του Καραθεοδωρή (From Carathéodory's Life). Αιών του Ατόμου (Century of the Atom) Δ´/2–3:326.

Hondros D (1950) Ο Κ. Καραθεοδωρή ως φυσικός (Carathéodory as a Physicist). Αιών του Ατόμου (Century of the Atom) Δ´/2–3:333–334.

Kappos D: Generalized Probability with Applications to the Quantum Mechanics. In: Pana-yotopoulos, A. (Ed.), op. cit., pp. 253–270.

Kritikos N (1950) Το μαθηματικό έργο του Κ. Καραθεοδωρή (Carathéodory's Mathematical Work). Αιών του Ατόμου (Century of the Atom) Δ'/2–3:328–332 and Δ'/4:378.

Landé A (1926) Axiomatische Begründung der Thermodynamik durch Carathéodory (Axiomatic Foundation of Thermodynamics by Carathéodory). Handb. d. Phys. 9:281–300.

Litten F (1994) Die Carathéodory-Nachfolge in München 1938–1944 (Succession to Carathéodory in Munich). Centaurus 37:154–172.

Mylonas V (undated) Constantin Carathéodory. Dissertation submitted to the Institut für Geschichte der Naturwissenschaften und Technik (Institute for the History of Natural Sciences and Technology) of the University of Stuttgart.

Orlandos A, Zervos P, Ioakimoglou G, Kalitsounakis I and Maltezos K: speeches at the Academy of Athens on the occasion of Carathéodory's death. Πρακτικά Ακαδημίας Αθηνών (Minutes of the Academy of Athens), session of 23 February 1950.

Pesch H J, Bulirsch R (1994) Historical Paper – The Maximum Principle, Bellman's Equation and Carathéodory's Work. Journal of Optimization Theory and Applications 80/2:199–225.

Perron O (1952) Constantin Carathéodory. Jber. Deutsch. Math.-Verein. 55:39—51.

Riesz F (1915) Über ein Problem des Herrn Carathéodory (On a Problem of Herr Carathéodory). Journal für die reine und angewandte Mathematik 146:83–87.

Rund H (1974) The Hamilton–Jacobi Theory of the Geodesic Fields of Carathéodory in the Calculus of Variations of Multiple Integrals. In: Panayotopoulos, A. (Ed.), op. cit., pp. 496–536.

Schmidt E (1943) Constantin Carathéodory zum 70. Geburtstag (To Constantin Carathéodory on the Occasion of his 70th Birthday). Forsch. u. Fortschr. 23/24:249–250.

Shields A (1988) Carathéodory and Conformal Mapping. The Mathematical Intelligencer 10/1:18–22.

Thiele R (1997) On Some Contributions to Field Theory in The Calculus of Variations from Beltrami to Carathéodory. Historia Mathematica 24:281–300.

Tietze H: Dem Andenken an C. Carathéodory (In Memory of C. Carathéodory). In: Sitzungsber. Bayer. Akad. d. Wiss. München, math.-nat. Kl. 1950, pp. 85–101.

Tietze H: Carathéodory, Constantin. In: Historische Kommission bei der Bayerischen Akademie der Wissenschaften (Ed.): Neue Deutsche Biographie, vol. III (Berlin, 1957), pp. 136–137.

Varopoulos Th: Κωνσταντίνος Καραθεοδωρή (1873–1950) (Constantin Carathéodory (1873–1950)). Speech delivered at the scientific event organised in memory of Carathéodory by the Faculty of Natural and Mathematical Sciences, University of Thessaloniki on 2 February 1951. (Thessaloniki, separate print, 1951).

Vovolinis (Ed.): Μέγα Ελληνικόν Βιογραφικόν Λεξικόν (Great Greek Biographical Dictionary), vol. 5 (Athens, Edn "Βιομηχανική Επιθεώρησις" 1962), pp. 469–542: Καραθεοδωρή Κωνσταντίνος (Carathéodory Constantin).

Books and Papers on Carathéodory's Ancestors

Georgiadou M (2000/2001) Vom ersten zum zweiten Phanar und die Carathéodorys (From the First to the Second Phanar and the Carathéodorys). Südost-Forschungen 59/60:164–217.

Georgiadou M: Expert knowledge between tradition and reform – The Carathéodorys: a Neo-Phanariot Family in 19th Century Constantinople. In: Institut Français d'études Anatoliennes (ed) Médecins et ingénieurs ottomans à l'âge des nationalismes (Ottoman Medical Doctors and Engineers in the Age of Nationalism) (Paris, Maisonneuve & Larose, 2003), pp. 243–294.

Mavrogenis, Spyridon: Βίος Στεφάνου Καραθεοδωρή (The Life of Constantin Carathéodory) (Paris, Gauthier-Villars, 1885).

Tantalidis, Ilias: Βίος Στεφάνου Καραθεοδωρή (The Life of Stephanos Carathéodory) (Constantinople, Ελληνικός Φιλολογικός Σύλλογος Κωνσταντινουπόλεως, 1868).

Tsonidis, Takis: Το Γένος Καραθεοδωρή (The Carathéodory Family) (Orestias, Cultural Association of Nea Vyssa "Stephanos Carathéodory", 1989).

Ülman, Yeşim Işıl: Gazette médicale de Constantinople'un –Tıp Tarihimizdeki Önemi (The Importance of the 'Gazette médicale de Constantinople' with Respect to Medical History) (Istanbul, unpublished doctoral thesis, 1999), pp. 79–81: Konstantin Karateodori (Edirne, 1802 – İstanbul, 28 Eylül 1879) (Constantin Carathéodory (Adrianople 1802 – Istanbul, 28 September 1879)).

Name Index

Geographic Index

Page numbers listed under the entry e.g. "Athens" etc. do not include pages which refer e.g. to "Technical University of Athens" or "University of Athens" etc. In those cases please check the "Index of Academic Organisations and Institutions", where you will find the respective entries like "Technical University of Athens" or "University of Athens" etc.

Subject Index

Index of Mathematical and Physical Subjects

Index of Academic Organisations and Institutions

Some Views of Munich
and
Ludwig-Maximilian University

The aerial photograph of Munich, 1932, on p. 648 is reproduced by permission of Stadtarchiv München.

The lithograph on the top of p. 649 shows the Munich-University square and is reproduced by permission of Münchner Stadtmuseum.

The photographs on pp. 650 and 651 are reproduced from Heinz Thiersch: *German Bestelmeyer – Sein Leben und Wirken für die Baukunst* (His Life and Work for Architecture). (München, Verlag Georg D.W. Callwey, 1961).

Munich-University square.
Lithograph: Münchner Stadtmuseum.

Middle wing of Munich University building today.
Photograph: S. Georgiadis.

University of Munich: light-well from upstairs.

University of Munich: light-well.
Vault next to Friedrich von Gärtner's old building. Ornamental lattices,
mosaics of the vault and floor by Wilhelm Koeppen.
Portrait sculptures of King Ludwig I and Prince Regent Luitpold by Bernhard Bleeker.

Printing: Mercedes-Druck, Berlin
Binding: Stein+Lehmann, Berlin